# Microscale Organic Laboratory

# Microscale Organic Laboratory

## with Multistep and Multiscale Syntheses

### THIRD EDITION

**Dana W. Mayo**   Charles Weston Pickard
Research Professor of Chemistry
Bowdoin College

**Ronald M. Pike**   Professor of Chemistry, Emeritus
Merrimack College

**Peter K. Trumper**   Professor of Chemistry
University of Southern Maine

John Wiley & Sons, Inc.
New York   Chichester   Brisbane   Toronto   Singapore

Acquisitions Editor    Nedah Rose
Marketing Manager    Catherine Faduska
Senior Production Editor    Marcia Craig
Text Designer    Eileen Burke
Cover Designer    Karin Gerdes Kincheloe
Manufacturing Manager    Andrea Price
Illustration    Sigmund Malinowski
Cover Illustration: De-ethyloxodeltaline    Roy Wiemann

This book was set in 10/12 Palatino by Bi-Comp, Inc. and
printed and bound by Hamilton Printing. The cover was printed by Lehigh Press.

All experiments contained herein have been performed several times by students in college
laboratories under supervision of the authors. If performed with the materials and equipment
specified in this text, in accordance with the methods developed in this text, the authors
believe the experiments to be a safe and valuable educational experience. However, all
duplication or performance of these experiments is conducted at one's own risk. The authors do
not warrant or guarantee the safety of individuals performing these experiments. The authors
hereby disclaim any liability for any loss or damage claimed to have resulted from or related
in any way to the experiments, regardless of the form of action.

"Permission for the publication herein of Sadtler Standard Spectra® has been granted, and all
rights reserved, by Sadtler Research Laboratories, Division of Bio-Rad Laboratories, Inc."

Recognizing the importance of preserving what has been written, it is a
policy of John Wiley & Sons, Inc. to have books of enduring value published
in the United States printed on acid-free paper, and we exert our best
efforts to that end.

*Library of Congress Cataloging in Publication Data:*

Mayo, Dana W.
    Microscale organic laboratory : with multistep and multiscale
syntheses / Dana W. Mayo, Ronald M. Pike, Peter K. Trumper. — 3rd
ed.
        p.  cm.
    Includes index.
    ISBN 0-471-57505-4 (acid-free)
    1. Chemistry, Organic—Laboratory manuals.  I. Pike, Ronald M.
II. Trumper, Peter K., 1955-    III. Title.
QD261.M38   1994
547'.0078—dc20
                                                    93-43051
                                                       CIP

ISBN 0-471-57505-4

Printed in the United States of America

10  9  8  7  6  5  4  3  2  1

To Jeanne d'Arc, Marilyn, and Susan

# Preface

## to the Third Edition

The Microscale Organic Laboratory, which began over a decade ago, is no longer a fledgling program. The germination of those early seeds has produced a bumper crop of conversions to this highly efficient method of laboratory instruction.

Over the eight years following the release of the *Preliminary - 1* edition, we have conducted numerous workshops and held seven, one-week, summer institutes at Bowdoin; these now continue at Merrimac College. These programs have been centered primarily on acquainting organic faculty at other institutions with the experimental techniques involved in successfully operating the microscale laboratory. In 1990 we published a separate book, *Microscale Techniques for the Organic Laboratory*, which was entirely devoted to describing the techniques required to operate an instructional program at this scale.

Based on the wide experience we have accumulated during this time, we have now undertaken a major revision of the material contained in the two earlier editions of *Microscale Organic Laboratory*. Our third edition, **MOL-3**, is a very different text.

The third edition of MOL represents a major reorganization of nearly the entire manuscript. We list here these modifications and the rationale for these changes, which significantly improve the effectiveness of the text in students' hands.

- We have moved the discussions of microscale techniques to their own chapter. No experimental exercises are to be found in this section of the text. Our approach in earlier editions placed representative experiments directly within the technique discussions. Many instructors prefer the option of employing other laboratory exercises to introduce particular techniques. As a result, students have found the unused experimental material surrounding a particular technique discussion to be confusing rather than helpful.
- To further help facilitate the student's introduction to a particular

technique, we have reorganized our earlier two-hour video program on the microscale laboratory into nine separate tapes; each cassette covers a particular technique. The technique videos range from approximately five to fifteen minutes in length. We use these video demonstrations routinely in the early stages of our own programs, and we have found them to be particularly effective during a student's initial contact with a laboratory procedure. *It is anticipated that the nine technique tapes will be available to all institutions that adopt the third edition.*

- Our technique discussions were updated and expanded during the preparation of the techniques manual in 1990. We have retained the majority of those discussions for the techniques chapter in MOL-3. However, since we have expanded our nuclear magnetic resonance (NMR) and ultraviolet-visible (UV-vis) spectroscopy sections, these discussions combined with the infrared spectroscopy (IR) section were moved to an easily identifiable individual chapter, Chapter 9.
- The NMR section now includes $^{13}C$ as well as $^1H$ spectroscopy, an expanded discussion of spectral interpretation, and introductions to two-dimensional and other multiple-pulse techniques. The expanded UV-vis discussion helps to support increased coverage of photochemical reactions in the experimental section of the text.

It is the experimental section of MOL-3, however, which has undergone the most significant revision.

- *All* experimental procedures have been rewritten and changed from the passive voice of the scientific literature to the active voice of conventional directions. Our logic in earlier editions was to acquaint the student with literature terminology and conventions early in the laboratory experience, but clearly this approach has also made the introduction of a difficult subject a bit more remote. As the vast majority of undergraduate organic students will never have to come to grips with the *Journal of Organic Chemistry*, we have moved to make the experimental discussions more "user friendly."
- In addition to revising the experimental discussions, we have significantly expanded the introductory discussions to a large majority of the experiments. Most of the classic reactions now carry sufficient background information concerning the mechanism and associated chemistry involved to become free-standing assignments, with little or no dependence on lecture text material. We have come to appreciate that in many organic programs, close correlation between the laboratory and the lecture material is often difficult to achieve. Thus, the efficient development of the experiment requires significant background chemistry present in the laboratory text. We feel that the convenience of having the complete discussion gathered together in one area more than offsets the increased size of MOL-3.
- During the process of expanding our discussions, which in earlier editions had been consciously kept to a minimum, we also have taken the opportunity to update, correct, and expand the mechanistic section of each experiment.
- Many of the reactions suggested for study by microscale techniques have important relationships with biochemical reactions. We have therefore expanded the discussions of several of these transformations to include the biological role played by these reactions.
- To give the student an historical perspective to the chemistry being studied, we have added numerous biographical sketches of famous

chemists who discovered or investigated the reactions. Specific attention has been placed on those cases where the name of a reaction has become associated with a particular individual. In a few cases where fundamental mechanistic examples are introduced, we have also included historical discussions of the development of the theory surrounding these transformations and the scientists responsible for elucidating the chemistry.

- In the experimental sections we have retained the format that was used in earlier editions. Headers that identify the various stages of the experiment, margin graphics that are a guide to assembling apparatus, clear tables of physical properties and quantities of starting materials and reagents, estimated time for students to complete the experimental procedures, and convenient stopping points in those exercises expected to require more than a single laboratory period are noted. **Warnings** and **Cautions**, as well as advice regarding use of the **hood**, are in **bold type** to bring this advice to the attention of the students, teaching assistants, and instructors. References to the literature and numerous questions are given at the end of each experiment. The latter have been expanded and reorganized to correlate effectively with the experimental sections. We have dropped tables of environmental data from each experiment. While we still view this information as important data for the student, our reviewers felt that this material was readily available from other sources and was an unnecessary redundancy. We have also deleted, where possible, the physical properties of the products in order to encourage the students to locate the data in common, readily available handbooks. In a number of experiments we have retained the detailed interpretations of infrared spectra of starting materials and products. These discussions can be used as illustrative examples of how this type of spectroscopic data can be used to more deeply characterize the reaction under study.
- We have added two new sections to the experimental format:

  First, at the beginning of each exercise a description is given of the *Purpose* for which the experimental work is designed.

  Second, an assigned *Prior Reading* list follows the statement of *Purpose*. This list refers to those technique sections that apply to the experimental procedures to be utilized during the particular experiment under study.
- As in earlier editions, we continue to emphasize several very powerful experimental methods modified by us for the purification of microscale and semimicroscale quantities of liquid reaction products. A number of the experiments introduce preparative gas chromatographic (prep-GC) routines for high-resolution separation and purification of liquids. For semimicroscale liquid separations, two powerful spinning band distillation columns are introduced.
- A number of new experiments have been added and a few less attractive ones have been dropped. **Ninety-two** individual experiments are available for study.

The experiments are organized into three chapters, rather than all being lumped into a single massive section.

The first experimental chapter (Chapter 6), *Microscale Organic Laboratory Experiments*, contains thirty-five microscale experiments that have detailed and explicit directions.

The second experimental chapter (Chapter 7), *Advanced Microscale Organic Laboratory Experiments*, consists of seven experiments. These exercises utilize more esoteric reactions and/or reagents. In this set of reactions, the conditions must be more tightly controlled than in the reactions described in Chapter 6 in order to effect successful conversions. The chemistry involved in these transformations may not normally be covered in the introductory organic lecture section. Nevertheless, these are particularly interesting and challenging experiments for study.

The third experimental chapter (Chapter 8), *Sequential Syntheses: The Transition from Macro to Micro*, contains the majority of the "new" chemistry in MOL-3. The chapter consists of six synthetic sequences employing twenty-four separate reactions. The experiments range from macro–semimicroscale conversions early in each synthetic sequence, to microscale reactions for the final transformations. The sequences are graded in difficulty, and each has a target molecule with interesting properties.

The experiments in the two advanced chapters (7 and 8) can be used as special student exercises late in the second semester or may also be used to form the core collection of experiments in an advanced third-semester laboratory course.

We strongly urge that the multistep sequences be assigned toward the end of the second semester, after the students have completely mastered microscale techniques. We are then confident that both the student and the instructor will be impressed with the ease with which these rather complex sequences can be successfully carried out.

- While the exercises in the introductory experimental chapter (Chapter 6) have quite detailed instructions, the two advanced chapters have less explicit instructions. It is quite possible that in future editions we may revert to literature style experimental details in Chapter 7 if our audience is interested in this change.
- As described, all but fifteen of the reactions can be utilized as standalone microscale exercises; downscaling instructions are also supplied within many of the scaled-up procedures. Alternatively, full macro scaleup directions are also contained in the introductory microscale chapter for 6 of these experiments. In addition, there are numerous scaled-up reactions present in the sequential reactions found in Chapter 8.

  In a single laboratory text, MOL-3 presents an unparalleled variety of experimentation for use at the undergraduate and graduate levels of instruction.

  The very large majority of these experiments have now been successfully tested by several generations of students in our laboratories. We have considerable confidence, therefore, that they can be reproduced provided *uncontaminated starting materials* are employed and the instructions are *carefully* followed.
- To further improve the quality of the spectral data contained in the text we have entered into a collaboration with Terry Grim of the Sadtler Research Laboratories. Thanks to his efforts we have obtained an excellent collection of infrared spectra that replace many of the spectra of the earlier editions that were used in group frequency analysis (Chapter 9).
- We have also included in MOL-3 a glossary of important chemical terms as a further aid to the student's entrance to the Microscale Organic Laboratory.

As the third edition of MOL is about to go to press we recall David Brooks' outrageous prediction made in a letter written to us following the biennial ChemEd meeting of 1984. He made the prediction that "within ten years 80% of all undergraduate organic students would be employing the microscale approach." With the conversions of the University of Michigan, New York University, North Carolina State, the University of North Carolina, Duke University, Purdue University, University of California at Los Angeles, the entire University of Wisconsin system (23 campuses), Rice University, the United States Military Academy, and the University of Texas to name just a few of the larger programs, it is now evident that David's vision was considerably clearer than our own. By the end of 1994 the number may not reach 80%, but certainly a majority of undergraduate organic chemistry instruction will be at the microscale level or close to it. It is no longer a question of, "Is this the best pedagogic approach?," but, "How soon will the conversion occur?" In fact, a recent advertisement for an organic chemistry laboratory instructor listed experience with microscale organic laboratory techniques as a primary job requirement.

With the introduction of *the microscale laboratory* feature column edited by Arden P. Zipp in the *Journal of Chemical Education*, the need for our *Smaller is Better Newsletter* has declined. It is now circulated only on an irregular basis. The authors are only too happy to see the journal absorb this role.

The authors wish to acknowledge a number of very helpful contributions and comments that have been incorporated into the third edition: the development of the improved separation scheme for the isomers of 4-*tert*-butylcyclohexanol by T. J. Dwyer and S. Jones of the University of California, San Diego; the development of the Sequence C experiments by S. Danishefsky of Yale University; and the Sequence F experiments by R. Marshall Wilson and D. L. Lieberman of the University of Cincinnati.

In addition, we are grateful to the colleagues listed below whose careful reviews, helpful suggestions, and thoughtful criticisms of the manuscript have been of such great value to us in developing this Third Edition.

Mark Midland
*University of California, Riverside*

Peggy Alley
*Amarillo College*

Deborah Lieberman
*University of Cincinnati*

Dale Ledford
*University of South Alabama*

Donald Slavin
*Community College of Philadelphia*

Doris Kimbrough
*University of Colorado-Denver*

Allen M. Schoffstall
*University of Colorado-Colorado Springs*

Warren Sherman
*Chicago State University*

We continue to acknowledge the outstanding contributions of the early pioneers of instructional microscale programs and techniques such as F. Emich and F. Pregl in Austria; N. D. Cheronis, L. Craig, R. C. Fuson, E. H. Huntress, T. S. Ma, A. A. Morton, F. L. Schneider, and R. L. Shriner, in the United States; and J. T. Stock in both England and the United States. Clearly, these educators laid the foundation on which we were able to fashion much of the current introductory program.

We have been very pleased to see significant reductions in the cost of the microscale glassware since the last edition and we applaud the efforts of J. Ryan and Larry Riley of the ACE Glass Company to make this educational equipment more accessible to institutions on very tight instructional budgets. Both Ryan and Riley, together with Robert Stevens of J. J. Stevens, have helped to make further significant improvements in the performance of the already powerful spinning band distillation systems that we have developed in collaboration with Robert Hinkle (Bowdoin

1986, Ph.D., University of Utah 1993). We are greatly indebted to the efforts of Robert Mathieu of GOW-MAC in providing information that will assist institutions to obtain funding for instructional gas chromatographic instrumentation. We wish to add our deep appreciation of the patience, understanding, and support that our editor at Wiley, Nedah Rose, has given us over the past three years. This major revision could not have materialized without her thoughtful advice and counsel.

The development of Chapter 8 is largely the work of a collection of outstanding Bowdoin students: Marlene L. Castro, Jodie K. Chin, Helen E. Counts, Jonathan M. Dugan, Patricia A. Ernst, Lenore R. Menger-Anderson, Joanne M. Holland, and Jessica B. Radin. We are particularly grateful for the tenacity with which Joanne attacked the chemistry leading to the photosensitive azomethine ylide in Sequence F. A number of our close associates continue to play vital roles in the evolution of the microscale program and we are pleased to acknowledge them: Lauren Bartlett (now in the Department of Environmental Engineering, Duke University), Judith Foster, Henry Horner, and Samuel Butcher. Sam, as our resident expert on air chemistry and laboratory atmosphere contamination, has retired from active participation in the further evolution of the microscale instructional program.

We wish to thank both the National Science Foundation and the PEW Charitable Trusts for continued support of our Microscale Summer Institutes. We are further indebted to the PEW Charitable Trusts for major support that has allowed us to develop and construct a new microscale organic laboratory at Bowdoin College. This new laboratory, which is now in operation, has the potential to become the prototype laboratory design for new instructional facilities at other institutions. We also acknowledge the PEW Charitable Trusts for support of the development of our second generation video programs and add our thanks to Barbara J. Kaster, Harrison King McCann Professor Emerita of Oral Communication, and her assistant, biochemistry major Anthony Gosselin, for their time, effort, enthusiasm, and dedication to detail, which have made these Technique Tapes such a success.

We are more than pleased to acknowledge the unparalleled contributions of Paulette Fickett, Laboratory Instructor, to the microscale program. Paulette lead us through a three-year conversion period in the early 1980s when the program was in its infancy, and she remains in charge of making the microscale program function effectively on the day-to-day basis at Bowdoin in the early 1990s. Without exaggeration, Paulette has had more experience at operating microscale instructional laboratories than any person in history. Her dedication to this program, her friendly advice to hundreds of calls for help from students in Brunswick and faculty halfway around the world, her starring role in the videotapes, and her organization of the laboratory program for the summer institutes, have all been vital to the success of Microscale Organic Laboratory. For all of these endless contributions we are most grateful.

The revolution is over. The microscale era has arrived! We are confident that future students now will be assured the opportunity to experience the thrill of coaxing one complex molecular structure into another and to bring about these transformations on quantities of material that are just visible to the human eye!

Brunswick, Maine
March 22, 1993

**DANA W. MAYO**
**RONALD M. PIKE**
**PETER K. TRUMPER**

# Preface
## to the Second Edition

In the three years since 1985, when bound xerox copies of *Microscale Organic Laboratory—1 Preliminary*, the manuscript of the later *Microscale Organic Chemistry*, first became available, nearly three hundred institutions have moved to convert to this style of program. The magnitude and rapidity of adoptions really does mean that we are now in the midst of an educational revolution. Most impressive to us, however, is that of all the institutions initiating trial microscale programs, none to our knowledge has later dropped the microscale approach. Those areas in need of further refinement are being attacked by an ever-widening and skillful group of enthusiastic evangelists. Indeed, it is our view that we have just skimmed the surface of fruitful micro techniques and experiments, and we look forward in the next several years to bringing into sharper focus the direction in which this exciting approach will be heading. It is particularly gratifying to see the program being explored by a cross section of institutions ranging from small community colleges to large research-oriented universities. This variety of interest has in turn led to a wide range of useful and ingenious comments and suggestions, many of which we are pleased to be able to incorporate in this latest edition.

In response to favorable reviews, we have retained the basic format of the first edition. Thus, we continue to integrate illustrative experimental examples in the techniques section; to incorporate keyed marginal illustrations of apparatus setups throughout most of the experimental section; and to describe the experiments in formal journal style prose. We have extended our energies during the last few years to refining a number of key experimental techniques and to improving the reliability, flexibility, and variety of reactions to be studied. For a number of experiments in which clarity needed to be improved, we have rewritten procedures.

We are now convinced that by the end of the coming decade the great majority of chemical educational programs in the United States will have converted to the microscale approach. The question is not whether to miniaturize but how best to go about the job. We feel that *flexibility* in program design is vitally important in order to accommodate the wide variety of local educational environments. In our first edition we sent the message "Smaller is better," and all the experiments utilized approximately 150 mg or less of starting material. In this second edition we feel that it is now appropriate to build more flexibility of scale into the experimental design. We have, therefore spent a good deal of time modifying ten experiments and three of our four Sequential Sequences to include *Optional Scaleup* procedures. These range from twofold to two hundredfold above the microscale level. We have two reasons for this change in content. First, although many students have the opportunity to experience multigram reaction scales in upper-level laboratories or independent study, a significant number depend on a single year of experimental organic chemistry to gain acquaintance with the field. We agree that this second category of students, and indeed perhaps many of the first, should experience the excitement of undertaking a set of sequential steps (starting with multigram quantities) that clearly emulates a research laboratory synthesis. These experiments form a fitting conclusion to the end of the microscale laboratory. Second, for programs that are interested in "going micro," but need time and equipment to effect the transition, these optional experiments offer a temporary bridge to begin moving in the micro direction. In this regard we should emphasize that for programs that have reached the microscale level, the recommended time for incorporation of the Optional Upscale procedures is late in the second semester.

The rapid, almost breathtaking speed of technique innovation taking place in the microscale laboratory program is best exemplified by the changes found in the completely rewritten section on fractional distillation. The development of bottom-driven, low-cost micro spinning band distillation columns in the authors' laboratories over the past three years has resulted in a major breakthrough in the technique for illustrating classic fractional distillation. These systems are unquestionably the most powerful stills ever to become available as instructional equipment at the introductory level. In particular, the Hinkle modification of the classic Hickman still will clearly have a major impact on the entire field of micro distillation, including the advanced research laboratory.

Developments in chemical instrumentation also continue to accelerate. The availability of low-cost Fourier transform-infrared systems has broken the jam of making infrared data available to students in real time and has provoked us into adding new detailed interpretations of the FT–IR spectra of reactants and products in fourteen experiments. We hope that these discussions can be used as a means of leading the student into a deeper appreciation of the value of these spectroscopic data. Fourier transform–nuclear magnetic resonance ($^1H$, $^{13}C$) and ultraviolet–visible data have been introduced for the first time in modest fashion where they are appropriate for the characterization of reaction products. It is clear that the instrumental and cost barriers to the utilization of experimentally derived NMR data in the introductory organic laboratory are finally beginning to be bridged. Although earlier we had significant reservations about incorporating NMR data in the microscale experiments, we now expect their use in these programs to expand. The ability of the student to collect and interpret spectroscopic data is an essential aspect of "characterization science," which is one of the most vital elements of the modern-day academic and industrial laboratory.

To further the flexibility of the approach, we have added keyed references in each experiment to the appropriate qualitative identification tests and derivative preparations in Chapter 7. These operations can be effectively utilized by programs looking to expand the amount of microscale chemistry involved in a particular experiment or as an alternative to instrumental characterization.

The very successful utilization of the chromatographic techniques given in the first edition has lead us to expand the coverage and application of these powerful experimental routines. Gas chromatography and thin-layer chromatography are effectively utilized in a number of new locations.

Several other areas of the text have been improved. For example, the section on acid-base extraction has been expanded and in addition includes a sequence that illustrates the separation of a three-component mixture. The variety of alternate methods to illustrate a particular reaction continues to increase. For example, an improved extraction technique for the isolation of caffeine, the Horner–Emmons modification of the Wittig reaction, and a new nitrating agent prepared from silica gel and nitric acid are now included in the reaction selections.

The authors wish to acknowledge the many helpful suggestions from Professors Charles E. Sundin, University of Wisconsin–Platteville, Chaim N. Sukenik, Case Western Reserve University, and Bruce Ronald of Idaho State University. We are also indebted to all those who have attended the Bowdoin College Microscale Summer Institutes and our workshops across the country. Your enthusiasm, interest, and insight have led to many of the improvements in procedures and techniques now incorporated into the program. The many contributors to the newsletter *Smaller Is Better* have significantly helped to sustain the momentum of the program. We especially thank Paulette Fickett and Lauren Bartlett, Laboratory Instructors at Bowdoin, for their dedication, hard work, and eternal optimism. The microscale program could not have evolved in such a successful fashion without the continued contributions of John Ryan of Ace Glass, Stephen Cantor of Pfaltz and Bauer, Robert Stevens of J. J. Stevens, and of Henry Horner and Thomas Tarrant, all of whom have given encouragement and helpful guidance at crucial stages along the way. Judy Foster again waved her magic wand and this time transformed a first-edition Apple Writer manuscript into Microsoft Word. We continue to marvel at the patience of Dennis Sawicki, our Chemistry Editor. We would like to acknowledge continued support from the Surdna Foundation in developing the microscale program. The Sloane Foundation and the National Science Foundation have supported two of the summer institutes.

We are particularly indebted to our colleague Peter Trumper for injecting his expertize in FT-NMR into the microscale program. His enthusiastic application of high-field NMR to the introductory organic program promises future exciting developments in this area.

As we stand on the threshold of the second decade of the microscale revolution, we wish to thank our students. Especially for them do we feel the program has meaning: a cleaner atmosphere in which to work, an awareness of the toxic factors associated with the chemical workplace, and a commitment to detail not hitherto fostered in the introductory laboratory. For their perseverance and eagerness to learn and adapt to new ideas, for the freshness that each new class brings, and for their willingness to grow, we are indeed most grateful.

**DANA W. MAYO**
**RONALD M. PIKE**
Brunswick, Maine
**SAMUEL S. BUTCHER**
October 30, 1988

# Preface
## to the First Edition

This introductory organic laboratory textbook is a major departure from all other modern texts dealing with this subject. For a number of very cogent reasons, we have chosen to introduce experimental organic chemistry at the microscale level. Currently, beginning students perform the large majority of their experimental work at least two orders of magnitude above that described in this manual. Although contraction from the multigram to the milligram scale is the most obvious change, there are a number of other very unique aspects to our approach.

1. The laboratory environment has been made a distinct part of the experimental process. The student is given the means of easily determining his or her exposure to all volatile substances employed in the experiments. These calculations can be carried out for any laboratory by utilizing instructor-supplied ventilation rates.

2. Chemical instrumentation is given very high priority in the laboratory. The text avoids emphasizing data not directly determined by the student. Product characterization by infrared spectroscopy is routine, and detailed means for interpreting such data are provided.

3. Modern separation and purification techniques, including preparative gas chromatography, thin-layer chromatography, and column chromatography, are extensively utilized in product workups.

4. Over a third of the 82 reaction products are new to the undergraduate laboratory. Many of the reactions involve reagents or substrates that would present potential safety problems or entail exorbitant costs in a macroscale laboratory program; nevertheless, the use of these materials becomes safe and practical for experimentation at the microscale level. Reagents such as anhydrous nitric

acid, diborane, chloroplatinic acid, "instant ylids," silver persulfate, chromium trioxide resin, tetrabutylammonium bromide, and triflic acid are representative of the materials that the student will encounter during the year.

Why are we committed to the goal of attempting to reduce significantly the scale of starting materials in the introductory organic laboratory? The academic community has become increasingly aware of the necessity of improving air quality in instructional laboratories. The standard solution to this problem, a costly upgrading of the ventilation system, has generally been considered the most reasonable answer.

Our study of the problem led us to the following conclusions. First, although current organic laboratory texts are filled with details of product characterization employing the latest spectroscopic methods, the descriptions of the techniques of preparing compounds have changed very little from those of a century ago. In particular, the scale of synthesis has changed very little over this period. (Indeed, the quantities of materials employed have only decreased modestly, and in some cases have actually increased!) Clearly, the strategy of introducing the student to organic chemical laboratory techniques, originally and still today, is centered on the multigram level (see Table 1).

**TABLE 1   Starting Materials Employed in Classical Organic Laboratory Syntheses, 1902–1980**

| | | Acetanilide | 4-Bromoacetanilide | Benzoin |
|---|---|---|---|---|
| | | Starting Materials Required (grams) | | |
| Date | Author | Aniline | Acetanilide | Benzaldehyde |
| 1902 | Levy, 4th ed. | 46.2 | — | 50.0 |
| 1915 | Cohen, 3rd ed. | 25.0 | 5.0 | 25.0 |
| 1933 | Adkins | 25.0 | 13.5 | 10.0 |
| 1941 | Fieser, 2nd ed. | 18.2 | 13.5 | 25.0 |
| 1963 | Adams | 20.0 | 13.5 | 16.0 |
| 1980 | Durst | 10.0 | 5.2 | 10.0 |

We seriously question the wisdom of maintaining the introduction of laboratory work at conventional levels. Is this approach relevant when one considers the quantities of materials commonly used in natural products, pharmaceutics, biochemistry, and other fields of modern research in which costly substances are employed?

Second, it is now fully recognized that there has been a very serious decrease in undergraduate laboratory contact time over the past two decades. It seems to us that an increasingly important question should be asked in evaluating the introductory organic program: At what scale of introductory laboratory work can the student gain the maximum ability to handle organic materials within the shortest period of time?

We now firmly believe that the microscale laboratory approach resolves both these concerns and, in addition, affords a number of significant bonuses. The immediate result of "going micro" is that there is a change in the laboratory air quality that can be described only as spectacular! Of greater importance, and a point that we have come to appreciate in retrospect, is that many reactions and operations carried out at the micro level require far less time to reach completion. Indeed, *many of the time-consuming aspects of current instructional experiments are dramatically shortened.* We believe that the concomitant advantage of the microscale approach, the substantial increase in the number of manipulations possible per laboratory period, will have a major pedagogic impact. Herein lies the significant advantage of this approach to teaching laboratory technique.

We also see the increase in use of chemical instrumentation as a further pedagogic advantage of operating at the microscale level. Routine use of gas chromatography may be avoided in macro experiments; however, if it is not employed at the microscale level, successful experimentation with liquid substances is limited. The capital investment that would allow routine use of this equipment in many undergraduate laboratories might appear to elevate the microscale program out of the reach of institutions with very limited budgets. For many institutions, however the very substantial savings in chemical costs (75–90% of current expenditures), if carefully managed, can offer a rapid payback period (2–4 years) for expanded gas chromatography capacity and for other pieces of equipment. The end result of conversion to microscale is a far more effective integration of modern instrumentation into the organic laboratory program.

Once the microscale approach is initiated, the advantages are endless. We will not dwell on them here, as many of the following points vary in importance with each institution: (1) major reduction in cost of chemicals, (2) elimination of fire or explosion danger, (3) elimination of chemical waste disposal costs, (4) expansion in variety and sophistication of experiments, (5) elimination of dependence on commercially available starting materials, and (6) more durable and less expensive glassware.

These advantages are highly compelling reasons for advocating the microscale approach. Initially, however, we had serious reservations. First, the microscale approach appeared to involve a sophisticated set of techniques and manipulations too advanced for sophomore undergraduates to master. Second, a student having experience only at the microscale level might be expected to encounter problems when larger-scale preparations were undertaken. Third, significantly less organic chemistry would be covered during the year because increased attention would have to be paid to the development of microtechniques. Fourth, certain classical procedures such as fractional distillation would have to be abandoned.

As of the completion of this text, we have conducted eight semesters of microorganic chemistry laboratory assessment. The experience with the test laboratory groups (made up of a cross section of volunteers) has been a revelation. It is clear from our observations that the entire range of the class achieves significantly better results on micro experiments. Better yields are realized and, in particular, the class appears to master experimental details and procedures more effectively. It is felt that these results arise from several causes, one of which is the increased attention to detail required in the laboratory. We also sense a synergistic influence with the analytical chemistry laboratory, which is often scheduled for the sophomore year concurrently with the organic course. Analytical chemistry is carried out at a scale not unlike that employed in the microscale organic laboratory.

At present there is no indication that the learning of micro techniques during the introduction to organic chemistry causes any adverse effects when scaleup work is introduced in advanced laboratory courses or in research areas. To the contrary, our students appear to be performing significantly better in upper-level work.

The development of the experimental section of the text, which contains eighty reactions, has been a major effort. The conditions for a large proportion of these reactions have been optimized to maximize yields at the micro level. We have chosen to describe the experimental work in language similar to formal journal style. At first, this impersonal construction would appear to be "user unfriendly" and to make the text a bit boring and difficult to read. On field testing with students, however, we have found that they quickly adapt to the style and soon come to appreciate the use of precise routine terminology. This introduction to formal style pays substantial dividends in upper-level courses where students are expected to consult the original literature.

The philosophy of this text is to focus the student, to a very large extent, on the *experimental* aspects of organic chemistry. We have purposely attempted to keep to a minumum the theoretical discussions, supplying only sufficient background material to cover potential discontinuities between lecture and laboratory. We want the student to become comfortable, and to develop a substantial degree of independence with the use of chemical instrumentation. To meet this objective, we have centered attention specifically on infrared spectroscopy and gas chromatography. We do this because we feel that these basic instrumental techniques can be utilized to a much larger extent by sophomores than the more expensive techniques.

We recognize the dominant role of nuclear magnetic resonance in modern organic chemistry. The current real-life situation is, however, that only in a very limited number of cases do second-year students ever have the opportunity to generate their own data with this instrumentation. Because we are concerned primarily with focusing student excitement and interest on gathering actual laboratory data, we feel that artificially incorporating material from outside the laboratory as a means of including the nuclear magnetic resonance or mass spectroscopy experiment may not enhance but divert attention away from the laboratory experience. We are seriously exploring ways to overcome this particular problem and hope to address this issue successfully in future editions.

The six-year road to completion of this text has been a long, but rewarding one. Many individuals have made vital contributions. Arnold Brossi, David Brooks, Miles Pickering, Lea Clapp, Eugene Cordes, and Henry Horner provided encouragement and sound advice at crucial points along the way.

We gratefully acknowledge that the large majority of the infrared and nuclear magnetic resonance spectra not recorded at Bowdoin College were obtained from the Aldrich Libraries of FT-IR and NMR Spectra through the courtesy of Dr. Charles Pouchert and the Aldrich Chemical Company.

John Ryan, Larry Reilly, Don Sellars, and Hugh Bowie of Ace Glass are primarily responsible for the development of the novel microglassware. They exhibited considerable patience throughout the ordeal. Ed Hollenbach and Lyle Phifer of Chem Services saw the advantage of making available small quantities of high-purity reagents and starting materials as a way of ensuring success at the microscale level.

The vast amount of experimental development was shared by Teaching Research Fellows Janet Hotham, David Butcher, Paulette Fickett, and Caroline Foote, plus a number of Bowdoin students—Mark Bowie, Sandy Hebert, Rob Hinkle, Marcia Meredith, and Gregory Merklin. The experimental ground covered by this group, much of which required ingenious solutions, is remarkable. The breadth of experiments available is a tribute to their dedication. Janet deserves a special thanks. A Merrimack College graduate adopted by Bowdoin College, she has been with the program almost from the beginning. Her thoughtful suggestions based on experience in the trenches and her willing contributions in any area of need at any time are most gratefully remembered. The enthusiasm of the sophomore organic students who volunteered for the initial pilot sections at Bowdoin and the field testing sections at Merrimack played a key role in encouraging us to continue these efforts.

Judy Foster's talent with a Macintosh enabled her to generate the majority of the illustrations and reaction schemes. Judy's efforts have greatly enhanced our descriptions of the techniques involved in microscale work. Her ingenuity also led to the pictorial keying of the equipment setups. The patience, understanding, and thoughtful advice of Dennis Sawicki, Chemistry Editor for Wiley, has been particularly valuable.

We thank Dean Alfred Fuchs for his constant encouragement of this program during its development. The initial exploratory work was supported by a grant from Bowdoin College and a department grant from the du Pont Company. A semester leave given to D.W.M. (spring 1984) was funded by a grant from the ARCO Corporation. A semester sabbatic leave was granted to R.M.P. by Merrimack College (spring 1985), as was an appointment as Visiting Charles Weston Pickard Professor of Chemistry at Bowdoin College (1980–1981, spring 1984). The Surdna Foundation awarded two major grants which allowed the complete development and implementation of the program at Bowdoin College and the field testing of experiments at Merrimack College. The support and faith in this educational concept by these institutions is gratefully acknowledged.

**DANA W. MAYO**

March 1985
Brunswick, Maine

**RONALD M. PIKE**

**SAMUEL S. BUTCHER**

# About
the Authors

**Dana W. Mayo** holds the Charles Weston Pickard Research Professor of Chemistry Chair at Bowdoin College. A former Fellow of the School for Advanced Study at MIT and a Special Fellow of the National Institute of Health at the University of Maryland, he received his Ph.D. in Chemistry from Indiana University. Professor Mayo is Director of the Bowdoin College Summer Course in Infrared Spectroscopy. His research interests include the application of vibrational spectroscopy to molecular structure determination, natural products chemistry, and environmental studies of oil pollution.

**Ronald M. Pike** is Professor Emeritus of Chemistry at Merrimack College and is currently Director of the National Microscale Chemistry Center (NMC²) at that institution. He received his Ph.D. from Massachusetts Institute of Technology. He is the author of numerous papers and patents in the area of silicone chemistry, is coauthor (with Szafran and Singh) of *Microscale Inorganic Chemistry: A Comprehensive Laboratory Experience* and (with Szafran and Foster) of *Microscale General Chemistry With Selected Macroscale Experiments*. He was previously associated with Union Carbide Corporation and the Lowell Technological Institute. He has been a Visiting Charles Weston Pickard Professor of Chemistry at Bowdoin College and a Visiting Professor at the U.S. Military Academy, West Point, NY.

**Peter K. Trumper** is an Assistant Professor of Chemistry at the University of Southern Maine. He received his B.A. from St. Olaf College and his Ph.D. from the University of Minnesota. He held an NIH postdoctoral fellowship at the University of Pennsylvania.

**Dana Mayo** and **Ronald Pike** jointly received the James Flack Norris Award (1988) for outstanding achievement in the teaching of chemistry, and the

John A. Timm Award (1987) for their work in developing the microscale instructional program. They also received (DM, 1989; RP, 1990) the Catalyst National Award of the Chemical Manufacturers Association for excellence in chemistry teaching. Together with Samuel Butcher, they were corecipients of the first Charles A. Dana Foundation Award (1986) for pioneering Achievement in Health and Higher Education and the American Chemical Society Division of Chemical Health and Safety Award (1987). They are also the coauthors together with Samuel Butcher and Peter Trumper of *Microscale Techniques For The Organic Laboratory*.

# Contents

CHAPTER **6**
MICROSCALE ORGANIC
LABORATORY EXPERIMENTS
119

**EXPERIMENT [7$_{adv}$]    Preparation of an Enol Acetate:
Cholesta-3,5-dien-3-ol Acetate**                                          453

CHAPTER  **8**
SEQUENTIAL SYNTHESES:
THE TRANSITION FROM
MACRO TO MICRO
459

**CHAPTER 9**
**SPECTROSCOPIC
IDENTIFICATION OF
ORGANIC COMPOUNDS**
585

CHAPTER **10**
QUALITATIVE
IDENTIFICATION OF
ORGANIC COMPOUNDS
693

APPENDIX **A**
# TABLES OF DERIVATIVES
729

APPENDIX **B**
# CHAPTERS 6, 7 AND 8: EXPERIMENTS CLASSIFIED BY MECHANISM
743

# GLOSSARY
751

# INDEX
755

# Contents of Biographies and Essays

# 1

# Introduction

**CH$_4$, Methane**
a substance of natural origin, known
as Marsh Gas to the alchemists.

You are breaking new ground in the organic chemistry laboratory!

Your course is going to be quite different from the conventional manner in which this laboratory has been taught. You will be learning the experimental side of organic chemistry from the microscale level. Surprising as it may seem, because you will be working with very small amounts of materials, you will be able to observe and learn more organic chemistry in one year than many of your predecessors did in nearly two years of laboratory work. You will find this laboratory an exciting and interesting place to be. While we cannot guarantee it for you individually, the majority of students who have been through the program during its development have found the microscale organic laboratory to be a pleasant adventure.

This text is centered on helping you develop skills in microscale organic laboratory techniques. Its focus is twofold. For those of you in the academic environment and involved with the introductory organic laboratory, it allows the flexibility of developing your own scaling sequence without being tied to a prescribed set of quantities. For those of you working in a research environment centered at the advanced undergraduate or graduate level or in the industrial area, this text will provide the foundation from which you can develop a solid expertise in microscale techniques directly applicable to your work. Working at the microscale level, though similar in a number of ways, is substantially different from the conventional operations carried out in the organic laboratory.

During the last decade, the experimental side of organic chemistry has begun a transformation to the microscale level. This conversion has been spurred on by the rapidly accelerating cost of chemical waste disposal. As we have said, surprising as it may seem, because you will be working with very small amounts of materials, the techniques that you will learn and experience will allow you to accomplish more organic chemistry in the long run than many of your predecessors.

1

At the very beginning, we want to acquaint you with the organization and contents of the text. We will then give you a few words of advice, which, if they are heeded, will allow you to avoid many of the sand traps you will find as you develop microscale laboratory techniques. Finally, we will wax philosophical and attempt to describe what we think you should derive from this experience.

After this brief introduction, the second chapter is concerned with safety in the laboratory. This chapter supplies information that will allow you to estimate your maximum possible exposure to volatile chemicals employed in the microscale laboratory. Chapter 2 also discusses general safety protocol for the laboratory. It is vitally important that you become familiar with the details of the material contained in this chapter; your health and safety depend on this knowledge.

The next three chapters are concerned primarily with the development of experimental techniques. Chapter 3 describes in detail the glassware employed in microscale organic chemistry: the logic behind its construction, tips on its usage, the common arrangements of equipment, and various other laboratory manipulations, including techniques for transferring microquantities of materials. Suggestions for the organization of your laboratory notebook are presented at the end of this chapter.

Chapter 4 deals with equipment and techniques for determining a number of physical properties of microscale samples. Chapter 5 is divided into nine technique sections. Detailed discussions develop the major areas of experimental technique that are used in the microscale organic laboratory.

Chapters 6–8 contain the main experimental sections of this text. Chapter 6 is focused on preparative organic chemistry at the microscale level and consists of 35 experiments. Additional selections of individual experiments can be drawn from those experiments presented in Chapter 8. Chapter 7 contains a series of seven experiments of a more sophisticated nature. A number of the experiments contained in Chapters 6 and 7 are of optional scale so that you may also have the opportunity to gain some experience with experimentation at the semimicroscale level. Chapter 8 consists of a set of six sequential experiments that are essentially identical to the type of problems tackled by research chemists involved in synthetic organic chemistry. A number of these multistep procedures begin the first step in the experiment with large scale, multigram quantities of starting material, but require microscale techniques to complete the final step or two. The use of this chapter is most appropriate in the final stages of the course, for example, the latter part of the second semester of a two-semester sequence.

Chapter 9 develops the characterization of organic materials at the microscale level by spectroscopic techniques. A detailed discussion of the interpretation of infrared (IR) group frequencies and nuclear magnetic resonance (NMR) spectral data make up a significant portion of this chapter, which also includes a brief discussion of ultraviolet-visible (UV–vis) spectroscopy. An introduction to the theoretical basis for these spectroscopic techniques is also presented.

Chapter 10 develops the characterization of organic materials at the microscale level by the use of classical organic reactions to form solid derivatives. Tables of derivative data for use in compound identification by these techniques are discussed, and are included in Appendix A. A list of experiments grouped by reaction mechanism is presented in Appendix B.

The organization of the experimental procedures given in Chapters 6–8 is arranged in the following fashion. A short opening statement describing the reaction to be studied is followed by the reaction scheme.

Generally, a brief discussion of the reaction follows, including a mechanistic interpretation. In a few cases of particularly important reactions, or where the experiment is likely to precede presentation of the topic in the classroom, a more detailed description is given. The estimated time needed to complete the work, and a table of reactant data come next. For ease in organizing your laboratory time, the experimental section is divided into **four** subsections: *reagents and equipment, reaction conditions, isolation of product,* and *purification and characterization.*

We then introduce a series of questions and problems designed to enhance and focus your understanding of the chemistry and the experimental procedures involved in a particular laboratory exercise. Finally, a bibliography offering a list of literature references is given. Although this list comes at the end of the experimental section, we view it as a very important part of the text. The discussion of the chemistry involved in each experiment is necessarily brief. We hope that you will take time to read and expand your knowledge about the particular experiment that you are conducting. You may, in fact, find that some of these references become assigned reading.

The experimental apparatus and materials involved at important stages of an experiment are indicated by a prompt (■) in the text and are shown in the margin. Important comments are italicized in the text, and **Warnings** and **Cautions** are given in boxes.

## GENERAL RULES FOR THE MICROSCALE LABORATORY

**1.** *Study the experiment before you come to lab.* This rule is a historical plea from all laboratory instructors. In the microscale laboratory it takes on a more important meaning. You will not survive if you do not prepare ahead of time. In microscale experiments, operations happen much more quickly than in the macroscale laboratory. Your laboratory time will be overflowing with many more events. If you are not familiar with the sequences you are to follow, you will be in deep trouble. Although the techniques employed at the microscale level are not particularly difficult to acquire, they do demand a significant amount of attention. For you to reach a successful and happy conclusion, you cannot afford to have the focus of your concentration broken by having to constantly refer to the text during the experiment. Disaster is ever present for the unprepared.

**2.** *ALWAYS work with clean equipment.* You must take the time to scrupulously clean your equipment before you start any experiment. Contaminated glassware will ultimately cost you additional time, and you will face the frustration of experiencing inconsistent results and lower yields. Dirty equipment is the primary cause of reaction failure at the microscale level.

**3.** *CAREFULLY measure the quantities of materials to be used in the experiments.* A little extra time at the beginning of the laboratory can speed you on your way at the end of the session. A great deal of time has been spent optimizing the conditions employed in these experiments in order to maximize yields. Many organic reactions are very sensitive to the relative quantities of substrate (the material on which the reaction is taking place) and reagent (the reactive substance or substances that bring about the change in the substrate). After equipment contamination, the second-largest cause of failed reactions is attempting to run a reaction with incorrect quantities of the reactants present. Do not be hurried or careless at the balance.

**4.** *Clean means DRY.* Water or cleaning solution can be as detrimental to the success of a reaction as dirt or sludge in the system. You often will be working with very small quantities of moisture-sensitive reagents. The glass surface areas with which these reagents come in contact, however, are relatively large. A slightly damp piece of glassware can rapidly deactivate a critical reagent and result in reaction failure. *This rule must be strictly followed.*

**5.** *ALWAYS work on a clean laboratory bench surface, preferably glass!*

**6.** *ALWAYS protect the reaction product* that you are working with from a disastrous spill by carrying out all solution or solvent transfers over a crystallizing dish.

**7.** *ALWAYS place reaction vials or flasks in a clean beaker* when standing them on the laboratory bench. Then, when a spill occurs the material is more likely to be contained in the beaker and less likely to be found on the laboratory bench or floor.

**8.** *NEVER use cork rings to support round-bottom flasks*, particularly if they contain liquids. You are inviting disaster to be a guest at your laboratory bench.

**9.** *ALWAYS think through the next step* you are going to perform *before* starting it. Once you have added the wrong reagent, it is back to square one.

**10.** *ALWAYS save everything* you have generated in an experiment until it is successfully completed. You can retrieve a mislabeled chromatographic fraction from your locker, but not from the waste container!

## THE ORGANIC CHEMISTRY LABORATORY

The confidence gained by mastering the microscale techniques described here will pay big dividends as you progress into modern-day experimental chemistry. Historically, the organic laboratory has had a reputation of being smelly, long, tedious, and pockmarked with fires and explosions. Modern organic chemistry still has trouble shaking this image, but present-day organic chemistry is undergoing a revolution at the laboratory bench. New techniques are sweeping away many of the old complaints, as an increasing fraction of industrial and academic research is being carried out at the microscale level.

This book allows the interested participant to rapidly develop the skills needed to slice more deeply into organic chemistry than ever before. The attendant benefits are greater confidence and independence in acquired laboratory techniques. The happy result is that in a microscale-based organic chemistry laboratory, you are more likely to have a satisfying encounter with the experimental side of this fascinating field of knowledge.

# 2

# Safety

**C₂H₄, Ethylene**
a substance of natural origin,
released by ripening fruit.

Research laboratories vary widely with respect to facilities and support given to safety. Large laboratories may have several hundred chemists and an extensive network of co-workers, supervisors, safety officers, and hazardous waste managers. In small laboratories, the chemist may be left pretty much alone to work things out. Large laboratories may have an extensive set of safety procedures and detailed practices for the storage of hazardous wastes. The individual chemist in the small laboratory may have to take care of all of those things. Some laboratories may routinely deal with very hazardous materials and may conduct all chemistry in hoods. Others may deal mainly with relatively innocuous compounds and have very limited hood facilities.

Our approach is to raise some questions for the individual to think about and to suggest places to look for further information. This chapter will not present a large list of safety precautions for use in all situations. We do present a list of very basic precautionary measures. A bibliography offers a list of selected references at the end of the chapter. Many laboratories may have safety guidelines that will supersede this very cursory treatment. This chapter is no more than a starting point.

Murphy's law states in brief, "If anything can go wrong, it will." Although it is often taken to be a silly law, it is not. Murphy's law means that if sparking switches are present in areas that contain flammable vapors, sooner or later there will be a fire. If the glass container can move to the edge of the shelf as items are moved around or because the building vibrates, it will come crashing to the floor at some time. If the pipet can become contaminated, then the mouth pipetter is going to ingest a contaminant sometime.

**MAKING THE LABORATORY
A SAFER PLACE**

We cannot revoke Murphy's law, but we can do a lot to minimize the damage. We can reduce the incidence of sparks and flames and flammable vapors. We can make sure that if the accident does occur, we have the means to contain the damage and take care of any injuries that result. All this means thinking about the laboratory environment. Think about what can go wrong and then do what can be done to minimize the chance of an accident and be prepared to respond if one does occur.

## Nature of Hazards

The chemistry laboratory presents a wide assortment of risks. These risks are outlined briefly here so that you can begin to think about the steps necessary to make the laboratory safer.

**1.** *Physical hazards.* Injuries resulting from flames, explosions, and equipment (cuts from glass, electrical shock from faulty instrumentation, or improper use of instruments).

**2.** *External exposure to chemicals.* Injuries to skin and eyes resulting from contact with chemicals that have splashed or have been left on the bench top or on equipment.

**3.** *Internal exposure.* Longer term (usually) health effects resulting from breathing hazardous vapors or ingesting chemicals.

## Reduction of Risks

Many things can be done to reduce risks. The rules below may be absolute in some laboratories. In others, the nature of the materials and apparatus used may justify the relaxation of some of these rules or the addition of others.

**1.** *Stick to the procedures described by your supervisor.* This attention to detail is particularly important for the chemist with limited experience. In other cases, variation of the reagents and techniques may be part of the work.

**2.** *Wear approved safety goggles.* We can often recover quickly from injuries affecting only a few square millimeters on our bodies, *unless* that area happens to be in our eyes. Often, larger industrial laboratories require that laboratory work clothes and safety shoes be worn. *Wear them,* if requested.

**3.** *Do not put anything in your mouth while in the laboratory.* This rule includes food, drinks, and pipets. There are countless ways that surfaces can become contaminated in the laboratory. Since there are substances that must *never* be pipetted by mouth, one should get into the habit of *not* mouth pipetting anything.

**4.** *Be cautious with flames and flammable solvents.* Remember that the flame at one end of the bench can ignite the flammable liquid at the other end in the event of a spill or improper disposal. Flames must never be used when certain liquids are present in the laboratory, and flames must always be used with care.

**5.** *Be sure that you have the proper chemicals for your reaction.* Check labels carefully, and return unused chemicals to the proper place for storage or disposal.

**6.** *Minimize the loss of chemicals to air or water and dispose of waste properly.* Some water-soluble materials may be safely disposed of in the water drains. Other wastes should go into special receptacles. Pay attention to the labels on these receptacles.

**7.** *Minimize skin contact with any chemicals.* Use impermeable gloves, when necessary, and wash any chemical off your body promptly. If you have to wash something off with water, use lots of it. Be sure that you know where the nearest water spray device is located.

**8.** *Tie back, or otherwise confine, long hair and loose items of clothing.* You do not want them falling into a reagent or getting near flames.

**9.** *Do not work alone.* Too many things can happen to a person working alone that might leave him/her unable to obtain assistance. In the rare event that you *are* working alone, be sure that someone checks on you at regular intervals.

**10.** *Exercise care in assembling glass and electrical apparatus.* Separating standard taper glassware, as well as other operations with glass, all involve the risk that the glass may break and that lacerations or punctures may result. Seek help or advice with glassware, if necessary. Electrical shock can result in many ways. When making electrical connections, make sure that your hands, the laboratory bench, and the floor are all dry and that *you* do not complete an electrical path to ground. Be sure that electrical equipment is properly grounded and insulated.

**11.** *Report any injury or accident to the appropriate person.* Reporting injuries and accidents is important so that medical assistance can be obtained, if necessary. It is also essential for others to be made aware of any safety problems. These problems may be correctable.

**12.** *Keep things clean.* Put unused apparatus away. Immediately wipe up or care for spills on the bench top or on the floor.

**13.** *Attend safety programs* as requested. Many laboratories offer excellent seminars and lectures on a wide variety of safety topics. *Pay careful attention* to the advice and counsel of the *safety officer.*

## Precautionary Measures

Locate the nearest

- Fire extinguisher
- Eye wash
- Spray shower
- Fire blanket
- Exit

Know who to call (have the numbers posted) for

- Fire
- Medical emergency
- Spill or accidental release of chemicals

Know where to go

- In case of injury
- To evacuate the building

## THINKING ABOUT THE RISKS IN USING CHEMICALS

The smaller quantities used in the microscale laboratory carry with them a reduction in hazards caused by fires and explosions. Hazards associated with skin contact are also reduced; however, care must be exercised when working in close proximity to even the small quantities involved.

There is a great potential for reducing the exposure to chemical vapors, but these reductions will be realized only if everyone in the laboratory is careful. One characteristic of vapors emitted outside hoods is that they mix rapidly throughout the lab and will quickly reach the nose of the person on the other side of the room. For this reason some operations may have to be performed in hoods. In some laboratories, the majority of the reactions may be carried out in hoods. When reactions are carried out in the open laboratory, each experimenter becomes a "polluter" whose emissions affect the people nearby the most, but these emissions become added to the laboratory air and to the burden each of us must bear.

The concentration of vapor in the general laboratory air space depends on the vapor pressure of the liquids, the area of the solid or liquid exposed, the nature of air currents near the sources, and the ventilation characteristics of the laboratory. One factor over which each individual has control is evaporation, which can be reduced by the following practices.

- Certain liquids must remain in hoods.
- Reagent bottles must be recapped when not in use.
- Spills must be quickly cleaned up and the waste discarded.

## Storage of Chemicals

Chemicals must be properly stored when not in use. Some balance must be struck between the convenience of having the compound in the laboratory where you can easily put your hands on it and the safety of having the compound in a properly ventilated and fire-safe storage room. Policies for storing chemicals will vary from place to place. There are limits to the amounts of flammable liquids that should be stored in glass containers, and fire-resistant cabinets must be used for storage of large amounts of flammable liquids. Chemicals that react with one another should not be stored in close proximity. There are plans for sorting chemicals by general reactivity classes in storerooms; for instance, Flinn Scientific Company (1989) includes a description of a storage system with their chemical catalog.

## Disposal of Chemicals

Chemicals must also be segregated into categories for disposal. The categories used will depend on the disposal services available, and upon federal, state, and local regulations. For example, some organic wastes are readily incinerated, while those containing chlorine may require much more costly treatment. Other wastes may have to be buried. For safety and economic reasons, it is important to place waste material in the appropriate container.

## Material Safety Data Sheets

Although risks are associated with the use of most chemicals, the magnitudes of these risks vary greatly. A short description of the risks is provided by a Material Safety Data Sheet, commonly referred to as an **MSDS**. These sheets are normally provided by the manufacturer or vendor of the chemical, and users are required to keep files of the MSDS of each material stored or used.

As an example, the MSDS for acetone is shown here. This sheet is provided by the J. T. Baker Chemical Company. Sheets from other sources will be very similar. Much of the information on these sheets is self-explanatory, but let us review the major sections of the acetone example.

Section I provides identification numbers and codes for the compound and includes a summary of the risks associated with the use of acetone. Because these sheets are available for many thousands of compounds and mixtures, there must be a means of unambiguously identifying the substance. A standard reference number for chemists is the Chemical Abstract Service Number, or CAS No.

J. T. BAKER CHEMICAL CO.  222 RED SCHOOL LANE, PHILLIPSBURG, NJ  08865
M A T E R I A L   S A F E T Y   D A T A   S H E E T
24-HOUR EMERGENCY TELEPHONE — (201) 859-2151
CHEMTREC # (800) 424-9300 — NATIONAL RESPONSE CENTER # (800) 424-8802

A0446 –01                          ACETONE                          PAGE: 1
EFFECTIVE: 10/11/85                                          ISSUED: 01/23/86

---

### SECTION I – PRODUCT IDENTIFICATION

PRODUCT NAME:    ACETONE
FORMULA:         (CH3)2CO
FORMULA WT:      58.08
CAS NO.:         00067-64-1
NIOSH/RTECS NO.: AL3150000
COMMON SYNONYMS: DIMETHYL KETONE;  METHYL KETONE;  2-PROPANONE
PRODUCT CODES:   9010,9006,9002,9254,9009,9001,9004,5356,A134,9007,9005,9008

### PRECAUTIONARY LABELLING

BAKER SAF-T-DATA(TM) SYSTEM

              HEALTH       –  1
              FLAMMABILITY –  3  (FLAMMABLE)
              REACTIVITY   –  2
              CONTACT      –  1

LABORATORY PROTECTIVE EQUIPMENT

SAFETY GLASSES; LAB COAT; VENT HOOD; PROPER GLOVES; CLASS B EXTINGUISHER

PRECAUTIONARY LABEL STATEMENTS

                    DANGER
              EXTREMELY FLAMMABLE
          HARMFUL IF SWALLOWED OR INHALED
                CAUSES IRRITATION
KEEP AWAY FROM HEAT, SPARKS, FLAME.  AVOID CONTACT WITH EYES, SKIN, CLOTHING.
AVOID BREATHING VAPOR.  KEEP IN TIGHTLY CLOSED CONTAINER.  USE WITH ADEQUATE
VENTILATION.  WASH THOROUGHLY AFTER HANDLING.  IN CASE OF FIRE, USE WATER SPRAY,
ALCOHOL FOAM, DRY CHEMICAL, OR CARBON DIOXIDE.  FLUSH SPILL AREA WITH WATER
SPRAY.

### SECTION II – HAZARDOUS COMPONENTS

| COMPONENT | % | CAS NO. |
|---|---|---|
| ACETONE | 90-100 | 67-64-1 |

### SECTION III – PHYSICAL DATA

| | | |
|---|---|---|
| BOILING POINT: | 56 C (  133 F) | VAPOR PRESSURE(MM HG):  181 |
| MELTING POINT: | –95 C (  –139 F) | VAPOR DENSITY(AIR=1):  2 |
| SPECIFIC GRAVITY: 0.79 (H2O=1) | | EVAPORATION RATE:  5.6 (BUTYL ACETATE=1) |

SOLUBILITY(H2O):    COMPLETE (IN ALL PROPORTIONS) % VOLATILES BY VOLUME: 100

APPEARANCE & ODOR:  CLEAR, COLORLESS LIQUID WITH FRAGRANT SWEET ODOR.

### SECTION IV – FIRE AND EXPLOSION HAZARD DATA

FLASH POINT:       –18 C (   0 F)    NFPA 704M RATING:  1-3-0

FLAMMABLE LIMITS:  UPPER –  13 %     LOWER –  2 %

FIRE EXTINGUISHING MEDIA
      USE ALCOHOL FOAM, DRY CHEMICAL OR CARBON DIOXIDE.
      (WATER MAY BE INEFFECTIVE.)

SPECIAL FIRE-FIGHTING PROCEDURES
      FIREFIGHTERS SHOULD WEAR PROPER PROTECTIVE EQUIPMENT AND SELF-CONTAINED
      (POSITIVE PRESSURE IF AVAILABLE) BREATHING APPARATUS WITH FULL FACEPIECE.
      MOVE EXPOSED CONTAINERS FROM FIRE AREA IF IT CAN BE DONE WITHOUT RISK.
      USE WATER TO KEEP FIRE-EXPOSED CONTAINERS COOL.

UNUSUAL FIRE & EXPLOSION HAZARDS
      VAPORS MAY FLOW ALONG SURFACES TO DISTANT IGNITION SOURCES AND FLASH BACK.
      CLOSED CONTAINERS EXPOSED TO HEAT MAY EXPLODE. CONTACT WITH STRONG
      OXIDIZERS MAY CAUSE FIRE.

### SECTION V – HEALTH HAZARD DATA

THRESHOLD LIMIT VALUE (TLV/TWA):  1780 MG/M3 (   750 PPM)

SHORT-TERM EXPOSURE LIMIT (STEL): 2375 MG/M3 (  1000 PPM)

TOXICITY:    LD50 (ORAL-RAT)(MG/KG)      –  9750
             LD50 (IPR-MOUSE)(G/KG)      –  1297

EFFECTS OF OVEREXPOSURE
CONTACT WITH SKIN HAS A DEFATTING EFFECT, CAUSING DRYING AND IRRITATION.
OVEREXPOSURE TO VAPORS MAY CAUSE IRRITATION OF MUCOUS MEMBRANES, DRYNESS
OF MOUTH AND THROAT, HEADACHE, NAUSEA AND DIZZINESS.

EMERGENCY AND FIRST AID PROCEDURES
CALL A PHYSICIAN.
IF SWALLOWED, IF CONSCIOUS, IMMEDIATELY INDUCE VOMITING.
IF INHALED, REMOVE TO FRESH AIR.  IF NOT BREATHING, GIVE ARTIFICIAL
RESPIRATION.  IF BREATHING IS DIFFICULT, GIVE OXYGEN.
IN CASE OF CONTACT, IMMEDIATELY FLUSH EYES WITH PLENTY OF WATER FOR AT
LEAST 15 MINUTES.  FLUSH SKIN WITH WATER.

### SECTION VI – REACTIVITY DATA

STABILITY: STABLE                                     HAZARDOUS POLYMERIZATION:   WILL NOT OCCUR

CONDITIONS TO AVOID:      HEAT, FLAME, SOURCES OF IGNITION

INCOMPATIBLES:             SULFURIC ACID, NITRIC ACID, STRONG OXIDIZING AGENTS

### SECTION VII – SPILL AND DISPOSAL PROCEDURES

STEPS TO BE TAKEN IN THE EVENT OF A SPILL OR DISCHARGE
WEAR SUITABLE PROTECTIVE CLOTHING.  SHUT OFF IGNITION SOURCES; NO FLARES,
SMOKING, OR FLAMES IN AREA.  STOP LEAK IF YOU CAN DO SO WITHOUT RISK.  USE
WATER SPRAY TO REDUCE VAPORS.  TAKE UP WITH SAND OR OTHER NON-COMBUSTIBLE
ABSORBENT MATERIAL AND PLACE INTO CONTAINER FOR LATER DISPOSAL.  FLUSH
AREA WITH WATER.

J. T. BAKER SOLUSORB(R) SOLVENT ADSORBENT IS RECOMMENDED
FOR SPILLS OF THIS PRODUCT.

DISPOSAL PROCEDURE
DISPOSE IN ACCORDANCE WITH ALL APPLICABLE FEDERAL, STATE, AND LOCAL
ENVIRONMENTAL REGULATIONS.

EPA HAZARDOUS WASTE NUMBER:          U002 (TOXIC WASTE)

### SECTION VIII – PROTECTIVE EQUIPMENT

VENTILATION:                USE GENERAL OR LOCAL EXHAUST VENTILATION TO MEET
TLV REQUIREMENTS.

RESPIRATORY PROTECTION:    RESPIRATORY PROTECTION REQUIRED IF AIRBORNE
CONCENTRATION EXCEEDS TLV.  AT CONCENTRATIONS UP
TO 5000 PPM, A GAS MASK WITH ORGANIC VAPOR
CANNISTER IS RECOMMENDED.  ABOVE THIS LEVEL, A
SELF-CONTAINED BREATHING APPARATUS WITH FULL FACE
SHIELD IS ADVISED.

EYE/SKIN PROTECTION:       SAFETY GLASSES WITH SIDESHIELDS, POLYVINYL ACETATE
GLOVES ARE RECOMMENDED.

### SECTION IX – STORAGE AND HANDLING PRECAUTIONS

SAF-T-DATA(TM) STORAGE COLOR CODE:     RED

SPECIAL PRECAUTIONS
BOND AND GROUND CONTAINERS WHEN TRANSFERRING LIQUID.  KEEP CONTAINER
TIGHTLY CLOSED.  STORE IN A COOL, DRY, WELL-VENTILATED, FLAMMABLE LIQUID
STORAGE AREA.

### SECTION X – TRANSPORTATION DATA AND ADDITIONAL INFORMATION

DOMESTIC (D.O.T.)

PROPER SHIPPING NAME     ACETONE
HAZARD CLASS             FLAMMABLE LIQUID
UN/NA                    UN1090
LABELS                   FLAMMABLE LIQUID

INTERNATIONAL (I.M.O.)

PROPER SHIPPING NAME     ACETONE
HAZARD CLASS             3.1
UN/NA                    UN1090
LABELS                   FLAMMABLE LIQUID

(TM) AND (R) DESIGNATE TRADEMARKS.
N/A = NOT APPLICABLE OR NOT AVAILABLE

THE INFORMATION PUBLISHED IN THIS MATERIAL SAFETY DATA SHEET HAS BEEN COMPILED
FROM OUR EXPERIENCE AND DATA PRESENTED IN VARIOUS TECHNICAL PUBLICATIONS. IT IS
THE USER'S RESPONSIBILITY TO DETERMINE THE SUITABILITY OF THIS INFORMATION FOR
THE ADOPTION OF NECESSARY SAFETY PRECAUTIONS. WE RESERVE THE RIGHT TO REVISE
MATERIAL SAFETY DATA SHEETS PERIODICALLY AS NEW INFORMATION BECOMES AVAILABLE.

A quick review of the degree of risks is given by the numerical scale under Precautionary Labeling. This particular scale is a proprietary scale that ranges from 0 (very little or nonexistent risk) to 4 (extremely high risk). The National Fire Protection Association (NFPA) uses a similar scale, but the risks considered are different. Other systems may use different scales, and there are some that represent low risks by the highest number! Be sure that you understand the scale being used. Perhaps some day, one scale will become standard.

Section II covers risks from mixtures. Because a mixture is not considered here, the section is empty. Selected physical data are described in Section III. Section IV contains fire and explosion data, including a description of the toxic gases produced when the compound is exposed to a fire. The MSDSs are routinely made available to fire departments that may be faced with fighting a fire in a building where large amounts of chemicals are stored.

Health hazards are described in Section V. The entries of most significance for evaluating risks from vapors are the Threshold Limit Value (or TLV) and the Short-Term Exposure Limit (STEL). The TLV is a term used by the American Conference of Governmental Industrial Hygienists (ACGIH). This organization examines the toxicity literature for a compound and establishes the TLV. This standard is designed to protect the health of workers exposed to the vapor 8 hours/day, 5 days a week. The Occupational Safety and Health Administration (OSHA) adopts a value to protect the safety of workplaces in the United States. Their value is termed the Time Weighted Average (TWA) and in many cases is numerically equal to the TLV. The STEL is a value not to be exceeded for even a 15-min averaging time. The TLV/TWA and STEL values for many chemicals are also summarized in a small handbook available from the ACGIH; they are also collected in the *CRC Handbook of Chemistry and Physics*.

The toxicity of acetone is also described in terms of the toxic oral dose. In this case, the $LD_{50}$ is the dose that will cause the death of 50% of the mice or rats given that dose. The dose is expressed as milligrams of acetone per kilogram of body weight of the subject animal. The figures for small animals are often used to estimate the effects on humans. If, for example, we used the mouse figure of 1297 mg/kg and applied it to a 60-kg chemist, a dose of 77,820 mg ($\sim$ 98.5 mL) would kill 50% of the subjects receiving that dose. As a further example, chloroform has an $LD_{50}$ of 80 mg/kg. For our 60-kg chemist, a dose of 4800 mg ($\sim$ 3 mL) would be fatal in 50% of these cases. The effects of exposure of skin to the liquid and vapor are also described.

Section VI describes the reactivity of acetone and the classes of compounds with which it should not come in contact. For example, sodium metal reacts violently with a number of substances (including water) and should not come in contact with them. Strong oxidizing agents (such as nitric acid) should not be mixed with organic compounds (among other things). The final sections (Sections VII–X) are self-explanatory.

## Estimating the Risks from Vapors

Other things (availability and suitability) being equal, one would, of course, choose the least toxic chemical for a given reaction. Some very toxic chemicals play very important roles in synthetic organic chemistry, and the toxicity of the chemicals in common use varies greatly. Bromine and benzene have TLVs of 0.7 and 30 mg/m$^3$, respectively, and are at the more toxic end of the spectrum of routinely used chemicals. Acetone has a TLV of 1780 mg/m$^3$. These representative figures do not mean that acetone is "harmless" or that bromine cannot be used. In general, one should exercise care at all times (make a habit of good laboratory practice) and should take special precautions when working with highly toxic materials.

The TLV provides a simple means to evaluate the relative risk of exposure to the vapor of any substance used in the laboratory. If the quantity of the material evaporated is represented by $m$ (in milligrams per hour) and the TLV is expressed by $L$ (in milligrams per cubic meter), a measure of relative risk to the vapor is given by $m/L$. This quantity actually represents the volume of clean air required to dilute the emissions to the TLV. As an example, the emission of 1 g of bromine and 10 g of acetone in 1 hour lead to the values of $m/L$ of 1400 m$^3$/hour for the bromine and 5.6 m$^3$/hour for acetone. These numbers provide a direct handle on the *relative* risks from these two vapors. It is difficult to assess the absolute risk to these vapors without a lot of information about the ventilation characteristics of the laboratory. If these releases occur within a properly operated hood, the threat to the worker in the laboratory is probably very small. (However, consideration must be given to the hood exhaust.)

Exposure in the general laboratory environment can be assessed if we assume that the emissions are reasonably well mixed before they are inhaled, and if we know something about the room ventilation rate. The ventilation rate of the room can be measured in a number of ways (Butcher et al.). Given the ventilation rate, it might be safe to assume that only 30% of that air is available for diluting the emissions. (This accounts for imperfect mixing in the room.) The effective amount of air available for dilution can then be compared with the amount of air required to dilute the chemical to the TLV.

Let us continue our example. Suppose that the laboratory has a volume of 75 m$^3$ and an air exchange rate of two air changes per hour. This value means that (75 m$^3$)(2/hour)(0.3) = 45 m$^3$/hour are available to dilute the pollutants. There may be enough margin for error to reduce the acetone concentration to a low level (5.6 m$^3$/hour are required to reach the TLV); use of bromine should be restricted to the hood. An assessment of the accumulative risk of several chemicals is obtained by adding the individual $m/L$ $\left(\frac{\text{mg/hour}}{\text{mg/m}^3}\right)$ values.

The $m/L$ figures may also be used to assess the relative risk of performing the experiment outside a hood. As $m/L$ represents the volume of air required for *each* student, this may be compared with the volume of air actually available for each student. If the ventilation rate for the entire laboratory is $Q$, (in cubic meters per min) for a section of $n$ students meeting for $t$ min, the volume for each student is $kQt/n$ cubic meters. Here $k$ is a mixing factor that allows for the fact that the ventilation air will not be perfectly mixed in the laboratory before it is exhausted. In a reasonable worst-case mixing situation a $k$ value of 0.3 seems reasonable. Laboratories with modest ventilation rates supplied by 15–20 linear feet of hoods can be expected to provide 30–100 m$^3$ per student over a 3-hour laboratory period if the hoods are working properly. Let us take the figure of 50 m$^3$ per student as an illustration. If the value of $m/L$ for a compound (or a group of compounds in a reaction) is substantially less than 50 m$^3$, it may be safe to do that series of operations in the open laboratory. If $m/L$ is comparable to or greater than 50 m$^3$, a number of options are available. (1) Steps using that compound may be restricted to a hood. (2) The instructional staff may satisfy themselves that much less than the assumed value is actually evaporated under conditions present in their laboratory. (3) The number of individual repetitions of this experiment may be reduced. The size of the laboratory section can be reduced or the experiment may be done in pairs or trios.

Conducting reactions in a hood does not automatically convey a stamp of safety. Hoods are designed to keep evaporating chemicals from entering the general laboratory space. For hoods to do their job, there must be an adequate flow of air into the hood, and this air flow must not be disturbed by turbulence at the hood face. A frequently used figure of merit for hood operation is the face velocity of 100 ft/min. This rate is an average velocity of air entering the hood opening. (There are instruments available for measuring this flow rate in the catalogs of major equipment suppliers. Prices range from less than $50 to several hundred dollars.) Even with a face velocity of 100 ft/min, vapors can be drawn out of an improperly designed hood simply by people walking by the opening, or by drafts from open windows.

Hood performance should be checked at regular intervals. The face velocity will increase as the front hood opening is decreased. If an adequate face velocity cannot be maintained with a front opening height of 15 cm, use of the hood for carrying out reactions will be limited. A low face velocity may indicate that the fans and ductwork need cleaning, that the exhaust system leaks (if it operates under lower than ambient pressure), or that the supply of make-up air is not adequate. When the hood system is properly maintained, the height of the hood opening required to provide an adequate face velocity is often indicated with a sticker.

Hoods are often used for storage of volatile compounds. A danger in this practice is that the hood space can become quickly cluttered, making work in the hood difficult, and the air flow may be disturbed. Of course, hoods used for storage should not be turned off.

**Concluding Thoughts**

This brief chapter only touches a few of the important points concerning laboratory safety. The risk from vapor exposure is discussed in some detail, but other risks are treated only briefly. Applications in some laboratories may involve reactions with a risk from radiation or infection or may involve compounds that are unstable with respect to explosion. The chemist must be aware of the potential risks and must be prepared to go to an appropriate and detailed source of information, as needed.

**QUESTIONS**

**2-1.** Think about what you would do in the following cases. (Note that you may need more information for some of these problems.)
  a. A hot solution "bumps," splashing your face.
  b. A beaker of solvent catches fire.
  c. A reagent bottle falls, spilling concentrated sulfuric acid ($H_2SO_4$).
  d. You hear a sizzle as you pick up a hot test tube.

**2-2.** A laboratory has four hoods each of which is 39 in. wide. When the hood door is open to a height of 8 in. and the hoods are operating, the average air velocity through the hood face is 170 ft/min.
  a. Evaluate the total ventilation rate for this room assuming that there are no other exhausts.
  b. The laboratory is designed for use by 30 students. Evaluate the air available per student if the mixing factor is 0.3, and experiments last for 3 hours.
  c. An experiment is considered in which each student would be required to evaporate 7 mL of methylene chloride ($CH_2Cl_2$). Estimate the average concentration of methylene chloride. Look up the TLV

or the TWA for methylene chloride and consider how the evaporation might be performed.

**2-3.**  A laboratory has a ventilation system that provides 20 m$^3$ for each student during the laboratory period. (This figure includes the mixing factor.) An experiment is considered for this program that uses the following quantities of materials A, B, and C. The TLV is also listed for each compound.

| Substance | Quantity (mg) | TLV (mg/m$^3$) |
|-----------|---------------|----------------|
| A | 400 | 1200 |
| B | 500 | 200 |
| C | 200 | 5 |

Assess the relative risks of these three compounds. Is there a likely need for operations to be conducted in a hood if the compounds are assumed to be entirely evaporated?

**2-4.**  An experiment is considered in which 1 mL of diethylamine would be used by each student. The ventilation rate for the laboratory is 5 m$^3$/min. Look up the TLV (or TWA) for diethylamine [$(C_2H_5)_2NH$]. What restrictions might be placed on the laboratory to keep the average concentration, over a 3-hour period, less than one third of the TWA? Assume a mixing factor of 0.3.

## GENERAL SAFETY REFERENCES

Committee on Chemical Safety: *Safety in Academic Chemical Laboratories*, 3rd ed.; American Chemical Society: Washington, DC, 1979.

Furr, A. K. Jr., Ed. *Handbook of Laboratory Safety*, 3rd ed.; CRC Press: Boca Raton, FL, 1990.

Green, M. E.; Turk, A. *Safety in Working with Chemicals*; Macmillan: New York, 1978.

National Research Council: *Prudent Practices for Handling Chemicals in Laboratories*; National Academy Press: Washington, DC, 1981.

National Research Council: *Prudent Practices for Disposal of Chemicals from Laboratories*; National Academy Press; Washington, DC, 1983.

## BIBLIOGRAPHY

ACGIH *Threshold Limit Values for Chemical Substances and Physical Agents in the Work Environment*; Cincinnati, OH; 1984. Available from ACGIH; 6500 Glenway Avenue, Building D-5, Cincinnati, OH 45211.

Butcher, S. S.; Mayo, D. W.; Hebert, S. M.; Pike, R. M. *Laboratory Air Quality, Part I, J. Chem. Educ.* **1985**, *62*, A238 and *Laboratory Air Quality, Part II, J. Chem. Educ.* **1985**, *62*, A261.

Flinn Scientific Company, *Chemical Catalog/Reference Manual (1989)*. Available from Flinn Scientific Co., P. O. Box 219, Batavia, IL 60510.

Lide, D.R., Ed. *CRC Handbook of Chemistry and Physics*, 72nd ed., CRC Press: Boca Raton, FL, 1991–1992.

# 3

# Introduction to Microscale Organic Laboratory Equipment and Techniques

**C₃H₄, Cyclopropene**
Demyanov and Doyarenko (1922).

We begin this chapter with a description of the standard pieces of glassware that are employed in our microscale laboratory, many of which will be used to perform the experiments carried out in your laboratory assignments. This equipment, in one form or another, is now available from at least seven different scientific glassware manufacturers. Two principal suppliers are ACE Glass of Vineland, New Jersey and Corning Glass of Corning, New York. The following descriptions are helpful when considering what supplementary equipment may be required to carry out the experiments selected by your instructor, and what type of application is of specific interest. We next consider a series of standard experimental apparatus setups that utilize this equipment. At the end of each of these short discussions, you will find a listing of arrangements of the particular setup presented.

It should be pointed out that most of the equipment listed below is glassware that was originally developed and tested by students at Bowdoin and Merrimack Colleges for use in the undergraduate microscale organic laboratory programs at these institutions. It is believed that this particular collection of glassware offers significant advantages over earlier versions commercially available to both student and research laboratories. This conclusion is our opinion, but it is only an opinion. As you more deeply explore the experimental side of the microscale laboratory, you will find in the techniques you choose to employ a component that is very much *artistic* in nature. Two chemists may do exactly the same experiment and describe the experimental procedures in very nearly the same terms, but in fact, in the laboratory they may go about the work in different ways. This aspect of the experimental process is the *art* of doing chemical research. The esthetics of laboratory experimentation touch chemists in very personal ways. Thus, the equipment you find described in this text may be identical, similar, or only vaguely reminiscent of the components found in other texts on the subject. Historically, the equipment employed in the undergraduate instructional laboratory became an extension of that used in the chemical research laboratory. The development of the microscale instructional programs has reversed this trend by actually developing various types of instructional laboratory equipment that will be particularly useful to the modern organic research chemist. As in any laboratory work, the microscale laboratory is in large part an extension of our own personal research laboratories. We hope you find that you agree with us, and that in using this design, you will be comfortable learning the art of experimental microscale organic chemistry (see Figs. 3.1–3.7).

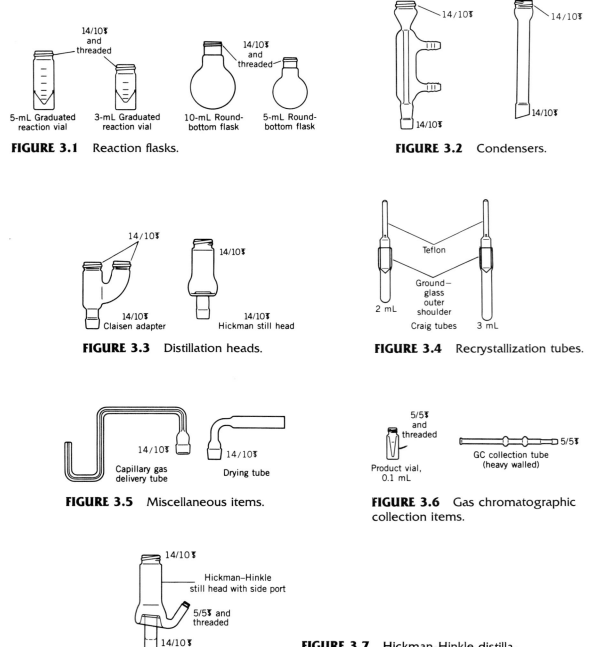

**FIGURE 3.1**   Reaction flasks.

**FIGURE 3.2**   Condensers.

**FIGURE 3.3**   Distillation heads.

**FIGURE 3.4**   Recrystallization tubes.

**FIGURE 3.5**   Miscellaneous items.

**FIGURE 3.6**   Gas chromatographic collection items.

**FIGURE 3.7**   Hickman-Hinkle distillation column.

Listed below are the items and quantities of glassware that we have installed in our laboratories over the last few years. We list these mainly as a guide to those who are considering conversion to this type of program. This collection obviously is not meant to imply that it is the only type and quantity of apparatus that must be present on the laboratory bench. It represents, in our current view, the minimum essential content when balancing cost against experimental breadth.

# MICROGLASSWARE EQUIPMENT

*Thirteen Glassware Items*

2   Conical vials, 5 mL, 14/10⏥ and threaded (one thin-walled)
2   Conical vials, 3 mL, 14/10⏥ and threaded
1   Round-bottom flask, 5 or 10 mL, 14/10⏥
1   Water-jacketed reflux condenser, male 14/10⏥, female 14/10⏥
1   Air condenser, male 14/10⏥, female 14/10⏥
1   Claisen head, male 14/10⏥ threaded, female 14/10⏥
1   Craig tube, 3 mL, with Teflon® head (optional)
1   Craig tube, 2 mL, accepts above head
1   Drying tube, male 14/10⏥
1   Capillary gas delivery tube, male 14/10⏥
1   Hickman still head, male 14/10⏥, female 14/10⏥, or Hickman–Hinkle Spinning Band Head, male 14–10/10⏥, female 14/10⏥ with side-arm collection port

*Four Gas Chromatographic Collection Items*

2   Gas chromatographic collection tubes, male 5/5⏥
2   Conical vials, 0.1 mL, 5/5⏥ and threaded

Standard taper (⏥) ground-glass joints are the common mechanism for assembling all conventional research equipment in the organic laboratory. The symbol ⏥ is commonly used to indicate the presence of this type of connector. Normally, ⏥ is either followed or preceded by #/#. The first # refers to the maximum inside diameter of a female (outer) joint or the maximum outside diameter of a male (inner) joint, measured in millimeters. The second number corresponds to the total length of the ground surface of the joint (Fig. 3.8). The advantage of this type of connection is that if the joint surfaces are lightly greased, a vacuum seal is achieved. One of the drawbacks in employing these joints, however, is that contamination of the reacting system readily occurs through extraction of the grease by the solvents present in the reaction vessel. In carrying out microscale reactions, this is particularly troublesome.

An alternative to using greased joints as connectors is to employ either outside or inside screw-threaded glass systems. This type of joint utilizes plastic Teflon-lined screw cap connectors. These threaded glass joints are commercially available; they originally suffered from the same problem as the ungreased standard taper joint, that is, they were not gas tight (at least not in the hands of most undergraduates). The more recent adaptations of these seals have minimized this problem to a reasonable extent. The glassware joints developed in the microscale experimental organic laboratory programs at Bowdoin and Merrimack, however, have the ease of assembly and physical integrity of research-grade standard taper ground-glass joints along with a number of extremely important added features. The dimensions usually are ⏥ 14/10, male or female (⏥ 7/10 systems are optional for some setups). The conical vials in which the large majority of reactions are carried out also employ this alternative type of sealing system. Note that in addition to being ground to a standard taper

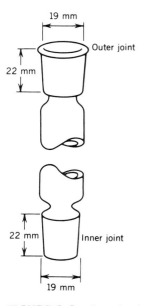

**FIGURE 3.8** Standard taper joints [⏥]. *From Zubric, James W. The Organic Chem Lab Survival Manual, 3rd ed.; Wiley: New York, 1992. (Reprinted by permission of John Wiley & Sons, Inc., New York.)*

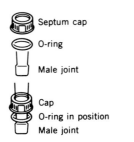

**FIGURE 3.9**   Threaded female joint.

Septum cap

O-ring

Male joint

Cap
O-ring in position
Male joint

**FIGURE 3.10**   Male joint with septum cap and O-ring.

on the inside surface, these vials also possess a screw thread on the outside surface (Fig. 3.9). This arrangement allows a standard taper male joint to be sealed to the reaction flask by a septum-type (open) plastic screw cap. The screw cap applies compression to a silicone rubber retaining O-ring positioned on the shoulder of the male joint (Fig. 3.10). The compression of the O-ring thereby achieves a *greaseless* gas-tight seal on the joint seam, while at the same time clamping the two pieces of equipment together. The ground joint provides both protection from intimate solvent contact with the O-ring, and mechanical stability to the connection. This type of joint (now available on all joints of the ACE Glass microscale line of glassware) provides a quick, easy, and reliable mechanism for assembling the glassware required for carrying out the microscale reactions. The use of this type of connector leads to a further bonus during construction of an experimental setup. Because the individual sections are small, light, and firmly sealed together, the entire arrangement often can be mounted on the support rack by a single clamp. In conventional systems, it is necessary in the majority of cases, to employ at least two clamps. This latter arrangement can easily lead to points of high strain in the glass components unless considerable care is taken in the assembly process. Clamp strain is one of the major sources of experimental glassware breakage. The ability to single-clamp essentially all microscale setups effectively eliminates this problem.

**It should be emphasized that the ground-glass joint surfaces are grease free; therefore, it is important to disconnect joints soon after use or they may become locked or "frozen" together.**

Joints of the size employed in these microscale experiments, however, seldom are a problem to separate, if given proper care (keep them clean!).

A complete set of glassware would involve 14 different components and a total of 17 items grouped according to function as follows (see Figs. 3.1–3.7):

*Five Reaction Flasks*

2   Conical vials, 5 mL, 14/10꜠ and threaded (one thin-walled)
2   Conical vials, 3 mL, 14/10꜠ and threaded
1   Round-bottom flask, 10 mL, 14/10꜠
1   Round-bottom flask, 5 mL, 14/10꜠ (optional)

Both the conical vials and the round-bottom flasks are designed to be connected via an O-ring compression cap installed on the male joint of the adjacent part of the system (see Fig. 3.1).

*Two Condensers*

1   Water-jacketed reflux condenser, male 14/10꜠, female 14/10꜠
1   Air condenser, male 14/10꜠, female 14/10꜠ (optional: The water-jacketed condenser can function as an air condenser)

These items form two sets of condensers for use with 14/10꜠ jointed reaction flasks. The female joints allow connection of the condenser to the 14/10꜠ drying tube and the 14/10 capillary gas delivery tube (see Fig. 3.2).

*Two Distillation Heads*

1   Hickman still head, male 14/10꜠, female 14/10꜠ or
1   Hickman–Hinkle Spinning Band Head, male 14–10/10꜠, female 14/10꜠ with side-arm collection port (optional)

The simple Hickman still is used with an O-ring compression cap to carry out semimicro simple or crude fractional distillations. The newly developed Hickman–Hinkle Spinning Band Head is a powerful modification of the simple system. The 3-cm fractionating column of this still routinely develops between five and six theoretical plates. The Hickman–Hinkle is currently available with 14/10$ joints and can be conveniently operated with the 14/10$, 3- and 5-mL conical vials (see Figs. 3.3 and 3.7). The still head is also available with an optional side-arm collection port.

*Two Recrystallization Tubes*

1   Craig tube, 2 mL
1   Craig tube, 3 mL

Craig tubes are a particularly effective method for recrystallizing small quantities of reaction products. These tubes possess a nonuniform ground joint in the outer section. The Teflon head modification (earlier models used glass inner sections) has made these systems more durable, and much less susceptible to breakage during centrifugation (see Fig. 3.4).

*Three Miscellaneous Items*

1   Claisen head, male 14/10$, female 14/10$
1   Drying tube, male 14/10$
1   Capillary gas delivery tube, 14/10$

The Claisen head (see Fig. 3.3) is often employed to facilitate the syringe addition of reagents to closed moisture-sensitive systems (such as Grignard reactions) via a septum seal in the vertical upper joint. This joint can also function to position the thermometer in the well of a Hickman–Hinkle still. The Claisen adapter is also used to mount the drying tube in a protected position remote from the reaction chamber. The drying tube, in turn, is used to protect moisture-sensitive reaction components from atmospheric water vapor, while allowing a reacting system to remain unsealed. The capillary gas delivery tube is employed in transferring gases, formed during reactions, to storage containers (see Figs. 3.5 and 3.28).

*Four Gas Chromatographic Collection Items*

2   Gas chromatographic collection tubes, male 5/5$
2   Conical vials, 0.1 mL, 5/5$ and threaded

The collection tube is connected directly to the exit port of the GC detector through a stainless steel adapter for fraction collection. The collected sample is then transferred to a 0.1-mL conical vial for storage. The system is conveniently employed in the resolution and isolation of two-component mixtures (see Fig. 3.6).

---

It is important to be able to carry out microscale experiments at accurately determined temperatures. Very often, transformations are successful, in part, because of the ability to maintain precise temperature control. In addition, many reactions require that reactants be intimately mixed to obtain a substantial yield of product. Therefore, the majority of the reactions you perform in this laboratory will be conducted with rapid stirring of the reaction mixture.

## STANDARD EXPERIMENTAL APPARATUS

### Heating and Stirring Arrangements

## Sand Bath Technique

The most convenient piece of equipment for heating or stirring or for performing both operations simultaneously on a microscale level is the hot plate magnetic stirrer. Heat transfer from the hot surface to the reaction flask is generally accomplished by employing a crystallizing dish containing a **shallow** layer of sand that can conform to the size and shape of the particular vessel employed. The temperature of the system is most conveniently monitored by embedding a thermometer in the sand near the reaction vessel. A successful procedure for determining the actual temperature inside the vial relative to the bath temperature, is to mount a second thermometer in the vial after first adding 2 mL of high-boiling silicone oil. The vial temperature is then measured at various sand bath temperatures, and the values entered on a graph of temperature versus hot plate setting. Recording the weight of sand used will help to make the values more reproducible.

The high sides of the crystallizing dish act to protect the apparatus from air drafts, and so the dish also operates somewhat as a hot-air bath. Heating can be made even more uniform by covering the crystallizing dish with aluminum foil (Figs. 3.11 and 3.12). This procedure works well, but is a bit awkward and is required in only a few instances.

Finally, it should be pointed out that in the use of sand baths, the insulating properties of the sand give rise to a readily available variable heat source. This results because the temperature of the sand rises steeply as one descends deeper into the bath. In this connection it should again be stressed that the depth of sand used in the bath is exceedingly important. **The depth should always be kept to a minimum, in the range of 10–15 mm**. Finally, sand baths offer a significant safety advantage over oil baths, which preceded them in the laboratory. Individual grains of sand are so small that they have little heat capacity and thus are less likely to burn the chemist in the event of a spill.

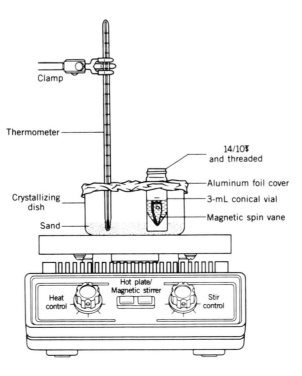

**FIGURE 3.11** Hot plate-magnetic stirrer with sand bath and reaction vial.

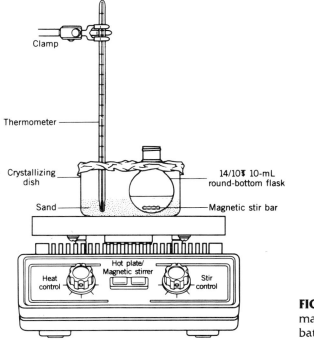

**FIGURE 3.12** Hot plate-magnetic stirrer with sand bath and reaction flask.

Labels in figure: Clamp; Thermometer; Crystallizing dish; Sand; 14/10₹ 10-mL round-bottom flask; Magnetic stir bar; Hot plate/Magnetic stirrer; Heat control; Stir control

## Metal Heat-Transfer Devices

During the early development of the microscale laboratory program, the use of drilled aluminum blocks that either directly contained the heat source (soldering iron element), or were used as the heat-transfer agent, was explored.

A number of nagging problems are associated with the use of heat blocks. First, observation of the reaction system is significantly restricted by the opaque metal surface. Slight color changes, the initial stages of reflux, the precipitation of small quantities of solids, and the early trends of decomposition all are more difficult to detect within cylindrical vials positioned in block heaters, as compared to baths in which sand is employed as the heat-transfer agent. Second, it is often necessary to gently swirl or slightly shake the reaction vial. This agitation can be readily accomplished while the system remains submerged in a malleable sand bath, but agitation is at best awkward without removing a system completely from a rigid aluminum block. The latter maneuver results in rapid cooling of the reacting system, which can produce unfavorable complications. Finally, it was noticed that there is enough variability in vial diameters so that occasionally vials will not fit blocks, or that as a result of uneven thermal expansion, breakage occurs.

Because of the obvious heat-transfer problems through sand, and the difficulties associated with running high-temperature reactions in which lower contact temperatures are desirable, considerable development work has taken place in the design of metal devices over the past few years. A detailed report of the use of aluminum block systems including a number of new designs has been made by Lodwig.[1] Other metal systems have recently been examined with rather promising results. To gain the advantages of both systems (sand and metal) the use of aluminum and zinc powders has been explored. Thirty-mesh zinc baths possess thermal characteristics very similar to solid aluminum blocks (see Fig. 3.13).

---

[1] Lodwig, Siegfried N. *J. Chem. Educ*, **1989**, *66*, 77.

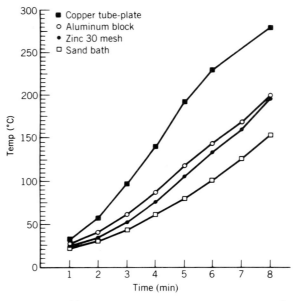

**FIGURE 3.13** Temperature rise characteristics for several thermal devices.

An evaluation of copper devices also has been undertaken, as copper has nearly double the thermal conductivity of aluminum. The configuration found most attractive in this series is the **copper tube–plate device**. It requires relatively little copper, as it utilizes a 3-mm plate and 15-mm lengths of standard tubing (see Fig. 3.14). The fabrication cost should be close to that of the aluminum block systems currently on the market. As shown in Figure 3.13, compared to all other systems, the copper plate is far superior in temperature rise from room temperature. Thus, it is found at nearly 200 °C at the 5-min mark, whereas the aluminum block is barely above 100 °C at this time. It is also easier to achieve equilibrium conditions (see Fig. 3.15) compared to the aluminum block. An equilibrium temperature of 150 °C could be reached in 9 min with copper, whereas it took 19 min with aluminum under the same conditions. The copper device also cools at nearly double the rate of the aluminum block. For example, to drop from 200 to 150 °C takes the aluminum block 8.1 min, whereas the copper system reaches the lower temperature in only 4.2 min. This characteristic is important in controlling temperature overruns.

Of particular significance is the observation that when distilling or refluxing organic materials, the block temperatures when employing the copper device are measurably lower. For example, 1 mL of *p*-xylene (bp 138 °C) contained in a 5-mL reaction vial refluxes at 175 °C in the copper bath. The aluminum block requires 188 °C under the same conditions and the sand bath ranks a distant third at 208 °C.

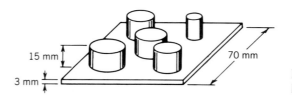

**FIGURE 3.14** Copper tube heat-transfer plate.

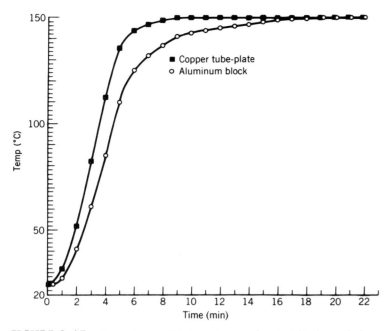

**FIGURE 3.15** Time to equilibrium (control set at high until the temperature reaches 85 °C, then heat control adjusted to a lower setting to give 150 °C at equilibrium).

## Stirring

Stirring the reaction mixture in a conical vial is carried out with Teflon-coated magnetic spin vanes, and in round-bottom flasks with Teflon-coated magnetic stirring bars (see Figs. 3.11 and 3.12).

It is important that the settings on the controls that adjust the current to the heating element, and to the motor that spins the magnet, are roughly calibrated to block temperatures and spin rates, respectively. You can make a rough graph of heat control setting versus temperature (Fig. 3.16) for your particular hot plate system (see sections on the Sand Bath Technique (p. 20) and Metal Heat-Transfer Devices (p. 21)). These data will save considerable time when you bring a reaction system to operating temperature. When you first enter the laboratory, it is advisable to adjust the temperature setting on the hot plate stirrer with the heating device, or bath, in place. The setting is determined from your control setting–temperature calibration curve. This procedure will allow the heated bath to reach a relatively constant temperature by the time it is required. You will then be able to make small final adjustments more quickly, if necessary.

**NOTE.** *If a sand bath is used, it should again be emphasized that heavy layers of sand act as an insulator on the hot plate surface. This insulation can result in damage to the heating element at high-temperature settings. Therefore, when temperatures over 150 °C are required, it is especially important to remember to use the minimum amount of sand.*

It is important to position the reaction flask as close to the bottom surface of the crystallizing dish as possible in those experiments that depend on magnetic stirring. This arrangement is good practice, in general, as it leads to use of the minimum amount of sand when this type of bath is employed.

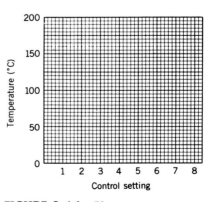

**FIGURE 3.16** Plot your temperature (°C) versus hot plate control setting.

If the reaction does not require elevated temperatures, but needs only to be stirred, the system is assembled without the heating device. Some stirred reactions, on the other hand, require cooling with a crystallizing dish filled with ice water, or with ice water and salt if lower temperatures are required.

## Reflux Apparatus

We often find that to bring about a successful reaction between two substances, it is necessary to intimately mix the materials together and to maintain a specific temperature. The mixing operation is conveniently achieved by dissolution of the materials in a solvent in which they are mutually soluble. If the reaction is carried out in solution under reflux conditions, the choice of solvent can be used to control the temperature of the reaction. Many organic reactions involve the use of a reflux apparatus in one arrangement or another.

What do we mean by *reflux?* The term means to "return," or "run back." This return is exactly how the reflux apparatus functions. When the temperature of the reaction system is raised to the solvent's boiling point (constant temperature), *all* vapors are condensed and returned to the reaction flask or vial; this operation is not a distillation and the liquid phase remains at a stable maximum temperature. In microscale reactions, two basic types of reflux condensers are utilized: the air-cooled condenser, or *air condenser*, and the *water-jacketed condenser*. The air condenser operates, as its name implies, by condensing solvent vapors on the cool vertical wall of an extended glass tube that dissipates the heat by contact with laboratory room air. This arrangement functions quite effectively with liquids boiling above 150 °C. Indeed, a simple test tube can act as a reaction chamber and air condenser all in one unit, and many simple reactions can be most easily carried out in test tubes. Air condensers can occasionally be used with lower boiling systems; however, the water-jacketed condenser, which employs flowing cold water to remove heat from the vertical column and thus facilitate vapor condensation, is more often employed in these situations. The latter apparatus is highly efficient at condensing vapor from low-boiling liquids. Both styles of condensers accommodate various sizes of reaction flasks and are available in standard taper joint size 14/10 Ŧ. The tops of both condenser columns possess the female 14/10 Ŧ joint.

In refluxing systems that do not require significant mixing or agitation, the stirrer (magnetic spin vane or bar) usually is replaced by a "boiling stone." These sharp-edged stones possess highly fractured surfaces that are very efficient at initiating bubble formation as the reacting medium approaches the boiling point. The boiling stone acts to protect the system from disastrous boil-overs and also reduces "bumping." (Boiling stones should be used only once and **must never** be added to a hot solution. In the first case, the vapor cavities become filled with liquid upon cooling, and thus a boiling stone becomes less effective after its first use. In the second case, **adding the boiling stone to the hot solution may initiate uncontrollable vapor eruption leading to catastrophic boil-over.**)

*Various Arrangements of Reflux Apparatus (see Figs. 3.17–3.19)*

Air condenser with a 3- or 5-mL conical vial, arranged for heating and magnetic stirring.

Water-jacketed condenser with a 3- or 5-mL conical vial, arranged for heating and magnetic stirring.

Water-jacketed condenser with a 10-mL round-bottom flask, arranged for heating and stirring.

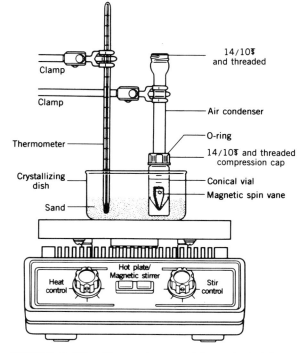

**FIGURE 3.17** Air condenser with conical vial, arranged for heating and magnetic stirring.

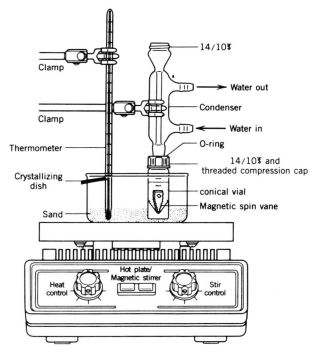

**FIGURE 3.18** Water-jacketed condenser with conical vial, arranged for heating and magnetic stirring.

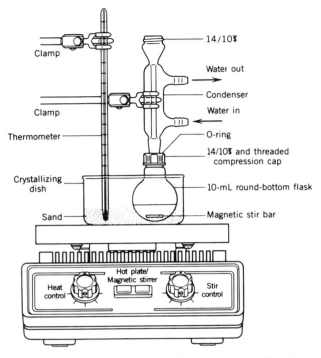

**FIGURE 3.19**   Water-jacketed condenser with 10-mL round-bottom flask, arranged for heating and magnetic stirring.

**Distillation Apparatus**

Distillation is a laboratory operation used to separate substances that have different boiling points. The mixture is heated, vaporized, and then condensed, with the more volatile component being enriched in the early fractions of condensate. Unlike the reflux operation, in distillations none, or only a portion, of the condensate is returned to the flask where vaporization is taking place. Many distillation apparatus have been designed to carry out this basic operation. These apparatus differ mainly in small features that are used to solve particular types of separation problems. In microscale experiments, a number of simple *semimicroscale* distillations are often required. For these distillations, the Hickman still head (Fig. 3.20) is ideally suited. This system has a 14/10₮ male joint for connection to either the 3- or 5-mL conical vials or the 5- or 10-mL round-bottom (RB) flasks. The still head functions as both an air condenser and a condensate trap. For a detailed discussion of this piece of equipment see Distillation Experiments [3A] and [3B]. In 1987, a new modification of the basic Hickman still was developed, the Hickman–Hinkle spinning band still (Fig. 3.21), and named for the Bowdoin College undergraduate student who first proposed this arrangement. The modified still (plain or side-arm types) functions in much the same way as the simple Hickman still, except that a Teflon spinning band element is mounted in the slightly extended section between the male joint and the collection collar. In addition, this system has a built-in thermometer well that allows fairly accurate measurement of vapor temperature. When the band is spun at 1500 rpm by a magnetic stirring hot

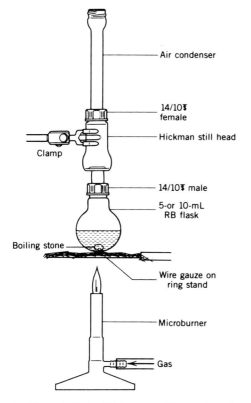

**FIGURE 3.20** Hickman still head and air condenser with 5-mL round-bottom flask, arranged for microburner heating.

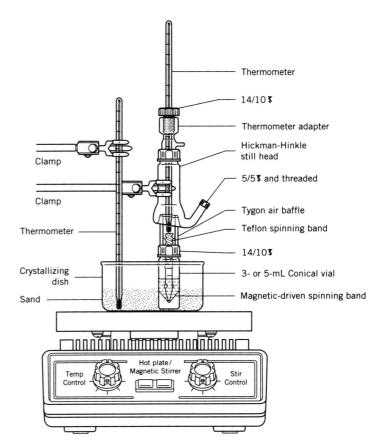

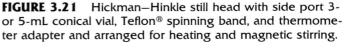

**FIGURE 3.21** Hickman–Hinkle still head with side port 3- or 5-mL conical vial, Teflon® spinning band, and thermometer adapter and arranged for heating and magnetic stirring.

plate, this still is transformed into a very effective short-path fractional distillation column (see Distillation, Experiment [3D]).

The most powerful system currently available for both the instructional and research laboratories, however, is the 2.5-in. vacuum-jacketed microscale spinning band distillation column (see Fig. 3.22*a* and Experiment [3C] for description and details). This still utilizes the same type of band drive as the Hickman–Hinkle columns, but in addition the system is designed for conventional downward distillate collection, nonstopcock reflux control, and reasonably accurate temperature sensing. The column, which can develop nearly 12 theoretical plates, possesses the highest resolution of any standard distillation apparatus presently in use in the introductory organic laboratory.

The high-performance low-cost atmospheric pressure 2.5-in. micro-spinning band distillation column has been modified to accommodate reduced pressure fractional distillation (Fig. 3.22*b*) by: (a) replacing the air condenser and suspended thermometer with a 14/10Ŧ, vacuum-tight, threaded thermometer adapter; (b) replacing the heavy-walled 3-mL conical collection vial with a thin-walled, 3-mL conical vial, which has, mounted near the bottom, a side arm with a threaded 5/5 Ŧ joint (a septum cap and silicone septum form a vacuum-tight seal on the side arm); and (c)

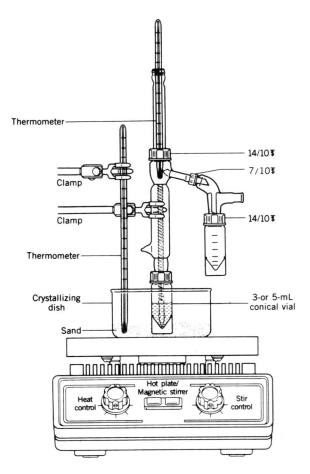

**FIGURE 3.22a**  Micro spinning band distillation column (2.5 in.).

a vacuum tubing nipple replacing the Teflon stopper (7/10$) used to establish a vapor lock on the collection side of the system in the atmospheric still.

The system is evacuated via the vacuum tubing nipple. Fractions are efficiently collected with a gas-tight syringe and needle inserted through a septum mounted in the side arm of the 3-mL collection vial. The collection vial may be cooled externally during collection. This arrangement allows convenient collection of distillate fractions down to pressures of approximately 100 torr (successful collections have been made at pressures as low as 10 torr).

The system continues to function very effectively under reduced pressure, even though the vapor lock has been removed. Data indicate that height equivalent/theoretical plate values will remain near 0.25 in. per plate in these columns.

*Various Distillation Apparatus (see Figs. 3.20 and 3.21)*

Hickman still head and air condenser with 5- and 10-mL round-bottom flasks, or 3- and 5-mL conical vials arranged for microburner or hot plate heating (see Experiments [3A] and [3B]).

Hickman–Hinkle still head with 5-mL conical vial (thin-walled), arranged for heating and magnetic spinning (see Experiment [3D].)

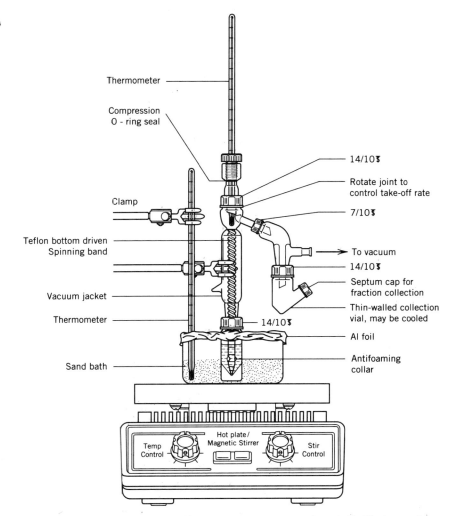

**FIGURE 3.22b** Reduced pressure microspinning band distillation column.

**Moisture-Protected Reaction Apparatus**

Many organic reagents react rapidly and preferentially with water. *The success or failure of many experiments depends to a large degree on how well atmospheric moisture is excluded from the reaction system.* The "drying tube," which is packed with a desiccant, such as anhydrous calcium chloride, is a handy way to carry out a reaction in apparatus that is not totally closed to the atmosphere, but which is reasonably well protected from water vapor. The microscale apparatus described here are designed to be used with the 14/10Ƭ drying tube. The reflux condensers discussed earlier are constructed with female 14/10Ƭ joints at the top of the column, which allows convenient connection of the drying tube, if the refluxing system is moisture sensitive.

Because many reactions are highly sensitive to moisture, successful operation at the microscale level can be rather challenging. If anhydrous reagents are to be added after an apparatus has been dried and assembled, it is important to be able to introduce these reagents without exposing the system to the atmosphere, particularly when operating under humid atmospheric conditions. In reactions not requiring reflux conditions and conducted at room temperature, this addition procedure is best accomplished by use of the microscale Claisen head adapter. The adapter has a vertical screw-threaded standard taper joint that will accept a septum cap. The septum seal allows syringe addition of reagents and avoids the necessity of opening the apparatus to the laboratory atmosphere.

In a few instances reactions that are unusually moisture sensitive will be encountered. In this situation, reactions are best carried out in completely sealed systems that are scrupulously dry. The use of the Claisen head adapter with a balloon substituted for the drying tube provides a satisfactory solution to the problem. Occasionally, it becomes important to maintain dry conditions during a distillation. The Hickman stills are constructed with a 14/10$ joint at the top of the head that readily accepts the drying tube (or Claisen head plus drying tube)

*Various Moisture-Protected Reaction Apparatus (see Figs. 3.23–3.26)*

Moisture-protected water-jacketed condenser with a 3- or 5-mL conical vial, arranged for heating and stirring.

Moisture-protected Claisen head with a 3- or 5-mL conical vial, arranged for syringe addition and magnetic stirring.

Sealed Claisen head with a 3- or 5-mL conical vial, arranged for nitrogen ($N_2$) flushing, heating, and magnetic stirring.

Moisture-protected Hickman still head with a 10-mL round-bottom flask, arranged for heating and stirring.

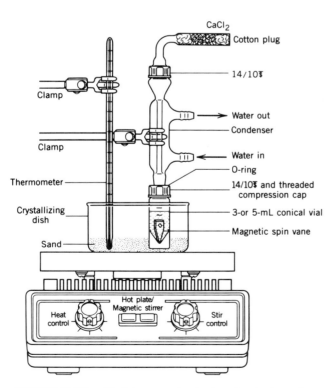

**FIGURE 3.23** Moisture-protected water-jacketed condenser with 3- or 5-mL conical vial, arranged for heating and magnetic stirring.

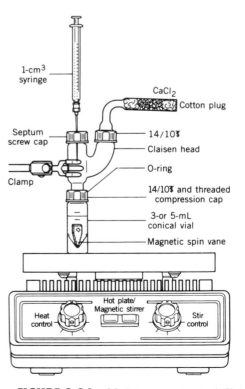

**FIGURE 3.24** Moisture-protected Claisen head with 3- or 5-mL conical vial, arranged for syringe addition and magnetic stirring.

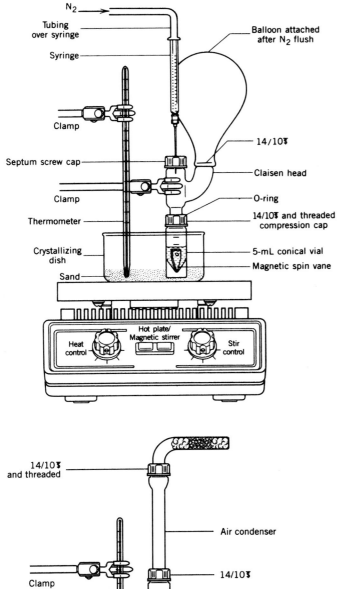

N₂
Tubing over syringe
Syringe
Clamp
Septum screw cap
Clamp
Thermometer
Crystallizing dish
Sand

Balloon attached after N₂ flush
14/10$
Claisen head
O-ring
14/10$ and threaded compression cap
5-mL conical vial
Magnetic spin vane

Hot plate/ Magnetic stirrer
Heat control
Stir control

**FIGURE 3.25** Sealed Claisen head with 3- or 5-mL conical vial, arranged for N₂ flushing, heating, and magnetic stirring.

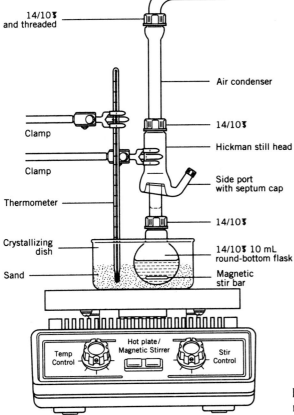

14/10$ and threaded
Clamp
Clamp
Thermometer
Crystallizing dish
Sand

Air condenser
14/10$
Hickman still head
Side port with septum cap
14/10$
14/10$ 10 mL round-bottom flask
Magnetic stir bar

Hot plate/ Magnetic Stirrer
Temp Control
Stir Control

**FIGURE 3.26** Moisture-protected Hickman still head with 10-mL round-bottom flask, arranged for heating and stirring.

## Specialized Pieces of Equipment

### Collection of Gaseous Products

Some experiments lead to gaseous products. The collection, or trapping, of gases is conveniently carried out by using the capillary gas delivery tube. This item is designed to be attached directly to a 1- or 3-mL conical vial, or to the female 14/10 \$ joint of a condenser connected to a reaction flask or vial. The tube leads to the collection system, which may be a simple, inverted, graduated cylinder; a blank-threaded septum joint; or an air condenser filled with water (where the gaseous product, or products, are not soluble in the aqueous phase). The 0.1-mm capillary bore considerably reduces dead volume, and increases the efficiency of product transfer (see Figs. 3.27 and 3.28).

### Collection of Gas Chromatographic Effluents

The trapping and collection of gas chromatographic liquid fractions becomes a particularly important exercise at the microscale level of experi-

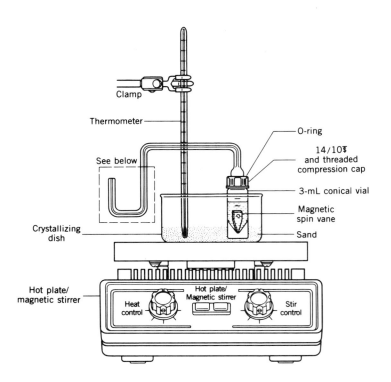

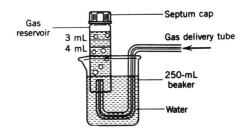

**FIGURE 3.27**   Conical vial (3-mL) and capillary gas delivery tube arranged for heating and magnetic stirring.

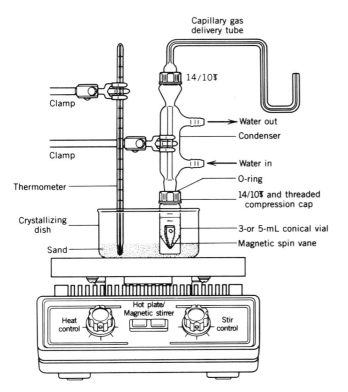

**FIGURE 3.28** Water-jacketed condenser with 3- or 5-mL conical vial and capillary gas delivery tube, arranged for heating and stirring.

mentation. When liquid product yields drop below 100 $\mu$L, conventional distillation, even utilizing microtechniques, becomes impractical. In this case, preparative gas chromatography replaces conventional distillation as the route of choice to product purification. A number of the reaction products in the experimental section of the text depend on this approach for successful purification and isolation. The ease and efficiency of carrying out this operation is greatly facilitated by employing the 5/5$\mathbb{T}$ collection tube and the 0.1-mL 5/5$\mathbb{T}$ conical collection vial (see Fig. 3.29).

*Various Sample Collection Apparatus (see Figs. 3.27–3.29)*

Conical vial (1-mL) and capillary gas delivery tube, arranged for heating and stirring.

Water-jacketed condenser with 3- or 5-mL conical vial and capillary gas delivery tube, arranged for heating and stirring.

Gas chromatographic collection tube and 0.1-mL conical vial.

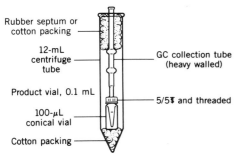

**FIGURE 3.29** Gas chromatographic collection tube and 0.1 mL conical vial.

## MICROSCALE LAWS

### Rules of the Trade for Handling Organic Materials at the Microscale Level

Now that we have briefly looked at the equipment we will be using to carry out microscale organic reactions, let us examine the specific techniques that are used to deal with the small quantities of material involved. Microscale synthetic organic reactions, as defined by Cheronis,[2] start with 15–150 mg of the limiting reagent. These quantities sound small, and they are. Although 150 mg of a light, powdery material will fill one-half of a 1-mL conical vial, you will have a hard time observing 15 mg of a clear liquid in the same container, even with magnification. On the other hand, this volume of liquid is reasonably easy to observe when placed in a 0.1-mL conical vial. A vital part of the game of working with small amounts of materials is to become familiar with microscale techniques and to practice them as much as possible in the laboratory.

### Rules for Working with Liquids at the Microscale Level

1. *Liquids are never poured at the microscale level.* Liquid substances are transferred by pipet or syringe. As we are working with small, easy-to-hold glassware at the microscale level, the best technique for transfer is to hold both containers with the fingers of one hand, with the mouths as close together as possible. The free hand is then used to operate the pipet (syringe) to withdraw the liquid and make the transfer. This approach reduces to a minimum the time that the open tip is not in, or over, the reservoir or the reaction flask. We employ three different pipets and two standard syringes for most experiments in which liquids are involved. *This equipment can be a prime source of contamination.* Be very careful to thoroughly clean the pipets and syringes after each use.

   a. *Pasteur pipet,* often called a capillary pipet, is a simple glass tube with the end drawn to a fine capillary. These pipets can hold several milliliters of liquid (Fig. 3.30*a*), and are filled using a small rubber bulb or one of the very handy, commercially available pipet pumps. Since many transfers are made using the Pasteur pipet, it is suggested that several of them be calibrated for approximate delivery of 0.5, 1.0, 1.5, and 2.0 mL of liquid. This calibration is easily done by drawing the measured amount of a liquid from a 10-mL graduated cylinder and marking the level of the liquid in the pipet. This mark can be made with transparent tape, or by scratching with a file. Indicate the level with a marking pen before trying to tape or file the pipet.

   b. *Pasteur filter pipet.* A very handy adaptation of the Pasteur pipet is a filter pipet. This pipet is constructed by taking a small cotton ball and placing it in the large open end of the standard Pasteur pipet. Hold the pipet vertically and tap it gently to position the cotton ball in the drawn section of the tube (Fig. 3.30*b*). Now form a plug in the capillary section by pushing the cotton ball down the pipet with a piece of copper wire (Fig. 3.30*c*). Finish by seating the plug flush with the end of the capillary (Fig. 3.30*d*). The optimum-size plug will allow easy movement along the capillary while it is being positioned by the copper wire. Compression of the cotton will build enough pressure against the walls of the capillary (once the plug is in position) to prevent plug slippage while the pipet is filled with liquid. If the ball is too big, it will wedge in the capillary

[2] Cheronis, N. D., *Semimicro Experimental Organic Chemistry,* Hadrion Press, Inc., New York, 1958.

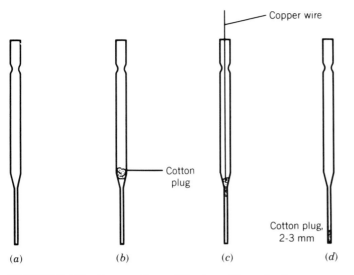

Copper wire

Cotton
plug

Cotton plug,
2-3 mm

(a)          (b)          (c)          (d)

**FIGURE 3.30**   Preparation of Pasteur filter pipet.

before the end is reached, and wall pressure will be so great that liquid flow will be shut off. Even some plugs that are loose enough to be positioned at the end of the capillary will still have developed sufficient lateral pressure to make the filling rate unacceptably slow. If the cotton filter, however, is positioned too loosely, it may be easily dislodged from the pipet by the solvent flow. These plugs can be quickly and easily inserted with a little practice. Once in place, the plug is rinsed with 1 mL of methanol, 1 mL of hexane, and dried before use.

The purpose of placing the cotton plug in the pipet is twofold. First, a particular problem with the transfer of volatile liquids via the standard Pasteur pipet is the rapid build up of back pressure from solvent vapors in the rubber bulb. This pressure quickly tends to force the liquid back out of the pipet and can cause valuable product to drip on the bench top. The cotton plug tends to resist this back pressure and allows much easier control of the solution once it is in the pipet. The time-delay factor becomes particularly important when the Pasteur filter pipet is employed as a microseparatory funnel (see the discussion on extraction techniques in Technique 4, p. 73).

Second, each time a transfer of material is made, the material is automatically filtered. This process effectively removes dust and lint, which are a constant problem when working at the microscale level with unfiltered room air. A second stage of filtration may be obtained by employing a disposable filter tip on the original Pasteur filter pipet as described by Rothchild.[3]

c. *Automatic pipet (considered the Mercedes-Benz of pipets).* Automatic pipets quickly, safely, and reproducibly measure and dispense specific volumes of liquids. These pipets are particularly valuable at the microscale level, as they generate the precise and accurate liquid measurements that are absolutely necessary when handling microliter volumes of reagent. The automatic pipet adds consider-

---

[3] Rothchild, R. *J. Chem. Educ.* **1990**, *67*, 425.

able insurance for the success of an experiment, as any liquid can be efficiently measured, transferred, and delivered to the reaction flask. Automatic pipets have become almost an essential instrument in the active microscale laboratory.

The automatic pipet system consists of a calibrated piston pipet with a specially designed disposable plastic tip. It is possible to encounter any one of three pipet styles: single volume, multirange, or continuously adjustable (see Fig. 3.31). The first type is calibrated to deliver only a single volume. The second type is adjustable to two or three predetermined delivery volumes. The third type is the most versatile and can be user set to deliver any volume within the range of the pipet. Obviously, the price of these valuable laboratory tools goes up with increasing features. These pipets are expensive, and often must be shared in the laboratory. Treat them with respect!

The automatic pipet is designed so that the liquid comes in contact only with the disposable tip. Never load the pipet without the tip in place. Never immerse the tip completely in the liquid that is being pipetted. Always keep the pipet vertical when the tip is attached. If these three rules are followed, most automatic pipets will give many years of reliable service. A few general rules for improving reproducibility with an automatic pipet should also be followed.

- Try to effect the same uptake and delivery motion for all samples. Smooth depression and release of the piston will give the most consistent results. Never allow the piston to snap back.
- *Always* depress the piston to the first stop before inserting the tip into the liquid. If the piston is depressed after submersion, formation of an air bubble in the tip becomes likely, and bubble formation will result in a filling error.
- *Never* insert the tip more than 5 mm into the liquid. It is good practice not to allow the body of the pipet to contact any surface, or bottle neck, that might be wet with a chemical.
- If an air bubble forms in the tip during uptake, return the fluid, discard the tip, and repeat the sampling process.

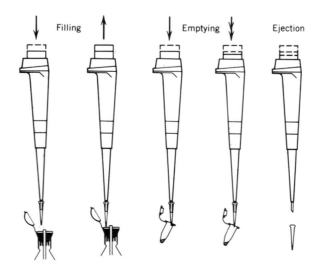

**FIGURE 3.31** Operation of automatic delivery pipet. *(Courtesy of Brinkmann Instruments Co., Westbury, NY)*

d. *Syringes.* Syringes are particularly helpful pieces of equipment when transferring liquid reagents or solutions to sealed reaction systems from sealed reagent or solvent reservoirs. Syringe needles can be inserted through a septum, which avoids opening the apparatus to the atmosphere. Syringes are also routinely employed in the determination of ultramicro boiling points (10-$\mu$L GC syringe). It is critically important to clean the syringe needle after each use. Effective cleaning of a syringe requires as many as a dozen flushes. For many transfers, the microscale laboratory utilizes a low-cost glass 1-mL insulin syringe in which the rubber plunger seal is replaced with a Teflon seal (ACE Glass). For preparative GC injections, the standard 50- or 100-$\mu$L syringes are preferred (see Technique 1).

**2.** *Liquid volumes may be converted easily to mass measures by the following relationship:*

$$\text{Volume (mL)} = \frac{\text{mass (g)}}{\text{density (g/mL)}}$$

**3.** *Work with liquids in conical vials,* and work in a vial whose capacity is approximately twice the volume of the material. The trick here is to reduce the surface area of the flask in contact with the sample to an absolute minimum. Conical systems are far superior to the spherical surface of the conventional round-bottom flask.

Liquids may also be weighed directly. A tared container (vial) should be used. After addition of the liquid, the vial should be kept capped throughout the weighing operation. This procedure prevents loss of the liquid material by evaporation. If the density of the liquid is known, the approximate volume of the liquid should be transferred to the container using an automatic delivery pipet or a calibrated Pasteur pipet. Use the above expression relating density, mass, and volume to calculate the volume required by the measured mass. Adjustment of the mass to give the desired value can then be made by adding or removing small amounts of liquid from the container by Pasteur pipet.

**NOTE.** *Before you leave the balance area, be sure to replace all caps on reagent bottles and clean up any spills. A modern balance is a precision instrument that can easily be damaged by contamination.*

## Rules for Working with Solids at the Microscale Level

**1.** *General considerations.* Working with a crystalline solid is much easier than working with the equivalent quantity of a liquid. Unless the solid is in solution, a spill on a clean glass working surface usually can be recovered quickly and efficiently. *Be careful, however, when working in solution. ALWAYS use the same precautions that you would use if you were handling a pure liquid.*

**2.** *The transfer of solids.* Solids are normally transferred with a microspatula, a technique that is not difficult to develop.

**3.** *Weighing solids at the milligram level.* The current generation of single-pan electronic balances has removed much of the drudgery from weighing solids. These systems can automatically tare an empty vial. Once the vial is tared, the reagent is added in small portions. The weight of each addition is instantly registered; material is added until the desired quantity has been transferred.

Most solids are weighed in glass containers (vials or beakers), in plastic or aluminum weighing boats, or on glazed weighing paper. Filter paper or other absorbing materials are not the best choice for this measurement, as small quantities of the weighed material will often adhere to the fibers of the paper.

## THE LABORATORY NOTEBOOK

Written communication is the most important method by which chemists transmit their work to the scientific community. It begins with the record kept in a laboratory notebook. The reduction to practice of an experiment originally recorded in the laboratory notebook is the source of information used to prepare scientific papers published in journals or presented at meetings. For the industrial chemist, it is especially critical in obtaining patent coverage.

It is important that potential scientists, whatever the field, learn to keep a detailed account of their work. A laboratory notebook has several key components. Note how each component is incorporated into the example that follows.

*Key Components of a Laboratory Experiment Writeup*

1. Date experiment was conducted.
2. Title of experiment.
3. Purpose for running the reaction.
4. Reaction scheme.
5. Table of reagents and products.
6. Details of procedure used.
7. Characteristics of the product(s).
8. References to product or procedure (if any).
9. Analytical and spectral data.
10. Signature of person performing the experiment and that of a witness, if required.

In reference to point 6, it is the obligation of the person doing the work to list the equipment, the amounts of reagents, the experimental conditions, and the method used to isolate the product. Any color or temperature changes should be carefully noted and recorded.

Several additional points can be made about the proper maintenance of a laboratory record.

11. A hardbound, permanent notebook is essential.
12. Each page of the notebook should be numbered in consecutive order. For convenience, an index at the beginning or end of the book is recommended and blank pages should be retained for this purpose.
13. If a page is not completely filled, an "X" should be used to show that no further entry was made.
14. Always record your data in ink. If a mistake is made, draw a neat line through the word or words so that they remain legible. Data are always recorded directly into the notebook, *never* on scrap paper!
15. Make the record clear and unambiguous. Pay attention to grammar and spelling.
16. In industrial research laboratories, your signature, as well as that of a witness, is required, because the notebook may be used as a legal document.
17. Always write and organize your work so that someone else could come into the laboratory and repeat your directions without confusion or uncertainty. *Completeness* and *legibility* are key factors.

For those of you undertaking study in the organic laboratory for the first time, it is likely that the reactions you will be performing have been worked out and checked in detail. Because of this, your instructor may not require you to keep your notebook in such a meticulous fashion. For example, when you describe the procedure (item 6), it may be acceptable to make a clear reference to the material in the laboratory manual and to note any modifications or deviations from the prescribed procedure. In some cases, it may be more practical to use an outline method. In any event, the following example should be studied carefully. It may be used as a reference when detailed records are important in your work.

**NOTE.** *Because of its length, the example is presented typed. Normally, the entry in the notebook is handwritten. However, now that computers are gaining wide acceptance in laboratory work, many chemists are using this means to record their data.*

---

*The circled numbers refer to list on page 38*

# EXAMPLE OF A LABORATORY NOTEBOOK ENTRY

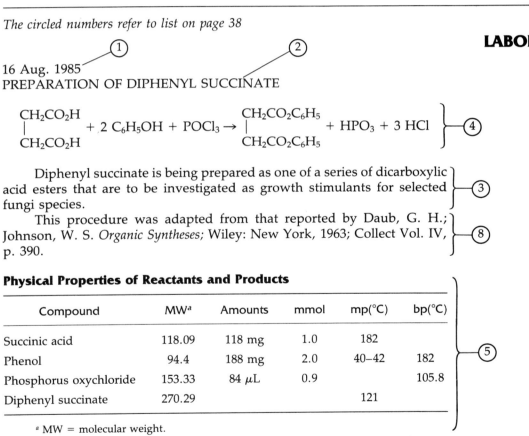

16 Aug. 1985 ①

PREPARATION OF DIPHENYL SUCCINATE ②

$$\begin{matrix} CH_2CO_2H \\ | \\ CH_2CO_2H \end{matrix} + 2\ C_6H_5OH + POCl_3 \rightarrow \begin{matrix} CH_2CO_2C_6H_5 \\ | \\ CH_2CO_2C_6H_5 \end{matrix} + HPO_3 + 3\ HCl \quad ④$$

Diphenyl succinate is being prepared as one of a series of dicarboxylic acid esters that are to be investigated as growth stimulants for selected fungi species. ③

This procedure was adapted from that reported by Daub, G. H.; Johnson, W. S. *Organic Syntheses*; Wiley: New York, 1963; Collect Vol. IV, p. 390. ⑧

**Physical Properties of Reactants and Products**

| Compound | MW[a] | Amounts | mmol | mp(°C) | bp(°C) |
|----------|-------|---------|------|--------|--------|
| Succinic acid | 118.09 | 118 mg | 1.0 | 182 | |
| Phenol | 94.4 | 188 mg | 2.0 | 40–42 | 182 |
| Phosphorus oxychloride | 153.33 | 84 µL | 0.9 | | 105.8 |
| Diphenyl succinate | 270.29 | | | 121 | |

⑤

[a] MW = molecular weight.

In a 3.0-mL conical vial containing a magnetic spin vane and equipped with a reflux condenser protected by a calcium chloride drying tube were placed succinic acid (118 mg, 1.0 mmol), phenol (188 mg, 2.0 mmol), and phosphorous oxychloride (84 µL, 0.9 mmol). The reaction mixture was heated with stirring at 115 °C in a sand bath in the **hood** for 1.25 hours. It was necessary to conduct the reaction in the **hood**, as hydrogen chloride (HCl) gas evolved during the course of the reaction. The drying tube was removed, toluene (0.5 mL) was added through the top of ⑥ the condenser using a Pasteur pipet, and the drying tube replaced. The mixture was then heated for an additional 1 hour at 115 °C.

The hot toluene solution was separated from the red syrupy residue of phosphoric acid using a Pasteur pipet. The toluene extract was filtered by gravity using a fast-grade filter paper and the filtrate collected in a 10-mL Erlenmeyer flask. The phosphoric acid residue was then extracted with two additional 1.0-mL portions of hot toluene. These extracts were also separated using the Pasteur pipet and filtered, and the filtrate was collected in the same Erlenmeyer flask. The combined toluene solutions were concentrated to a volume of approximately 0.6 mL by warming them in a sand bath under a gentle stream of nitrogen ($N_2$) gas in the **hood**. The pale yellow liquid residue was then allowed to cool to room temperature; the diphenyl succinate precipitated as colorless crystals. The solid was collected by vacuum filtration using a Hirsch funnel and the filter cake washed with three 0.5-mL portions of cold diethyl ether. The product was dried in a vacuum oven at 30 °C (3 mm) for 30 min.

There was obtained 181 mg (67%) of diphenyl succinate having a mp = 120–121 °C (lit. value 121 °C: *CRC Handbook of Chemistry and Physics*, 72nd ed.; CRC Press: Boca Raton, FL, 1991–1992; no. 13559, p. 3–471).

The IR spectrum exhibits the expected peaks for the compound. [*At this point, the data may be listed, or the spectrum pasted on a separate page of the notebook.*]

*Marilyn C. Waris*

*witnessed by*
*O. Jeanne d'Arc Mailhiot*        *16/Aug./1985*

Almost without exception, in each of the experiments presented in this text, you are requested to calculate the percentage yield. For any reaction, it is always important for the chemist to know how much of a product is *actually* produced (experimental) in relation to the *theoretical* amount (maximum) that could have been formed. The percentage yield is calculated on the basis of the relationship

$$\% \text{ yield} = \frac{\text{actual yield (experimental)}}{\text{theoretical yield (calculated maximum)}} \times 100$$

The percentage yield is generally calculated on a weight (gram or milligram) or on a mole basis. In the present text, the calculations are made using milligrams.

Several steps are involved in calculation of the percentage yield.

**1.** Write a *balanced* equation for the reaction. For example, consider Experiment [22A], the Wiliamson synthesis of propyl *p*-tolyl ether.

**2.** Identify the *limiting* reactant. The ratio of reactants is calculated on a millimole (or mole) basis. In the example, 0.78 mmol of *p*-cresol is used compared with 0.77 mmol of propyl iodide. Note that the sodium hydroxide is used as a reagent but *does not* appear in the product, propyl *p*-tolyl ether. Therefore, it is not considered in the calculations. Nor is the quaternary salt considered; it is used as a phase transfer catalyst. The reaction is run essentially on a 1:1 molar ratio and, thus, calculation of the theoretical yield can be based on the *p*-cresol or the propyl iodide.

**3.** Calculate the *theoretical* (maximum) amount of the product that could be obtained for the conversion, based on the limiting reactant. In the present case, referring to the balanced equation, one mole of propyl iodide affords one mole of the ether product. Therefore, if we start with 0.77 mmol of the propyl iodide, the maximum amount of propyl *p*-tolyl ether that can be produced is 0.77 mmol, or 115.7 mg.

**4.** Determine the *actual* (experimental) yield (milligrams) of product isolated in the reaction. This amount is invariably less than the theoretical quantity, unless the material is impure (one common contaminant is water). For example, student yields for the preparation of propyl *p*-tolyl ether average 70 mg.

**5.** Calculate the *percentage yield* using the weights determined in steps 3 and 4. The percentage yield is then

$$\% \text{ yield} = \frac{70 \text{ mg (actual)}}{115.7 \text{ mg (theoretical)}} \times 100 = 60.5\%$$

As you carry out each reaction in the laboratory, strive to obtain as high a percentage yield of product as possible. The reaction conditions have been carefully developed and, therefore, it is essential that you master the microscale techniques concerned with transfer of reagents and the isolation of products as soon as possible.

# 4

# Determination of Physical Properties

**C$_4$H$_6$, Bicyclo[1.1.0]butane**
Lemal, Menger and Clark (1963).

Determination of physical properties is important for substance identification, and as an indication of material purity. Historically, the physical constants of prime interest include: the boiling point, density, and refractive index in liquids and the melting point in solids. In special cases, optical rotation and molecular weight determinations may be required. Today, with the widespread availability of spectroscopic instrumentation, powerful new techniques may be applied to the direct identification and characterization of materials, including the analysis of individual components of very small quantities of complex mixtures. The sequential measurement of the infrared (IR) and mass spectrometric (MS) characteristics of a substance resolved "on the fly" by capillary gas chromatography (GC) can be quickly determined and interpreted. This particular combination (GC-IR-MS), which stands out among a number of "hyphenated" techniques just becoming available, is perhaps the most powerful system yet developed for molecular identification. The rapid development of high-field multinuclear nuclear magnetic resonance (NMR) spectrometers also has added another powerful dimension to identification techniques. The NMR sensitivities, however, are still considerably lower than either IR or MS. The IR spectrum alone, obtained with one data point per wavenumber, can add more than 4000 measurements to the few classically determined properties. *Indeed, even compared to high-resolution MS, and pulsed $^1$H and $^{13}$C NMR, the IR spectrum of a material still remains a powerful set of physical properties (transmission elements) available to the organic chemist for the identification of an unknown compound.*[1]

---

[1] Griffiths, P. R.; de Haseth, J. A.; Azaraga, L. V. *Anal. Chem.* **1983,** *55*, 1361A.

At the present time, simple physical constants are determined mainly to assist in establishing the purity of *known* materials. As the boiling point or the melting point of a material can be very sensitive to small quantities of impurities, these data can be particularly helpful in determining whether or not a starting material needs further purification, or whether a product has been isolated in acceptable purity. Gas (GC), high-performance liquid (HPLC), and thin-layer (TLC) chromatography, however, now provide considerably more powerful purity information when such data are required. When a new composition of matter has been formed, an elemental (combustion) analysis is normally reported, if sufficient material is available for this destructive analysis. For new substances, we are, of course, interested in establishing not only the identity, but also the molecular structure of the materials. In this situation, other modern techniques, such as $^1H$ and $^{13}C$ NMR spectroscopy, high-resolution MS, and single-crystal X-ray diffraction, can provide sensitive structural information.

When comparisons are made between experimental data and values obtained from the literature, it is essential that the latter information be obtained from the most reliable sources available. Certainly, judgment, which improves with experience, must be exercised in accepting any value as a standard. The known classical properties of a large number of compounds are found in the *CRC Handbook of Chemistry and Physics*. This reference work is a valuable source that lists inorganic, organic, and organometallic compounds. The handbook is kept current, and a new edition is published each year.

# LIQUIDS

## ULTRAMICRO-BOILING POINT

Upon heating, the vapor pressure of a liquid increases, though in a nonlinear fashion. When the pressure reaches the point where it matches the local pressure, the liquid boils. That is, it spontaneously begins to form vapor bubbles, which rapidly rise to the surface. If heating is continued, both the vapor pressure and the temperature of the liquid will remain constant until the substance has been completely vaporized (Fig. 4.1).

Since microscale preparations generally yield quantities of liquid products in the range 30–70 $\mu$L, the allocation of 5 $\mu$L or less of material to boiling point measurements becomes highly desirable. The modification of the Wiegand ultramicro boiling point procedure described here has established that reproducible and reasonably accurate ($\pm 1$ °C) boiling points can be observed on 3–4 $\mu$L of most liquids[2] thermally stable at the required temperatures.

## Procedure

Ultramicro boiling points can be conveniently determined in standard (90-mm length) Pyrex capillary melting point tubes. The melting point tube replaces the conventional 3–4 mm (o.d.) tubing used in the Siwoloboff procedure.[3] The sample (3–4 $\mu$L) is loaded into the melting point capillary via a 10-$\mu$L syringe and centrifuged to the bottom if necessary.

A small glass bell, which replaces the conventional melting point tube as the bubble generator in microboiling point determinations, is formed by

---

[2] Wiegand, C. *Angew. Chem.* **1955**, *67*, 77. Mayo, D. W.; Pike, R. M.; Butcher, S. S.; Meredith, M. L. *J. Chem. Educ.* **1985**, *62*, 1114.

[3] Siwoloboff, A. *Chem. Ber.* **1886**, *19*, 795.

heating 3-mm (o.d.) Pyrex tubing with a microburner and drawing it out to a diameter small enough to be readily accepted by the melting point capillary. A section of the drawn capillary is fused and then cut to yield two small glass bells approximately 5 mm long (Fig. 4.2a). It is important that the fused section be reasonably large. This section is more than just a seal. The fused glass must add sufficient weight to the bell so that it will firmly seat itself in the bottom of the melting point tube.

An alternate technique to prepare the glass bells is as follows: heat the midsection of an open-ended melting point capillary tube and then draw the glass to form a smaller capillary section. This section is then broken approximately in the middle and each end of the open end sealed. The appropriate length for the bell is then broken off. Thus two bells are obtained, one from each section. The sealing process (be sure that a significant section of glass is fused during the tube closure in order to give sufficient weight to the bell) can be repeated on each remaining glass section and thus a series of bells can be prepared in a relatively short period.

One of the glass bells is inserted into the loaded melting point capillary, open end first (down), and allowed to fall (centrifuged if necessary) to the bottom. The assembled system (Fig. 4.2b) is then inserted onto the stage of a Thomas-Hoover Uni-Melt® Capillary Melting Point apparatus (Fig. 4.3)[4] or similar system (such as Mel-Temp®).

The temperature is rapidly raised to 15–20 °C below the expected boiling point (the temperature should be monitored carefully in the case of unknown substances), and then is adjusted to a maximum 2 °C/min rise rate until a fine stream of bubbles is emitted from the glass bell. The heat control is then adjusted to drop the temperature. The boiling point is recorded at the point where the last escaping bubble collapses (i.e., when the vapor pressure of the substance equals the atmospheric pressure). The heater is then rapidly adjusted to again raise the temperature at 2 °C/min and induce a second stream of bubbles. This procedure may then be repeated several times. *It should be emphasized that a precise and sensitive temperature control system is essential to the successful application of this cycling technique. Although this control is a desirable feature it is not, however, essential for obtaining satisfactory boiling point data.*

Utilization of the conventional melting point capillary as the "boiler" tube has the particular advantage that the system is ideally suited for observation in a conventional melting point apparatus. The illumination and magnification available make the observation of rate changes in the bubble stream readily apparent. Economical GC injection syringes (10 μL) appear to be the most successful instrument for dealing with the small quantities of liquids involved in these transfers. The 3-in. needles normally supplied with the 10-μL barrels will not reach the bottom of the capillary; however, liquid samples deposited on the walls of the tube are easily and efficiently moved to the bottom by centrifugation. After the sample is packed in the bottom of the capillary tube, the glass bell is introduced. Use of the glass bell is necessary because, if a conventional Siwoloboff fused capillary insert is employed (it would extend beyond the top of the melting point tube), capillary action between the "boiler" tube wall and the capillary insert will draw the majority of the sample from the bottom of the tube up onto the walls. This effect often precludes the formation of the requisite bubble stream.

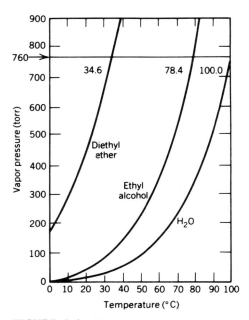

**FIGURE 4.1** Vapor-pressure curves. *From Brady, J. E.; Humiston, G. E.* General Chemistry, *3rd ed.; Wiley: New York, 1982.* (Reprinted by permission of John Wiley & Sons, New York)

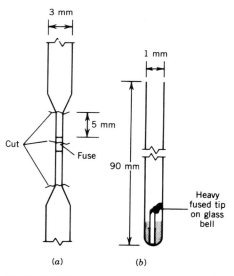

**FIGURE 4.2** (a) Preparation of small glass bell for ultramicro boiling point determination. (b) Ultramicro boiling point assembly. *From Mayo, D. W.; Pike, R. M.; Butcher, S. S.; Meredith, M. L.* J. Chem. Educ. *1985, 62, 1114.*

4 Thomas Scientific, 99 High Hill Road, P.O. Box 99, Swedesboro, NJ 08085.

**FIGURE 4.3**  Thomas-Hoover melting point determination device. *(Courtesy of Thomas Scientific, Swedesboro, NJ.)*

Little loss of low-boiling liquids occurs (see Table 4.1). Furthermore, if the boiling point is overrun and the sample is suddenly evaporated from the bottom section of the "boiler" capillary, it rapidly will condense on the upper (cooler) sections of the tube, which extend above the heat-transfer fluid. The sample can easily be recentrifuged to the bottom of the tube, and a new determination of the boiling point begun. It should be pointed out that if the bell cavity fills completely during the cooling point of a cycle, it is often difficult to reinitiate the bubble stream without first emptying the entire cavity by overrunning the boiling point.

Observed boiling points for a series of compounds, which boil over a wide range of temperatures, are summarized in Table 4.1.

**TABLE 4.1  Observed Boiling Points (°C)[a]**

| Compound | Observed | Literature Value | Reference |
|---|---|---|---|
| Methyl iodide | 42.5 | 42.4 | b |
| Isopropyl alcohol | 82.3 | 82.4 | c |
| 2,2-Dimethoxypropane | 80.0 | 83.0 | d |
| 2-Heptanone | 149–150 | 151.4 | e |
| Cumene | 151–153 | 152.4 | f |
| Mesitylene | 163 | 164.7 | g |
| p-Cymene | 175–178 | 177.1 | h |
| Benzyl alcohol | 203 | 205.3 | i |
| Diphenylmethane | 263–265 | 264.3 | j |

*REFERENCE(S)*

[a] Observed values are uncorrected for changes in atmospheric pressure (corrections all estimated to be less than ±0.5 °C).

[b] *CRC Handbook of Chemistry and Physics*, 72nd ed.; CRC Press: Boca Raton, FL, 1991–1992; no. 9082, p. 3–320.

[c] *CRC Handbook of Chemistry and Physics*, 72nd ed.; CRC Press: Boca Raton, FL, 1991–1992; no. 11972, p. 3–417.

[d] *Dictionary of Organic Compounds*, 4th ed.; Oxford University Press: London, 1965; Vol. I, p. 11.

[e] *CRC Handbook of Chemistry and Physics*, 72nd ed.; CRC Press: Boca Raton, FL, 1991–1992; no. 7627, p. 3–268.

[f] *CRC Handbook of Chemistry and Physics*, 72nd ed.; CRC Press: Boca Raton, FL, 1991–1992; no. 5394, p. 3–191.

[g] *CRC Handbook of Chemistry and Physics*, 72nd ed.; CRC Press: Boca Raton, FL, 1991–1992; no. 8987, p. 3–317.

[h] *CRC Handbook of Chemistry and Physics*, 72nd ed.; CRC Press: Boca Raton, FL, 1991–1992; no. 2192, p. 3–85.

[i] *CRC Handbook of Chemistry and Physics*, 72nd ed.; CRC Press: Boca Raton, FL, 1991–1992; no. 3160, p. 3–116.

[j] *CRC Handbook of Chemistry and Physics*, 72nd ed.; CRC Press: Boca Raton, FL, 1991–1992; no. 9074, p. 3–320.

Materials that are thermally stable at their boiling point will give identical values on repeat determinations. Substances that begin to decompose will give values that slowly drift after the first few measurements. The observation of color and/or viscosity changes, together with a variable boiling point, all signal the need for caution in making repetitive measurements.

Comparison of the boiling points obtained experimentally at various atmospheric pressures with reference boiling points at 760 torr is greatly

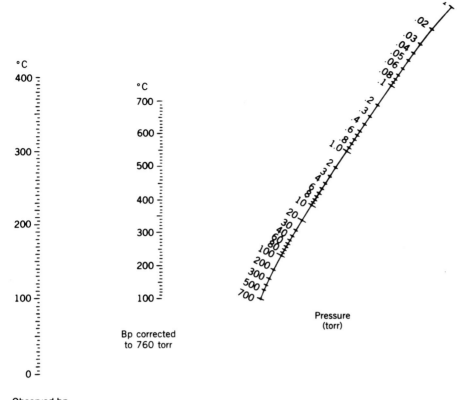

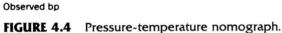

**FIGURE 4.4**  Pressure-temperature nomograph.

facilitated by the use of pressure–temperature nomographs, such as that shown in Figure 4.4. A straight line from the observed boiling point to the observed pressure will pass through the corrected boiling point value. These values can be of practical importance when carrying out reduced-pressure distillations.

Density, defined as mass per unit volume, is generally expressed as grams per milliliter (g/mL) or grams per cubic centimeter (g/cm³) for liquids. Accurate procedures have been developed for the measurement of this physical constant at the microscale level. A micropycnometer (density meter), developed by Clemo and McQuillen,[5] requires approximately 2 μL of liquid (Fig. 4.5). This very accurate device determines the density to three significant figures. The system is self-filling, and the fine capillary ends do not need to be capped while coming to temperature equilibrium, or while weighing (the measured values tend to degrade for substances boiling under 100 °C and when room temperatures are much above 20 °C). In addition, the apparatus must first be tared, filled, and then reweighed on an *analytical* balance. A technique that results in less precise densities (good to about two significant figures), but one that is far easier to use is simply to substitute a 50- or 100-μL syringe for the pycnometer. The method simply

**DENSITY**

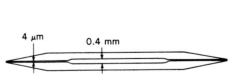

**FIGURE 4.5**  Pycnometer of Clemo and McQuillen. *From Schneider, F. L. Monographien aus dem Gebiete der qualitativen Mikroanalyse, Qualitative Organic Microanalysis, Vol. II; Benedetti-Pichler, A. A., Ed.; Springer-Verlag: Vienna, Austria, 1964.*

[5] Clemo, G. R.; McQuillen, A. *J. Chem. Soc.* **1935**, 1220.

requires weighing the syringe before and after filling it to a measured volume as in the conventional technique. With the volume and the weight of the liquid known, the density can be calculated. A further advantage of the syringe technique is that the pycnometer is not limited to a fixed volume. Although much larger samples are required, it is not inconvenient to utilize the entire sample obtained in the reaction for this measurement, as the material can be efficiently recovered from the syringe for additional characterization studies. Since density changes with temperature, these measurements should be obtained at a constant temperature.

An alternative to the syringe method is to use *Drummond Disposable Microcaps* as pycnometers. These precision-bore capillary tubes, calibrated to contain the stated volume from end to end (accuracy ±1%), are available from a number of supply houses.[6] These tubes are filled by capillary action or by suction using a vented rubber bulb (provided). The pipets can be obtained in various sizes, but as with the syringe, volumes of 50, 75, or 100 μL are recommended. When using this method, be sure to handle the micropipet with forceps and not with your fingers (heat). The empty tube is first *tared,* and then filled and weighed again. The difference in these values is the weight of liquid in the pipet. For convenience, the pipet may be placed in a small container (10-mL beaker or Erlenmeyer flask) when the weighing procedure is carried out.

## REFRACTIVE INDEX

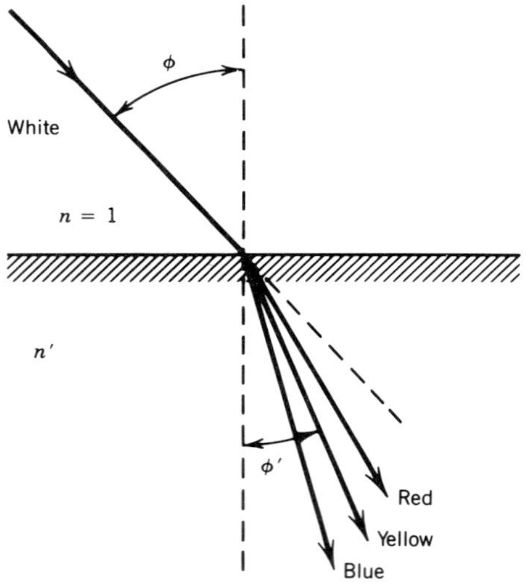

**FIGURE 4.6** Upon refraction, white light is spread out into a spectrum. This is called dispersion.

It is commonly observed that a beam of light "bends" as it passes from one medium to another. For example, an oar looks bent as one views (from the air) the portion under the water. This effect is a consequence of the refraction of light. It results from the change in velocity of the light at the interface of the media, and the angle of refraction, ($\phi'$). It is related to the velocity change as follows (see Fig. 4.6):

$$\frac{\sin \phi}{\sin \phi'} = \frac{\text{velocity in vacuum}}{\text{velocity in sample}} = n \text{ (refractive index)}$$

*where $\phi$ is the angle of incidence between the beam of light and the interface.*

Since the velocity of light in a medium must be less than that in a vacuum, the index of refraction ($n$) will always be greater than 1. In practice, $n$ is taken as the ratio of the velocity of light in air relative to the medium being measured. The refractive index is wavelength dependent.

The wavelength dependence gives rise to the effect of dispersion or the spreading of white light into its component colors. When we measure $n$, therefore, we must specify the wavelength at which the measurement is made. The standard wavelength for refractive index determinations has become the (yellow) sodium 589-nm emission, the sodium D line. Sodium, unfortunately, is a poor choice of wavelength for these measurements with organic substances, but as the sodium lamp represented one of the easiest-to-obtain monochromatic sources of light in the past, it has become widely used. Because the density of the medium is sensitive to temperature, the velocity of radiation also changes with temperature, and therefore, refractive index measurements must be made at constant temperatures. Many values in the literature are reported at 20 °C. The refractive index can be

---

[6] Drummond Disposable Microcaps are available from Thomas Scientific, 99 High Hill Road, P.O. Box 99, Swedesboro, NJ 08085; VWR Scientific, P. O. Box 626, Bridgeport, NJ 08014.

measured optically quite accurately to four decimal places. Since this measurement is particularly sensitive to the presence of impurities, the refractive index can be a valuable physical constant for tracking the purification of liquid samples.

For example, the measurement is reported as

$$n_D^{20} = 1.4628$$

In the Abbe-3L refractometer (Fig. 4.7), white light is used as the source, but compensating prisms give indexes for the D line. This refractometer is commonly used in many organic laboratories.

Samples (~10 $\mu$L) are applied between the horizontal surfaces of a pair of hinged prisms (Fig. 4.8). A sampling procedure recently developed by Ronald[7] significantly reduces the amount of sample required and allows accurate measurements on highly volatile materials. The technique involves placing a small precut 6-mm disk of good-quality lens paper at the center of the bottom prism. The sample is loaded onto the disk with a micro Pasteur pipet or a microliter syringe (see Table 4.2).

---

CAUTION:  Do not touch the prisms with the Pasteur pipet or syringe tip as they may be easily and permanently marred or scratched, and the refractometer will, from then on, give erroneous results.

---

The refractometer is adjusted so that the field of view has a well-defined light and dark split image (see your instrument manual for the correct routine for making adjustments on your particular refractometer).

**FIGURE 4.7**  Abbe-3L refractometer. *(Courtesy of Milton Roy Co., Rochester, NY.)*

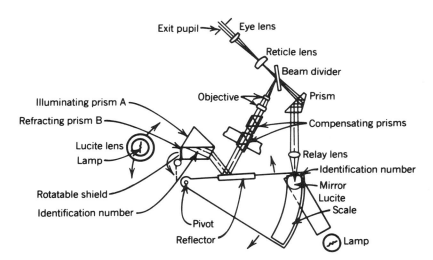

**FIGURE 4.8**  Diagram of a typical refractometer. *(Courtesy of Milton Roy Co., Rochester, NY.)*

[7] Ronald, B. P. Department of Chemistry, Idaho State University, Pocatello, ID (personal communication).

**TABLE 4.2   Refractive Index Measurements Utilizing Lens Paper Disk Technique**

| Substance | T(°C) | $n^t$ (normal) 100 $\mu$L | $n^t$ (microdisk) 2–4 $\mu$L |
|-----------|-------|---------------------------|------------------------------|
| Water | 24.5 | 1.3224 | 1.3226 |
| Diethyl ether | 24 | 1.3508 | 1.3505 |
| Chlorobenzene | 24.5 | 1.5225 | 1.5219 |
| Iodobenzene | 24.5 | 1.6151 | 1.6151 |

When using the refractometer, always clean the prisms with alcohol and lens paper before and after use. Record the temperature at which the reading is taken. A reasonably good extrapolation of temperature effects can be obtained by assuming that the index of refraction changes 0.0004 unit per degree Celsius, and that it varies inversely with temperature.

## SOLIDS

### MELTING POINTS

In general, the crystalline lattice forces holding organic solids together are distributed over a relatively narrow energy range. The melting points of organic compounds, therefore, are usually relatively sharp, that is, less than 2 °C. The range and maximum temperature of the melting point, however, are very sensitive to impurities. Small amounts of sample contamination by soluble impurities will nearly always result in melting point depressions.

The drop in melting point is usually accompanied by an expansion of the melting point range. Thus, in addition to the melting point acting as a useful guide in identification, it also can be a particularly effective indication of sample purity.

### Procedure

In the microscale laboratory, two different types of melting point determinations are carried out: (1) simple capillary melting points and (2) evacuated melting points.

#### Simple Capillary Melting Point

Because the microscale laboratory utilizes the Thomas-Hoover Uni-Melt apparatus or a similar system for determining boiling points, melting points are conveniently obtained on the same apparatus. The Uni-Melt system utilizes an electrically heated and stirred silicone oil bath. The temperature readings require no correction in this case as the depth of immersion is held constant. (This assumes, of course, that the thermometer is calibrated to the operational immersion depth.) Melting points are determined in the same capillaries as boiling points. The capillary is loaded by introducing about 1 mg of material into the open end. The sample is then tightly packed (~2 mm) into the closed end by dropping the capillary down a length of glass tubing held vertically to the bench top. The melting point tube is then ready for mounting in the metal stage, which is immersed in the silicone oil bath of the apparatus. If the melting point of the substance is expected to occur in a certain range, the temperature can be rapidly raised to about 20 °C below the expected value. At that point, the

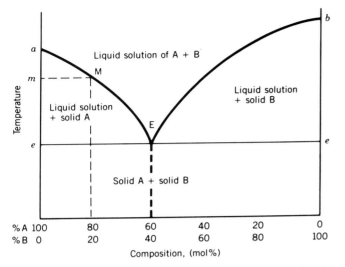

**FIGURE 4.9** Melting point composition diagram for the binary mixture, A + B. In this diagram, a is the melting point of the solid A, b of solid B, e of eutectic mixture E, and m of the 80% A:20% B mixture, M.

temperature rise should be adjusted to a maximum of 2 °C/min, which is the standard rate of change at which the reference determinations are obtained. The melting point range is recorded from the temperature at which the first drop of liquid forms (point e on Fig. 4.9) to that at which the last crystal melts (point m on Fig. 4.9).

### Evacuated Melting Points

Many organic compounds begin to decompose at their melting points. This decomposition often begins as the melting point is approached and may adversely affect the values measured. The decomposition can be invariably traced to reaction with oxygen at elevated temperatures. If the melting point is obtained in an evacuated tube, therefore, much more accurate melting points can be obtained. These more reliable values arise not only from increased sample stability, but because several repeat determinations can often be made on the same sample. The multiple measurements may then be averaged to provide more accurate data.

Evacuated melting points are quickly and easily obtained with a little practice. The procedure is as follows: Shorten the capillary portion of a Pasteur pipet to approximately the same length as a normal melting point tube (see Fig. 4.10a). Seal the capillary end by rotating in a microburner flame. Touch the pipet only to the very edge of the flame, and keep the large end at an angle below the end being sealed (see Fig. 4.10b). This technique will prevent water from the flame being carried into the tube where it will condense in the cooler sections. Then load 1–2 mg of sample into the drawn section of the pipet with a microspatula (see Fig. 4.10c). Tap the pipet gently to seat the solid powder as far down the capillary as it can be worked (see Fig. 4.10d). Then push the majority of the sample partially down the capillary with the same diameter copper wire that you used to seat the cotton plug in constructing the Pasteur filter pipet (see Fig. 4.10e). Next, connect the pipet to a mechanical high-vacuum system with a piece of vacuum tubing. Turn on the vacuum and evacuate the pipet for 30 seconds (see Fig. 4.10f). With a microburner, gently warm the surface of

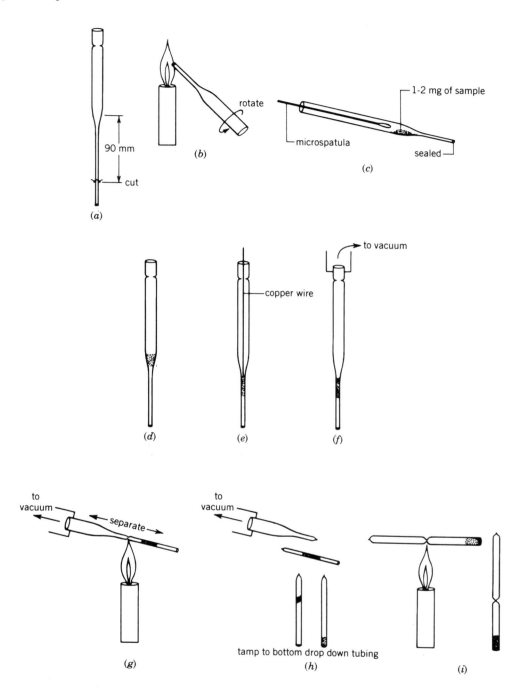

**FIGURE 4.10**   Procedure for obtaining evacuated melting point capillaries.

the capillary tubing just below the drawn section. On warming, the remaining fragments of the sample (the majority of which has been forced further down in the tube) will sublime in either direction away from the hot section. Once the traces of sample have been "chased" away, the heating is increased, and the capillary tube collapsed, fused, and separated from the shank, which remains connected to the vacuum system (see Fig. 4.10g). The vacuum system is then vented and the shank is discarded. The sample is tightly packed into the initially sealed end of the evacuated capillary by dropping down a section of glass tubing, as in the case of packing open melting point samples. After the sample is packed (~2 mm in length, see Fig. 4.10h), a section of the evacuated capillary about 10–15 mm above the sample is once more gently heated and collapsed by the microburner flame (see Fig. 4.10*i*).

This procedure is required to trap the sample below the surface of the heated silicone oil in the melting point bath, and thus avoid sublimation up the tube to cooler sections during measurement of the melting point. The operation is a little tricky and should be practiced a few times. It is very important that complete fusion of the tubing take place. Now the sample is ready to be placed in the melting point apparatus. The procedure beyond this point is the same as in the open capillary case, except that after the sample melts, it can be cooled, allowed to crystallize, remelted several times, and the average value of the range reported. If these latter values begin to drift downward, the sample can be considered to be decomposing even under evacuated, deoxygenated conditions. In this case, the first value observed should be recorded as the melting point, and the decomposition noted (mp = xx dec, where dec = decompose).

---

## MIXTURE MELTING POINTS

Additional information can often be extracted from the sensitivity of the melting point to the presence of impurities. Where two different substances possess identical melting points (not an uncommon occurrence), it would be impossible to identify an unknown sample as either material based on the melting point alone. If reference standards of the two compounds are available, however, then mixtures of the unknown and the two standards can be prepared. It is important to prepare several mixtures of varying concentrations for melting point comparisons, as the point of maximum depression need not occur on the phase diagram at the 50:50 ratio (see Fig. 4.9). The melting points of the unknown and the mixed samples are conveniently obtained simultaneously (the Uni-Melt stage will accept up to seven capillaries at one time). The unknown sample and the mixture of the unknown with the correct reference will have identical values, but the mixture of the reference with a different substance will give a depressed melting point. This procedure is a classical step in the positive identification of crystalline solids. Only very rarely do mixtures of two different compounds fail to exhibit mixture melting point depression.

# 5

# Development of Microscale Techniques

**C₅H₆, Propellane**
Wiberg and Walker (1982).

This chapter introduces the microscale organic laboratory techniques employed throughout the experimental sections of the text. These operations must be mastered in order to be successful when working at this scale. Detailed discussions are given for each individual experimental technique. At the end of each of these discussions, you will find a note listing the experiments described in Chapters 6–8 in which that particular technique plays a significant role. These lists should prove useful to instructors assembling experiments to be covered in the laboratory part of a course. The lists also will be handy for students who wish to examine the application of a particular technique to other experiments not covered in their laboratory sequence.

One of the principal hurdles in dealing with experimental chemistry is the isolation of materials in their pure state. Characterization of a substance requires a pure sample of the material. In organic chemistry, this is a particularly difficult demand since most organic reactions generate several products. We are generally satisfied if the desired product is the major component of the mixture obtained. Thus, in this chapter a heavy emphasis is placed on separation techniques.

## TECHNIQUE 1

### Microscale Separation of Liquid Mixtures by Preparative Gas Chromatography

**INTRODUCTION**

Technique 1 begins the discussion of the resolution (separation) of microliter quantities of liquid mixtures via preparative gas chromatography. Techniques 2 and 3 deal with semimicro adaptations of classical distillation routines that focus on the separation of liquid mixtures involving one to several milliliters of material.

The methods of chromatography have revolutionized experimental organic chemistry over the past 30-odd years. These methods are by far the most powerful of the techniques for separating mixtures and isolating pure substances, either solids or liquids. Chromatography can be defined as the resolution of a multicomponent mixture (several hundred in some cases) by distribution between two phases, one stationary and one mobile. The various methods of chromatography are categorized by the phases involved: column, thin-layer, and paper (solid–liquid); partition (liquid–liquid); and vapor phase (gas–liquid). The principal mechanism on which these separations depend is differential solubility, or adsorbtivity, of the mixture components with respect to the two phases involved. That is, the components must exhibit different partition coefficients.

Gas chromatography (GC, or sometimes termed vapor-phase chromatography, VPC) is an extraordinarily powerful technique for the separation of mixtures. In this case, the stationary phase is a high-boiling liquid and the mobile phase is a gas (the carrier gas). The GC systems develop resolutions vastly superior to those obtained via distillation techniques.

**Prep-GC** separations, which involve processing greater than submicroliter quantities of materials, require relatively sophisticated instrumentation and, even then, considerable time may be expended to develop the appropriate operating conditions.

On the other hand, **analytical-GC** separations, which require small quantities of material (often $<0.1 \mu L$), are extremely powerful. The necessity of working with small quantities of materials in GC separations, however, becomes advantageous at the microscale level. Note that this analytical mode is utilized in assaying distillation fractions in Experiments [3C] and [3D].

The instrumentation required to carry out GC can range from straightforward, and relatively simple systems, to those containing highly automated and relatively expensive components. A diagram of a common and simple instructional laboratory GC is given in Figure 5.1.

**Injection port:**   The analysis begins in a heated injection port. The sample mixture is introduced by syringe through a septum into the high-temperature chamber (injection port) through which the inert carrier gas (the mobile phase) is flowing. Helium and nitrogen are commonly used as carrier gases. The solubility of the sample in the carrier gas depends to a large extent on the vapor pressure of the substances in the mixture. Thus, heating the injection port helps to insure vaporization of less volatile samples. These requirements place two major constraints on GC: (1) the sample must be stable at the temperature required to cause vaporization, and (2) the sample must possess sufficient vapor pressure to be completely soluble in the carrier gas at the column operating temperatures.

NOTE.   *When injecting a sample, remember to always have your thumb positioned over the syringe plunger. This technique prevents a blow-back of the sample by the carrier gas pressure in the injection port.*

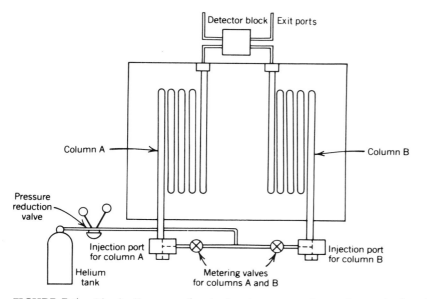

**FIGURE 5.1** Block diagram of a dual-column gas chromatograph showing essential parts. *(Courtesy of GOW-MAC Instrument Co., Bound Brook, NJ.)*

**Column:** The vaporized mixture is swept by the carrier gas from the injection port onto the column. Bringing the sample mixture into intimate contact with the column constitutes the key stage in the separation process. The stationary liquid phase, in which the sample will dissolve and partition with the mobile gas phase, is physically and/or chemically bonded to inert packing (often referred to as the "support") material contained in the column. Gas chromatographic columns are available from manufacturers in a variety of sizes and shapes. In the diagram of the GOW-MAC instrument (Fig. 5.1), two parallel coiled columns are mounted in a well-insulated oven. Considerable oven space can be saved, and better temperature regulation achieved, if the columns are coiled. Temperature regulation is particularly important, as column resolution degrades rapidly if the entire column is not at the same temperature, and most liquid mixtures will require oven temperatures above ambient to maintain reasonable vapor pressures during the course of the separation.

Resolution of the mixture occurs as the carrier gas sweeps the sample through the column. Most columns are constructed of stainless steel, glass, or fused silica. The diameter and length of the column are critical factors in determining how the internal part of the column is designed to achieve maximum separation of the sample mixture.

*Packed Columns:* In these columns the surface area of the liquid phase, in contact with the sample contained in the moving gas phase, is maximized by coating a finely divided inert support with a nonvolatile liquid material (stationary phase). The coated support is carefully loaded into the column to avoid void channels. Columns prepared in this fashion are termed "packed columns." Packed columns are usually 1/4 or 1/8 in. in diameter and range from 4 to 12 ft long. These columns are particularly attractive for use in the microscale laboratory as they can function in both the analytical and preparative modes. Numerous examples are known where simple mixtures ranging from 20 to 80 $\mu$L can be resolved into their pure components and collected at the exit port of the detector. On the other hand, samples in the 0.2–2-$\mu$L range will exhibit quite good analytical resolution.

*Capillary Columns:* Research grade columns are also available that have no packing; the liquid phase is simply applied directly to the walls of the column. These columns are referred to as wall-coated or open-tubular columns. The reduction in surface area is compensated for by making the diameter very small (0.1 mm) and the length very long (100 m would not be uncommon). Termed capillary columns, they are the most efficient columns employed for analytical separations. Mixtures of several hundred compounds can be completely resolved in a single pass through one of these systems. Capillary columns, generally, require a more sophisticated and expensive operating system, and, in addition, they are restricted to very small sample loading (0.1 $\mu$L or less). Thus, capillary columns cannot be used for preparative separations.

**Liquid phase:** Once the sample is introduced on the column (in the carrier gas), it will undergo partition with the liquid phase. The choice of the liquid phase is particularly important since it directly affects the relative distribution coefficients.

In general, the stationary liquid phase controls the partitioning of the sample by two criteria: (1) If little or no interaction occurs between the sample components and the stationary phase, the boiling point of the materials will determine the order of elution. Under these conditions, the highest boiling species will be the last to elute. (2) The functionality of the components may interact directly with the stationary phase to establish different partition coefficients. Elution in this latter case will depend on the particular binding properties of the sample components.

Some typical materials employed as stationary phases are shown below.

| Name | Stationary Phase[a] | Maximum Temperature (°C) | Mechanism of Interaction |
|---|---|---|---|
| Silicone oil DC 710, etc. | $R_3Si[OSiR_2]_nOSiR_3$ | 250 | According to boiling point |
| Polyethylene glycol (Carbowax®) | $HO[CH_2CH_2O]_nCH_2CH_2OH$ | 150 | Relatively selective toward polar compounds |
| Diisodecyl phthalate | $o\text{-}C_6H_4[CO_2\text{-isodecyl}]_2$ | 175 | According to boiling point |

[a] R = alkyl substituent

**Oven temperature:** The temperature of the column will also affect the separation. In general, the elution period of the sample components will decrease as the temperature is increased. That is, the retention times on the column are shorter at higher temperatures. Higher boiling components tend to undergo diffusion broadening at low column temperatures because of the increase in retention times. Programmed oven temperature increases, which speed up elution of the higher boiling components, tend to suppress peak broadening, and therefore, increase resolution. If the oven temperature is too high, however, equilibration with the stationary phase will not be established and the component mixture may elute together or, at best, undergo incomplete resolution. Obviously, temperature-programming capabilities require more costly ovens and controllers.

**Flow rate:** The flow rate of the carrier gas constitutes another important parameter. The rate must be slow enough to allow equilibration between the phases, but sufficiently rapid to ensure that diffusion will not overcome resolution of the components.

**Column Length:**   As noted, the length of the column also is an important factor in separation performance. As we will see in the distillation discussions, distillation column efficiency is proportional to column height, as it determines the number of evaporation–condensation cycles. In a similar manner, increasing the length of the GC column allows more partition cycles to occur. Difficult-to-separate mixtures, such as the xylenes (very similar boiling points, *o*-xylene, 144.4 °C, *m*-xylene, 139.1 °C, and *p*-xylene, 138.3 °C), will have a better chance of being resolved on longer columns. In fact, both GC and distillation resolution data are described using the same term, **theoretical plates** (see Technique 3 and Experiments [3C] and [3D]).

**Detector and Exit Port:**   A successfully resolved mixture will elute as individual components, sequentially with time, at the exit port (also temperature controlled) of the instrument. To monitor the exiting vapors, a detector is placed in the gas stream just prior to the above outlet (Fig. 5.1). After passing through the detector, the carrier gas and the separated sample components are then vented.

A widely used detector is the nondestructive, thermal conductivity sensor, often referred to as a hot-wire detector. A heated element in the gas stream changes electrical resistance when a substance dilutes the carrier gas and changes its thermal conductivity. Helium possesses a higher thermal conductivity than most organic substances. When samples other than helium are present, the conductivity of the gas stream decreases and the resistance of the heated wire changes. The change in resistance is measured by differences (Wheatstone bridge), with a reference detector mounted on a second (parallel) gas stream. This signal is plotted by a recorder. On the plot, the horizontal axis is time and the vertical axis is the magnitude of the resistance difference. The plot of resistance difference versus time is referred to as the **chromatogram**. Retention time ($t_R$) is defined as the time from sample injection to the time of maximum peak intensity. The baseline width ($W_b$) of a peak is defined as the distance between two points where tangents to the points of inflection cross the baseline (Fig. 5.2).

**Theoretical Plates:**   It is possible to estimate the number of theoretical plates (directly related to the number of distribution cycles) present in a column for a particular substance. The parameters are given in the relationship[1]

$$n = 16[t_R/W_b]^2$$

where the units of retention time ($t_R$) and baseline width ($W_b$) are the same (minutes, seconds, or centimeters). As in distillation columns, the larger the number of theoretical plates, *n*, the higher the resolution of the column.

The efficiency of a system may also be expressed as the *height equivalent theoretical plate* (HETP) in centimeters (or inches)/plate. This parameter is related to the number of theoretical plates *n* by

$$\text{HETP} = \frac{L}{n}$$

[1] Berg, E. W. *Physical and Chemical Methods of Separation*; McGraw-Hill: New York, 1963, p. 111.

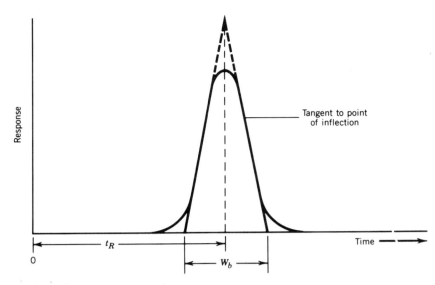

**FIGURE 5.2** Schematic chromatogram.

where $L$ is the length of the column, usually reported in centimeters. The smaller the HETP, the more efficient the column.

The number of theoretical plates available in fractional distillation columns is limited by column holdup (see Techniques 2A and 2B). Thus, distillations of less than 500 $\mu$L are not practical. Gas chromatographic columns, on the other hand, operate **most efficiently at the microscale or submicroscale levels**, where 500 $\mu$L would be an order (even 3–5 orders in the case of capillary columns) of magnitude too large.

**Fraction Collection:** Sequential collection of the separated materials can be made by attaching suitable sample condensing tubes to the exit port (see Fig. 3.6).

**Procedure for Preparative Collection**

The collection tube (oven dried until 5 min before use) is attached to the heated exit port by the metal 5/5₮ joint. Sample collection is initiated 30 s prior to detection of the expected peak on the recorder (time based on previously determined retention values)[2] and continued until 30 s following the return to baseline. After the collection tube is detached, the sample is transferred to the 0.1-mL conical GC collection vial. The transfer is facilitated by the 5/5₮ joint on the conical vial. After the collection tube is joined to the vial (preweighed with cap), the system is centrifuged (see Fig. 5.3). The collection tube is then removed, and the vial capped and reweighed.

The efficiency of collection can exceed 90% with most materials, even relatively low-boiling substances. In the latter case, the collection tube, after attachment to the instrument, is wrapped with a paper tissue. As the tube is being wrapped, it is also being flushed by the carrier gas. The wrapping is then saturated with liquid nitrogen.

The role of preparative GC in the microscale laboratory is that of a powerful replacement for the technique of fractional distillation in product purification. As all liquid product mixtures in the microscale laboratory are less than 500 $\mu$L, distillation techniques are impractical in most cases.

---

[2] Refer to your local laboratory instructions.

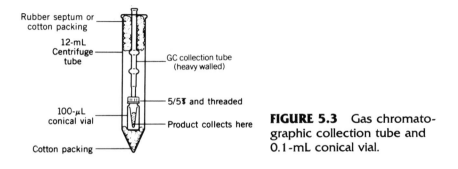

Rubber septum or
cotton packing

12-mL
Centrifuge
tube

GC collection tube
(heavy walled)

5/5ℑ and threaded

100-µL
conical vial

Product collects here

Cotton packing

**FIGURE 5.3** Gas chromatographic collection tube and 0.1-mL conical vial.

Refer to Experiment [2] for specific experimental details on preparative GC as applied to the separation of a number of binary mixtures. These are designed as practice examples to give you experience at sample collection.

**NOTE.** *Gas chromatographic purification of reaction products is suggested in the following list of experiments: Experiments [2], [3C], [3D], [5A], [5B], [8C], [9], [10], [13], [17], and [32].*

## Semimicroscale Distillation

### INTRODUCTION

Distillation is the process of heating a liquid to the boiling point, condensing the heated vapor by cooling, and returning either a portion of, or none of, the condensed vapors to the distillation vessel. Distillation varies from the process of reflux (see p. 24) only in that at least some of the condensate is removed from the boiling system. Distillations in which a fraction of the condensed vapors are returned to the boiler are often referred to as being under "partial reflux." Three types of distillations will be described under the headings Techniques 2A, 2B and 3.

### Distillation Theory

Distillation techniques often can be used for separating two or more components on the basis of differences in their vapor pressures. Separation can be accomplished by taking advantage of the fact that the vapor phase is generally richer in the more volatile (lower boiling) component of the liquid mixture. Molecules in a liquid are in constant motion and possess a range of kinetic energies. Those with higher energies (a larger fraction for the lower boiling component) moving near the surface have a greater tendency to escape into the vapor (gas) phase. If a pure liquid (e.g., hexane) is in a closed container, eventually hexane molecules in the vapor phase will reach equilibrium with hexane molecules in the liquid phase. The pressure exerted by the hexane vapor molecules at a given temperature is called the *vapor pressure* and is represented by the symbol $P_H^\circ$ where the superscript $^\circ$ indicates a pure component. For any pure component A, the vapor pressure would be $P_A^\circ$. Suppose a second component (e.g., toluene) is added to the hexane. The total vapor pressure ($P_{total}$) is then the sum of the individual component *partial vapor pressures* ($P_H$, $P_T$), where $P_H$ is the partial pressure of hexane, and $P_T$ is the partial pressure of toluene as given by *Dalton's law*.

$$P_{total} = P_H + P_T$$

or in general

$$P_{total} = P_A + P_B + P_C + \ldots + P_n$$

Assuming that the vapors are ideal, the mole fraction of hexane in the **vapor phase** is given by

$$Y_H = P_H/P_{total} \tag{5.1}$$

It is important to realize that the vapor pressure $(P_A^\circ)$ and the partial vapor pressure $(P_A)$ are not equivalent, since the presence of a second component in the liquid system has an effect on the vapor pressure of the first component. If the solution is ideal, the partial vapor pressure of hexane is given by Raoult's law

$$P_H = P_H^\circ X_H \tag{5.2}$$

where $X_H$ is the mole fraction of hexane in the **liquid system**.

For ideal solutions, Eqs. 5.1 and 5.2 may be combined to obtain the phase diagram shown in Figure 5.4. In this figure and elsewhere we will drop the subscripts from $X_H$ and $Y_H$. Here $X$ and $Y$ will represent the mole fractions of the <u>m</u>ost <u>v</u>olatile <u>c</u>omponent (MVC) (hexane in this case) in the liquid and vapor phases, respectively.

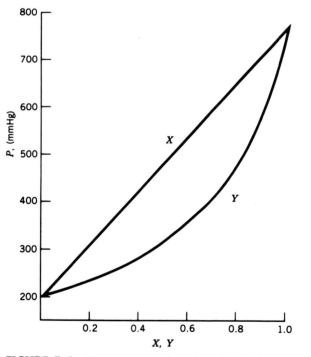

**FIGURE 5.4**  Pressure as a function of liquid composition ($X$) and vapor composition ($Y$), for hexane and toluene (temperature held constant at 69 °C).

Figure 5.4 describes hexane and toluene mixtures at a fixed temperature. For the region above the $X$ curve, there will be only liquid present. For the region below the $Y$ curve, there will be only vapor. In the area between (the sloping, lens-shaped region), liquid and vapor will be present in equilibrium. This area is the only region of interest to us in examining the distillation process.

To understand what the phase diagram tells us about the composition of the liquid and vapor phases, let us imagine that the total pressure of the system is 500 torr, shown by the horizontal line in Figure 5.5. At this pressure a liquid of composition $X_1$ will be in equilibrium with a vapor of composition $Y_1$. These two points are defined by the intersection of the constant pressure line with the $X$ and $Y$ curves. It is important to note here that $Y$ will be greater than $X$ for the equilibrium system. That is, the vapor in equilibrium with a given liquid will be richer in the more volatile component than in the liquid.

Diagrams such as Figure 5.5 are not very useful in describing the distillation process. We need a phase diagram for the mixture at constant pressure instead of constant temperature. Figure 5.4 may be transformed to the desired diagram if we know how $P_H^o$ and $P_T^o$ depend on $T$. This information may be supplied by the Clausius–Clapeyron equation or by appropriate experimental data.

We will obtain a qualitative diagram for temperature as a function of composition by the following reasoning: (1) The substance having the *higher* vapor pressure at a given temperature will have the *lower* boiling point at a given pressure. (2) At *low* temperatures, only the liquid phase will be present, and at *high* temperatures only the vapor phase will be

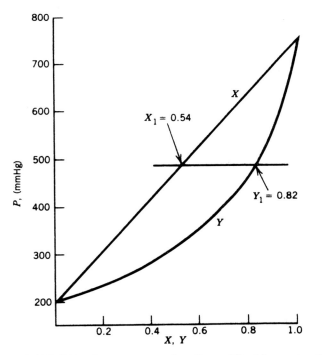

**FIGURE 5.5** Pressure as a function of liquid composition ($X$) and vapor composition ($Y$), for hexane and toluene (temperature held constant at 69 °C).

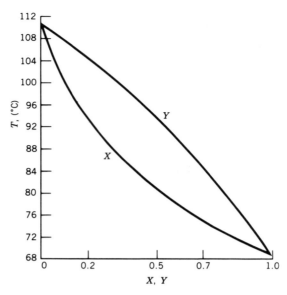

**FIGURE 5.6** Temperature as a function of liquid composition (X) and vapor composition (Y).

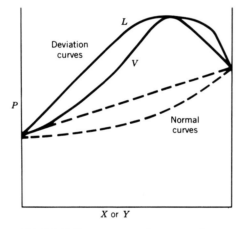

**FIGURE 5.7** Positive deviation from Raoult's law; L = liquid, V = vapor.

present. Thus, the temperature-composition diagram is shown by Figure 5.6. Note that this figure may also be obtained (qualitatively) by turning Figure 5.4 upside down.

Many pairs of liquids do not obey Raoult's law. Often, pairs of liquids encountered in organic chemistry exhibit a positive deviation from Raoult's law. The *positive deviation* means that the pressure above the solution is *greater than* would be predicted by Raoult's law. If this deviation is large, the pressure composition curve may exhibit a maximum, as shown in Figure 5.7. Here the curves for normal Raoult's law behavior are shown as dashed lines for reference. Mixtures in which one of the components is polar and the other component is at least partly nonpolar often exhibit positive deviations from Raoult's law.

The temperature–composition diagram for systems showing a positive deviation from Raoult's law is again obtained by the simple inversion process. Such a diagram is shown in Figure 5.8. In this diagram at a temperature of $T_1$ a liquid of composition $X_1$ will be in equilibrium with a vapor

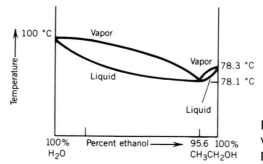

**FIGURE 5.8** Ethanol–water minimum boiling point phase diagram.

of $Y_1$. At a temperature of $T_{az}$, however, the composition of the liquid and vapor will be the same. This mixture is an *azeotropic* or constant-boiling mixture. Water ($H_2O$) and ethanol ($CH_3CH_2OH$) form one of the more familiar azeotropic systems. This mixture exhibits a positive deviation from Raoult's law and has a minimum boiling azeotrope at 78.1 °C, which consists of 95.6% ethanol by volume.

## Technique 2A  Simple Distillation at the Semimicroscale Level

### Process

Simple distillation involves the use of the distillation process to separate a liquid from minor components that are nonvolatile, or that have boiling points at least 30–40 °C above that of the major component. A typical setup for a macroscale distillation of this type is shown in Figure 5.9. At the microscale level, when one is working with volumes smaller than 500 $\mu L$, GC techniques (see Technique 1) have replaced conventional microdistillation processes.[3] Semimicroscale simple distillation in the volume range 0.5–2 mL still remains an effective separation technique. Apparatus have been developed that achieve effective separation of mixture samples in this

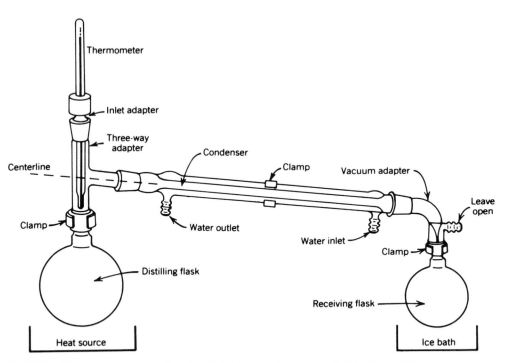

**FIGURE 5.9**  A complete simple distillation setup. *From Zubric, James W.* The Organic Chem Lab Survival Manual, *3rd ed.; Wiley: New York, 1992. (Reprinted by permission of John Wiley & Sons, Inc., New York.)*

[3] Schneider, F. L. *Monographien aus dem Gebiete der qualitativen Mikroanalyse*, Vol. II: *Qualitative Organic Microanalysis*; A. A. Benedetti-Pichler, Ed.; Springer-Verlag: Vienna, Austria, 1964; p. 31.

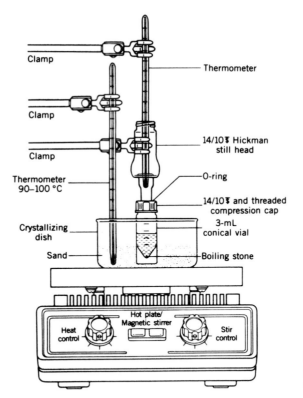

**FIGURE 5.10**  Hickman still [14/10₺ with conical vial (3 mL)].

range. One of the most significant of these designs is the classic Hickman still, shown in Figure 5.10.

This still is employed in several modes in microscale experiments described in Chapters 6–8; such as the purification of solvents, carrying out reactions, and concentration of solutions for recrystallization. An introduction to the use of the Hickman still is given in Experiment [3].

In a distillation where a liquid is separated from a nonvolatile solute, the vapor pressure of the liquid is lowered by the presence of the solute, but the vapor phase consists only of one component. Thus, except for the incidental transfer of nonvolatile material by splashing, the material condensed should consist only of the volatile component.

We can understand what is going on in a simple distillation of two volatile components by referring to the phase diagrams shown in Figures 5.11 and 5.12. Figure 5.11 is the phase diagram for hexane and toluene. The boiling points of these liquids are separated by 42 °C. Figure 5.12 is the phase diagram for methylcyclohexane and toluene. Here the boiling points are separated by only 9.7 °C.

Imagine a simple distillation of the hexane–toluene pair in which the liquid in the pot is 50% hexane. In Figure 5.11, when the liquid reaches 80.8 °C it will be in equilibrium with a vapor having a composition of 77% hexane. This result is indicated by the line *A–B*. If this vapor is condensed to a liquid of the same composition, as shown by line *B–C*, we will have achieved a significant enrichment of the condensate with respect to hexane. This change in composition is referred to as a simple distillation. The process of evaporation and condensation is achieved by the theoretical construct known as a *theoretical plate*. When this distillation is actually done with a Hickman still, some of the mixture will go through one evaporation and condensation cycle, some will go through two of these cycles, and some may be splashed more directly into the collar. A resolution of between one and two theoretical plates is generally obtained.

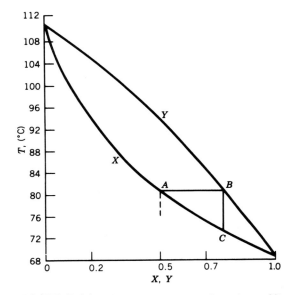

**FIGURE 5.11**  Temperature as a function of liquid composition (*X*) and vapor composition (*Y*): hexane and toluene.

Referring to Figure 5.12, if we consider the same process for a 50% mixture of methylcyclohexane and toluene, the methylcyclohexane composition will increase to 58% for one theoretical plate. Thus, simple distillation may provide adequate enrichment of the MVC if the boiling points of the two liquids are reasonably well separated, as they are for hexane and toluene. If the boiling points are close together, as they are for methylcyclohexane and toluene, the simple distillation will not provide much improvement.

As we continue the distillation process and remove some of the MVC by condensing it, the residue in the heated flask becomes less rich in the MVC. This means that the next few drops of condensate will be less rich in the MVC. As the distillation is continued the condensate becomes less and less rich in the MVC.

We can improve upon simple distillation by repeating the process. For example, we could collect the condensate until about one third is obtained. Then we could collect a second one-third aliquot in a separate container. Our original mixture would then be separated into three fractions. The first one third would be richest in the MVC and the final one

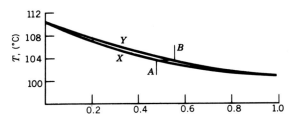

**FIGURE 5.12**  Temperature as a function of liquid composition (*X*) and vapor composition (*Y*): methylcyclohexane and toluene.

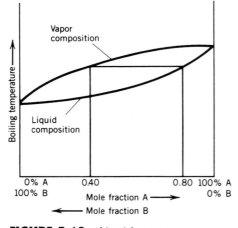

**FIGURE 5.13**   Liquid–vapor composition curve.

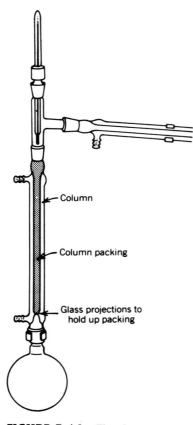

**FIGURE 5.14**   The fractional distillation setup. *From Zubric, James W. The Organic Chem Lab Survival Manual, 3rd ed.; Wiley: New York, 1992. (Reprinted by permission of John Wiley & Sons, Inc., New York.)*

third in the pot would be richest in the least volatile component. If the MVC were the compound of interest we could redistill the first fraction collected (in a clean flask!) and collect the first one third of the material condensing in that process. This simplest of all fractional distillation strategies is employed in Experiment [3B].

## Technique 2B   Fractional Distillation at the Semimicroscale Level

### Process

Fractional distillation is the application of a distillation system containing more than one theoretical plate. This process must be used if fairly complete separation is desired when the boiling points of the components differ by less than 30–40 °C. In this situation a fractionating column is required to increase the efficiency of the separation. As discussed previously, it may be seen from a liquid–vapor composition curve (Fig. 5.13) that the lower boiling component of a binary mixture makes a larger contribution to the vapor composition than does the higher boiling component. On condensation, the liquid formed will be richer in the lower boiling component. This condensate will not be pure, however, and, in the case of closely boiling components, it may show only slight enrichment. If the condensate is volatilized a second time, the vapor in equilibrium with this liquid will show a further enrichment in the lower boiling component. Thus, the trick to separating liquids that possess similar boiling points is to repeat the vaporization–condensation cycle many times. Each cycle is one *theoretical plate*. A number of different column designs are available for use at the macro level, which achieve varying numbers of theoretical plates (see Fig. 5.14).

In most distillation columns, the design is such that increased fractionation efficiency is dependent on a very large increase in the surface area in contact with the vapor phase. This increased surface area is normally accomplished by packing the fractionating column with wire gauze or glass beads. Unfortunately, a large volume of liquid must be distributed over the column surface in equilibrium with the vapor. Furthermore, the longer the column the more efficient it becomes, but longer columns also require additional liquid phase. The column requirement of the liquid phase is termed *column holdup*. Column holdup is defined as the amount of liquid distributed over the column packing required to maintain the system in equilibrium. This material is essentially lost from the liquid phase held in the distillation pot. The amount of column holdup can be large compared with the total volume of material available for the distillation. With mixtures of less than 2 mL, column holdup precludes the use of the most common fractionation columns. Microfractionating columns constructed of rapidly spinning bands of metal gauze or Teflon have very low column holdup and have a large number of plates relative to their height (Fig. 5.15). These columns are, however, rather expensive and normally are available only for research purposes.

In the development of the microscale laboratory, several new distillation systems have been designed. The microscale spinning band distillation apparatus (Fig. 5.16) can achieve nearly 12 theoretical plates and operates simply enough to be used in the instructional laboratory. This system contains a Teflon band that fits rather closely inside an insulated glass tube. The Teflon band has spiral grooves which, when the band is spun, rapidly return condensed vapor to the distillation pot. The Teflon band is rotated at 1000–1500 rpm when the column is fully operational. A powerful extension of this type of instructional apparatus uses a short spinning band inside a modified Hickman still head (see Fig. 3.21). These stills are referred to as Hickman–Hinkle stills, and 4-cm Hickman–Hinkle columns have been rated in excess of 10 theoretical plates! The commercially available 2.5-cm model is rated at 6 plates. Distillation Experiments [3C] and [3D] involve fractional distillation with spinning band columns.

An alternative to the microscale spinning band distillation column is the concentric-tube column. In these columns the fractionating section is constructed of two concentric tubes in which the vapor–liquid equilibrium is established within the annular space between the two columns. The resolution of the concentric tube system is inversely proportional to the thickness of the annular ring. Columns of this type can achieve very good separations with the number of theoretical plates approaching three per centimeter. In addition, column holdup can be close to 10 $\mu$L per theoretical plate. The major constraint in the use of these columns is the very low throughput, which can be as little as 100 $\mu$L/hour. Concentric tube columns also have the nasty habit of flooding even in experienced hands. These factors cause long residence times at elevated temperatures for the liquid components. When time and thermal stability are not a problem, the concentric-tube column can be a powerful system for mixture separations.

Experiment [3B] utilizes the Hickman still in a simple demonstration of fractional distillation. The system is arranged so that the thermometer is positioned directly down the center of the still column with the bulb ex-

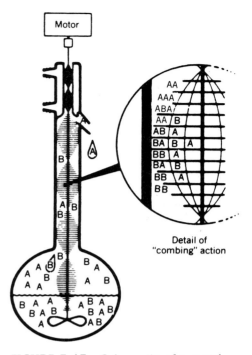

**FIGURE 5.15** Schematic of a metal mesh spinning band still.

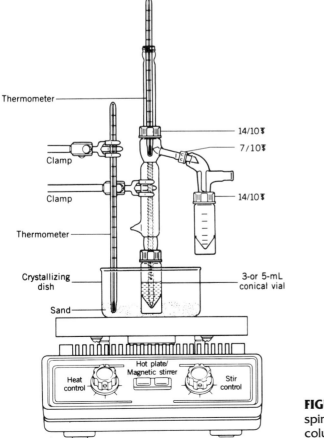

**FIGURE 5.16** Micro-spinning band distillation column (3 in.).

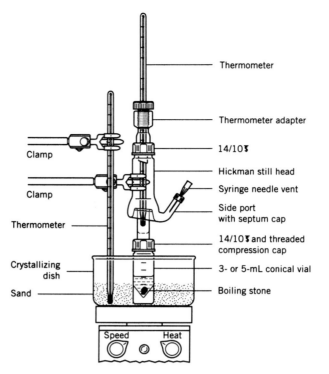

**FIGURE 5.17** Hickman still with thermometer adapter.

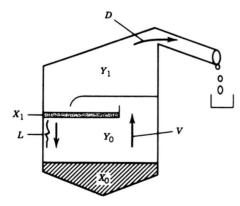

**FIGURE 5.18** Model for fractional distillation with partial reflux.

tending just to the bottom of the well. As assembled, the system functions as a very rough concentric tube fractionating column. Thus, it is very important to position both the still and the thermometer as close to the vertical as possible, and in no circumstances should the two elements come into direct contact (see Fig. 5.17). A successful, two-theoretical-plate distillation is obtained with a two-component mixture by carrying out two *sequential* fractional distillations with this system.

To understand how the spinning band improves the performance of the column, and to see some of the important characteristics of these fractionating systems, we will analyze them a bit further. Figure 5.18 shows a very simple distillation column possessing just one plate. Overall, however, this system has two theoretical plates.

The compositions of the liquid in the pot and of the infinitesimal amount of liquid at the first plate are $X_0$ and $X_1$, respectively. Both $Y_0$ and $Y_1$ are compositions of the vapor in equilibrium with each of the liquids, $X_0$ and $X_1$, respectively. Since we are talking about the MVC, $Y_0 > X_0$, $Y_1 > X_1$, and we may assume that $Y_1 > Y_0$ (i.e., we have already established that the vapor will be enriched in the MVC). The parameter $V$ is the *rate* at which vapor is transported upward, $L$ is the rate of downward transport of liquid back to the pot, and $D$ is the rate at which material is distilled. The parameters $L$, $D$, and $V$ are related by $V = L + D$. (What goes up, must come down.)

We will first look at the process qualitatively. If $D$ and $V$ are comparable ($L$ is small), as vapor at composition $Y_1$ is removed, $X_1$ becomes smaller—the liquid in the first plate becomes less rich in the MVC as the MVC rich vapor is removed. If, on the other hand, $L$ and $V$ are both large compared with $D$, the composition at plate 1 will be maintained at $Y_0$ by the large supply of incoming vapor with composition $Y_0$. This relationship may also be seen in the following quantitative discussion.

The first relationship we have already seen:

$$V = L + D \qquad (5.3)$$

In addition, if an insignificant amount of material is held up at plate 1, the moles of MVC going into plate 1 must equal the moles of MVC leaving plate 1;

$$V \cdot Y_0 = L \cdot X_1 + D \cdot Y_1 \qquad (5.4)$$

We may eliminate $V = (L + D)$ from Eq. 5.4 and rearrange to get

$$X_1 = Y_0 - (D/L) \cdot (Y_1 - Y_0) \qquad (5.5)$$

Now, in Eq. 5.5, $Y_1 - Y_0$ will always be greater than 0 under ideal conditions (since they both represent the MVC). Both $D$ and $L$ are positive. Therefore, the *upper limit* for $X_1$ is $Y_0$. This situation would be the case when $D/L = 0$ (and nothing is being distilled). In the real world something is being distilled ($D > 0$) and $X_1$ will be less than $Y_0$ (and down the line the condensate will be less rich in the MVC; there will be less separation). You should be able to generalize this result for one plate for the situations in which we have $n$ plates. When $D/L$ is small the composition of the first plate is at a maximum; therefore, $Y_1$ is also a maximum and, furthermore, $X_2$ will be a maximum, and so on.

The purpose of the spinning band is to make $D/L$ as small as possible so that $X_1$ approaches its upper limit of $Y_0$. The spinning band ensures that the optimum separation of a pair of liquids will be achieved. We will have a lot of vapor going up, a lot of liquid being returned by the spinning action of the spiral band, and a relatively small amount of material actually being distilled. This result also implies that if we want to achieve a high degree of separation, the distillation rate should be low. There is, of course, a compromise between the low rate of distillation required to obtain maximum separation and a rate that will allow someone else to use the apparatus and you to get on to other things.

**NOTE.** *The following list of experiments utilize Technique 2A: Experiments [3A], [3B], [11C], [29], [32], and [3A_adv]. The following Experiments utilize Technique 2B: Experiments [3C] and [3D].*

---

## Steam Distillation

**TECHNIQUE 3**

---

If two substances are immiscible (as the term "steam distillation" implies, one of the substances will be water), the **total vapor pressure** ($P_t$) above the two-phase liquid mixture is equal to the **sum of the vapor pressures** of the pure individual components ($P_i$) according to Dalton's law.

**THEORY**

$$P_t = P_1 + P_2 + \ldots + P_i$$

In other words, with an *immiscible* pair of liquids, neither component lowers the vapor pressure of the other. The two liquids would exert the same vapor pressure (at a given temperature) even if they were in separate containers. If a mixture of these two liquids is heated, **boiling** will occur when the **combined vapor pressures** of the liquids equals atmospheric pressure. At this point, distillation will commence. **Condensation** of the vapors gives a **two-phase mixture (condensate** or **distillate)** of the organic species and water. The composition of this distillate is determined by the vapor pressure and molecular weight of the compounds.

To illustrate this concept let us take an insoluble mixture of bromobenzene (bp = 156 °C) and water. At 30 °C the vapor pressure of bromobenzene is 6 torr; water is 32 torr. Therefore, the vapor pressure of the mixture is **the sum** of 6 + 32 or 38 torr, at this temperature. At 95 °C, the vapor pressure of bromobenzene is 120 torr; water is 640 torr. It follows that the **sum** is now equal to 760 torr or 1 atm. That is, $P_A^{\circ} + P_B^{\circ} = 1$ atm.

As a result, the mixture boils and the bromobenzene and water distill together. A further example is the isolation of cyclohexanone from nonvolatile reaction byproducts by steam distillation. The normal boiling point of pure cyclohexanone is 156 °C and that for water is 100 °C. At about 94.5 °C, the vapor pressure of cyclohexanone is 112 torr and the vapor pressure of water is 648 torr. The vapor pressures of cyclohexanone and water add up to 760 torr. Thus the two compounds steam distill at 94.5 °C.

Several important aspects of steam distillation are summarized below:

**1.** The mixture **boils below** the boiling point of either pure component. In this regard the technique is similar to reduced pressure distillation in that the liquid distills and condenses at a temperature below its normal boiling point. This result occurs because compounds that are immiscible in water have very large, positive deviations from Raoult's law.

**2.** The **boiling point** of the mixture will **hold constant** as long as both substances are present to saturate the vapor volume.

**3.** The molar ratio of the two species in the **distillate** remains constant as long as aspect 2 holds.

From Dalton's law, the condensate in a steam distillation will consist of water and the compound in the same *molar* ratio as the ratio of their vapor pressures (P°) at the steam distillation temperature. The relationship is

$$\frac{\text{Moles water}}{\text{Moles of organic species}} = \frac{P^{\circ} \text{ water}}{P^{\circ} \text{ organic species}}$$

This relationship may be altered to show the weight relationship of the organic substance to that of water.

Substituting grams per molecular weight (g/MW) for moles we obtain

$$\frac{\text{g/MW water}}{\text{g/MW organic species}} = \frac{P^{\circ} \text{ water}}{P^{\circ} \text{ organic species}}$$

Transposition and rearrangement of this expression leads to

$$\frac{\text{g water}}{\text{g organic species}} = \frac{P^{\circ} \text{ water} \times \text{MW water}}{P^{\circ} \text{ organic species} \times \text{MW organic species}}$$

Based on this relationship, the weight of water required per weight of organic species can be calculated. Note that if the vapor pressure of water is known at the boiling temperature of the mixture that is being steam distilled, the vapor pressure of the organic substance is 760 torr $- P^{\circ}_{\text{water}}$.

**NOTE.**   *These two experiments utilize Technique 3: Experiments [11C] and [32].*

## Solvent Extraction

Solvent extraction is a technique frequently used in the organic laboratory to separate or isolate a desired species from a mixture of compounds or from impurities. Solvent extraction methods are readily adapted to microscale work, since small quantities are easily manipulated in solution. This method is based on the solubility characteristics of the organic substances involved in relation to the solvents used in a particular separation procedure.

Substances vary greatly in their solubility in various solvents, but based on observations, many of which were made in the very early days, a useful principle has evolved that allows the chemist to predict rather accurately the solubility of a particular substance.[4] It is generally true that *a substance tends to dissolve in a solvent that is chemically similar to itself. In other words, like dissolves like.*

   Thus, for a particular substance to exhibit solubility in water requires that species to possess some of the characteristics of water. For example, an important class of compounds, the organic alcohols, have the hydroxyl group (—OH) bonded to a hydrocarbon chain or framework (R—OH). The hydroxyl group can be viewed as being effectively one half a water ($H_2O$) molecule, and it has a similar polarity to that of water. This results from a charge separation arising from the difference in electronegativity between the hydrogen and oxygen atoms. The O—H bond, therefore, is considered to have *partial ionic character*.

$$-\overset{\delta^-}{\underset{..}{\ddot{O}}}-\overset{\delta^+}{H}$$

Partial ionic character of the hydroxyl group

   This *polar*, or partial ionic, character leads to relatively strong hydrogen bond formation between molecules having this entity. Strong hydrogen bonding is evident in molecules that contain a hydrogen atom attached to an oxygen, nitrogen, or fluorine atom, as shown here for the (ethanol–water) system. This polar nature of a functional group is present when there are sufficient differences in electronegativity between the atoms making up the group.

$$CH_3-CH_2-\overset{..}{\underset{..}{O}}:\underset{\overset{|}{\underset{H}{}}}{\overset{\delta^-}{}}\overset{\delta^+}{}$$

Ethanol                    Hydrogen bond formation

   In ethanol ($CH_3CH_2OH$), it is apparent that the hydroxyl end of the molecule is very similar to water (HOH). When ethanol is added to water, therefore, they are miscible in all proportions. That is, ethanol is completely soluble in water and water is completely soluble in ethanol. This solubility results because the attractive forces set up between the two molecules are nearly as strong as between two water molecules; however, the

---

[4] This section is repeated in Experiment [4], and is presented here for the sake of continuity.

attraction in the first case is somewhat weakened by the presence of the nonpolar alkyl ethyl group, $CH_3CH_2—$. Hydrocarbon groups attract each other only weakly, as evidenced by their low melting and boiling points. Three examples of the contrast in boiling points between compounds of different structure, but similar molecular weight, are summarized in Table 5.1. Clearly, those molecules that attract each other weakly have lower boiling points.

**TABLE 5.1    Comparison of Boiling Point Data**

| Name | Formula | MW | bp(°C) |
|---|---|---|---|
| Ethanol | $CH_3CH_2OH$ | 46 | 78.3 |
| Propane | $CH_3CH_2CH_3$ | 44 | −42.2 |
| Methyl acetate | $CH_3CO_2CH_3$ | 74 | 54 |
| Diethyl ether | $(CH_3CH_2)_2O$ | 74 | 34.6 |
| Ethylene | $CH_2{=}CH_2$ | 28 | −102 |
| Methylamine | $CH_3NH_2$ | 31 | −6 |

When we compare the water solubility of ethanol (a two-carbon ($C_2$) alcohol that, as we have seen, is completely miscible with water) with that of octanol (a straight-chain eight-carbon ($C_8$) alcohol), we find that the solubility of octanol is less than 2% in water. Why the difference in solubilities between these two alcohols? The answer lies in the fact that the dominant structural feature of octanol has become the nonpolar nature of its alkyl group.

$$CH_3—CH_2—CH_2—CH_2—CH_2—CH_2—CH_2—CH_2—\overset{\delta-}{\ddot{O}}\underset{H}{\overset{}{\diagdown}}{}_{\delta+} \qquad\qquad CH_3—CH_2—\ddot{O}—CH_2—CH_3$$

Octanol                                                                                      Diethyl ether

As the bulk of the hydrocarbon section of the alcohol molecule increases, the intramolecular attraction between the polar hydroxyl groups of the alcohol and the water molecules is no longer sufficiently strong to overcome the *hydrophobic character* (lack of attraction to $H_2O$) of the nonpolar hydrocarbon section of the alcohol. On the other hand, octanol has a large nonpolar hydrocarbon group as its dominant structural feature. We might, therefore, expect octanol to exhibit enhanced solubility in less polar solvents. In fact, octanol is found to be completely miscible with diethyl ether. Ethers are solvents of weak polarity (here the polarity depends on the magnitude of C—O bond dipoles, which are considerably less than those present in O—H bonds). Since the nonpolar characteristics are significant in both molecules, mutual solubility is observed. In general, it has been empirically demonstrated that if a compound has both polar and nonpolar groups present in its structure, those compounds having five or more carbon atoms in the hydrocarbon portion of the molecule will be more soluble in nonpolar solvents, such as pentane, diethyl ether, or methylene chloride. Figure 5.19 summarizes the solubilities of a number of straight-chain alcohols, carboxylic acids, and hydrocarbons in water. As expected, those compounds with more than 5 carbon atoms are shown to possess solubilities similar to those of the hydrocarbons.

Several additional relationships between solubility and structure have been observed and are pertinent to the discussion.

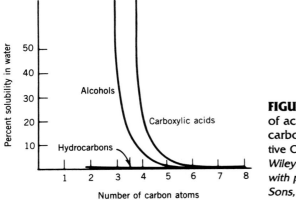

**FIGURE 5.19** Solubility curve of acids, alcohols, and hydrocarbons. *From Kamm, O.* Qualitative Organic Analysis, *2nd ed.; Wiley: New York, 1932. (Reprinted with permission of John Wiley & Sons, New York.)*

**1.** Branched-chain compounds have greater water solubility than their straight-chain counterparts, as illustrated in Table 5.2 with a series of alcohols.

**TABLE 5.2   Water Solubility of Alcohols[a]**

| Name | Formula | Solubility (g/100 g $H_2O$) |
|------|---------|------------------------------|
| Pentanol | $CH_3(CH_2)_3CH_2OH$ | 4.0 |
| 2-Pentanol | $CH_3(CH_2)_2CH(OH)CH_3$ | 4.9 |
| 2-Methyl-2-butanol | $(CH_3)_2C(OH)CH_2CH_3$ | 12.5 |

[a] Data at 20°C.

**2.** The presence of more than one polar group in a compound will increase that compound's solubility in water and decrease its solubility in nonpolar solvents. For example, high molecular weight sugars, such as cellobiose, that contain multiple hydroxyl and/or acetal groups, are water soluble, and ether insoluble—whereas cholesterol ($C_{27}$), which possesses only a single hydroxyl group, is water insoluble and ether soluble.

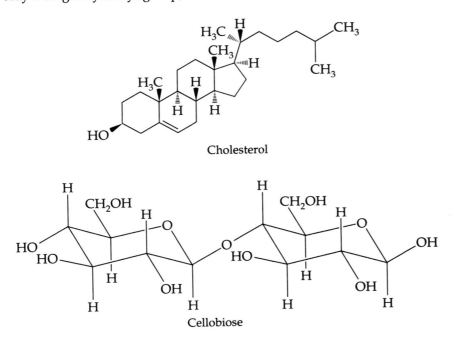

Cholesterol

Cellobiose

**3.** The presence of a chlorine atom, even though it lends some partial ionic character to the mostly covalent C—Cl bond, does not normally impart water solubility to a compound. In fact, such compounds as methylene chloride ($CH_2Cl_2$), chloroform ($CHCl_3$), and carbon tetrachloride ($CCl_4$) have long been used as solvents for the extraction of aqueous solutions. It should be noted that use of the latter two solvents is no longer recommended, unless strict safety precautions are exercised, because of their potential carcinogenic nature.

**4.** Most functional groups that are capable of forming a hydrogen bond with water, if it constitutes the dominant structural feature of a molecule, will impart increased water solubility characteristics to a substance (the five-carbon rule obviously applies here in determining just what is a dominant feature). For example, certain alkyl amines (organic relatives of ammonia) might be expected to have significant water solubility. This finding is indeed the case, and the water-solubility data for a series of amines is summarized in Table 5.3.

**TABLE 5.3   Water Solubility of Amines[a]**

| Name | Formula | Solubility (g/100 g $H_2O$) |
|---|---|---|
| Ethylamine | $CH_3CH_2NH_2$ | $\infty$ |
| Diethylamine | $(CH_3CH_2)_2NH$ | $\infty$ |
| Trimethylamine | $(CH_3)_3N$ | 91 |
| Triethylamine | $(CH_3CH_2)_3N$ | 14 |
| Aniline | $C_6H_5NH_2$ | 3.7 |
| p-Phenylenediamine | $H_2NC_6H_4NH_2$ | 3.8 |

[a] Data at 25 °C.

It is important to realize that the solubility characteristics of any given compound will uniquely govern that substance's distribution (*partition*) between the phases of two immiscible solvents (in which the material has been dissolved) when these phases are intimately mixed. In this experiment we determine the partition coefficient (distribution coefficient) of benzoic acid between two immiscible solvents, methylene chloride, and water.

## PARTITION COEFFICIENT

A given substance, if placed in a mixture of two immiscible solvents, will distribute (partition) itself in a manner that is a function of its relative solubility in the two solvents. For example, a solute X will be distributed between two immiscible solvents according to the following equilibrium distribution.

$$X_{solvent\,1} \Longleftrightarrow X_{solvent\,2}$$

Then,

$$K_{eq} = \frac{[X_{solvent\,2}]}{[X_{solvent\,1}]}$$

The equilibrium constant ($K_{eq}$) is the *ratio* of the concentrations of the species in each solvent for a given system at a given temperature. This particular equilibrium constant is designated the *partition coefficient* (also referred to as the *distribution coefficient*). This coefficient is similar to the partitioning of a species that occurs in chromatographic separations.

The basic equation used to express the coefficient $K$ is

$$K = \frac{(g/100\ mL)_{organic\ layer}}{(g/100\ mL)_{water\ layer}}$$

This expression uses grams per 100 mL or grams per deciliter (g/dL), but grams per liter (g/L), parts per million (ppm), and molarity (M) are also valid. The partition coefficient is dimensionless so that any concentration units may be used, provided the units are the same for both phases. If equal volumes of both solvents are used, the equation reduces to the ratio of the weights of the given species in the two solvents.

$$K = \frac{\delta_{organic\ layer}}{\delta_{water\ layer}}$$

Determination of the partition coefficient for a particular compound in various immiscible-solvent combinations often can give information valuable to the isolation and purification of the species using extraction techniques.

Let us now look at a typical calculation for the extraction of an organic compound P from an aqueous solution using diethyl ether. We will assume that the $K_{ether/water}$ value (*partition coefficient* of P between ether and water) is 3.5 at 20 °C.

If an aqueous solution containing 100 mg of P in 300 $\mu$L of water is extracted at 20 °C with 300 $\mu$L of ether, the following expression holds:

$$K_{ether/water} = \frac{C_e}{C_w} = \frac{W_e/300\ \mu L}{W_w/300\ \mu L}$$

where

    $W_e$ = weight of P in the ether layer
    $W_w$ = weight of P in the water layer
    $C_e$ = concentration of P in the ether layer
    $C_w$ = concentration of P in the water layer

Since $W_w = 100 - W_e$, the preceding relationship can be written as

$$K_{ether/water} = \frac{W_e/300\ \mu L}{(100 - W_e)/300\ \mu L} = 3.5$$

If we solve for the value of $W_e$, we obtain 77.8 mg; the value for $W_w$ = 22.2 mg. Thus, we see that after one extraction with 300 $\mu$L of ether, 77.8 mg of P (77.8% of the total) is removed by the ether and 22.2 mg (22.2% of the total) remains in the water layer. The question often comes up whether it is preferable to make a single extraction with the total quantity of solvent available, or to make multiple extractions with portions of the solvent. The second method is usually preferable in terms of efficiency of extraction. To illustrate, let us consider the following.

In relation to the foregoing example, let us now extract the 100 mg of P in 300 $\mu$L of water with **two** 150-$\mu$L portions of ether **instead of one** 300-$\mu$L portion as previously done.

For the first 150-$\mu$L extraction,

$$\frac{W_e/150\ \mu L}{W_w/300\ \mu L} = \frac{W_e/150\ \mu L}{(100 - W_e)/300\ \mu L}$$

Solving for the value of $W_e$, we obtain 63.6 mg. The amount of P remaining in the water layer ($W_w$) is then 36.4 mg. The aqueous solution is now extracted with the second portion of ether (150 $\mu$L). We then have

$$\frac{W_e/150\ \mu L}{(36.4 - W_e)/300\ \mu L} = 3.5$$

As before, by solving for $W_e$, we obtain 23.2 mg for the amount of P in the ether layer; $W_w = 13.2$ mg.

The two extractions, each with 150 $\mu$L of ether, removed a total of 63.6 mg + 23.2 mg = 86.8 mg of P (86.8% of the total). The P left in the water layer is then 100 − 86.8 or 13.2 mg (13.2% of the total).

Based on the preceding calculations, it can be seen that the multiple extraction technique is the more efficient. Whereas the single extraction removed 77.8% of P, the double extraction increased this to 86.8%. To extend this relationship, multiple extractions with one third of the total quantity of the ether solvent in three portions would be even more efficient. You might wish to calculate this extension to prove the point. Of course, there is a practical limit to the number of extractions that can be performed based on time and the degree of efficiency realized.

## EXTRACTION

The two major types of extractions utilized in the organic laboratory are the solid–liquid and liquid–liquid methods.

### Solid–Liquid Extraction

The simplest form of solid–liquid extraction is the treatment of a solid with a given solvent in a beaker or Erlenmeyer flask followed by the decantation or filtration of the solvent extract from the solid sample. An example of this technique (see Experiment [11A]) is the extraction of usnic acid from its native lichen using acetone as the extraction solvent. This approach is most useful when only one main component of the solid phase has appreciable solubility in the solvent. The extraction of caffeine from tea (see Experiment [11B]) is a further example of this method and is accomplished by heating the tea in an aqueous solution of sodium carbonate. This approach works well because the solubility characteristics of the compounds involved are improved, and also because the water swells the tea leaves and allows the caffeine to be removed more readily.

The usual method of choice, however, is to carry out solid–liquid extractions on a continuous basis employing a *countercurrent* process. Various types of apparatus have been developed over the years, but perhaps the best known is the "Soxhlet" extractor, first described in 1879. It is pictured in Figure 5.20. The solid sample is placed in a porous thimble. The extraction-solvent vapor, generated by refluxing the extraction solvent contained in the distilling pot, passes up through the vertical side tube into the condenser.

The liquid condensate then drips onto the solid, which is extracted. The extraction solution passes through the pores of the thimble, eventually filling the center section of the Soxhlet. The siphon tube also fills with this extraction solution and when the liquid level reaches the top of the tube, the siphoning action commences and the extract is returned to the distillation pot. The cycle is automatically repeated numerous times. In this manner the desired species is concentrated in the distillation pot. Equilibrium is not generally established in the system, since the extraction is usually a slow process. However, the rate of extraction can be influenced by choices

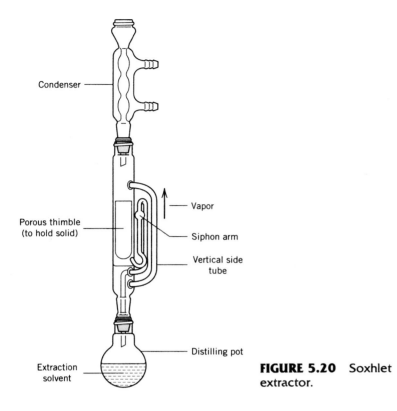

Condenser

Porous thimble
(to hold solid)

Vapor

Siphon arm

Vertical side
tube

Distilling pot

Extraction
solvent

**FIGURE 5.20**  Soxhlet
extractor.

of solvent, temperature, and so on. The method is classified as a discontinuous-infusion process, since the thimble containing the solid must fill with solvent before the extraction solution returns to the distillation pot. The extraction solution collected in the pot is then concentrated to isolate the desired compound.

Soxhlet extractors are available from many supply houses and can be purchased in various sizes. Of particular interest to us is the micro variety, which is effective for small amounts of material. Numerous designs for microextractors have been reported in the literature. Two examples are shown in Figure 5.21a and 5.21b. The apparatus developed by Garner (Fig. 5.21a) consists of a small cold-finger condenser inserted into a test tube. The test tube has indentations near the bottom to support a small funnel. The sample is carefully wrapped in filter paper and placed in the funnel. The Blount extractor (Fig. 5.21b) is designed for the simultaneous extraction and recrystallization of small amounts of material. In this arrangement, the sample is placed in a small fritted-glass filter crucible that is hung from the bottom of the condenser. Soxhlet microextractors also have been described, and are now commercially available.[5]

Recently, a new apparatus has become commercially available. It is reported to be much more efficient than the traditional Soxhlet method. Named the "Soxtec," it has been demonstrated to be applicable for extraction of petroleum, food, textiles, plastics, chemicals, and so on.[6] Enhanced safety, the claim of shorter extraction times, a built-in solvent recovery system, and indirect heating by circulating oil from an electronically controlled service unit are some of the advantages of this system. The three

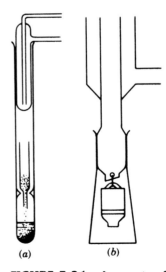

(a)        (b)

**FIGURE 5.21**  Apparatus for continuous extraction of a solid. (a) Garner microextractor.(b) Blount microextractor. *From Schneider, F. L.* Monographien aus dem Gebiete der qualitativen Mikroanalyse, Qualitative Organic Microanalysis, Vol. II; Benedetti-Pichler, A. A., Ed.; Springer-Verlag: Vienna, Austria, 1964.

[5] Microscale Soxhlet equipment is available from Wheaton, 1501 North Tenth St., Millville, NJ 08332.
[6] The "Soxtec" extractor is available from Tecator, Inc., P.O. Box 405, Herndon, VA 22070.

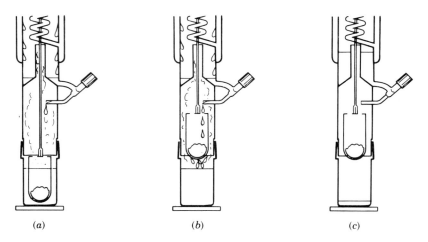

**FIGURE 5.22** The Soxtec principle. (*a*) Boiling. Rapid pre-extraction in hot solvent. (*b*) Rinsing. Condensed solvent washes the last traces of soluble matter from the sample. (*c*) Recovery. Solvent is evaporated, condensed, and collected. (*Courtesy of Tecator, a Perstorp Analytical Co., 2875C Towerview Road, Herndon, VA 22071.*)

stages of the process: boiling, rinsing, and solvent recovery, are shown in Figure 5.22.

## Liquid–Liquid Extraction

The more common type of procedure, liquid–liquid extraction, is used extensively in the laboratory. It is a very powerful method for the separation and isolation of materials encountered at the microscale level.

In the majority of extractions, a capped centrifuge tube, or a 10 × 75-mm test tube, or a conical vial is used as the container. In any extraction technique, it is essential that complete mixing of the two immiscible solvents be realized.

Let us consider a practical example. Benzanilide, prepared by the in situ rearrangement of benzophenone oxime in acid solution, is separated from the product mixture by extraction with three 1.0-mL portions of methylene chloride solvent.

**NOTE.** *This wording is the accepted manner of indicating that three successive extractions are performed, each using 1.0 mL of methylene chloride.*

At the microscale level, the extraction process consists of two parts: (1) mixing of the two immiscible solutions, and (2) separation of the two layers after the mixing process.

**1.** *Mixing.* In the experimental procedure for the isolation of the benzanilide product, methylene chloride solvent (1.0 mL) is added to the aqueous reaction mixture contained in a 5.0-mL conical vial. The extraction procedure is outlined in the following steps.

a. The vial is capped.
b. The vial is shaken gently to thoroughly mix the two phases.
c. The vial is carefully vented by loosening the cap to release any pressure that may develop.
d. The vial is allowed to stand on a level surface to permit the two phases to separate. A sharp phase boundary should be evident.

**NOTE.** *The mixing stage may be carried out using a Vortex® mixer.*

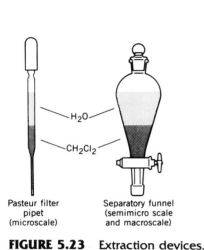

**FIGURE 5.23** Extraction devices.

Pasteur filter pipet (microscale)

Separatory funnel (semimicro scale and macroscale)

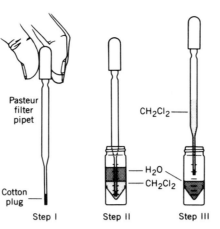

**FIGURE 5.24** Pasteur filter pipet separation of two immiscible liquid phases, with the denser layer containing the product.

**2.** *Separation.* At the microscale level the two phases are separated with a Pasteur filter pipet (a simple Pasteur pipet can be used in some situations), which acts as a miniature separatory funnel. The separation of the phases is shown in Figure 5.23.

A major difference between macro and micro techniques is that at the microscale level, as just discussed, the mixing and separation are done in two parts (1, 2), whereas at the macroscale level with the separatory funnel, the mixing and separation are both done in the funnel in one step. It is important to note that the separatory funnel is an effective device for extractions at the semimicroscale or macroscale level, but at the microscale level it is not practical to use because the volumes are so small. The recommended procedures are diagrammed in Figures 5.24 and 5.25. Continuing the example, it is known that benzanilide is more soluble in methylene chloride than in water. Multiple extractions are performed to ensure complete removal of the benzanilide from the aqueous phase. The methylene chloride solution is the lower layer since it is more dense than water. The following Steps (I–IV) outline the general method (refer to Fig. 5.24).

**I.** Squeeze the pipet bulb to force air from the pipet.

**II.** Insert the pipet into the vial until close to the bottom. Be sure to hold the pipet in a vertical position.

**III.** Carefully allow the bulb to expand, drawing only the lower methylene chloride layer into the pipet. This procedure is done in a smooth, steady manner so as not to disturb the boundary between the layers. With practice, one can judge the amount that the bulb must be squeezed to just separate the layers.

**IV.** (Step IV is not shown in the figure). Holding the pipet in a vertical position, place it over an empty vial and gently squeeze the bulb to transfer the methylene chloride solution into the vial. A second extraction can now be performed after addition of another portion of methylene chloride to the original vial. The identical procedure is repeated. In this manner, multiple extractions can be performed, with each methylene chloride extract being transferred to the same vial; that is, the extracts are combined. The reaction product has now been transferred from the aqueous layer to the methylene chloride layer, and the phases have been separated.

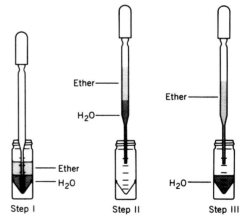

**FIGURE 5.25** Pasteur filter pipet separation of two immiscible liquid phases, with the less dense layer containing the product.

In a diethyl ether–water extraction, the ether layer is less dense and thus is the top phase. Generally, the reaction product dissolves in the ether layer and is thus separated from byproducts and other impurities. The procedure followed to separate the water–ether phases is identical to that outlined earlier, except that it is the top layer that is transferred to the new container. The Steps (I–III) are shown in Figure 5.25.

**I.** Draw both phases into the pipet as outlined before in Steps I and II. Try not to allow air to be sucked into the pipet, since this will tend to mix the phases in the pipet. If mixing does occur, allow time for the boundary to reform.

**II.** Return the bottom aqueous layer to the original container by gently squeezing the pipet bulb.

**III.** Transfer the separated ether layer to the new vial.

## Separatory Funnel Extraction

As previously mentioned, the separatory funnel (Fig. 5.23) is an effective device for extractions carried out at the semimicroscale and macroscale levels. With the use of this funnel, the mixing and separation are done in the funnel itself in one step. Many of you may be familiar with this device from your work in the general chemistry laboratory.

The solution to be extracted is added to the funnel, after first making sure that the stopcock is closed. The funnel is generally supported in an iron ring attached to a ring stand or the lab bench. The proper amount of extraction solvent is now added (about one third of the volume of the solution to be extracted is a good rule of thumb) and the stopper placed on the funnel.

**NOTE.** *The size of the funnel should be such that the total volume of solution is less than three fourths the total volume of the funnel. If the funnel is constructed with a ground-glass stopcock and/or stopper, the ground-glass surfaces must be lightly greased to prevent sticking, leaking, or freezing. If Teflon® stoppers and stopcocks are used, grease is not necessary since they are self-lubricating.*

The funnel is removed from the ring stand, the stopper rested against the index finger of one hand, and the funnel held in the other with the fingers positioned so as to operate the stopcock (Fig. 5.26a). The funnel is carefully inverted, the liquid is allowed to drain away from the stopcock,

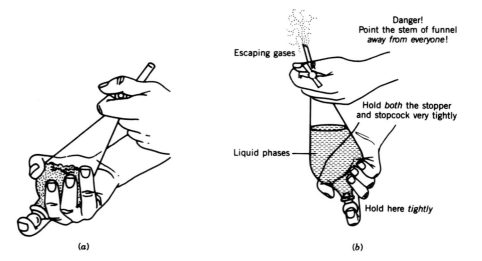

**FIGURE 5.26** (a) Correct position for holding a separatory funnel while shaking. (b) Correct method for venting a separatory funnel.

and then the stopcock is opened slowly to release any built-up pressure (Fig. 5.26b).

**NOTE.** *Make sure the stem of the funnel is pointing up and that it does not point at anyone.*

The stopcock is closed, the funnel shaken for several seconds, the funnel positioned for venting, and the stopcock opened to release built-up pressure. This process is repeated several more times. After the final sequence, the stopcock is closed and the funnel returned upright to the iron ring.

The layers are allowed to separate, the stopper is removed, the stopcock is opened gradually, and the bottom layer is drained into a suitable container; the upper layer is removed by pouring from the top of the funnel.

When aqueous solutions are extracted with a *less dense solvent*, such as ether, the bottom, aqueous layer is drained *into the original container* from which it was poured into the funnel. Once the top ether layer is removed from the funnel, the aqueous layer can then be returned for further extraction. Losses can be minimized by rinsing this original container with a small portion of the extraction solvent, which is then added to the funnel. When the extraction solvent is denser than the aqueous phase (e.g., methylene chloride) the aqueous phase is the top layer, and therefore is retained in the funnel for subsequent extractions.

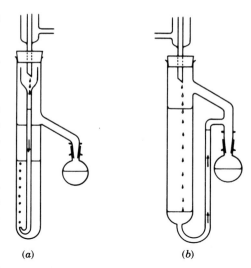

**FIGURE 5.27** Early designs for single stage extractors. (a) Kutscher–Steudel extractor. (b) Wehrli extractor.

Continuous extraction of liquid–liquid systems is also possible. Figure 5.27a illustrates a single-stage type of extractor developed originally by Kutscher and Steudel for use with less-dense extraction solvents, and Figure 5.27b, one developed by Wehrli for use with more-dense solvents. A modification of these early designs, which combines the utility of both, is now commercially available (Fig. 5.28).

By using this device (Fig. 5.28), it is possible to extract various species from aqueous solutions using a less-dense or more-dense immiscible organic solvent. In both cases, the important aspect is that the extraction can be carried out with a limited amount of solvent. Furthermore, the number of individual extractions that would have to be performed to accomplish the same task would be prohibitive.

When one uses the convertible liquid–liquid continuous extraction apparatus (Fig. 5.28) with an extraction solvent that is heavier than water, the stopcock is open and the insert A is not used. The apparatus is fitted with a condenser at B and a distilling pot at C. The extraction solvent is distilled from the pot at C and the condensate from the condenser is allowed to pass down through the aqueous solution containing the desired compound as shown in Figure 5.27b. The more-dense extraction solution eventually returns to the distillation pot. In this manner the removed material is concentrated in the pot and fresh extraction solvent is continually passed through the aqueous solution.

When extraction solvents lighter than water are used, the stopcock is closed and the insert, A, is placed in the apparatus (Fig. 5.28). The extraction is carried out by allowing the condensate of the extraction solvent, as it forms on the condenser on continuous distillation, to drop through the inner tube and to percolate up through the solution containing the material to be extracted (see Fig. 5.27a). This inner tube usually contains a sintered glass plug on its end, which generates smaller droplets of the solvent and thus increases the efficiency of the procedure. The extraction solution is

## Continuous Liquid– Liquid Extraction

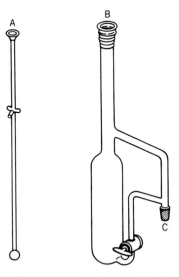

**FIGURE 5.28** Convertible liquid– liquid continuous extractors. (*Courtesy of Aldrich Chemical Co., Inc., Milwaukee, WI.*)

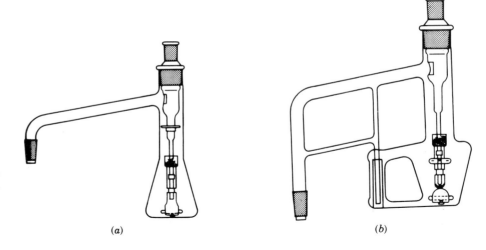

**FIGURE 5.29** Normag liquid–liquid extractors. (*a*) For extraction with solvents of lower density. (*b*) For extraction with solvents of higher density. *(Courtesy of Aldrich Chemical Co., Inc., Milwaukee, WI.)*

(*a*)          (*b*)

then returned to the original distilling flask. Eventually, in this manner, the desired material, extracted in small increments, is collected in the boiling flask and can then be isolated by concentration of the collected solution. This method works on the premise that fresh portions of the less-dense phase are continuously introduced into the system and is often used in those instances where the organic material to be isolated has an appreciable solubility in water. Continuous liquid–liquid extraction is useful for removal of extractable components from those having partition ratios that approach zero. It should be realized that this is a method that requires a long period of time.

Extractors have been developed to further increase the efficiency of the extraction process. One such technique is to insert a mechanical or magnetic rotating distributor in the extractor vessel. The latter type are commercially available. One, the "Normag" (Fig. 5.29), is arranged so that the extraction solvent is presented to the solution being extracted as fine droplets.

This results in optimum conditions for the extraction and thus the rate of the process is increased dramatically. It is available in models using solvents of both higher and lower density.

An alternate design for a liquid–liquid extractor has recently become available (Fig. 5.30). It is used with a solvent more dense than water, which is placed in flask A. Distillation of the solvent produces a condensate, which drips through a frit in the joint at C. The small droplets from the frit fall through the aqueous phase in the chamber D. The solution of extractant–solvent then returns to the distilling flask through tube E.

Microextractors of the lighter and heavier solvent type have also been developed. Two designs are illustrated in Figure 5.31.

These microextractors work on the same principle that we discussed previously. An alternate design has been reported by Gould in which the Soxhlet apparatus is modified by using various adapters, depending on whether the solvent is heavier or lighter than the aqueous solution to be extracted. A novel technique for removing medium-polar solvents by continuous extraction with water has been reported by Uzar.[7]

Countercurrent multiple fractional extraction, an automated technique, has been developed to allow 200 or more successive extractions to

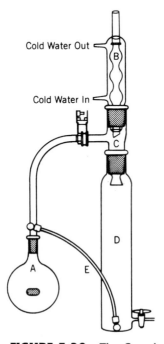

Cold Water Out

B

Cold Water In

C

D

A     E

**FIGURE 5.30** The Supelco liquid–liquid extractor. *(Courtesy of Supelco Inc., Supelco Park, Bellefonte, PA 16823.)*

[7] Uzar, H. C. *J. Chem Educ.* **1990** *67,* 349.

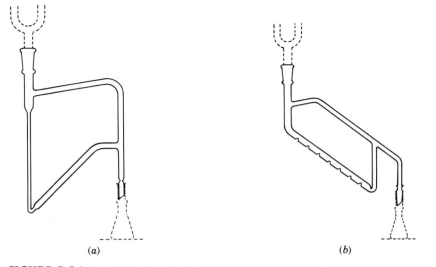

**FIGURE 5.31** Micro liquid—liquid extractors. (*a*) For extraction with solvents of lower density. (*b*) For extraction with solvents of higher density.

be performed.[8] The basic type of extractor consists of a series of extraction tubes through which the upper phase is moved. The extraction tubes (Fig. 5.32) are shaken to equilibrate the phases. After each equilibration, the less-dense phase is transferred to the next tube by a reservoir; the more-dense phase remains in the original tube. Fresh extraction solvent is added after each extraction. The material to be extracted is placed in the first tube at the start of the extraction. As the process proceeds, the different components being extracted progress through the series of extraction tubes at

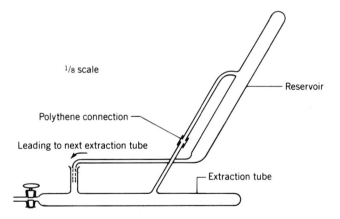

**FIGURE 5.32** A single unit of the multiple extraction apparatus. *From Elvidge, J. A.; Summer, P. G. A Course in Modern Techniques of Organic Chemistry 2nd ed., 1966. (By permission of the publishers, Butterworth & Co. (Publishers) Ltd. ©, London, England).*

---

[8] For a complete discussion of the continuous liquid—liquid extraction method, see Craig, L. C.; Craig, D. in *Technique of Organic Chemistry*, 2nd ed., A. Weissberger, Ed.; Interscience: New York, 1956; Vol. III, pp. 149–332. A shorter presentation is given in Elvidge, J. A.; Sammes, P. G. *A Course in Modern Techniques of Organic Chemistry*, 2nd. ed.; Butterworths: London, 1966; Chapter 7, p. 57.

different rates, depending on their partition coefficients. This technique permits separation of various species having nearly the same partition coefficients. Craig has shown that a series of 10 amino acids which have partition coefficients that differ by less than 0.1, can be separated by this approach.

## Drying of the Wet Organic Layer

It is important to realize that organic extracts separated from aqueous phases usually contain traces of water. Before evaporation of the solvent to isolate the desired species, or before further purification steps can be taken, the organic extract must be dried to remove any residual water. This condition is conveniently achieved with an inorganic anhydrous salt, such as magnesium, sodium, or calcium sulfate. These materials readily form insoluble hydrates, thus removing the water from the wet organic phase. The hydrated solid can then be removed from the dried solution by filtration or decantation. An ideal drying agent should have a short drying time, high capacity for water, and a high degree of absorption. Table 5.4 summarizes the properties of some of the more common drying agents used in the laboratory.

There are two basic requirements for an effective solid drying agent: (1) it should not react with the material in the system, and (2) it must be easily and completely separated from the dried liquid phase. The amount of drying agent used depends on the amount of water present, on the capacity of the solid desiccant to absorb water, and on its particle size. If the solution is wet, the first amount of drying agent will clump (molecular

**TABLE 5.4   Properties of Common Drying Agents**

| Drying Agent | Formula of Hydrate | Comments |
|---|---|---|
| Sodium sulfate | $Na_2SO_4 \cdot 10\ H_2O$ | Slow in absorbing water and is inefficient but is inexpensive and has a high capacity. Loses water above 32 °C. Granular form available |
| Magnesium sulfate | $MgSO_4 \cdot 7\ H_2O$ | One of the best. Can be used with nearly all organic solvents. Usually in powder form |
| Calcium chloride | $CaCl_2 \cdot 6\ H_2O$ | Relatively fast drying agent. However, it reacts with many oxygen and nitrogen containing compounds. Usually in granular form |
| Calcium sulfate | $CaSO_4 \cdot \frac{1}{2}\ H_2O$ | Very fast and efficient. However, notice that it has a low dehydration capacity |
| Silica gel | $(SiO_2)_m \cdot n\ H_2O$ | High capacity and efficient. Commercially available t.h.e.® $SiO_2$ drying agent is excellent[a] |
| Molecular sieves | $[Na_{12}(Al_{12}Si_{12}O_{48})] \cdot 27\ H_2O$ | High capacity and efficient. Use the 4-Å size[b] |

[a] Available from EM Science, Cherry Hill, NJ 08034.
[b] Available from Aldrich Chemical Co. Inc, 940 West Saint Paul Ave., Milwaukee, WI 53233.

sieves and t.h.e.® SiO₂ are exceptions). In this case, additional drying agent is added until the agent appears mobile on swirling the container. Swirling the contents of the container increases the rate of drying, since it aids in establishment of the equilibrium for hydration.

$$\text{Drying agent} + n\,H_2O = \text{Drying agent} \cdot n\,H_2O$$
(anhydrous solid)           (solid hydrate)

Most drying agents attain approximately 80% of their drying capacity within 15 min, so longer drying times are generally not required.

As mentioned above, particle size plays a role in the effectiveness of the drying agent used. Magnesium sulfate is a good all-around drying agent. However, it is supplied as a fine powder and thus has a high surface area. The disadvantage is that the desired product can become trapped on the surface of the magnesium sulfate particles. If the drying agent, after separation, is not thoroughly washed, precious product may be lost. Furthermore, removal of a finely powdered solid agent is generally more difficult since it may pass through the filter paper (if used) or clog the pores of a fine porous filter. It is advisable to use a drying agent that is available in a granular form. Anhydrous sodium sulfate is a good example. A smaller surface area translates into less adsorption of product on the surface and easier separation from the dried solution.

The drying agent may be added directly to the container containing the organic extract or the extract may be passed through a Pasteur filter pipet packed with the drying agent. A funnel fitted with a cotton, glass wool, or polyester plug to hold the drying agent may also be used.

## Separation of Acids and Bases

The separation of organic acids and bases constitutes another important and extensive use of the extraction method. An organic acid reacts with dilute aqueous sodium hydroxide solution to form a salt. The salt, having an ionic charge, dissolves in the more polar aqueous phase (e.g. Experiment [7]).

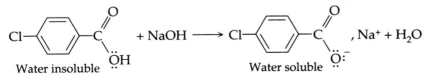

The reaction reverses the solubility characteristics of a water insoluble acid. The water phase may then be extracted with an immiscible organic solvent to remove any impurities, leaving the acid salt in the water phase. Neutralization of the soluble salt with hydrochloric acid causes precipitation of the insoluble organic acid in a relatively pure state. In a similar fashion, organic bases, such as amines, can be rendered completely water soluble by treatment with dilute hydrochloric acid to form hydrochloride salts (e.g. Experiment [23]).

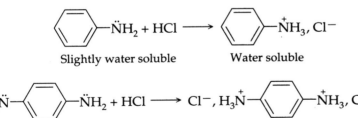

Extraction procedures are also used in the separation of mixtures of solids. For example, the separation of a mixture made up of an aromatic (Ar) organic acid (ArCO$_2$H), base (ArNH$_2$), and neutral compound (ArH). A flow chart is given below that diagrams the sequence.

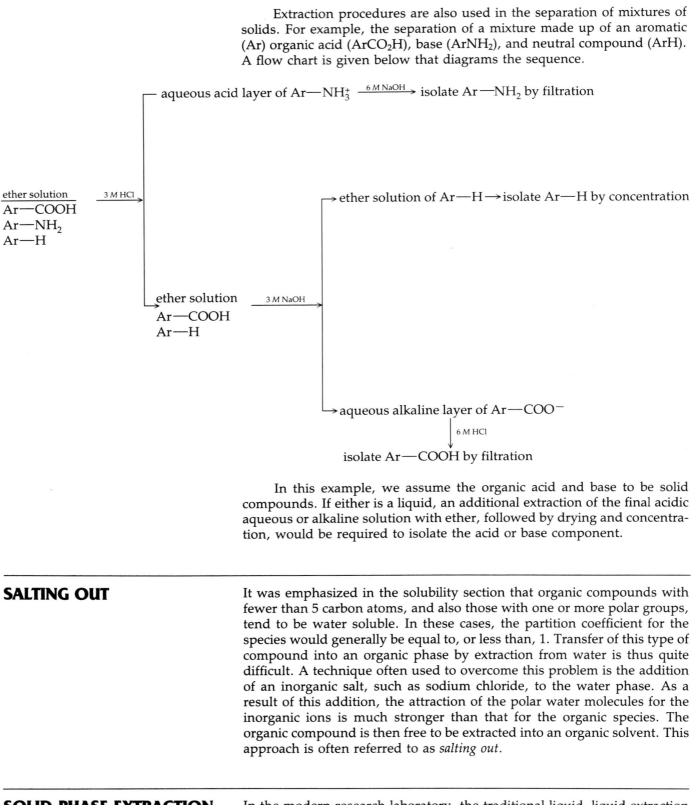

In this example, we assume the organic acid and base to be solid compounds. If either is a liquid, an additional extraction of the final acidic aqueous or alkaline solution with ether, followed by drying and concentration, would be required to isolate the acid or base component.

## SALTING OUT

It was emphasized in the solubility section that organic compounds with fewer than 5 carbon atoms, and also those with one or more polar groups, tend to be water soluble. In these cases, the partition coefficient for the species would generally be equal to, or less than, 1. Transfer of this type of compound into an organic phase by extraction from water is thus quite difficult. A technique often used to overcome this problem is the addition of an inorganic salt, such as sodium chloride, to the water phase. As a result of this addition, the attraction of the polar water molecules for the inorganic ions is much stronger than that for the organic species. The organic compound is then free to be extracted into an organic solvent. This approach is often referred to as *salting out*.

## SOLID-PHASE EXTRACTION

In the modern research laboratory, the traditional liquid–liquid extraction technique may be replaced by the **solid-phase extraction** method.[9] The advantage of this new approach is that it is rapid, small volumes of solvent

[9] For a description of this method see Zief, M.; Kiser, R. *Am. Lab.* **1990**, 70 and Zief, M. *NEACT J.* **1990**, *8*, 38.

are used, emulsion formation is avoided, isolated solvent extracts do not require a further drying stage, and it is ideal for working at the microscale level. This technique is finding wide acceptance in the food industry and in the environmental and clinical area, and it is becoming the accepted procedure for the rapid isolation of drugs of abuse and their metabolites from urine.

Solid-phase extraction is accomplished using prepackaged, disposable, extraction columns. A typical column is shown in Figure 5.33. The columns are available from several commercial sources.[10]

The polypropylene columns can be obtained packed with 100, 200, 500, or 1000 mg of 40-$\mu$m sorbent sandwiched between two 20-$\mu$m polyethylene frits. The columns are typically 57 mm long. Sample volumes generally average 1–6 mL.

The adsorbents used in the columns are silica-gel based with a chemically bonded nonpolar stationary phase. In fact, they are the same nonpolar adsorbents used extensively in the reversed-phase high performance liquid chromatography (HPLC) technique. More specifically, the adsorbents are derivatized silica gel where the —OH groups of the silica gel have been replaced with siloxane units by treatment with the appropriate organochlorosilanes.

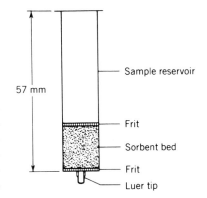

**FIGURE 5.33** Polyethylene solid-phase extraction column.

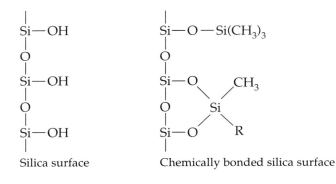

Silica surface          Chemically bonded silica surface

Two of the most popular nonpolar packings are those containing R groups consisting of octadecyl ($C_{18}H_{37}$—) or phenyl ($C_6H_5$—) units. These systems can adsorb nonpolar (like attracts like) organic material from aqueous solutions. This material is then eluted from the column, using a solvent strong enough to displace it. Solvents such as methylene chloride or hexane are used extensively. In many instances, methanol appears to be an effective solvent for eluting analytes from the octadecyl bonded-phase columns. The analyte capacity of bonded silica gels has been estimated to be 10–20 mg of analyte per gram of packing.

It is imperative to understand that for this extraction technique to be effective, information on the structure of the material to be separated, the number and type of functional groups present, the polarity, the molecular weight, the solubility characteristics, and so on, is essential, if good separation is to be obtained. Furthermore, the elution solvent must be evaluated, and it must be established that adsorption of the desired species will occur on the column and that it also can be removed.

An example of a typical solid-phase extraction is the determination of the amount of caffeine in coffee. A 1-mL capacity column is used containing 100 mg of octadecyl-bonded silica. The column is conditioned by aspirating 2 mL of methanol followed by 2 mL of water through the column. We then draw 1 mL of a coffee solution containing approximately 0.75 mg

[10] These columns are available from Analytichem International, J. T. Baker, Inc., Supelco, Inc., and Waters Associates.

of caffeine/mL through the sorbent at a flow rate of 1 mL/min. The column is washed with 1 mL of water and air dried under vacuum for 10 min. The adsorbed caffeine is eluted using two 500-$\mu$L portions of chloroform. High-performance liquid chromatographic analysis indicates that 94% of the caffeine was isolated from the coffee sample.

## BIBLIOGRAPHY

For overviews on extraction methods see the following general references:

Blount, B. *Mikrochemie* **1936**, *19*, 162.

Colegrave, E. B. *Analyst* **1935**, *60*, 90.

Craig, L. C. *Anal. Chem.* **1950**, *22*, 1346.

Craig, L. C.; Craig, D. In *Technique of Organic Chemistry*, 2nd ed., A. Weissberger, Ed.; Interscience: New York, 1956; Vol. III, Part I, Chapter 2.

Garner, W. *Ind. Chem.* **1928**, *4*, 332.

Gould, B. S. *Science* **1943**, *98*, 546.

Hanson, C., Ed., *Recent Advances in Liquid–Liquid Extraction*; Pergamon: Oxford, 1971.

Hartland, S. *Countercurrent Extraction*; Pergamon: Oxford, 1970.

Jubermann, O. in *Houben–Weyl Methoden der Organischen Chemie*, 4th ed.; Verlag: Stuttgart, 1958, Vol. I, p. 223.

*Kirk–Othmer Encyclopedia of Chemical Technology* 3rd ed.; Wiley: New York, 1980; Vol. 9, p. 672.

Kutscher, K; Steudel, H. *Z. Physiol. Chem.* **1903**, *39*, 474.

Schneider, Frank L. *Qualitative Organic Microanalysis*; Vol. II of *Monographien aus dem Gebiete der qualitativen Mikroanalyse*, A. A. Benedetti-Pichler, Ed.; Springer-Verlag: Vienna, Austria, 1964; p. 61.

Soxhlet, F; Szombathy *Dinglers Polytech. J.* **1879**, *232*, 461.

Thorpe, J. F.; Whiteley, M. A. *Thorpe's Dictionary of Applied Chemistry*, 4th ed.; Longman: New York, 1940; Vol. IV, p. 575.

Wehrli, S. *Helv. Chim. Acta* **1937**, *20*, 927.

### References to Solid-Phase Extractions:

Baker-10 spe Applications Guide, Vol. 1, p 182; J. T. Baker Chemical Co., Phillipsburg, NJ.

Cooke, N. H. C.; Olsen, K. *Amer. Lab.* **1979**, *11*, 45.

Dorsey, J.; Dill, K. A. *Chem. Rev.* **1989**, *89*, 331.

Nawrocki, J.; Baszewski, B. *J. Chromatogr.* **1988**, *559*, 1.

**NOTE.**    *The following list of experiments utilize Technique 4: Experiments [4A], [4B], [5A], [5B], [7], [8A], [8B], [8C], [11A], [11B], [11C], [12], [13], [16], [17], [19A], [19B], [19C], [19D], [22A], [22B], [23], [27], [30], [32], [34A], [34B], [3A$_{adv}$], [4$_{adv}$], [6$_{adv}$], [A1$_b$], [D3], [E3], [F1], [F2], [F3], and [F4].*

## TECHNIQUE 5

# Crystallization

This discussion introduces the basic strategy involved in achieving the purification of solid organic substances by crystallization. The technique of crystallizing an organic compound is one of fundamental importance, and one that must be mastered in order to deal successfully with the purification of these materials. *It is not an easy art to acquire.* Organic solids tend not to crystallize with the ease of inorganic substances.

Indeed, in earlier times, an organic chemist occasionally would resist an invitation to leave a well-worn laboratory for new quarters. This concern arose from the suspicion that the older facility (in which many crystallizations had been carried out) harbored seed crystals for a large variety of substances in which the resident investigator had an interest. Carried by dust from the earlier work, this trace of material presumably aided the successful initiation of crystallization of reluctant materials. Further support for this argument/legend was gained by the often quoted (but not substantiated) observation that after a material was first crystallized in a particular laboratory, subsequent crystallizations of the material, regardless of its purity or origin, were always easier to carry out.

A reaction could very often have been viewed as a failure if an amorphous sludge could not be enticed to become a collection of beautiful white crystals. The melting point of an amorphous substance is ill-defined, and if this material is mixed with a crystalline reference compound, large melting point depressions usually result.

In several areas of organic chemistry, particularly those dealing with natural products, the success or failure of an investigation can depend to a large extent on the ability of the research chemist to isolate tiny quantities of crystalline substances. Often, the compounds of interest must be extracted from enormous amounts of extraneous material. In one of the more spectacular examples, Reed et al. in 1953 isolated 30 mg of the crystalline coenzyme, lipoic acid, from 10 tons of beef liver residue.[11]

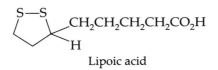

Lipoic acid

---

## PROCEDURE

The essentials of this purification technique are outlined as follows: (1) dissolve the material (primarily made up of the compound of interest along with smaller quantities of contaminating substances) in a warm solvent; (2) once the solid mixture is fully dissolved, filter the heated solution, and then bring it to the point of saturation by evaporating a portion of the solvent; (3) cool the warm saturated solution to cause a drop in solubility of the dissolved substance. This results in precipitation of the solid material; and (4) isolate the precipitate by filtration, and then remove the last traces of solvent.

The technique is considered successful if the solid is recovered in good yield and is obtained in a higher state of purity than that before the procedure. This cycle, from the solid state to solution and back to the solid state, is termed *recrystallization*, if both the initial and final solid materials are crystalline.

Although the technique sounds fairly simple, in reality it is demanding. The successful purification of microscale quantities of solids will require your utmost attention.

The first major problem to be faced is the choice of solvent system. To achieve high recoveries, the compound to be crystallized would ideally be very soluble in the solvent of choice at elevated temperatures, but nearly insoluble when it is cold. If the crystallization is to increase the purity of the

[11] Reed, L. J.; Gunsalus, I. C.; Schnakenberg, G. H. F.; Soper, Q. F.; Boaz, H. E.; Kem, S. F.; Parke, T. V. *J. Am. Chem. Soc.* **1953,** *75,* 1267.

compound, however, the impurities should be either very soluble in the solvent at all temperatures, or not soluble at any temperature. In addition, the solvent should possess as low a boiling point as possible so that traces can be easily removed from the crystals after filtration.

Thus, the choice of solvent is critical to a good crystallization. Table 5.5 is a list of common solvents used in the purification of most organic solids. (The list has contracted significantly in the past few years as health concerns about these very volatile compounds have arisen.)

**TABLE 5.5 Common Solvents**

| Solvent | bp (°C) |
|---|---|
| Water | 100 |
| Methanol | 65 |
| Ethanol, 95% | 78 |
| Ligroin | 60–90 |
| Acetone | 56 |
| Diethyl ether | 35 |
| Methylene chloride | 41 |
| Petroleum ether | 30–60 |

Seldom are the solubility relationships ideal for crystallization and most often a compromise is made. If there is no suitable single solvent available, it is possible to employ a mixture of two solvents, termed a solvent pair. In this situation, a solvent is chosen that will readily dissolve the solid. After dissolution, the system is filtered. A second solvent, miscible with the first, in which the solute has a lower solubility, is then added dropwise to the hot solution to achieve saturation. In general, polar organic molecules have higher solubilities in polar solvents, and nonpolar materials are more soluble in nonpolar solvents (like dissolves like). Considerable time can be spent in the laboratory working out an appropriate solvent system for a particular reaction product. In most instances, with known compounds, the optimum solvent system has been established.

Because many impurities have solubilities similar to those of the compounds of interest, most crystallizations are not very efficient. Recoveries of 50–70% are not uncommon. It is important that the purest possible material be isolated prior to recrystallization.

A number of microscale crystallization routines are available.

## SIMPLE CRYSTALLIZATION

Simple crystallization works well with large quantities of material (100 mg and up), and it is essentially identical to that of the macroscale technique.

1. Place the solid in a small Erlenmeyer flask or test tube.
2. Add a minimum of solvent and heat the mixture to the solvent's boiling point in a sand bath.
3. Stir, and add solvent dropwise with continued heating, until all of the material has dissolved.
4. Add a decolorizing agent, if necessary (powdered charcoal, or better, charcoal pellets), to remove colored minor impurities and resinous byproducts.

5. Filter the hot solution into a second Erlenmeyer flask (preheat the funnel with hot solvent). This operation removes the decolorizing agent and any insoluble material initially present in the sample.

6. Evaporate enough solvent to reach saturation.

7. Cool to allow crystallization (crystal formation will be better if this step takes place slowly).

8. Collect the crystals by filtration.

9. Wash (rinse) the crystals.

10. Dry the crystals.

---

The standard filtration system for collecting products purified by recrystallization in the microscale laboratory is vacuum filtration with an 11-mm Hirsch funnel. In addition, many reaction products that do not require recrystallization are collected directly by this technique. The Hirsch funnel, as shown in Figure 5.34a, is composed of a ceramic cone with a circular flat bed perforated with small holes. The diameter of the bed is covered by a flat piece of filter paper of the same diameter. The funnel is sealed into a filter flask with a Neoprene adapter (see Fig. 5.34b).

The filter flask, which is heavy walled and especially designed to operate under vacuum, is constructed with a side arm (they are often called "side-arm pressure flasks"; see Fig. 5.34c).

The side arm is connected with heavy-walled rubber vacuum tubing to a water aspirator, or water pump. The water pump utilizes a very simple and inexpensive aspirator. The system is based on the Venturi effect in which water is forced through a constricted throat in the pump (see Fig. 5.35).

If water, as an incompressible liquid, completely fills the water pump at all points, the same volume of liquid must pass every cross section in any given time as it is discharged out of the aspirator. Hence,

$$Q = Av$$

where $Q$ = the discharge rate (cm$^3$/s)
$A$ = the cross section of the discharge
$v$ = the velocity of the discharge

and the *equation of continuity* follows

$$Q = Av = A_1v_1 = \text{constant}$$

where $A_1$ = the cross section of the throat
$v_1$ = the velocity in the throat

A consequence of this relationship is that the velocity is greatest at points where the cross section is least, and vice versa. Hence, the water velocity through the throat of the aspirator is much faster than the discharge velocity. Air is drawn into the low-pressure water rushing through the constricted portion (Bernoulli's principle).

Two pressure effects are involved in the operation of this type of pump. First, the pressure is limited to the vapor pressure of water at the temperature of the water supply reservoir. Second, a drop in water pressure (caused by a laboratory neighbor turning on a tap directly connected to your water line) will often result in a backup of water through the hose connection and into the filter flask, as the pressure momentarily is lowest

## FILTRATION TECHNIQUES

### Use of the Hirsch Funnel

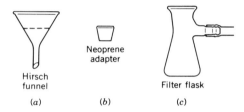

Hirsch funnel    Neoprene adapter    Filter flask

(a)    (b)    (c)

**FIGURE 5.34** Component parts for vacuum filtration (see parts a–c).

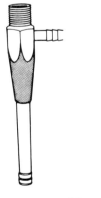

**FIGURE 5.35** Aspirator pump.

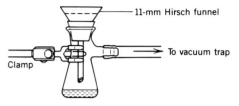

**FIGURE 5.36** Vacuum filtration apparatus.

on that side of the system. It is, therefore, essential that safety traps (see Fig. 5.39) be mounted between the aspirator and the filtration apparatus.

When water is running through the aspirator, a partial vacuum is formed, which creates a flow of air down the vacuum tubing from the filter flask. With the rubber adapter in place, the entering air is forced through the filter paper, which is held flat by suction. The mother liquors of the crystallization are rapidly forced into the filter flask, while the crystals retained by the filter are quickly dried by the stream of air passing through them (Fig. 5.36).

In some situations, the substance collected on the Hirsch funnel is not highly crystalline. If this is true, the compact filter cake may be difficult to dry using the method outlined above. In these situations, a thin, flexible rubber sheet or a piece of plastic food wrap is placed over the mouth of the funnel such that the suction generated from the aspirator (or vacuum pump) pulls the sheeting down onto the filter cake. This process creates pressure on the cake. This pressure assists in forcing a large portion of the remaining solvent from the collected material and thus further drying is achieved. It is advisable to use a section of sheeting large enough that, when it is placed over the funnel, the entire filter cake is covered. If this is not done, a vacuum will not be created and thus drying will not occur.

In some instances, substances may retain water or other solvents with great tenacity. To dry these materials, a **desiccator** is often used. A desiccator is generally a glass or plastic container in which a material (desiccant) having the capacity to absorb water is placed. The substance to be dried, held in a suitable open container, is then placed on a support above the desiccant layer. This technique is widely used in the quantitative analysis laboratory for drying collected precipitates. Vacuum desiccators are available if required (see Fig. 5.37a). If this method of drying is still insufficient, a **drying pistol** is then employed. These pistols (Abderhalden vacuum-drying apparatus, see Fig. 5.37b) are commercially available from most supply houses. The sample, in an open container (vial) is placed in the apparatus, which is then evacuated. The pistol has a pocket in which a strong absorbing agent, such as $P_4O_{10}$(for water), NaOH or KOH (for acidic gases), or paraffin wax (for organic solvents) is placed. The temperature of the pistol is controlled by refluxing vapors that surround the barrel. A simple alternative to this method is the use of a side-armed test tube as shown in Figure 5.38.

(a)

**FIGURE 5.37a** Vacuum desiccator

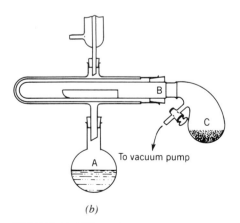

(b)

**FIGURE 5.37b** Abderhalden vacuum drying apparatus. A, refluxing heating liquid; B, vacuum drying chamber; C, desiccant.

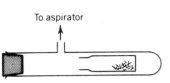

**FIGURE 5.38** Side arm test tube as a vacuum drying apparatus.

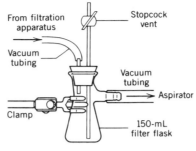

**FIGURE 5.39** Vacuum trap.

*As mentioned above, when you are using a water pump, it is very important to have a safety trap mounted in the vacuum line leading from the filter flask. Any drop in water pressure (easily created by one or two other students on the same water line turning on their aspirators at the same time) can result in the backup of water into the system as the flow through the aspirator decreases (see Fig. 5.39).*

The Craig tube is commonly used for microscale crystallizations in the range of 10–100 mg of material (see Fig. 5.40). The process consists of the following steps.

**1.** The sample is placed in a small test tube (10 x 75 mm).

**2.** The solvent (0.5–2 mL) of choice is added, and the sample dissolved by heating in the sand bath. Rapid stirring with a microspatula (roll the spatula rod between your fingers) greatly aids the dissolution and protects against boil-over. A modest excess of solvent is added after the sample is completely dissolved. It will be easy to remove this excess at a later stage, as the volumes involved are very small. The additional solvent ensures that the solute will stay in solution during the hot transfer.

**3.** The heated solution is transferred to the Craig tube by Pasteur filter pipet (the pipet is preheated with hot solvent). This transfer automatically filters the solution (if decolorizing charcoal powder has been added, two filtrations by the pipet may be required. The second filtration can almost always be avoided by the use of charcoal pellets.)

**4.** The hot, filtered solution is then concentrated to saturation by gentle boiling in the sand bath. Constant agitation of the solution with a microspatula during this short period will avoid the use of a boiling stone, and guarantees that a boil-over will not occur. Ready crystallization on the microspatula, just above the solvent surface, serves as a good indication that saturation is close at hand.

**5.** The upper section of the Craig tube is set in place, and the system is allowed to cool in a safe place. As cooling commences, seed crystals, if necessary, may be added by crushing them against the side of the Craig tube with a microspatula just above the solvent line. A good routine, if the time is available, is to place the assembly in a small Erlenmeyer, then place the Erlenmeyer in a beaker, and finally cover the first beaker with a second inverted beaker. This procedure will ensure slow cooling, which will enhance good crystal growth (Fig. 5.41). A Dewar flask may be used when very slow cooling and/or larger crystals are required (as in X-ray crystallography).

## Craig Tube Crystallizations

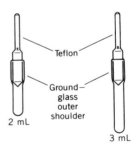

**FIGURE 5.40** Craig tubes.

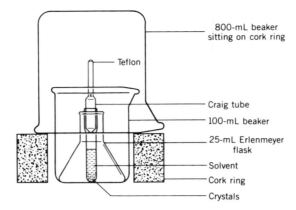
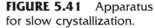

FIGURE 5.41 Apparatus for slow crystallization.

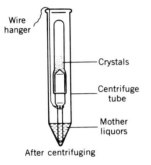

FIGURE 5.42 Crystal collection with a Craig tube.

**6.** After the system reaches room temperature, cooling in an ice bath will further improve the yield.

**7.** The solvent is now removed by inverting the Craig tube assembly into a centrifuge tube and spinning the mother liquors away from the crystals (Fig. 5.42). This operation takes the place of the usual filtration step in simple crystallizations. It avoids another transfer of material and also avoids product contact with filter paper.

**8.** After removal from the centrifuge, the Craig tube is disassembled and any crystalline product clinging to the upper section is scraped into the lower section. If the lower section is tared, it can be left to air-dry to constant weight, or it can be placed in a warm vacuum oven (wrap a piece of filter paper over the open end secured by a rubber band to prevent dust from collecting on the product while drying). The yield can then be directly calculated.

The cardinal rule in carrying out the purification of small quantities of solids is: *Keep the number of transfers to an absolute minimum!* The Craig tube is very helpful in this regard.

The preceding routine will maximize the crystallization yield. If time is important, the process can be shortened considerably. Shortcuts, however, invariably lead to a corresponding drop in yield.

## RECRYSTALLIZATION PIPET

Landgrebe has recently described an alternative to the Craig tube method.[12] This approach utilizes a modified Pasteur pipet as a recrystallization tube. The method works well for 10–100-mg quantities, as long as the volume of solvent used in the recrystallization does not exceed 1.5 mL, the capacity of a Pasteur pipet. A description of the sequence follows:

**1.** Prepare a recrystallization tube (Fig. 5.43) by pushing a plug of cotton (copper wire is used) into the Pasteur pipet so that the cotton resides 1–2 cm below the wider bore of the pipet.

**2.** Seal the (lower) part of the tube below the cotton plug with a microburner. Pull the glass so that a very narrow tip is formed. This procedure allows the tip to be broken easily at a later stage of the operation.

**3.** Place the solid into the tared tube, reweigh to determine the weight of solid, and then clamp the tube near the top in a vertical position.

[12] Landgrebe, J. A. *J. Chem. Educ.* **1988**, *65*, 460.

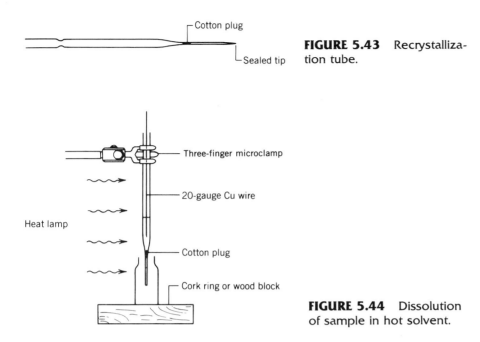

**FIGURE 5.43** Recrystalliza-tion tube.

**FIGURE 5.44** Dissolution of sample in hot solvent.

Arrange a tared vial so that the bottom tip of the recrystallization tube protrudes about 1 cm into it. (Fig. 5.44).

**4.** Add an appropriate amount of solvent to the tube using a Pasteur pipet. Stir the suspension with a copper wire, and arrange a heating lamp approximately 6–8 cm from the tube.

**5.** When the solid has dissolved, remove the vial, snap the tip off the tube, and quickly replace the vial under it. Continue to warm the solution being filtered (do not boil). If the filtration is too slow, gently apply pressure using a pipet bulb.

**6.** After crystallization is complete, the mother liquor may be removed using a Pasteur filter pipet. Cold, fresh solvent may be added to wash the crystals, and the wash solvent then can be removed by using the Pasteur filter pipet as before. Dry the crystals as discussed above (see Fig. 5.37 or 5.38).

**NOTE.** *The following list of experiments utilize Technique 5: Experiments [6], [7], [15], [16], [18], [19A], [19B], [19C], [19D], [20], [23A], [23B], [24A], [24B], [25A], [25B], [26], [28], [29A], [29B], [29C], [29D], [30], [31], [33A], [33B], [34A], [34B], [2<sub>adv</sub>], [3A<sub>adv</sub>], [3B<sub>adv</sub>], [5<sub>adv</sub>], [6<sub>adv</sub>], [7<sub>adv</sub>], [A1<sub>a</sub>], [A2<sub>a</sub>], [A3<sub>a</sub>], [A1<sub>b</sub>], [A2<sub>b</sub>], [A3<sub>b</sub>], [A4<sub>ab</sub>], [B1], [C2], [C3], [D1], [D2], [E1], [E2], [F1], [F2], and [F4].*

---

## Chromatography                                    **TECHNIQUE 6**

---

### Technique 6A   Column, High Performance Liquid, and Thin-Layer Chromatography

The theory of chromatography is defined in Technique 1 during the development of gas-phase separations. The term is derived from the Greek word for color, *chromatos*. Tswett discovered the technique (1903) during studies that were centered on the separation of mixtures of natural plant pigments.[13] The chromatographic zones were detected simply by observing the color bands. Thus, as originally applied, the name was not an inconsistent use of terminology. Today, however, most mixtures that are chromatographed are colorless. The separated zones in these cases are established by other methods. Coincidentally, "tswett" means "color" in Russian (the discoverer's nationality).

In this section, two additional chromatographic techniques are explored. Both of these procedures depend on adsorption and distribution between a stationary solid phase and a moving liquid phase. The first to be discussed is "column chromatography," a very powerful technique that is used extensively throughout organic chemistry. It was one of the earliest of the modern chromatographic methods to be applied to the separation of organic mixtures by Tswett. The second procedure, thin-layer chromatography (TLC), was developed in the late 1950s. Thin-layer chromatography is particularly effective in rapid assays of sample purity. It is also effective when employed as a preparative technique for obtaining high-quality material for analytical data (ideal conditions for the microscale laboratory).

## COLUMN CHROMATOGRAPHY

The term *column chromatography* is derived from the use of a column packed with a solid stationary phase, a relationship similar to the liquid phase of GC. The mobile liquid phase descends, by gravity or applied pressure, through the column.

A wide variety of substances have been employed as the stationary phase in this technique. In practice, however, two materials have become dominant in this type of separation chemistry. Finely ground (100–200 mesh) alumina (aluminum oxide, $Al_2O_3$) and silica gel (silicic acid) are by far the most useful of the known adsorbents. The liquids, which act as the moving phase and elute (wash) sample materials through the column, are many of the common organic solvents. Table 5.6 lists the better known column packings and elution solvents.

**TABLE 5.6  Column Chromatography Materials**

| Stationary Phase | | | Moving Phase | | |
|---|---|---|---|---|---|
| Alumina | ↑ | *Increasing* | Water | ↑ | *Increasing* |
| Silica gel | | *adsorption* | Methanol | | *solvation* |
| Magnesium sulfate | | *of polar* | Ethanol | | *of polar* |
| Cellulose paper | | *materials* | Acetone | | *materials* |
| | | | Ethyl acetate | | |
| | | | Diethyl ether | | |
| | | | Methylene chloride | | |
| | | | Cyclohexane | | |
| | | | Pentane | | |

Silica gel impregnated with silver nitrate, $AgNO_3$, (usually 5–10%) is an attractive solid-phase adsorbent. The silver salt selectively binds to un-

[13] Tswett, M. *Ber. Deut. Botan. Ges.* **1906**, 24, 235.

saturated sites via a silver ion $\pi$ complex. Traces of alkene materials are easily removed from saturated reaction products by chromatography with this system (see also Experiment [12]). This adsorbent, however, must be protected from light until used, or the mixture will rapidly darken and become ineffective (photooxidation).

Column chromatography is usually carried out according to the procedures discussed in the following five sections.

**Packing the Column**

The quantity of stationary phase required is determined by the sample size. A common rule of thumb is to use a weight of packing material 30–100 times the weight of the sample to be chromatographed. The size of the column is chosen to give roughly a 10:1 ratio of height to diameter, appropriately sized for the amount of adsorbent required.

In the microscale laboratory, two standard chromatographic columns are employed.

**1.** A Pasteur pipet, modified by shortening the capillary tip, is used for the separation of smaller mixtures (10–100 mg). Approximately 0.5–2.0 g of packing is used in the pipet column (Fig. 5.45a).

**2.** A 50-mL titration buret modified by reducing the length of the column (to 10 cm above the stopcock) is used for the larger sample mixtures (50–200 mg) and for the difficult-to-separate mixtures. Approximately 5–20 g of packing is employed in the buret column (Fig. 5.45b).

Both columns are prepared by first clamping the empty column in a vertical position, and then seating a small cotton, or glass wool, plug at the bottom. The cotton is covered with a thin layer of sand, in the case of the buret. The Pasteur pipets are loaded by adding the adsorbent with gentle tapping, "dry packing." The column is then premoistened just prior to use. The burets are packed by a slurry technique. In this procedure the column is filled part way with solvent; then the stopcock is opened slightly, and as the solvent slowly drains from the column, a slurry of the adsorbent–solvent is poured into the top of the column. The column should be gently tapped while the slurry is added. The solvent is then

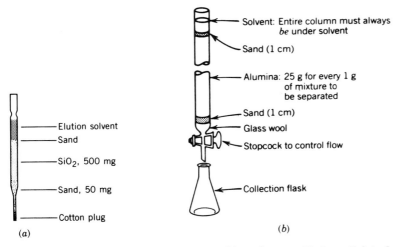

**FIGURE 5.45**  (a & b) Chromatographic columns. (b) *From Zubric, James W. The Organic Chem Lab Survival Manual, 3rd ed.; Wiley: New York, 1992. (Reprinted by permission of John Wiley & Sons, Inc., New York.)*

drained to the top of the adsorbent level, and held at that level until used. Alternatively, the wet-packed column can be loaded by sedimentation techniques rather than using a slurry. One such routine is to initially fill the column with the least-polar solvent to be employed in the intended chromatographic separation. Then the solid phase is slowly added with gentle tapping, which helps to avoid subsequent channeling. As the solid phase is added, the solvent is slowly drained from the buret at the same time. After the adsorbent has been fully loaded, the solvent level is then lowered to the top of the packing as in the slurry technique.

## Sample Application

The sample is applied in a minimum amount of solvent (usually the least-polar solvent in which the material is readily soluble) to the top of the column by Pasteur pipet. The pipet is rinsed, and the rinses are added to the column just as the sample solution drains to the top of the adsorbent layer.

## Elution of the Column

The critical step in resolving the sample mixture is eluting the column. Once the sample has been applied to the top of the column, the elution begins (a small layer of sand can be added to the top of the buret column after addition of the first portion of elution solvent).

**NOTE.**   *It is very important not to let the column run dry because instant cavitation can occur, which leads to extensive channeling of the column.*

The Pasteur pipet is free flowing (the flow rate is controlled by the size of the capillary tip), and once the sample is on the column, the chromatography will require constant attention. Buret column flow is controlled by the stopcock. The flow rate should be set to allow time for equilibrium to be established between the two phases. The choice of solvent is dictated by a number of factors. A balance between the adsorption power of the stationary phase and the solvation power of the elution solvent will govern the rate of travel of the material descending through the stationary phase. If the material travels rapidly down the column, then too few adsorption–elution cycles will occur and the materials will elute together in one fraction. If the sample travels too slowly, diffusion broadening takes over, and resolution is degraded. In the latter situation, samples then elute over many fractions with overlapping broad bands. Ideal solvent and elution rates strike a balance between these two situations and maximizes the separation. It can take considerable time to develop a solvent or mixture of solvents that produces a satisfactory separation of a particular mixture.

## Fraction Collection

As the solvent elutes from the column, it is collected in a series of "fractions" using small Erlenmeyer flasks or vials. Under ideal conditions, as the mixture of material travels down the column, it will separate into several individual bands of pure substances. By careful collection of the fractions, these bands can be separated as they sequentially elute from the column (similar to the collection of GC fractions in the example described in Technique 1). The bands of material being eluted can be detected by a number of techniques (weighing fraction residues, visible absorption bands, TLC, etc.).

Column chromatography is a powerful technique for the purification of organic materials. In general, it is significantly more efficient than crystallization procedures. Thus, this technique is used extensively in micro-scale laboratory experiments. Recrystallization is avoided until the last stages of purification, where it will be most efficient, relying upon chromatography to do most of the separation.

**NOTE.** *One major advantage of working with small amounts of product is that the chromatographic times are shortened dramatically.*

Column chromatography of a few milligrams of product usually takes no more than 30 min, but to chromatograph 10 g of product might take a whole afternoon, or even the better part of a day. Large scale chromatography (50–100 g) may even take several days to complete using this type of equipment.

Flash chromatography, first described by Still, Kahn, and Mitra in 1978,[14] has rapidly developed as a standard method for the separation and purification of nonvolatile mixtures of organic compounds. The technique is rapid, easy to perform, and relatively inexpensive. Recovery of material is high, since band-tailing is minimized. At present, many laboratories routinely use the technique to separate mixtures weighing 0.01–10 g in 10–15 min.

This moderate-resolution, preparative technique was originally developed using silica gel, with an optimum particle size of 40–63 $\mu$m. More recently, it has been demonstrated that bonded-phase silica gel of a larger particle size can also be used effectively. The columns are generally packed dry, using approximately 6 in. of the silica. Thin-layer chromatography is an efficient technique to employ as an aid in establishing experimental parameters for flash chromatography. It was found that a solvent resulting in TLC differential retardation factor ($DR_f$) values greater than or equal to 0.15 between the mixture components gives effective separation with flash chromatography. Table 5.7 lists typical experimental parameters for various sample sizes, as a guide to separations using flash chromatography. In general, a mixture of organic compounds that can be separated by TLC can be separated preparatively using flash chromatography.

## Flash Chromatography

**TABLE 5.7  Typical Experimental Parameters**[a]

| Column Diameter (mm) | Total Volume of Eluent (mL)[b] | Typical Sample Loading (mg) | | Typical Fraction (mL) |
|---|---|---|---|---|
| | | $DR_{f>0.2}$ | $DR_{f>0.1}$ | |
| 10 | 100–150 | 100 | 40 | 5 |
| 20 | 200–250 | 400 | 160 | 10 |
| 30 | 400–450 | 900 | 360 | 20 |
| 40 | 500–650 | 1600 | 600 | 30 |
| 50 | 1000–1200 | 2500 | 1000 | 50 |

[a] Data from Majors, R. E.; Enzweiler, T. *LC, GC* **1988**, *6*, 1046.
[b] Required for both packing and elution.

[14] Still, W. C.; Kahn, M.; Mitra, A. *J. Org. Chem.* **1978**, *43*, 2923.

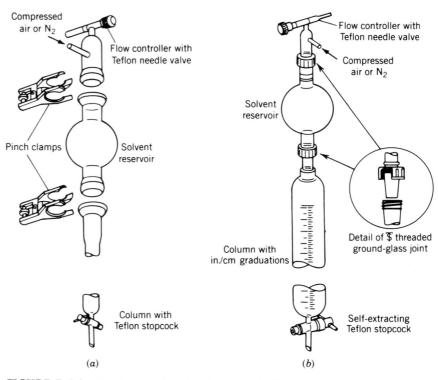

**FIGURE 5.46** (*a*) Conventional column and (*b*) screw-threaded column.

The chromatographic apparatus generally consists of a 20-mm i.d. glass column, modified so that a positive pressure of compressed air or nitrogen can be applied to the top of the column. A typical arrangement, which is commercially available, is shown in Figure 5.46.[15]

Generally, a 20–25% solution of the sample in the elution solvent is recommended, as is a flow rate of about 2 in./min. It is important that the column be conditioned before the sample is applied. This process is accomplished by flushing the column with the mobile phase (under pressure) to drive out all the air that may be trapped in the stationary phase, and to also equilibrate the packing material and solvent.

Several modifications of the basic arrangement have been reported, especially in regard to the adaptation of the technique to the instructional laboratory. These involve inexpensive pressure control valves, use of a "vibrator" air pump, the adaptation of a balloon reservoir as a supply of pressurized gas and, at the microscale level, the adaptation of a pipet bulb or pump to supply pressure on the column.

A recent method, utilizing a capillary Pasteur pipet for introducing the sample onto the chromatographic column, has been reported that approximately doubles the effectiveness of the column in terms of theoretical plates.[16]

---

[15] A complete line of glass columns, reservoirs, clamps, and packing materials for flash chromatography is offered by Aldrich Chemical Co., 1001 St. Paul Ave., Milwaukee, WI. Silica gels for use in this technique are also available from Amicon, Danvers, MA; J. T. Baker, Phillipsburg, NJ; EM Science/Merck, Gibbstown, NJ; ICN Biomedicals, Inc., Cleveland, OH; Universal Solvents, Atlanta, GA; Whatman, Clifton, NJ.
[16] Pivnitsky, K. K. *Aldrichimica Acta* **1989**, *22*, 30.

Thin-layer chromatography (TLC) is another solid–liquid partition technique of more recent development. It is a close relative to column chromatography, in that the phases used in both techniques are essentially identical. That is, alumina and silica gel are typically used as stationary phases, and the mobile phases are the usual solvents. There are, however, some distinct operational differences between TLC and column chromatography. While the mobile phase *descends* in column chromatography; in TLC, the mobile phase *ascends*. The column of stationary-phase material used in column chromatography is replaced, in TLC, by a very thin layer (100 $\mu$m) of the material spread over a flat surface. The technique has some distinct advantages at the microscale level. It is very rapid (2–10 min), and it employs *very* small quantities of material (2–20 $\mu$g). The chief disadvantage of this type of chromatography is that it is not very amenable to preparative scale work. Even when large surfaces and thicker layers are used, separations are most often restricted to the 5–10-mg level unless sophisticated research equipment is available.

The sequence of operations for TLC is as follows:

**1.** A piece of window glass, a microscope slide, or a sheet of plastic can be used as a support for a thin layer of adsorbent spread over the surface. It is possible to locally prepare the glass plates, but plastic-backed thin-layer plates are only commercially available. The plastic-backed plates are particularly attractive because they possess very uniform coatings and are highly reproducible in operation. Another convenient feature of the plastic-backed plates is that they can be cut with scissors into very economical 1 × 3-in. strips (even smaller sizes can be satisfactory). The latter style is used extensively in the microscale laboratory.

**2.** A pencil line is drawn parallel to the short side of the plate, 1.0 cm from the edge. One or two points, evenly spaced, are marked on the line. The sample to be analyzed (1 mg or less) is placed in a 100-$\mu$L conical vial and a few drops of solvent are added. A micropipet (prepared by the same technique used for constructing the capillary insert in the ultramicro-boiling point determination; see Chapter 4) is used to apply a small fraction of the solution from the vial to the plate (Fig. 5.47).

**3.** The chromatography is carried out by placing the spotted thin-layer plate in a screw-capped, wide-mouth jar, or in a beaker with a watch glass cover, containing a small amount of elution solvent (Fig. 5.48). The material spot on the TLC plate must initially be positioned above the solvent level. The jar is quickly recapped, or the watch glass replaced, to maintain an atmosphere saturated with the elution solvent. The elution solvent rapidly ascends the plate by capillary action. The choice of solvent

## THIN-LAYER CHROMATOGRAPHY

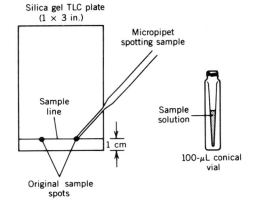

**FIGURE 5.47** Sample application to a TLC plate.

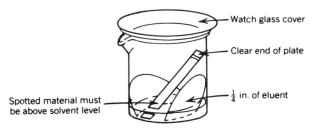

**FIGURE 5.48** Development of a TLC plate. *From Zubric, James W. The Organic Chem Lab Survival Manual, 3rd ed.; Wiley: New York, 1992. (Reprinted by permission of John Wiley & Sons, Inc., New York.)*

will be similar to that used in column chromatography, but need not be identical. The spotted material is eluted vertically up the plate. Resolution of mixtures into individual spots along the vertical axis occurs by precisely the same mechanism as in column chromatography. Elution is stopped, when the solvent line nears the top of the plate, by removing the plate from the jar or beaker, and the position of the solvent front should be quickly marked on the plate (before the solvent evaporates) when the chromatography is terminated.

**4.** Visualization of colorless, separated components can be achieved by placing the plate in an iodine-vapor chamber (a sealed jar containing solid $I_2$) for several seconds. Iodine forms a reversible complex with most organic substances. Thus, dark spots will develop in those areas containing sample material. On removal from the iodine chamber, the spots should be marked by pencil, because they may fade rather rapidly. TLC plates are commonly prepared with an ultraviolet (UV) activated fluorescent indicator mixed in with the silica gel. Sample spots can be detected with a hand-held UV lamp as the sample quenches the fluorescence induced by the lamp.

**5.** The elution characteristics are reported as $R_f$ values. The $R_f$ value is a measure of the travel of a substance up the plate relative to the solvent movement. This value is defined as the distance traveled by the substance divided by the distance traveled by the solvent front (the position of the solvent front should be quickly marked on the plate when the chromatogram is terminated; see Fig. 5.49).

Thin-layer chromatography is used in a number of applications. The speed of the technique makes it quite useful for monitoring large scale column chromatography. Analysis of fractions can guide decisions on the solvent elution sequence. The TLC analysis of column-derived fractions can also give an indication of how best to combine collected fractions. Following the progress of a reaction by periodically removing small aliquots for TLC analysis is an extremely useful application of thin-layer chromatography.

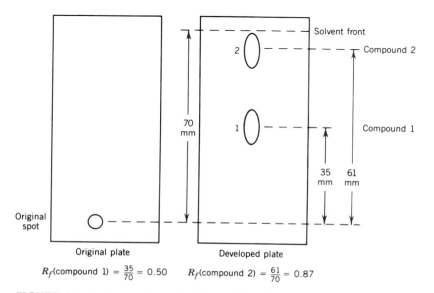

$$R_f(\text{compound 1}) = \frac{35}{70} = 0.50 \qquad R_f(\text{compound 2}) = \frac{61}{70} = 0.87$$

**FIGURE 5.49**   A sample calculation of $R_f$ values.

Although GC is the chromatographic method of choice in many instances, it is limited to compounds that have a significant vapor pressure at temperatures up to about 200 °C. Thus, compounds of high molecular weight and/ or polarity are not amenable to separation by GC, and high-performance liquid chromatography (HPLC) becomes an attractive chromatographic method.

Both GC and HPLC are somewhat similar, in the instrumental sense, in that the analyte is partitioned between a stationary and a mobile phase that travels down a column. Whereas the mobile phase in GC is a gas, the mobile phase in HPLC is a liquid. As shown schematically in Figure 5.50, the mobile phase, or solvent, is delivered to the system by a pump capable of pressures up to about 6000 psi. The sample is introduced by the injection of a solution into an injection loop. The injection loop is brought in line between the pump and the column by turning a switch, and the sample then flows down the column, is partitioned, and then flows into a detector.

The solid phase in HPLC columns used for organic monomers is usually based on silica gel. "Normal" HPLC refers to chromatography using a solid phase (usually silica gel) which is more polar than the liquid phase, or solvent, so that less polar compounds elute earlier. Typical solvents include ethyl acetate, hexane, acetone, low molecular weight alcohols, chloroform, and acetonitrile. For extremely polar compounds, such as amino acids, "reversed-phase" HPLC is used. Here, the liquid phase is more polar than the stationary phase, and the more polar compounds elute more quickly. The mobile phase is usually a mixture of water and some water-miscible organic solvent, such as acetonitrile, dioxane, methanol, 2-propanol, or acetone. The stationary phase is usually a derivatized silica gel where the —OH groups of the silica gel have been replaced by —OSiOR groups where R is typically a linear $C_{18}$ alkyl chain. These so-called bonded-phase columns are not capable of handling as much analyte as normal silica gel columns, and are thus easily overloaded and are less useful for preparative work. For further discussion see the section on Solid-Phase Extraction in Technique 4.

## HIGH-PERFORMANCE LIQUID CHROMATOGRAPHY

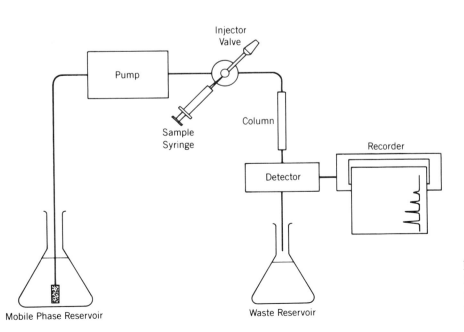

**FIGURE 5.50** High-Performance Liquid Chromatography system block diagram. *(Courtesy of the Perkin-Elmer Corp., Norwalk, CT.)*

A wide variety of detection systems are available for HPLC. A common, inexpensive and sensitive system is UV detection. The solvent flowing off the column is sent through a small cell, and the UV absorbance is recorded with respect to time. Many detectors are capable of variable wavelength operation so the detector can be set to the wavelength most suitable to the compound or compounds being analyzed. Photodiode array detectors are available that can obtain a full UV spectrum in a fraction of a second, so that more information can be obtained on each component of a mixture. For compounds that absorb light in the visible (vis) spectrum, many UV detectors can also be tuned to visible wavelengths. The principal shortcoming of UV–vis detection is that the compounds being studied must possess a chromophore, such as an aromatic ring or other conjugated $\pi$ system.

For compounds that lack a UV–vis chromophore, refractive index (RI) detection is a common substitute. An RI detector measures the difference in refractive index between the eluant and a reference cell filled with the elution solvent. Refractive index detection suffers in that it is significantly less sensitive than UV–vis detection, and the detector is quite sensitive to temperature changes during the chromatographic run.

More sophisticated HPLC instruments offer the ability to mix two or three different solvents and to use solvent gradients by changing the solvent composition as the chromatographic run progresses. This procedure allows the simultaneous analysis of compounds that differ greatly in their polarity. For example, a silica gel column might begin elution with a very nonpolar solvent, such as hexane, while the solvent is continuously delivered with an increasing solvent polarity by adding in ethyl acetate until the run is ended by eluting with pure ethyl acetate. This effect is directly analogous to temperature programming in GC.

For analytical work, typical HPLC columns are about 5 mm in diameter and about 25 cm in length. The maximum amount of analyte for columns such as this is generally less than 1 mg, and the minimum amount is determined by the detection system. High-performance liquid chromatography can thus be used to obtain small amounts of purified compounds for infrared (IR), nuclear magnetic resonance (NMR) or mass spectrometric (MS) analysis. Larger "semipreparative columns" that can handle up to about 20 mg of material without significant overloading, are available and are useful for obtaining material for $^{13}$C NMR spectroscopy or for further synthetic work.

**NOTE.**  *The following experiments utilize Technique 6A: Experiments [8A], [8B], [8C], [11C], [12], [13], [16], [17], [19A], [19B], [19C], [19D], [22A], [22B], [27], [29A], [29B], [29C], [29D], [30], [33A], [33B], [35], [1A$_{adv}$], [1B$_{adv}$], [4$_{adv}$], [7$_{adv}$], [A2$_a$], [A1$_b$], [E1], and [E3].*

### Technique 6B   Concentration of Solutions

The solvent can be removed from the chromatographic fractions by a number of different methods.

---

**DISTILLATION**

Concentration of solvent by distillation is straightforward, and the standard routine is described in Technique 2. This approach allows for high recovery of volatile solvents and often can be done outside a hood. The Hickman still head and the 5- or 10-mL round-bottom flask are useful for this purpose. Distillation should be used primarily for concentration of the solution, followed by transfer of the concentrate with a Pasteur filter pipet to a vial for final isolation.

## EVAPORATION WITH NITROGEN GAS

A very convenient method for removal of final solvent traces is the concentration of the last 0.5 mL of solution by evaporation with a gentle stream of nitrogen gas while the sample is warmed in a sand bath. This process is usually done at a hood station where several Pasteur pipets can be attached to a manifold leading to a tank of the compressed gas. Gas flow to the individual pipets is controlled by needle valves. *Always test the gas flow with a blank vial of solvent.* This evaporation technique is a hand-held operation. The sample vial will cool rapidly on evaporation of the solvent, and gentle warming of the vial with agitation will thus aid removal of the last traces of the volatile material. This procedure avoids possible moisture condensation on the sample residue. *Do not leave the heated vial in the gas flow after the solvent is removed!* This precaution is particularly important in the isolation of liquids. Remember to tare the vial before loading the solution to be concentrated, because achievement of constant weight is the best indication of total solvent removal.

## REMOVAL OF SOLVENT UNDER REDUCED PRESSURE

Concentration of solvent under reduced pressure is very efficient. It reduces the time of solvent removal in microscale experiments to a few seconds, or at most, a few minutes. In contrast, distillation or evaporation procedures require many minutes for even relatively small volumes. Vacuum-concentration techniques, however, can be tricky and should be practiced prior to committing a hard-won reaction product to this test. The procedure is most beneficial when applied to fairly large chromatographic fractions (5–10 mL).

The sequence of operations is as follows (see also Fig. 5.51):

**1.** Transfer the chromatographic fraction to the 25-mL filter flask.
**2.** Insert the 11-mm Hirsch funnel and rubber adapter into the flask.
**3.** Turn on the water pump (with trap) and connect the vacuum tubing to the pressure flask side arm while holding the flask in one hand.
**4.** Place the thumb of the hand holding the filter flask over the Hirsch funnel filter bed to shut off the air flow through the system (Fig. 5.51). This step will result in an immediate drop in pressure. The volatile solvent will rapidly come to a boil at room temperature. Thumb pressure adjusts air leakage through the Hirsch funnel and thereby controls the pressure in the system. It is also good practice to learn to manipulate the pressure so that the liquid does not foam up into the side arm of the filter flask.

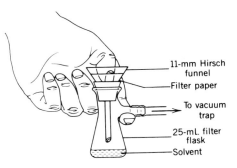

**FIGURE 5.51**  Removal of solvent under reduced pressure.

It is essential that the filter flask be warmed by the sand bath during this operation, for rapid evaporation of the solvent will quickly cool the solution. The air leak used to control the pressure results in a stream of moist laboratory air being rapidly drawn over the surface of the solution. If the evaporating liquid becomes cold, water will condense over the interior of the filter flask and eventually will contaminate the isolated residue. Warming the flask while the evaporation process is being carried out will avoid this problem and help to speed solvent removal. The temperature of the flask should be checked from time to time by touching it with the palm of the free hand. The flask is kept slightly above room temperature by adjusting the heating and evaporation rates. It is best to practice this operation a few times with solvent blanks to see whether you can avoid boilovers and the accumulation of water residue in the flask.

In the majority of research laboratories, the most efficient way to concentrate a solution under reduced pressure is to use a **rotary evaporator** (Fig. 5.52). The use of this commercially available device makes it possible to recover the solvent removed during the operation.

The rotary evaporator is a motor-driven device that rotates the flask containing the solution to be concentrated under reduced pressure. The rotary motion continuously exposes a thin film of the solution for evaporation. This process is very rapid, even well below the boiling point of the solvent being removed. Since the walls of the flask are constantly rewetted by the solution as it rotates, bumping and superheating are minimized. The rotating flask may be warmed in a water or oil bath, or other suitable device that controls the rate of evaporation. Where very large volumes (> 2 L) must be concentrated, a constant feed device is available for use. It is recommended that a suitable adapter be used on the rotary evaporator to guard against splashing and sudden ebullition, which may lead to loss, and to contamination of both the product and apparatus.

In microscale work, it is important to remember never to pour the recovered product, if a **liquid**, from the rotary flask. Always use a Pasteur pipet to accomplish this transfer. We should note that rotary evaporators can also be used for final solvent removal and vacuum drying of powders and solids.

**NOTE.** *The following experiments utilize Technique 6B: Experiments [11C], [12], [13], [27], [29A], [29B], [29C], [29D], [30], [33A], [33B], [1A$_{adv}$], [1B$_{adv}$], [4$_{adv}$], [7$_{adv}$], [A2a], [C1], [D3], [E1], [E3], and [F2].*

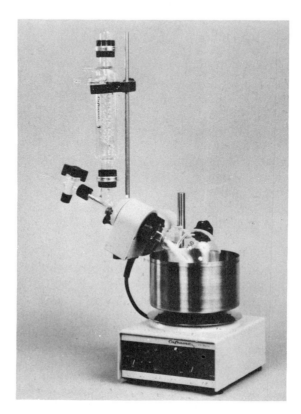

**FIGURE 5.52** Heidolph microrotary evaporator. *(Courtesy of Caframo, Ltd., Wiarton, Ontario, Canada.)*

## Collection or Control of Gaseous Products                    **TECHNIQUE 7**

Numerous organic reactions lead to the formation of gaseous products. If the gas is insoluble in water, collection is easily accomplished by the displacement of water from a collection tube. A typical experimental setup for the collection of gases is shown in Figure 5.53.

As illustrated, the glass capillary efficiently transfers the evolved gas to the collection tube. The delivery system need not be glass, as small diameter polyethylene or polypropylene tubing may also serve this purpose. In this latter arrangement, a syringe needle is inserted through a septum to accommodate the plastic tubing as shown in Figure 5.54. An alternative to the use of this connector is to employ a shortened Pasteur pipet inserted through a thermometer adapter (also shown in Fig. 5.54 )

An example of a reaction leading to gaseous products that can utilize this collection technique is the dehydration of 2-butanol with an acid catalyst as described in Experiment [9]. The products of this reaction are a mixture of the alkenes, 1-butene, *trans*-2-butene, and *cis*-2-butene having boiling points of $-6.3$, 0.9, and 3.7 °C, respectively.

In Figure 5.53, the gas collection tube is capped with a rubber septum. This arrangement allows for convenient removal of the collected gaseous butenes using a gas-tight syringe as shown in Figure 5.55.

In this particular reaction, the mixture of gaseous products is conveniently analyzed by ambient temperature GC (see Technique 1).

Some organic reactions release poisonous or irritating gases as byproducts. For example, hydrogen chloride, ammonia, or sulfur dioxide are typical byproducts in organic reactions. In these cases, the reaction is generally run in a **hood**. In addition, a **gas trap** may or may not be employed to prevent the gases from being released into the laboratory atmosphere. If the volatile material evolved is water soluble, the trap technique is convenient at the microscale level. The evolved gas is directed from the reaction

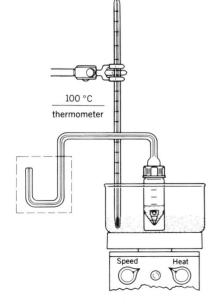

100 °C
thermometer

Speed     Heat

2-Butanol, 100 µL, +
concd $H_2SO_4$, 20 µL

3 mL
4 mL

**FIGURE 5.53** Microscale gas collection apparatus.

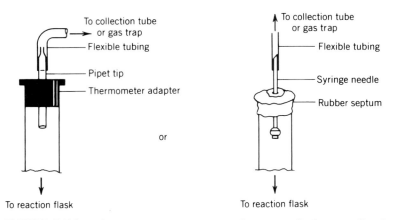

**FIGURE 5.54** Alternative arrangements for controlled gas collection.

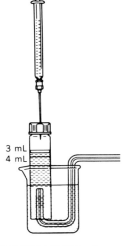

**FIGURE 5.55** Removal of collected gases.

vessel to a container of water or other aqueous solution, wherein it dissolves (reacts). For example, a dilute solution of sodium or ammonium hydroxide is suitable for acidic gases (such as HCl) or a dilute solution of sulfuric or hydrochloric acid for basic gases (such as $NH_3$ or low molecular weight amines).

Various designs are available for the gas trap arrangement. A simple one that is easily assembled is shown in Figure 5.56.

This arrangement is used for a gas that is very soluble in water. Note that the funnel is not immersed in the water. If the funnel is held below the surface of the water and a large quantity of gas is absorbed or dissolved, the water easily could be drawn back into the reaction assembly. If the gas to be collected is *not* very soluble, the funnel may be immersed just below the surface of the water.

When working at the microscale level, small volumes of these gases are evolved, and the funnel may not be necessary. Three alternatives are available in this situation.

**1.** In the first situation, the beaker (100 mL) in Figure 5.56 is filled with fine, moistened, glass wool and the gas delivery tube is led directly into the wool.

**2.** The second approach is to place moistened glass wool in a drying tube, which is then attached to the reaction apparatus as shown in Figures 3.23 or 3.26. However, be careful that the added moisture is not allowed to drip into the reaction vessel. It is advisable to place a small section of **dry** glass wool in the tube before the moist section is added.

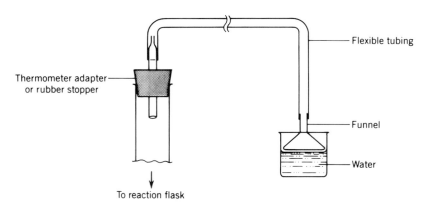

**FIGURE 5.56** Trapping of a water-soluble gas.

**3.** The third alternative is to use a water aspirator. An inverted funnel is placed over the apparatus opening where the evolved gas is escaping (usually the top of a condenser) and connected with flexible tubing to a water trap, and hence to the aspirator. In some cases, a Pasteur pipet can be substituted for the funnel when the volume of evolved gas is relatively small.

**NOTE.** *The following experiments utilize Technique 7: Experiments [9], [10], [14], [A2ₐ], and [B2].*

---

## Measurement of Specific Rotation

TECHNIQUE 8

Solutions of optically active substances, when placed in the path of a beam of polarized light, may rotate the plane of the polarized light clockwise or counterclockwise. The observed optical rotation is measured using a *polarimeter*. This technique is one of the oldest instrumental procedures used to characterize chemical compounds. It is a technique applicable to a wide range of analytical problems that vary from purity control to the analysis of natural and synthetic products in the medicinal and biological fields. The results obtained from the measurement of the observed angle of rotation, $\alpha$, are generally expressed in terms of *specific rotation* $[\alpha]$.

---

Ordinary light behaves as though it is composed of electromagnetic waves in which the oscillating electric field vectors may take up all possible orientations around the direction of propagation (see Fig. 5.57).

OPTICAL ROTATION THEORY

**NOTE.** *A beam of light behaves as though it is composed of two, mutually perpendicular, oscillating fields: an electrical field and a magnetic field. The oscillating magnetic field is not considered in the following discussion.*

The planes in which the electrical fields oscillate are perpendicular to the direction of propagation of the light beam. If one separates one particular plane of oscillation from all other planes by passing the beam of light through a polarizer, the resulting radiation is said to be plane-polarized (see Fig. 5.58).

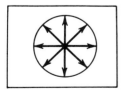

**FIGURE 5.57** Oscillation of the electric field of ordinary light occurs in all possible planes perpendicular to the direction of propagation. *From Solomons, T. W. Graham* Organic Chemistry, *5th ed., Wiley: New York, 1992. (Reprinted by permission of John Wiley & Sons, Inc. New York.)*

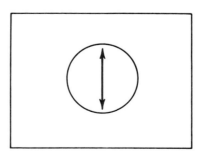

**FIGURE 5.58** The plane of oscillation of the electric field of plane polarized light. In this example the plane of polarization is vertical. *From Solomons, T. W. Graham* Organic Chemistry, *5th ed., Wiley: New York, 1992. (Reprinted by permission of John Wiley & Sons, Inc. New York.)*

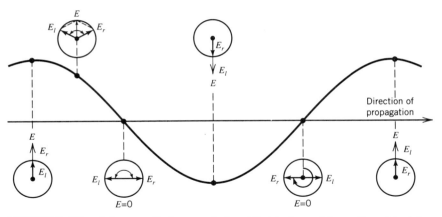

**FIGURE 5.59**   A beam of plane-polarized light viewed from the side (sine wave) and along the direction of propagation at specific times (circles) where the resultant vector $E$ and the circularly polarized components $E_l$ and $E_r$ are shown. *From Douglas, Bodie; McDaniel, Darl H.; Alexander, John J.* Concepts and Models of Inorganic Chemistry, *2nd ed., Wiley: New York, 1983. (Reprinted by permission of John Wiley & Sons, Inc., New York.)*

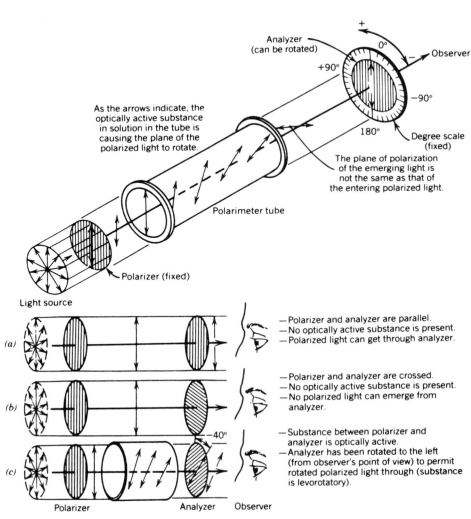

**FIGURE 5.60**   Operation of a polarimeter. *From Solomons, T. W. Graham* Organic Chemistry, *5th ed., Wiley: New York, 1992. (Reprinted by permission of John Wiley & Sons, Inc. New York.)*

In the interaction of light with matter, this plane-polarized radiation is represented as the vector sum of two circularly polarized waves. The electric vector of one of the waves moves in a clockwise direction, whereas the other moves in a counterclockwise direction, both waves having the same amplitude (see Fig. 5.59). These two components add vectorially to produce plane-polarized light.

If the passage of plane-polarized light through a material results in the velocity of one of the circularly polarized components being decreased more than the other by interaction with bonding and nonbonding electrons, the transmitted beam of radiation has the plane of polarization rotated from its *original* position (Figs. 5.60 and 5.61). A **polarimeter** is used to measure this angle of rotation.

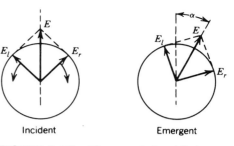

**FIGURE 5.61**    Plane-polarized light before entering and after emerging from an optically active substance. *From Douglas, Bodie; McDaniel, Darl H.; Alexander, John J.* Concepts and Models of Inorganic Chemistry, *2nd ed., Wiley: New York, 1983. (Reprinted by permission of John Wiley & Sons, Inc., New York.)*

## THE POLARIMETER

The *polarimeter* measures the amount of rotation caused by an optically active compound (in solution) placed in the beam of the plane-polarized light. The principal parts of the instrument are diagrammed in Figure 5.60. Two Nicol prisms are used in the instrument. The first prism, which polarizes the original light source, is called the polarizer. The second prism, called the analyzer, is used to examine the polarized light after it passes through a solution of the optically active species.

When the axes of the analyzer and polarizer prisms are parallel and no optically active substance is present, the maximum amount of light is transmitted and the instrument dial is set to 0°. However, if the axes of the analyzer and polarizer are at right angles to each other, no transmission of light is observed and the field is dark. The introduction of a solution of a nonracemic chiral substance into the path of the plane-polarized light causes one of the circularly polarized components, through dissymmetric interaction, to be slowed more than the other. The refractive indexes are, therefore, different in the two circularly polarized beams. Figure 5.60 represents a case in which the left-hand component has been affected the most.

**NOTE.**    *In this simplified figure, the effect on only one of the circularly polarized waves is diagrammed. See Figure 5.61 for a more accurate description (view from behind the figure).*

As seen, this results in a tilt of the plane of polarization. The analyzer prism must be rotated to the left to maximize the transmission of radiation. If rotation is counterclockwise, the angle of rotation is defined as (−) and the enantiomer that caused the effect is termed levorotatory (*l*). Conversely, clockwise rotation is defined as (+), and the enantiomer is said to be dextrorotatory (*d*). It is important to note that if a solution of equal amounts of a *d* and an *l* enantiomeric pair is placed in the beam path of the polarimeter, no rotation is observed. Such a solution is said to be *racemic* if, as in this case, it is an equimolar mixture of enantiomers.

The magnitude of optical rotation depends on several factors: (1) the nature of the substance, (2) the path length through which the light passes, (3) the wavelength of light used as a source, and (4) the temperature. It also depends on the concentration and the solvent of the solution of the optically active material.

The results obtained from the measurement of the observed angle of rotation, $\alpha_{obs}$, are generally expressed in terms of *specific rotation* $[\alpha]$.[17] The sign and magnitude of $[\alpha]$ are dependent on the specific molecule and are determined by complex features of molecular structure and conformation, and thus cannot be easily explained or predicted. The relationship of $[\alpha]$ to $\alpha_{obs}$ is as follows:

$$[\alpha]_{\lambda}^{T} = \frac{\alpha_{obs}}{l \cdot c}$$

where $T$ = the temperature of the sample **in degrees Celsius** (°C),
      $l$ = the length of the polarimeter cell **in decimeters** (1 dm = 0.1 m = 10 cm)
      $c$ = the concentration of the sample **in grams per milliliter** (g/mL)
      $\lambda$ = the wavelength of the light **in nanometers** (nm) used in the polarimeter.

These units are traditional, though most are esoteric by contemporary standards. The specific rotation for a given compound is dependent on both the concentration and the solvent, and thus both the solvent and concentration used must be specified. For example: $[\alpha]_D^{25}$ ($c$ = 0.4, $CHCl_3$) = 12.3° implies that the measurement was recorded in a $CHCl_3$ solution of 0.4 g/mL at 25 °C using the sodium D line (589 nm) as the light source.

For increased sensitivity, most lower cost polarimeters are equipped with an optical device that divides the viewed field into three adjacent parts (triple-shadow polarimeter; Fig. 5.62). A very slight rotation of the analyzer will cause one portion to become dimmer and the other lighter (Figs. 5.62a and 5.62c). The angle of rotation reading ($\alpha$) is recorded when the sections of the fields all have the same intensity. An accuracy of ±0.1° is easily obtained using this technique.

High-performance polarimeters are now available with a digital read-out and an accuracy of ± 0.001°. These instruments are based on an automatic optical-null balance principle that operates within a rotary range of ±80°. The schematic diagram for two such polarimeters is shown in Figure 5.63. Model 241 (Fig. 5.63a) has two spectral line sources (Hg and Na) and a filter wheel (F) containing five optical filters. A rotatable mirror (M) is coupled to this shaft so that when a desired wavelength is selected, the beam of the corresponding lamp is automatically reflected along the optical path.

The Model 241 MC, (Fig. 5.63b) is equipped with a high-resolution grating monochrometer (M) plus deuterium ($D_2$) and quartz–iodine lamps (continuous sources). The desired wavelength(s) can be selected by the monochrometer.

In both instruments, the monochromatic light passes through the polarizer (P), the sample cell (K), the analyzer (A), on to the photomultiplier tube detector (PM). The polarizer and analyzer are rotatable Glan prisms of calcite. The polarizer (and thus the plane of the plane-polarized light) oscillates at an amplitude of about ±0.7° at the line frequency about the optical longitudinal axis. In an unbalanced condition, the photomultiplier tube receives a frequency signal that is amplified and fed to the servomotor (SM) with the corresponding polarity (the latter gives the direction of rotation). The servomotor drives the analyzer until this signal is

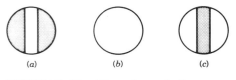

(a)        (b)        (c)

**FIGURE 5.62** View through the eyepiece of the polarimeter. The analyzer should be set so that the intensity in all parts of the field is the same (b). When the analyzer is displaced to one side or the other, the field will appear as in (a) or (c).

[17] The following discussion is also contained in Experiment [11A]. It is repeated here for the sake of continuity.

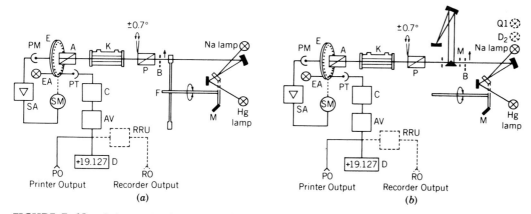

**FIGURE 5.63** Schematic diagrams of Perkin-Elmer polarimeters: (*a*) Model 241, (*b*) Model 241 MC. *(Courtesy of the Perkin-Elmer Corp., Newton Centre, MA.)*

reduced to null (this gives the magnitude of the rotation). This condition constitutes a balanced system (optical null).

When an optically active substance is placed in the sample cell (K), the plane of the polarized light is changed based on the optical rotation of the sample. The analyzer is rotated by the servomotor to the new null balance position. The difference between the original and new balance position corresponds to the optical rotation of the sample.

Optical rotary dispersion, the dependence of the rotation of an optically active species versus wavelength, is a natural extension of the basic polarimetric method. The newer instruments (Fig. 5.63*b* , Model 241 MC) are readily adapted to this technique. Information on the degree of coiling of protein helices and the establishment of the configurations of chiral molecules indicate the wide utility of this approach.

An interesting area for measuring optical activity is in the case of natural products. An excellent example is the lichen metabolite, usnic acid, which can be easily isolated from its native source, "Old Man's Beard," as golden yellow crystals, see Experiment [11A].

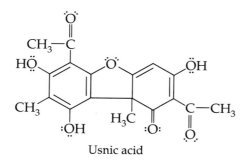

Usnic acid

Usnic acid contains a single chiral center and, therefore, has the possibility of existing as an enantiomeric pair of isomers. Generally, in a given lichen, only one of the isomers (*R* or *S*) is present. Usnic acid has a very high specific rotation ($\sim \pm 460°$), and for this reason is an ideal candidate to measure rotation at the microscale level.

**NOTE.** *The following experiment utilizes Technique 8: Experiment [11A].*

## TECHNIQUE 9

## Sublimation

Sublimation is a technique that is especially suitable for the purification of solid substances at the microscale level. It is particularly advantageous when the impurities present in the sample are nonvolatile under the conditions employed. Sublimation is a relatively straightforward method in that the impure solid need only be heated and mechanical losses are easily kept to a minimum.

The technique has additional advantages: (1) it can be the technique of choice for purifying sensitive materials, as it can be carried out under very high vacuum and thus it is effective at low temperatures; (2) solvents are not involved and indeed, final traces of solvents are effectively removed (see point 4); (3) impurities most likely to be separated are those having lower vapor pressures than the desired substance and, often therefore, lower solubilities, exactly those materials very likely to be contaminants in attempts at recrystallization; (4) solvated materials tend to desolvate during the process; and (5) in the specific case of water of solvation, it is very effective even with those substances that are deliquescent. The main disadvantage of the technique is that it may not be as selective as recrystallization. This nonselectivity occurs when the vapor pressure of the materials being sublimed are similar.

Materials sublime when heated below their melting points under reduced pressure. Substances that are candidates for purification by sublimation are those that do not have strong intermolecular attractive forces. Naphthalene, ferrocene, and *p*-dichlorobenzene are examples of compounds that meet these requirements.

## SUBLIMATION THEORY

The processes of *sublimation* and *distillation* are closely related. Crystals of a solid substance that sublimes, when placed in an evacuated container, will gradually generate molecules in the vapor state by the process of *evaporation* (i.e., the solid exhibits a vapor pressure). Occasionally, one of the vaporized molecules will strike the crystal surface or the walls of the container and be held by attractive forces. This latter process is the reverse of evaporation and is termed *condensation*.

*Sublimation* is the complete process of *evaporation* from the solid phase to *condensation* from the gas phase to form crystals *directly* without passing through the liquid state.

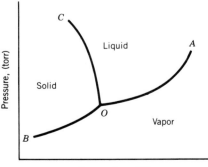

**FIGURE 5.64** Single-component phase diagram.

**FIGURE 5.65** Vacuum sublimator. *(Courtesy of ACE Glass Inc., Vineland, NJ)*

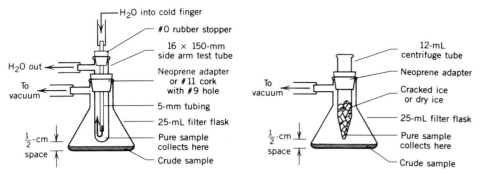

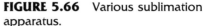

**FIGURE 5.66**  Various sublimation apparatus.

A typical single-component phase diagram is shown in Figure 5.64, which relates the solid, liquid, and vapor phases of a substance to temperature and pressure. Where two of the areas (solid, liquid, or vapor) touch, there is a line, and along each line, the two phases exist in *equilibrium*. Line BO is the sublimation–vapor pressure curve of the substance in question, and only along line BO can solid and vapor exist together in equilibrium. At temperatures and pressures along the BO curve, the liquid state is thermodynamically unstable. Where the three lines representing pairs of phases intersect, all three phases exist together in equilibrium. This point is called the *triple point*.

Many solid substances have a sufficiently high vapor pressure near their melting point and they thus can be sublimed easily *under reduced pressure* in the laboratory. Sublimation occurs when the vapor pressure of the solid equals the applied pressure.

Heating the sample with a microburner or a sand bath to just below the melting point of the solid causes sublimation to occur. The vapors condense on the cold-finger surface, whereas any less volatile residue will remain at the bottom of the flask. Apparatus suitable for sublimation of small quantities are now commercially available (Fig. 5.65).

A simple apparatus suitable for sublimation of small quantities of material in the microscale organic laboratory are shown in Figure 5.66.

An example of the purification of a natural product, where the sublimation technique at the microscale level is effective, is in the case of the alkaloid caffeine. This substance can be isolated by extraction from tea (see Experiment [11B]).

**NOTE.**  *The following experiments utilize Technique 9: Experiments [11B], [25A], and [25B].*

# 6

# Microscale Organic Laboratory Experiments

$C_6H_6$, Prismane
Katz and Acton (1973).

**INTRODUCTION**

This chapter contains the experimental details of a collection of famous organic reactions that are at the heart of this field of chemistry. One of the great triumphs of the human intellect has been our ability to rationalize the physical transformations of organic materials. The experimental laboratory is the source of information out of which predictive organic theory has been fashioned. The **microscale organic laboratory** is designed to give you the opportunity to experience, first hand, how organic reactions occur. This program will allow you to see how experimental data are directly related to the development of the structural and mechanistic theory surrounding these transformations. If successful, the **microscale organic laboratory** should bring the lecture portion of your course to life. What you have been studying in two dimensions in lecture, now becomes alive in three dimensions in the laboratory.

The 35 experiments that make up this chapter focus largely on studying some of the most important of the fundamental organic reactions that have been discovered over the last two centuries. As the application of these reactions to synthesis has been extensive, the microscale laboratory program offers a broad and practical introduction to organic chemistry. Six of the experiments include optional scale-up procedures to provide laboratory experience at the semimicroscale level, if an alternative scale of experimental work is desired. Studying one or two expanded-scale experiments at the beginning of the second semester can be helpful, particularly if some of the multistep syntheses covered in Chapter **8** are planned for study later in that semester. These latter sequences make extensive use of semimicroscale experimentation.

The microscale experiments are designed to enhance your ability to master the miniaturized experimental techniques employed. A **Prior Reading** section highlights which techniques are to be used in a particular experiment, and outlines the pages to refer to in the Techniques section (see Chapter 5). At the beginning of the semester it is important to *review these sections before proceeding* (see the advice in Chapter 1) with the assigned experiment. After examining the written material, it would be particularly helpful if you could see the videotape program that we have prepared. It covers the techniques in detail and can be viewed either in your spare time, or in a prelab lecture period, depending on local arrangements. As the basic techniques are used repeatedly, you will rapidly gain a firm grasp of the manipulations required and become comfortable working at this scale. You should also take advantage of the large number of graphics in the text margin that detail how the experimental apparatus are assembled. There are prompt signs (■) in the running text near the point where the equipment is to be employed.

The important role of spectroscopic techniques in the modern organic laboratory is emphasized in the **microscale organic laboratory**. This information is set aside as a separate chapter to be treated as reference material to be used while undertaking the experimental sections in Chapters 6–8. Numerous cases of detailed analyses of the spectra are found in the *Purification and Characterization* sections of the experiments. As infrared spectroscopy continues to play a major role in the characterization of reaction products in the introductory laboratory, and as this technique is currently given only a cursory treatment in most lecture texts, we have included a fairly detailed qualitative introduction to the theory of the effect, and the instrumentation used in obtaining these observations, in Chapter 9. *While the IR part of the spectroscopic section may cover more ground than is normally found in many introductory laboratories, we feel it is exceedingly important to overcome the black-box attitude which students can rapidly develop toward complex chemical instrumentation when they are turned loose on these powerful instruments with very limited knowledge.* Thus, we hope to be able to accomplish this transformation of attitude by offering the student the essential details which will allow them to gain a command of the logic and the mechanics of obtaining and interpreting infrared spectral data. In addition, as mentioned earlier, we have also included, in a number of the experiments, illustrative examples of the detailed spectral analysis that can be used to examine starting materials, follow the progress of the reaction, and finally to ultimately assess the character and purity of the products.

Each experiment is preceded by a Discussion section. In a number of cases, especially when named reactions are involved, a brief biographical sketch of the individual so honored is included. At this point we also introduce pertinent information concerning the reaction mechanism, often in considerable detail. When appropriate, the relevance of a reaction to the life sciences and the chemical industry is explored.

Note that *Safety* and *Warning* indicators are highlighted or boxed in the experiments. We urge you to *always* adhere strictly to the safety precautions listed.

The nomenclature of organic compounds is often extremely confusing to the beginning student and even occasionally to the experienced research chemist! In an attempt to ease your introduction to the name game, the *common name* (sometimes referred to as the *trivial name*) of the compound to be synthesized is also listed at the beginning of each experiment. In addition, the *Chemical Abstracts* (CA) number and CA index name are also given.

Good luck! Enjoy your adventure in transforming small quantities of a large number of organic materials.

## Measurement of Physical Properties

### Purpose

To become acquainted with the experimental techniques used to measure certain classic physical properties of organic substances. These properties include the boiling point, density, and refractive index for liquids, and the melting point for solids. The procedures are outlined in Chapter 4.

A further objective is to study sampling techniques for obtaining the spectral characteristics of organic materials. You will observe that absorption spectra of organic substances contain, without question, the most important collection of physical constants of a material available to the investigator. Spectral information can lead to the elucidation of molecular structure and the rapid identification of unknown substances.

To learn how to locate the literature values for these measured properties using various chemical handbooks.

*Prior Reading*

Chapter 4:   Determination of Physical Properties
  Ultramicro-Boiling Point (see pp. 44–47)
  Density (see pp. 47–48)
  Refractive Index (see pp. 48–50)
  Melting Point
    Capillary Method (see pp. 50–51)
    Evacuated Technique (see pp. 51–52)
    Mixture Melting Points (see p. 53)

Organic compounds have a number of physical properties that allow their precise characterization. These include the classical physical constants: boiling point, density, and refractive index for liquids and melting point for solids. The rapid development of modern chemical instrumentation, however, has also made easily accessible many of the spectral properties of these materials. Spectroscopy, in particular, provides information that is extremely powerful for establishing the structure of unknown molecular systems on the one hand, and the rapid identification of known materials on the other.

**DISCUSSION**

Not only are physical properties used to characterize a specific organic compound, but they are often employed to compare one compound to another. Examples of this approach are illustrated in Chapter 10, Qualitative Identification of Organic Compounds. The route to identification of an unknown organic species has become increasingly dependent on the measurement of the physical properties of the pure substance.

The various classical physical constants of a large number of organic substances appear in the *CRC Handbook of Chemistry and Physics*, *Lange's Handbook*, and the *Aldrich Catalog Handbook of Fine Chemicals*. The latter collection also contains references to published infrared (IR) and nuclear magnetic resonance (NMR) spectral data for many of the compounds. The *CRC Atlas of Spectral Data and Physical Constants for Organic Substances* contains IR and ultraviolet (UV) peak positions. Collections of spectra may be located in the Aldrich Libraries of both IR and NMR spectra, and in the Sadtler Library.

A detailed discussion of the spectral properties of organic compounds is given in Chapter 9.

## EXPERIMENTAL PROCEDURE

**NOTE.** *Your instructor will select which of the physical properties you are to measure. The length of your laboratory period, the size of your section, and the number of instruments available will all play a role in determining how many of these properties will be suggested.*

Estimated time for the total experiment: 4.5 hours.

### Melting Point

**1.** Using the melting point apparatus (*your instructor will provide you with the experimental details concerning the operation of the particular instrument to be used in your laboratory*), determine the melting point of acetanilide and compare your result with those reported in the CRC and Aldrich references (see pp. 50–51 for the experimental details on how to proceed with this measurement).

Determined Value _____

CRC: lit. value _____ ; Ed._____ ; Page _____ ;
Compound No. _____

Aldrich: lit. value _____ ; Year _____ ; Page _____ ;
Compound No. _____

**2.** Now determine an *evacuated* melting point (see pp. 51–52 for the experimental details on how to proceed with this measurement) of caffeine and compare it to the values in the CRC and Aldrich references. *Your instructor will provide you with the experimental details concerning the operation of the particular instrument to be used in your laboratory.*

First, determine the melting point of caffeine in an unevacuated melting point tube. After observing the melting point, allow the temperature to drop below the melting point and observe if the sample crystallizes again. If crystallization occurs, observe a second melting point, and then repeat this procedure a third time. Follow the same routine with an evacuated sample, and compare the results of the two sets of melting points. Do you observe any differences between these data sets?

| *Unevacuated* | | *Evacuated* | |
|---|---|---|---|
| 1st Determined value | _____ | 1st Determined value | _____ |
| 2nd Determined value | _____ | 2nd Determined value | _____ |
| 3rd Determined value | _____ | 3rd Determined value | _____ |

CRC: lit. value _____ ; Ed. _____ ; Page _____ ;
Compound No. _____

Aldrich: lit. value _____ ; Year _____ ; Page _____ ;
Compound No. _____

**3.** Why is a second seal required on the evacuated melting point tube?

**4.** When are evacuated melting points necessary?

**5.** Next consider the technique of *mixture melting points* (see p. 53 for a discussion of the details of this procedure). This technique is the classical approach for establishing a positive identification of a substance when pure reference standards are available in the laboratory. For two examples where this type of measurement is applied in the microscale organic laboratory, see Experiments [6] and [34A].

**6.** In the present experiment, you have obtained the melting points of acetanilide and of caffeine. Using these values as reference standards observe the melting points of two mixtures (1) caffeine 75%–acetanilide 25% and (2) caffeine 25%–acetanilide 75%. Should these observations be made in evacuated capillaries or not? (*Your instructor will provide you with the experimental details concerning the operation of the particular instrument to be used in your laboratory.*)

(1) caffeine 75%–acetanilide 25%      (2) caffeine 25%–acetanilide 75%
   Determined value _____               Determined value _____

**7.** Is it possible to detect the presence of impurities by melting point measurements? Why?

**8.** Would it be possible to establish the composition of an unknown binary mixture of two substances from mixture melting point data? Explain.

## Ultramicro-Boiling Point

**1.** Make several (5) glass bells (p. 45). Your instructor will demonstrate the procedure, or you can view the procedure on video, if available. Place the bells in a small glass vial and store them in your micro kit.

**2.** Use the technique for determining ultramicro-boiling points on a *melting point apparatus hot stage* (very likely this will be the same apparatus you used in the melting point section of the experiment) as discussed in Chapter 4 (p. 45), and complete the following table (*your instructor will provide you with the experimental details concerning the operation of the particular instrument to be used in your laboratory*).

|         | Propane | Butane | Pentane | Hexane | Heptane |
|---------|---------|--------|---------|--------|---------|
| bp (°C) | −42.1   | −0.5   | 36.1    | 67     | ?       |

**3.** Compare your measured value for heptane to those listed in the CRC handbook and the Aldrich catalog.

Determined value _____; Barometric pressure _____ torr
Corrected value _____

CRC: lit. value _____; Ed. ____; Page ____;
Compound No. ____

Aldrich: lit. value _____; Year ____; Page ____;
Compound No. ____

**4.** Prepare, and attach, a graph of molecular weight (MW) vs. boiling point (bp) using the above data.

**5.** Explain the trend in boiling point as the molecular weight increases.

## Density

**1.** Use the syringe method of measuring the density outlined in Chapter 4 (p. 47), and 0.5 mL of liquid, to complete the following table and compare your value to those listed in the CRC and Aldrich references. (*Your instructor will provide you with the experimental details concerning the* *operation of the particular equipment to be used in this experiment. Pay close attention to the details concerning the operation of the balances.)*

|  | Methylene Chloride ($CH_2Cl_2$) | Octane $C_8H_{18}$ |
|---|---|---|
| Density (g/mL) | 1.33 | ? |
| Determined value _____ | | |

CRC: lit. value _____; Ed. _____; Page _____;
Compound No. _____

Aldrich: lit. value _____; Year _____; Page _____;
Compound No. _____

**2.** Methylene chloride is often used as a solvent in the laboratory to extract an *organic species* from an aqueous solution. Will the $CH_2Cl_2$ normally form the top or bottom layer?

**3.** If octane were used to extract an aqueous phase, would it form the top or bottom layer? How did you arrive at this answer?

## Refractive Index

**1.** Using the lens paper disk technique outlined in Chapter 4 (pp. 49–50) determine the refractive index of methanol and compare the value you have obtained to that found in the CRC and Aldrich data collections. *(Your instructor will provide you with the experimental details concerning the operation of the particular instrument used in your laboratory.)*

Determined value _____

CRC: lit. value _____; Ed. _____; Page _____;
Compound No. _____

Aldrich: lit. value _____; Year _____; Page _____;
Compound No. _____

**2.** Correct your measured value to 20 °C.

Temperature of your measurement _____.
Temperature of literature value _____.
Your measured value corrected to 20 °C. _____.

Calculations:

**3.** Based on your observations, how does refractive index vary with temperature?

## Infrared Spectroscopy Sampling Procedures
### Sampling of Liquids

**1.** Use the technique for obtaining an IR spectrum of a thin-liquid film as described on pages 634–635 *(your instructor will provide you with the experimental details concerning the operation of the particular instrument to be used in your laboratory and the type of windows on which your sample will be mounted)*.

**2.** Obtain the spectra of *n*-octane and 1-octanol [use a single scan with Fourier-transform (FT) instruments]. Compare your data with the Aldrich or Sadtler reference collections and compare the two spectra to each other.

**3.** What differences do you observe between the two spectra? Can you associate differences in molecular structure to the differences in the spectra (see Infrared Discussions in Chapter 9 (pp. 613–614) relating to the IR spectra of *n*-hexane and 1-hexanol).

**4.** Occasionally, when an IR spectrum is obtained, some of the very strong bands will appear with flattened peaks, as if they were totally absorbing the energy at those wavenumber values. The flat bottom of the band, however, does not correspond to 0% transmission on the scale, but will indicate an energy transmission of 5–10% or even higher. Can you explain the odd shape of the band? Is the energy being totally absorbed or not? Explain your answers.

### Sampling of Solids

**1.** Use the technique for obtaining an IR spectrum of a solid melting above 100 °C as described in Chapter 9 (pp. 635–636) (*your instructor will provide you with the experimental details concerning the operation of the particular instrument to be used in your laboratory and the type of KBr die in which your sample will be pressed*).

**2.** Obtain the IR spectrum of caffeine (use four scans with FT instruments). Compare your data with the Aldrich or Sadtler reference collections, and with the data given in Experiment [11B]. Your sample may be saved by taping it to a file card with your name, and stored by your instructor in a desiccator. If, later in the year, you isolate caffeine from its natural source (Experiment [11B]), you will be able to compare, at that time, the material you have extracted and purified from the plant, with your own authentic reference spectrum.

**3.** Occasionally, the spectral region from 4000–2000 $cm^{-1}$ in solid samples is steeply sloping to lower transmission values at higher wavenumber values. Is this drop in transmission an absorption of the radiation? If so, to what process can the absorption be ascribed, and if not, what is the cause of the decreased transmission?

**4.** Compare the spectra of caffeine obtained in your laboratory section. Are the spectra all identical? Where do they differ? To what effect can you ascribe the differences, if there are any? (This is a good open question for the entire lab section.)

### Nuclear Magnetic Resonance Spectroscopy Sampling Procedures

For NMR sampling procedures and examples see Chapter 9 and Experiments [5B], [22A], and [28].

**6-1.** An unknown carboxylic acid has a boiling point of 100 °C at 25 torr. Using the pressure–temperature nomograph on page 47, determine its boiling point at 760 torr. Identify the acid from the list in Appendix A, Table A.1 (p. 729).

**QUESTIONS**

**6-2.** Discuss several factors that might lead you to obtain an incorrect value of the melting point for your solid sample.

**6-3.** The mass of a certain volume of an unknown liquid at 25 °C was determined to be 234 mg. The mass of an equal volume of water at the same temperature was measured as 201 mg. Calculate the density of the unknown liquid at 25 °C.

**6-4.** The melting point of pure *trans*-cinnamic acid is 133 °C, and that of 2-acetoxybenzoic acid (aspirin) is 135 °C (Appendix A, Table A.2, p. 730). Given pure reference standards of both acids, describe a melting point procedure by which you could identify if an unknown sample melting in this range could be assigned to either structure, or to neither one.

**6-5.** Why is filter paper a poor choice of surface on which to powder or crush a solid crystalline sample before placing it in a capillary melting point tube?

**6-6.** Why is the ultramicro-boiling point determined precisely at the point when the last vapor bubble has escaped and the liquid phase begins to rise in the bell cavity?

---

**EXPERIMENT [2]**

## The Separation of a 25-$\mu$L Mixture of Heptanal (bp 153 °C) and Cyclohexanol (bp 160 °C) by Gas Chromatography

### Purpose
This experiment illustrates the separation of a 25-$\mu$L mixture, consisting of heptanal and cyclohexanol, into the pure components. The volume of the mixture is approximately that of a single drop, and the materials boil within 7 °C of each other. This mixture would be difficult, if not impossible, to separate by the best distillation techniques available. The purity of the fractions collected from the gas chromatograph (GC) can be assessed by boiling points, refractive indexes, or infrared (IR) spectra.

*Prior Reading*

| | |
|---|---|
| *Technique 1:* | Microscale Separation of Liquid Mixtures by Preparative Gas Chromatography (pp. 56–61). |
| *Chapter 4:* | Determination of Physical Properties Ultramicro-Boiling Point (pp. 44–47). Refractive Index (pp. 48–50). |
| *Experiment [1]:* | Measurement of Physical Properties Ultramicro-Boiling Point (p. 123). |
| *Chapter 9:* | Introduction to Infrared Spectroscopy (pp. 585–594, 613–614 |

---

**DISCUSSION**

The efficacy of GC separations is highly dependent on the experimental conditions. For example, two sets of experimental data on the heptanal–cyclohexanol mixture are given below to demonstrate the effects of variations in oven temperature on retention times.

In Data Set A, the oven temperature was allowed to rise slowly from 160 to about 170 °C during a series of sample collections. The retention time of the first component, heptanal, dropped from about 3 min to close to 2 min, whereas the retention time of the second component, cyclohexanol, was reduced from about 5.5 min to nearly 4 min. The significant decrease in resolution over this series of collections is reflected in the number of theoretical plates calculated, which was over 300 for heptanal and about 500 for cyclohexanol in the first trial, but declined to below 200 for both compounds toward the last run (see Data Set A).

**Data Set A**

| | Heptanal | | | | Cyclohexanol | | | |
|---|---|---|---|---|---|---|---|---|
| Trial No. | Retention Time (min) | Baseline Width (min) | Number of Theoretical Plates | Recovery (mg) | Retention Time (min) | Baseline Width (min) | Number of Theoretical Plates | Recovery (mg) |
| 1 | 3.1 | 0.7 | 314 | 8.0 | 5.6 | 1.0 | 502 | 8.0 |
| 2 | 2.9 | 0.7 | 275 | 8.0 | 5.3 | 1.0 | 449 | 8.0 |
| 3 | 3.0 | 0.7 | 294 | 7.0 | 5.7 | 1.0 | 520 | 8.0 |
| 4 | 2.8 | 0.7 | 256 | 8.0 | 5.1 | 1.1 | 344 | 8.0 |
| 5 | 2.5 | 0.6 | 278 | 8.0 | 4.3 | 1.1 | 244 | 9.0 |
| 6 | 2.7 | 0.5 | 467 | 7.0 | 4.6 | 1.0 | 339 | 10.0 |
| 7 | 2.5 | 0.6 | 278 | 10.0 | 4.2 | 1.0 | 282 | 8.0 |
| 8 | 2.2 | 0.5 | 310 | 9.0 | 3.5 | 1.0 | 196 | 8.0 |
| 9 | 1.8 | 0.5 | 207 | 8.0 | 3.0 | 1.0 | 144 | 8.0 |
| 10 | 2.3 | 0.7 | 173 | 8.0 | 3.9 | 1.0 | 243 | 8.0 |
| Av | $2.6 \pm 0.4$ | $0.6 \pm 0.09$ | $285 \pm 7$ | $8.1 \pm 0.8$ | $4.5 \pm 0.9$ | $1.0 \pm 0.05$ | $326 \pm 129$ | $8.3 \pm 0.7$ |

## COLLECTION YIELD

### Cyclohexanol

Density of cyclohexanol = 0.963 mg/$\mu$L.
In 25 $\mu$L of 1:1 cyclohexanol–heptanal, we have 12.5 $\mu$L of cyclohexanol.
Therefore, 12.5 $\mu$L $\times$ 0.963 mg/$\mu$L = 12 mg of cyclohexanol injected.
Percent recovered = (8.3 mg/12.0 mg) x 100 = 69% cyclohexanol collected.

### Heptanal

Density of heptanal = 0.850 mg/$\mu$L.
Therefore, 12.5 $\mu$L $\times$ 0.85 mg/$\mu$L = 10.6 mg of heptanal injected.
Percent recovered = (8.1 mg/10.6 mg) $\times$ 100 = 76% heptanal.

In the Data Set B collections, stable oven temperatures and flow rates were maintained, and the data exhibit excellent reproducibility. Oven temperature was held at 155 °C throughout the sampling process. The retention time of heptanal was observed to be slightly longer than 3 min, with a variance of 6 s, whereas the cyclohexanol retention time was found to be slightly longer than 6 min, with a variance of 12 s. The resolution remained essentially constant throughout the series, and the number of theoretical plates calculated was about 350 for heptanal and about 500 for cyclohexanol (see Data Set B).

## DATA SET B

| | Heptanal | | | | Cyclohexanol | | | |
|---|---|---|---|---|---|---|---|---|
| Trial No. | Retention Time (min) | Baseline Width (min) | Number of Theoretical Plates | Recovery (mg) | Retention Time (min) | Baseline Width (min) | Number of Theoretical Plates | Recovery (mg) |
| 1 | 3.5 | 0.7 | 400 | 8.0 | 6.6 | 1.1 | 576 | 8.0 |
| 2 | 3.2 | 0.7 | 334 | 9.0 | 6.0 | 1.1 | 476 | 7.0 |
| 3 | 3.5 | 0.7 | 400 | 7.0 | 6.6 | 1.2 | 484 | 10.0 |
| 4 | 3.2 | 0.7 | 334 | 9.0 | 6.1 | 1.0 | 595 | 9.0 |
| 5 | 3.1 | 0.6 | 427 | 8.0 | 6.0 | 1.1 | 476 | 8.0 |
| 6 | 3.2 | 0.7 | 334 | 9.0 | 6.0 | 1.1 | 476 | 9.0 |
| 7 | 3.3 | 0.8 | 272 | 9.0 | 6.1 | 1.1 | 492 | 8.0 |
| 8 | 3.1 | 0.7 | 313 | 8.0 | 6.0 | 1.1 | 476 | 10.0 |
| 9 | 3.2 | 0.7 | 334 | 8.0 | 6.1 | 1.1 | 492 | 8.0 |
| 10 | 3.2 | 0.7 | 334 | 8.0 | 6.2 | 1.1 | 508 | 8.0 |
| Av. | 3.2 ± 0.1 | 0.7 ± 0.05 | 348 ± 47 | 8.3 ± 0.7 | 6.2 ± 0.2 | 1.1 ± 0.05 | 505 ± 44 | 8.5 ± 1.0 |

## COLLECTION YIELD

Cyclohexanol

Density of cyclohexanol = 0.963 mg/$\mu$L.
In 25 $\mu$L of 1:1 cyclohexanol–heptanal, there are 12.5 $\mu$L of cyclohexanol.
Therefore, 12.5 $\mu$L × 0.963 mg/$\mu$L = 12 mg of cyclohexanol injected.
Percent recovered = (8.5 mg/12.0 mg) × 100 = 71% cyclohexanol collected.

Heptanal

Density of heptanal = 0.850 mg/$\mu$L.
Therefore, 12.5 $\mu$L × 0.85 mg/$\mu$L = 10.6 mg of heptanal injected.
Percent recovered = (8.3 mg/10.6 mg) × 100 = 78% heptanal.

The results just described demonstrate that the resolution of gas chromatographic (GC) peaks may be very sensitive to changes in retention time resulting from instability in oven temperatures. Since the number of theoretical plates is related to resolution values, significant degradation in column plate values can occur with variations in oven temperatures. When you compare the time and effort required to obtain a two-plate fractional distillation on a 2-mL mixture (see Experiment [3B] and Technique 3), with the speed and ease used to obtain a 500 plate separation on 12.5 $\mu$L of cyclohexanol in this experiment, it is hard not to be impressed with the enormous power of this technique.

## COMPONENTS

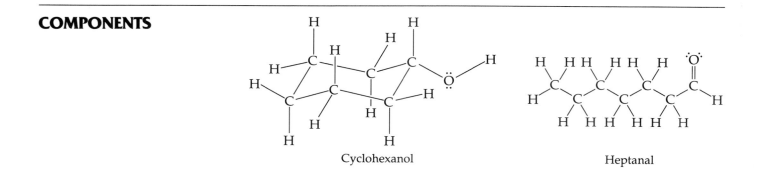

Cyclohexanol                    Heptanal

Estimated time for the experiment: 2.0 hours

## EXPERIMENTAL PROCEDURE

### Physical Properties of Components

| Compound | MW | Amount | bp(°C) | Density (d) | $n_D$ |
|---|---|---|---|---|---|
| Heptanal | 114.19 | 12.5 $\mu$L | 153 | 0.85 | 1.4113 |
| Cyclohexanol | 100.16 | 12.5 $\mu$L | 160 | 0.96 | 1.4641 |

### Reagents and Equipment

The procedure involves injection of a 25-$\mu$L mixture of heptanal–cyclohexanol 1:1 (v/v) onto a 1/4-in. × 8-ft stainless steel column packed with 10% Carbowax 80/100 20M PAW-DMS. Experimental conditions (GOW-MAC series No. 350) are He flow rate, 50 mL/min; chart speed, 1 cm/min; oven temperature, 155 °C.

### Procedure for Preparative Collection

The liquid effluents are collected in an uncooled, 4-mm diameter collection tube (double reservoirs; overall tube length 40–50 mm, see Fig. 6.1)

The collection tube (oven dried until 5 min before use) is attached to the heated exit port by the 5/5$ joint. Sample collection is initiated 0.5 min prior to detection on the recorder of the expected peak (time based on previously determined retention values)[1] and continued until 0.5 min following return to baseline. After the collection tube is detached, the sample is transferred to the 0.1-mL conical GC collection vial. The transfer is facilitated by the 5/5$ joint on the conical vial. After the collection tube is joined to the vial (preweighed with stopper), the system is centrifuged (see Fig. 6.1). The collection tube is then removed, and the vial stoppered and reweighed.

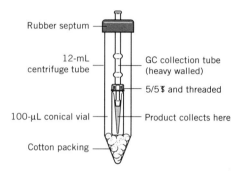

Rubber septum

12-mL centrifuge tube

100-$\mu$L conical vial

Cotton packing

GC collection tube (heavy walled)

5/5$ and threaded

Product collects here

**FIGURE 6.1** Gas chromatographic collection tube and 0.1 mL conical vial.

### Characterization

Calculate the percent recovery. These amounts should range between 7 and 10 mg. Determine the boiling point of each fraction and obtain the refractive index or infrared (IR) spectrum, if time permits. These latter measurements will require most, if not all, of the sample not used for boiling point determination.

Assess the purity and efficiency of the separation from your tabulated data and the GC chromatogram.

### Alternative Mixture Pairs for Preparative Collection

(all mixtures are 1:1 v/v)

### Mixture

a. *Separation* of a 40-$\mu$L mixture of[2]
(1S)-(−)-$\alpha$-pinene (bp 156 °C, $n_D$ = 1.4650, d = 0.855) and
(1S)-(−)-$\beta$-pinene (bp 165 °C, $n_D$ = 1.4782, d = 0.859)

---

[1] Refer to your local laboratory instructions.
[2] Refractive index at D line of sodium = $n_D$ and density = d.

## Components

(1S)-(−)-α-Pinene          (1S)-(−)-β-Pinene

*Chromatographic Parameters*

A 40-μL injection
Flow rate: 50 mL/min
Column temperature: 120 °C
Column: 20% Carbowax

Elution time
   α-Pinene: ~8 min
   β-Pinene: ~12 min

Average recovery
   α-Pinene: 8.3 μL
   β-Pinene: 10.6 μL

## Mixture

**b.** *Separation* of a 40-μL mixture of
   2-Heptanone (bp 149–150 °C, $n_D$ = 1.4085, $d$ = 0.820) and
   Cyclohexanol (bp 160–161 °C, $n_D$ = 1.4641, $d$ = 0.963)

## Components

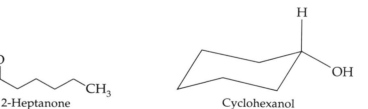

2-Heptanone                    Cyclohexanol

*Chromatographic Parameters*

A 40-μL injection
Flow rate: 50 mL/min
Column temperature: 145 °C
Column: 20% Carbowax

Elution time
   2-Heptanone: ~5.5 min
   Cyclohexanol: ~10.0 min

Average recovery
   2-Heptanone: 8.1 μL (41%)
   Cyclohexanol: 11.4 μL (57%)

## Mixture

**c.** *Separation* of a 40-μL mixture of
   d-Limonene (bp 175–176 °C, $n_D$ = 1.4743, $d$ = 0.8402) and
   Cyclohexyl acetate (bp 173 °C, $n_D$ = 1.4401, $d$ = 0.9698)

## Components

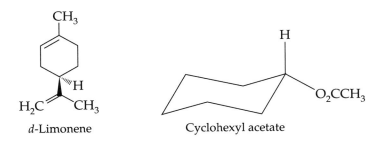

d-Limonene          Cyclohexyl acetate

*Chromatographic Parameters*

A 40-$\mu$L injection
Flow rate: 50 mL/min
Column temperature: 170 °C
Column: 20% Carbowax

Elution time
    d-Limonene: ~5.5 min
    Cyclohexyl acetate: ~7.5 min

Average recovery
    d-Limonene: 8.7 $\mu$L (44%)
    Cyclohexyl acetate: 10.0 $\mu$L (50%)

**6-7.** Based on the data presented in the Data Set A chromatographic separation, can you explain why there is such a steep decline in column efficiency with temperature change?

**6-8.** Consider the following gas chromatogram for a mixture of analytes X and Y:

**QUESTIONS**

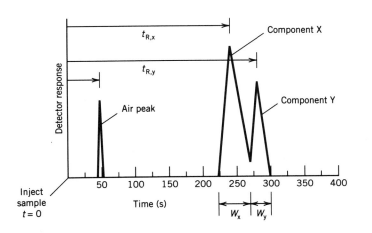

a. Calculate the number of theoretical plates for the column in reference to the peaks of each component (X and Y).
b. If the column is 12 m long, calculate the height equivalent theoretical plate (HETP) (in plates per cm) for this column.

**6-9.**   The number of theoretical plates a column has is important, but the crucial factor is the ability to separate two or more substances. That is, how well resolved are the peaks? The resolution of two peaks depends not only on how far apart they are ($t_R$) but also on the peak width (W). Baseline resolution (R) is defined by the following equation:

$$R = \frac{2\,\Delta t_R}{(W_x + W_y)}$$

Because of the tailing of most species on the column, a value of 1.5 is required to give baseline resolution.
a. Calculate the resolution for the peaks in Question 6-8.
b. Do you think a quantitative separation of the mixture is possible based on your answer?
c. Has baseline resolution been achieved?

**6-10.**   Discuss at least two techniques you might employ to increase the resolution of the column in Question 6-9 (without changing the column).

**6-11.**   Retention times for several organic compounds separated on a gas chromatographic column are given below.

| Compound | $t_R$ (seconds) |
| --- | --- |
| Air | 75 |
| Pentane | 190 |
| Heptane | 350 |
| 2-Pentene | 275 |

a. Calculate the relative retention of 2-pentene with respect to pentane.
b. Calculate the relative retention of heptane with respect to pentane.

---

**EXPERIMENT [3]**

## Distillation

In the following set of experiments, we will examine the applications of a variety of distillation techniques to the purification of liquid mixtures. In Experiments [3A] and [3B] you will conduct simple distillations. In Experiment [3A] a volatile liquid component is separated from a nonvolatile solid. Experiment [3B] illustrates the use of the Hickman still in the separation of hexane and toluene, which have boiling points 42 °C apart. The composition of the fractions is analyzed by refractive index and boiling point. Experiments [3C] and [3D] introduce the use of micro spinning band distillation columns for the separation of cyclohexane (bp 80.7 °C) and 2-methylpentane (bp 60.3 °C). The composition of the distillate fractions are determined by gas chromatography. The number of theoretical plates is determined for the spinning band column used. In Experiment [3D] you will be introduced to one of the simplest yet most efficient and powerful distillation techniques for the separation of liquid mixtures at the semimicroscale level, the Hickman–Hinkle still.

## Experiment [3A]   Simple Distillation at the Semimicroscale Level: Separation of Ethyl Acetate from *trans*-1,2-Dibenzoylethylene

### Purpose

To examine *simple distillation*, which involves the use of the distillation process to separate a liquid ester from minor components that are nonvolatile or that have boiling points much greater (>100 °C) than that of the major component.

### Prior Reading

Techniques 2 and 3:   Distillation (see pp. 61–72).
   Distillation Theory (see pp. 61–65).
   Simple Distillation at the Semimicroscale
   Level (see pp. 65–68).

Chapter 4:   Determination of Physical Properties
   Ultramicro-Boiling Point (see pp. 44–47).
   Density (see pp. 47–48).
   Refractive Index, (see pp. 48–50).
   Evacuated Melting Point (see pp. 51–52).

**DISCUSSION**

Semimicroscale simple distillation can be an effective separation technique with volumes from 0.5 to 2 mL. Apparatus have been developed that achieve effective separation of mixture samples in this range. One of the most significant of these designs is the classic Hickman still, shown in Figure 5.10. In this experiment you will effect the separation of a two component mixture by the use of this still.

The Hickman still is employed in several of the microscale experiments to purify solvents, carry out reactions, and concentrate solutions for recrystallization. An introduction to the use of the Hickman still is given in this experiment.

In a distillation where a liquid is separated from a nonvolatile solute, the vapor pressure of the liquid is lowered by the presence of the solute, but the vapor phase consists only of the pure liquid component. Thus, except for the transfer of nonvolatile material by incidental splashing, the material condensed in the collar of the Hickman still should consist only of the volatile component. In the present experiment this component is ethyl acetate. The temperature of the vial and contents being distilled will rise during the distillation process since the concentration of the impurity is increasing as the volatile liquid is removed. This effect lowers the vapor pressure of the liquid. However, the boiling point of the liquid remains constant, since only the pure liquid component is being vaporized.

## COMPONENTS

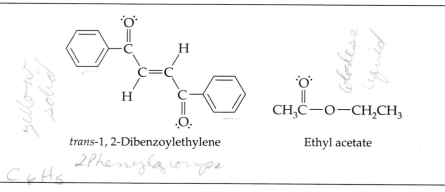

*trans*-1, 2-Dibenzoylethylene

Ethyl acetate

## EXPERIMENTAL PROCEDURE

Estimated time for the experiment: 2.0 hours.

### Physical Properties of Components

| Compound | MW | Amount | mp(°C) | bp(°C) | d | $n_D$ |
|---|---|---|---|---|---|---|
| Ethyl acetate | 88.12 | 1.0 mL | | 77 | 0.90 | 1.3723 |
| *trans*-1,2-Dibenzoylethylene | 236.27 | 50 mg | 111 | | | |

### Reagents and Equipment

Transfer 1 mL of the yellow stock solution (*trans*-1,2-dibenzoylethylene/ ethyl acetate, 50 mg/mL) to a 3-mL conical vial by automatic delivery pipet (remember to place the vial in a small beaker to prevent tipping during the transfer). Place a boiling stone in the vial and assemble the Hickman still head. The still assembly is mounted in a sand bath on a hot plate (see Fig. 6.2).

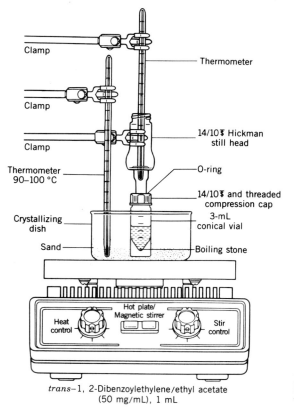

*trans*–1, 2-Dibenzoylethylene/ethyl acetate (50 mg/mL), 1 mL

**FIGURE 6.2** Hickman still [14/10 with conical vial (3 mL)].

## Experimental Conditions

The temperature of the bath is raised to 90–100 °C at a rate of 5 °C/min.

---

**CAUTION:** Do not let the temperature of the still rise too rapidly.

---

Once boiling commences, the rate of heating should be lowered to the point where the temperature increases at 2–3 °C/min. A slow distillation rate is very important in establishing equilibrium between the vapor and liquid components in the mixture. Follow the course of the distillation by the rise of condensate on the sides of the Hickman column. When the condensate reaches the trap, adjust the bath temperature so that liquid is removed from the column slowly (~100 $\mu$L/3 min). A smooth, slow distillation will provide a cleaner separation of the components, and will also avoid mechanical transfer of nonvolatile components via splattering to the condensate trap (if the condensate appears yellow, contamination has occurred).

Collect approximately 50–150 $\mu$L of the ester in the collar of the still (the first fraction collected is often referred to as the forerun; give the temperature range). As the distillation continues, remove the forerun with a Pasteur pipet having a slightly bent tip (microburner). Place the fraction in a clean, dry 1-dram screw-capped vial (use an aluminum foil liner to avoid cap contamination). Label the fraction with a marking pen. Collect a second fraction of ester (400–500 $\mu$L, which may require combining two, or even three, collections from the collar; give the temperature range), which should be clear and colorless. Remove and store as before. Discontinue the distillation. Allow the distilling flask to cool slowly by leaving it in the warm sand bath while measuring the physical properties of the distillate fractions.

## Characterization

Three physical properties of the ester will be measured to establish the identity and purity of the compound by comparison with known literature values.

Determine the refractive index (see Chapter 4) of the two fractions collected. Compare the experimental values to those found in the literature for ethyl acetate. If the values are within 0.0010 unit of each other, the fractions can be considered to have the same constitution. Are the values for the two fractions the same? If not, which one deviates the most from the reference data? Attempt to explain the result.

Determine the density (see Chapter 4) of the ester, using material contained in the second fraction. This measurement is nondestructive and the material used may be recovered for use in further tests. Compare your results with those values found in the literature.

Determine the boiling point of the second fraction by the ultramicro-boiling point procedure (see Chapter 4). Compare your result with the literature value. Does this fraction appear to be pure ethyl acetate?

In the next step, disconnect and cool the 3-mL conical vial in an ice water bath for 10 min. *trans*-1,2-Dibenzoylethylene will crystallize from the concentrated solution. Remove the remaining solvent from the distillation vial with a Pasteur filter pipet and place the crystals on a porous clay plate to air dry. The melting point of the crystalline material is obtained by the evacuated capillary method and compared with the literature value.

Reference values of the physical constants are available in the *CRC Handbook of Chemistry and Physics.* Submit a copy of the table prepared in your laboratory notebook to the instructor, after first tabulating the experimentally measured values of the physical properties, in addition to those reported in the literature for ethyl acetate (see *acetic acid, ethyl ester* if necessary).

## Experiment [3B]   Fractional Semimicroscale Distillation: Separation of Hexane and Toluene

### Purpose
To effect the separation of a binary liquid mixture composed of liquids having boiling points that are relatively far apart, greater than 30 °C. To develop the skills to operate a semimicrodistillation apparatus so that purifications required in later experiments can be successfully carried out.

### Prior Reading

Technique 2:   Distillation (see pp. 61–72).
Distillation Theory (see pp. 61–65).
Simple Distillation at the Semimicroscale Level (see pp. 65–68).
Fractional Semimicroscale Distillation (see pp. 68–71).

Chapter 4:   Determination of Physical Properties
Ultramicro-Boiling Point (see pp. 44–47).
Refractive Index (see pp. 48–50).

## DISCUSSION

Hexane and toluene are liquid hydrocarbons that have boiling points approximately 40 °C apart. The liquid–vapor composition curve in Figure 5.11 represents this system; therefore, it is apparent that a two-plate distillation should yield nearly pure components. The procedure to be outlined consists of two parts. The first deals with the initial distillation (first plate), which separates the liquid mixture into three separate fractions. The second deals with *redistillation* of the first and third fractions (second plate). Exercising careful technique during the first distillation should provide a fraction rich in the lower boiling component, a middle fraction, and a fraction rich in the higher boiling component. Then careful *redistillation* of these fractions can be expected to complete the separation of the two components and to produce fractions of relatively pure hexane and toluene. The Hickman still employed in the microscale laboratory is a simple, short-path column, and, therefore, one would not expect complete separation of the hexane and toluene in one cycle.

## COMPONENTS

$$CH_3CH_2CH_2CH_2CH_2CH_3$$

Hexane

Toluene

Estimated time for the experiment: 2.0 hours.

**EXPERIMENTAL PROCEDURE**

**Physical Properties of Components**

| Compound | MW | Amount | bp(°C) | $d$ | $n_D$ |
|----------|------|--------|--------|------|--------|
| Hexane | 86.18 | 1.0 mL | 69 | 0.66 | 1.3751 |
| Toluene | 92.15 | 1.0 mL | 111 | 0.87 | 1.4961 |

## Reagents and Equipment

In a clean, dry, stoppered 5-mL conical vial are placed 1.0 mL of hexane and 1.0 mL of toluene by using an automatic delivery pipet.

*Place the vial in a small beaker to prevent tipping.* A boiling stone is added, the Hickman still is assembled with the thermometer positioned directly down the center of the column (see previous discussion), and the system is mounted in a sand bath (see Fig. 6.3).

## Experimental Conditions

The temperature of the sand bath is raised to 80–90 °C, at a maximum rate of 5 °C/min (> 70 °C, at 3 °C/min), using a hot plate.

---

CAUTION:  Do not let the temperature of the still rise too rapidly.

---

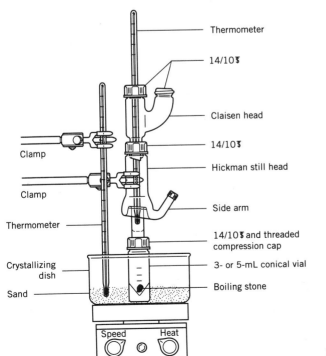

**FIGURE 6.3**  Hickman still with Claisen head adapter.

Once gentle boiling begins, the heating rate should be lowered to a maximum of 2 °C/min. It is ***absolutely crucial*** that the distillation rate be kept below 100 $\mu$L/3 min to achieve the necessary fraction enrichment that will permit good separation during the second stage of the experiment. The distillate is collected in *three fractions* over the temperature ranges (**1**) 65–85 °C (bath temperature ~95–110 °C); (**2**) 85–105 °C (bath temperature ~140 °C); and (3) 105–110 °C (bath temperature ~170 °C) in amounts of approximately 800, 400, and 800 $\mu$L, respectively. Remove each fraction from the still with a bent-tip Pasteur pipet. Store the liquid condensate (fractions) in clean, dry, 1-dram, screw-cap vials. *Remember to number the vials in order and use an aluminum foil cap liner.*

## Characterization of Crude Fractions

For each of the three fractions, record the refractive index. Fraction 1 has been enriched in one of the two components. Which one? Does the refractive index agree with that found in the literature? Fraction 3 has been enriched in the other component. Does the refractive index of that fraction support your first conclusion? If partial enrichment has been achieved, proceed to the second phase of the distillation.

## Redistillation of Fraction 1

Redistill fraction 1 in a clean Hickman still with a thermometer arranged as before (Fig. 6.3), using a 3-mL conical vial and the procedure just outlined. Collect an initial fraction over the boiling range 68–71 °C (~100–200 $\mu$L). Remove it from the collar, using the Pasteur pipet, and place it in a 1-dram screw-cap vial.

## Characterization of Fraction 1

Determine the ultramicro-boiling point and the refractive index of this lower boiling fraction. Compare the experimental values obtained with those of pure hexane reported in the literature.

## Redistillation of Fraction 3

Fraction **3** is placed in a clean Hickman still, using a thermometer and a 3-mL conical vial (Fig. 6.3), and redistilled using the procedure outlined. Collect an initial fraction over the boiling range 95–108 °C (~500 $\mu$L), and transfer this fraction by Pasteur pipet to a screw-cap vial. Collect a final fraction at 108–110 °C (~250 $\mu$L), and transfer the material to a second vial. *This second fraction is the highest boiling fraction to be collected in the three distillations and should be the richest in the high-boiling component.*

## Characterization of Fraction 3

Determine the refractive index and boiling point of the second fraction, and compare your results with those found in the literature for toluene. *Determine the refractive index and boiling point of pure toluene for comparison purposes.*

## Experiment [3C]   Fractional Semimicroscale Distillation: Separation of 2-Methylpentane and Cyclohexane Using a Spinning Band Column

### Purpose
To separate two liquids possessing boiling points that are relatively similar: less than 20 °C apart. To learn the operation of a high-performance spinning band distillation column. To develop the skills for purifying small quantities of liquid mixtures.

### Prior Reading
    *Technique 1:*   Microscale Separation of Liquid Mixtures by GC (see
                        pp. 56–61).
    *Technique 2:*   Distillation (see pp. 61–72).
                        Distillation Theory (see pp. 61–65).
                        Simple Distillation at the Semimicroscale Level (see
                           pp. 65–68).
                        Fractional Semimicroscale Distillation (see pp.
                           68–71).
    *Chapter 4:*   Determination of Physical Properties
                        Ultramicro Boiling Point (see pp. 44–47).
                        Refractive Index (see pp. 48–50).

**DISCUSSION**

In this experiment, the separation of a 2-mL mixture of 2-methylpentane and cyclohexane using a 2.5-in. spinning band distillation column will be described. The purity of the fractions will be determined by gas chromatography and by measurement of the refractive index. Finally, the number of theoretical plates will be estimated. You will separate a 50:50 mixture of 2-methylpentane and cyclohexane using the spinning band distillation column shown in Figure 6.4.

**COMPONENTS**

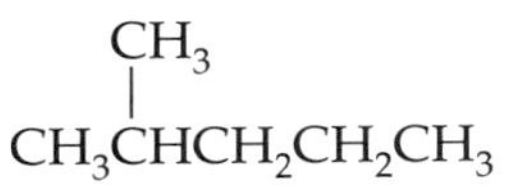

    CH₃
    |
CH₃CHCH₂CH₂CH₃

    2-Methylpentane 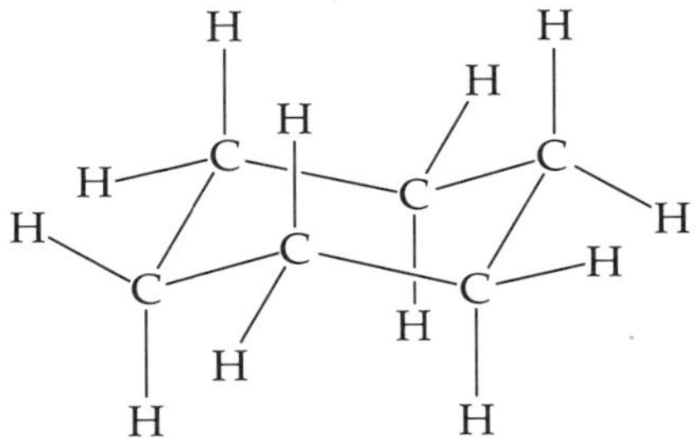                   Cyclohexane

**EXPERIMENTAL PROCEDURE**

Estimated time for the experiment: 3.0 hours.

### Physical Properties of Components

| Compound | MW | Amount | bp(°C) | $n_D$ |
|---|---|---|---|---|
| 2-Methylpentane | 86.18 | 1.0 mL | 60.3 | 1.3715 |
| Cyclohexane | 84.16 | 1.0 mL | 80.7 | 1.4266 |

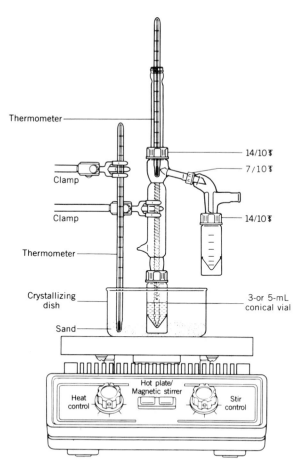

Thermometer

Clamp

Clamp

Thermometer

Crystallizing dish

Sand

14/10 ƒ

7/10 ƒ

14/10 ƒ

3-or 5-mL conical vial

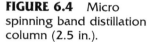

Heat control

Hot plate/ Magnetic stirrer

Stir control

**FIGURE 6.4** Micro spinning band distillation column (2.5 in.).

## Reagents and Equipment

Assemble the system as shown in Figure 6.4. In the process, make sure that the Teflon band is aligned as straight as possible in the column. In particular, the pointed section extending into the vial must be straightened to minimize vibration during spinning of the band.

Place a pipet bulb on the side arm of the collection adapter. This bulb plays a very important function in the operation of the column. Attachment of the bulb creates a closed system. Suspension of the thermometer with a septum on the top of the condenser can act to release any build up of pressure. **THE SYSTEM MUST BE ABLE TO VENT AT THE THERMOMETER DURING OPERATION!**

Once the spinning band has been tested and rotates freely, place 1.0 mL of 2-methylpentane and 1.0 mL of cyclohexane in the vial (to be delivered with a Pasteur pipet or an automatic delivery pipet). Reassemble the system and lower the column into the sand bath or copper tube block. It is important to note that the beveled edge on the air condenser should be rotated 180° from the collection arm.

**NOTE.** *It is important to make an aluminum foil cover for the sand bath; this cover will reflect the heat and hot air away from the collection vial.*

## Experimental Conditions

Gently heat (copper-tube block, see Fig. 3.14) the vial until boiling occurs. The magnetic stirrer is turned to a low-spin rate when heating commences. When reflux is observed at the base of the column, the magnetic stirrer is adjusted to intermediate spin rate. Once liquid begins to enter the column the spin rate is increased to the maximum (1000–1500 rpm).

**NOTE.** *It is absolutely critical that the temperature of the vial be adjusted so that vapors in the column rise* very slowly. *It is possible for overheated vapors to be forced through the air condenser.*

When the vapors slowly arrive in the unjacketed section of the column head, the condenser joint acts as a vapor shroud to effectively remove vapors from the receiver-cup area. During this total reflux period, maximum separation of the components is achieved. Once vapor reflux occurs in the head of the column, the system is left for 20–30 min to reach thermal equilibrium. During this period of total reflux, the head thermometer should read about 57–60 °C (at least for most of the equilibration time).

Following the equilibration period, collection of the resolved components may begin. Rotate the air condenser 180° so that the beveled edge is over the collection duct. At this point, manipulation of the pipet bulb allows drainage of the condensate from the side arm. (This procedure is repeated occasionally to continue drainage from the side arm.) Collect six drops (~0.30 mL). After removing the collection vial, transfer the contents into a covered vial using a Pasteur pipet. Label all fractions. Collect two 0.6-mL fractions (the pipet bulb may be removed during collection of these latter fractions); then turn off the heat and stirring motor, and remove the vial from the sand bath. Transfer the material remaining in the vial, using a Pasteur pipet, to a fourth covered vial.

## Characterization of the Fractions

The composition of each of the fractions may be determined by gas chromatographic analysis and measurement of the refractive index.

A GOW-MAC gas chromatograph should be set up as follows:

| Column | DC 710 |
|---|---|
| Injection | 10 $\mu$L |
| Temperature | 80 °C |
| Flow rate (He) | 55 mL/min |
| Chart speed | 1 cm/min |

If we assume that the refractive index is a linear function of the volume fraction, the following relationship gives us the volume fraction of 2-methylpentane in a mixture. The volume fraction is $X$ and the measured refractive index is $n_D$.

$$X = (1.4266 - n_D)/(1.4266 - 1.3715)$$

The curve shown below in Figure 6.5 may be used to estimate the number of theoretical plates from the composition (mole fraction) of the *first* 0.30-mL fraction. For example, if the composition of the *first* 0.30 mL is 0.89, we would infer that the system had a resolution equivalent to about

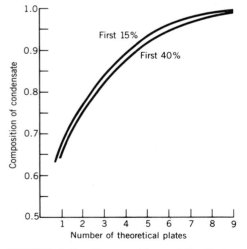

**FIGURE 6.5** Composition of the first 15% and the first 40% of the volume collected in the distillation of a 50% (v/v) mixture of 2-methylpentane and cyclohexane.

four theoretical plates. Note that the number of plates cannot be determined with confidence if the composition is greater than about 0.97. If we really wanted to determine the number of theoretical plates for a system with more than five plates, we could start with a mixture containing only 10 or 20% of <u>m</u>ost <u>v</u>olatile <u>c</u>omponent (MVC), rather than the 50% used in this example.

### Experiment [3D]   Fractional Semimicroscale Distillation: The Separation of 2-Methylpentane and Cyclohexane Using a Spinning Band in a Hickman–Hinkle Still

#### Purpose
To become familiar with a powerful modification of the classic Hickman still: the Hickman–Hinkle spinning band distillation apparatus. This small still is one of the most efficient techniques developed for the purification of small quantities of liquids. To develop the skills for handling small quantities of liquids and their purification by distillation. To become familiar with these techniques, so that they may be employed in the purification of reaction products formed in later experiments.

## DISCUSSION

This distillation separates the same two compounds used in Experiment [3C]. The distillate can be analyzed to determine the number of theoretical plates. If careful attention is given to the procedure, the spinning Hickman–Hinkle is capable of more than six theoretical plates.

As in Experiment [3C] the separation of a 2-mL mixture of 2-methylpentane and cyclohexane is achieved. The purity of the fractions can be determined by gas chromatography and by measurement of the refractive index.

## COMPONENTS

2-Methylpentane

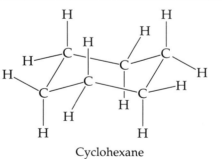

Cyclohexane

## EXPERIMENTAL PROCEDURE

Estimated time for the experiment: 3.0 hours.

#### Physical Properties of Components

| Compound | MW | Amount | bp(°C) | $n_D$ |
|---|---|---|---|---|
| 2-Methylpentane | 86.18 | 1.0 mL | 60.3 | 1.3715 |
| Cyclohexane | 84.16 | 1.0 mL | 80.7 | 1.4266 |

## Reagents and Equipment

The system is assembled as shown in Figure 3.21. In the process, make sure that the Teflon band is aligned as straight as possible in the column. In particular, the pointed section extending into the vial must be straightened to minimize vibration during spinning of the band.

Once the spinning band has been tested and rotates freely, place 1.0 mL of 2-methylpentane and 1.0 mL of cyclohexane in the vial (to be delivered with a Pasteur pipet or an automatic delivery pipet). Reassemble the system and lower the column into the sand bath.

*Cover the sand bath with aluminum foil during the distillation in order to prevent the collar of the still from overheating.* It is easier to regulate the temperature of the bath when it is covered. However, for distillations, a more efficient heating technique is the recently developed copper-tube block (see Fig. 3.14).

## Experimental Conditions

Gently heat the vial until boiling occurs. When heating commences, turn on the magnetic stirrer at a low setting. When reflux commences at the base of the column the magnetic stirrer is raised to intermediate settings. Once liquid begins to enter the column, the spin rate is increased to the maximum (1000–1500 rpm).

**NOTE.** *It is extremely important that careful temperature control be exercised at this stage so that the condensing vapors ascend the column very slowly.*

Vapor-phase enrichment by the most volatile component is limited mainly to this period, as fraction collection commences immediately on arrival of the vapor column at the annular ring. Once condensation occurs, fractions are collected by the same technique used in Experiment [3B]. Characterization of the fractions, however, follows the procedure given in Experiment [3C].

## Characterization of the Fractions

The composition of each of the fractions may be determined by gas chromatography, the refractive index, or both. See Experiment [3C] for details.

An alternate approach to the procedures discussed in Experiment [3C] is to establish the fraction volume by weight. The curves shown in Figure 6.5 may again be used to estimate the number of theoretical plates. The volume of the first fraction can be estimated, or determined more accurately by weighing the fraction in a tared screw-cap vial. The composition of this fraction then may be determined and the fraction of the total represented by this portion calculated. If, for example, the first fraction has a volume of 0.4 mL (20% of the total) and has a composition 0.89 by volume of 2-methylpentane, we would infer that the system had a resolution equivalent to about four theoretical plates.

**QUESTIONS**

**6-12.** The boiling point of a liquid is affected by several factors. What effect does each of the following conditions have on the boiling point of a given liquid?

a. The pressure of the atmosphere.
b. Use of an uncalibrated thermometer.
c. Rate of heating of the liquid in a distillation flask.

**6-13.** Calculate the vapor pressure of a solution containing 30 mol% hexane and 70 mol% octane at 90 °C assuming Raoult's law is obeyed.

*Given:* vapor pressure of the pure compounds at 90 °C. Vapor pressure of hexane = 1390 torr; vapor pressure of octane = 253 torr.

**6-14.** In any distillation for maximum efficiency of the column, the distilling flask should be approximately one-half full of liquid. Comment on this fact in terms of (a) a flask that is too full and (b) a flask that is nearly empty.

**6-15.** Occasionally during a distillation, a solution will foam rather than boil. A way of avoiding this problem is to add a *surfactant* to the solution.
a. What is a surfactant?
b. What is the chemical constitution of a surfactant?
c. How does a surfactant reduce the foaming problem?

**6-16.** Explain why packed and spinning band fractional distillation columns are more efficient at separating two liquids having close boiling points than are unpacked columns.

**6-17.** Explain what effect each of the following mistakes would have had on the simple distillation carried out in this experiment.
a. You did not add a boiling stone. *make boil more evenly*
b. You heated the distillation flask at too rapid a rate. *boil over, catch fire*

**6-18.** In the ultramicro-boiling point determination, why is the boiling point taken just as bubbles cease emerging from the bell?

**6-19.** Define each of the following terms, which are related to the distillation process.
a. Distillate.
b. Normal boiling point.
c. Forerun.

**6-20.** How does the refractive index of a liquid vary with temperature? What corrective factor is often used to determine the value at a specific temperature, for example, 20 °C, if the measurement were made at 25 °C?

---

## EXPERIMENT [4]    Solvent Extraction

---

### Experiment [4A]   Determination of a Partition Coefficient for the System Benzoic Acid, Methylene Chloride, and Water

#### Purpose
This exercise illustrates the general procedures that are used to determine a partition coefficient at the microscale level. Experience in weighing milligram quantities of materials on an electronic balance, the use of automatic delivery pipets for accurately dispensing microliter quantities of liquids, the transfer of microliter volumes of solutions with the Pasteur filter pipet, and the use of a Vortex mixer, are techniques encountered in this experiment.

*Prior Reading*
    *Chapter 3:*  Experimental Apparatus
                Pasteur Filter Pipet (See pp. 34–35).
                Automatic Delivery Pipet (see pp. 35–36).
                Weighing of Solids in Milligram Quantities (see pp.
                    37–38).
   *Technique 4:*  Solvent Extraction
                Liquid–Liquid Extraction (see pp. 80–82).
                Partition Coefficient Calculations (see pp. 76–78).
                Drying of the Wet Organic Layer (see pp. 86–87).
                Separation of Acids and Bases (see pp. 87–88).

## Solubility

**DISCUSSION**

Substances vary greatly in their solubility in various solvents, but based on observations, many of which were made in the very early days, a useful principle has evolved that allows the chemist to predict rather accurately the solubility of a particular substance. It is generally true that *a substance tends to dissolve in a solvent that is chemically similar to itself. In other words, like dissolves like.*

    Thus, for a particular substance to exhibit solubility in water requires that species to possess some of the characteristics of water. For example, an important class of compounds, the organic alcohols, have the hydroxyl group (—OH) bonded to a hydrocarbon chain or framework (R—OH). The hydroxyl group can be viewed as being effectively one-half a water (HOH) molecule, and it has a similar polarity to that of water. This results from a charge separation arising from the difference in electronegativity between the hydrogen and oxygen atoms. The O—H bond, therefore, is considered to have *partial ionic character.*

$$\overset{\delta^-}{\ddot{\text{O}}}\,\overset{\delta^+}{\text{—H}}$$

Partial ionic character of the hydroxyl group

    This *polar,* or partial ionic, character leads to relatively strong hydrogen bond formation between molecules having this entity. Strong hydrogen bonding is evident in molecules that contain a hydrogen atom attached to an oxygen, nitrogen, or fluorine atom, as shown here for the ethanol–water system. This polar nature of a functional group is present when there are sufficient differences in electronegativity between the atoms making up the group.

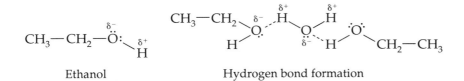

    Ethanol               Hydrogen bond formation

    In ethanol ($CH_3CH_2OH$), it is apparent that the hydroxyl end of the molecule is very similar to water (HOH). When ethanol is added to water, therefore, they are miscible in all proportions. That is, ethanol is completely soluble in water and water is completely soluble in ethanol. This solubility results because the attractive forces set up between the two molecules ($CH_3CH_2OH$ and $H_2O$) are nearly as strong as between two water molecules; however, the attraction in the first case is somewhat weakened by the presence of the nonpolar alkyl ethyl group, $CH_3CH_2—$. Hydrocarbon groups attract each other only weakly, as evidenced by their low melting and boiling points. Three examples of the contrast in boiling points between compounds of different structure, but similar molecular weight, are summarized in Table 6.1. Clearly, those molecules that attract each other weakly have lower boiling points.

**TABLE 6.1    Comparison of Boiling Point Data**

| Name | Formula | MW | bp(°C) |
|------|---------|-----|--------|
| Ethanol | $CH_3CH_2OH$ | 46 | 78.3 |
| Propane | $CH_3CH_2CH_3$ | 44 | −42.2 |
| Methyl acetate | $CH_3CO_2CH_3$ | 74 | 54 |
| Diethyl ether | $(CH_3CH_2)_2O$ | 74 | 34.6 |
| Ethylene | $CH_2{=}CH_2$ | 28 | −102 |
| Methylamine | $CH_3NH_2$ | 31 | −6 |

When we compare the water solubility of ethanol (a two-carbon ($C_2$) alcohol that, as we have seen is completely miscible with water) with that of octanol (a straight-chain eight-carbon ($C_8$) alcohol), we find that the solubility of octanol is less than 2% in water. Why the difference in solubilities between these two alcohols? The answer lies in the fact that the dominant structural feature of octanol has become the nonpolar nature of its alkyl group.

$$CH_3—CH_2—CH_2—CH_2—CH_2—CH_2—CH_2—CH_2—\overset{\delta^-}{\ddot{O}{:}}$$

Octanol
$$\underset{H}{\overset{\delta^+}{\diagdown}}$$

$$CH_3—CH_2—\ddot{O}—CH_2—CH_3$$
Diethyl ether

As the bulk of the hydrocarbon section of the alcohol molecule increases, the intramolecular attraction between the polar hydroxyl groups of the alcohol and the water molecules is no longer sufficiently strong to overcome the *hydrophobic character* (lack of attraction to water) of the nonpolar hydrocarbon section of the alcohol. On the other hand, octanol has a large nonpolar hydrocarbon group as its dominant structural feature. We might, therefore, expect octanol to exhibit enhanced solubility in less polar solvents. In fact, octanol is found to be completely miscible with diethyl ether. Ethers are solvents of weak polarity. Since the nonpolar characteristics are significant in both molecules, mutual solubility is observed. In general, it has been empirically demonstrated that if a compound has both polar and nonpolar groups present in its structure, those compounds having five or more carbon atoms in the hydrocarbon portion of the molecule will be more soluble in nonpolar solvents, such as pentane, diethyl ether, or methylene chloride. Figure 5.19 summarizes the solubilities of a number of straight-chain alcohols, carboxylic acids, and hydrocarbons in water. As expected, those compounds with more than 5 carbon atoms are shown to possess solubilities similar to those of the hydrocarbons.

Several additional relationships between solubility and structure have been observed and are pertinent to the discussion.

**1.** Branched-chain compounds have greater water solubility than their straight-chain counterparts, as illustrated in Table 6.2 with a series of alcohols.

**2.** The presence of more than one polar group in a compound will increase that compound's solubility in water and decrease its solubility in nonpolar solvents. For example, high molecular weight sugars, such as cellobiose, that contain multiple hydroxyl and/or acetal groups, are water soluble, and ether insoluble; whereas cholesterol ($C_{27}$), which possesses only a single hydroxyl group, is water insoluble and ether soluble.

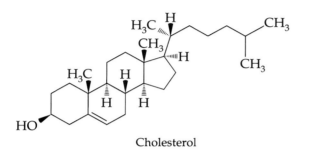

Cholesterol

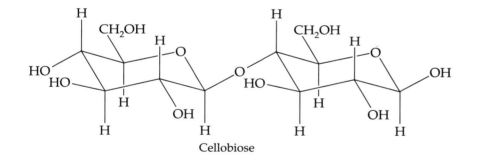

Cellobiose

**TABLE 6.2   Water Solubility of Alcohols[a]**

| Name | Formula | Solubility (g/100 g $H_2O$) |
|---|---|---|
| Pentanol | $CH_3(CH_2)_3CH_2OH$ | 4.0 |
| 2-Pentanol | $CH_3(CH_2)_2CH(OH)CH_3$ | 4.9 |
| 2-Methyl-2-butanol | $(CH_3)_2C(OH)CH_2CH_3$ | 12.5 |

[a] Data at 20 °C.

**3.** The presence of a chlorine atom, even though it lends some partial ionic character to the covalent C—Cl bond, does not normally impart water solubility to a compound. In fact, such compounds as methylene chloride ($CH_2Cl_2$), chloroform ($CHCl_3$), and carbon tetrachloride ($CCl_4$) have long been used as solvents for the extraction of aqueous solutions. It should be noted that use of the latter two solvents is no longer recommended, unless strict safety precautions are exercised, because of their potential carcinogenic nature.

**4.** Most functional groups that are capable of forming a hydrogen bond with water, if it constitutes the dominant structural feature of a molecule, will impart increased water solubility characteristics to a substance (the five-carbon rule obviously applies here in determining just what is a dominant feature). For example, certain alkyl amines (organic relatives of ammonia) might be expected to have significant water solubility. This finding is indeed the case, and the water-solubility data for a series of amines is summarized in Table 6.3.

**TABLE 6.3 Water Solubility of Amines[a]**

| Name | Formula | Solubility (g/100 g $H_2O$) |
|---|---|---|
| Ethylamine | $CH_3CH_2NH_2$ | $\infty$ |
| Diethylamine | $(CH_3CH_2)_2NH$ | $\infty$ |
| Trimethylamine | $(CH_3)_3N$ | 91 |
| Triethylamine | $(CH_3CH_2)_3N$ | 14 |
| Aniline | $C_6H_5NH_2$ | 3.7 |
| p-Phenylenediamine | $H_2NC_6H_4NH_2$ | 3.8 |

[a] Data at 25 °C.

It is important to realize that the solubility characteristics of any given compound will uniquely govern that substance's distribution (*partition*) between the phases of two immiscible solvents (in which the material has been dissolved) when these phases are intimately mixed. In this experiment we determine the partition coefficient (distribution coefficient) of benzoic acid between two immiscible solvents, methylene chloride and water.

## COMPONENTS

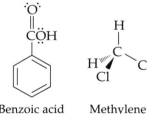

Benzoic acid          Methylene          Water
                      chloride

## EXPERIMENTAL PROCEDURE

Estimated time of experiment: 1.5 hours.

**Physical Properties of Components**

| Compound | MW | Amount | mmol | mp(°C) | bp(°C) | d |
|---|---|---|---|---|---|---|
| Benzoic acid | 122.13 | 50 mg | 0.41 | 122 | | |
| Methylene chloride | | 1.20 mL | | | 40 | 1.33 |
| Water | | 600 μL | | | 100 | 1.00 |

## Equipment Setup and Addition of Reagents

Weigh and add to a 5.0-mL conical vial fitted with a screw cap, 50 mg (0.41 mmol) of benzoic acid. Now add 600 $\mu$L of methylene chloride followed by 600 $\mu$L of water.

*The solvents are transferred to the vial with automatic delivery pipets (use a different pipet or pipet tip for each solvent). The methylene chloride addition should be carried out in the **hood**.*

HOOD

## Procedure for Establishing Equilibrium Distribution

Cap the vial and shake (or use a Vortex mixer) until the benzoic acid dissolves and the two phases have been thoroughly mixed. Vent the vial (by releasing the Cap-seal®) and then allow the two layers to separate.

Carefully draw the lower methylene chloride layer into a Pasteur filter pipet and transfer it to a 5-mL conical vial containing 100 mg of anhydrous, granular sodium sulfate. Recap the vial.

**NOTE.** *If the volume of the methylene chloride layer is so large that it cannot be transferred completely in one operation, a second transfer may be required. Be careful not to overfill the pipet to insure that solvent does not come in contact with the rubber bulb.*

*The technique of removing the last traces of water from the methylene chloride solution is often referred to as* drying the solution. *It can involve any one of a number of insoluble anhydrous salts, which convert the moisture retained in the organic phase to water of crystallization. In this case, we are employing sodium sulfate.*

## Isolation of the Benzoic Acid

After drying the methylene chloride solution for 10–15 min, transfer the anhydrous solution to a previously **tared** (the term **tare** means to preweigh the empty vial) vial, using a Pasteur filter pipet (the use of the filter pipet is a convenient way of separating the solid hydrated sodium sulfate from the dried solution). Rinse the sodium sulfate with an additional 600 $\mu$L of methylene chloride and combine the rinse with the solution in the tared 5.0-mL conical vial. Evaporate the solvent under a gentle stream of nitrogen gas in a warm sand bath in the **hood**. (It is important to warm the solution while evaporating the solvent; otherwise the heat of vaporization will rapidly cool the solution. In this latter case, as the cold, solid acid precipitates from the saturated solution, moisture will condense from the air entrained in the evaporation process, and contaminate the surface of the recovered material.)

HOOD

**NOTE.** *If a hot sand bath is used, a boiling stone is placed in the vial before it is tared. The boiling stone will help to avoid explosive, sudden boiling of the solvent when the vial is placed in the sand bath.*

## Weight Data and Calculations

Weigh the vial and determine the weight of benzoic acid that remains following removal of the methylene chloride. Break up the hard cake of precipitated benzoic acid with a microspatula and *briefly* reheat the vial and contents in a sand bath to remove the last traces of solvent and any water that remains in the system. Cool and reweigh. Repeat this operation until a constant weight is obtained. This weight represents the benzoic acid that dissolved in the *methylene chloride layer.*

The *original weight* of benzoic acid used, minus the amount of benzoic acid recovered in the methylene chloride layer, equals the weight of the benzoic acid that dissolved and still remains in the *water layer*.

Since equal volumes of both solvents were used, the partition coefficient may be simply determined from the ratio of the weight of benzoic acid in the methylene chloride solvent to the weight of benzoic acid in the water layer.

Calculate the partition coefficient for benzoic acid in the solvent pair used in this exercise.

## Experiment [4B]   Solvent Extraction I: The System; Benzoic Acid, Methylene Chloride, and 10% Sodium Bicarbonate Solution; An Example of Acid–Base Extraction Techniques

### Purpose
This exercise illustrates a further important, and extensively used, extraction technique in which a reversible reaction is used to alter the solubility characteristics of the substance of interest.

*Prior Reading*

|  |  |
|---|---|
| Chapter 3: | Experimental Apparatus |
|  | Pasteur Filter Pipet (see pp. 34–35). |
|  | Automatic Delivery Pipet (see pp. 35–36). |
|  | Weighing of Solids in Milligram Quantities (see pp. 37–38). |
| Technique 4: | Solvent Extraction |
|  | Liquid-Liquid Extraction (see pp. 80–82). |
|  | Partition Coefficient Calculations (see pp. 76–78). |
|  | Drying of the Wet Organic Layer (see pp. 86–87). |
|  | Separation of Acids and Bases (see pp. 87–88). |

**REACTION**

$$C_6H_5\overset{\overset{\displaystyle \cdot\cdot O\cdot\cdot}{\|}}{C}-\overset{\cdot\cdot}{O}H + Na^+HCO_3^- \rightleftharpoons C_6H_5\overset{\overset{\displaystyle \cdot\cdot O\cdot\cdot}{\|}}{C}-\overset{\cdot\cdot}{O}{:}^-\,Na^+ + H_2CO_3$$

Benzoic acid                               Sodium benzoate      Carbonic acid

$$H_2CO_3 \rightleftharpoons CO_2 + H_2O$$

**DISCUSSION**

Benzoic acid reacts readily with sodium bicarbonate to form sodium benzoate, carbon dioxide, and water. The sodium salt of benzoic acid has highly ionic characteristics and thus, unlike the free acid, the salt is very soluble in water and nearly insoluble in methylene chloride. This salt is characterized by a full ionic bond between the carboxylic acid group of the acid and the sodium ion. It is, therefore, a new substance exhibiting many of the solubility properties commonly associated with inorganic ionic salts.

Estimated time of experiment: 1.0 hour.                    **EXPERIMENTAL PROCEDURE**

**Physical Properties of Reactants**

| Compound | MW | Amount | mmol | mp(°C) | bp(°C) | d |
|----------|-----|--------|------|--------|--------|-----|
| Benzoic acid | 122.13 | 50 mg | 0.41 | 122 | | |
| Methylene chloride | | 1.20 mL | | | 40 | 1.33 |
| Sodium bicarbonate (10% solution) | | 600 μL | | | | |

### Procedure for Establishing Distribution

**Repeat the identical procedures** carried out in Experiment [4A], but replace the 600 μL of water with 600 μL of 10% sodium bicarbonate solution. A good estimate of the efficiency of the conversion of benzoic acid to the sodium salt of the acid, which because of its ionic character is found almost exclusively in the aqueous phase, can be made by recovering any unreacted acid from the methylene chloride layer and employing the distribution coefficient established in Experiment [4A]. Also, be sure to obtain a melting point of any recovered residue (assumed above to be benzoic acid) from the organic phase, as contamination of free acid by the acid salt can be detected by this measurement. Sodium benzoate has a melting point above 300 °C, whereas benzoic acid melts near 122 °C.

### Test for a Carboxylic Acid

As illustrated in the above reaction, when a carboxylic acid comes in contact with a solution containing bicarbonate ion, carbon dioxide is generated. Once saturation of the solution by carbon dioxide occurs, bubbles of carbon dioxide gas are observed to form in the liquid phase. This effervescence may be used as a qualitative test for the presence of the carboxylic acid functional group in an unknown substance.

Place 1–2 mL of 10% sodium or potassium bicarbonate on a small watch glass. Add the pure acid, one drop from a Pasteur pipet if the sample is a liquid (~5 mg if it is a solid), to the bicarbonate solution. Evolution of bubbles ($CO_2$) from the mixture indicate the presence of an acid.

Perform the above test for the presence of carboxyl groups on several organic acids, such as acetic, benzoic, propanoic, or chloroacetic acid.

### Experiment [4C]   Solvent Extraction II: A Three Component Mixture; An Example of the Separation of an Acid, a Base, and a Neutral Substance

#### Purpose
To investigate how solvent extraction techniques can be applied effectively to problems that require the separation of mixtures of organic acids, bases, and neutral compounds in the research or industrial laboratory.

*Prior Reading*

Chapter 3:   Experimental Apparatus
Pasteur Filter Pipet (see pp. 34–35).
Automatic Delivery Pipet (see pp. 35–36).
Weighing of Solids in Milligram Quantities (see pp. 37–38).

Technique 4:   Solvent Extraction
Liquid–Liquid Extraction (see pp. 80–82).
Drying of the Wet Organic Layer (see pp. 86–87).
Separation of Acids and Bases (see pp. 87–88).
Salting Out (see p. 88).

## DISCUSSION

As implied in the discussions of Experiments [4A] and [4B], the solubility characteristics of organic acids in water can be shown to be highly dependent on the pH of the solution. By extending this extraction approach to include organic bases, it has been possible to develop a general procedure for the separation of mixtures of organic acids, bases, and neutral substances.

**NOTE.**   *Refer to Technique 4, p. 88 for a chart outlining the procedure.*

The components of the mixture to be separated in this experiment are benzoic acid, ethyl 4-aminobenzoate (a base), and 9-fluorenone, a neutral compound (may be prepared in Experiment [33A]).

## COMPONENTS

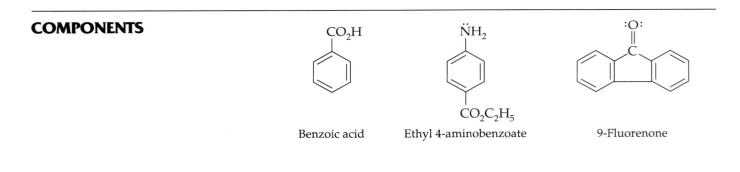

Benzoic acid            Ethyl 4-aminobenzoate            9-Fluorenone

## EXPERIMENTAL PROCEDURE

Estimated time of experiment: 2.5 hours.

### Physical Properties of Reactants

| Compound | MW | Amount | mmol | mp(°C) | bp(°C) | $d$ |
|---|---|---|---|---|---|---|
| Benzoic acid | 122.13 | 50 mg | 0.41 | 122 | | |
| Ethyl 4-aminobenzoate | 165.19 | 50 mg | 0.31 | 89 | | |
| 9-Fluorenone | 180.22 | 50 mg | 0.27 | 84 | | |
| Diethyl ether | 74.12 | 4 mL | | | 35 | 0.7184 |
| 3 *M* HCl | | 4 mL | | | | |
| 3 *M* NaOH | | 4 mL | | | | |
| 6 *M* HCl | | | | | | |
| 6 *M* NaOH | | | | | | |

**NOTE.** *In carrying out the separation, you should keep a record or flow chart of your procedure (as suggested in the prior reading assignment) in your laboratory notebook. You should also be particularly careful to label all flasks.*

### Reagents and Equipment

Weigh and add to a stoppered or capped 15-mL centrifuge tube the following: 50 mg (0.41 mmol) of benzoic acid, 50 mg (0.31 mmol) of ethyl 4-aminobenzoate, and 50 mg (0.27 mmol) of 9-fluorenone. Now add 4 mL of diethyl ether using a 10-mL graduated cylinder [HOOD] for the transfer. HOOD
The solids may be dissolved by either stirring with a glass rod or mixing on a Vortex mixer (capped vial).

### Separation of the Basic Component

Cool the solution in an ice bath. Now, using a calibrated Pasteur pipet, add 2 mL of 3 M HCl dropwise to the cooled solution with swirling. Cap and thoroughly mix the resulting two-phase system for several minutes (a Vortex mixer works well). Vent carefully, and after the layers have separated, remove the bottom (aqueous) layer using a Pasteur filter pipet and transfer this phase to a **labeled,** 10-mL Erlenmeyer flask.

Repeat this step with an additional 2 mL of the 3 M acid solution. As before, transfer the aqueous layer to the same labeled Erlenmeyer flask. Stopper or cap this flask. *Save the ether solution.* The aqueous acidic solution is to be used in the next step.

**NOTE.** *A small amount of crystalline material may form at the interface between the layers. A second extraction generally dissolves this material.*

### Isolation of Ethyl 4-aminobenzoate: The Basic Component

To the aqueous acidic solution, separated and set aside in the previous step, add 6 M NaOH dropwise until the solution is distinctly alkaline to litmus paper. Cool the flask in an ice bath for about 10–15 min. Collect the solid precipitate that forms in the basic solution by reduced-pressure filtration using a Hirsch funnel. Wash the precipitate with two 1-mL portions of distilled water. Air-dry the washed microcrystals by spreading them on a clay plate, filter paper, or in a vacuum drying oven. Weigh the material and calculate the percent recovery. Obtain the melting point of the dry ethyl 4-aminobenzoate and compare your result to the literature value. It is of interest to note that this material is used as a topical anesthetic.

### Separation of the Acidic Component

Add 2 mL of 3 M NaOH to the ether solution that was set aside earlier in the experiment. At this point, if necessary, add additional ether (~1–2 mL) so that the volume of the organic layer is at least equal to, or somewhat larger than, that of the aqueous phase. This adjustment in volume should allow an efficient distribution to take place when the two phases are mixed. Then carry out the extraction as before, allow the layers to separate, and finally transfer the bottom aqueous basic layer to a **labeled,** 10-mL Erlenmeyer flask.

Repeat this routine and again remove the aqueous layer and transfer it to the same Erlenmeyer flask. Stopper this flask (that contains the extracted aqueous basic phase) and set it aside for later use.

## Separation of the Neutral Component

Wash (extract) the remaining ether solution contained in the centrifuge tube with two 1-mL portions of water. Separate the lower aqueous layer in each sequence. *Save the aqueous wash layer temporarily, as it will be discarded at the very end of the experiment. (It is good practice to never discard any layer until you have recovered or accounted for all of the material.)* Now add about 300 mg of anhydrous granular sodium sulfate to the wet ether (this is ether saturated with water) solution. Cap the tube and set it aside while working up the alkaline extraction solution. This procedure will allow sufficient time for the traces of moisture to be removed from the ether solution by hydration of the insoluble drying agent. If the sodium sulfate initially forms large clumps, you may add a further quantity of the anhydrous salt.

## Isolation of Benzoic Acid: The Acidic Component

Add 6 M HCl dropwise to the aqueous alkaline solution, which was separated and set aside earlier, until the solution becomes distinctly acidic to litmus paper. Then cool the flask in an ice bath for about 10 min. If only a small amount (10–25 mg) of precipitate is obtained on acidification, add a small amount of a saturated aqueous solution of sodium chloride (salting out effect; see Prior Reading assignment) to help promote further precipitation of the benzoic acid. Collect the precipitated benzoic acid by filtration under reduced pressure using a Hirsch funnel. Wash (rinse) the filter cake (precipitated acid) with two 1-mL portions of cold distilled water. Dry the solid product using one of the techniques described earlier for ethyl 4-aminobenzoate. Weigh the dry benzoic acid and calculate your percent recovery.

Obtain a melting point of this material and compare your result to the literature value.

## Isolation of 9-Fluorenone: The Neutral Component

Transfer the dried ether solution, which had been collected earlier, by use of a Pasteur filter pipet, to a tared 10-mL Erlenmeyer flask containing a boiling stone. Rinse the drying agent with an additional 1 mL of ether and combine the ether wash with the anhydrous organic phase.

HOOD

Concentrate the ether solution on a warm sand bath using a *slow* stream of nitrogen gas [HOOD]. Obtain the weight of the residue (9-fluorenone) after removal of the solvent and calculate the percent recovery. Obtain a melting point of the material and compare your result to the literature value.

## QUESTIONS

**6-21.** Explain why diethyl ether would be expected to be a satisfactory solvent for the straight-chain hydrocarbons, hexane and heptane.

**6-22.** The solubility of p-dibromobenzene in benzene is 80 $\mu$g/100 $\mu$L at 25 °C. Would you predict the solubility of this compound to be greater, less, or approximately the same in acetone solvent at this temperature? Explain.

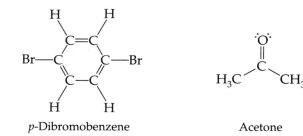

p-Dibromobenzene                      Acetone

**6-23.** Each of the solvents listed below are used in experiments in this text to extract organic compounds from aqueous solutions.
   a. Methylene chloride
   b. Pentane
   c. Toluene
   d. Diethyl ether

Will the organic phase be the upper or lower layer when each of these solvents is mixed with water? Explain your answer for each case.

**6-24.** A 36-mg sample of an organic compound (MW = 84) is dissolved in 10 mL of water. This aqueous solution is extracted with 5.0 mL of hexane. Separation and analysis of the aqueous phase shows that it now contains 12 mg of the organic compound. Calculate the partition coefficient for the compound.

**6-25.** A qualitative test often used to determine whether an organic compound contains oxygen is to test its solubility in concentrated sulfuric acid. Almost all oxygen-containing compounds are soluble in this acid. Explain.

**6-26.** In the discussion section, which covers multiple extractions (see p. 77), it was suggested that you might extend the relationship in the example given to the next step by using one third of the total quantity of the ether solvent in three portions. The reason for increasing the number of extractions was to determine whether this expansion would increase the efficiency of the process even further. To determine if this next step is worth the effort, perform the calculations for the extraction of 100 mg of P in 300 $\mu$L of water with three 100-$\mu$L portions of ether. Assume the partition coefficient is 3.5 (as before).

Compare the amounts of P extracted from the water layer using one, two, or three extractions.

Do you think that the additional amount of P extracted from the water layer using three extractions is justified? Might it be justified if P were valuable and you were working on the industrial scale of 100 kg of P in 3000 L of water?

---

## Reduction of Ketones Using a Metal Hydride Reagent: Cyclohexanol and *cis-* and *trans-*4-*tert-*Butylcyclohexanol

**EXPERIMENT 5**

---

Common name: Cyclohexanol
CA number: [108-95-0]
CA name as indexed: Cyclohexanol

Common name: 4-*tert*-Butylcyclohexanol
CA number: [98-52-2]
CA name as indexed: Cyclohexanol, 4-(1,1-dimethylethyl)-

## Purpose

To carry out the reduction of a ketone carbonyl to the corresponding alcohol by use of sodium borohydride, a commercially available metal-hydride reducing agent. The alcoholic reaction products are isolated by extraction techniques and purified by preparative gas chromatography. Cis and trans diastereomers are formed in the reduction of the 4-*tert*-butylcyclohexanone. These diastereomeric products can be separated during the preparative GC isolation. The stereochemistry of the structures can be deduced, once the mixture is separated into its pure components, using either NMR or IR spectroscopy.

### *Prior Reading*

*Technique 1:*  Gas Chromatography (pp. 56–61).
*Technique 4:*  Solvent Extraction
    Liquid-Liquid Extraction (pp. 80–82).
    Drying of a Wet Organic Layer (pp. 86–87).
    Concentration of Solutions (pp. 106–108).
*Chapter 9:*  Infrared Spectroscopy (pp. 613–614).
    Nuclear Magnetic Resonance Spectroscopy
    (pp. 657–663).

## REACTION (Experiment [5A])

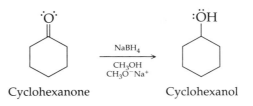

Cyclohexanone          Cyclohexanol

## DISCUSSION

An important route for the synthesis of primary and secondary alcohols is the reduction of aldehydes and ketones, respectively. Reduction involves the addition of the equivalent of molecular hydrogen (H—H) to the carbonyl group (C=O).

There are a variety of pathways that have been discovered to accomplish this conversion, but the most common method used in the research laboratory involves the use of complex metal-hydride reagents. Two reagents that enjoy wide application are lithium aluminum hydride ($LiAlH_4$) and sodium borohydride ($NaBH_4$).

Lithium aluminum hydride is a powerful reducing agent that reacts not only with aldehydes and ketones, but many other carbonyl containing functional groups as well. It will attack esters, lactones, carboxylic acids, anhydrides, and amides. It has also been found to reduce noncarbonyl systems, such as, alkyl halides, alkyl azides, alkyl isocyanates, and nitriles. It is particularly important to note that *$LiAlH_4$ can only be used safely in aprotic solvents* [a solvent that does not contain an ionizable (acidic) proton], such as diethyl ether or tetrahydrofuran (THF). In protic solvents, lithium aluminum hydride reacts *violently* with the acidic hydrogen of the solvent to rapidly generate hydrogen gas.

$$LiAlH_4 + 4\ CH_3OH \rightarrow LiAl(OCH_3)_4 + 4\ H_2$$

*It is not uncommon for the hydrogen gas to ignite. This particular hydride reagent should not be employed unless specific instructions are made available for its proper use under anhydrous conditions.*

Sodium borohydride is a much more selective reducing reagent. That is, it is a much milder reagent compared to LiAlH$_4$. It is for this reason that it is usually employed for the reduction of aldehydes and ketones. It does not react with the vast majority of the less reactive organic functional groups, such as C=C, C≡C, nitro, cyano, and even some carbonyl containing systems, such as, amides, and carboxylic acids. Sodium borohydride does react at an appreciable rate with water, but only slowly with aqueous alkaline solution (no available protons), methanol, α,β-unsaturated ketones, and esters. For small scale reactions an excess of reagent is generally used to compensate for the amount of borohydride that reacts with the protic solvent (methanol). This approach is preferred to that of using a solvent in which the sodium borohydride is less soluble (it is insoluble in ether), as the reaction is driven more rapidly to completion under the former conditions. On the other hand, sodium borohydride can react rapidly with strong acids to generate hydrogen gas. This latter reaction may be used to advantage as a source of in situ hydrogen for the reduction of C=C bonds (see Experiment [12]). The relatively high cost of the metal hydride reducing agents is offset by their low molecular weight (more moles per gram) and the fact that 1 mol of reducing agent reduces 4 mol of aldehyde or ketone.

The key step in the reduction of a carbonyl group by sodium borohydride is the transfer of a hydride ion (:H$^-$) from boron to the carbon atom of the polarized carbonyl group

$$\underset{\diagup}{\overset{\diagdown}{C}} \overset{\delta+}{\cdots} \overset{\delta-}{\ddot{O}}:$$

In the reaction, the electron-rich hydride ion is acting as a *nucleophile* (nucleus seeking), which attacks the *electrophilic* (electron seeking) carbon atom of the carbonyl group.

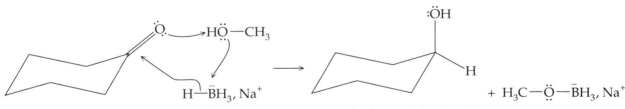

The hydride may make an equatorial (e) or axial (a) attack
depending on steric factors.

The overall reduction process requires two hydrogen atoms, but only one comes directly from the borohydride reagent. The other hydrogen atom is derived from the protic solvent (methanol).

In the 4-*tert*-butylcyclohexanone example, the steric environment is different on either face of the carbonyl group. In this case, the hydride reducing agent attacks more rapidly from the axial direction, and thus the equatorial alcohol (axial H) is found to be the major product. This reaction pathway is found to be preferred with the relatively small sodium borohydride and lithium aluminum hydride reagents. This type of reaction, where one stereoisomeric product is preferentially formed, is termed a *stereoselective* reaction.

Sterically large hydride reducing reagents, such as lithium tri-*sec*-butylborohydride, are forced to make an equatorial attack, due to steric factors (these hydride reagents run into the 1,3-diaxial hydrogen atoms), and thus, the cis isomer of 4-*tert*-butylcyclohexanol is the major product. The data are summarized in Table 6.4 which relates the stereochemistry of the reduction to the metal hydride reagent employed.

**TABLE 6.4  Reduction of 4-*tert*-Butylcyclohexanone**

| Reagent | trans (%) | cis (%) |
|---|---|---|
| Sodium borohydride | 80 | 20 |
| Lithium aluminum hydride | 92 | 8 |
| Lithium tri-*sec*-butylborohydride | 7 | 93 |

## Experiment [5A]   Cyclohexanol

The reaction is shown above.

## EXPERIMENTAL PROCEDURE

Estimated time for the experiment: 1.5 hours.
For the GC analysis, 15 min per student.

### Physical Properties of Reactants

| Compound | MW | Amount | mmol | bp(°C) | d | $n_D$ |
|---|---|---|---|---|---|---|
| Cyclohexanone | 98.15 | 100 μL | 0.97 | 156 | 0.95 | 1.4507 |
| Methanol | 32.04 | 250 μL | 10.3 | 65 | 0.79 | 1.3288 |
| Sodium borohydride, reducing solution | | 300 μL | | | | |

### Reagents and Equipment

With the aid of an automatic delivery pipet, place 100 μL (95 mg, 0.97 mmol) of cyclohexanone in an oven dried and tared (preweighed) 5.0-mL conical vial equipped with an air condenser. Now add 250 μL of methanol and gently shake the vial to obtain a homogeneous solution. (■)

HOOD    To the solution of the ketone add 300 μL of sodium borohydride reducing solution dropwise, with swirling **[HOOD]**.

**NOTE.**  *The cyclohexanone, methanol, and sodium borohydride solutions are dispensed using automatic delivery pipets. It is suggested that the cyclohexanone be weighed after delivery so as to get an accurate weight for the yield calculations.*

*The stock reducing solution should be prepared just prior to conducting the experiment.*

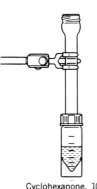

Cyclohexanone, 100 μL
+ CH₃OH, 250 μL +
NaBH₄ solution, 300 μL

## Instructor Preparation

*In a 10-mL Erlenmeyer flask place 50 mg of anhydrous sodium methoxide and 2.5 mL of methanol. To this solution add 100 mg of sodium borohydride. Stopper the flask tightly and swirl the contents gently to dissolve the solid phase (100 $\mu$L of this solution provides ~2.0 mg of $NaOCH_3$ and 4.0 mg of $NaBH_4$).*

**NOTE.** *Test for activity of the reducing solution: Add 1–2 drops of the freshly prepared reducing solution to about 200 $\mu$L of concentrated hydrochloric acid. Generation of hydrogen gas bubbles is a positive test.*

## Reaction Conditions

Allow the resulting solution to stand at room temperature for a period of 15 min.

## Isolation of Product

By use of a calibrated Pasteur pipet, add dropwise 1.0 mL of cold dilute hydrochloric acid. Extract the aqueous mixture with three 0.5-mL portions of methylene chloride. Upon each addition of methylene chloride, cap the vial, shake it gently, and then carefully vent it by loosening the cap (a Vortex mixer may be used if available). After the layers have separated, remove the bottom methylene chloride layer using a Pasteur filter pipet and transfer it to a Pasteur filter pipet containing about 500 mg of anhydrous sodium sulfate.

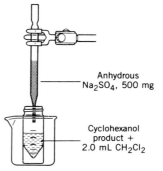

Anhydrous
$Na_2SO_4$, 500 mg

Cyclohexanol
product +
2.0 mL $CH_2Cl_2$

Collect the dried eluate in a tared 5.0-mL conical vial containing a boiling stone. Use an additional 0.5 mL of methylene chloride to rinse the sodium sulfate and collect the rinse in the same conical vial. (■)

*An additional rinse of the sodium sulfate may be made if desired.* Remove the methylene chloride solvent by careful evaporation in the **HOOD** by gentle warming in a sand bath (constantly agitate the surface of the solution with a microspatula to prevent superheating and subsequent boil-over). *In this instance, do not use a stream of nitrogen gas to hasten the evaporation. The volatility of the product alcohol is such that a substantial loss of product will occur if this latter technique is employed.*

HOOD

## Purification and Characterization

The crude cyclohexanol reaction product remaining after evaporation of the methylene chloride solvent is usually of sufficient purity for direct characterization.

Determine the weight of the liquid residue and calculate the percent crude yield. Determine the refractive index (3 $\mu$L, optional) and boiling point (4 $\mu$L) of the material. Compare your values with data given in the literature.

Obtain the infrared spectrum of the crude (dry) cyclohexanol product by the capillary film sampling technique. Compare the spectrum of the starting ketone in Figure 6.6 to that of your isolated material. Is there evidence of the unreacted starting material in your product? The spectrum of cyclohexanol crude product is shown in Figure 6.7.

**NOTE.** *It should be noted that the majority of the infrared spectra referred to in the experimental analysis sections are Fourier-transform derived and have been plotted on a slightly different scale than the other spectra presented in the text. These spectra utilize a 12.5 cm$^{-1}$/division format below 2000 cm$^{-1}$ and undergo a 2 : 1 compression above 2000 cm$^{-1}$ (25 cm$^{-1}$/division).*

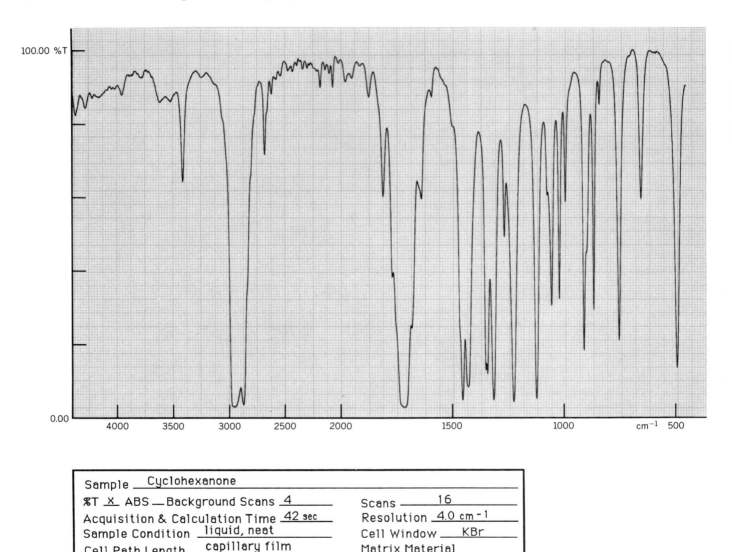

| Sample | Cyclohexanone | | |
| --- | --- | --- | --- |

Sample ___Cyclohexanone_____

%T _X_ ABS __Background Scans _4_____    Scans _____16_____

Acquisition & Calculation Time _42 sec___    Resolution _4.0 cm⁻¹___

Sample Condition _liquid, neat_____    Cell Window ____KBr_____

Cell Path Length __capillary film____    Matrix Material _____

**FIGURE 6.6**  IR spectrum: cyclohexanone.

## Infrared Analysis: A Comparison of Reactant and Product

The key absorption bands to examine in the spectrum of cyclohexanone occur at 3420, 3000–2850, 1715, and 1425 cm⁻¹. The lack of significant absorption between 3100–3000 and 1400–1350 cm⁻¹ also should be noted. The sharp weak band at 3420 cm⁻¹ is not a fundamental vibration (not O—H or N—H stretching), but arises from the first overtone of the very intense carbonyl stretching mode found at 1715 cm⁻¹. Note that the overtone does not fall exactly at double the frequency of the fundamental, but usually occurs slightly below that value because of anharmonic effects (see Chapter 9). The lack of absorption in the region near 3100–3000 cm⁻¹ and the presence of a series of very strong absorption bands at 3000–2850 cm⁻¹ indicates that the only C—H stretching modes present are part of $sp^3$ systems. Thus, the spectrum is typical of an aliphatic ketone. The occurrence of a 1425-wavenumber band suggests the presence of at least one methylene group adjacent to the carbonyl group while the 1450-wavenumber band requires other methylene groups more remote to the C=O group. The lack of absorption in the 1400–1375-wavenumber region indicates the absence of any methyl groups (a good indication of a simple aliphatic ring system) and further suggests that the absorption at 1450

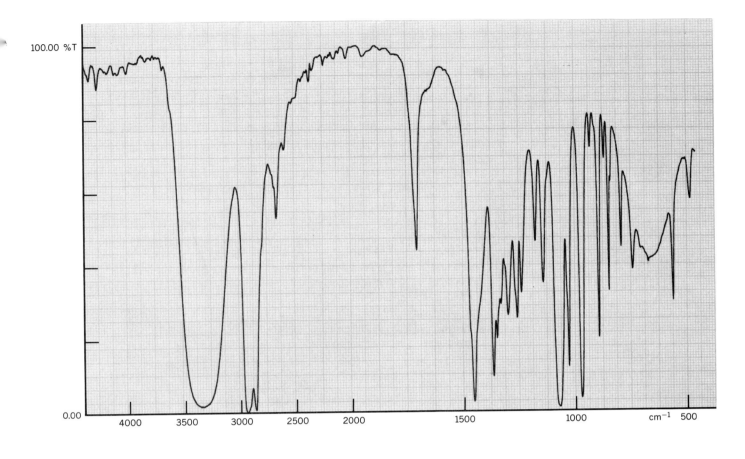

| Sample __Cyclohexanol (crude product)__ |
| --- |

%T _X_ ABS __ Background Scans _4___          Scans ___16_____

Acquisition & Calculation Time _42 sec__   Resolution _4.0 cm⁻¹___

Sample Condition _liquid, neat___          Cell Window ___KBr_____

Cell Path Length _capillary film___        Matrix Material _____

**FIGURE 6.7**   IR spectrum: cyclohexanol (crude product).

cm$^{-1}$ must arise entirely from methylene scissoring modes. The value of 1715 cm$^{-1}$ for the C=O stretch supports the presence of a six-membered ring.

Now examine the spectrum of your reaction product (a typical example is given in Fig. 6.7). The spectrum is rather different from that of the starting material. The major changes are a new very strong broad band occurring between 3500 and 3100 cm$^{-1}$ and the large drop in the intensity of the band found at 1715 cm$^{-1}$. These changes indicate the reductive formation of an alcohol group from the carbonyl system. The band centered near 3300 cm$^{-1}$ results from the single polarized O—H stretching mode. The drop in intensity of the 1715-wavenumber band indicates the loss of the carbonyl function. The exact amount of cyclohexanone remaining can be determined by carrying out a Beer's law type analysis, but in this case we will use gas chromatographic techniques to determine this value. Other bands of interest in the spectrum of cyclohexanol occur at 1069 and 1031 cm$^{-1}$. These bands can be assigned, respectively, to the equatorial and axial C—O stretching of the rotational conformers of this alicyclic secondary alcohol. A broad band (width ~300 cm$^{-1}$) can be found near 670 cm$^{-1}$. This absorption arises from an O—H bending, out-of-plane mode, of the associated alcohol. This band is generally identified only in neat samples where extensive hydrogen-bonding occurs. Also note that the band at 1425 cm$^{-1}$ has vanished as there are no methylene groups alpha to carbonyl systems in the product.

Now proceed with purification of the reaction product by preparative gas chromatography. Use the following conditions and refer to Experiment [2] for the collection technique. If time permits, or perhaps in a later laboratory period, determine the infrared spectrum of the purified product. Describe and explain the changes observed in the new spectrum compared to that of the crude product.

## Separation of Small Quantities of Cyclohexanone from Cyclohexanol

*Example*

9:1 (v/v) cyclohexanol/cyclohexanone
10% Carbowax 20M (stationary phase)
Injection volume: 15 $\mu$L
Temperature: 130 °C
He flow rate: 50 mL/min
Column: 1/4-in. × 8-ft stainless steel
Chart speed = 1 cm/minute

| Run | Cyclohexanol Retention Time (min) | Cyclohexanol Yield (mg) | Cyclohexanone Retention Time (min) |
|---|---|---|---|
| 1 | 15.3 | 6.5 | 11.6 |
| 2 | 17.3 | 7.3 | 12.5 |
| 3 | 17.2 | 8.5 | 12.6 |
| 4 | 16.0 | 9.6 | 12.0 |
| 5 | 14.6 | 7.3 | 11.2 |
| 6 | 14.5 | 8.7 | 11.2 |
| 7 | 15.5 | 8.9 | 11.7 |
| 8 | 15.5 | 8.9 | 11.8 |
| 9 | 16.4 | 8.8 | 12.3 |
| 10 | 15.4 | 8.4 | 12.7 |
| Av | 15.8 ± 1 | 8.3 ± 0.9 | 12.0 ± 0.6 |

$$\text{Cyclohexanol injected} = 0.9(15 \ \mu L)(0.963 \ mg/\mu L) = 13.0 \ mg$$
$$\text{Percent yield} = 8.3/13.0 \times 100 = 63.8\%$$

**NOTE.** *Collection efficiencies approaching 90% can be obtained by cooling the collection tube. Liquid nitrogen soaked tissues work best, but ice water gives a measurable improvement.*

## Chemical Tests

Several chemical tests (see Chapter 10) may also be used to establish that an alcohol has been formed by the reduction of a ketone. Perform the ceric nitrate and 2,4-dinitrophenylhydrazine test on **both** the starting ketone and the alcohol product. Do your results demonstrate that an alcohol was obtained? You may also wish to prepare a phenyl- or $\alpha$-naphthylurethane derivative of the cyclohexanol. Before the development of chemical instrumentation, the formation of solid derivatives was used extensively to identify reaction products.

## Experiment [5B]   *cis-* and *trans-4-tert-*Butylcyclohexanol

**REACTION**

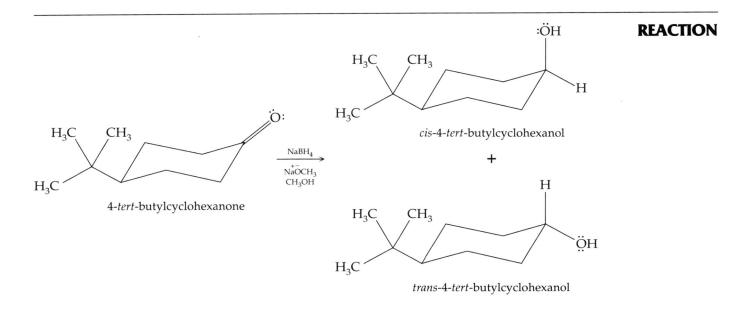

4-*tert*-butylcyclohexanone

*cis*-4-*tert*-butylcyclohexanol

+

*trans*-4-*tert*-butylcyclohexanol

## EXPERIMENTAL PROCEDURE

Estimated time of the experiment: 2.0 hours.
For the GC analysis, 15 min per student.

### Physical Properties of Reactants

| Compound | MW | Amount | mmol | mp(°C) | bp(°C) |
|---|---|---|---|---|---|
| 4-*tert*-Butylcyclohexanone | 154.25 | 50 mg | 0.33 | 47–50 | |
| Methanol | 32.04 | 50 $\mu$L | | | 65 |
| Sodium borohydride reducing solution | | 100 $\mu$L | | | |

### Reagents and Equipment

In a tared 3.0-mL conical vial equipped with an air condenser weigh and place 50 mg (0.33 mmol) of 4-*tert*-butylcyclohexanone followed by 50 $\mu$L of methanol. (■) Gently shake the vial to obtain a homogeneous solution.

Now slowly add 100 $\mu$L of the sodium borohydride reducing solution while swirling.

**NOTE.** *The liquid reagents are dispensed by use of automatic delivery pipets. The preparation of the reducing solution is given in Experiment [5A], Reagents and Equipment (see p. 159).*

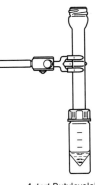

4-*tert*-Butylcyclohexanone,
50 mg + CH$_3$OH, 50 $\mu$L +
NaBH$_4$ solution, 100 $\mu$L

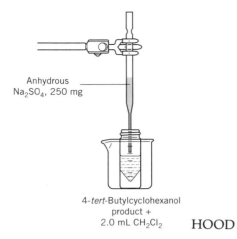

Anhydrous
Na$_2$SO$_4$, 250 mg

4-*tert*-Butylcyclohexanol
product +
2.0 mL CH$_2$Cl$_2$      HOOD

## Reaction Conditions

Allow the solution to stand at room temperature for 10 min.

## Isolation of Product

Work up the resulting solution using the procedure described in Experiment [5A], Isolation of Product (see p. 159), with the exception that 250 mg of sodium sulfate is placed in the Pasteur filter pipet. (■)

Remove the dried methylene chloride solvent from the final solution by directing a gentle stream of nitrogen gas onto the liquid surface while at the same time externally warming the vial in a sand bath. The use of a heating bath will help to avoid moisture condensation on the residue during solvent evaporation [**HOOD**].

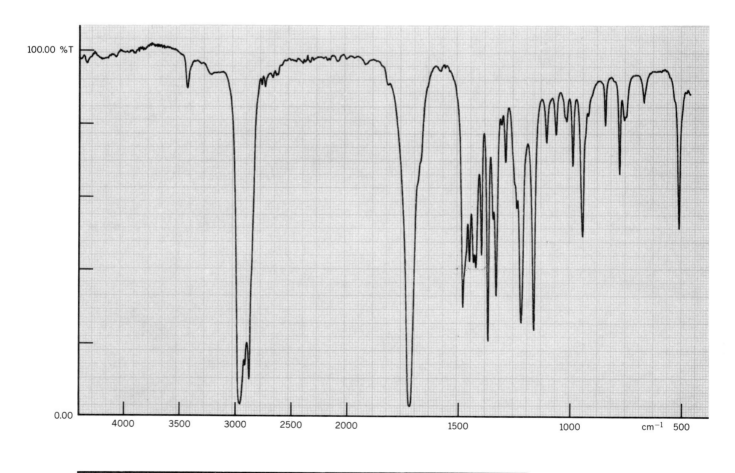

| Sample __4-tert-Butylcyclohexanone__ | |
|---|---|
| %T _X_ ABS _ Background Scans _4_ | Scans ___16___ |
| Acquisition & Calculation Time _42 sec_ | Resolution _4.0 cm⁻¹_ |
| Sample Condition __solid – melt__ | Cell Window ___KBr___ |
| Cell Path Length __capillary film__ | Matrix Material _____ |

**FIGURE 6.8** IR spectrum: 4-*tert*-butylcyclohexanone.

## Purification and Characterization

The product mixture remaining after removal of the methylene chloride is normally of sufficient purity for direct characterization. Weigh the solid product and calculate the percent yield. Determine the melting point of your material. 4-*tert*-Butylcyclohexanol (mixed isomers) has a melting point of 62–70 °C.

Obtain the IR and NMR spectra of the crude mixture of isomers. Infrared sampling in this instance is best accomplished by the capillary film-melt (use the heat lamp) technique (see Chapter 9, IR discussions).

## Infrared Analysis

Refer to the discussion in Experiment [5A] for an interpretation of the absorption bands found at 3435, 3000–2850, 1717, and 1425 cm$^{-1}$ in the starting material (Fig. 6.8), and at 3250, 3000–2850 (1717 variable relative intensity – may be quite weak – why?), 1069, and 1031 cm$^{-1}$ in the crude alcohol (Fig. 6.9). In addition, the ketone

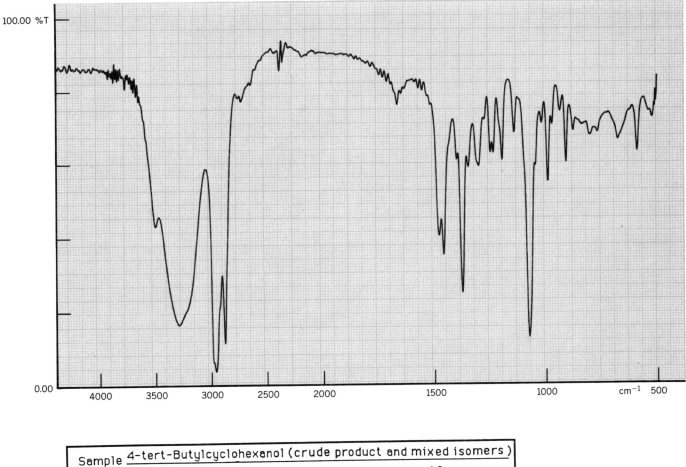

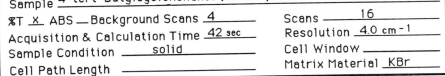

**FIGURE 6.9**   IR spectrum: 4-*tert*-butylcyclohexanol (crude product & mixed isomers).

has bands at 1396 (weak) and 1369 (strong) , and the alcohol has bands at 1399 (weak) and 1375 (strong) cm$^{-1}$. These two pairs of bands establish the presence of the tertiary butyl group in these compounds.

Note that (a) a weak band (3495 cm$^{-1}$) is present on the high wavenumber side of the 3250 cm$^{-1}$ O—H stretching mode and (b) that even in neat samples of the tertiary butyl derivative, the 670-wavenumber band, clearly evident in cyclohexanol, is difficult to observe.

The mixture of two diastereomeric alcohols that have been synthesized provides an ideal opportunity to introduce you to nuclear magnetic resonance (NMR) spectroscopy. This technique is an extremely powerful tool for the discrimination and characterization of diastereomeric compounds. As you will see, the two diastereomers have quite different NMR spectra. An interpretation of these spectra will allow you to determine the ratio of the two isomers as well as to make an unambiguous assignment of their relative stereochemistry.

This experiment presents the option for students to obtain and interpret NMR data if the local opportunity exists. The two diastereomeric alcohols exhibit different splitting patterns for the proton on the carbon bearing the —OH group. Integration of these signals allows one to determine the ratio of the diastereomeric alcohols in the sample.

### Nuclear Magnetic Resonance Analysis

Refer to the expanded NMR spectrum in Figure 6.10. The signals at about 4.04 ppm and 3.52 ppm correspond to the proton on the carbon bearing the —OH group in the two diastereomers of 4-*tert*-butylcyclohexanol shown. On closer inspection, the downfield signal (4.04 ppm) is a pentet and the upfield signal (3.52 ppm) is a triplet of triplets. The pentet implies that the proton in question is coupled with equal coupling constants (*J*) to four adjacent protons. The triplet of triplets implies that the proton in question is coupled to two adjacent protons with a large coupling constant and to two other adjacent protons with a smaller coupling constant. Specifically, the proton in the first case must be equatorial and the proton in the second case must be axial, because the dihedral angle between an equatorial proton and each of the four adjacent protons is the same, about 60°, *J* = 2–3 Hz. When a proton is axial, the dihedral angle to the two adjacent equatorial protons is about 60° (*J* ≈ 3 Hz) and the dihedral angle to the two adjacent axial protons is about 180° (*J* ≈ 13 Hz), thus producing a triplet of triplets.

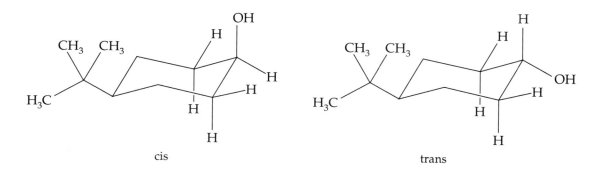

cis                                    trans

### Gas Chromatographic Analysis

The cis and trans isomers of 4-*tert*-butylcyclohexanol may be separated by gas chromatography using a 1/4-in. x 8-ft, 10% FFAP column at 110 °C. Prepare a methylene chloride solution of the alcohol mixture having a

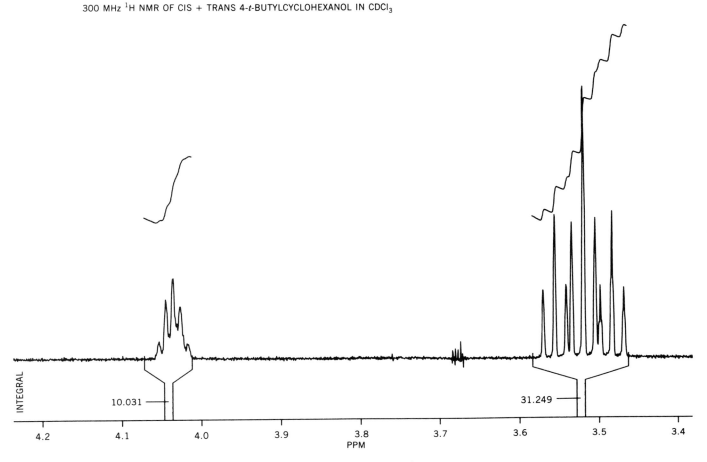

300 MHz $^1$H NMR OF CIS + TRANS 4-*t*-BUTYLCYCLOHEXANOL IN CDCl$_3$

INTEGRAL

10.031

31.249

4.2    4.1    4.0    3.9    3.8    3.7    3.6    3.5    3.4
PPM

**FIGURE 6.10**    NMR spectrum: *cis-* and *trans-4-tert*-butylcyclohexanol.

concentration of 0.5 mg/μL and inject a 5.0-μL sample into the GC apparatus. At a flow rate of 50 mL/min (He), the retention time of the cis isomer is 13 min, and the trans isomer is 16 min.

Determine the percentage of each isomer present in the sample by determining the area under each curve.

**NOTE.** *If a Carbowax column (170 °C) is used, the starting ketone has a retention time similar to that of the cis alcohol. Therefore, if the reaction does not go to completion, the apparent cis/trans ratio may not be accurate. It has recently been demonstrated that 10% FFAP columns will resolve all components present in the product mixture. Thus, the starting ketone and the cis isomer concentrations may be effectively established in addition to the trans isomer.*

The latter separation scheme was developed by T. J. Dwyer and S. Jones at the University of California, San Diego.

**NOTE.** *Calculation of the Area Under a Curve: There are several techniques that may be used, but the following method is satisfactory for your needs and gives reproducible results of ±3–4%.*

*Multiplication of the peak height (mm) by the width at one half height (mm), measured from the base line of the curve, yields the area (mm²) under the curve.*

**QUESTIONS**

**6-27.** Suggest a chemical test that would allow you to distinguish between *tert*-butyl alcohol and 1-butanol, both of which give a positive ceric nitrate test.

**6-28.** Which of the isomeric butanols ($C_4H_{10}O$) can be prepared by reduction of a ketone with sodium borohydride?

**6-29.** Why are there axial and equatorial hydroxyl isomers for 4-*tert*-butylcyclohexanol, but not for cyclohexanol itself?

**6-30.** What aldehyde or ketone would you reduce to prepare the following alcohols?

Benzyl alcohol     3,3-Dimethyl-2-butanol     2,2-Dimethyl-1-pentanol

**6-31.** The *cis*- and *trans*-4-*tert*-butylcyclohexanol prepared in Experiment [5B] each have a plane of symmetry. Draw this symmetry element for each of the diastereomers.

**6-32.** In the spectrum of the crude product obtained from the reduction of 4-*tert*-butylcyclohexanone, the fingerprint region appears to possess bandwidths that are slightly broader than those found in cyclohexanol. Explain.

**6-33.** The reduction of 4-*tert*-butylcyclohexanone is a stereoselective reaction. Which isomer predominates? If you do not have NMR data available, it is still possible to arrive at a rough estimate of the product ratio. How would you go about this measurement? Suggest a value.

**6-34.** In the spectrum of the crude 4-*tert*-butylcyclohexanols one observes:
(a) A weak band (3495 $cm^{-1}$) located on the high side of the 3250 $cm^{-1}$ O—H stretching mode.
(b) Even in neat samples of this alcohol, the 670-$cm^{-1}$ band, clearly evident in cyclohexanol, is difficult to observe.
Explain these observations. Is the same effect operating in both cases?

**6-35.** Sketch the NMR spectrum you would expect to observe for the following compounds.
(a) Acetone
(b) 1,1,2-Tribromoethane
(c) Ethyl chloride
(d) 2-Iodopropane
(e) 1-Bromo-4-methoxybenzene

**BIBLIOGRAPHY**

General references on metal hydride reduction:

House, H. O. *Modern Synthetic Reactions*; Benjamin: Reading, MA, 1972.
Walker, E. R. H. *Chem. Soc. Rev.* **1976,** 5, 23.

Sodium borohydride as a reducing agent:

Brown, H. C. *Boranes in Organic Chemistry*; Cornell Univ. Press: Ithaca, NY, 1972.
Cragg, G. M. W. *Organoboranes in Organic Synthesis*; Marcel Dekker: New York, 1973.

Fieser, L. F.; Fieser, M. *Reagents for Organic Synthesis*; Wiley: New York, 1967; Vol. I, p. 1050 and subsequent volumes.

Todd, D. *J. Chem. Educ.* **1979**, *56*, 540.

Lithium aluminum hydride as a reducing agent:

Brown, W. G. *Org. React.* **1951**, *6*, 469.

Fieser, L. F.; Fieser, M. *Reagents for Organic Synthesis*; Wiley: New York, 1967; Vol. I, p. 581 and subsequent volumes.

*trans*-4-*tert*-Butylcyclohexanol has been prepared from the ketone using LiAlH₄ as the reducing agent:

Eliel, E. L.; Martin, R. J. L.; Nasipuri, D. *Organic Syntheses*; Wiley: New York, 1973; Collect. Vol. V, p. 175.

---

# Photochemical Isomerization of an Alkene: *cis*-1,2-Dibenzoylethylene

**EXPERIMENT [6]**

Common names: *cis*-1,2-Dibenzoylethylene
              *cis*-1,4-Diphenyl-2-butene-1,4-dione

CA number: [959-27-3]
CA name as indexed: 2-Butene-1,4-dione, 1,4-diphenyl-, (Z)-

## Purpose

To illustrate the ease of cis–trans isomerization in organic molecules and, specifically in this case, to demonstrate the isomerization of a trans alkene to the corresponding cis isomer via photochemical excitation.

---

A number of very important biochemical reactions are promoted by the adsorption of UV–vis radiation. Vitamin D₃, which regulates calcium deposition in bones, is biosynthesized in just such a photochemical reaction. This vitamin is formed when the pro-vitamin, 7-dehydrocholesterol, is carried through fine blood capillaries just beneath the surface of the skin and exposed to sunlight. The amount of radiation exposure, which is critical for the regulation of the concentration of this vitamin in the blood stream, is controlled by skin pigmentation and geographic latitude. Thus, the color of human skin is an evolutionary response to control the formation of vitamin D₃ via a photochemical reaction.

**BIOLOGICALLY IMPORTANT PHOTOCHEMICAL REACTIONS**

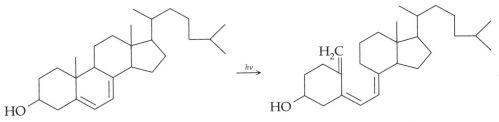

7-dehydrocholesterol        Vitamin D₃

Another set of significant photochemical reactions in human biochemistry is one contained in the chemistry of vision. These reactions involve vitamin A₁ (retinol) which is a $C_{20}$ compound belonging to a class of compounds known as diterpenes. These compounds are molecules formally constructed by the biopolymerization of four isoprene, $CH_2{=}C(CH_3){-}CH{=}CH_2$, molecules. Retinol (an all-trans pentaene) is first oxidized via liver enzymes (biological catalysts) to vitamin A aldehyde (*trans*-retinal). The *trans*-retinal, which is present in the light sensitive cells (the retina) of the eye, undergoes further enzymatic transformation (retinal isomerase) to give *cis*-retinal (a second form of vitamin A aldehyde) in which one of the double bonds of the all-trans compound is isomerized.

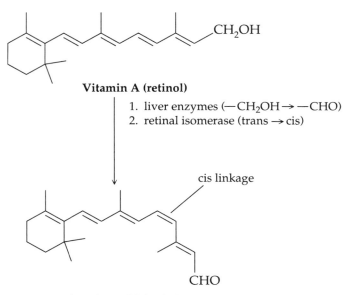

**Vitamin A (retinol)**

1. liver enzymes $(-CH_2OH \rightarrow -CHO)$
2. retinal isomerase (trans $\rightarrow$ cis)

cis linkage

**Vitamin A aldehyde (*cis*-retinal)**

The cis isomer of vitamin A aldehyde (retinal) possesses exactly the correct dimensions to become coupled to opsin, a large protein molecule (MW ~38,000) [coupling involves a reaction of the retinal aldehyde group, $-CH{=}O$, with an amine group ($-NH_2$) of the protein to form an imine linkage ($RCH{=}NR$)], to generate a light-sensitive substance, rhodopsin. This material is located in the rodlike structures of the retina. When the protonated form of rhodopsin [$-CH{=}\overset{+}{N}HR$], which absorbs in the blue-green region of the visible spectrum near 500 nm, is exposed to radiation of this wavelength, isomerization of the lone cis double bond of the diterpene group occurs and *trans*-rhodopsin is formed.

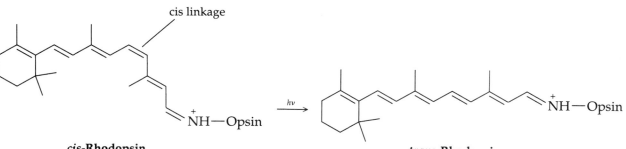

cis linkage

*cis*-Rhodopsin

*trans*-Rhodopsin

This photoreaction (a fast reaction, $10^{-12}$ s) involves a significant change in the geometry of the diterpene group, which eventually ($10^{-9}$ s) results in both a nerve impulse and the separation of *trans*-retinal from the opsin. The trans isomer is then enzymatically reisomerized back to the cis compound, which then starts the initial step of the visual cycle over again.

There are two interesting facts about this reaction: (1) This reaction is incredibly sensitive. A single photon will cause the visual nerve to fire. (2) All known visual systems employ *cis*-retinal, regardless of their evolutionary trail.

The photoreaction that we will study next is very similar to the cis–trans double-bond isomerism found in the vitamin A visual pigments. The only difference is that in our case we will be photochemically converting a trans double-bond isomer to the cis isomer.

### Prior Reading

*Technique 5:* Crystallization
Introduction (pp. 90–92).
Use of the Hirsch Funnel (pp. 93–95).
Craig Tube Crystallization (pp. 95–96).
Recrystallization Pipet (pp. 96–97).

**REACTION**

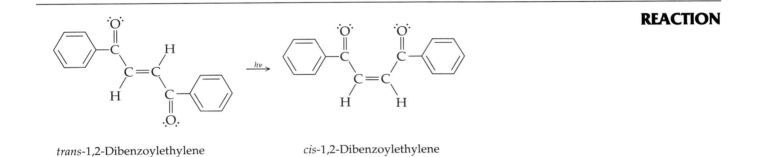

*trans*-1,2-Dibenzoylethylene          *cis*-1,2-Dibenzoylethylene

**DISCUSSION**

The $\pi$ bond of an alkene (C=C) is created by overlap of the $sp^2$ hybridized carbons' $p$ orbitals. Rotation about the axis of the C=C requires a good deal of energy because it destroys the $p$ orbital overlap, and thus the $\pi$ bond. Unless the material is irradiated with light of the appropriate wavelength, absorption of the radiation does not take place, and the isomers do not interconvert unless sufficient thermal energy (60–65 kcal/mol, typically > 200 °C ) is supplied to break the $\pi$ bond. If one of the electrons in the $\pi$ bond ($\pi$ molecular orbital) is photochemically excited into an antibonding, $\pi^*$, molecular orbital, as occurs in this experiment, the $\pi$ bond is weakened significantly. Rotation about this bond can then occur rapidly at room temperature.

It is the high-energy barrier to rotation about the C=C that gives rise to the possibility of alkene stereoisomers. The cis and trans isomers of an alkene system are termed diastereoisomers, or diastereomers. Like all stereoisomers, these isomers differ only in the arrangement of the atoms in space. These isomers have all the same atoms bonded to each other. Diastereomers are not mirror images of each other and, of course, are not superimposable (identical). This particular type of diastereomer often was referred to in the older literature as a "geometric isomer." Diastereomers, therefore, would be expected to possess different physical properties, such as melting points, boiling points, dipole moments, densities, and solubilities, as well as different spectroscopic properties. Because of these differences in physical properties, diastereomers are amenable to separation by chromatography, distillation, crystallization, and other separation techniques. We will employ both chromatographic and crystallization methods in the present experiment.

The course of the isomerization may be followed using *thin-layer chromatography* (Experiment [6B]) or alternatively, by spectroscopic techniques employing NMR analysis (Experiment [6C]).

The photochemical isomerization of a system containing an —N=N— linkage is illustrated in Experiment [35]. The photochemical isomerization of a diazabicyclohexene system is presented in Experiment [F4].

## Experiment [6A] Purification of *trans*-1,2-Dibenzoylethylene

---

**EXPERIMENTAL PROCEDURE**

Estimated time to complete the purification: 0.75 hours of laboratory time.

### Physical Properties of Components

| Compound | MW | Amount | mmol | mp(°C) | bp(°C) |
|---|---|---|---|---|---|
| *trans*-1,2-Dibenzoylethylene | 236.27 | 100 mg | 0.42 | 111 | |
| Ethanol (95%) | | 6.0 mL | | | 78.5 |
| Methylene chloride | | 4.0 mL | | | 40 |

### Purification Conditions

Purify the starting alkene by recrystallization.

Weigh and add to a 10-mL Erlenmeyer flask 100 mg (0.42 mmol) of *trans*-1,2-dibenzoylethylene and 3.0 mL of methylene chloride.

**NOTE.** *If the melting point of the alkene was not supplied to you, set aside a small sample (1–2 mg) of the weighed sample so that the evacuated melting point of this initial material can be obtained later.*

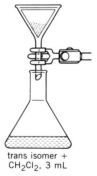

trans isomer +
CH$_2$Cl$_2$, 3 mL

Add decolorizing charcoal pellets (10 mg) to this solution and swirl the mixture gently for several minutes. Use a Pasteur filter pipet to transfer the methylene chloride solution to a second 10-mL Erlenmeyer flask containing a boiling stone (remember to hold the necks of the two flasks close together with the fingers of one hand during the transfer).

**NOTE.** *If powdered charcoal is used instead of pellets, filter the solution by gravity through a fast grade filter paper. Rinse the filter paper with an additional 1-mL of methylene chloride and collect this rinse in the same flask.* (■)

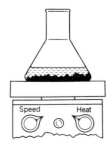

trans-1, 2-Dibenzoylethylene,
150 mg +CH$_2$Cl$_2$, 3.0 mL,
charcoal pellets

HOOD

Concentrate the filtered solution to dryness in a warm (50 °C) sand bath under a slow stream of nitrogen gas in the **hood**.

Now add 3 mL of 95% ethanol to the flask and dissolve the yellow solid residue by warming in a sand bath, with magnetic stirring, until a homogeneous solution is obtained. (■)

Allow the solution to cool slowly to room temperature over a period of 15 min and then place it in an ice bath for an additional 10 min. Collect the yellow needles by vacuum filtration using a Hirsch funnel (■) and then air-dry them on a porous clay plate or on filter paper.

Weigh the product and calculate the percent recovery. Determine the evacuated melting point and compare your result with both the literature value and that obtained with the material saved prior to recrystallization. If Experiment [3A] has been completed, compare the melting point of that sample of the trans alkene, which was obtained by concentration, via distillation, of an ethyl acetate solution. In the latter experiment, a simple crystallization was performed without the aid of decolorizing charcoal.

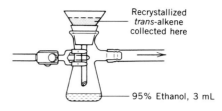

Recrystallized
*trans*-alkene
collected here

95% Ethanol, 3 mL

## Experiment [6B]   Isomerization of an Alkene: Thin-Layer Chromatographic Analysis

Estimated time to complete the experiment: 2.5 hours of laboratory time. The reaction requires approximately 1 hour of irradiation; the actual time to completion is quite sensitive to both radiation flux and temperature. These factors are largely determined by the distance the reaction vessel is positioned from the source of radiation.

**EXPERIMENTAL PROCEDURE**

### Physical Properties of Reactants

| Compound | MW | Amount | mmol | mp(°C) | bp(°C) |
|---|---|---|---|---|---|
| *trans*-1,2-Dibenzoylethylene | 236.27 | 25 mg | 0.11 | 111 | |
| Ethanol (95%) | | 3.5 mL | | | 78.5 |

### Reagents and Equipment

To a 13 × 100-mm test tube weigh and add 25 mg (0.11 mmol) of recrystallized *trans*-1,2-dibenzoylethylene and 3.0 mL of 95% ethanol.

### Reaction Conditions

Use a sand bath to warm the mixture [**GENTLY**] until a homogeneous solution is obtained. Stopper the test tube *loosely*, or cover it with filter paper held in place by a rubber band, and then place it approximately 2–4 in. from a 275-W sun lamp. Irradiate the solution for approximately 1 hour. (■)

GENTLY

**NOTE.** *If a lower wattage lamp is used, longer irradiation times or shorter distances will be necessary. In either case, solvent evaporation can be significantly reduced by directing a flow of cool air (fan) over the reaction tubes. An alternative procedure is to allow the tube to stand in sunlight at room temperature for several days.*

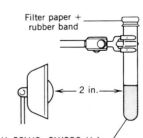

Filter paper +
rubber band

2 in.

$(C_6H_5CO)HC=CH(COC_6H_5)$,
25 mg + 95% ethanol, 3 mL

The progress of the isomerization may be followed by TLC analysis.

**INFORMATION.** *The TLC analysis is carried out using Eastman Kodak silica gel-polyethylene terephthalate plates with a fluorescent indicator. **Activate the plates at an oven temperature of 100 °C for 30 min** and then place them in a* *desiccator to cool until needed. After spotting, elute the plates using a 1:1 mixture of hexane–methylene chloride as the solvent. Visualize the spots with a UV lamp. The course of the reaction is followed by removing small samples (2–3 drops) of*

HOT

*solution from the test tube [HOT] at set time intervals with a Pasteur pipet and placing them in separate 1/2-dram vials. See Technique 6A for the method of TLC analysis and the determination of $R_f$ values.*

*Approximate $R_f$ values: trans = 0.3; cis = 0.7.*

### Isolation of Product

HOT    Remove the test tube from the light source [**HOT**] and allow the solution to cool to room temperature.

Place the resulting mixture in an ice bath to complete crystallization of the *colorless cis*-1,2-dibenzoylethylene product. Collect the crystals by vacuum filtration using a Hirsch funnel, wash them with 0.5 mL of cold 95% ethanol and then air-dry them on a porous clay plate or on filter paper.

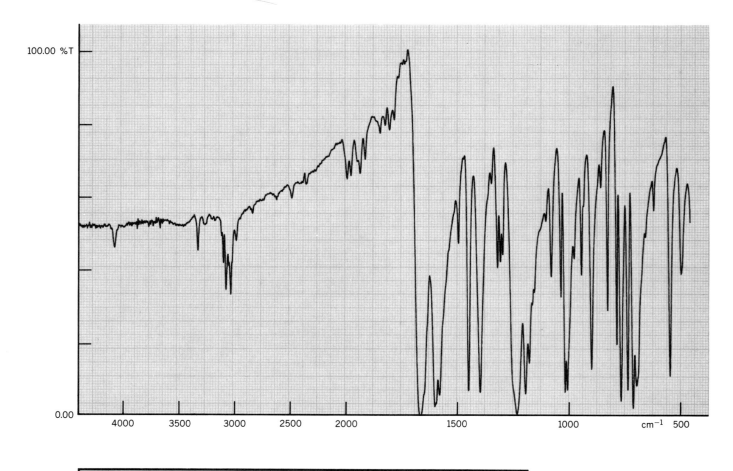

**FIGURE 6.11**   IR spectrum: *cis*-dibenzoylethylene.

## Purification and Characterization

Weigh the dried product and calculate the percent yield. Determine the melting point (evacuated) and compare your result with the literature value. The purity of the crude isolated product may be further determined by TLC analysis (if not used above). Finally, if necessary, further purify a portion of the isolated product by recrystallization from 95% ethanol using a Craig tube.

Obtain IR spectra (KBr pellet technique) of the cis and trans isomers and compare them to Figures 6.11 and 6.12. Alternatively, or in addition to the IR analysis, the UV–vis spectra may be observed in methanol solution and the results compared to Figures 6.13 and 6.14.

## Ultraviolet–Visible Analysis

The bright yellow color of the *trans*-dibenzoylethylene rapidly fades as the conversion to the colorless cis compound progresses under irradiation. This visual observation may be supported by an examination of the absorp-

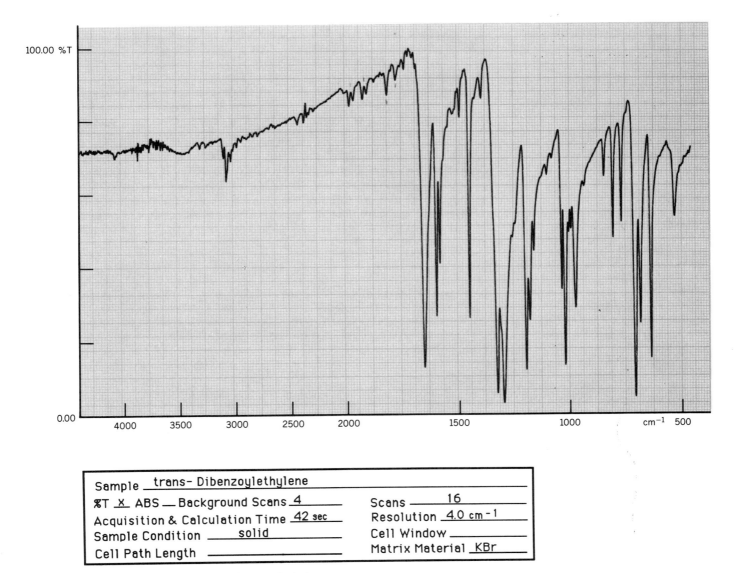

**FIGURE 6.12**    IR spectrum: *trans*-dibenzoylethylene.

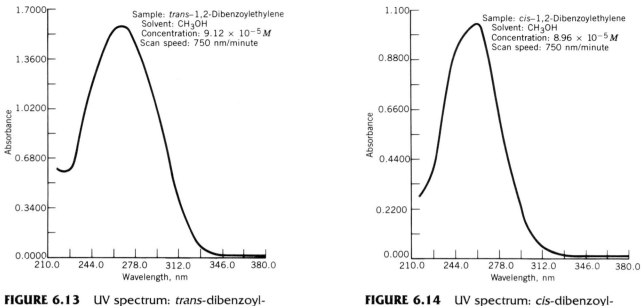

**FIGURE 6.13** UV spectrum: *trans*-dibenzoyl-ethylene.

**FIGURE 6.14** UV spectrum: *cis*-dibenzoyl-ethylene.

tion spectra of the isomers in a methanol solution from 225 to 400 nm (see Figs. 6.13 and 6.14). The $\lambda_{max}$ of the trans isomer drops from 268 to 259 nm in the cis compound. This shift to shorter wavelengths is just enough to move the long-wavelength end of the absorption band in the trans isomer out of the visible region and into the near-ultraviolet (thus, the cis compound does not absorb light to which the eye is sensitive and the compound appears colorless). This observation is consistent with the theory that the spatial contraction of extended $\pi$ systems moves the associated electronic transitions to higher energy gaps (higher frequencies or shorter wavelengths) which is exactly what takes place during the trans–cis isomerization.

### Infrared Analysis

The infrared spectral changes are consistent with the proposed reaction product. Consider the spectrum of the trans starting material (Fig. 6.12). It contains the following.

    **1.** The macro group frequency train (see Strategies for Interpreting Infrared Spectra, p. xx) for conjugated aromatic ketones: 3080 and 3030 (C—H, aromatic), 1652 (doubly conjugated carbonyl), 1599 and 1581 ($\nu_{8a}$, $\nu_{8b}$ degenerate ring stretch, strong intensity of 1581-wavenumber peak confirms ring conjugation), 1495 and 1450 ($\nu_{19a}$, $\nu_{19b}$ degenerate ring stretch, $\nu_{19a}$ weak) cm$^{-1}$.

    **2.** The monosubstituted phenyl ring macro group frequency train: 1980(d), 1920(d), 1820, 1780 ( mono combination-band pattern), 708 (C—H out-of-plane bend, C=O conjugated), 686 (ring puckering) cm$^{-1}$.

    **3.** The presence of the trans double bond is indicated by the single medium intensity 970-cm$^{-1}$ band, as the C—H stretching region is overlapped by the aromatic ring C—H stretches.

    The spectrum of the cis photo product (Fig. 6.11) possesses the same macro group frequencies as the starting material.

    **1.** The conjugated aromatic ketone frequency train is as follows: 3335 (overtone of C=O stretch), 3080 and 3040, 1667, 1601, 1581, 1498 (weak), and 1450 cm$^{-1}$.

**2.** The monosubstituted phenyl ring macro frequency train is assigned the following peaks: 1990, 1920, 1830, 1795, 710, and 695 cm$^{-1}$.

**3.** The cis double bond is clearly present and utilizes the following macro frequency train: 3030 (overlapped by aromatic C—H stretch), 1650 (C=C stretch, shoulder on the low wavenumber side of conjugated carbonyl; resolved in spectra run on thin samples), 1403 (=C—H out-of-phase, in-plane bending mode, strong band not present in trans compound), 970 (trans, in-phase, out-of-plane bend missing), 820 (cis, in-phase out-of-plane bend; band not found in spectrum of trans isomer) cm$^{-1}$.

Discuss the similarities and differences of the experimentally derived spectral data to the reference spectra (Figs. 6.11 and 6.12).

## Optional Isolation of the Thermodynamically *Most* Stable Reaction Product Under the Conditions Used to Carry Out the Above Reaction

If the photoreaction exposure is continued, the cis isomer, which is formed quickly, slowly undergoes conversion to a more stable, new product, ethyl 4-phenyl-4-phenoxy-3-butenoate. This conversion of the intermediate cis isomer may be followed conveniently by thin-layer chromatography. Maximum yields are obtained over approximately 24 h under the above reaction conditions. The product has R$_f$ = 0.8 in 1:1 hexane/methylene chloride. Evaporation of the solvent yields a yellow oil.

The infrared spectrum of this new material (Fig. 6.15) indicates that significant changes have occurred in the structure of the material. The proposed structure involves a rather spectacular molecular rearrangement of the cis-unsaturated ketone to yield an ethyl ester containing a phenoxy-substituted double bond. Can you rationalize the data to fit this structure? Suggest possible macro group frequencies that are present in the IR spectrum of the rearranged product.

## Mixture Melting Point Measurements

It is interesting to observe a series of evacuated mixture melting points (see Chapter 4) with isomer ratios of 75:25; 50:50; 25:75. These values will allow you to *estimate* the eutectic temperature of this system. This same technique is used in Experiment [29A] to aid in identifying the isolated product, 2,5-dichloronitrobenzene, which has a melting point only 3 °C higher than 1,4-dichlorobenzene, the starting material in this nitration reaction.

## Experiment [6C]   Isomerization of an Alkene: Nuclear Magnetic Resonance Analysis

Estimated time of experiment: 0.5 hour.

**EXPERIMENTAL PROCEDURE**

### Physical Properties of Reactants

| Compound | MW | Amount | mmol | mp(°C) | bp(°C) |
|---|---|---|---|---|---|
| *trans*-1,2-Dibenzoylethylene | 236.27 | 5 mg | 0.02 | 111 | |
| Chloroform-*d* | 120.39 | 500 μL | | | 62 |

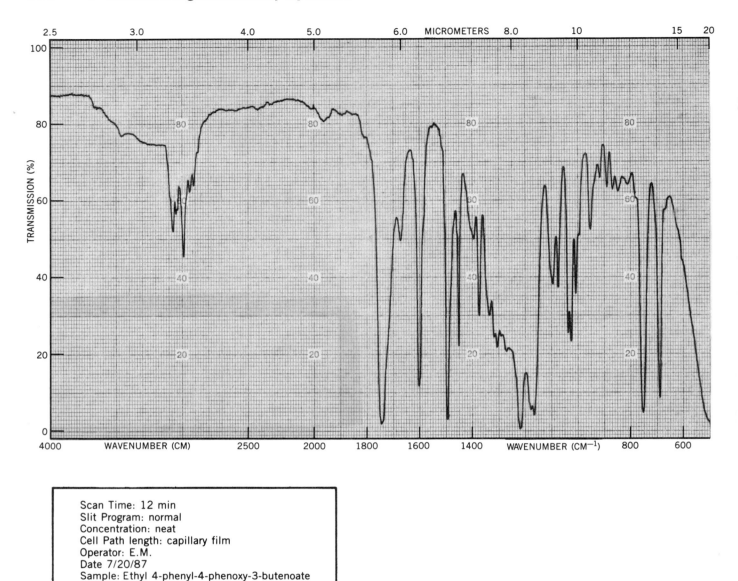

Scan Time: 12 min
Slit Program: normal
Concentration: neat
Cell Path length: capillary film
Operator: E.M.
Date 7/20/87
Sample: Ethyl 4-phenyl-4-phenoxy-3-butenoate

**FIGURE 6.15**   Ethyl 4-phenyl-4-phenoxy-3-butenoate.

### Reagents and Equipment

Prepare a sample of 4–5 mg of recrystallized *trans*-1,2-dibenzoylethylene in about 500 μL of $CDCl_3$ in an NMR tube. Using a fine capillary, spot a silica gel TLC plate (see Experiment [6B]) with a sample of this solution. Use TLC to track the results of the NMR experiment.

Obtain an NMR spectrum of this solution.

### Reaction Conditions

**NOTE.** *A 15-min exposure is normally adequate but, as can be seen by TLC analysis, complete isomerization of this alkene may not occur in this short a time.*

Clamp the NMR tube 3–4 in. (see comments above, Experiment [6A]) from a 275-W sun lamp for 15–20 minutes (see Experiment [6B]).

## Analysis of the Results

After irradiation, spot the original TLC plate with this irradiated solution and elute the plate with methylene chloride/hexane (1:1) solvent (see Experiment [6B]). Obtain an $^1H$ NMR spectrum of the irradiated solution.

Does the TLC analysis correlate with the NMR data? Is there evidence that the isomerization did occur?

The data in the following table were taken on a high-field (300 MHz) NMR instrument. Which signal in each isomer comes from the hydrogen atoms attached to the C=C group?

### $^1H$ Chemical Shift Data

| trans Isomer (ppm) | cis Isomer (ppm) |
|---|---|
| 8.06 doublet | 7.9 doublet |
| 8.01 singlet | 7.55 triplet |
| 7.63 triplet | 7.43 triplet |
| 7.52 triplet | 7.14 singlet |

QUESTIONS

**6-36.** What properties should an ideal recrystallization solvent have?

**6-37.** What is meant by the term *solvent pair* when it is employed in reference to the recrystallization of solids?

**6-38.** What is the purpose of adding powdered charcoal to the solvent system during a recrystallization sequence?

**6-39.** Why do we obtain larger crystals if a solution of an organic compound is allowed to cool slowly?

**6-40.** List several advantages of using the Craig tube to purify crystalline products in contrast to using the Hirsch funnel filtration method.

**6-41.** The stereochemistry of the more highly substituted alkenes is difficult to define using the cis and trans designations. Therefore, a more systematic manner of indicating stereochemistry in these systems has been developed that uses an *E* and *Z* nomenclature. Draw the structures of the *E* and *Z* stereoisomers of 1,4-diphenyl-2-butene-1,4-dione used in this experiment. In this case, which is cis and which is trans?

**6-42.** The cis H—C=C—H in-phase out-of-plane bending frequency normally is found in the range 740–680 cm$^{-1}$ and is broad. In addition, this vibrational mode can be quite variable in intensity. On conjugation with C=O groups, the band rises to near 820 cm$^{-1}$.
   a. Can you explain the underlying cause of this observation? (*Hint*: See the discussion of cis double-bond group frequencies in Chapter 9.)
   b. Can you explain why a broad band near 690 cm$^{-1}$ assigned to this out-of-plane mode is found in the cis product obtained in this reaction?

**6-43.** In *trans*-1,2-dibenzoylethylene, even though the conjugated C=O group frequency coincides directly with the expected C=C stretching frequency, we would not have expected to observe the latter stretching mode in the infrared spectrum. Why?

**6-44.** Explain how this photochemical isomerization allows the production of the thermodynamically *less* stable cis isomer. In other words, why is the trans isomer exclusively converted to the cis isomer during short reaction periods and not vice versa? Is it possible, under these conditions, that the trans and cis isomers are in equilibrium with one another?

| | |
|---|---|
| **BIBLIOGRAPHY** | The photochemical isomerization was adapted from the following references: |

Pasto, D. J.; Ducan, J. A.; Silversmith, E. F. *J. Chem. Educ.* **1974**, *51*, 277.

Silversmith, E. F.; Dunsun, F. C. *J. Chem. Educ.* **1973**, *50* , 568.

Reviews on photochemical isomerization reactions may be found in the following references.

Coyle, J. D. *Introduction to Organic Photochemistry*; Wiley: New York, 1986.

Crombie, L. *Q. Rev.* **1952**, *6*, 101.

DeMayo, P. *Adv. Org. Chem.*,**1960**, *2*, 367.

Fonken, G. L. In *Organic Photochemistry*; O. L. Chapman, Ed.; Marcel Dekker: New York, 1967; Vol. I, p. 197.

Saltiel, J.; D'Agostine, J.; Megarity, E. D.; Metts, L.; Neuberger, K. R.; Wrighton, M.; Zafiriou, O. C. In *Organic Photochemistry*; O. L. Chapman, Ed.; Marcel Dekker: New York, 1973; Vol. III, p. 1.

| | |
|---|---|
| **EXPERIMENT [7]** | ## The Cannizzaro Reaction with 4-Chlorobenzaldehyde: 4-Chlorobenzoic Acid and 4-Chlorobenzyl Alcohol[3] |

Common name: 4-Chlorobenzoic acid
CA number: [74-11-3]
CA name as indexed: 4-Chlorobenzoic acid

Common name: 4-Chlorobenzyl alcohol
CA number: [873-76-7]
CA name as indexed: Benzenemethanol, 4-chloro-

**Stanislao Cannizzaro** (1826–1910): This famous reaction is named for Stanislao Cannizzaro. He was born in Palermo, Italy, where he became interested in chemistry at an early age and studied with Professor Piria at the University of Pisa. He eventually arrived in Paris following participation in the 1847 Sicilian uprising. By 1851 he had moved to Egypt where he became Professor of Chemistry at the National School of Alexandria. It was here that he discovered the reaction that carries his name. In the process he also was the first person to identify benzyl alcohol, which he obtained from the reaction of potash (KOH) with benzaldehyde. It is interesting to note that benzyl alcohol was the first alcohol of the aromatic series to be isolated and characterized. Cannizzaro devoted extensive work to the study of aromatic alcohols and he was the first to propose the term *hydroxyl* for the —OH group. Later he demonstrated the conversion of aromatic alcohols of the benzyl class into their corresponding halides and further to their

[3] Portions of this experiment were previously published. Mayo, D. W.; Butcher, S. S.; Pike, R. M.; Foote, C. M.; Hotham, J. R.; Page, D. S. *J. Chem. Educ.* **1985**, *62*, 149.

phenyl acetic acid derivatives. In 1855 he moved to Genoa and in 1861 he moved back to his birthplace to become Professor of Chemistry at Palermo. Finally, 10 years later at the young age of 45, he assumed the chemistry chair at the University of Rome, which he held for 39 years! Upon his move to Rome his old political interests surfaced, as he became a member of the Italian Senate that same year. He later carried the honor of serving as vice president of that august body. Cannizzaro was also honored by the Royal Society of London in 1891 when he was awarded the Copley Medal for his investigations of atomic and molecular weights. He was quick to recognize the importance of Avogadro's hypothesis and his studies played a key role in placing this hypothesis on a sound scientific basis, which ultimately was the spark that ushered chemistry into the modern era.[4]

## Purpose

This experiment illustrates the simultaneous oxidation and reduction of an aromatic aldehyde to form the corresponding benzoic acid and benzyl alcohol. The experiment further demonstrates the techniques for separation of a carboxylic acid from a neutral alcohol. For a detailed discussion of this extraction procedure, see Experiment [4C].

### Prior Reading

Technique 4:   Solvent Extraction
                     Liquid–Liquid extraction (see pp. 80–82).
                     Separation of acids and bases (see pp. 87–88).
Technique 5:   Crystallization.
                     Use of the Hirsch funnel (see pp. 93–95).
                     Craig Tube Crystallization (see pp. 95–96).
Chapter 9:   Infrared Spectroscopy (see pp. 613–614, 619–620).

---

**REACTION**

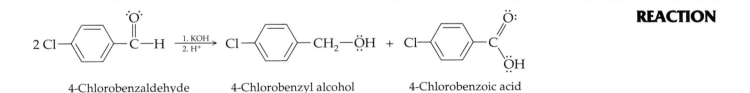

4-Chlorobenzaldehyde          4-Chlorobenzyl alcohol          4-Chlorobenzoic acid

---

**DISCUSSION**

The carbonyl group of an aldehyde represents the intermediate stage of oxidation (ox) [or reduction (red)] between an alcohol and a carboxylic acid.

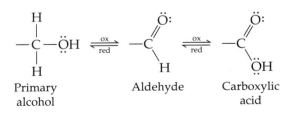

Primary                 Aldehyde              Carboxylic
alcohol                                                 acid

[4] See Newell, L. C. *J. Chem. Educ.* **1926**, *3*, 1361; Parravano, N. *ibid.*, **1927**, *4*, 836; Tilden, W. A. *J. Chem. Soc.* **1912**, 1677; *Dictionary of Scientific Biography* Gillespie, C. C., Ed.; Scribner's: New York, 1971, Vol. III, p. 45; Surrey, A. R. *Name Reactions in Organic Chemistry*; Academic Press: New York, NY, 1954; p. 27.

It is not surprising then, to find a reaction in which a specific type of aldehyde will undergo an oxidation–reduction sequence to form the corresponding alcohol and carboxylic acid. Such a reaction is the **Cannizzaro reaction**. In the presence of hydroxide ion, aldehydes that lack (acidic) *alpha (α)-hydrogen* atoms undergo a self-oxidation–reduction reaction. Thus, under the influence of strong base, one molecule of the aldehyde reduces a second molecule of aldehyde to the primary alcohol, and, in the process, is itself oxidized to the corresponding carboxylate anion. Aldehydes with α-hydrogen atoms, however, do not undergo this reaction because in the presence of base these acidic hydrogens are deprotonated. The resulting enolate generally leads to an aldol reaction (see Experiments [20] and [A3a]).

The first step in the mechanism is the nucleophilic attack of the hydroxide anion on the carbonyl group of the aldehyde. This attack is followed by the key step in the reaction, the transfer of a hydrogen atom with its pair of electrons (a hydride ion) to the carbonyl group of a second molecule of aldehyde. This sequence is shown here.

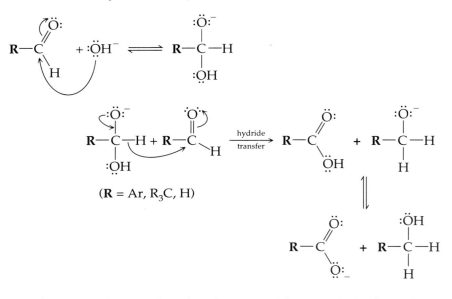

$(\mathbf{R} = \mathrm{Ar}, \mathrm{R_3C}, \mathrm{H})$

The strong electron-donating character of the negatively charged oxygen atom in the first intermediate anion greatly facilitates the ability of the aldehydic hydrogen to be transferred as a hydride ion, that is, *with* its pair of electrons (:H$^-$), to a second molecule of aldehyde. As seen in the preceding diagram, this nucleophilic attack on the carbon of a carbonyl group leads to the formation of a carboxylic acid and an alkoxide anion. The equilibrium established in the final step lies far to the right, and involves a fast acid–base reaction that yields both an alcohol and a carboxylate anion. *Thus, even though the Cannizzaro reaction is an equilibrium reaction, it proceeds nearly to completion.* The proposed mechanism is supported by evidence obtained by carrying out the reaction in D$_2$O. It was found that under these conditions the product alcohol did not contain any α-deuterium substitution. This result suggests that the transferred hydride ion must come from the aldehyde group and not from the solvent.

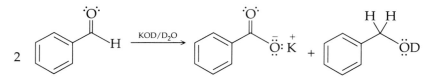

Since few aldehydes lack α-hydrogen atoms, the Cannizzaro reaction is of limited use in modern synthetic sequences, although a variation called the *crossed* Cannizzaro is occasionally employed for the reduction of aromatic aldehydes. In this latter reaction excess formaldehyde is mixed with the aromatic aldehyde and becomes preferentially activated by base since it is present in excess. Thus, the aldehyde present to the lesser extent is reduced to the primary alcohol in high yield. For example:

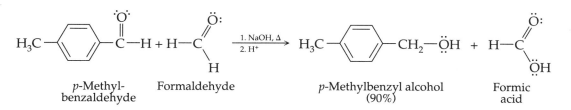

| *p*-Methyl-benzaldehyde | Formaldehyde | *p*-Methylbenzyl alcohol (90%) | Formic acid |

In addition, it is known that α-ketoaldehydes undergo an internal Cannizzaro reaction to yield α-hydroxy carboxylic acids.

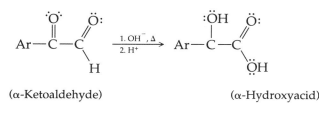

(α-Ketoaldehyde)                    (α-Hydroxyacid)

---

Estimated time to complete the experiment: two laboratory periods.

**NOTE.** *This reaction may be carried out in a centrifuge tube containing a boiling stone. The tube should be covered by a piece of filter paper held in place by a rubber band.*

### Physical Properties of Reactants

| Compound | MW | Amount | mmol | mp(°C) | bp(°C) | $n_D$ |
|---|---|---|---|---|---|---|
| 4-Chlorobenzaldehyde | 140.57 | 150 mg | 1.1 | 47.5 | | |
| Methanol | 32.04 | 400 μL | | | 65 | 1.3288 |
| Potassium hydroxide (11 *M* ) | 56.11 | 400 μL | | | | |

### Reagents and Equipment

Weigh and add to a 5.0-mL conical vial containing a magnetic spin vane and equipped with a reflux condenser, 150 mg (1.1 mmol) of 4-chlorobenzaldehyde followed by 0.4 mL of methanol. (■) With gentle swirling, add 0.4 mL of an 11 *M* aqueous solution of potassium hydroxide.

**NOTE.** *It is convenient to dispense the methanol and KOH solution using automatic delivery pipets. The glass equipment should NOT be rinsed or cleaned with acetone followed by air-drying. The residual acetone undergoes an aldol reaction that forms high-boiling contaminants, which interfere with the isolation of the desired products.*

## EXPERIMENTAL PROCEDURE

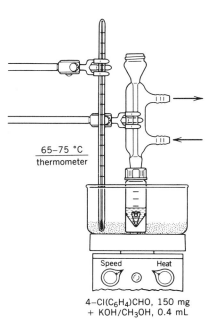

65–75 °C thermometer

4–Cl(C₆H₄)CHO, 150 mg + KOH/CH₃OH, 0.4 mL

> CAUTION: The concentrated methanolic KOH solution is very caustic. Do not allow it to come into contact with the skin or eyes.

### Reaction Conditions

Stir the reaction mixture in a sand bath at 65–75 °C for 1 hour.

### Isolation and Purification

Cool the reaction mixture to room temperature, and add 2.0 mL of chilled distilled water. Extract the resulting solution with three 0.5-mL portions of methylene chloride, using a Pasteur filter pipet to transfer the extracts to a 3.0-mL conical vial. On each addition of methylene chloride, cap the vial, shake gently, and then carefully vent by loosening the cap. A Vortex mixer, if available, is convenient for this extraction step. After separation of the layers, remove the lower methylene chloride layer using a Pasteur filter pipet.

**IMPORTANT.** *Save both the alkaline and methylene chloride phases for further workup.*

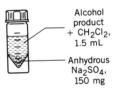

Alcohol product + CH$_2$Cl$_2$, 1.5 mL

Anhydrous Na$_2$SO$_4$, 150 mg

**1.** *4-Chlorobenzyl alcohol.* Wash the combined methylene chloride extracts with two 0.25-mL portions of saturated sodium bicarbonate solution followed by one 0.5-mL portion of distilled water. Remove the aqueous upper phase (Pasteur filter pipet) used in each washing step and save this combined material in a separate 10-mL Erlenmeyer flask [*this material will be discarded at the end of the experiment (see: Rule 10 for the Microscale Laboratory, Chapter 1, p. 4)*]. Dry the methylene chloride layer over 150 mg of granular anhydrous sodium sulfate. (■)

After drying the solution, transfer it by use of a Pasteur filter pipet to a tared 3.0-mL conical vial. Be particularly careful not to let any granules of the hydrated sodium sulfate adhere to the surface of the pipet and become transferred with the dried organic phase. Rinse the sodium sulfate drying agent with 0.3 mL of fresh methylene chloride, and combine the rinse with the dried organic phase. Evaporate the methylene chloride using a stream of dry nitrogen gas in a warm sand bath in the **hood.** If formation of the alcohol took place during the reaction, crude 4-chlorobenzyl alcohol will remain in the vial following removal of the solvent.

HOOD

To purify the crude alcohol, recrystallize the residue using a solution of 4% acetone in hexane (0.25 mL). Collect the product under reduced pressure using a Hirsch funnel, and wash the filter cake (*product crystals packed on the funnel are often referred to as filter cake*) with 0.2 mL of ice cold hexane to give the desired 4-chlorobenzyl alcohol. Air-dry the product on a porous clay plate or on filter paper.

Weigh the 4-chlorobenzyl alcohol and calculate the percent yield. Determine the melting point and compare your value with that found in the literature. Obtain the IR spectrum and compare it with that in Figure 6.16.

*This point in the procedure is a logical place to divide the experiment into two laboratory sessions, if this seems appropriate. A second melting point may be obtained at the beginning of the next laboratory period after the sample of aromatic alcohol has had additional time to air-dry. A convenient way to let the sample dry for several days is to transfer the crystals from the porous plate and spread them*

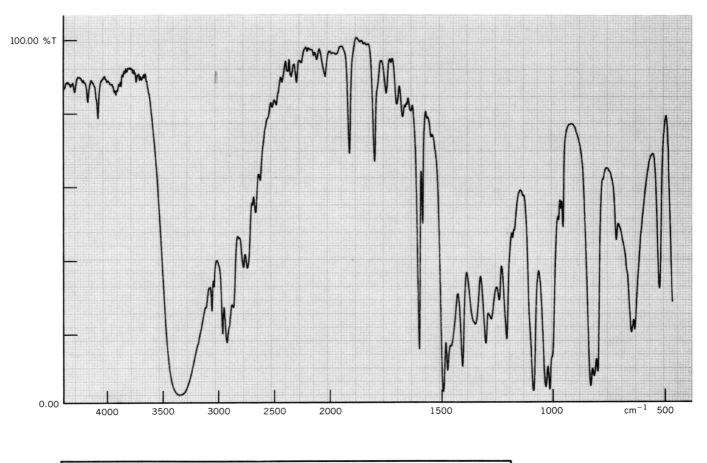

| | |
|---|---|
| Sample __4-Chlorobenzyl alcohol__ | |
| %T _X_ ABS __ Background Scans _4_ | Scans _____16_____ |
| Acquisition & Calculation Time _42 sec_ | Resolution _4.0 cm⁻¹_ |
| Sample Condition __solid – melt__ | Cell Window ____KBr____ |
| Cell Path Length __capillary film__ | Matrix Material _____ |

**FIGURE 6.16**   IR spectrum: 4-chlorobenzyl alcohol.

*over the bottom of a small Erlenmeyer flask (either 5 or 10 mL). Then, cover the
mouth of the flask with filter paper held in place by a rubber band. This procedure
will allow the last traces of moisture and solvent to evaporate, and at the same time
protect the sample from dirt and dust particles. This procedure is a good technique
for drying most crystalline samples. If moisture is a particularly difficult problem,
the sample stored in this fashion can be placed in a desiccator in the presence of a
hygroscopic material, such as calcium chloride.*

**2.** *4-Chlorobenzoic acid.* Dilute the alkaline phase obtained and saved
during the original extraction procedure, by adding 2.0 mL of water. The
dilute aqueous phase is then acidified by the addition of 0.4 mL of concentrated hydrochloric acid. Collect the voluminous white precipitate of the
product that forms on addition of the acid, under reduced pressure by use
of a Hirsch funnel. Rinse the filter cake with 2.0 mL of distilled water. (■)

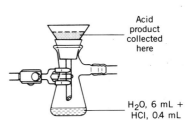

Air-dry the solid product on a porous clay plate or on filter paper to
obtain the crude 4-chlorobenzoic acid product. If a white precipitate is not
obtained upon acidification, add a small amount of a saturated sodium
chloride solution [salting out technique (see Technique 4, p. 88)] to aid the
process.

To purify the crude acid, recrystallize with methanol in a Craig tube. Weigh the dried material and calculate the percent yield. Determine the melting point and compare your value to that found in the literature. Obtain the IR spectrum in a potassium bromide disk and compare your spectrum with that in Figure 6.17.

### Infrared Analysis

The molecule is 4-chlorobenzaldehyde. The infrared spectrum of this aromatic aldehyde (Fig. 6.18) is rich and interesting. The aromatic aldehyde macro group frequency train (see: Strategies for Interpreting Infrared Spectra) consists of peaks near 3090, 3070, 2830 and 2750, 1706, 1602, 1589, and 1396 cm$^{-1}$.

    **a.** 3070 cm$^{-1}$: A C—H stretch on $sp^2$ carbon.

    **b.** 2840 and 2740 cm$^{-1}$: This pair of bands is a famous example of powerful

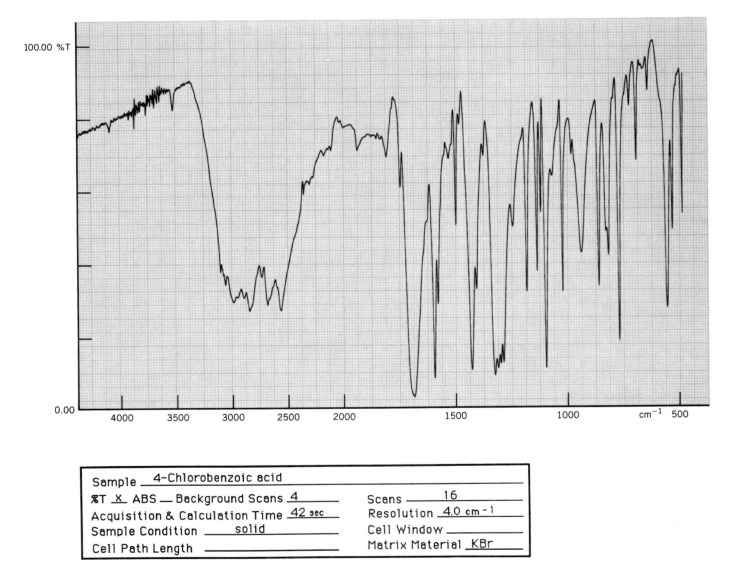

**FIGURE 6.17** IR spectrum: 4-chlorobenzoic acid.

Fermi coupling (see Chapter 9, IR discussions). The unperturbed C—H stretching of the aldehyde C—H group would be expected to occur near 2790 cm$^{-1}$ (the C—H stretch on an $sp^2$ carbon would have been expected to be found at higher values, but the oxygen system in this case significantly lowers the observed values). The in-plane bending mode of this C—H group occurs at 1390 cm$^{-1}$. Thus, the first harmonic should fall near 2780 cm$^{-1}$, very close (~10 cm$^{-1}$) to the stretching frequency of this oscillator. The essential conditions are, therefore, satisfied and strong Fermi coupling occurs and gives rise to the split peaks at 2840 and 2740 cm$^{-1}$. The latter band is moved well away from the normal $sp^3$ C—H symmetric stretching modes. Thus, the lower wavenumber component of the Fermi-coupled aldehyde C—H mode leads to easy identification of an aldehyde, even when it represents a very small fraction of an aliphatic system. In the present case where no aliphatic C—H oscillators are present, both components are observed.

   **c.** 1704 cm$^{-1}$: The carbonyl stretch of the aldehyde group. The frequency observed in aliphatic aldehydes falls in the range 1735–1720 cm$^{-1}$, but when conjugated, the value drops 15–25 cm$^{-1}$ (see Chapter 9, IR discussions) and is found in the range 1720–1700 cm$^{-1}$.

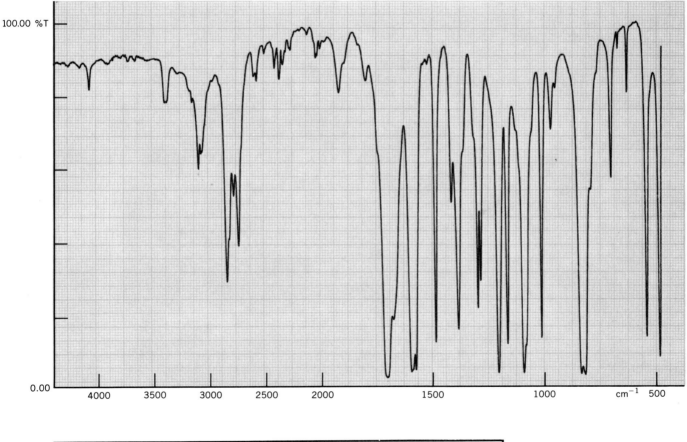

**FIGURE 6.18**   IR spectrum: 4-chlorobenzaldehyde.

**d.** 1601 and 1578 cm$^{-1}$: This pair of bands is related to the degenerate ring stretching vibrations, $\nu_{8a}$ and $\nu_{8b}$, of benzene (also see Infrared discussions in Experiment [1B$_{adv}$]).

**e.** 1390 cm$^{-1}$: The aldehyde C—H in-plane bending vibration. The first harmonic of this vibration is Fermi coupled to the aldehyde C—H stretching mode, as discussed earlier.

This compound also possesses a second powerful macro group frequency train that assesses the substitution pattern of the aromatic ring system (see Chapter 9, IR discussions). The para-disubstituted benzene ring macro group frequency train requires peaks in the following regions: 1950, 1880, 1800, 1730, 750, and 690 cm$^{-1}$.

**a.** 1905, 1795 cm$^{-1}$: This pair of two weak bands, with the higher wavenumber band more intense than the lower member, arises from combination bands (see Chapter 9, IR discussions) which involve the out-of-plane bending frequencies of the ring C—H bonds (see below). The exact wavenumber positions are not very important, but the overall shape of the pattern can be used to determine the ring substitution pattern (see Chapter 9, IR discussions).

**b.** 832 cm$^{-1}$: This strong band is very characteristic of para disubstituted benzene rings. The 832-cm$^{-1}$ peak arises from the in-phase out-of-plane bending vibration of the two pairs of C—H groups on opposite sides of the six membered ring (see Chapter 9, IR discussions).

The C—Cl stretching mode when substituted on an aromatic ring often becomes heavily coupled with ring vibrations, and as in this case, does not give rise to an identifiable group frequency. The presence of this group must be determined by other methods, such as a Beilstein or sodium fusion test (see Chapter 10, pp. 696–702). The presence of a hidden group (very likely halogen), however, is strongly indicated. Because the system must be para substituted (see above paragraphs), one of the substituents on the ring must exhibit very little identifiable absorption from 4000 to 500 cm$^{-1}$.

The reaction involves the formation of two products, one neutral and one acidic, both of which incorporate large portions of the substrate molecule. These materials can be characterized by their IR spectra. The neutral product is proposed to be 4-chlorobenzyl alcohol. An examination of the spectrum (Fig. 6.16) supports the presence of two macro group frequency trains. One (a) is the same one found in the starting aldehyde, the para-disubstituted benzene ring frequency train. The other (b) is a primary aliphatic alcohol macro frequency train. Macro (a) has been expanded to include all aromatic ring specific group frequencies, as the alcoholic side chain is decoupled from the ring and the carbon–chlorine frequencies fall outside the range of the instrumentation normally used in these measurements.

The macro group frequencies for 4-chlorobenzyl alcohol are as follows:

**a.** 3055 (aromatic C—H stretch), 1906 and 1793 (para combination band pattern), 1596 and 1583 (ring stretch degenerate pair, $\nu_{8a}$ and $\nu_{8b}$; 1583 intensity indicates weak conjugation with the ring), 1495 and 1475 (ring stretch degenerate pair, $\nu_{19a}$ and $\nu_{19b}$), 834 (C—H, out-of-plane bend) cm$^{-1}$.

**b.** 3340 (broad, O—H stretch), 2965–2925 (C—H, aliphatic $sp^3$), 1450–1300 (broad, O—H bend, associated), 1015 (C—O, stretch, primary alcohol), 630 (broad, weak O—H bend, associated) cm$^{-1}$.

The acidic product generated in the reaction is assumed to be 4-chlorobenzoic acid. This material also is a para-substituted aromatic compound; thus, the spectrum of this compound (Fig. 6.17) possesses a macro frequency train (a), similar to those of the aldehyde and alcohol. In addition, this benzoic acid derivative exhibits an extended aromatic acid macro group frequency train (b). The macro frequencies are as follows:

**a.** 1935 and 1795 (para combination band pattern), 852 (ring C—H, in-phase, out-of-plane bend) cm$^{-1}$.

**b.** 3400–2200 (very broad, very strong, O—H stretch, associated acid), 1683 (C=O, stretch, out-of-phase, associated acid dimer), 1596 and 1578 (degenerate ring stretch), 1500 and 1432 (degenerate ring stretch), 1320–1280 (C—O, stretch), 885 (O—H, out-of-plane, ring dimer bend) cm⁻¹.

The carbon–chlorine stretch is not observed.

Examine the spectra of the reaction products you have obtained, in a potassium bromide matrix. Discuss the similarities and differences of the experimentally derived spectral data to the reference spectra (Figs. 6.16–6.18).

It is a useful exercise to compare the results of organic qualitative analysis reaction tests, which were used historically to classify the preceding compounds, with the data now available from modern spectroscopic instrumentation.

### Chemical Tests

Perform each of the following tests (see Chapter 10). Do the results confirm that you have isolated an aromatic carboxylic acid and an aromatic alcohol?

1. The ignition test.
2. The Beilstein or the sodium fusion test for the presence of halogen.
3. The ceric nitrate and/or the Jones oxidation test for the alcohol.
4. The solubility of the acid in sodium bicarbonate and sodium hydroxide solutions. Is carbon dioxide evolved in the bicarbonate test?

If you were to prepare a derivative for the alcohol and acid products, which one would you choose? See Chapter 10, Preparation of Derivatives.

**QUESTIONS**

**6-45.** The discussion mentions that the crossed Cannizzaro reaction can be realized when one of the components is formaldehyde. Predict the product(s) of the reaction below and give a suitable name to the reactants and products.

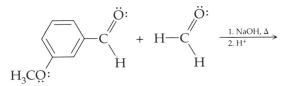

**6-46.** One group of investigators has suggested that a dianion might be the source of hydride in the Cannizzaro reaction. Explain why this species would be a better source of hydride in comparison to the species depicted in the mechanism presented in the discussion section.

**6-47.** The Cannizzaro reaction is an oxidation–reduction sequence. What type of reagent is formaldehyde acting as in Question 6.45?

**6-48.** Propose a mechanism for the internal Cannizzaro reaction depicted below.

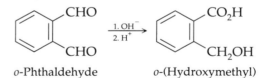

*o*-Phthaldehyde　　　　　*o*-(Hydroxymethyl) benzoic acid

---

**BIBLIOGRAPHY**

Cannizzaro, S. *Annalen* **1853**, *88*, 129.

Reviews on the Cannizzaro reaction:

Geissman, T. A. *Org. React.* **1944**, *2*, 94.

Swain, C. G.; Powell, A. L.; Sheppard, L. A.; Morgan, C. R. *J. Am. Chem. Soc.* **1979**, *101*, 3576.

Examples of the Cannizzaro Reaction:

Cheney, L. C. *J. Am. Chem. Soc.* **1951**, *73*, 685.

Davidson, D.; Weiss, M. *Organic Syntheses*; Wiley: New York, 1943; Collect. Vol. II, p. 590.

Wilson, W. C. *Organic Syntheses*; Wiley: New York, 1941; Collect. Vol. I, p. 256.

---

**EXPERIMENT [8]**

## The Esterification Reaction: Ethyl Laurate, Isopentyl Acetate, and the Use of Acidic Resins

Common names: Ethyl laurate, ethyl dodecanoate
CA number: [106-33-2]
CA name as indexed: Dodecanoic acid, ethyl ester

Common names: isopentyl acetate, isoamyl acetate
CA number: [123-92-2]
CA name as indexed: 1-Butanol, 3-methyl-, acetate

Common name: Butyl acetate
CA number: [123-86-4]
CA name as indexed: Acetic acid, butyl ester

Common names: Pentyl acetate, amyl acetate
CA number: [628-63-7]
CA name as indexed: Acetic acid, pentyl ester

Common name: Hexyl acetate
CA number: [142-92-7]
CA name as indexed: Acetic acid, hexyl ester

Common name: Octyl acetate
CA number: [112-14-1]
CA name as indexed: Acetic acid, octyl ester

***Prior Reading***
   *Technique 4:*   Solvent Extraction
                   Liquid–Liquid Extraction (see pp. 80–82).
                   Drying of the Wet Organic Layer (see pp. 86–87).
   *Technique 6:*   Chromatography
                   Packing the Column (see pp. 99–100).
                   Elution of the Column (see p. 100).

## Purpose

An exploration of the classic reactions of carboxylic acids ($RCO_2H$) with alcohols ($R'OH$), in the presence of acid catalyst, to yield esters ($RCO_2R'$) plus water ($H_2O$, a small stable molecule). To examine the physical properties of these esterification products and to apply the techniques of distillation and column chromatography to the purification of these materials.

---

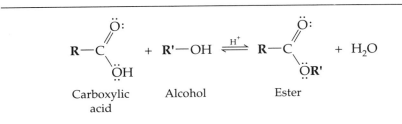

Carboxylic        Alcohol        Ester
acid

**REACTION**

---

## DISCUSSION

Esters are among the most important of the carboxylic acid (and alcohol) derivatives. Substances possessing this functional group are widely distributed in nature in the form of waxes, essential oils, fatty acid esters, and aromas. The ester functionality plays a significant role in biochemistry, both in primary metabolism and in a variety of substances exhibiting remarkable physiological activity in humans (hormones and neurotransmitters). Esters find extensive use in commercial products from fingernail polish remover and artificial sweeteners, to polymeric fibers, plasticizers, and surfactants.

## Biosynthesis of Esters

The biosynthesis of fatty acids (fatty acids are naturally occurring, long, straight-chain, $C_{12}$–$C_{40}$ carboxylic acids; most contain an even number of carbon atoms) provides an important and interesting example of a primary metabolic pathway in which a special type of ester is the essential link between the enzyme and the substrate (acetic acid). In this instance, the enzyme-bound substrate grows by repeated addition of two-carbon ($C_2$) units and, when eventually released from the enzyme, has undergone an extension of the fatty acid hydrocarbon chain.

   The first step in fatty acid biosynthesis involves the formation of a **thiolester**, acetyl coenzyme A (acetyl CoA), from acetic acid (present in the primary metabolic pool) and the thiol group (mercapto, or —SH group) of the **coenzyme** (HSCoA). A thiolester is an ester in which the single-bonded oxygen (from the alcohol component) is replaced by a sulfur atom,

$$\overset{\overset{\displaystyle \cdot\overset{..}{O}\cdot}{\|}}{-C} - \overset{..}{\underset{..}{S}} - \overset{}{C} -$$ , from the coenzyme. Coenzymes are loosely bound, nonprotein factors attached to the enzyme that play an important role in the catalytic function of the enzyme. These coenzymes are distinguished, in an ill-defined manner, from **prosthetic groups**, which are intimately attached to active sites of enzymes. Part of the role of the CoA is to facilitate the transfer of the substrate (the $C_2$ unit) to a new thiol group of the enzyme (protein), where the next stage of the biosynthesis takes place.

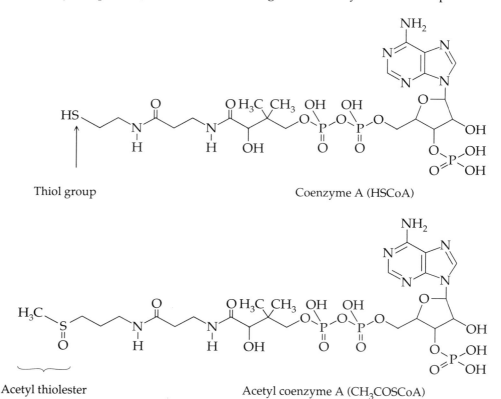

Thiol group               Coenzyme A (HSCoA)

Acetyl thiolester             Acetyl coenzyme A (CH$_3$COSCoA)

This reactive thiolester is capable of undergoing **aldol**-type (see Experiment [20]) condensations under physiological conditions. AcetylCoA is first carboxylated with the help of the enzyme, acetyl CoA-carboxylase, to yield a **thiolmalonyl** derivative. The resulting intermediate possesses an

activated methylene group, $-\overset{\overset{\displaystyle \cdot\overset{..}{O}\cdot}{\|}}{C} - CH_2 - \overset{\overset{\displaystyle \cdot\overset{..}{O}\cdot}{\|}}{C} - \overset{..}{\underset{..}{S}}-$, which with further enzymatic support undergoes a **Claisen** condensation (see Experiment [3A$_{adv}$]) with an acetyl group that has been also transferred via acetylCoA to an appropriate acyl-carrier protein, fatty acid synthetase. Reduction, dehydration, further reduction, and finally hydrolysis of the thiolester, yields the fatty acid extended by a $C_2$ hydrocarbon group.

$$CH_3 - \overset{\overset{\displaystyle \cdot\overset{..}{O}\cdot}{\|}}{C} - \overset{..}{\underset{..}{O}}H + HSCoA \longrightarrow CH_3 - \overset{\overset{\displaystyle \cdot\overset{..}{O}\cdot}{\|}}{C} - SCoA + H\overset{..}{\underset{..}{O}}H$$

CoA                        Acetyl CoA

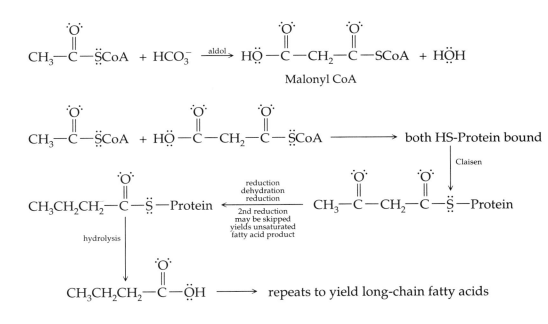

Malonyl CoA

both HS-Protein bound

Claisen

reduction
dehydration
reduction
2nd reduction
may be skipped
yields unsaturated
fatty acid product

hydrolysis

repeats to yield long-chain fatty acids

---

## Oils, Fats, Waxes, and Aromas                                    LIPIDS

Fatty acids derived from primary metabolism play a key role in the formation of naturally occurring oils, fats, and waxes. Fats and oils are esters of
these acids with a triol, glycerol ($HOCH_2CHOHCH_2OH$). The common fats
and oils are formed from mixtures of $C_4$–$C_{26}$ saturated fatty acids with the
vast majority derived from $C_{12}$–$C_{18}$ acids. The oils are more likely to include
significant contributions from mixtures of unsaturated fatty acids.

Since fats and oils are triesters of glycerol, they are generally called
**triglycerides**. In plants and animals, triglycerides function as energy re-
serves that can be used in primary metabolism when food (energy) is not
available to the organism.

Although the fats have high molecular weights, they are generally
found to be very low-melting solids, particularly if they contain unsaturated fatty acids. The bent chains, which result from incorporating cis-
alkene (C=C) groups into the chain, prevent close packing in a solid and,
as a result, such molecules exhibit lower melting points. For example,
compare **oleic acid** ($C_{18}H_{34}O_2$, mp 4 °C) to its saturated analogue, **stearic
acid** ($C_{18}H_{36}O_2$, mp 69–70 °C). The former melts more than 60 °C lower
because of the cis carbon–carbon double bond.

Oleic acid is the simplest of the unsaturated $C_{18}$ fatty acids, as it has a
single C=C group located in the middle of the chain. This unsaturated
fatty acid is the most widely distributed of all fatty acids. It is the dominant
component (76–86%) of the triglycerides in olive oil. Highly saturated fats,
on the other hand, are generally solids at room temperature because the
straight-chain fatty acids pack together well (see Fig. 6.19).

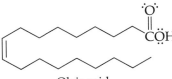

Oleic acid
[(Z)-9-octadecenoic acid]

Stearic acid
[Octadecanoic acid]

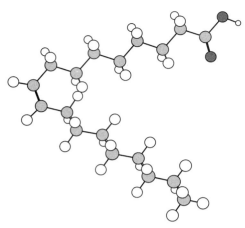

Oleic acid — ball and stick model

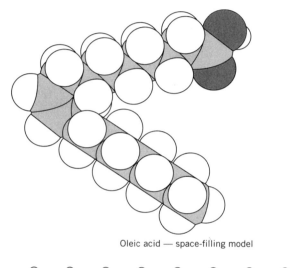

Oleic acid — space-filling model

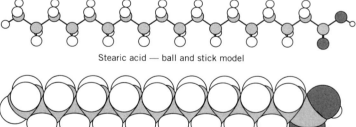

Stearic acid — ball and stick model

**FIGURE 6.19** Molecular models of fatty acids.

Stearic acid — space-filling model

Many vegetable and fish oils are liquid triglycerides. Because these organisms operate at ambient temperatures, evolution dictated that low-melting fats were required to avoid solidification. In warm-blooded animals, higher melting fats can be tolerated and are used.

The cheap and plentiful unsaturated oils can be converted to solid fats by hydrogenation of the alkene groups, which gives straight-chain alkyl groups. As consumers historically have desired to cook with solid, white, and creamy fats (such as lard) derived from animal triglycerides (low in unsaturation), hydrogenation of vegetable oils, such as peanut, soybean, and cottonseed oils, has been carried out on a large scale (this process is referred to as **hardening** the fat).

Unfortunately, the relationship between saturated fats in the human diet and the formation of cholesterol (a **simple lipid**, see below) plaque and coronary heart disease has been established. The dietary switch to less saturated fats is currently underway.

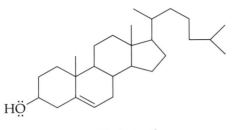

Cholesterol

The triglycerides obtained from animal fats have been used for a very long time as a source of **soap**. When fats are boiled with lye (sodium hydroxide) the ester linkages are cleaved by a process known as **saponification** (the term originates from the Latin word for soap, *sapon*, as does the modern French word for soap, *savon*) to yield the sodium salt of the fatty acid and the esterifying alcohol (glycerol).

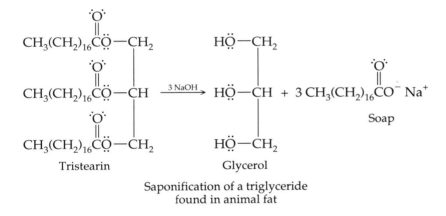

Saponification of a triglyceride
found in animal fat

Salts of fatty acids function effectively as soaps because one end of the straight-chain system has the highly polar carboxylate ion and is readily solvated in water. The rest of the fatty acid molecule has all the characteristics of a nonpolar hydrocarbon and readily dissolves in hydrocarbons, such as greases and oils. We refer to the polar end (head) as being **hydrophilic** (attracted to water) and the hydrocarbon end (tail) as being **lipophilic** (attracted to oils). When dispersed in an aqueous solution, fatty acids tend to form **micelles** (spherical clusters of molecules). The lipophilic ends of the fatty acids occupy the interior of the cluster, while the polar ends, which are heavily solvated by water molecules, form the outer surface of the spherical micelle. Micelles absorb the hydrocarbon chains of the triglycerides, and thus soaps break up and help to dissolve the fats and oils that tend to coat skin, clothes, and the surfaces of eating and cooking utensils (see Figs. 6.20 and 6.21).

**Waxes** are naturally occurring esters of fatty acids (in waxes, chain lengths can reach as high as $C_{36}$) and a variety of other alcohols that often possess relatively complicated structures (steroid alcohols) and/or long chains. For example, *n*-octacosanol, $CH_3(CH_2)_{26}CH_2OH$, has been isolated from the esters in wheat waxes, and a component of carnauba wax (traditionally an automobile wax) has 62 carbons, $CH_3(CH_2)_{33}CO_2(CH_2)_{26}CH_3$.

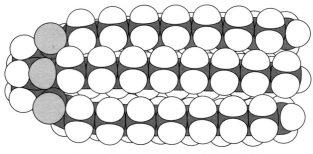

A space-filling model of a saturated triglyceride.

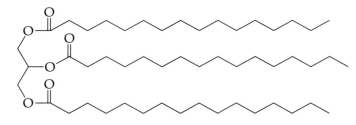

A saturated triglyceride

**FIGURE 6.20**  A space-filling model of a saturated triglyceride.

The biological role of carnauba wax is as a leaf coating involved in the conservation of plant moisture. Animal waxes include cetyl palmitate (spermaceti) found in sperm whales and beeswax [one constituent of which has been identified as $CH_3(CH_2)_{29}CO_2(CH_2)_{29}CH_3$, which is used in the construction of the honeycomb].

Lower molecular weight, naturally occurring esters make major contributions to the pleasant **aromas** of fruits and flowers. These odors have been shown to generally be composed of complex mixtures of materials that have been separated only since the development of modern chemical instrumentation. Single components, however, may play a dominant role in an individual plant or animal. Propyl acetate (pears), ethyl butyrate (pineapples), and 3-methylbutyl acetate (bananas) are examples of simple esters responsible for a particular plant odor. Odors derived from esters are not limited just to esters of straight-chain carboxylic acids, as is demonstrated by oil of wintergreen, **methyl salicylate**.

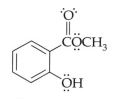

Oil of wintergreen

## Phospholipids

**Lipid** is a term applied to those natural substances that are more soluble in nonpolar solvents than in water. In its most general sense, it is a broad definition that includes fats, waxes, hydrocarbons, and so on. In biochemistry, lipids are more narrowly defined as substances that yield fatty acids upon hydrolysis.

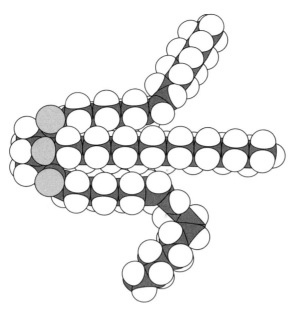

A space-filling model of an unsaturated triglyceride.

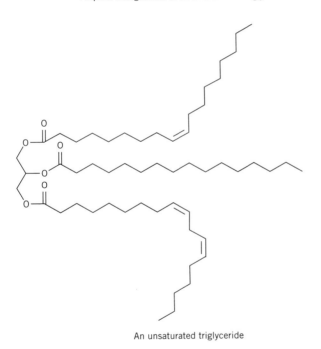

An unsaturated triglyceride

**FIGURE 6.21**   A space-filling model of an unsaturated triglyceride.

Another class of glycerides are those substances in which one of the fatty acid groups has been replaced by a phosphoric acid residue: the phospholipids, or more accurately, the **phosphoglycerides**. The phosphate group is almost always further esterified, usually with a biological amino alcohol, such as choline (the **lecithins**) or ethanolamine (the **cephalins**).

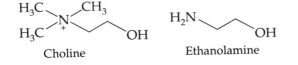

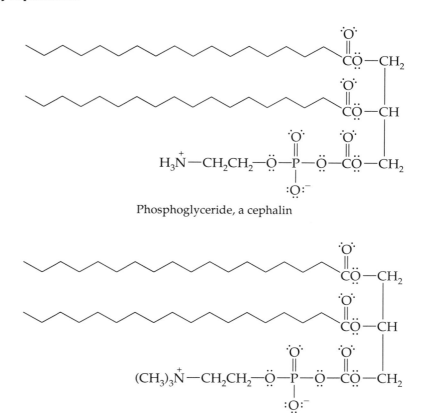

Phosphoglyceride, a cephalin

Phosphoglyceride, a lecithin

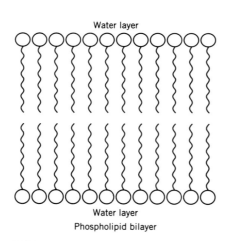

Water layer

Water layer
Phospholipid bilayer

**FIGURE 6.22**   Phospholipid bilayer.

These latter groups significantly increase the polarity of the glycerol section of the molecule so that phosphoglycerides undergo strong self-association. In aqueous solutions, this intermolecular attraction can lead to **lipid bilayer** formation (Fig. 6.22). In a lipid bilayer, the molecules organize themselves to form sheets that contain a double layer of the molecules formed by tail-to-tail association within the interior of the sheet; the outer surface of the lipid bilayer contains the polar heads, which are heavily solvated by water molecules. This association of phosphoglycerides is the key feature in the construction of cell membranes. Thus, esters must have played a vital role at the very earliest stages as cell structures evolved in the development of living systems.

## Preparation of Esters

Esters are generally synthesized by one of four fundamental routes.

1. Esterification of a carboxylic acid with an alcohol in the presence of an acid catalyst.
2. Alcoholysis of acid chlorides, anhydrides, or nitriles.
3. Reaction of a carboxylate salt with an alkyl halide or sulfate.
4. Transesterification reactions.

The first of these pathways, known as Fischer esterification, is the method used for the preparation of ethyl laurate in Experiment [8A] and of isopentyl acetate in Experiment [8B]. A modern variation of the Fischer esterification is used in Experiment [8C]. The development of this esterification reaction represents just one of a number of major discoveries in organic chemistry by Emil Fischer.

Emil Fischer (1852–1919, German).[5] In 1874 Fischer obtained his Ph.D. from the University of Strasbourg, studying with Adolf von Baeyer. He later had appointments as Professor of Chemistry at Erlangen, Würzburg and Berlin Universities.

In 1875, at the age of 23, and 1 year after completing his graduate studies, he synthesized phenyl hydrazine ($C_6H_5$—$NHNH_2$) for the first time. This highly reactive reagent later played a key role in Fischer's work on elucidating structures of a large majority of the sugars (carbohydrates), an entire class of important and complex organic molecules. Sugars, or carbohydrates, represent the prime pathway for the storage of radiant energy from the sun, through photosynthesis, as chemical energy. In the short period from 1891 to 1894, Fischer established not only the basic structures, but also the configurations of all the known sugars. In addition, he predicted all the theoretically possible isomers and, in the process, developed a method of representing the three-dimensional molecular structures in two-dimensional drawings that became known as Fisher projection formulas. These representations are still in use today, and have been widely applied beyond sugar chemistry. This work by Fischer led directly to proving the existence of the asymmetric carbon atom, a concept proposed by Van't Hoff and LeBel in 1874.

Fischer was also active in the area of protein chemistry. He demonstrated that amino acids are the basic subunits from which proteins are constructed. Fischer also devised methods for the synthesis of many of the known amino acids. Perhaps his most ingenious contribution was the "lock and key" hypothesis of how proteins bind with substrates of complementary shapes. This work ultimately led to our understanding of how enzymes, the catalysts of biochemical reactions, function.

Fischer carried out extensive work on the chemistry of purine and on those compounds containing its nucleus. Purine is one of the two nitrogen base ring systems present in DNA. He synthesized approximately 150 members of this class of heterocyclic compounds (including the first synthesis of the alkaloid, **caffeine** (see Experiment [11B]), uric acid, and the **xanthines**, also see Experiment [11B]). He developed a general synthesis of another nitrogen heterocycle, **indole**, which was so effective that it has become one of the classic synthetic methods of organic chemistry and is known today as the "Fischer Indole synthesis."

Indole

Fischer's work essentially laid the foundation of modern biochemistry. Regarded as the greatest organic chemist of his time, Fischer became the second chemist to receive the Nobel Prize (1902). Depressed by the loss of his young wife at the age of 33, by the loss of two of his three sons (one by suicide, the other in World War I), and suffering from the advanced stages of intestinal cancer, and saddened by the socioeconomic conditions of post-war Germany, Fischer committed suicide.

[5] See *Chem. Ind.* **1919**, 42, 269; Darmstaedter, L.; Oester, R. E. *J. Chem. Educ.* **1928**, 5, 37; Ratman, C. V. *ibid.* **1942**, 38, 93; Kauffman, G. B.; Priebe, P. M. *ibid.* **1990**, 67, 93; *Chem. Eng. News* **1992**, June 8th, p. 25. Recommended reading, *The Emil Fischer–William Ramsay Friendship: The Tragedy of Scientists in War*; *J. Chem. Educ.* **1990**, 67, 451.

## Mechanism of the Fischer Esterification Reaction

The Fischer esterification proceeds by nucleophilic attack of the alcohol on the *protonated* carbonyl group of the carboxylic acid to form a tetrahedral intermediate. Collapse of the tetrahedral intermediate regenerates the carbonyl group and produces the ester and water. The overall sequence is outlined here.

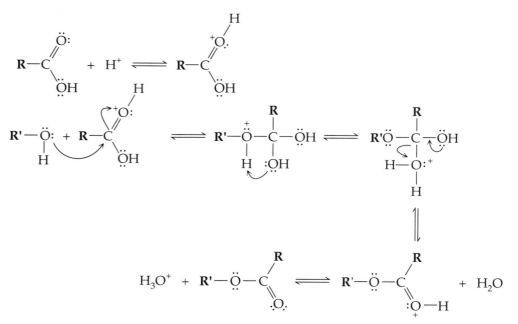

In the Fischer esterification with primary alcohols, the products are only slightly favored by the equilibrium and therefore, in order to obtain substantial yields of the ester, the equilibrium must be shifted toward the products. This result can be accomplished in a number of ways. For example, an excess of the starting alcohol can be used to shift the position of equilibrium towards the products. This technique is used in the preparation of ethyl laurate (Experiment [8A]). An analogous alternative is to use an excess of the carboxylic acid. A third option to drive the reaction is the removal of one or both of the products (the ester or water) as they are formed during the reaction. The preparation of isopentyl acetate, synthesized in Experiment [8B], depends on two of these strategies: (1) the reaction is run in an excess of the carboxylic acid (it doubles as the solvent) and (2) the water generated as one of the products is removed by a drying agent. The acid catalyst employed in Fischer esterifications is generally dry hydrogen chloride, concentrated sulfuric acid, or a strong organic acid, such as *p*-toluenesulfonic acid.

When the carboxyl group and hydroxyl group are present in the *same* molecule, an intramolecular esterification may occur and a cyclic ester (called a lactone) may be formed. Lactonization requires an acceptable conformation (i.e., the two groups must be close and spatially positioned

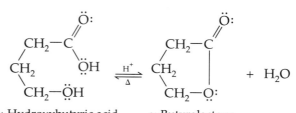

γ-Hydroxybutyric acid        γ-Butyrolactone

to react). Ring closure is especially favorable if lactone formation yields five- or six-membered (stable and rapidly formed) ring systems.

As noted earlier, Fischer esterification is an equilibrium reaction and is thus reversible. Thus, heating an ester in aqueous solution, in the presence of an acid catalyst, regenerates the corresponding carboxylic acid and alcohol. This latter reaction is termed *acid hydrolysis* of an ester. The rate-determining step in both the forward esterification reaction and the reverse reaction, acid hydrolysis, is the formation of the tetrahedral intermediate. It is, therefore, evident that the rate of the reaction will be determined by the ease with which the nucleophile (alcohol on esterification and water on hydrolysis) approaches the carbonyl group. Steric and electronic factors have been shown to have large effects on the rate of esterification. An increase in the number of bulky substituents substituted on the $\alpha$ and $\beta$ positions of the carbonyl-containing compound decreases the rate (steric effects). Electron-withdrawing groups near the carbonyl group, on the other hand, tend to increase the rate, as this increases the electrophilicity (partial positive charge) of the carbonyl carbon (electronic effects). Conversely, electron-donating groups act to retard the rate of esterification (electronic effects).

You are to synthesize the ethyl ester of lauric acid in Experiment [8A]. Lauric acid, $CH_3(CH_2)_{10}CO_2H$ (dodecanoic acid), is one of the four most common fatty acids found in naturally occurring triglycerides. It is named for the laurel botanical family from which it was first isolated in 1842. It is the most abundant of the fatty acids isolated from the vegetable oils of palm kernel oil (52%), the seed fat of *Elaeis guineensis*, of coconut oil (48%), *Cocos nucifera*, and of babassu oil (46%), *Attalea funifera*.

In the preparation given for ethyl laurate (Experiment [8A]), acetyl chloride is used to generate the HCl catalyst in situ as shown below. Notice that the other product of this step is a molecule of the desired ester.

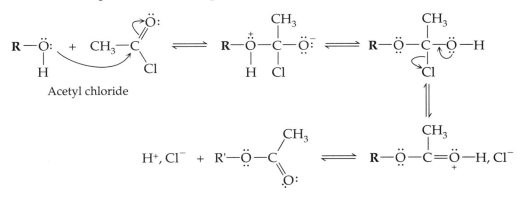

Acetyl chloride

In Experiment [8B] you will be synthesizing the isopentyl alcohol (3-methylbutanol) ester of acetic acid (the basic building block of the fatty acids), isopentyl acetate (isoamyl acetate). This low molecular weight ester has a distinct banana or pear-like odor, and the liquid product is often referred to as banana oil or pear oil (see above). Isopentyl acetate has a wide variety of uses: as a flavoring agent in mineral waters and syrups; a solvent for oil paints, tannins, nitrocellulose, lacquers, and a number of other commercial products; a perfume ingredient in shoe polish; and in the manufacture of artificial silk, leather, and pearls. You are very likely to find this experiment to be a pleasant olfactory experience!

In Experiment [8C] you will use a polymer-bound acid reagent to catalyze the esterification reaction. Polymer-bound reagents are becoming increasingly useful in organic synthesis; both in the research laboratory and in industrial-scale reactions.

## Experiment [8A]   Ethyl Laurate

## REACTION

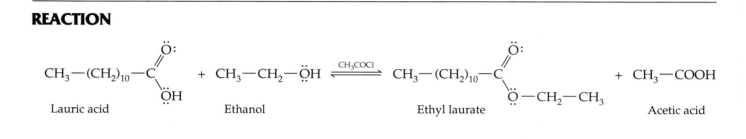

## EXPERIMENTAL PROCEDURE

Estimated time to complete the experiment: 3.0 hours.

### Physical Properties of Reactants

| Compound | MW | Amount | mmol | mp(°C) | bp(°C) | d | $n_D$ |
|----------|-----|--------|------|--------|--------|---|-------|
| Lauric acid | 200.33 | 70 mg | 0.35 | 44 | | | |
| Ethanol | 46.07 | 1.0 mL | 17 | | 78.5 | 0.71 | 1.3611 |
| Acetyl chloride | 78.50 | 30 μL | 0.42 | | 50.9 | 1.11 | 1.3898 |

### Reagents and Equipment

Weigh and add 70 mg (0.35 mmol) of lauric acid to a 3.0-mL conical vial containing a magnetic spin vane. Then equip the reaction vial with a reflux condenser that is protected by a calcium chloride drying tube. (■) Next use a graduated 1.0-mL pipet to add 1.0 mL of absolute ethanol followed by 30 μL (0.43 mmol) of acetyl chloride (automatic delivery pipet **[HOOD]**) to the reaction vial. (Remember not to turn on the cooling water in the condenser until the system is completely assembled. Otherwise, the cold inner surface will condense moisture from the laboratory atmosphere. It takes only a very small amount of water on the condenser walls to completely deactivate your acid chloride catalyst.)

HOOD

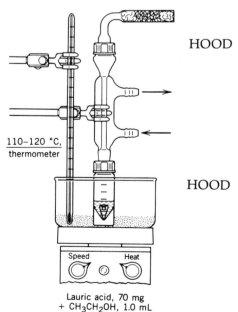

110–120 °C, thermometer

Lauric acid, 70 mg
+ CH₃CH₂OH, 1.0 mL
+ CH₃COCl, 30 μL

HOOD

**WARNING:** Acetyl chloride is an irritant. Dispense this reagent in the *hood* using an automatic delivery pipet. (Be sure to quickly reassemble the reaction vial and condenser following addition of the acid chloride, as this reagent will rapidly react with moist laboratory air and lose its activity.)

### Reaction Conditions

Heat the reaction mixture for 1 hour at gentle reflux with stirring, using a sand bath temperature of 110–120 °C. Cool the resulting mixture to room temperature and then remove the spin vane with forceps.

### Isolation of Product

Add a boiling stone to the cooled vial and then concentrate the reaction solution to a volume of about 0.25 mL by warming in a sand bath in the HOOD  **hood.** Use a 1.0-mL graduated pipet to add 0.5 mL of diethyl ether and 0.25

mL of 5% sodium bicarbonate solution to the concentrated product mixture. Cap the conical vial and shake gently (or mix on a Vortex mixer). Loosen the cap carefully to vent the two-phase mixture. Remove the bottom aqueous layer using a Pasteur filter pipet and set it aside in a labeled Erlenmeyer flask. Extract the ether phase with three additional 0.25-mL portions of 5% sodium bicarbonate solution. Save the aqueous wash after each extraction (you may combine them with the initial aqueous basic phase in the Erlenmeyer) and do not discard them until you have successfully purified and characterized the ethyl laurate.

## Purification and Characterization

Dry and purify the wet, crude ether solution of ethyl laurate by column chromatography. In a Pasteur filter pipet, place 500 mg of activated silica gel followed by 500 mg of anhydrous sodium sulfate. (■) Wet the column first with 0.5 mL of methylene chloride and then transfer the crude ether solution of ethyl laurate to the column using a Pasteur filter pipet. Use a **tared** 5-mL conical vial containing a boiling stone as a collection flask for the column eluant. Rinse the reaction vial with two 0.5-mL portions of methylene chloride and transfer each rinse to the column using the same pipet. Add an additional 1.0 mL of methylene chloride directly to the column to ensure complete elution of the ester.

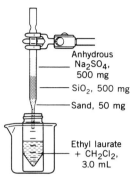

Anhydrous
Na$_2$SO$_4$,
500 mg

SiO$_2$, 500 mg

Sand, 50 mg

Ethyl laurate
+ CH$_2$Cl$_2$,
3.0 mL

HOOD

Remove the ether–methylene chloride solvent by evaporation in the **hood** by using a stream of nitrogen gas with gentle warming in a sand bath. Make sure the vial remains warm to your fingers during the evaporation process, this will insure that condensation of moisture on the liquid product is avoided during solvent concentration.

The recovered ethyl laurate, is a clear, viscous, pleasant-smelling ester, and is usually fully characterized without further purification. Weigh the product and calculate the percent yield. Determine the refractive index (optional) and boiling point and compare your data to the literature values. Obtain an IR spectrum of the ester using the capillary film technique and compare it to the spectrum of an authentic sample or to the one in the *Aldrich Library of IR spectra.*

## Chemical Tests
Does this ester give a positive hydroxamate test (see Chapter 10)? Check the solubility of ethyl laurate in water. Would you have predicted the result? Is the ester soluble in 85% phosphoric acid or in concentrated sulfuric acid?

## Experiment [8B]    Isopentyl Acetate: Semimicroscale Preparation

Isopentyl acetate is prepared by the following procedure.

**REACTION**

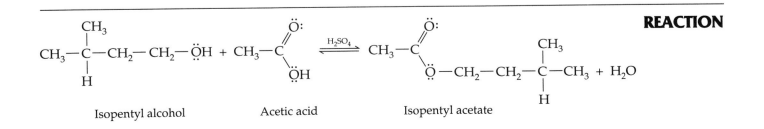

Isopentyl alcohol            Acetic acid            Isopentyl acetate

## EXPERIMENTAL PROCEDURE

Estimated time to complete the experiment: 3.0 hours.

### Physical Properties of Reactants

| Compound | MW | Amount | mmol | bp(°C) | d | $n_D$ |
|---|---|---|---|---|---|---|
| Isopentyl alcohol | 88.15 | 800 μL | 7.4 | 132 | 0.81 | 1.4053 |
| Acetic acid | 60.1 | 1.5 mL | 26.2 | 139.6 | 1.05 | 1.3720 |
| Sulfuric acid, concd | | 4 Drops | | | | |

### Reagents and Equipment

In a 5.0-mL conical vial containing a magnetic spin vane and equipped with a reflux condenser protected by a calcium chloride drying tube, place 1 mL (800 μL, 7.4 mmol) of isopentyl alcohol, 1.5 mL (26.2 mmol) of glacial acetic acid, 4 drops (Pasteur pipet) of concentrated sulfuric acid, and approximately 100 mg of silica gel beads. (■)

---

**CAUTION:  Cap the vial immediately after addition of each reagent. Dispense the reagents in the *hood* using automatic delivery pipets. Concentrated acetic and sulfuric acids are *corrosive***

---

**NOTE.**  *The silica gel beads (t.h.e. desiccant) are used to absorb the water as it is generated in the reaction*

### Reaction Conditions

Heat and stir the reaction mixture using a sand bath temperature of 160–180 °C for 1 hour. Cool the resulting mixture to room temperature and remove the spin vane with forceps. Add 0.5 mL of diethyl ether to increase the volume of the organic phase.

### Isolation of Product

Extract the crude organic product with three 2-mL portions of 5% sodium bicarbonate solution, followed by 1 mL of water. During each extraction, cap the vial, shake gently, vent carefully, and then allow it to stand so the layers may separate. A Vortex mixer may be used to good advantage in this sequence. Remove the aqueous layer and be sure not to discard it until the final product is purified and characterized.

Following the aqueous extraction, dry the organic phase by adding anhydrous sodium sulfate to the vial. Transfer the anhydrous solution, using a Pasteur filter pipet, to a clean, dry 3-mL conical vial containing a boiling stone. Remove the added ether by evaporation in the *hood* by using a gentle stream of nitrogen gas with gentle warming in a sand bath. Next, attach the vial containing the crude ester residue to a Hickman still equipped with an air condenser and arranged in a sand bath for distillation. (■) Cover the sand bath with aluminum foil during the procedure.

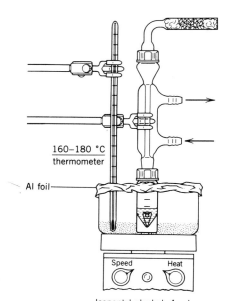

160–180 °C
thermometer

Al foil

Isopentyl alcohol, 1 mL
+ CH₃CO₂H, 0.55 mL
+ 4 drops concd H₂SO₄
+ silica gel beads, 100 mg

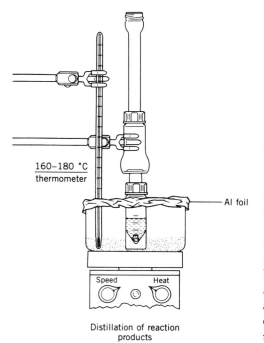

160–180 °C
thermometer

Al foil

Distillation of reaction
products

## Purification and Characterization

Distill the isopentyl acetate product at a sand bath temperature of 160–180°C. As the material collects in the collar of the still, transfer it by Pasteur pipet (9 in.) to a **tared** $\frac{1}{2}$-dram vial.

Weigh the clear, viscous, pleasant-smelling isopentyl acetate, and calculate the percent yield. Determine the refractive index (optional) and boiling point, and compare your values with those found in the literature.

Obtain an IR spectrum of the ester using the capillary film technique and compare it with that shown in Figure 6.23.

## Infrared Analysis

The conversion of a branched-chain aliphatic primary alcohol to an acetate ester results in significant changes in the infrared spectrum of the molecule. These changes are similar to those observed in straight-chain systems. The two macro

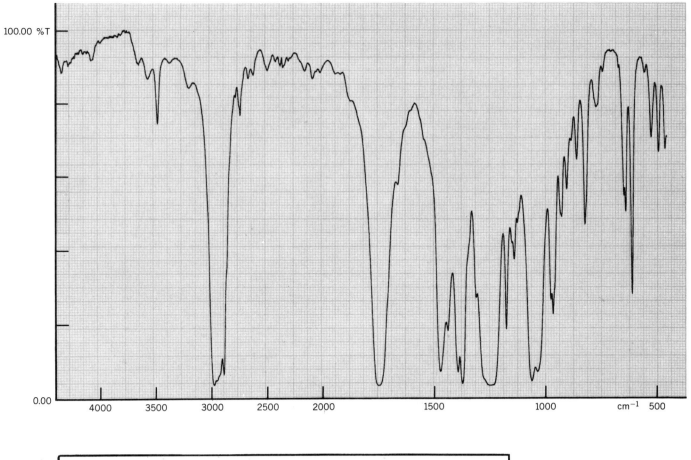

| Sample | Isopentyl acetate | | |
|---|---|---|---|
| %T X ABS __ Background Scans 4 | | Scans 16 | |
| Acquisition & Calculation Time 42 sec | | Resolution 4.0 cm⁻¹ | |
| Sample Condition liquid, neat | | Cell Window KBr | |
| Cell Path Length capillary film | | Matrix Material | |

**FIGURE 6.23**   IR spectrum: isopentyl acetate.

group frequency trains in the present example are

     **1.** Isopentyl alcohol (Fig. 6.24): 3350 (broad), 3000–2850, 1460–1300, 1060, and 660 cm$^{-1}$.
     **2.** Isopentyl acetate (Fig. 6.23): 3490 (weak), 3000–2850, 1746 (strong), 1367 (broad), 1250, and 1040 cm$^{-1}$.

     If the macro group frequencies are further constrained to include only those systems in which the chain branching involves the presence of an isopropyl group, two additional bands near 1385 and 1365 cm$^{-1}$ are required. These peaks, which are present in both the alcohol and the ester, arise from the spatially coupled and split symmetric methyl bending vibrations of the aliphatic backbone. In both isopentyl alcohol and isopentyl acetate, this pair of bands are found at identical locations, 1386 and 1367 cm$^{-1}$.

## Chemical Tests

Does the product give a positive hydroxamate test for an ester (Chapter 10)? Check the solubility of this ester in water, ether, and 85% H$_3$PO$_4$. In which solubility group (see Chapter 10, p. 702) does isopentyl acetate fall?

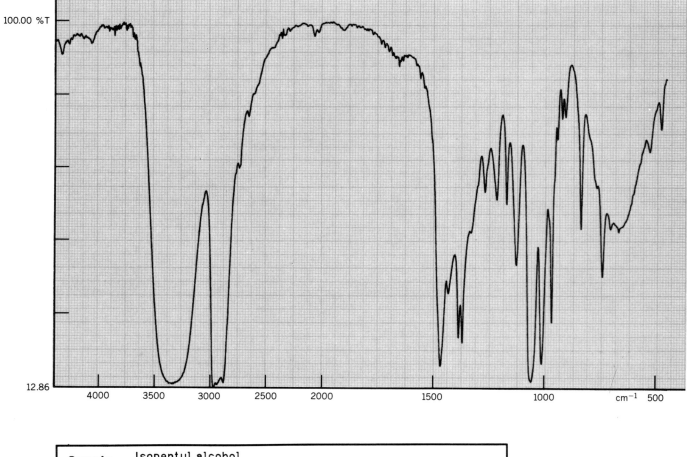

**FIGURE 6.24**    IR spectrum: isopentyl alcohol.

# Experiment [8C]   Esterification by Use of Acidic Resins: Semimicroscale Preparations

Modern synthetic reactions are making increased use of reagents that are heterogeneous in character. The use of resins as the support material for reactive compounds has become very popular because of their ease of removal and reliability. In this experiment you will utilize one of the recent additions to the arsenal of resin catalysts, Nafion® 417. This catalyst is a powerfully acidic resin. It can approach the acidities of 100% sulfuric acid and of trifluoromethanesulfonic acid in trifluoroacetic anhydride solution. The "superacidity" of the sulfonic acid group in Nafion 417 is attributable to the electron-withdrawing ability of the perfluorocarbon backbone of the resin to which it is attached.

$$-[CF_2CF_2]_n[CFCF_2]_x-$$
$$(OCF_2CF)_mOCF_2CF_2SO_3H$$
$$CF_3$$

Nafion 417 resin with an equivalent weight of approximately 1200 contains tetrafluoroethylene (the monomer used to make Teflon) and perfluorovinyl ether units in a ratio of 7:1. These resins have been used successfully by Olah,[6] as in this experiment, for the esterification of carboxylic acids.

**REACTION**

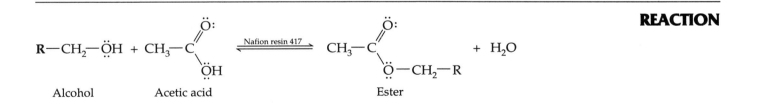

| Alcohol | Acetic acid | Ester |

Estimated time to complete the experiment: 1.5 hours.

**EXPERIMENTAL PROCEDURE**

### Physical Properties of Reactants

| Compound | MW | Amount | mmol | bp(°C) | $d$ | $n_D$ |
|---|---|---|---|---|---|---|
| 1-Butanol | 74 | 365 μL | 4.0 | 118 | 0.810 | 1.3990 |
| Isopentyl alcohol | 88 | 500 μL | 4.6 | 132 | 0.81 | 1.4053 |
| 1-Pentanol | 88 | 500 μL | 4.6 | 138 | 0.811 | 1.4090 |
| 1-Hexanol | 102 | 570 μL | 4.55 | 156 | 0.814 | 1.4180 |
| 1-Octanol | 130 | 700 μL | 4.45 | 196 | 0.827 | 1.4290 |
| Acetic acid | 60 | 275 μL | 4.8 | 116 | 1.049 | 1.3720 |
| Nafion 417 resin | | Equivalent weight 1100 | | 1.5 × 0.5-cm strip | | |

[6] Olah, G. A.; Keumi, T.; Meidar, D. *Synthesis* **1978**, 929.

## Reagents and Equipment

In a 5.0-mL conical vial containing a magnetic spin vane and equipped with a reflux condenser protected by a calcium chloride drying tube, place (*see table for correct molar quantities for your particular alcohol*), 275 $\mu$L (4.8 mmol) of glacial acetic acid, and a strip of Nafion 417 resin (1.5 × 0.5 cm). (If the strip of resin is placed so as to encircle the spin vane, mixing of the reaction solution is particularly effective.)

**NOTE.**   *The Nafion 417 resin has a significant number of hydrophilic centers that function in exactly the same way as the silica gel does in Experiment [8B]. Thus, the resin helps to drive the esterification to completion (i.e., Nafion 417 in addition to functioning as an acid catalyst also acts as a desiccant) by absorbing water as it is generated in the reaction.*

## Reaction Conditions

The reaction mixture is heated with stirring at a sand bath temperature of 160–180 °C for 30 min. The mixture is then cooled to room temperature.

## Isolation of Product

Remove the Nafion strip and the spin vane (with forceps) from the cool reaction mixture and carefully (to avoid foaming) rinse them with 1 mL of 5% $NaHCO_3$. The rinse is added to the vial (to form a two-phase system) and swirled with the reaction mixture. The bicarbonate wash is then separated and the aqueous phase transferred to a 10-mL Erlenmeyer flask for temporary storage. The reaction mixture is washed with a further 1-mL portion of 5% $NaHCO_3$, and then by two 1-mL distilled water washes (2 × 1 mL). All the aqueous washes are transferred to the storage Erlenmeyer. Dry the crude and wet reaction product by passing it down a Pasteur pipet containing a cotton plug and 1 g of anhydrous granular sodium sulfate ($Na_2SO_4$). Collect the product residue from the column in a tared 3-mL conical vial and weigh it to determine the crude yield. Obtain an IR spectrum and measure the refractive index (optional) of the crude material. Does the IR spectrum indicate the presence of any unreacted starting material in the crude product? What would be the most likely contaminant?

## Purification and Characterization

A 25 $\mu$L sample of the crude ester is purified by prep-GC (see GC conditions for each product mixture and Experiment [2] for details on the collection procedure). Following collection of the purified ester, obtain an IR spectrum, boiling point and refractive index and compare these with the data obtained on the crude residue. The ester is the first compound to elute from the column. It may be followed by a small amount of a second component. You can calculate the purity of the crude product by measuring the area under the two elution bands (only in the case of the butyl acetate synthesis does the workup procedure distort the apparent crude yields: Can you explain why this occurs? If you want to identify the contaminant (second band) directly, it may be possible to also collect a second fraction from the GC if the contaminant comprises a reasonable fraction of the product mixture (5–10%). It may, however, require a second injection of 30–40 $\mu$L.

## Gas Chromatographic Conditions

### General

Stainless steel columns 8 ft × 1/4 in. packed with 20% Carbowax

Flow rates: 50 mL/min (He); sample size: 25 $\mu$L.

### Specific

| Ester 1 | Ester 2 |
|---|---|
| *Butyl Acetate* | *Pentyl Acetate* |
| Column temperature: 120 °C | Column temperature: 130 °C |
| Retention time:<br>    ester: 8.5 min<br>    impurity: 11 min | Retention time:<br>    ester: 8.5 min<br>    impurity: 11 min |

| Ester 3 | Ester 4 |
|---|---|
| *Isopentyl Acetate* | *Hexyl Acetate* |
| Column temperature: 140 °C | Column temperature: 160 °C |
| Retention time:<br>    ester: 6 min<br>    impurity: 8 min | Retention time:<br>    ester: 10 min<br>    impurity: 12 min |

| Ester 5 |
|---|
| *Octyl Acetate* |
| Column temperature: 185 °C |
| Retention time:<br>    ester: 10 min<br>    impurity: 13 min |

**QUESTIONS**

**6-49.** In the preparation of the esters given in this experiment, the reaction product was extracted with 5% sodium bicarbonate solution ($NaHCO_3$) in the isolation step. Why? What gas was evolved during this washing step? Write a balanced equation for the reaction that produced it.

**6-50.** Why is a large excess of acetic acid used in the preparation of isopentyl acetate?

**6-51.** Concentrated sulfuric acid is used as a catalyst for the esterification of acetic acid in the preparation of isopentyl acetate. Why is the sulfuric acid needed if another acid, acetic acid, is already present?

**6-52.** Fatty acids are long-chain carboxylic acids, usually of 12 or more carbon atoms, isolated from saponification of fats and oils (esters of glycerol). Draw the structure of each of the fatty acids named below and also determine its common name.

Hexadecanoic acid    (Z)-9-octadecenoic acid
Octadecanoic acid    (Z,Z)-9,12-octadecadienoic acid

**6-53.** Write a mechanism for the acid-catalyzed transesterification reaction of ethyl acetate with 1-butanol, which gives butyl acetate.

**6-54.** In the infrared spectra of acetates, two intense bands are usually observed in the 1270–1000-wavenumber region. These peaks are related to the stretching vibrations of the two C—O bonds of the ester group. Should we expect to be able to assign the C—O bonds to individual peaks, and if so, which mode is associated with which band? Why? (*Hint:* The peak that occurs at higher wavenumbers is consistently close to 1250 cm$^{-1}$.)

## BIBLIOGRAPHY

These references are selected from the large number of examples of esterification given in *Organic Syntheses* :

Bailey, D. M.; Johnson, R. E.; Albertson, N. F. *Organic Syntheses*; Wiley: New York, 1988; Collect. Vol. VI, p. 618.

Bowden, E. *Organic Syntheses*; Wiley: New York, 1943; Collect. Vol. II, p. 414.

Eliel, E. L.; Fisk, M. T. *Organic Syntheses*; Wiley: New York, 1963; Collect. Vol. IV, p. 169.

Emerson, W. S.; Longely, R. I., Jr. *Organic Syntheses*; Wiley: New York, 1963; Collect. Vol. IV, p. 302.

Fuson, R. C.; Wojcik, B. H. *Organic Syntheses*; Wiley: New York, 1943; Collect. Vol. II, p. 260.

McCutcheon, J. W. *Organic Syntheses*; Wiley: New York, 1955; Collect. Vol. III, p. 526.

Mic'ovic', V. M. *Organic Syntheses*; Wiley: New York, 1943; Collect. Vol. II, p. 264.

Peterson, P. E.; Dunham, M. *Organic Syntheses*; Wiley: New York, 1988; Collect. Vol. VI, p. 273.

Weissberger, A.; Kibler, C. J. *Organic Syntheses*; Wiley: New York, 1955; Collect. Vol. III, p. 610.

## EXPERIMENT [9]

## The E1 Elimination Reaction: Dehydration of 2-Butanol to Yield 1-Butene, *trans*-2-Butene, and *cis* 2-Butene

Common name: 1-butene
CA number: [06-98-9]
CA name as indexed: 1-Butene

Common name: *trans*-2-Butene
CA number: [624-64-6]
CA name as indexed: 2-Butene, (*E*)-

Common name: *cis*-2-Butene
CA number: [590-18-1]
CA name as indexed: 2-Butene, (*Z*)-

### Purpose

This experiment illustrates the variety of pathways that are available to acid-catalyzed elimination reactions of secondary (2°) alcohols via carbocation intermediates. The dehydration of 2-butanol forms a mixture of gaseous alkene products. The alkenes formed in this reaction are separated and identified by employing one of the most powerful instrumental techniques available to the modern research chemist for the separation of complex mixtures: gas chromatography (GC).

---

The dehydration reaction that you are about to study is representative of the large collection of reactions that are classified as E1 elimination reactions. These reactions all form an intermediate in which one of the carbon atoms bears, if not a *full* positive charge, at least a significant fractional positive charge. It is this fleeting, high-energy intermediate, an aliphatic carbocation, which makes these reactions so interesting.

The development of bonding theory in organic chemistry during the late nineteenth and early twentieth century did not accept the existence of carbocations, except in a few esoteric instances. This position was reasonable to take, as aliphatic compounds show little ionic character. The first proposal that these nonpolar substances might actually form cations came from Julius Stieglitz (1867–1937), of the University of Chicago, in a paper published in 1899. Eight years later, James F. Norris (1871–1940) at Massachusetts Institute of Technology produced rather compelling evidence that these substances might be intermediates in reactions of certain *tert*-butyl halides.

These two papers were the origin of the concept that organic carbon cations, which in those days became known as *carbonium ions*, were far more widespread in organic reactions than previously anticipated. Indeed, these early investigations caused more controversy and more experimental work to unambiguously prove the existence of what are now termed *carbocations*, than any other single problem in American chemistry.

It was, however, the English chemists Arthur Lapworth, Sir Robert Robinson, C. K. Ingold (University of London) and E. D. Hughes who, from 1920–1940, undertook a massive effort to develop the experimental and theoretical data to place these early postulates on a solid scientific foundation. Between 1920 and 1922, Hans Meerwein in Bonn, Germany, demonstrated that carbon rearrangements in the camphene series could be best explained by postulating the presence of carbocation intermediates. Perhaps the most important contribution to the entire subject was published in 1932 by Frank C. Whitmore of Pennsylvania State College, where he was Dean of the School of Chemistry and Physics. His paper, *The Common Basis of Intramolecular Rearrangements*, brought together a vast array of data in a beautifully consistent interpretation that essentially cemented the carbocation into contemporary organic chemical theory.

A reaction not too distant (in fact, rather close if you overheat your own!) from the one that is to be carried out below, but one that also included a rearrangement along with the dehydration, was studied by Dorothy Bateman and C. S. "Speed" Marvel at the University of Illinois as early as 1927.

Within 2 years of the publication of the Whitmore paper, Robinson proposed the formation of the steroids (including cholesterol) from squalene (a $C_{30}$ polyunsaturated polyisoprene molecule) via an incredible series of intermediates and rearrangements. Later, following the elucidation of the structure of lanosterol, R. B. Woodward and K. Bloch made a brilliant proposal that at once rationalized the biosynthetic origin of both lanosterol and cholesterol and implicated lanosterol as an intermediate in cholesterol biosynthesis. Their mechanism involved the concerted (bonds made and broken simultaneously) cyclization of four rings, as well as four

## HISTORY OF THE DEVELOPMENT OF CARBOCATION THEORY

rearrangements following the generation of the initial carbocation intermediate, to ultimately yield lanosterol.[7]

The elucidation of these biochemical pathways is further evidence of the major impact that our understanding of carbocation chemistry has had on related fields, such as biochemistry.

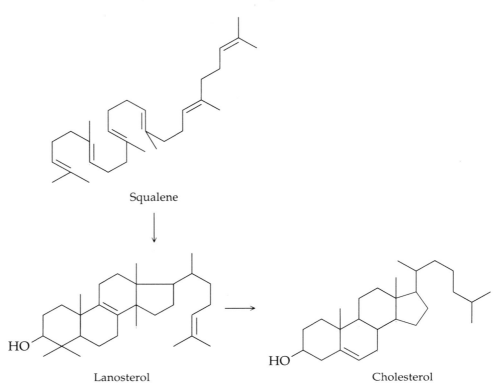

Squalene

Lanosterol

Cholesterol

***Prior Reading***

> *Technique 1:*   Gas Chromatography (see pp. 56–61).
> *Technique 7:*   Collection or Control of Gaseous Products (see pp. 109–111).

---

## REACTION

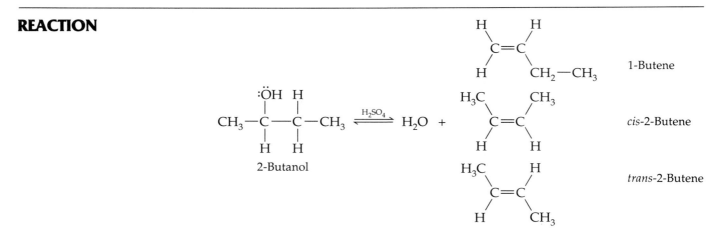

[7] Stieglitz, J. *Am. Chem. J.* **1899**, *21*, 101. Norris, J. F. *Am. Chem. J.* **1907**, *38*, 627. Meerwein, H.; van Emster, K. *Berichte* **1922**, *55*, 2500. Whitmore, F. C. *J. Am. Chem. Soc.* **1932**, *54*, 3274. Bateman, D. E.; Marvel, C. S. *J. Am. Chem. Soc.* **1927**, *49*, 2914. Robinson, R. *Chem. Ind.* **1934**, *53*, 1062. Woodward, R. B.; Bloch, K. *J. Am. Chem. Soc.* **1953**, *75*, 2023. See also: Tarbell, D. S.; Tarbell, T. *The History of Organic Chemistry in the United States, 1875–1955*; Folio: Nashville, TN, 1986.

The formation of an alkene (or alkyne) frequently involves loss of a proton and a leaving group from adjacent carbon atoms. The generalized reaction scheme is shown below.

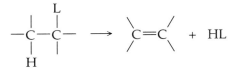

Such a reaction is termed an *elimination reaction*, as a small molecule (HL) is *eliminated* from the organic molecule. If the molecule eliminated is water, the reaction may also be referred to as a *dehydration*. One of the common synthetic routes to alkenes is the dehydration of an alcohol.

Dehydration of alcohols is an acid-catalyzed elimination reaction. Experimental evidence shows that alcohols react in the order tertiary (3°) > secondary (2°) > primary (1°); this reactivity relates directly to the stability of the carbocation intermediate formed in the reaction. Generally, sulfuric or phosphoric acid is used as the catalyst in the research laboratory. A Lewis acid, such as aluminum oxide or silica gel, is usually the catalyst of choice at the fairly high temperatures used in industrial scale reactions.

The mechanism for this reaction is classified as E1 (elimination, unimolecular). The elimination, or dehydration reaction, proceeds in several steps, as outlined below.

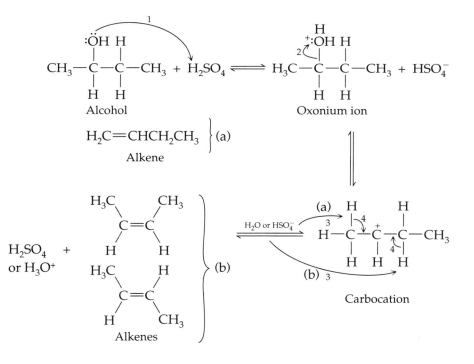

The first step (1) involves the very rapid, though reversible, protonation of the oxygen atom of the alcohol to form an *oxonium ion* (an oxygen cation with a full octet of electrons). This protonation step is important because it produces a good leaving group, water. Without acid, the only available leaving group is hydroxide ion (HO$^-$) which, as a strong base, is a poor leaving group. This finding is the reason why acid plays such an important role in the mechanism of this dehydration. The second step (2) of the reaction is the dissociation of the oxonium ion to form an intermediate *carbocation* and water. This step is the rate-determining (and therefore the slowest) step of the reaction. In the third step (3), the carbocation is deprotonated by a ubiquitous water molecule (or other base, such as bisulfate ion, present in the system) in another rapid equilibrium. The carbocation gains stability (lower energy) by releasing a proton ($H^+$) from a carbon atom adjacent ($\alpha$) to the carbocation (route a or b shown) to the attacking base. Thus, in Step 4 the catalyst is regenerated as a protonated molecule of water ($H_3O^+$), and the electron pair, previously comprising the C—H bond adjacent to the carbocation, flows toward the positive charge, generating a stable and neutral alkene. A variety of isomeric products may be formed, as different protons adjacent to the carbocation may be removed, and different conformations of the intermediate carbocation are also possible [see routes (a) and (b) above and further discussion below].

E1 elimination reactions, as we have just seen, involve equilibrium conditions and thus, to maximize the yield (drive the reaction to completion), the alkene is usually removed from the reaction while it is in progress. A convenient technique for accomplishing this task is distillation, which is often used because alkenes *always* have a lower boiling point than the corresponding alcohol. In the present reaction, the alkenes are gases and, therefore, are easily removed and collected as described in the experimental section.

It is important to note that many 1° (primary) alcohols also undergo dehydration, but primarily by a different route, the E2 (elimination, bimolecular) mechanism. This step is governed mainly by the fact that the 1° carbocation that would be required in an E1 process is a relatively unstable (very high energy) intermediate. In this case, attack by the base occurs directly on the oxonium ion, as shown below.

$$CH_3CH_2CH_2\ddot{O}H \; \underset{}{\overset{H_2SO_4}{\rightleftharpoons}} \; CH_3CH_2CH_2\overset{+}{\underset{\cdot\cdot}{O}}H_2 \; + \; HSO_4^-$$

Oxonium ion

$$HSO_4^- + \; CH_3 - \underset{\underset{H}{|}}{\overset{\overset{H}{|}}{C}} - CH_2 - \overset{+}{\underset{\cdot\cdot}{O}}H_2 \; \rightleftharpoons \; CH_3CH\!=\!CH_2 \; + \; H_2SO_4 \; + \; H_2O$$

Oxonium ion

E1 elimination is usually accompanied by a competing $S_N1$ substitution reaction that involves the *same* carbocation intermediate. Since both reaction mechanisms are reversible in this case, and since the alkene product gases are easily removed from the reaction, the competing substitution reaction (which predominantly regenerates the starting alcohol by attack of water, as a nucleophile, at the carbocation) is not troublesome and the equilibrium eventually leads to gas evolution.

As discussed earlier, many alcohols can dehydrate to yield more than one isomeric alkene. In the present reaction involving the dehydration of 2-butanol, at least three alkenes are usually formed: 1-butene, *trans*-2-butene, and *cis*-2-butene.

It would not be unreasonable to expect that the alkene generated in the largest amount is that possessing the highest degree of substitution, as this is the most stable product (lowest energy). This finding is exactly what is observed: more than 90% of the products are isomers of 2-butene. As trans alkenes are thermodynamically more stable than their cis counterparts, and since the reaction is reversible, one might expect the dominant isomer in the 2-butene mixture to be trans. Indeed, nearly twice as much trans as cis isomer is formed. An empirical rule, originally formulated by the Russian chemist Alexander Zaitsev (or Saytzeff), for base catalyzed E2

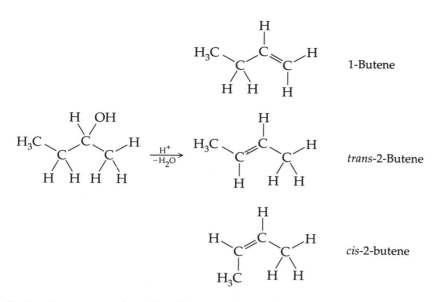

1-Butene

trans-2-Butene

cis-2-butene

eliminations states that the alkene with the largest number of alkyl substituents on the double bond will be the major product. This rule also is able to correctly predict the relative ratio of substituted alkenes to be expected from a given E1 elimination reaction, and it is obvious that it correctly applies in the case of 2-butanol.

Rearrangements of alkyl groups (such as methide, $^-$:$CH_3$; ethide, $^-$:$CH_2CH_3$) and hydrogen (as hydride, :$H^-$) are often observed during the dehydration of alcohols, especially in the presence of very strong acid where carbocations can exist for longer periods of time. For example, when 3,3-dimethyl-2-butanol is treated with sulfuric acid, the elimination reaction yields the mixture of alkenes shown below. Can you predict which alkene is the principal product?

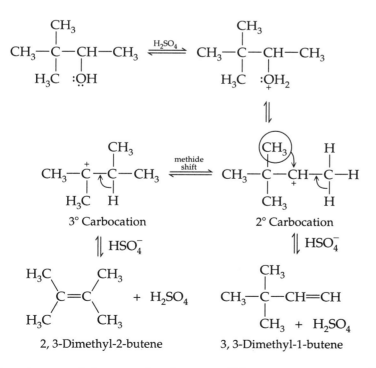

2, 3-Dimethyl-2-butene          3, 3-Dimethyl-1-butene

Reactions carried out under these conditions are very susceptible to alkide or hydride shifts if a more stable carbocation intermediate can be formed. In the example above, a 2° carbocation rearranges into the more stable 3° carbocation. This intramolecular rearrangement involves the transfer of an entire alkyl group (in this case a methyl substituent), together with its bonding pair of electrons, to an adjacent carbon atom. This migration, or shift, of the methyl group to an adjacent carbocation is called a *1,2-methide shift*. It commonly occurs in aliphatic systems involving carbocation intermediates that have alkyl substituents adjacent to the cation. Hydride shifts are also frequently observed and actually appear in a wider range of molecules.

## EXPERIMENTAL PROCEDURE

Estimated time to complete the experiment: 2 hours.

**NOTE.**   *It is suggested that the starting times of the reaction be staggered to allow more students easier access to the GC instrument when the product gases are analyzed.*

### Physical Properties of Reactants

| Compound | MW | Amount | mmol | bp(°C) | d | $n_D$ |
|---|---|---|---|---|---|---|
| 2-Butanol | 74.12 | 100 $\mu$L | 1.1 | 99.5 | 0.81 | 1.3978 |
| Concd sulfuric acid | 98.08 | 50 $\mu$L | | | | |

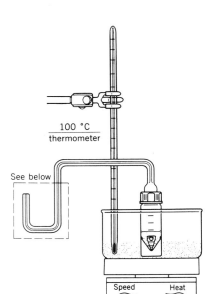

2-Butanol, 100 $\mu$L, + concd H$_2$SO$_4$, 50 $\mu$L

### Reagents and Equipment

Assemble the gas collection apparatus shown in the figure *before* the reactants are mixed. (■)

The capacity of the gas collection reservoir is determined by the following procedure. Seal the collection tube with a septum cap and invert it. Then add 3.0 mL of water and mark the 3.0-mL level. Finally, add an additional 1.0 mL, and also mark the 4.0-mL level.

**NOTE.**   *Use of a Teflon-lined septum cap on the gas collection reservoir is necessary to prevent loss of the collected butene gases by permeation.*

To position the gas collection reservoir, carry out the following steps: (1) fill the reservoir with water; (2) place your finger (index finger is usually used) over the open end of the reservoir; (3) invert it; (4) place it, with the open end down, into a beaker (250 mL) filled with water. When your finger is removed, the column of water should remain in the reservoir.

Place 100 $\mu$L (81 mg, 1.1 mmol) of 2-butanol and 50 $\mu$L of concentrated sulfuric acid in a **clean**, dry, 1.0-mL conical vial containing a magnetic spin vane.

3 mL
4 mL

---

**CAUTION:   Sulfuric acid is a strong, corrosive material. Contact with the skin or eyes can cause severe burns. It is convenient to dispense the reactants using automatic delivery pipets.**

---

Cap-seal the 1-mL vial to the gas delivery tube and position the tube under water into the open end of the gas collection reservoir as shown in the figure. Clamp the reservoir in place.

## Reaction Conditions

Heat the reaction mixture, with gentle stirring, using a sand bath until the evolution of gas takes place (sand bath temperature ~110–120 °C ). The mixture should be warmed slowly through the upper temperature range to prevent foaming.

## Isolation of Product

Collect about 3–4 mL of gas in the collection reservoir. Remove the gas delivery tube from the gas collection reservoir, and **then** from the water bath, **before** removing the reaction vial from the heat. Use this sequence of steps in shutting down the reaction to prevent water from being sucked back into the hot reaction flask while it is cooling down.

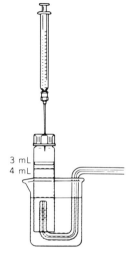

---

CAUTION:  If water is drawn back into hot concentrated sulfuric acid, a very dangerous situation can occur, particularly if larger quantities of the reagents than recommended above are employed in the experiment.

---

NOTE.  *Do **not** remove the gas collection reservoir from the water bath. The beaker containing the reservoir is carried to the gas chromatograph for analysis of the collected gas.*

With the aid of a gas-tight syringe, withdraw a 0.5-cm³ sample through the rubber septum for GC analysis. (■)

## Purification and Characterization

Analyze the collected gas by GC without further purification.

*Gas Chromatographic Conditions*

Column: 0.25 in. x 8 ft packed with 20% silicone DC 710
Room temperature
Flow rate: 20 mL/min (He gas)
Sample size: 0.5 cm³ of collected gas

Record the literature values of the physical properties of the products.

The butenes have been determined to elute from the DC 710 column in the following order: 1-butene, *trans*-2-butene, and *cis*-2-butene. *If the reaction mixture is heated above the recommended temperatures, a rearrangement can occur to yield isobutene, which will be detected as an additional isomeric product.*

If we make the assumption that the amount of each substance in the gaseous mixture is proportional to the area under its corresponding GC peak, determining the relative amounts of the three components of the gas sample becomes a straightforward calculation. The accuracy is, of course, dependent on how well the three peaks are resolved in the gas chromatogram.

NOTE.  *Calculation of relative quantities of alkenes formed in the reaction.*
*There are several techniques that may be used to quantitatively determine the composition of the gas mixture if an integrating recorder is not available. Two methods are described here.*

1. *Determination of the areas under the peaks gives reproducible results of ± 3–4% when these areas are assumed equal to the peak heights (mm) × the peak widths at half-height (mm), measured from the baseline of the curve.*

2. *An alternate method for determination of these areas is to cut out the peaks from the chromatogram and weigh them on an analytical balance (sensitivity to 0.1 mg). The weights of the peaks are directly proportional to the relative amount of each compound in the gas sample.[8]*

## QUESTIONS

**6-55.** Gas chromatographic analysis of a mixture of organic compounds gave the following peak areas ($cm^2$): hexane = 2.7; heptane = 1.6; hexanol = 1.8; toluene = 0.5.

    a. Calculate the mole percent composition of the mixture. Assume that the response of the detector (area per mole) is the same for each component.

    b. Calculate the weight percent composition of the mixture, using the same assumptions as in part a.

**6-56.** It is noted at the end of the experiment that if the mixture is heated strongly, rearrangement can occur and isobutene (2-methyl-2-propene) is also formed. Suggest a mechanism to account for the formation of this compound.

**6-57.** When *tert*-pentyl bromide is treated with 80% ethanol, the following amounts of alkene products are detected on analysis.

$$CH_3CH_2-\overset{\overset{\displaystyle CH_3}{|}}{\underset{\underset{\displaystyle CH_3}{|}}{C}}-Br \xrightarrow{C_2H_5OH(80\%)} \underset{\underset{(32\%)}{\mathbf{I}}}{CH_3CH=C(CH_3)_2} + \underset{\underset{(8\%)}{\mathbf{II} \;\; CH_3}}{CH_3CH_2C=CH_2}$$

$$+ \begin{cases} \textit{tert-Pentyl alcohol} \\ \textit{tert-Pentyl ethyl ether} \end{cases}$$

$$(60\%)$$

Offer an explanation of why Compound I is formed in far greater amount than the terminal alkene.

**6-58.** The $-\overset{+}{S}R_2$ group is easily removed in elimination reactions, but the $-SR$ group is not. Explain.

**6-59.** Why is sulfuric acid, rather then hydrochloric acid, used to catalyze the dehydration of alcohols?

## BIBLIOGRAPHY

Several dehydration reactions of secondary alcohols using sulfuric acid as the catalyst are given in *Organic Syntheses*:

Adkins, H.; Zartman, W. *Organic Syntheses*; Wiley: New York, 1943; Collect. Vol. II, p. 606.

Bruce, W. F. *Organic Syntheses*; Wiley: New York, 1943; Collect. Vol. II, p. 12.

Coleman, G. H.; Johnstone, H. F. *Organic Syntheses*; Wiley: New York, 1941; Collect. Vol. I, p. 183.

[8] See Pecsok, R. L. *Principles and Practices of Gas Chromatography;* Wiley: New York, 1961; p. 145.

Grummitt, O.; Becker, E. I. *Organic Syntheses*; Wiley: New York, 1963; Collect. Vol. IV, p. 771.

Norris J. F. *Organic Syntheses*; Wiley: New York, 1941; Collect. Vol. I, p. 430.

Wiley, R. H.; Waddey, W. E. *Organic Syntheses*; Wiley: New York, 1955; Collect. Vol. III, p. 560.

Experiment [9] is adapted from the method given by:

Helmkamp, G. K.; Johnson, H. W. Jr. *Selected Experiments in Organic Chemistry*, 3rd ed., W. H. Freeman: New York, 1983; p. 99.

## The E2 Elimination Reaction: Dehydrohalogenation of 2-Bromobutane to Yield 1-Butene, *trans*-2-Butene, and *cis*-2-Butene

Common name: 1-Butene
CA number: [06-98-9]
CA name as indexed: 1-Butene

Common name: *trans*-2-Butene
CA number: [624-64-6]
CA name as indexed: 2-Butene, (*E*)-

Common name: *cis*-2-butene
CA number: [590-18-1]
CA name as indexed: 2-Butene, (*Z*)-

### Purpose
This reaction illustrates the base-induced dehydrohalogenation of alkyl halides with strong base. This reaction is used extensively for the preparation of alkenes. The stereo- and regiochemical effect of the size of the base is investigated, and the product mixture is analyzed by the use of gas chromatography.

### *Prior Reading*
Technique 1: Gas Chromatography (see pp. 56–61).
Technique 7: Collection or Control of Gaseous Products (see pp. 109–111).

**REACTIONS**

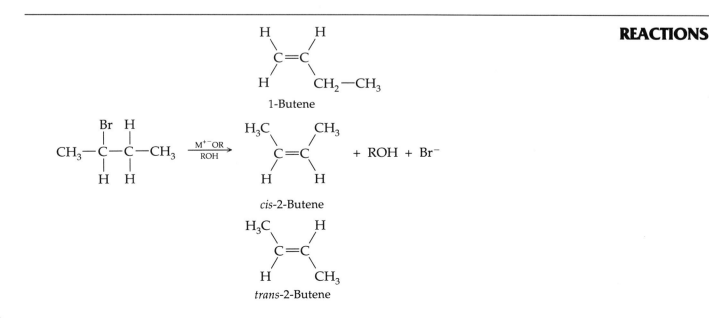

**DISCUSSION**

Base-induced elimination (dehydrohalogenation) of alkyl halides is a general reaction and is an excellent method for preparing alkenes. This process is often referred to as $\beta$ *elimination*, since a hydrogen atom is always removed $\beta$ to the halide (leaving group).

$$
\text{B:} \rightarrow \text{H} \quad \overset{\text{B:}\rightarrow\text{H}}{\underset{\beta}{-}\text{C}} - \overset{\alpha}{\text{C}} - \quad \xrightarrow{E_2} \quad \text{C}=\text{C} \quad + \quad \text{B:H} \quad + \quad \text{:X}^-
$$

A high concentration of a strong base in a relatively nonpolar solvent is used to carry out the dehydrohalogenation reaction. Such combinations as sodium methoxide in methanol, sodium ethoxide in ethanol, potassium isopropoxide in isopropanol, and potassium *tert*-butoxide in *tert*-butanol or dimethyl sulfoxide (DMSO) are often employed.

Elimination reactions almost always yield an isomeric mixture of alkenes, where this is possible. Under the reaction conditions, the elimination is *regioselective* and follows the Zaitsev rule when more than one route is available for the elimination of HX from an unsymmetrical alkyl halide. That is, the reaction proceeds in the direction that yields the most highly substituted alkene. For example,

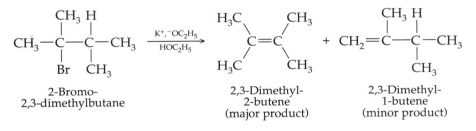

2-Bromo-
2,3-dimethylbutane

2,3-Dimethyl-
2-butene
(major product)

2,3-Dimethyl-
1-butene
(minor product)

In cases where cis or trans alkenes can be formed, the reaction exhibits *stereoselectivity*, and the more stable trans isomer is the major product.

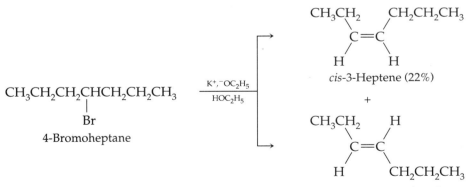

CH₃CH₂CH₂CHCH₂CH₂CH₃
|
Br
4-Bromoheptane

*cis*-3-Heptene (22%)

+

*trans*-3-Heptene (78%)

Experimental evidence indicates that the five atoms involved in the E2 elimination reaction must lie in the same plane, the anti-periplanar conformation is preferred. This conformation is necessary for the orbital overlap that must occur for the $\pi$ bond to be generated in the alkene. The $sp^3$-hybridized atomic orbitals on carbon that comprise the C—H and C—X $\sigma$ bonds broken in the reaction develop into the $p$ orbitals comprising the $\pi$ bond of the alkene formed.

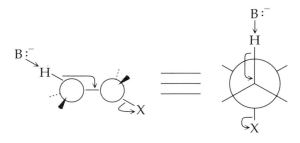

Anti-periplanar conformation

There is a smooth transition between reactant and product. Analogous to the $S_N2$ reaction, no intermediate has been isolated or detected. Furthermore, no rearrangements occur under E2 conditions. This situation is in marked contrast to E1 elimination reactions, where carbocation intermediates are generated and rearrangements are frequently observed (see Experiment [9]).

Since the alkyl halide adopts the anti-periplanar conformation in the transition state, if the size of the base is increased experimental evidence demonstrates that it must be difficult for the large base to abstract an internal $\beta$-hydrogen atom. In such cases, the base removes a less hindered $\beta$-hydrogen atom leading to a predominance of the thermodynamically **less stable (terminal) alkene** in the product mixture. This type of result is often referred to as **anti-Zaitsev or Hofmann elimination**. Thus, in the reaction of 2-bromo-2,3-dimethylbutane given above, the 2,3-dimethyl-1-butene would be the major product (anti-Zaitsev) if the conditions used employed a bulkier base. The anti-periplanar arrangements are illustrated in the Newman projections below. This phenomenon will be explored in the present experiment.

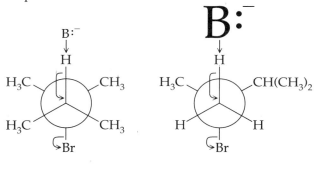

Zaitsev product
internal alkene

Anti-Zaitsev product
terminal alkene

Two anti-periplanar conformations
of 2-bromo-2,3-dimethylbutane

Dehydrohalogenation of alkyl halides in the presence of strong base (E2) is often accompanied by the formation of substitution ($S_N2$) products. The extent of the competitive substitution reaction depends on the structure of the alkyl halide. Primary alkyl halides give predominantly substitution products (the corresponding ether) secondary alkyl halides give predominantly elimination products, and tertiary alkyl halides give exclusively elimination products. For example, the reaction of 2-bromopropane with sodium ethoxide proceeds as follows:

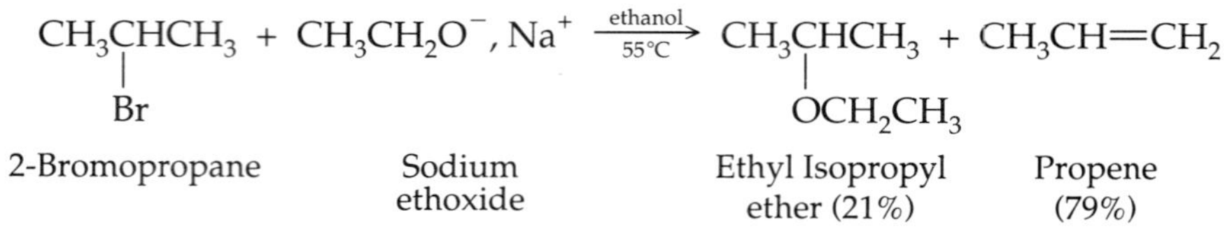

$$\underset{\substack{|\\ Br}}{CH_3CHCH_3} + CH_3CH_2O^-, Na^+ \xrightarrow[55°C]{ethanol} \underset{\substack{|\\ OCH_2CH_3}}{CH_3CHCH_3} + CH_3CH=CH_2$$

| | | | |
|---|---|---|---|
| 2-Bromopropane | Sodium ethoxide | Ethyl Isopropyl ether (21%) | Propene (79%) |

In general, for the reaction of alkyl halides with strong base:

$$1° \quad 2° \quad 3°$$

$$\xleftarrow{\hspace{3cm}}$$
ease of $S_N2$ reaction

$$\xrightarrow{\hspace{3cm}}$$
ease of $E_2$ reaction

---

**EXPERIMENTAL PROCEDURE**

Estimated time of experiment: 2.5 hours, if two reactions are run by each student.

### Physical Properties of Reactants

| Compound | MW | Amount | mmol | bp(°C) | d | $n_D$ |
|---|---|---|---|---|---|---|
| 2-Bromobutane | 137.03 | 100 μL | 0.92 | 91.2 | 1.2556 | 1.4366 |
| Methanol | 32.04 | 3.5 mL | | 64.9 | 0.7914 | 1.3288 |
| 2-Propanol | 60.09 | 3.5 mL | | 82.4 | 0.7851 | 1.3776 |
| 2-Methyl-2-propanol (*tert*-butanol) | 74.12 | 3.5 mL | | 82–83 | 0.7856 | 1.3838 |
| 3-Ethyl-3-pentanol | 116.20 | 3.5 mL | | 140–142 | 0.8389 | 1.4266 |
| Sodium | 22.98 | 60 mg | 2.6 | 883 | 0.97 | |
| Potassium | 39.10 | 60 mg | 1.5 | 760 | 0.86 | |

### Reagents and Equipment

The following combinations of reagents (Table 6.5) may be used to prepare the alkoxide base. It is recommended that students compare results so that a total picture of the effect be observed.

### TABLE 6.5   Reagent Combinations

| Alcohol Solvent | Metal | Alkoxide Base Produced |
|---|---|---|
| Methanol | Sodium | Sodium methoxide |
| 2-Propanol | Potassium | Potassium 2-propoxide |
| 2-Methyl-2-propanol (*tert*-butanol) | Potassium | Potassium 2-methyl-2-propoxide (potassium *tert*-butoxide) |
| 3-Ethyl-3-pentanol | Potassium | Potassium 3-ethyl-3-pentoxide |

## Preparation of the Alkoxide Base

Measure and add to a 5.0 mL conical vial containing a magnetic spin vane 3–3.5 mL of the anhydrous alcohol to be used (see Table 6.5). Add a 60-mg piece of potassium (or sodium) metal and immediately attach the vial to a reflux condenser protected by a calcium chloride drying tube. Place the arrangement in a sand bath and with stirring heat the mixture gently (~50 °C). (■)

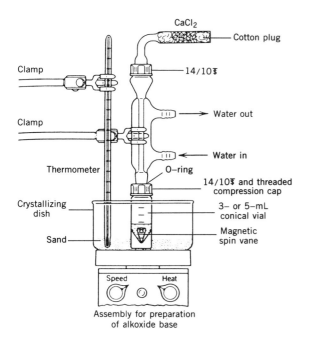

Assembly for preparation
of alkoxide base

**NOTE.** *If the sodium/methanol combination is used, it is **not** necessary to heat the mixture. A fairly vigorous reaction occurs at room temperature. It is recommended that the instructor cut the Na/K metal before commencing the laboratory.*

*Handle sodium and potassium with care. These metals react vigorously with moisture and are kept under paraffin oil or xylene. Remove the small piece of the metal from the oil using a pair of forceps or tongs (**NEVER USE YOUR FINGERS**), dry it quickly by pressing with filter paper (to soak up the oil), and immediately add the metal to the alcohol in the reaction vial. Any residual pieces of sodium/potassium should be stored in a bottle marked "sodium/potassium residues." **NEVER** throw small pieces of these metals in the sink or in water. To destroy, add small amounts of the metal to absolute ethanol [HOOD].*

NEVER

HOOD

When all the metal has reacted, remove the assembly from the sand bath and cool to near room temperature (do not remove the drying tube.).

## Reaction Conditions

Remove the drying tube from the condenser and using a calibrated Pasteur pipet introduce 100 μL of 2-bromobutane down through the condenser into the vial. Replace the drying tube and place the assembly in the **preheated** sand bath (see Table 6.6). Remove the drying tube and attach the gas delivery tube to the top of the condenser so that the open end of the

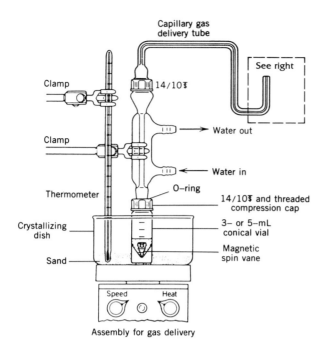

Assembly for gas delivery

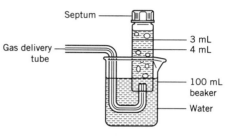

Arrangement for gas collection

tube is beneath the water level of the reservoir (■). If the connection to the top of the condenser is not made with an O-ring cap seal connection, lightly grease the ground-glass joint to insure a gas-tight seal. After about 10–15 air bubbles emerge, place the water-filled gas collection tube over the open end of the gas delivery tube.

**TABLE 6.6  Temperature Conditions**

| Base | Temperature (°C) |
|------|------------------|
| $NaOCH_3$ | 100–110 |
| $KOCH(CH_3)_2$ | 130–140 |
| $KOC(CH_3)_3$ | 140–150 |
| $KOC(CH_2CH_3)_3$ | 175–180 |

## Isolation of Product

Collect about 6–7 mL of gas in the collection reservoir and then, using a hypodermic syringe, withdraw a 0.7–0.8-mL sample through the rubber septum for GC analysis.

**NOTE.**  *Remove the gas delivery tube from the collecting reservoir and then from the water before discontinuing the heat on the reaction vial. This order of events prevents water from being sucked back into the reaction flask.*

## Purification and Characterization

The collected gas is analyzed by gas chromatography without further purification.

*Gas Chromatographic Conditions*

Column: ¼ in. × 8 ft packed with 20% DC 710
Room temperature

Flow rate (He gas): 30 mL/min
Sample size: 0.7–0.8 mL of collected gas
Chart speed: 1 cm/min

Assuming the amount of each substance in the gas is proportional to the areas of its corresponding peak, determine the ratio of the three components in the gas sample.

### Area Under a Curve

*Several techniques may be used. The following method gives reproducible results of ±3–4%:* Area = peak height (mm) × width at half-height (mm), measured from the baseline of the curve.

The order of elution of the butenes is 1-butene, *trans*-2-butene, and *cis*-2-butene. Record the literature values of their physical properties.

**6-60.** Outline a complete mechanistic sequence for the reaction of 2-bromobutane with potassium 2-methyl-2-propoxide in 2-methyl-2-propanol solvent to form the three alkenes generated in the reaction (1-butene, *trans*-2-butene, and *cis*-2-butene). Include a clear drawing of the anti-periplanar transition state for the formation of each alkene.

**6-61.** Does the mixture of gases collected in this experiment consist only of alkenes? If not, what other gases might be present?

**6-62.** Predict the predominant alkene product that would form when 2-bromo-2-methylpentane is treated with sodium methoxide in methanol. If the base were changed to $\overset{+}{K}\,\overset{-}{O}$—C(CH$_2$CH$_3$)$_3$ would the same alkene predominate? If not, why? What would be the structure of this alternate product, if it formed?

**6-63.** Predict the more stable alkene of each of the following pairs:
  a. 1-Hexene or *trans*-3-hexene
  b. *trans*-3-hexene or *cis*-3-hexene
  c. 2-Methyl-2-hexene or 2,3-dimethyl-2-pentene

**6-64.** Starting with the appropriate alkyl halide and base–solvent combination, outline a synthesis that would yield each of the following alkenes as the major or only product.
  a. 1-Butene
  b. 3-Methyl-1-butene
  c. 2,3-Dimethyl-1-butene
  d. 4-Methylcyclohexene

**6-65.** When *cis*-1-bromo-4-*tert*-butylcyclohexane reacts with sodium ethoxide in ethanol, it reacts rapidly to yield 4-*tert*-butylcyclohexene. Under similar conditions, *trans*-1-bromo-4-*tert*-butylcyclohexane reacts very slowly. Using conformational structures, explain the difference in reactivity of these cis–trans isomers.

**QUESTIONS**

Several dehydrohalogenation reactions of alkyl halides using alkoxide bases are given in *Organic Syntheses:*

Allen, C. F.; Kalm, M. J. *Organic Syntheses;* Wiley: New York, 1963; Collect. Vol. IV, p. 398.

**BIBLIOGRAPHY**

McElvain, S. M.; Kundiger, D. *Organic Syntheses*; Wiley: New York, 1955; Collect. Vol. III, p. 506.

Paquette, L. A.; Barrett, J. H. Organic *Syntheses*; Wiley: New York, 1973; Collect. Vol. V, p. 467.

Schaefer, J. P.; Endres, L. *Organic Syntheses*; Wiley: New York, 1973; Collect. Vol. V, p. 285.

Experiment [10] is adapted from the method given by:
S. A. Leone; J. D. Davis *J. Chem. Educ.* **1992**, *69*, A175.

---

## EXPERIMENT [11]   The Isolation of Natural Products

---

## INTRODUCTION

These experiments are designed to acquaint you with the procedures used for the isolation of naturally occurring and often biologically active organic compounds. These substances are known as *natural products* because they are produced by living systems. The particular natural products you are going to study come from the plant kingdom.

At the end of the nineteenth century more than 80% of all medicines in the Western world were natural substances found in roots, barks, and leaves. There was a widespread belief at that time that in plants there existed cures for all diseases. As Kipling wrote, "Anything green that grew out of the mold/ Was an excellent herb to our fathers of old." Even as the power of synthetic organic chemistry has grown during this century, natural materials still constitute a significant fraction of the drugs employed in modern medicine. For example, in the mid-1960s when approximately 300 million new prescriptions were written each year, nearly half were for substances of natural origin. Furthermore, these materials have played a major role in successfully combating the worst of human illnesses, from malaria to high blood pressure; diseases that affect hundreds of millions of people.

Unfortunately, during the latter one-half of this century a number of very powerful natural products that subtly alter the chemistry of the brain have been used in vast quantities by our society. The ultimate impact on civilization is of grave concern. Evidence clearly demonstrates that these natural substances disrupt the exceedingly complex and delicate balance of biochemical reactions that lead to normal human consciousness. How well the brain is able to repair the damage from repetitive exposure is unknown. Unfortunately, we are, whether we like it or not, currently conducting the experiment to answer that question.

The natural products that you may isolate in the following experiments include a bright-yellow crystalline antibiotic (Experiment [11A]), a white crystalline alkaloid that acts as a stimulant in humans (Experiment [11B]), and an oily material with a pleasant odor and taste (Experiment [11C].

### Experiment [11A]   Isolation and Characterization of an Optically Active Natural Product: Usnic Acid

Common name: Usnic acid
CA number: [7562-61-0]
CA name as indexed: 1,3(2*H*,9b*H*)-Dibenzofurandione, 2,6-diacetyl-7,9-dihydroxy-8,9b-dimethyl-

**Purpose**

To extract the active principle, usnic acid, from one of the lichens that produce it. Usnic acid is a metabolite found in a variety of lichens. For this experiment we will utilize a local (in Maine) species of lichen, *Usnea hirta* (often referred to as *Old Man's Beard*), which is a fruticose lichen (a lichen that possesses erect, hanging, or branched structures). This experiment illustrates an extraction technique often employed to isolate natural products from their native sources (see also Experiment [11B] for another extraction strategy). As usnic acid possesses a single chiral center (stereocenter), and as only one of the enantiomers is produced in *Old Man's Beard*, this experiment also functions as an introduction to the methods used to measure the specific rotation of optically active substances.

## LICHENS AND NATURAL PRODUCTS

Lichens, of which there are estimated to be greater than 15,000 species, are an association between an algae and a fungus that live together in an intimate relationship. This association is often termed symbiosis. Symbiosis requires that two different organisms live together in both close structural proximity and interdependent physiological combination. It ordinarily is applied to situations where the relationship is advantageous, or even required, for one or both, but not harmful to either. In the case of lichens, the algae can be grown independently of the fungi that obtain nutrients from the algae cells. The fungi are, therefore, considered to be parasitic and their contribution to the union has been viewed historically only as an aid in the absorption and retention of water and perhaps to provide a protective structure for the algae. It appears, however, that the fungi may play a far more important role in the life of the lichen than earlier appreciated. The fungi appear to generate a metabolite, usnic acid, which is the most common substance found in these primitive systems. This acid can comprise up to 20% of the dry weight of some lichens! Even more intriguing is the original belief that usnic acid appears to have *no biological function* in these plants. Why would a living system channel huge amounts of its precious energy into making an apparently useless substance? Recently, with our increased understanding of the role of chemical communication substances in ecology, it has been recognized that usnic acid very likely makes a major symbiotic contribution as a *chemical defense agent*. Indeed, in 1945 Burkholder demonstrated that several New England lichens possessed antibiotic properties, and usnic acid was subsequently shown to be the active agent against several kinds of bacteria, including staphylococcus. The Finnish company, Lääke Oy Pharmaceutical, has prepared from reindeer lichen a broad-spectrum usnic acid antibiotic for treating tuberculosis and serious skin infections. There is, in fact, evidence that lichens were used in medicine by the ancient Egyptians, and from 1600–1800 AD these plants were considered an outstanding cure for tuberculosis. Usnic acid has been investigated for use as an antibiotic by the U.S. Public Health Service. It proved to be effective in dilutions between 1 part in 100,000 and 1 part in 1,000,000 against several Gram-positive organisms.

This widespread lichen metabolite is the material isolated in this experiment. Usnic acid was first isolated and identified in 1843 by Rochleder, but a molecule of this complexity was beyond the structural knowledge of organic chemistry in those days. The structure was finally determined in 1941 by Schöpf, and not too long after that (1956) it was synthesized in the laboratory by Sir D. H. R. Barton (Nobel Laureate). Barton's route involved a spectacular one-step dimerization of a simple precursor, a synthesis that very closely mimicked the actual biogenetic pathway (see chemistry). The key step was the one-electron (1 e⁻) oxidation of methylphloracetophenone, which leads directly to the dimerization. The mechanism of this reaction, both in the plant and in the laboratory synthesis, is essentially identical to the oxidative coupling of 2-naphthol to give 1,1′-bi-2-naphthol, which is explored in detail in Experiment [5_adv].

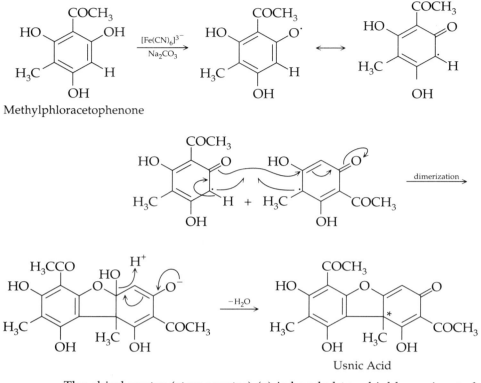

Methylphloracetophenone

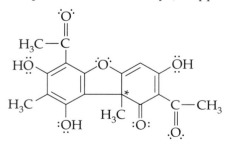

Usnic Acid

The chiral center (stereocenter) (*) is bonded to a highly conjugated aromatic ring system (see structure), which gives rise to a very large specific rotation. This enhanced interaction with polarized radiation makes this compound a particularly interesting molecule to examine for optical activity. The production of a single enantiomer in the natural product, which, as discussed above, is formed by an oxidative coupling process, implies that there must be an intimate association between the substrate and an enzyme (a biological catalyst that itself is optically active) during the crucial coupling process.[9]

### Prior Reading
Technique 8:   Measurement of Specific Rotation
Optical Rotation Theory (see pp. 111–115).

[9] Dean, F. M.; Halewood, P.; Mongkolsuk, S.; Roberston, A.; Whally, W. B. *J. Chem. Soc.* **1953**, 1250. Kreig, M. B. *Green Medicine*; Rand McNally: New York, 1964. Lewis, W.H.; Elvin-Lewis, M. P. F. *Medical Botany*; Wiley: New York, 1977. Hendrickson, J. B. *The Molecules of Nature*, W. A. Benjamin, New York, 1965. Richards, J. H.; Hendrickson, J. B. *The Biosynthesis of Steroids, Terpenes, and Acetogenins*; W. A. Benjamin, New York, 1964. Schöpf, C.; Ross, F. *Annalen* **1941**, 546, 1 (see further references cited in Experiment [5_adv]).

**DISCUSSION**

Nonracemic solutions of chiral substances, when placed in the path of a beam of polarized light, may rotate the plane of the polarized light clockwise or counterclockwise and are thus referred to as *optically active*. This angle of optical rotation is measured using a *polarimeter*. This technique is applicable to a wide range of analytical problems varying from purity control to the analysis of natural and synthetic products in the medicinal and biological fields. The results obtained from the measurement of the observed angle of rotation ($\alpha_{obs}$) are generally expressed in terms of *specific rotation* [$\alpha$]. The sign and magnitude of [$\alpha$] are dependent on the specific molecule and are determined by complex features of molecular structure and conformation, and thus cannot be easily explained or predicted. The relationship of [$\alpha$] to $\alpha_{obs}$ is as follows: $[\alpha]_\lambda^T = \dfrac{\alpha_{obs}}{l \cdot c}$ where $T$ is the temperature of the sample **in degrees Celsius** (°C), $l$ is the length of the polarimeter cell **in decimeters** (1 dm = 0.1 m = 10 cm), $c$ is the concentration of the sample **in grams per milliliter** (g/mL), and $\lambda$ is the wavelength of the light **in nanometers** (nm) used in the polarimeter. These units are traditional, though most are esoteric by contemporary standards. Thus, the specific rotation for a given compound is normally reported in terms of temperature, wavelength, concentration, and the nature of the solvent. For example: $[\alpha]_D^{25} = +12.3°$ ($c = 0.4$, CHCl$_3$) implies that the measurement was recorded in a CHCl$_3$ solution of 0.4 g/mL at 25 °C using the sodium D line (589 nm) as the light source. Unless indicated the pathlength is assumed to be 1 decimeter in these observations.

Usnic acid contains a single stereocenter (see structure), and therefore it can exist as a pair of enantiomers. In nature, however, only one of the enantiomers (*R* or *S*) would be expected to be present. Usnic acid has a very high specific rotation, $[\alpha]_D^{25} = +488°$ ($c = 0.4$, CHCl$_3$), which will give a large $\alpha_{obs}$ even at low concentrations, and for this reason it is an ideal candidate to measure rotation in a microscale experiment.

Racemic (equimolar amounts of each enantiomer) usnic acid has been resolved (separated into the individual enantiomers) through preparation and separation of the diastereomeric (−) brucine salts. This procedure was the route followed in order to obtain an authentic synthetic sample for comparison with the natural material. This separation was required because the dimerization step in the synthesis, which was carried out in the absence of enzymatic, or other chiral, influence, gave a racemic product.

A common method of extracting chemical constituents from natural sources is presented in this experiment. In this case, only one chemical compound, the usnic acid, is significantly soluble in the extraction solvent, acetone. For this reason, the isolation sequence is straightforward.

## Isolation of Usnic Acid

**EXPERIMENTAL PROCEDURE**

Estimated time for completion of the experiment: 2.5 hours.

### Physical Properties of Components

| Compound | MW | Amount | bp(°C) |
|---|---|---|---|
| Lichen | | 1.0 g | |
| Acetone | 58.08 | 15.0 mL | 56.2 |

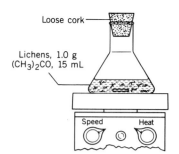

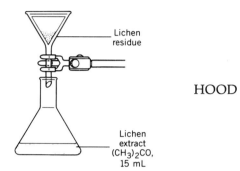

HOOD

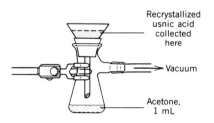

## Reagents and Equipment

Weigh and place about 1.0 g of oven dried (40 °C) crushed, or cut-up, lichens and 15.0 mL of acetone in a 50-mL Erlenmeyer flask containing a magnetic stirrer. Loosely cap the flask with a cork stopper. (■) The lichens used in this experiment are *Usnea hirta*.

## Reaction Conditions

Stir or occasionally swirl the mixture for about 30 min at room temperature. If necessary, periodically push the lichens below the surface of the acetone solvent using a glass rod.

## Isolation of Product

Filter the resulting mixture by gravity and collect the filtrate in a 25-mL Erlenmeyer flask. (■) A Pasteur filter pipet may be used to make this transfer, if desired. Remove the acetone solvent under a slow stream of air or nitrogen [HOOD] on a warm sand bath nearly to dryness. Allow the remainder of the acetone to evaporate at room temperature to obtain the crude bright yellow or orange usnic acid crystals.

## Purification and Characterization

Recrystallize the crude extract from acetone/95% ethanol (10:1). Dissolve the crystals in the minimum amount of hot acetone and add the appropriate volume of 95% ethanol. Allow the mixture to cool to room temperature and then place the flask in an ice bath to complete the recrystallization. Collect the golden-yellow crystals by vacuum filtration (■) and wash them with *cold* acetone. Dry the crystals on a porous clay plate or on a sheet of filter paper. As an alternative and more efficient procedure, the crude material may be recrystallized using a Craig tube, avoiding the filtration step with the Hirsch funnel.

Weigh the yellow needles of usnic acid and calculate the percent of the acid extracted from the dry lichen. Determine the melting point (use the evacuated melting point technique) and compare your value to that found in the literature. Obtain an IR spectrum and compare it with that of an authentic sample or that given in *The Aldrich Library of IR Spectra*.

## Chemical Tests

Chemical tests can assist in establishing the nature of the functional groups in usnic acid. Perform the 2,4-dinitrophenylhydrazine test and the ferric chloride test (see Chapter 10). Are the results significant?

## Determination of the Specific Rotation

Though usnic acid is an optically active compound with a very high specific rotation, a low-volume, long-path-length cell must be employed to successfully determine its specific rotation with microscale quantities.

Dissolve usnic acid (80 mg) in 4.0 mL of tetrahydrofuran (THF) solvent and transfer the solution to the polarimeter cell using a Pasteur pipet.

**NOTE.** *To obtain this quantity (80 mg) of usnic acid will very likely require pooling the recrystallized product of eight or nine students. Spectral grade THF* *should be used as the solvent. Many of the early specific rotation values on these substances were recorded with chloroform as the solvent, but, because it possesses some toxicity, it is now avoided if possible.*

Place the cell in the polarimeter and measure the angle of rotation. Calculate the specific rotation using the equation given in the discussion section.

**6-66.** Determine the correct *R* or *S* designation for each of the following molecules.

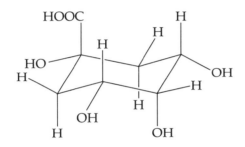

**6-67.** The structure originally proposed for cordycepic acid, which has [*α*] = +40.3°, was

Why is this not a plausible structure?

**6-68.** A sample of 150 mg of an organic compound is dissolved in 7.5 mL of water. The solution is placed in a 20-cm polarimeter tube and the rotation measured in a polarimeter. The rotation observed was +2.676°. Distilled water, in the same tube, gave a reading of +0.016°. Calculate the specific rotation for the compound.

**6-69.** Compound **A** is optically active and has the molecular formula $C_5H_{10}O$. On catalytic hydrogenation (addition of hydrogen) of **A**, Compound **B** is obtained. Compound **B** has the molecular formula $C_5H_{12}O$ and is optically inactive. Give the structure for Compounds **A** and **B**.

**6-70.** Which of the following compounds have a meso form?
2,3-Dibromopentane        2,4-Dibromopentane        2,3-Dibromobutane

---

**BIBLIOGRAPHY**

This experiment is adapted from that given by:

Todd, D. *Experimental Organic Chemistry*; Prentice-Hall: Englewood Cliffs, NJ, **1979**, p. 57.

Synthesis of usnic acid:

Barton, D. H. R.; DeFlorin, A. M.; Edwards, O. E. *J. Chem. Soc.* **1956**, 530.

Penttila, A., Fales, H. M. *Chem. Commun.* **1966**, 656.

A large scale method of isolation of usnic acid has been reported:

Stark, J .B.; Walter, E. D.; Owens, H. S. *J. Am. Chem. Soc.* **1950**, 72, 1819.

Optical, crystallographic and X-ray diffraction data have been reported for usnic acid:

Jones, F .T.; Palmer, K. J. *J. Am. Chem. Soc.* **1950**, *72*, 1820.

## Experiment [11B]   Isolation and Characterization of a Natural Product: Caffeine and Caffeine 5-Nitrosalicylate

### Product
Common names: Caffeine, 1,3,7-trimethyl-2,6-dioxopurine
CA number: [58-08-2]
CA name as indexed: 1*H*-Purine-2,6-dione, 3,7-dihydro-1,3,7-trimethyl-

### Purpose
To extract the active principle, an alkaloid, caffeine, from a native source, tea leaves. Caffeine is a metabolite (a product of the living system's biochemistry) found in a variety of plants. We will use ordinary tea bags as our source of raw material. This experiment illustrates an extraction technique often employed to isolate water-soluble, weakly basic natural products from their biological source (see also Experiment [11A] for another extraction strategy). The isolation of caffeine will also give you the opportunity to use sublimation as a purification technique, as caffeine is a crystalline alkaloid that possesses sufficient vapor pressure to make it a good candidate for this procedure. In addition, the preparation of a derivative of caffeine, its 5-nitrosalicylate salt, will be carried out. This latter conversion takes advantage of the weakly basic character of this natural product.

## ALKALOIDS

Caffeine belongs to a rather amorphous class of natural products called alkaloids. This collection of substances is unmatched in its variety of structures, biological response on nonhost organisms, and the biogenetic pathways to their formation.

The history of these fascinating organic substances begins at least 4000 years ago. While their therapeutic activity was incorporated into poultices, potions, poisons, and medicines, no attempt was made to isolate and identify the substances responsible for the physiological response until the very early 1800s.

The first alkaloid to be obtained in the pure crystalline state was morphine; isolated by Friedrich Wilhelm Sertürner (1783–1841) in 1805. He recognized that the material possessed basic character and he, therefore, classified it as a vegetable alkali (that is a base with its origin in the plant kingdom). Thus, compounds with similar properties ultimately became known as alkaloids. The term "alkaloid" was introduced for the first time by an apothecary, Meissner, in Halle in 1819.

Sertürner, also a pharmacist, lived in Hamelin, another city in Prussia. He isolated morphine from opium, the dried sap of the poppy. As the analgesic and narcotic effects of the crude resin had been known for centuries, it is not surprising that, with the emerging understanding of chemistry, the interest of Sertürner became focused on this drug, which is still medicine's major therapy for intolerable pain. He published his studies in detail in 1816 and very quickly two French professors, Pierre Joseph Pelletier (1788–1842) and Joseph Caventou (1795–1877) at the Ecole de Pharmacie in Paris recognized the enormous importance of Sertürner's work.

In the period from 1817–1820, these two men and their students isolated many of the alkaloids, which continue to be of major importance.

Included in that avalanche of purified natural products was caffeine, which they obtained from the coffee bean. This substance is the target compound that you will be isolating directly from the raw plant in this experiment. A little more than 75 years later, caffeine was first synthesized by Fischer in 1895 from dimethylurea and malonic acid.

---

These compounds are separated into three general classes of materials.

**THE CLASSIFICATION OF ALKALOIDS**

**1. True alkaloids:** these compounds contain nitrogen in a hetero-cyclic ring; are almost always basic (the lone-pair of the nitrogen is respon-sible for this basic character); are derived from amino acids in the biogene-sis of the alkaloid; invariably are toxic and possess a broad spectrum of pharmacological activity; are found in a rather limited number of plants (of the 10,000 known genera only 8.7% possess at least one alkaloid); and normally occur in a complex with an organic acid (this helps to make them rather soluble in aqueous media). As we will see, there are numerous exceptions to these rules. For example, there are several very well-known quaternary alkaloids. These are compounds in which the nitrogen has become tetravalent and positively charged (as in the ammonium ion). Thus, they are not actually basic.

**2. Protoalkaloids:** These compounds are simple amines, derived from amino acids, in which the basic nitrogen atom is not incorporated into a ring system; and they are often referred to as *biological amines*. An exam-ple of a protoalkaloid is mescaline.

**3. Pseudoalkaloids:** These compounds contain nitrogen atoms usu-ally *not* derived from amino acids. There are two main classes into which pseudoalkaloids are divided, the steroidal alkaloids and the *purines*. Caf-feine has been assigned to this latter class of alkaloids.

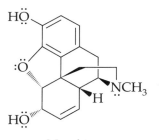

Morphine

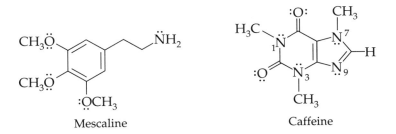

Mescaline                    Caffeine

*Prior Reading*
   *Technique 4:*   Solvent Extraction
                        Solid–Liquid extraction (see pp. 78–80).
                        Liquid–Liquid Extraction (see pp. 80–82).
   *Technique 9:*   Sublimation
                        Sublimation Theory (see pp. 116–117).

## DISCUSSION

Caffeine (1,3,7-trimethylxanthine) and its close relative theobromine (3,7-dimethylxanthine) both possess the oxidized purine skeleton (xanthine). These compounds are classified as pseudoalkaloids, as only the nitrogen atom at the 7 position can be traced to an amino group originally derived from an amino acid (in this case glycine). This classification emphasizes the rather murky problem of deciding just what naturally occurring nitrogen bases are *true* alkaloids. We will simply treat caffeine as an alkaloid.

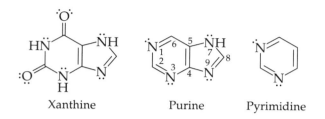

Xanthine          Purine          Pyrimidine

It should be noted that although the pyrimidine ring (present in caffeine's purine system) is a significant building block of nucleic acids it is rare elsewhere in nature.

These two methylated xanthines are found in quite a number of plants and have been extracted and widely used for centuries. Indeed, they very likely have been, and remain today, the predominant stimulant consumed by humans. Every time you make a cup of tea or coffee, you perform an aqueous extraction of plant material (tea leaves, *Camellia sinenis*, 1–4%, or coffee beans, *Coffea* spp., 1–2%) to obtain a dose of between 25–100 mg of caffeine. Caffeine is also the active substance (~2%) in maté (used in Paraguay as a tea) made from the leaves of *Ilex paraguensis*. In coffee and tea, caffeine is the dominant member of the pair, whereas in *Theobroma cacao*, from which we obtain cocoa, theobromine (1–3%) is the primary source of the biological response. Caffeine acts to stimulate the central nervous system with its main impact on the cerebral cortex, and as it makes one more alert, it is no surprise that it is the chief constituent in No-Doz® pills.

Caffeine is readily soluble in hot water (because the alkaloid is often bound in thermally labile, partially ionic complexes with naturally occurring organic acids, such as with 3-caffeoylquinic acid in the coffee bean), which allows for relatively easy separation from black tea leaves by aqueous extraction.

Other substances, mainly tannic acids, are also present in the tea leaves and they are also water soluble. The addition of sodium carbonate, a base, during the aqueous extraction helps to increase the water solubility of these acidic substances by forming ionic sodium salts and liberating the free base.

Subsequent extraction of the aqueous phase with methylene chloride, in which free caffeine has a moderate solubility, allows the transfer of the caffeine from the aqueous extract to the organic phase. At the same time, methylene chloride extraction leaves the water-soluble sodium salts of the organic acids behind in the aqueous phase.

Extraction of the tea leaves directly with nonpolar solvents (methylene chloride) to remove the caffeine gives very poor results—since, as we have seen, the caffeine is bound in the plant in a partially ionic complex that will not be very soluble in nonpolar solvents. Thus, water is the superior extraction solvent for this alkaloid. The water also swells the tea leaves and allows for easier transport across the solid–liquid interface.

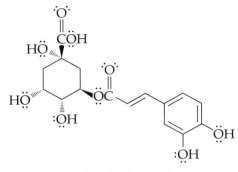

3-Caffeoylquinic acid

Following extraction and removal of the solvent, sublimation techniques are applied to the crude solid residues in order to purify the caffeine. This technique is especially suitable for the purification of solid substances at the microscale level, if they possess sufficient vapor pressure. Sublimation techniques are particularly advantageous when the impurities present in the sample are nonvolatile under the conditions employed.

Sublimation occurs when a substance goes directly from the solid phase to the gas phase upon heating, bypassing the liquid phase. Sublimation is technically a straightforward method for purification in that the materials need only be heated and therefore, mechanical losses can be kept to a minimum (the target substance must, of course, be thermally stable at the required temperatures). Materials sublime only when heated *below* their melting points, and reduced pressure is usually required to achieve acceptable sublimation rates. Obviously, substances that lend themselves best to purification by sublimation are those that do not possess strong intermolecular attractive forces. Caffeine, and ferrocene (the latter is used as a reactant in Experiment [27]) meet these criteria as they present large flat surfaces occupied predominantly with repulsive $\pi$ electrons. For other isolations, see the discussion of *solid-phase* extraction methods in Technique 4 for an example of the extraction of caffeine from coffee beans (see pp. 89–90).

---

**EXPERIMENTAL PROCEDURE**

Estimated time to complete the experiment: 2.5 hours.

**Physical Properties of Constituents**

| Compound | MW | Amount | mmol | mp(°C) |
|---|---|---|---|---|
| Tea | | 1.0 g | | |
| Water | | 10 mL | | |
| Sodium carbonate | 105.99 | 1.1 g | 10 | 851 |

## Reagents and Equipment

Carefully open a commercial tea bag (2.0–2.5 g of tea leaves) and empty the contents. Weigh out 1.0 g of tea leaves and place them back in the empty tea bag. Close and secure the bag with staples.

Weigh, and add to a 50-mL Erlenmeyer flask, 1.1 g (0.01 mol) of

Anhydrous Na$_2$CO$_3$, 1.1 g
+ H$_2$O, 10 mL + tea bag
with tea leaves, 1.0 g

anhydrous sodium carbonate followed by 10 mL of water. (■) The mixture is heated with occasional swirling on a hot plate to dissolve the solid. Now add the 1.0 g of tea leaves (in the tea bag) to the solution. Place the bag in the flask so that it lies flat across the bottom.

## Reaction Conditions

GENTLY

Place a small watch glass over the mouth of the Erlenmeyer flask and then heat the aqueous suspension to boiling **[GENTLY]** for 30 min using a sand bath.

## Isolation of Product

Cool the flask and contents to room temperature. Transfer the aqueous extract from the Erlenmeyer flask to a 12- or 15-mL centrifuge tube using a Pasteur filter pipet. In addition, gently squeeze the tea bag by pressing it against the side of the Erlenmeyer flask to recover as much of the basic extract as possible. Set aside the tea bag and its contents.

Extract the aqueous solution with 2.0 mL of methylene chloride.

**NOTE.** *The tea solution contains some constituents that may cause an emulsion. If you find that during the mixing of the aqueous and organic solvent layers (by shaking or using a Vortex mixer) an emulsion is obtained, it can be broken readily by centrifugation.*

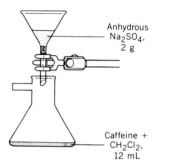

Anhydrous
Na$_2$SO$_4$,
2 g

Caffeine +
CH$_2$Cl$_2$,
12 mL

Separate the lower (methylene chloride) layer (check to make sure that the lower layer is, indeed, the organic layer by testing the solubility of a few drops of it in a test tube with distilled water) using a 9-in. Pasteur pipet. Drain the wet extracts through a filter funnel containing a small plug of cotton that is covered with about 2.0 g of anhydrous sodium sulfate, previously "moistened" with a small amount of methylene chloride. (■) (The organic phase will be saturated with water following the extraction, therefore it is referred to as "wet." It also may contain a few droplets of the aqueous phase, which become entrained during the phase separation; this can be particularly troublesome if an emulsion forms during the mixing.)

Collect the dried filtrate in a 25-mL filter flask. Extract the remaining aqueous phase with four additional 2.0-mL portions of methylene chloride (4 × 2 mL). Each extraction (an extraction is often referred to as a washing) is separated, and then dried as above, and transferred to the same filter flask. Finally, rinse the sodium sulfate with an additional 2.0 mL of methylene chloride and combine this wash with the earlier organic extracts.

HOOD

Add a boiling stone to the flask and concentrate the solution to dryness **[HOOD]** by warming the flask in a sand bath. The crude caffeine should be obtained as an off-white crystalline solid.

## Purification and Characterization

Purify the crude solid caffeine by sublimation.

Assemble a sublimation apparatus as shown in Figure 5.66; either arrangement is satisfactory. Using an aspirator, apply a vacuum to the system through the filter flask (remember to install a water-trap bottle between the side-arm flask and the aspirator). After the system is evacuated, run cold water gently through the cold finger or add ice to the centrifuge tube. By cooling the surface of the cold finger *after* the system has been evacuated, you will minimize the condensation of moisture on the area where the sublimed sample will collect.

Once the apparatus is evacuated and cooled, begin the sublimation by gently heating the flask with a microburner or sand bath. If you use a gas burner, always keep moving the flame back and forth around the bottom and sides of the flask.

**BE CAREFUL.** *Do not MELT the caffeine. If the sample does begin to melt, remove the flame for a few seconds before heating is resumed. Overheating the crude sample will lead to decomposition and the deposition of impurities on the cold finger. High temperatures are not necessary since the sublimation temperature of caffeine (and of all solids that sublime) is below the melting point. It is generally worthwhile to carry out sublimations as slowly as possible, as the purity of the material collected will be enhanced.*

When no more caffeine will sublime onto the cold finger, remove the heat, shut off the aspirator and the cooling water to the cold finger, and allow the apparatus to cool to room temperature under reduced pressure. Once cooled, carefully vent the vacuum and return the system to atmospheric pressure. *Carefully,* remove the cold finger from the apparatus.

**NOTE.** *If the removal of the cold finger is done carelessly, the sublimed crystals may be dislodged from the sides and bottom of the tube and drop back onto the residue left in the filter flask.*

Scrape the caffeine from the cold finger onto weighing paper using a microspatula and a sample brush. Weigh the purified caffeine and calculate its percent by weight in the original tea leaves. Determine the melting point and compare your value to that in the literature.

If your melting point apparatus uses capillary tubes to determine the melting point, an evacuated sealed tube is necessary since caffeine sublimes; the melting point is above the sublimation temperature (see Chapter 4). The melting point may be obtained using the Fisher–Johns apparatus without this precaution.

Obtain an IR spectrum as a KBr pressed disk and compare it with that of an authentic sample.

## Chemical Test

Does the soda lime or the sodium fusion test (see Chapter 10) confirm the presence of nitrogen in your caffeine product?

---

It is not completely surprising to find that caffeine in the coffee bean is bound in a thermally labile complex with acid, as this alkaloid is a weakly basic substance possessing a base strength somewhat greater than that of an aryl amide. As the purine ring system of caffeine has little reactive

**DERIVATIVE: CAFFEINE 5-NITROSALICYLATE**

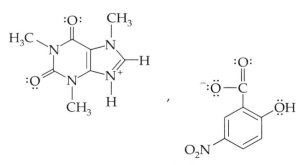

Caffeine 5-nitrosalicylate

functionality, the formation of simple chemical derivatives is limited. The ease of association with high-melting carboxylic acids, however, offers a route to a variety of materials with well-defined melting points.

One such acid is 5-nitrosalicylic acid, which is prepared by nitration of salicylic acid in Experiment [29]. The purified caffeine obtained in this experiment may be further characterized by preparation of the 5-nitrosalicylate complex. This association is similar to the natural one formed with 3-caffeoylquinic acid.

## EXPERIMENTAL PROCEDURE

Estimated time to complete the experiment: 0.5 hours.

### Physical Properties of Reactants and Product

| Compound | MW | Amount | mmol | mp(°C) | bp(°C) |
|---|---|---|---|---|---|
| Caffeine | 194.20 | 11 mg | 0.06 | 238 | |
| 5-Nitrosalicylic acid | 183.12 | 10 mg | 0.06 | 229–30 | |
| Petroleum ether (60–80 °C) | | 0.5 mL | | | |
| Ethyl acetate | 88.12 | 0.7 mL | | | 77 |
| Caffeine 5-nitrosalicylate | 377.32 | | | 180 | |

### Reagents and Equipment

Weigh and add to a 3.0-mL conical vial, containing a magnetic spin vane and equipped with an air condenser, 11 mg (0.06 mmol) of caffeine, 10.0 mg (0.06 mmol) of 5-nitrosalicylic acid, and 0.7 mL of ethyl acetate.

### Reaction Conditions

*Gently* warm the mixture on a hot plate, with stirring, to dissolve the solids. Add 0.5 mL of petroleum ether (bp. 60–80 °C) to the warm ethyl acetate solution, mix, and warm for several seconds. Remove the spin vane using forceps.

### Isolation of Product

Cool the mixture to room temperature and then place it in an ice bath for 10–15 min. Collect the crystals under reduced pressure using a Hirsch funnel, and then wash the filter cake with 0.5 mL of cold ethyl acetate. Dry the product on a porous clay plate or filter paper.

### Purification and Characterization

The product normally is sufficiently pure for direct characterization. Weigh the caffeine 5-nitrosalicylate and calculate the percent yield. Determine the melting point and compare your value with that reported above.

Obtain an IR spectrum by the KBr disk technique and compare it with those shown in Figures 6.25 and 6.26. The infrared spectrum reveals some of the details of the derivative formation (see below).

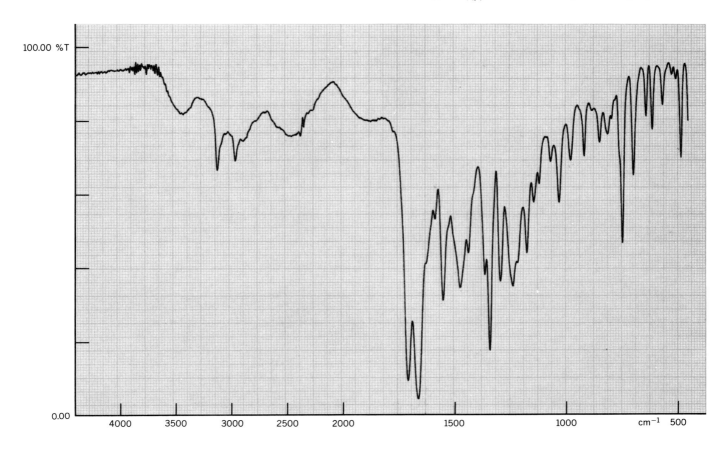

| | |
|---|---|
| Sample __Caffeine 5-nitrosalicylate__ | |
| %T _X_ ABS __ Background Scans __4__ | Scans _____16__ |
| Acquisition & Calculation Time __42 sec__ | Resolution __4.0 cm⁻¹__ |
| Sample Condition _____solid_____ | Cell Window _____ |
| Cell Path Length _____ | Matrix Material __KBr__ |

**FIGURE 6.25**    IR spectrum: caffeine 5-nitrosalicylate.


## Infrared Analysis

The infrared spectrum of 5-nitrosalicylic acid (Fig. 6.26) is characteristic of aromatic carboxylic acids (see discussion in Experiment [7]). Three points to note in the spectrum are (a) the substitution of the ring is revealed by the presence of the 1,2,4-combination band pattern with peaks at 1940, 1860, and 1815 cm$^{-1}$; (b) the strongest band in the spectrum below 1750 cm$^{-1}$ is assigned to the symmetric stretch of the —NO$_2$ group found at 1339 cm$^{-1}$; and (c) the conjugated carboxyl C=O stretch is located at 1675 cm$^{-1}$.

In the spectrum of the complex (Fig. 6.25) we do not find evidence for ionized carboxylate. This group, if present, would give rise to two very strong *broad* bands 1600–1550 and 1400–1330 cm$^{-1}$. What is observed is the carboxylate C=O stretch at 1665 cm$^{-1}$ overlapped with a caffeine band. Evidence for very strong hydrogen bonding, however, is indicated by the series of very broad bands extending from 3550–2000 cm$^{-1}$. The complex association does not, therefore, very likely involve a complete proton transfer as is indicated in the above simplified chemical structures of the complex. The unambiguous formation of the 5-nitrosalicylic acid complex is best ascertained by the identification of the presence of the —NO$_2$ symmetric stretching vibration from a strong band located at 1345 cm$^{-1}$. Caffeine does not possess a band in this region.

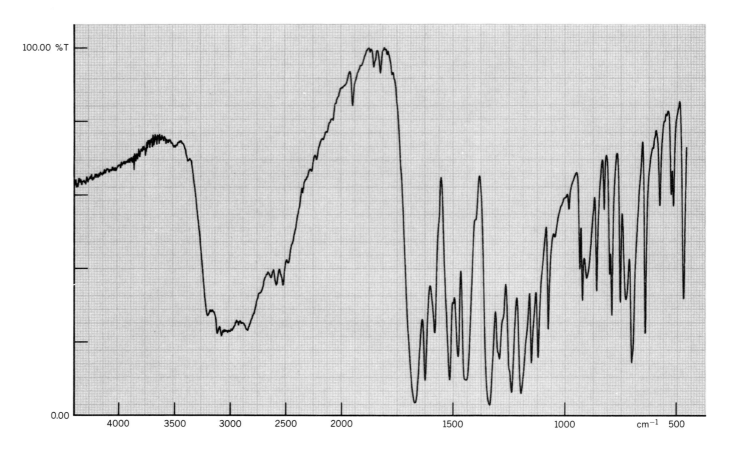

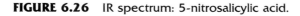

Sample __5-Nitrosalicylic acid__

%T _X_ ABS __ Background Scans _4_       Scans _____16_____

Acquisition & Calculation Time _42 sec_    Resolution _4.0 cm⁻¹_

Sample Condition __solid__           Cell Window _____

Cell Path Length _____     Matrix Material _KBr___

**FIGURE 6.26**   IR spectrum: 5-nitrosalicylic acid.

## QUESTIONS

**6-71.** Compounds such as naphthalene and 1,4-dichlorobenzene find use as mothballs since they sublime at a slow rate at atmospheric pressure. Explain this behavior in terms of the structure of the molecules.

**6-72.** a. How many peaks would you expect to find in the NMR spectrum of caffeine? b. What characteristic absorption bands would you expect to find in the infrared spectrum of caffeine?

**6-73.** The vapor pressures of 1,2-diphenylethane, $p$-dichlorobenzene, and 1,3,5-trichlorobenzene are 0.06, 11.2, and 1.4 torr, respectively, at their melting point (52–54 °C). Which compound is likely to be sublimed most rapidly at a reduced pressure of 15 torr and a temperature of 40 °C?

**6-74.** To color spots on TLC plates for easier visualization after elution with solvent, the plates can be "developed" in a sealed chamber containing solid iodine. Explain how the solid–vapor equilibrium operates in this instance.

**6-75.** Caffeine is soluble in ethyl acetate. Do you think that the purity of your product could be checked by TLC using ethyl acetate as an elution solvent? Explain.

**6-76.** The infrared spectrum of 5-nitrosalicylic acid (Fig. 6.26) possesses the typical broad medium band found in *acid dimers* (908 cm$^{-1}$). In the caffeine 5-nitrosalicylate complex, however, this band is missing. Suggest a reason why the 908-wavenumber peak vanishes.

**BIBLIOGRAPHY**

Several extraction procedures for isolating caffeine from tea, some designed for the introductory organic laboratory, have been reported.

Ault, A.; Kraig, R. *J. Chem. Educ.* **1969,** *46,* 767.

Fischer, E. *Ber.* **1895,** *28,* 2473, 3135.

Laswick, J. A.; Laswick, P. H. *J. Chem. Educ.* **1972,** *49,* 708.

Mitchell, R. H.; Scott, W. A.; West, P. R. *J. Chem. Educ.* **1974,** *51,* 69.

## Experiment [11C]  Isolation of a Natural Product by Steam Distillation: Cinnamaldehyde from Cinnamon

Common name: Cinnamaldehyde
CA number: [14371-10-9]
CA name as indexed: 2-Propenal, 3-phenyl-, (*E*)-

### Purpose

To extract oil of cinnamon from a native plant source, such as *Cinnamomum zeylanicum*, and to then purify the principal flavor and odor component of the oil, cinnamaldehyde. To demonstrate the steam distillation techniques (at the semimicro level) to the collection of essential oils.

### Prior Reading

Technique 2A:  Simple Distillation at the Semimicroscale Level (see pp. 65–68).
Technique 3:   Steam Distillation (see pp. 71–72).
Technique 4:   Solvent Extraction
   Liquid–Liquid Extraction (see pp. 80–82).
   Drying of the Wet Organic Layer (see pp. 86–87).
Technique 6:   Chromatography
   Concentration of Solutions with Nitrogen Gas (see p. 107).

**ESSENTIAL OILS**

Let us begin by defining what we mean by the term "metabolite." The metabolism of an organism is composed of the biochemical reactions and pathways in that living system. The products (most of them organic molecules) derived from this array of molecular transformations are the *metabolites*. This vast collection of substances that are generally referred to as *natural products* are, in fact, the metabolites of the natural living world.

*Natural products* are divided into two large families of compounds. Those metabolites that are *common* to the large majority of all organisms are known as the *primary metabolites*. In general, they have well-defined roles in the biochemistry of the system. For example, the amino acids are the building blocks for protein synthesis in all organisms. The second great category of *natural products* is known as the *secondary metabolites*. Individual secondary metabolites are far less widely distributed in nature and may be unique to single species (or even limited to a variety of a particular species). While the biochemical role of some of these compounds was established early and easily, the majority of these materials were believed to be of little importance to the functioning of the living system, and their presence was unexplained until very recently. With the development of chemical communication theory over the last few decades, however, the vitally important roles of many of the secondary metabolites in the life cycles of their particular host organisms have been revealed.

You may have had the opportunity to become acquainted with secondary metabolites in Experiments [11A] and [11B]. In Experiment [11A] an acetogenin (this term refers to the biochemical origin of this material from eight acetic acid residues), usnic acid, was isolated from a lichen where it can occur in dramatically high concentrations. Only recently has the role of usnic acid as a defense mechanism come to be fully appreciated.

In Experiment [11B], the alkaloid caffeine was obtained from tea. This compound is a very unusual example of a purine ring system in a secondary metabolite. The ecological significance of the presence of caffeine in both tea and coffee seeds has been established and it has been shown that caffeine acts against both predators and competitors.

We will now examine, in Experiment **[11C]**, a third class of secondary metabolites: the *essential oils*. The majority of these materials are high-boiling liquids that can be extracted from plant material via steam distillation techniques. The value of codistilling high-boiling substances was learned early in the days of alchemy, and as these oils very often gave pleasant odors and flavors, they were considered to be the "essence" of the original plant material. Eventually, they became known as *essential oils*.

The use of these materials became widespread as flavorings, perfumes, medicines, and as both insect repellents and attractants. By the early 1800s, as it became possible to establish the carbon/hydrogen ratio in organic substances, many of the oils possessing pine-type odors (the oil of turpentine) were shown to have identical C/H ratios. These materials ultimately became known as terpenes. The terpenes all have their origin in mevalonic acid, from which they utilize, as their building block, a branched five-carbon unit as in isoprene. The terpenes of the essential oils occur as $C_{10}$ (monoterpenes), $C_{15}$ (sesquiterpenes), $C_{20}$(diterpenes), $C_{30}$ (triterpenes), and $C_{40}$ (tetraterpenes) compounds. Today, this collection of substances represents a large fraction of the known secondary metabolites, including the steroids. Terpenes may be polymerized, extending to much higher molecular weights, with between 1000 and 5000 repeating isoprene units (MW = 60,000–350,000) to yield polymers known as the natural rubbers.

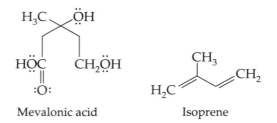

Mevalonic acid                    Isoprene

As we have seen, many of the compounds found in the essential oils possess pleasing properties of taste and odor, and we now know that many of these systems contain either ketone or aldehyde functional groups. Our senses of taste and smell, however, possess a wide range of responses to the shape and dimensions of the carbon skeleton supporting the main functionality that triggers the odor signal. Thus, our sense of odor may involve simultaneous multiple stimulations by many different molecular species or, as in a number of cases, the principal response may be to a single component. As the shape of the odor- or taste-inducing molecule plays a significant role in the effect, it is not surprising and quite interesting to find that chirality can have a dramatic impact on our perception of a particular odor.

One of the classic examples of this type of response is the case of the cyclic ketone carvone, which contains a single stereocenter (*). The *S* enantiomer is the principal odor and flavor component in caraway seed, whereas the *R* enantiomer gives rise to the odor and flavor of spearmint!

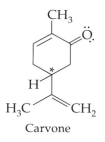
Carvone

What we find pleasant may be offensive to others. A constituent of the oil of lemon grass is citronellal, a $C_{10}$ unsaturated aldehyde. While we find this compound to have a pleasant fragrance, it is a potent alarm signal in ants that is shunned by many other insects. Thus, this terpene aldehyde has been employed effectively by both ants and humans as an insect repellent.

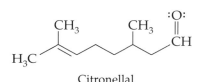
Citronellal

In this experiment we will isolate the principal component of the oil of cinnamon, another naturally occurring aldehyde, cinnamaldehyde. The oil is first extracted from the dried parts of the *Cinnamomum* plant by steam distillation. Although this aromatic aldehyde is a component of an essential oil, it is not formed from mevalonic acid and is not a terpene. Cinnamaldehyde is also not an acetogenin nor is it related to usnic acid. The origin of this fragrant material is shikimic acid, which is part of the plant's *primary metabolism*.

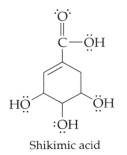
Shikimic acid

Cinnamaldehyde's formation from shikimic acid utilizes one of only two biogenetic routes in nature that lead to the aromatized benzene ring (the other pathway is found in the acetogenins and produces secondary metabolites like usnic acid; see Experiment [11A]). The shikimic acid route contributes to a class of metabolites called the phenylpropanes (Ph—C$_3$), of which cinnamaldehyde is one of a limited number of simple end products. Another close relative is, for example, eugenol from oil of cloves.

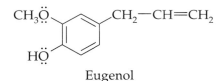

Eugenol

The principal metabolic fate of the phenylpropanes is the formation of lignin polymers that are the fundamental basis of the structural tissue in all plants. Thus, cinnamaldehyde itself is a relatively rare example of a primary metabolite which has been expressed in an *essential oil*.

## DISCUSSION

### Steam Distillation

The process of steam distillation can be a valuable technique in the laboratory for the separation of thermally labile, high-boiling substances from relatively nonvolatile materials. For steam distillation to be successful, however, the material to be isolated must be nearly immiscible with water.

Steam distillation is in essence the codistillation (or simultaneous distillation) of two immiscible liquid phases. By definition, one of these liquid phases is water and the other phase is usually a mixture of organic substances that have a low solubility in water. Though steam distillation is widely used as a separation technique for natural products, and occasionally for the isolation and/or purification of synthetic products that decompose at their normal boiling points, it has several limitations. For example, it is not the method of choice when a dry product is required or if the compound to be isolated reacts with water. Obviously, steam distillation is not feasible if the compound to be isolated decomposes upon contact with steam at 100 °C.

Although cinnamaldehyde decomposes at its normal boiling point, it may be extracted from the plant without degradation by boiling water. Steam distillation, therefore, is the method of choice for the isolation of this pleasant smelling and tasting aldehyde.

## COMPONENT

Cinnamaldehyde

## EXPERIMENTAL PROCEDURE

Estimated time to complete the experiment: 2.5 hours.

**Physical Properties of Components**

| Compound | MW | Amount |
|----------|-----|--------|
| Cinnamon |  | 1 g |
| Water | 18.02 | 4.0 mL |

## Components and Equipment

Place 1 g of chopped stick cinnamon (or powder) and 4 mL of water in a 10-mL round-bottom flask containing a boiling stone. Attach the flask to a Hickman still head equipped with an air condenser. (■)

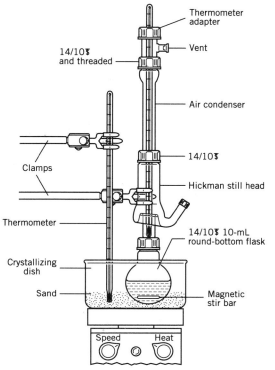

Thermometer adapter

Vent

14/10⸓ and threaded

Air condenser

Clamps

14/10⸓

Hickman still head

Thermometer

14/10⸓ 10-mL round-bottom flask

Crystallizing dish

Sand

Magnetic stir bar

Speed    Heat

Cinnamon stick + H₂O, 4 mL

## Distillation Conditions

Place the apparatus on a sand bath maintained at 150–160 °C. Use an aluminum foil shield or baffle (not shown in drawing) to cover the sand bath. This procedure will prevent the collection collar of the Hickman still from overheating.

**NOTE.**  *The mixture tends to foam during the distillation, so care must be taken to prevent contamination of the distillate by cinnamon particles, especially if powdered cinnamon is used.*

Position a thermometer in the throat of the Hickman still to record the distillation temperature, which should be very close to 100 °C. In early Hickman stills (without side-arm access) this thermometer makes pipetting of the condensate difficult. It is suggested that in the latter setup, the thermometer be removed after the initial distillation temperature is established and recorded.

## Isolation of Cinnamaldehyde

Remove the milky cinnamaldehyde–water (two-phase) distillate that collects in the collar of the still using a 9-in. Pasteur pipet (or 6-in. in the case of side-arm stills). Transfer this material to a 12- or 15-mL centrifuge tube.

Continue the distillation for approximately 1 h or until about 5–6 mL of distillate is collected in the centrifuge tube.

**NOTE.**   *Add additional water during the course of the distillation to maintain the original volume in the flask. Add this water using a 9-in. Pasteur pipet inserted down the neck of the Hickman still, after first removing the thermometer if it is still in place (the thermometer need not be replaced in the still following the addition).*

Extract the combined distillate fractions, which you collected in the centrifuge tube, with three successive 2-mL portions of methylene chloride (3 × 2 mL). Use the first portion of methylene chloride to rinse the collection collar of the Hickman still. After each extraction, transfer the lower methylene chloride layer (Pasteur filter pipet) to a 25-mL Erlenmeyer flask.

Dry the combined extracts over anhydrous sodium sulfate.

### Purification and Characterization

Transfer the dried methylene chloride solution **in at least two portions**, using a Pasteur filter pipet, to a **tared** 5-mL conical vial. Evaporate the solvent in a warm sand bath using a slow stream of nitrogen gas. After all the solution has been transferred and the solvent evaporated, rinse the sodium sulfate with an additional two 0.5-mL portions of methylene chloride. Transfer the rinses to the same vial and concentrate as before.

Weigh the flask and calculate the percentage of crude cinnamaldehyde extracted from the original sample of cinnamon.

Record the infrared spectrum and compare it to that reported in the *Aldrich Library of Infrared Spectra.*

## QUESTIONS

**6-77.**   List several advantages and disadvantages of steam distillation as a method of purification.

**6-78.**   Explain why the distillate collected from the steam distillation of cinnamon is cloudy.

**6-79.**   Calculate the weight of water required to steam distill 500 mg of bromobenzene at 95 °C. The vapor pressure of water at this temperature is 640 torr; bromobenzene, 120 torr.

**6-80.**   Steam distillation may be used to separate a mixture of *p*-nitrophenol and *o*-nitrophenol. The ortho isomer distills at 93 °C; the para isomer does not. Explain.

**6-81.**   A mixture of nitrobenzene and water steam distills at 99 °C. The vapor pressure of water at this temperature is 733.2 torr. What weight of water is required to steam distill 300 mg of nitrobenzene?

## EXPERIMENT [12]

### Reductive Catalytic Hydrogenation of an Alkene: Octane

Common name: Octane
CA number: [1111-65-9]
CA name as indexed: Octane

## Purpose

To reduce the carbon–carbon double bond of an alkene by addition of molecular hydrogen ($H_2$). To gain an understanding of the important role that metal catalysts play in the stereospecific reductions of alkenes (and alkynes), to form the corresponding alkanes, by the activation of molecular hydrogen. To observe the powerful influence on column chromatography of heavy metal ions, such as silver ($Ag^{1+}$), which lead to effective separation of mixtures of alkenes from alkanes. Finally, to appreciate the enormous importance and breadth of application of these reduction reactions in both industrial synthesis and basic biochemistry (for important examples see Experiment [8])

---

**REACTION**

$$CH_3—(CH_2)_5—CH{=}CH_2 \; \underset{\substack{NaBH_4 \\ C_2H_5OH \\ HCl(6\,M)}}{\overset{H_2PtCl_6}{\rightleftharpoons}} \; CH_3—(CH_2)_5—CH_2—CH_3$$

1-Octene                                         Octane

### Prior Reading

*Technique 4:*   Solvent Extraction
                      Liquid–Liquid Extraction (see pp. 80–82).
*Technique 6:*   Chromatography
                      Column Chromatography (see pp. 89–102).
                      Concentration of Solutions (see pp. 106–108).

---

**DISCUSSION**

The addition of hydrogen to an alkene (or to put it another way, the saturation of the double bond of an alkene with hydrogen) to produce an alkane is a very important reaction in organic chemistry. Alkanes are also called *saturated hydrocarbons*, as the carbon skeletons of alkanes contain the greatest possible number of hydrogen atoms permitted by tetravalent carbon atoms; alkenes are thus *unsaturated hydrocarbons*. Hydrogenation reactions have widespread use in industry. For example, we all consume vegetable fats *hardened* by partial hydrogenation (see Experiment [8]) of the polyunsaturated oils that contained several carbon–carbon double bonds per molecule as isolated from their original plant sources. Partially hydrogenated fats represented a consumer market of over 1.7 billion kg in 1977.

The hydrogenation reaction is exothermic; the energy released is approximately 125 kJ/mol for most alkenes, but on the kinetic side of the ledger, this reductive pathway requires a significant activation energy to reach the transition state. Thus, alkenes can be heated in the presence of hydrogen gas at high temperatures for long periods without any measurable evidence of alkane formation. However, when the reducing reagent ($H_2$) and the substrate (the alkene) are in intimate contact with each other in the presence of a finely divided metal catalyst, rapid reaction does occur at room temperature. Under these conditions, successful reduction is generally observed at pressures of 1–4 atm. For this reason, this reaction is often referred to as *low-pressure catalytic hydrogenation*. These reactions are termed *heterogeneous reactions*, as they occur at the boundary between two phases—in this case a solid and a liquid.

The main barrier to the forward progress of the reaction is the very strong H—H bond that must be broken. Molecular hydrogen, however, has been found to be adsorbed by a number of metals in substantial quantities, indeed in some instances the amount of hydrogen contained in the metal lattice can be greater than in an equivalent volume of pure liquid hydrogen! In this adsorption process the H—H bond is broken or severely weakened. [It is interesting to note that this adsorption process, which necessarily involves a large exchange of chemical energy between the metal lattice and the adsorbed hydrogen, may in some, as yet unexplained way, be related to the "cold fusion" problem in which palladium saturated with a heavy isotope of hydrogen (deuterium) exhibits apparent excess thermal energies on electrolysis.]

The $\pi$ system of the alkene is also susceptible to adsorption onto the metal surface and when this occurs the barrier to reaction between the alkene and the activated hydrogen drops dramatically.

Catalytic hydrogenations are also a representative example of a class of organic reactions known as *addition reactions,* which are reactions in which two new substituents are *added* to a molecule (the alkene substrate in this case) across a $\pi$ system. Usually addition is 1,2, but in extended $\pi$ systems such as 1,3 dienes, the addition may occur 1,4. In catalytic hydrogenations, formally, one hydrogen atom of a hydrogen molecule adds to each carbon of the alkene linkage, C═C. It is not at all clear that both hydrogen atoms must come from the same original hydrogen molecule even though they are added stereospecifically in syn (cis) fashion while both systems are coordinated with the metal surface. A representation of the stereochemistry of this addition is given below.

The metals most often used as catalysts in low-pressure (1–4 atm) hydrogenations in the laboratory are nickel, platinum, rhodium, ruthenium, and palladium. In industry, high-pressure large scale processes are more likely to be found. For example, Germany had little or no access to naturally formed petroleum deposits during WW II, but did possess large coal mines. The Germans mixed powdered coal with heavy tar (from previous production runs) and 5% iron oxide and heated this in the presence of $H_2$ at a pressure of 3000 lb./in$^2$, at about 500 °C for 2 hours to yield synthetic crude oil. Thirteen German plants operating in 1940 produced 24 million barrels that year, with an average of 1.5–2 tons of coal ultimately converted to about 1 ton of gasoline.

In the present experiment the metal catalyst, platinum, is generated in situ by the reaction of chloroplatinic acid with sodium borohydride. The reduced platinum metal is formed in a colloidal suspension, which provides an enormous surface area, and therefore excellent conditions, for heterogeneous catalysis. The molecular hydrogen necessary for the reduction can also be conveniently generated in situ by the reaction of sodium borohydride with hydrochloric acid.

$$4\ NaBH_4 + 2\ HCl + 7\ H_2O \rightarrow Na_2B_4O_7 + 2\ NaCl + 16\ H_2$$

This reduction technique does not require equipment capable of safely withstanding high pressures. The use of chloroplatinic acid, therefore, is particularly attractive for saturating easily reducible groups, such as unhindered alkenes or alkynes, in the laboratory. The potential limitation to the use of this reagent is that other reducible functional groups, such as aldehydes and ketones, normally inert to catalytic hydrogenations of alkenes and alkynes, may be reduced by the sodium borohydride. Thus, with chloroplatinic acid and sodium borohydride, we accept, as a compromise, a more limited set of potential reactants (substrates) for the convenience inherent in the reagent.

The platinum catalyst generated in the reaction medium adsorbs both the internally generated molecular hydrogen and the target alkene on its surface. The addition of the hydrogen molecule ($H_2$) (evidence strongly suggests that it is actually atomic hydrogen that attacks the alkene $\pi$ system) to the alkene system while they are both adsorbed on the metal surface results in the reduction of the substrate and the formation of an alkane. The addition is, as mentioned earlier, syn (or cis) as both hydrogen atoms add to the same face of the alkene plane. The mechanistic sequence is outlined below.

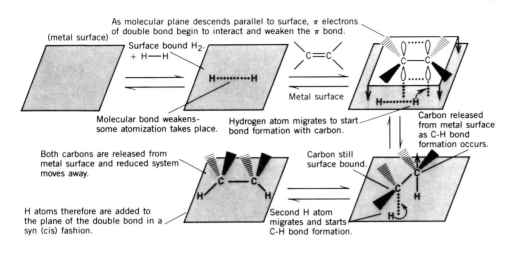

This experiment also provides an opportunity for you to study a powerful aspect of column chromatography in which heavy metal ions have a particularly important role in the purification of the product. Unreacted alkene has the potential to be a problem contaminant during the isolation and purification of the relatively low-boiling saturated reaction product. The successful removal of the remaining 1-octene from the desired *n*-octane in the product mixture is achieved by employing column chromatography with silver nitrate/silica gel as the stationary phase. Complex formation between the silver ion (on the silica gel surface) and the $\pi$ system of the unreacted alkene acts to retard the rate of elution of the alkene relative to that of the alkane.

The ability of alkenes to form coordination complexes with certain metal ions having nearly filled *d* orbitals was established some time ago. In the case of the silver ion complex with alkenes, the orbital nature of the bonding is believed to involve a $\sigma$ bond formed by overlap of the filled $\pi$ orbital of the alkene with the free *s* orbital of the silver ion plus a $\pi$ bond formed by overlap of the vacant antibonding $\pi^*$ orbitals of the alkene together with the filled *d* orbitals of the metal ion.

---

Estimated time to complete the experiment: 2.5 hours.

## EXPERIMENTAL PROCEDURE

### Physical Properties of Reactants

| Compound | MW | Amount | mmol | bp(°C) | d | $n_D$ |
|---|---|---|---|---|---|---|
| 1-Octene | 112.22 | 120 $\mu$L | 0.76 | 121.3 | 0.72 | 1.4087 |
| Ethanol (absolute) | 46.07 | 1.0 mL | | 78.5 | | |
| Chloroplatinic acid (0.2 *M*) | 517.92 | 50 $\mu$L | | | | |
| Sodium borohydride (1 *M*) | 37.83 | 125 $\mu$L | | | | |
| Dilute HCl (6 *M*) | | 100 $\mu$L | | | | |

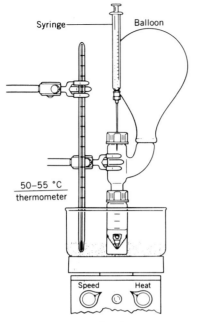

Syringe — Balloon

50–55 °C
thermometer

Speed   Heat

**Step I:**
0.2 *M* H$_2$PtCl$_6$ solution,
50 μL +
CH$_3$CH$_2$OH, 1.0 mL
+ NaBH$_4$ solution, 125 μL
**Step II:**
6 *M* HCl, 100 μL
+ 1-octene, 120 μL
+ NaBH$_4$ solution, 1.0 mL

## Reagents and Equipment

Equip a 5.0-mL conical vial containing a magnetic spin vane with a Claisen head fitted with a rubber balloon and Teflon-lined rubber septum-cap (good *GC septa* work best – this is an important point, as several injections through the septum are required and the seal must remain gas-tight). Mount the assembly in a sand bath on a magnetic stirring hot plate. (■)

**NOTE.** *1. No residual acetone (perhaps from cleaning the equipment) can be present since it reacts with the NaBH$_4$.*
*2. The balloon must make a gas-tight seal to the Claisen head, so be sure to secure it with copper wire or a rubber band.*

Remove the 5.0-mL vial from the Claisen head and add the following reagents. *Recap the vial after each addition.*

    a. Add 50 μL of a 0.2 *M* solution of chloroplatinic acid (H$_2$PtCl$_6$) (automatic delivery pipet).
    b. Add 1.0 mL of absolute ethanol (calibrated Pasteur pipet).
    c. Add 125 μL of the sodium borohydride reagent (automatic delivery pipet).

**IMPORTANT.** *Reattach the vial **IMMEDIATELY** to the Claisen head after the NaBH$_4$ solution is added.*

Stir the mixture vigorously. The solution should turn black immediately as the finely divided platinum catalyst is formed.

## Instructor Preparations

    **1.** *The 0.2 M H$_2$PtCl$_6$ solution is prepared by adding 41 mg (0.1 mmol) of the acid to 0.5 mL of deionized water.*
    **2.** *The sodium borohydride reagent is prepared by adding 0.38 g (0.01 mol) of NaBH$_4$ to a solution of 0.5 mL of 2.0 M aqueous NaOH in 9.5 mL of absolute ethanol.*

After 1 min, use a syringe to add 100 μL of 6 *M* HCl solution through the septum-cap. In a like manner using a fresh syringe, add *immediately* to the acid solution, a solution of 120 μL (86 mg, 0.76 mmol) of 1-octene dissolved in 250 μL of absolute ethanol. (This solution is conveniently prepared in a 1-mL conical vial; the reagents are best dispensed using automatic delivery pipets.)
Now add dropwise (clean syringe) 1.0 mL of the NaBH$_4$ reagent solution over a 2-min interval.

**IMPORTANT.** *At this point the balloon should inflate and remain inflated for at least 30 min. If it does not, the procedure must be repeated.*

## Reaction Conditions

Stir the reaction mixture vigorously at a sand bath temperature of 50 °C for 45 min.

## Isolation of Product

Cool the reaction to ambient temperature and dropwise add 1 mL of water. Extract the resulting mixture in the reaction vial with three 1.0-mL portions of pentane. Transfer each pentane extract to a stoppered 25-mL Erlenmeyer flask containing 0.5 g of anhydrous sodium sulfate.

The extraction of the reaction mixture is as follows:

Upon addition of each portion of pentane, cap the vial, shake, vent carefully and then allow the layers to separate. A Vortex mixer may be used if available. The transfers must be made using a Pasteur filter pipet as the pentane solvent is particularly volatile.

Using a Pasteur filter pipet, transfer the dried solution to a second 25-mL Erlenmeyer flask. Rinse the drying agent with an additional 1.0 mL of pentane (calibrated Pasteur pipet) and add the rinse to this second flask. Add a boiling stone and concentrate the solution to a volume of about 1.0–1.5 mL by warming gently in a sand bath [**HOOD**].

HOOD

## Purification and Characterization

The saturated product, *n*-octane, is purified by column chromatography. In a Pasteur filter pipet place about 50 mg of sand, 500 mg of 10% silver nitrate on activated silica gel (200 mesh) and then 50 mg of anhydrous sodium sulfate. (■)

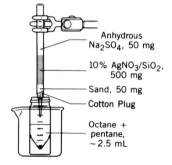

Anhydrous
$Na_2SO_4$, 50 mg

10% $AgNO_3/SiO_2$,
500 mg

Sand, 50 mg

Cotton Plug

Octane +
pentane,
~2.5 mL

## Instructor Preparation

The ~10% silver nitrate/silica gel used in this separation is commercially available.

---

**CAUTION:** Silver nitrate causes stains on the skin. Protective gloves should be worn during this operation.

---

Wet the column with 0.5 mL of pentane (calibrated Pasteur pipet) and then transfer the concentrated crude product, as obtained above, to the column by Pasteur filter pipet. Elute the octane from the column using 1.5 mL of pentane and collect the eluate in a tared 5.0-mL conical vial containing a boiling stone.

Fit the vial with an air condenser and then place the assembly in a sand bath maintained at a temperature of 90–100 °C to evaporate the pentane solvent [**HOOD**].

HOOD

**OPTIONAL.** *The evaporation is continued until a constant weight of product is obtained. This procedure is the best approach, but has to be done very carefully or a considerable amount of product can be lost.*

Record the weight of product and calculate the percent yield. Determine the boiling point, and refractive index (optional) of your material and compare your results to those reported in the literature for octane. Obtain an IR spectrum. Compare your results with those reported in the *Aldrich Library of IR Spectra*. Also, compare your IR spectrum to that of the 1-octene starting material. Can you establish from the above data if your sample is contaminated by traces of the pentane extraction solvent? If not, how would you go about determining the presence of this potential impurity?

## QUESTIONS

**6-82.** Squalene, first isolated from shark oil and a biological precursor of cholesterol, is a long-chain aliphatic alkene ($C_{30}H_{50}$). The compound undergoes catalytic hydrogenation to yield an alkane of molecular formula $C_{30}H_{62}$. How many double bonds does a molecule of squalene have?

**6-83.** A chiral carboxylic acid **A** ($C_5H_6O_2$) reacts with 1 mol of hydrogen gas on catalytic hydrogenation. The product is an achiral carboxylic acid **B** ($C_5H_8O_2$). What are the structures of Compounds **A** and **B**?

**6-84.** Two hydrocarbons, **A** and **B**, each contain six carbon atoms and one C=C. Compound **A** can exist as both *E* and *Z* isomers but Compound **B** cannot. However, both **A** and **B** on catalytic hydrogenation give only 3-methylpentane. Draw the structures and give a suitable name for Compounds **A** and **B**.

**6-85.** What chemical test would you use to distinguish between the 1-octene starting material and the octane product?

**6-86.** Give the structure and names of five alkenes having the molecular formula $C_6H_{12}$ that produce hexane on catalytic hydrogenation.

## BIBLIOGRAPHY

Selected references on catalytic hydrogenation:

Adkins, H.; Shriner, R. L. in Gilman, H. *Advanced Organic Chemistry*, 2nd ed.; Wiley: New York, 1943; Vol. I, p. 779.

Carruthers, W. *Some Modern Methods of Organic Synthesis*, 3rd ed.; Cambridge Univ. Press: London, 1986; p. 411.

March, J. *Advanced Organic Chemistry*, 4th ed.; Wiley: New York, 1992; p. 771.

Selected examples of catalytic hydrogenation of alkenes in *Organic Syntheses*:

Adams, R.; Kern, J. W.; Shriner, R. L. *Organic Syntheses*; Wiley: New York, 1941; Collect. Vol. I, p. 101.

Bruce, W. F.; Ralls, J. O. *Organic Syntheses*; Wiley: New York, 1943; Collect. Vol. II, p. 191.

Cope, A. C.; Herrick, E. C. *Organic Syntheses*; Wiley: New York, 1963; Collect. Vol. IV, p. 304.

Herbst, R. M.; Shemin, D. *Organic Syntheses*; Wiley: New York, 1943; Collect. Vol. II, p. 491.

Ireland, R. E.; Bey, P. *Organic Syntheses*; Wiley: New York, 1988; Collect. Vol. VI, p. 459.

McMurry, E. *Organic Syntheses*; Wiley: New York, 1988; Collect. Vol. VI, p. 781.

Meyers, A. I.; Beverung, W. N.; Gault, R. *Organic Syntheses*; Wiley: New York, 1988; Collect. Vol. VI, p. 371.

## EXPERIMENT [13]

## Hydroboration–Oxidation of an Alkene: Octanol

Common name: Octanol
CA number: [111-87-5]
CA name as indexed: 1-Octanol

### Purpose

To investigate the oxidation of an alkene to an alcohol via the in situ formation of the corresponding trialkylborane, followed by the oxidation of the carbon–boron bond with hydrogen peroxide. To explore the conditions required for hydroboration (a reduction) of unsaturated hydrocarbons. To recognize that alkylboranes are particularly useful synthetic intermediates for the preparation of alcohols. The example employed in this experiment is the conversion of 1-octene to 1-octanol in which an anti-Markovnikov addition to the double bond is required to yield the intermediate, trioctylborane. As it is this alkyl borane that subsequently undergoes oxidation to the alcohol, hydroboration offers a synthetic pathway for introducing substituents at centers of unsaturation that are not normally available to the anti-Markovnikov addition reactions that are based on radical intermediates.

### Prior Reading

*Technique 4:* Solvent Extraction
Liquid–Liquid Extraction (see pp. 80–82).
*Technique 6:* Chromatography
Column Chromatography (see pp. 98–109).
Concentration of Solutions (see pp. 106–108).

**REACTION**

$$3\ CH_3-(CH_2)_5-CH{=}CH_2 \xrightarrow{\text{THF·BH}_3} [CH_3(CH_2)_7]_3B \xrightarrow[\text{OH}^-]{\text{H}_2\text{O}_2} 3\ CH_3-(CH_2)_7-OH$$

1-Octene  Trioctylborane  1-Octanol

**DISCUSSION**

The course of this reaction depends (1) on the *stereospecific* reductive addition of diborane [$B_2H_6$, introduced as the borane•tetrahydrofuran complex ($BH_3$•THF)] to an alkene to form an intermediate trialkylborane and (2) on oxidation of the borane with alkaline hydrogen peroxide to yield the corresponding alcohol.

The first step in the reaction sequence is generally termed a *hydroboration*.[10] The addition of diborane has been found to be a rapid, quantitative, and general reaction for all alkenes (as well as alkynes) when carried out in a solvent that can act as a Lewis base. The ether solvation of the diborane, for example, is the key to the success of this reaction. In the absence of a Lewis base, borane ($BH_3$) exists as a dimer ($B_2H_6$) which is much less reactive than the monomer ($BH_3$). Borane, however, does exist in coordination with ether type solvents. It is the monomer ($BH_3$) that functions as the active reagent in the reductive addition.

As depicted in the following mechanism, the boron hydride rapidly adds successively to three molecules of the alkene to form a trialkylborane.

It is important to note that the transition state of this addition reaction is generally considered to be a *four-center* one, and that the 1-octene sub-

---

[10] For references relating to the use of diborane as a hydroboration reagent, see Experiment [1$_{adv}$].

strate is oriented such that the boron becomes bonded to the least-substituted carbon atom of the double bond. Thus, the reaction can be classified as *regioselective,* and it will be sensitive to substitution on the carbon–carbon double bond.

In the developing transition state, the alkene $\pi$ electrons (the least tightly held, and most nucleophilic) flow to the electron-deficient boron atom (the vacant *p* orbital is the electrophile). The formation of the *transition state* is controlled in large part by the polarization of the alkene $\pi$ system during the early stages of formation of the transition state. At this point a partial positive charge begins to form on the more highly substituted carbon (the more stable carbocation), and a partial negative charge on the least substituted carbon. The orientation of the polarization, therefore, is to a large extent controlled by the electron-releasing effects of the alkyl substituents on the alkene, which enables the more highly substituted of the *sp²* carbon atoms to better accommodate the positive charge. As the reaction proceeds, the boron acquires a partial negative charge in response to the incoming electron density. The ease of hydride ($:H^-$) transfer from the boron to the more highly substituted carbon atom of the alkene, therefore, increases. Thus, hydroboration involves simultaneous hydride release and boron–carbon bond formation, and is a concerted reaction. The reaction can be conveniently considered as passing through a four-centered transition state, wherein the atoms involved undergo simultaneous changes in bonding [i.e., electron redistribution (see below)].

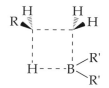

Hydroboration, as we have seen, can be classified as a **concerted addition reaction** in which no intermediate is formed. The mechanism is characteristic of a group of reactions called **pericyclic** (from the Greek, meaning *around the circle*) **reactions,** which involve a cyclic shift of electrons in and around the *transition state.* The mechanism proposed is further supported by the fact that rearrangements are not normally observed in hydroboration reactions, which implies that there are no carbocationic intermediates.

When alkenes with varying degrees of substitution undergo hydroboration, the boron ends up on the least substituted *sp²* carbon atom. While it might appear from the products that the regioselectivity is controlled by steric factors, this is probably too simplistic. Steric and electronic factors both favor, and are both likely responsible for, the observed regioselectivity in hydroboration reactions.

Accumulated evidence demonstrates that the reaction occurs by **syn** addition, which is a consequence of the four-centered transition state. Therefore, the new C—B and C—H bonds are necessarily formed on the same face of the C=C bond, as shown in the following example.

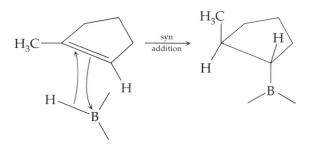

Organoboranes are important in organic synthesis as reactive intermediates. Reactions have been developed by which the boron atom may be replaced by a wide variety of functional groups, such as —H, —OH, —NH$_2$, —Br, and —I. The present experiment demonstrates the conversion of an organoborane to an alcohol by oxidation with alkaline hydrogen peroxide. It is not necessary to isolate the organoborane prior to its oxidation. This simplification is particularly fortuitous in this case as most alkylboranes, when not in solution, are pyrophoric (spontaneously flammable in air).

With regard to the second stage of the hydroxylation process, there is now conclusive evidence that oxidation of the C—B bond proceeds with retention of configuration at the carbon atom bearing the boron. That is, the hydroxyl group that replaces the boron atom has the identical orientation in the molecule as the boron.

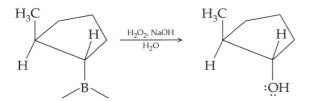

Thus, in unsymmetrical alkenes the hydroboration–oxidation sequence of reactions leads to the addition of the elements of H—OH to the original C=C in an anti-Markovnikov manner.

In the oxidation step a hydroperoxide anion (HOO$^-$) is generated in the alkaline medium. This species makes a nucleophilic attack on the boron atom to form a boron hydroperoxide. A 1,2 migration of an alkyl group from boron to oxygen occurs to yield a boron monoester (a borate). Hydrolysis of the boron triester, generated by successive rearrangement of all three alkyl groups, produces the desired alcohol. The mechanism of the oxidation sequence is given below.

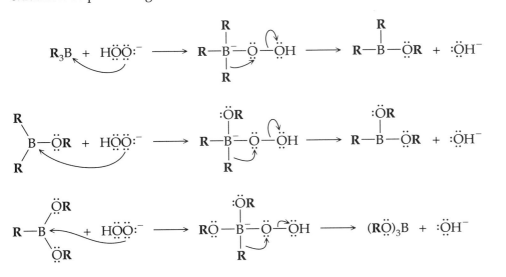

In the final step, alkaline hydrolysis of the trialkyl borate ester yields 3 mol of the alcohol.

$$B(OR)_3 + OH^- \xrightarrow{H_2O} 3\ ROH + B(OH)_4^-$$

The effective use of B$_2$H$_6$ in the hydroboration reaction was discovered in 1955 by Professor H. C. Brown, and is just one of the many important hydride reagents developed by Professor Brown and his co-workers at Purdue University.

**Herbert Charles Brown** (1912, American). Brown obtained his B.S. in chemistry from the University of Chicago (1936) and his Ph.D. from the same institution in 1938. He later became a Professor of Chemistry at Wayne State and Purdue Universities.

Working with H. I. Schlesinger at the University of Chicago, Brown developed practical routes for the synthesis of diborane ($B_2H_6$). He discovered that diborane reacted rapidly with LiH to produce lithium borohydride ($LiBH_4$), discovering and opening a synthetic route to the metal borohydrides. These compounds proved to be powerful reducing agents. Later, he developed an effective route to $NaBH_4$, which led to the commercial production of this material. Metal borohydrides, and particularly $LiAlH_4$ (developed by Schlesinger and Albert Finholt), have revolutionized how organic functional groups are reduced in both the research laboratory and the industrial plant.

In 1955, Brown discovered that alkenes can be converted to organoboranes by reaction with diborane (actually the monomer in ether solution) and with organoboranes containing a B—H bond (the hydroboration reaction). The organoboranes are valuable intermediates in organic synthesis because the boron substituent can be quickly and quantitatively replaced by groups such as —OH, —H, —$NH_2$, or —X (halogen). Thus organoboranes have become an attractive pathway for the preparation of alcohols, alkanes, amines, or organohalides.

Brown's investigations of the addition compounds of trimethyl borane, diborane, and boron trifluoride with amines has provided a quantitative estimation for steric strain effects in chemical reactions. He also investigated the role of steric effects in solvolytic, displacement, and in elimination reactions. His results demonstrate that steric effects can assist, as well as hinder, the rate of a chemical reaction.

Brown has published over 700 scientific papers and is the author of several texts. For his extensive work on organoboranes, Brown [with G. Wittig (organophosphorus compounds)] received the Nobel Prize in Chemistry in 1979.[11]

## EXPERIMENTAL PROCEDURE

Estimated time to complete the experiment: 4.0 hours.

### Physical Properties of Reactants

| Compound | MW | Amount | mmol | bp(°C) | $d$ | $n_D$ |
|---|---|---|---|---|---|---|
| 1-Octene | 112.22 | 210 $\mu$L | 1.34 | 121 | 0.72 | 1.4087 |
| Borane•THF (1 $M$) | | 500 $\mu$L | | | | |
| Sodium hydroxide (3 $M$) | 40.00 | 300 $\mu$L | | | | |
| Hydrogen peroxide (30%) | 34.01 | 300 $\mu$L | | | | |

---

[11] See *McGraw-Hill Modern Scientists and Engineers* S. P. Parker, Ed.; McGraw-Hill: New York, 1980, Vol. 1, p. 150.

## Reagents and Equipment

Equip a 5.0-mL conical vial, containing a spin vane, with a Claisen head fitted with a rubber septum and calcium chloride drying tube. (■) Through the rubber septum add 210 μL (150 mg, 1.34 mmol) of 1-octene (in one portion) with a 1.0-cm³ syringe.

**IMPORTANT.** *Dry the glassware and syringe in a 100 °C oven for at least 30 min before use. An alternate method is to "flame-out" the glassware with a microburner and a flow of dry nitrogen, and to then add the drying tube and caps.*

Cool the reaction vessel in an ice bath and, using the same syringe, add 500 μL (0.5 mmol) of the 1 M borane•THF solution through the septum over a 5-min period.

---

CAUTION:    The BH₃•THF reagent reacts *violently* with water.

---

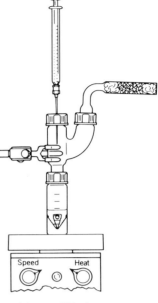

1-Octene, 210 μL
+ 1M BH₃•THF, 500 μL

## Reaction Conditions

Allow the reactants to warm to room temperature and then stir for 45 min. Using a Pasteur pipet, carefully add two drops of water to hydrolyze any unreacted borane complex.

**NOTE.** *At this stage of the procedure the vial may be removed from the Claisen head, capped, and allowed to stand until the next laboratory period.*

If the experiment is not interrupted, proceed by first removing the reaction vial from the Claisen head. Then use a graduated 1.0-mL pipet to add 300 μL of 3 M NaOH solution, followed by the dropwise addition of 300 μL of 30.0% hydrogen peroxide solution over a 10-min period using another graduated 1.0-mL pipet. Stir the reaction vial gently after each addition.

---

CAUTION:    Hydrogen peroxide blisters the skin.
Concentrated solutions of hydrogen peroxide can explode!

---

Attach the vial to a reflux condenser and warm the reaction mixture with stirring, for 1 hour in a sand bath at 40–50 °C. (■)
Cool the resulting two-phase mixture to room temperature and remove the spin vane using forceps. Add 0.5 mL of diethyl ether to establish a reasonable volume for extraction of the organic phase.

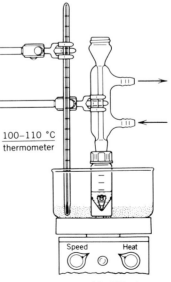

100–110 °C
thermometer

3 M NaOH, 300 μL +
30% H₂O₂, 300 μL +
[CH₃(CH₂)₇]₃B intermediate +
0.5 mL THF

## Isolation of Product

Using a Pasteur filter pipet, separate the bottom aqueous layer and transfer it to a 3.0-mL reaction vial. *Save* the organic phase in the 5.0-mL conical vial.

Extract the aqueous phase placed in the 3.0-mL conical vial with two 1.0-mL portions of diethyl ether. Upon the addition of each portion of ether, cap, shake, and carefully vent the vial and allow the layers to separate. The top ether layer is then separated using a Pasteur filter pipet and the ether extracts combined with the previously saved organic phase in the 5.0-mL conical vial. *If a solid forms during the extraction, add a few drops of 0.1 N HCl.*

Extract the combined organic phases with 750 $\mu$L of 0.1 N HCl solution, followed by extraction with several 0.5-mL portions of distilled water or until the aqueous extract is neutral to pH paper. Transfer the neutral organic phase to a 10-mL Erlenmeyer flask. Rinse the conical vial with a further 0.5 mL of ether and combine the rinse with the ether solution in the Erlenmeyer flask. Add granular anhydrous sodium sulfate (Na$_2$SO$_4$) (200 mg) to the combined organic phases and let it stand with occasional swirling for 20 min. If large clumps of drying agent form, add an additional 100 mg of Na$_2$SO$_4$. The solution should be clear at the end of the drying period. Then transfer the solution to a tared 3.0-mL conical vial in 1-mL aliquots, add a boiling stone to the vial, and concentrate by warming in a sand bath (60–65 °C) in the **hood** to yield the crude product residue.

HOOD

Weigh the vial and calculate the crude yield.

### Gas Chromatographic Analysis

This crude product is easily analyzed by gas chromatography. The procedure involves the injection of 10 $\mu$L of the liquid material onto a $\frac{1}{4}$ in. × 8-ft steel column packed with 10% Carbowax 80/100 20M PAW-DMS. Experimental conditions are: He flow rate, 50 mL /min; chart speed, 1 cm/min; temperature, 190 °C.

The liquid components elute in the order: unreacted 1-octene, a small amount of 2-octanol, and the major product, 1-octanol. Approximate retention times (conditions above) are 1.3, 3.1, and 4.2 min respectively.

Collect the eluted 1-octanol in an uncooled, 4-mm diameter collection tube. Transfer the collected material to a 0.1-mL conical vial (see Technique 1). This material may then be analyzed by IR spectroscopy and/or included in the following section on Purification and Characterization.

### Purification and Characterization

Pack a Pasteur filter pipet with 400 mg of 10% silver nitrate treated activated silica gel followed by 100 mg of anhydrous sodium sulfate.

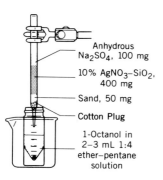

Anhydrous
Na$_2$SO$_4$, 100 mg

10% AgNO$_3$–SiO$_2$,
400 mg

Sand, 50 mg

Cotton Plug

1-Octanol in
2–3 mL 1:4
ether–pentane
solution

Dissolve the organic residue isolated above in 500 $\mu$L of pentane (spectral or HPLC grade) and then transfer this solution by Pasteur filter pipet to the column. (■) Elute the material from the column with 3 mL of a 1:4 diethyl ether–pentane solution. Collect the eluate in a tared 5.0-mL conical vial containing a boiling stone.

HOOD

Concentrate the collected eluate to a constant weight by warming on a sand bath (60–65 °C) in the **hood**. Weigh the octanol product and calculate the percent yield.

This product may again be analyzed by gas chromatography. Follow the procedure and experimental conditions outlined above. The 1-octene impurity should have been removed during the column chromatography step. The small percentage of 2-octanol byproduct, however, more than likely will still be detected. The actual percent composition of the mixture can be calculated by determination of the areas under the chromatographic peaks (see Technique 1).

Obtain an IR spectrum of the alcohol and compare your result to that recorded in the *Aldrich Library of IR Spectra*.

## Chemical Tests

A positive ceric nitrate test (Chapter 10) should confirm the presence of the alcohol grouping. The ignition test may be used to establish that the material is an aliphatic species. The phenyl or $\alpha$-naphthylurethane derivative may also be prepared to further characterize the alcohol (Chapter 10).

It might also be of interest to determine the solubility characteristics of this $C_8$ alcohol in water, ether, concentrated sulfuric acid, and 85% phosphoric acid (Chapter 10). Do your results agree with what you would predict for this alcohol?

What chemical tests would you perform to determine the difference between the starting alkene and the alcohol product?

**QUESTIONS**

**6-87.** Using the hydroboration reaction, outline a reaction sequence for each of the following conversions:
   a. 1-Pentene to 1-pentanol.
   b. 1-Methylcyclopentene to *trans*-2-methylcyclopentanol.
   c. 2-Phenylpropene to 2-phenyl-1-propanol.

**6-88.** When diborane ($B_2H_6$) dissociates in ether solvents, such as tetrahydrofuran (THF), a complex between borane ($BH_3$) and the ether is formed. For example,

$$B_2H_6 \; + \; 2 \; \ddot{O} \bigcirc \longrightarrow \; 2 \; \bigcirc \overset{+}{\ddot{O}} \text{—} \overset{-}{B}H_3$$

   a. In the Lewis sense, what is the function of $BH_3$ as it forms the complex? Explain.
   b. Write the Lewis structure for $BH_3$. Diagram its expected structure indicating the bond angles in the molecule.

**6-89.** In reference to question 6-88a:
   a. Explain why borane ($BH_3$) reacts readily with the $\pi$-electron system of an alkene.
   b. Explain why diborane ($B_2H_6$) reacts only very slowly with C=C groups.

**6-90.** In an unsymmetrical alkene, the boron atom adds predominantly to the least substituted carbon atom. For example, 2-methyl-2-butene gives the products indicated below.

$$\underset{CH_3}{\overset{CH_3}{CH_3\text{—}CH\text{=}CH\text{—}CH_3}} \xrightarrow[\text{diglyme}]{BH_3} \underset{\underset{2\%}{BH_2}}{\overset{CH_3}{CH_3\text{—}C\text{—}CH_2\text{—}CH_3}} + \underset{\underset{98\%}{BH_2}}{\overset{CH_3 \; H}{CH_3\text{—}CH\text{—}C\text{—}CH_3}}$$

(Diglyme: $CH_3\ddot{O}\text{—}CH_2CH_2\text{—}\ddot{O}\text{—}CH_2CH_2\text{—}\ddot{O}CH_3$, diethyleneglycol dimethyl ether)

Offer a reasonable explanation to account for the ratio obtained.

**NOTE.** *The above solvent (diglyme) has been shown to cause a significant increase in the number of miscarriages by workers who come in contact with it.*

**6-91.** An advantage of the hydroboration reaction is that rearrangement of the carbon skeleton does not occur. This lack of migration contrasts with results obtained upon the addition of hydrogen chloride to the double bond. For example,

$$CH_3-\underset{\underset{\displaystyle CH_3}{|}}{CH}-CH{=}CH_2 \xrightarrow{HCl} CH_3-\underset{\underset{\displaystyle CH_3}{|}}{CH}-\underset{\underset{\displaystyle Cl}{|}}{CH}-CH_3 \; + \; CH_3-\underset{\underset{\displaystyle Cl}{\overset{\overset{\displaystyle CH_3}{|}}{|}}}{C}-CH_2CH_3$$

$$CH_3-\underset{\underset{\displaystyle CH_3}{|}}{CH}-CH{=}CH_2 \xrightarrow{BH_3} CH_3-\underset{\underset{\displaystyle CH_3}{|}}{CH}-CH_2CH_2BH_2$$

Offer an explanation for the difference in these results.

## BIBLIOGRAPHY

Brown, H. C.; Subba, B. C. *J. Am. Chem. Soc.* **1956,** *78,* 5694.

Experiment [13] was based on the work reported by:

Kabalka, G. W. *J. Chem. Educ.* **1975,** *52,* 745.

## EXPERIMENT [14]

## Diels–Alder Reaction: 4-Cyclohexene-*cis*-1,2-dicarboxylic Acid Anhydride

Common name: 4-Cyclohexene-*cis*-1,2-dicarboxylic acid anhydride
CA number: [85-43-8]
CA name as indexed: 1,3-Isobenzofurandione, 3a,4,7,7a-tetrahydro-

### Purpose

To demonstrate the use of the Diels–Alder reaction in the preparation of six-membered carbocyclic rings. The cyclic products are obtained by reaction of a conjugated diene with an alkene. The illustration given here involves the treatment of 1,3-butadiene (generated in situ) with maleic anhydride to form the corresponding Diels–Alder product. These *addition* products are often termed *adducts*.

### *Prior Reading*

*Technique 7:* Collection or Control of Gaseous Products (see pp. 109–111).

*Chapter 9:* Infrared Spectroscopy (see pp. 599–602, 620–621).
Nuclear Magnetic Resonance Spectroscopy (see pp. 668–669).

**REACTION**

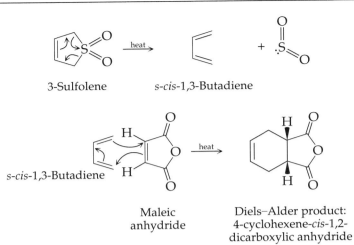

3-Sulfolene          *s-cis*-1,3-Butadiene

*s-cis*-1,3-Butadiene          Maleic anhydride          Diels–Alder product: 4-cyclohexene-*cis*-1,2-dicarboxylic anhydride

**Otto Paul Hermann Diels** (1876–1954, German, born in Hamburg).

Diels obtained his Ph.D. in 1899 while studying with Emil Fischer at the University of Berlin. He later became Associate Professor of Chemistry at the University of Berlin, and in 1916 he moved to the University of Kiel. In 1906 Diels discovered carbon suboxide gas ($C_3O_2$), obtained from the dehydration of malonic acid. He did extensive studies on saturated fats and fatty acids. Diels also developed the use of selenium as a mild dehydrogenation agent. This latter work led to the commercial production of polyunsaturated oils.

In the same year that he identified carbon suboxide, Diels began to investigate cholesterol with E. Abderhalden. The structure of this lipid had not yet been determined. He was the first to study the products of the selenium dehydrogenation of cholesterol and isolated a hydrocarbon ($C_{18}H_{16}$), which became known as "Diels' hydrocarbon." This substance proved to possess the basic steroidal ring structure; its subsequent synthesis by Diels in 1935 (almost 30 years after he started this work) led to the rapid elucidation of the structures of a vast array of steroidal sex hormones, saponins, cardiac glycosides, bile pigments, and adrenal cortical hormones, such as cortisone.

Diels, however, is best known for his discovery (with his student Kurt Alder) of the reaction that now bears his name (in modern terminology it is known as the *Diels–Alder cycloaddition reaction*), which was first published in 1928. This reaction involves the 1,4 addition of dienophile reagents to diene substrates to produce six-membered cycloalkenes. The reaction has found extensive application in the synthesis of terpenes and other natural products, as six-membered rings abound in the metabolites of living systems, and because for some time it was one of the few methods available for the synthesis of these cyclic structures.

Because of the impact of their work on the field of organic synthesis, Diels shared the 1950 Nobel Prize (in chemistry) with Alder. He also was the author of a popular textbook (*Einfuhrung in die organische Chemie*), first published in 1907, which went through 19 editions by 1962.[12]

[12] See Newett, L. C. *J. Chem. Educ.* **1931**, *8*, 1493; *Dictionary of Scientific Biography*, C.C. Gillespie, Ed.; Scribner's: New York, 1971, Vol. IV, p. 90; *McGraw-Hill Modern Scientists and Engineers* S. P. Parker, Ed.; McGraw-Hill: New York, 1980, Vol.1, p. 289.

## DISCUSSION

The Diels–Alder reaction is one of the most useful synthetic reactions in organic chemistry because, in a *single* step, it produces two new carbon–carbon bonds and up to four stereocenters. It is an example of a [4 + 2] cycloaddition reaction (4 $\pi$ electrons + 2 $\pi$ electrons) between a *conjugated* 1,3-diene and an alkene (dienophile; to have an affinity for dienes, from the Greek *philos,* meaning loving), which leads to the formation of cyclohexenes. Alkynes may also be used as dienophiles, in which case the reaction produces 1,4-cyclohexadienes.

The reaction proceeds faster if the dienophile bears electron-withdrawing groups and if the diene bears electron-donating groups. Thus, $\alpha,\beta$-unsaturated esters, ketones, nitriles, and so on, make excellent dienophiles, which are often used in the Diels–Alder reaction. By varying the nature of the diene and dienophile, a very large number of compounds can be prepared. Unsubstituted alkenes, such as ethylene, are poor dienophiles and react with 1,3-butadiene only at elevated temperatures and pressures. These high activation energies (slow reactions) pose a particular problem for the Diels–Alder reaction.

The Diels–Alder reaction is a reversible, equilibrium reaction that is not very exothermic. Since the equilibrium constant ($K_{eq}$) is temperature dependent [$K_{eq} = e^{-\Delta G/RT}$], $K_{eq}$ decreases with increasing temperature and eventually can become quite small at the high temperature needed for the reaction of an unactivated dienophile to proceed at a reasonable rate. Elevating the temperature will increase the rate of the reaction, but this will also reduce the amount of product formed, and therefore the lowest possible temperature must often be used.

The reaction is a thermal *cycloaddition* (a ring is formed), which occurs in one step and is thus a *concerted* reaction. Both new C—C single bonds and the new C=C $\pi$ bond are formed simultaneously, as the three $\pi$ bonds in the reactants break. The electron flow for the reaction is shown below. The reaction is thus classified as a *pericyclic* reaction (from the Greek *peri,* meaning *around the circle*).

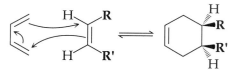

The diene component must be in the *s-cis* (the "s" refers to the conformation about a <u>s</u>ingle bond) conformation in order to yield the cyclic product with the cis C=C required by the six-membered ring. For this reason, cyclic dienes usually react more readily than acyclic species. For example, 1,3-cyclopentadiene, which is locked in the s-cis configuration, reacts with maleic anhydride about 1000 times faster than 1,3-butadiene, which prefers an s-trans conformation.

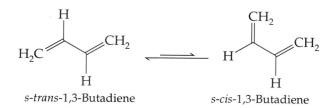

*s-trans*-1,3-Butadiene      *s-cis*-1,3-Butadiene

The reaction is highly stereospecific and the orientation of the groups on the dienophile are *retained* in the product, and thus the addition must be suprafacial–suprafacial. That is, both new bonds are formed on the same face of the diene and on the same face of the dienophile. Thus, two groups that are cis on the dienophile will be cis in the product (and trans will give a trans product).

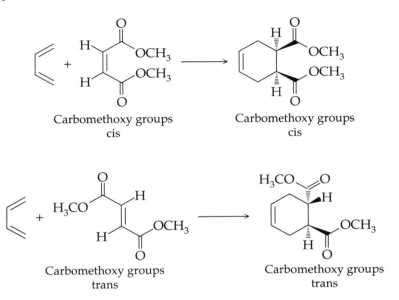

Carbomethoxy groups cis → Carbomethoxy groups cis

Carbomethoxy groups trans → Carbomethoxy groups trans

The reaction of cyclopentadiene with maleic anhydride demonstrates the further stereochemical consequence of the relative orientation of the reactants in Diels–Alder reactions. In this situation, there are two possible ways in which the reactants may bond. This reaction leads to the formation of two products: the endo and exo stereoisomers.

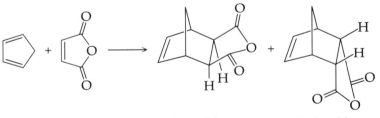

Exo adduct          Endo adduct

Generally, the endo form of the product predominates, but endo/exo ratios may vary depending on several steric and electronic factors, and with reaction conditions.

The diene used in this experiment, 1,3-butadiene, is a gas at room temperature (bp −5 °C), which makes it a difficult reagent to measure and handle in the laboratory. Fortunately, 1,3-butadiene can be generated in situ from a solid reagent that is easily handled, 3-sulfolene. In an example of a retro-cycloaddition reaction, 3-sulfolene decomposes at a moderate temperature to yield sulfur dioxide and 1,3-butadiene.

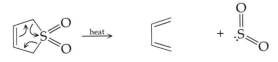

3-Sulfolene          s-cis-1,3-Butadiene

## EXPERIMENTAL PROCEDURE

Estimated time to complete the experiment: 1.5 hours.

### Physical Properties of Reactants

| Compound | MW | Amount | mmol | mp(°C) | bp(°C) |
|---|---|---|---|---|---|
| 3-Sulfolene | 118.15 | 170 mg | 1.42 | 66 | |
| Maleic anhydride | 98.06 | 90 mg | 0.92 | 60 | |
| Xylene | | 80 μL | | | 137–140 |

### Reagents and Equipment

Weigh and place 170 mg (1.42 mmol) of 3-sulfolene, 90 mg (0.92 mmol) of maleic anhydride, and 80 μL of xylene in a 3.0-mL conical vial equipped with an air condenser protected with a calcium chloride drying tube and containing a boiling stone. (■)

**NOTE.** *The maleic anhydride should be finely ground and protected from moisture to prevent hydrolysis to the corresponding acid. A mixture of xylenes in the boiling point range 137–140 °C will suffice. Use freshly distilled solvent or dry it over molecular sieves before use. Dispense the xylene in the **hood** using an automatic delivery pipet.*

*In large laboratory sections it is recommended that the evolved $SO_2$ be trapped (see Prior Reading).*

HOOD

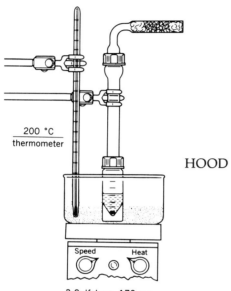

200 °C
thermometer

3-Sulfolene, 170 mg,
+ maleic anhydride, 90 mg
+ xylene, 80μL

### Reaction Conditions

Heat the reaction mixture at reflux, using a sand bath at 200 °C for 20 min.

---

**CAUTION: The reaction is exothermic. Avoid overheating. Sulfur dioxide is evolved in the process and adequate ventilation should be provided.**

---

HOT

Carefully remove the conical vial **[HOT]** from the sand bath and allow the contents to cool to room temperature.

### Isolation of Product

Add 0.5 mL of toluene to the cooled solution, and then add petroleum ether (60–80 °C) dropwise until a slight cloudiness persists. Roughly 0.25–0.35 mL of petroleum ether will be needed.

Reheat the solution until it becomes clear and then cool it in an ice bath. *During the recrystallization step, the sides of the vial may have to be scratched with a glass rod to induce crystallization.*

Collect the crystalline product by vacuum filtration and wash the filter cake on the Hirsch funnel with 0.5 mL of *cold* petroleum ether (60–80 °C). (■)

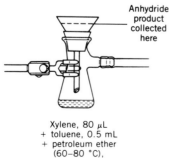

Anhydride
product
collected
here

Xylene, 80 μL
+ toluene, 0.5 mL
+ petroleum ether
(60–80 °C),
~0.8 mL

**BE CAREFUL.** *Do not wash with an excess of the petroleum ether, as a loss of product will result.*

## Purification and Characterization

The product is of sufficient purity for direct characterization. Weigh the material and calculate the percent yield. Determine the melting point and compare your result with the value found in the literature.

Obtain an IR spectrum of the material utilizing the KBr pellet technique.

The reaction involves two reactants (butadiene and maleic anhydride) which both contribute functional groups to the product. The infrared spectrum of the isolated material reflects this observation. Compare the infrared spectrum of your product with that shown in Figure 6.29.

## Infrared Analysis

The spectrum of one of the starting materials, 3-sulfolene (Fig. 6.27), is representative of an alkene sulfone. The macro group frequency train for an unconjugated five-membered ring alkene fits the data reasonably well: 3090 (=C—H, stretch),

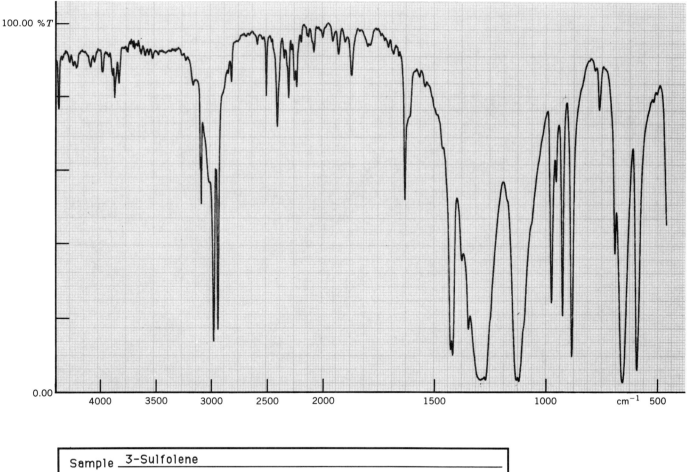

| Sample  3-Sulfolene | |
| --- | --- |
| %T _X_ ABS _ Background Scans _4_ | Scans _____16_____ |
| Acquisition & Calculation Time _42 sec_ | Resolution _4.0 cm⁻¹_ |
| Sample Condition ____solid____ | Cell Window _____ |
| Cell Path Length _____ | Matrix Material _KBr_ |

**FIGURE 6.27**  IR spectrum: 3-Sulfolene.

1635 (C=C, stretch), and 657 (H—C=C—H, cis, out-of-plane bend) cm$^{-1}$. The presence of the sulfone group is convincingly identified by the very strong bands at 1287 and 1127 cm$^{-1}$, which are assigned to the coupled in-phase and out-of-phase S—O stretching vibrations. The antisymmetric and symmetric C—H stretching vibrations of the methylene groups are observed at 2975 and 2910 cm$^{-1}$. Note that the influence of the heterocyclic five-membered ring system has raised all of the C—H stretching modes by 40–50 cm$^{-1}$.

The spectrum of the other reactant, maleic anhydride (Fig. 6.28), possesses all the peaks representative of a conjugated five-membered ring anhydride: 3110 (=C—H, stretch), 1858 (C=O, in-phase stretch, weak), 1777 (C=O, out-of-phase stretch, strong), 1595 (C=C, stretch, weak), 1060 (C—O, antisymmetric stretch), 899 (C—O, symmetric stretch), 835 (H—C=C—H, out-of-plane bend, normally near 700 cm$^{-1}$, but raised by conjugation to carbonyls). The in-phase stretch of the anhydride carbonyls is particularly weak as the five-membered ring system forces the two oscillators into nearly opposing positions.

The Diels–Alder product exhibits all of the spectral properties of a nonconjugated, alkenyl five-membered anhydride (Fig. 6.29). The macro group frequency train for the anhydride portion of the molecule involves the following peaks: 1844, 1772, 998 and 935 cm$^{-1}$.

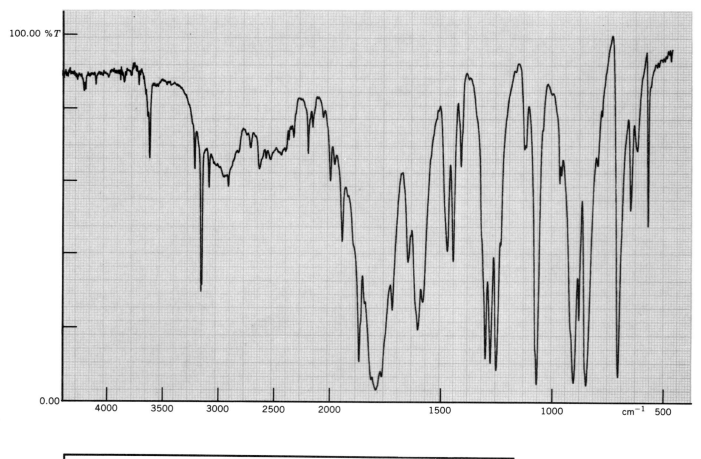

| Sample | Maleic anhydride | | |
|---|---|---|---|
| %T _X_ ABS __ Background Scans _4_ | | Scans ____16____ | |
| Acquisition & Calculation Time _42 sec_ | | Resolution _4.0 cm$^{-1}$_ | |
| Sample Condition ____solid____ | | Cell Window _____ | |
| Cell Path Length _____ | | Matrix Material _KBr_ | |

**FIGURE 6.28**   IR spectrum: Maleic anhydride.

**a.** 1844 cm$^{-1}$: This weak band is similar to the 1858-wavenumber band found in maleic anhydride, and involves the in-phase stretch of the two coupled carbonyl groups. The band is somewhat more intense in the case of this saturated five-membered ring example, as the ring system is more easily distorted from planarity. This band often exhibits some secondary splitting on the high wavenumber side (see discussion of the 935-wavenumber band below).

**b.** 1772 cm$^{-1}$: This band arises from the out-of-phase stretch of the anhydride carbonyls and is the most intense band in the spectrum. The separation, or splitting, between the two carbonyl stretching modes of anhydrides is relatively constant and falls in the range of 60–90 cm$^{-1}$.

**c.** 998 and 935 cm$^{-1}$: These two bands are identified with the antisymmetric and symmetric C—O—C stretching modes. It is the overtone of the symmetric stretch that is Fermi coupled (see Chapter 9) to the symmetric carbonyl stretch ($935 \times 2 = 1870$ cm$^{-1}$). The frequency match is variable, with the harmonic generally falling on the high wavenumber side of the fundamental.

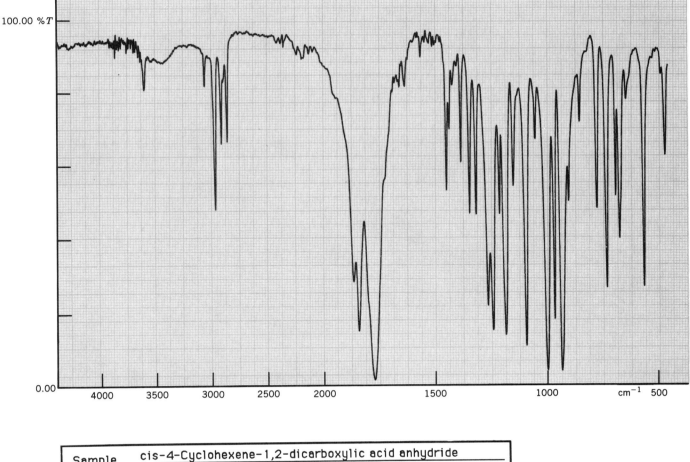

**FIGURE 6.29** IR spectrum: 4-Cyclohexene-*cis*-1,2-dicarboxylic acid anhydride.

The alkenyl section of the product possesses the following macro group frequency bands: 3065 (=C—H stretch), 1635 (C=C stretch, unsubstituted C=C, 5,6-membered fused ring system), and 685 (H—C=C—H, cis, out-of-plane bend) cm$^{-1}$.

The saturated C—H section of the cyclohexene ring possesses two identical methylene groups that contain the short macro frequency train: 2972, 2929 and 2860, 1447 cm$^{-1}$.

**a.** 2972 cm$^{-1}$: Antisymmetric C—H stretch of the —CH$_2$— group.

**b.** 2929 and 2860 cm$^{-1}$: Split symmetric C—H stretch of the —CH$_2$— group. This mode is split by Fermi coupling (see Chapter 9) with the overtone of the symmetric methylene scissoring vibration found at 1447 cm$^{-1}$ (1447 × 2 = 2896). The uncoupled fundamental mode would be expected to occur at or near 2890 cm$^{-1}$, and thus, falls very close to the predicted location of the overtone.

**c.** 1447 cm$^{-1}$: Symmetric deformation vibration (scissoring motion) of the methylene group present in the cyclohexene system.

Examine the spectrum of your reaction product, which you have obtained in a potassium bromide matrix. Discuss the similarities and differences of the experimentally derived spectral data to the reference spectra (Figs. 6.27–6.29).

300 MHz NMR OF 4-CYCLOHEXENE-CIS-1,2-DICARBOXYLIC ACID ANHYDRIDE

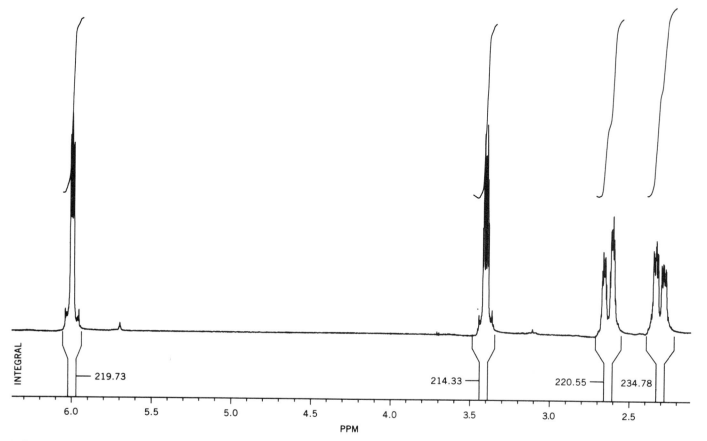

219.73

214.33

220.55

234.78

INTEGRAL

6.0    5.5    5.0    4.5    4.0    3.5    3.0    2.5

PPM

**FIGURE 6.30** $^1$H-NMR spectrum: 4-Cyclohexene-*cis*-1,2-dicarboxylic acid anhydride.

## Nuclear Magnetic Resonance Analysis

The 4-cyclohexene-*cis*-1,2-dicarboxylic acid anhydride provides an ideal example of the utility of $^{13}$C NMR when only limited information is available from the $^1$H NMR spectrum. The 300-MHz $^1$H spectrum is shown in Figure 6.30. Due in part to the presence of two stereocenters, as well as long-range coupling through the $\pi$ system of the alkene, the entire $^1$H spectrum is second order and no information is available from the coupling constants because the spectrum is too complex. Limited assignments to peaks could be made on the basis of chemical shift, but it would be difficult to make any statements regarding the purity of your sample based on the $^1$H NMR spectrum, as an impurity could well be hidden beneath any of the complex signals.

On the other hand, the fully $^1$H-decoupled $^{13}$C spectrum (Figure 6.31) of 4-cyclohexene-*cis*-1,2-dicarboxylic acid anhydride is much less complex. Because of the mirror plane of symmetry in the compound, there are only

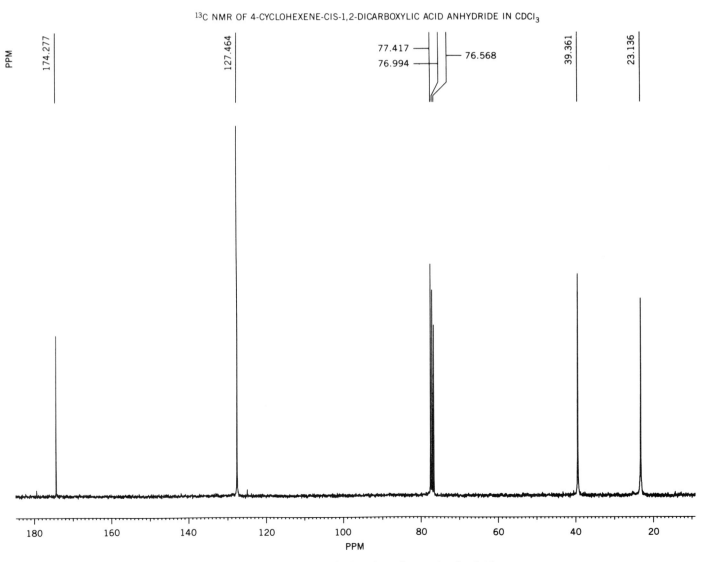

$^{13}$C NMR OF 4-CYCLOHEXENE-CIS-1,2-DICARBOXYLIC ACID ANHYDRIDE IN CDCl$_3$

**FIGURE 6.31**   $^{13}$C-NMR spectrum: 4-Cyclohexene-*cis*-1,2-dicarboxylic acid anhydride.

four unique carbon atoms and thus only four peaks are seen in the $^{13}$C NMR spectrum. The 1:1:1 triplet centered at 77 ppm is due to the solvent, CDCl$_3$. Although no $^1$H–$^{13}$C coupling is seen as a result of the $^1$H decoupling, $^2$H–$^{13}$C coupling is observed because $^1$H and $^2$H (D) resonate at different frequencies.

Sample preparation for $^{13}$C NMR is essentially the same as for $^1$H NMR spectroscopy except that significantly more material is required to obtain a $^{13}$C NMR spectrum in a reasonable amount of time. In this case, acceptable signal-to-noise levels can be obtained on 40 mg of material in about 10 min. Although tetramethylsilane (TMS) can be added as a reference, it is often more convenient to reference the spectrum relative to the known chemical shift of the solvent signal. Do not hesitate to use all of your material for the $^{13}$C spectrum; it can be easily recovered later by emptying the NMR tube into a small vial, rinsing once with solvent, and evaporating the solvent in a hood under a gentle stream of dry nitrogen.

### Chemical Tests

Selected chemical classification tests can also be used to aid in characterization of this compound.

Is the compound soluble in water? If so, does the aqueous solution turn blue litmus paper red? Is the compound soluble in 5% NaOH and 5% NaHCO$_3$? Is there evidence of CO$_2$ evolution with the bicarbonate solution? If so, what does this test indicate? Give the structure of the product formed when the material is added to the sodium hydroxide solution.

Perform the Baeyer test for unsaturation (Chapter 10). Is there evidence for the presence of a C=C bond?

## OPTIONAL SEMIMICROSCALE PREPARATION

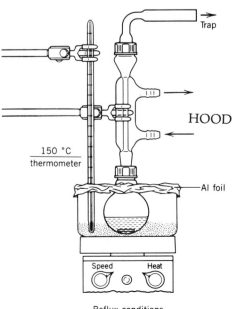

Reflux conditions
10-mL RB flask

The microscale Diels–Alder addition may be scaled up by a factor of 10. The procedure is similar to that outlined above, with the exceptions noted below.

**1.** Use a 10-mL round-bottom flask containing a magnetic stirrer, equipped with a reflux condenser protected by a calcium chloride drying tube. Place an insulating aluminum foil guard between the bottom of the reflux condenser and the round-bottom flask. (■)

---

**CAUTION:  At this scale, due to the generation of a considerably larger quantity of sulfur dioxide than at the microscale level, the reaction must be run in the *hood* or provisions made to trap the evolved gas.**

---

**2.** Increase the reagent and solvent amounts about 10-fold:

**Physical Properties of Reactants**

| Compound | MW | Amount | mmol | mp(°C) | bp(°C) |
|---|---|---|---|---|---|
| 3-Sulfolene | 118.15 | 850 mg | 7.2 | 66 | |
| Maleic anhydride | 98.06 | 450 mg | 4.6 | 60 | |
| Xylene | | 400 μL | | | 137–140 |

**3.** Heat the reaction mixture at 150–155 °C for 20 min. *Avoid overheating*. Then, cool the reaction to room temperature.

**4.** After cooling, add 5 mL of toluene and transfer the resulting clear solution to a 25-mL Erlenmeyer flask. The toluene solution may require heating and stirring to dissolve any crystals that may have formed. Now add petroleum ether (60–80 °C) (2.5–3.5 mL).

**5.** After isolation by vacuum filtration, wash the product with 5 mL of cold petroleum ether (60–80 °C).

**6.** Characterize the product as outlined in the above microscale procedure.

**QUESTIONS**

**6-92.** Predict the product in each of the following Diels–Alder reactions.

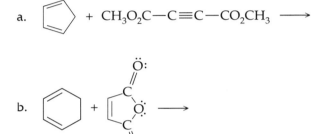

a.

b.

c.

**6-93.** Cyclopentadiene reacts as a diene in the Diels–Alder reaction a great deal faster than does 1,3-butadiene. Explain.

**6-94.** Predict the diene and dienophile that would lead to each of the following products.

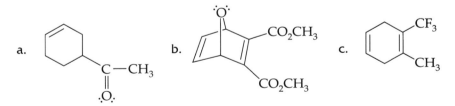

**6-95.** The decomposition of 3-sulfolene to form 1,3-butadiene generates 1 mole of sulfur dioxide gas per mole of 1,3-butadiene. Substantial quantities of $SO_2$ would be generated if this decomposition were carried out on a large scale. Suggest a method for trapping the gas to prevent its escape into the environment.

**6-96.** Two structural isomers are formed when 2-methyl-1,3-butadiene reacts with ethyl acrylate (ethyl 2-propenoate). Draw structures for these isomers.

**6-97.** In general, on going from the unsaturated fully conjugated anhydrides to the saturated systems, the two C—O stretching modes begin to coalesce, and in many cases only a single band is observed. In maleic anhydride, these modes occur at 1060 and 899 cm$^{-1}$, while in *cis*-4-cyclo-hexene-1,2-dicarboxylic acid anhydride they are found at 998 and 935 cm$^{-1}$. Explain this observation.

**6-98.** Acyclic anhydrides exhibit a reversal of the intensity relationship of the carbonyl stretching vibrations found in the cyclic anhydrides. That is, the lower wavenumber band is now the weaker member of the pair. Explain why this intensity exchange occurs.

**6-99.** If in Experiment [14] you had prepared the other possible diastereomer of the product, 4-cyclohexene-*trans*-1,2-dicarboxylic acid anhydride, how many lines would you expect to see in the $^{13}$C NMR spectrum? Note that this isomer does not have a mirror plane of symmetry.

**6-100.** Does the $^{13}$C NMR spectrum unambiguously demonstrate the position of the carbon–carbon double bond? How?

**6-101.** In $^1$H spectra, the relative intensities of lines within a triplet are 1:2:1. The signal for CDCl$_3$ is a triplet with relative intensities of 1:1:1. Why?

---

**BIBLIOGRAPHY**

The preparation of 4-cyclohexene-*cis*-1,2-dicarboxylic acid anhydride by the reaction of 1,3-butadiene with maleic anhydride is recorded:

Cope, A. C.; Herrick, E. C. *Organic Syntheses*; Wiley: New York, 1963; Collect. Vol. IV, p. 890.

Other Diels–Alder reactions reported in *Organic Syntheses* include:

Carlson, R. M.; Hill, R. K. *Organic Syntheses*; Wiley: New York, 1988; Collect. Vol. VI, p. 196.

Jung, M. E.; McCombs, C. A. *Organic Syntheses*; Wiley: New York, 1988; Collect. Vol. VI, p. 445.

Sample, T. E., Jr.; Hatch, L. F. *Organic Syntheses*; Wiley: New York, 1988; Collect. Vol. VI, p. 454.

The conditions of this reaction were adapted from those reported by:

Sample, T. E., Jr.; Hatch, L. F. *J. Chem. Educ.* **1968**, *45*, 55.

---

**EXPERIMENT [15]**

## Diels—Alder Reaction: 9,10-Dihydroanthracene-9,10-$\alpha$,$\beta$-succinic Acid Anhydride

Common name: 9,10-Dihydroanthracene-9,10-$\alpha$,$\beta$-succinic acid anhydride
CA number: [85-43-8]
CA name as indexed: 1,3-Isobenzofurandione, 3a,4,7,7a-tetrahydro-

### Purpose

To investigate the Diels–Alder reaction. To explore the role of an aromatic ring system as the diene substrate in this addition reaction. The reaction studied in this experiment is an example of a 1,4 addition by an activated alkene dienophile across the 9,10 positions of anthracene.

*Prior Reading*
*Technique 5:*   Crystallization
        Use of the Hirsch funnel (see pp. 93–95).

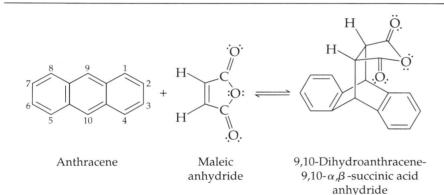

Anthracene          Maleic          9,10-Dihydroanthracene-
                    anhydride       9,10-$\alpha,\beta$-succinic acid
                                    anhydride

This famous class of reactions are named for **Otto Paul Hermann Diels** and **Kurt Alder**, who were primarily responsible for its development. Diels and Alder received the Nobel Prize in 1950 for this work. See Experiment [14] for a biography of Diels. A short biography of his student Kurt Alder follows:[13]

**Kurt Alder** (1902–1958, German). Alder obtained his Ph.D. while studying with Otto Diels in 1926 at the University of Kiel. His dissertation was titled *Causes of the Azoester Reaction*. He became Professor of Chemistry at the University of Kiel in 1934 and later became head of the Chemical Institute at the University of Cologne (1940). For a few years (1936–1940) he was Research Director of the Baeyer dye works.

Together with Diels, Alder was responsible for the development of what came to be known as the Diels–Alder reaction. This reaction typically involves the reaction of a 1,3-conjugated diene with an activated alkene (dienophile) to form a six-membered cycloalkene. While reactions of this type had been reported as early as 1893, Diels and Alder were the first to recognize their great versatility. Alder continued to focus his academic research in this area following his graduate work with Diels. Over a number of years, Alder carried out a systematic study of the reactivity of a large number of dienes and dienophiles and established the structure and stereochemistry of many new adducts. He also expanded his doctoral research, studying the condensation of azoesters with dienes to yield the corresponding heterocyclic adducts.

Alder demonstrated that successful addition required that the diene double bonds possess an s-cis conformation (s refers to the single bond connecting the two double bonds). Furthermore, he realized that the bridged ring adducts formed by using cyclic dienes were closely related to natural products, such as camphor, and that this reaction offered a powerful route for the synthesis of a wide variety of naturally occurring compounds, particularly the terpenes. The Diels–Alder reaction also has been invaluable in the industrial synthesis of thousands of new organic materials from insecticides and dyes to lubricating oils and pharmaceuticals.

Alder investigated autooxidation and polymerization processes particularly during his industrial years. For example, he was involved in an extensive study of polymerizations related to the formation of Buna-type synthetic rubbers.[14]

---

[13] For references to the Diels–Alder reaction, see Experiment [14].
[14] See Allen, C. F. H. *J. Chem. Educ.* **1933**, *10*, 494; *Dictionary of Scientific Biography* C.C. Gillespie, Ed.; Scribner's: New York, 1970, Vol. I, p. 105; *McGraw-Hill Modern Scientists and Engineers* S. P. Parker, Ed., McGraw-Hill: New York, 1980, Vol. 1, p. 8.

## DISCUSSION

This experiment is a further example of the Diels–Alder reaction. *For a discussion of the basic aspects of this reaction see Experiment [14].* In the present case, the central ring of anthracene is shown to possess the characteristic properties of a diene system. Thus, this aromatic compound reacts to form stable Diels–Alder adducts with many dienophiles at the 9 and 10 positions (the two positions on the central ring where new bonds can be made without destroying the aromaticity of the other two rings). Maleic anhydride, a very reactive dienophile, is used here in the reaction with anthracene. Note, that as this reaction is reversible, it is usually best carried out at the lowest possible temperatures consistent with an acceptable reaction rate (see Experiment [14]).

Higher molecular weight polynuclear aromatic hydrocarbons (PAHs) containing the anthracene nucleus have also been found to react with maleic anhydride. These ring systems, however, can differ widely in reaction rates. Typical examples of those systems that undergo the Diels–Alder reaction are 1,2,5,6-dibenzanthracene (**a**), 2,3,6,7-dibenzanthracene (pentacene) (**b**), and 9,10-diphenylanthracene (**c**).

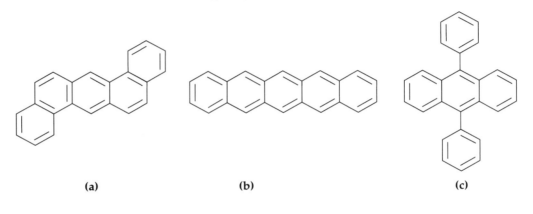

|      (a)      |      (b)      |      (c)      |

## EXPERIMENTAL PROCEDURE

Estimated time to complete the experiment: 1.5 hours.

### Physical Properties of Reactants and Product

| Compound | MW | Amount | mmol | mp(C°) | bp(°C) |
|---|---|---|---|---|---|
| Anthracene | 178.24 | 80 mg | 0.44 | 216 | |
| Maleic anhydride | 98.06 | 40 mg | 0.40 | 60 | |
| Xylenes | | 1.0 mL | | | 137–140 |
| 9,10-Dihydroanthracene-9,10-$\alpha$,$\beta$-succinic acid anhydride | 276 | | | 261–262 | |

### Reagents and Equipment

Weigh and place 80 mg (0.44 mmol) of anthracene and 40 mg (0.40 mmol) of maleic anhydride in a 3.0-mL conical vial containing a boiling stone, and equipped with an air condenser protected by a calcium chloride drying tube. Now add 1.0 mL of xylene to the solid mixture **[HOOD]** using an automatic delivery pipet. (■)

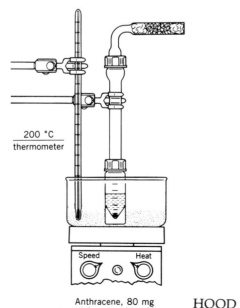

200 °C thermometer

Speed    Heat

Anthracene, 80 mg
+ maleic anhydride, 40 mg
+ xylene, 1.0 mL

HOOD

**NOTE.** *High-purity grades of anthracene and maleic anhydride are strongly recommended. Anthracene may be recrystallized from 95% ethanol. A mixture of xylenes with a boiling point range of 137–140 °C is sufficient, but the solvent (xylenes) should be dried over molecular sieves before use.*

## Reaction Conditions

Heat the reaction mixture at reflux for 30 min in a sand bath at about 200 °C. During this time the initial yellow color of the reaction mixture gradually disappears (Why?). Allow the resulting bleached solution to cool to room temperature and then place it in an ice bath for 10 min to complete the crystallization of the product.

## Isolation of Product

*3-0 mL.*

Collect the crystals by vacuum filtration using a Hirsch funnel and wash the filter cake with two 300-μL portions of cold ethyl acetate. (■) Partially dry the filter cake under suction using plastic food wrap (see Prior Reading). Transfer the filtered, washed, and partially dried product to a porous clay plate or filter paper to complete the drying.

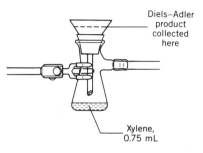

Diels–Adler product collected here

Xylene, 0.75 mL

## Purification and Characterization

The product is often of sufficient purity for direct characterization at this point. The adduct, however, may be further purified by recrystallization from ethyl acetate using a Craig tube. Weigh the dried Diels–Alder product and calculate the percent yield. Determine the melting point and compare your result with the value listed above.

**NOTE.** *It is suggested that a hot-stage melting point apparatus, such as the Fisher–Johns, be used due to the high melting point of the product.*

## Chemical Tests

The *ignition test* is often used as a preliminary method to categorize hydrocarbon materials (see Chapter 10). Aromatic compounds give a yellow, sooty flame. Perform this test on anthracene. Based on your results can anthracene be classified as aromatic?

Carry out the ignition test on the following compounds and determine whether they should be classified as aromatic materials.

Toluene    Octane    Isopropyl alcohol    Nitrobenzene    *trans*-Stilbene

---

This Diels–Alder reaction may be scaled up by a 5- or 50-fold increase in reactants. The scaled-up procedures are nearly identical in either case to that given for the microscale preparation; changes are noted below.

**OPTIONAL
SEMIMICROSCALE
PREPARATIONS**

1. **Fivefold Scaleup:**

   a. In place of the conical vial as the reaction vessel, use a 10-mL round-bottom flask containing a magnetic stirrer connected to the

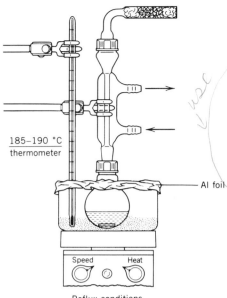

*no drying tube.*   *grease joint*

water-jacketed reflux condenser that is protected by a calcium chloride drying tube. (■)

b. The reagent and solvent quantities are given in the following table:

| Compound | MW | Amount | mmol | mp(°C) | bp(°C) |
|---|---|---|---|---|---|
| Anthracene | 178.24 | 400 mg ~0.4 g | 2.2 | 216 | |
| Maleic anhydride | 98.06 | 200 mg ~0.2 g | 2.0 | 60 | |
| Xylene | 106.16 | 5.0 mL | | | 137–140 |
| Ethyl acetate | 88.11 | 3.0 mL | | | 77 |

c. The reaction flask is **heated at 185–190 °C** for 30 min to obtain optimized yields.   *reflux for 30 min. at start of dripping*

2. **Fiftyfold Scaleup:**

a. In place of the conical vial as the reaction vessel, use a 100-mL round-bottom flask connected to a water-cooled reflux condenser that is fitted with a calcium chloride drying tube.

b. The required reagent and solvent quantities are given in the following table.

| Compound | MW | Amount | mmol | mp(°C) | bp(°C) |
|---|---|---|---|---|---|
| Anthracene | 178.24 | 4 g | 0.2 | 216 | |
| Maleic anhydride | 98.06 | 2 g | 0.2 | 60 | |
| Xylene | 106.16 | 50 mL | | | 137–140 |
| Ethyl acetate | 88.11 | 30 mL | | | 77 |

c. Heat the reaction mixture at **vigorous** reflux for **2 hours.**
d. Recrystallize the crude product from ethyl acetate.

## QUESTIONS

**6-102.**   Given the data tabulated below for the rate of reaction of maleic anhydride with a series of substituted 1,3-butadienes, offer a reasonable explanation to account for the trend in the rates.

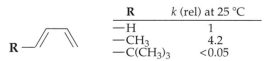

| R | $k$ (rel) at 25 °C |
|---|---|
| —H | 1 |
| —CH$_3$ | 4.2 |
| —C(CH$_3$)$_3$ | <0.05 |

**6-103.**   Predict the structure of the product formed in the following reactions.

a.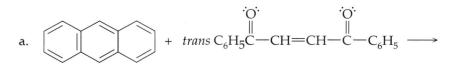

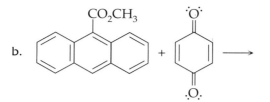

b.

**6-104.**  Offer an explanation of why anthracene preferentially forms a Diels–Alder adduct at the 9,10 positions.

**6-105.**  Experiment [31] demonstrates the monobromination of anthracene to yield one specific monosubstituted product. Other conditions may be used to form other monobrominated anthracenes. Draw the structures of, and name, the possible monobromo-substituted anthracenes that could be prepared in the laboratory.

**6-106.**  There are four reasonable resonance structures for anthracene. Draw them.

**6-107.**  A large number of polycyclic benzenoid aromatic hydrocarbons are known. One of these, benz[*a*]pyrene, is a powerful carcinogen found in tobacco smoke. From the literature, locate and then draw the structure of this hydrocarbon. Can you suggest other sources where this material might be expected to be present?

**6-108.**  Anthracene undergoes a *suprafacial* $[_\pi 4_s + _\pi 4_s]$ cycloaddition reaction at the 9,10 positions when subjected to irradiation to form a cyclic dimer. *Suprafacial* is a term used to describe in detail that the addition of the dienophile to the 1,3-$\pi$-system has occurred all from the same side (or face) of this planar structure. It comes from the Latin meaning above, or the dorsal side, and it is symbolized by the subscript *s* in the terminology: $[_\pi 4_s + _\pi 4_s]$. This is a theoretically forbidden thermal pericyclic reaction, but it can occur under photochemical conditions. Draw the structure of this product.

---

# Grignard Reaction with a Ketone: Triphenylmethanol

**EXPERIMENT [16]**

Common name: Triphenylmethanol
CA number: [76-84-6]
CA name as indexed: Benzenemethanol, $\alpha,\alpha$-diphenyl-

## Purpose
To develop the techniques required to prepare Grignard reagents. To investigate the reaction of these reagents with ketones to form *tertiary* alcohols. To experience, first hand, working with these highly air- and moisture-sensitive materials. To observe the formation of this famous reagent, in which magnesium metal is transformed (heterogeneous conditions) into organometallic salts of enormous value to the synthetic chemist.

This exercise is the first of a number of experiments available in this chapter in which the Grignard reaction is studied at the microscale level. Even if you do not actually do more than one of them, it is well worth the time to read through the other Experiments [17], [21], and [4$_{adv}$] to get a sense of the breadth of the applications of the reaction.

*Prior Reading*
Technique 4:  Solvent Extraction
              Liquid–Liquid Extraction (see pp. 80–82).
              Drying of the Wet Organic Layer (see pp. 86–87).
Technique 5:  Crystallization
              Craig Tube Crystallizations (see pp. 95–96).
Technique 6:  Chromatography
              Thin-Layer Chromatography (see pp. 103–104).

## REACTION

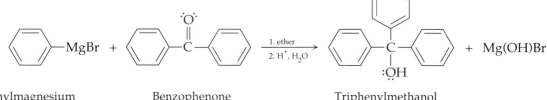

Phenylmagnesium          Benzophenone          Triphenylmethanol
    bromide

**François Auguste Victor Grignard** (1871–1935, French). Born in Cherbourg, Grignard was professor of Chemistry at the Universities of Lyons and Nancy. After studying for 1 year with Bouveault, Grignard became a graduate student of Phillippe Antoine Barbier, a professor at the University of Lyons. Barbier, who was working in the area of terpene chemistry, had found that magnesium could be used in place of zinc in the reaction of methyl iodide with an unsaturated ketone (methylheptenone) to yield the corresponding tertiary alcohol. This route was much preferred since the zinc reagents were difficult to work with as they were pyrophoric (spontaneously flammable in air). The use of magnesium in the formation of tertiary alcohols was reported in 1899. Barbier suggested to Grignard, that it might be interesting to further investigate the reaction of magnesium with alkyl halides. This study was to form the basis of Grignard's doctoral dissertation. Grignard discovered that treatment of alkyl iodides with magnesium in *diethyl ether* produced an alkylmagnesium iodide by a spontaneous reaction at ambient temperatures. His initial results, reported in 1900, were followed by seven papers the following year. His doctoral thesis on organomagnesium compounds and their application to synthetic organic chemistry was presented in 1901, when Grignard was 30 years old.

Grignard continued this work, having recognized the enormous potential of the alkyl magnesium halides in organic synthesis. These species are now known as *Grignard reagents* and when the reagent is employed in synthesis, the reaction is termed a *Grignard reaction*. These reagents have found great utility in the preparation of many kinds of organic compounds including alcohols, ketones, esters, and carboxylic acids. As mentioned above, these reagents contain a carbon–metal bond, and therefore they are classed in a large group of substances called *organometallic* compounds.

For this work, Grignard received the Nobel Prize in 1912. In 1919 Grignard returned to Lyons where he succeeded Barbier as chairman of the Department. By the end of his life, the scientific literature contained over 6000 papers dealing with Grignard reagents and their application.

Grignard also did extensive work in the areas of the terpenes, quantitative ozonolysis of alkenes, aldol reactions, catalytic hydrogenation, and dehydrogenation and cracking of hydrocarbons.[15]

[15] See Gordon, N. E. *J. Chem. Educ.* **1930**, *7*, 1487; Rheineoldt, H. *ibid.* **1950**, *27*, 476; Kauffman, G. B. *ibid.* **1990**, *67*, 569; Gilman H. *J. Am. Chem. Soc. (Proc. )* **1937**, *59*, 17; Gibson, C. S.; Pope, W. J. *J. Chem. Soc.* **1937**, 171; *Dictionary of Scientific Biography* C. C. Gillespie, Ed., Scribner's: New York, 1972, Vol. V, p. 540.

**DISCUSSION**

Grignard reagents possess significant nucleophilic character because of the highly polarized carbon–metal bond that results in considerable carbanionic character at carbon. Grignard discovered that these reactive materials readily attack the electrophilic carbon of a carbonyl group. It is this direct attack on carbon by a carbon nucleophile, resulting in carbon–carbon bond formation, that makes these such important reactions. Furthermore, as the carbonyl is the most ubiquitous functionality in all of organic chemistry, Grignard reagents have found great utility and widespread use in organic synthesis.

The formation of the organomagnesium halide (Grignard reagent) involves a heterogeneous reaction between magnesium metal and an alkyl, alkenyl, or aryl halide in ether solution. The solvent may be any one of a number of ethers, but diethyl ether and tetrahydrofuran are by far the most popular.

$$R(Ar)-X \; + \; Mg \xrightarrow{\text{ether}} R(Ar)-Mg-X$$
$$R = \text{alkyl, alkenyl}$$
$$\text{and } Ar = \text{aryl}$$

The reaction between an alkyl, alkenyl, or aryl halide and magnesium takes place on the surface of the metal and is an example of a *heterogeneous* (across two phases) reaction. The reactivity of the alkyl halides is in the order $Cl < Br < I$; fluorides do not generally react. Substituted alkyl halides react in the order $1° > 2° > 3°$; alkenyl and aromatic halides also form Grignard reagents to varying degrees.

It is important to understand the role of the ether solvent in the formation of the Grignard reagents. The reaction at the surface of the metal is essentially an oxidation–reduction. The metal is partially oxidized to the greater than 1+ state and the organohalide is reduced to halide ion and a highly polarized carbon–metal bond with the magnesium. The overall reaction can be viewed as forming the species, $R^{\delta-}-Mg^{+}X^{-}$ as the Grignard reagent. This highly polarized material is insoluble in most nonpolar organic solvents. The reaction will proceed at the surface of the metal until a layer of the insoluble organometallic reagent has formed. At this point, the surface reaction with the magnesium will immediately cease. If protic solvents are employed, they would instantly react with the highly basic Grignard reagent, R—MgX, to form the corresponding hydrocarbon, R—H. Thus, the use of either nonpolar or protic solvents does not lead to successful Grignard reagent formation. Why then is ether, a relatively nonpolar solvent, essential to the preparation of Grignard reagents?

The magnesium is essentially divalent, and electron deficient, when it reacts with the halide to form the RMgX species. A full octet around the metal atom requires two additional pairs of electrons. It is the energy gained by filling this octet that drives the coordination of the magnesium with two molecules of the ether solvent. This association in turn dramatically increases the solubility of the Grignard reagent in the relatively nonpolar ether solvent, and thus promotes further Grignard reagent formation.

$$
\overset{\overset{\displaystyle \delta^{+}}{}}{CH_3CH_2\ddot{O}CH_2CH_3}
$$
$$\downarrow$$
$$R-X \; + \; Mg \xrightarrow{\text{ether}} (R-Mg-X)^{\delta-}$$
$$\uparrow$$
$$\underset{\underset{\displaystyle \delta^{+}}{}}{CH_3CH_2\ddot{O}CH_2CH_3}$$

This interaction of RMgX with ether solvent also may be described as a Lewis acid–base interaction in which the coordinating solvent molecules are usually not written. When a Grignard reagent is described, it is important to remember that this vital solvation is always taking place.

The reactions of Grignard reagents with different types of carbonyl groups yield a number of important functional groups. For example, reaction with formaldehyde yields 1° alcohols; with higher aldehydes, 2° alcohols; with ketones, 3° alcohols; with esters, 3° alcohols; with acyl halides, ketones; with N,N-dialkylformamides, aldehydes; and with carbon dioxide, carboxylic acids.

In this experiment you will study the addition of the aryl Grignard reagent (phenylmagnesium bromide) to a diaryl ketone (benzophenone) to yield the corresponding tertiary (3°) alcohol. Because it is possible to vary both the structure of the Grignard reagent and the ketone, a wide variety of 3° alcohols may be obtained by this synthetic route.

The mechanism, as discussed above, can be thought of as involving rapid nucleophilic attack by the Grignard reagent at the carbon of the carbonyl group. Hydrolysis of the resulting alkoxide ion intermediate with dilute acid yields the desired alcohol. The reaction sequence is outlined here.

By using Grignard reagents, it is theoretically possible to synthesize a very large number of alcohols. Indeed, there is often more than one synthetic pathway open to a desired product. The choice of route is generally dictated by the availability of starting materials and the associated costs of these compounds.

---

**EXPERIMENTAL PROCEDURE**     Estimated time to complete the experiment: Two laboratory periods.

---

**CAUTION:  Ether is a flammable liquid. All flames must be extinguished during the time of this experiment.**

---

**Physical Properties of Reactants**

| Compound | MW | Amount | mmol | mp(°C) | bp(°C) | d | $n_D$ |
|---|---|---|---|---|---|---|---|
| Bromobenzene | 157.02 | 76 μL | 0.72 | | 156 | 1.50 | 1.5597 |
| Diethyl ether | 74.12 | 1.3 mL | | | 34.5 | 0.73 | |
| Magnesium | 24.3 | 17.5 mg | 0.73 | | | | |
| Iodine | 253.8 | 1 crystal | | | | | |
| Benzophenone | 182.21 | 105 mg | 0.58 | 48 | | | |

## Reagents and Equipment

### Preparation of Phenylmagnesium Bromide

**NOTE.**    *All the glassware used in the preparation of the Grignard reagent should be cleaned and dried in an oven at 110 °C for at least 30 min. After removal from the drying oven, the hot glassware should be placed in a desiccator and cooled before being assembled. Flame drying of the apparatus with a microburner annealing flame (high gas mixture) is an alternative, if ovens are not available. This latter method preferably should be carried out prior to assembly, and as with the oven-dried equipment, the hot glassware should be placed in a desiccator to cool. Care must be taken when flaming the reaction vial, as thermal shock can easily crack this heavy-walled vessel. If the flame drying procedure is performed on the assembled apparatus care also must be taken not to overheat the O-rings and plastic Capseals.*

In a 3.0-mL conical vial containing a magnetic spin vane and equipped with a Claisen head fitted with a calcium chloride drying tube and a rubber septum, weigh and place 18 mg (0.73 mmol) of polished magnesium ribbon, a small crystal of iodine, and 100 $\mu$L of anhydrous ether (automatic delivery pipet, **HOOD**). (■)                                    HOOD

**NOTE.**    *Scrape a 2–3-in. piece of magnesium ribbon clean of surface oxide (MgO) coating and then cut it into 1-mm long sections.*

### Alternate Procedure
Place the (polished) magnesium metal and an iodine crystal in the reaction vial and quickly assemble the apparatus. Warm the mixture (microburner or hot plate) *GENTLY* until evidence of purple iodine vapor is observed.          GENTLY

**NOTE.**    *If a microburner is used in this step, all flames must be extinguished before proceeding.*

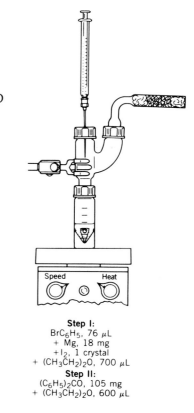

Now add 100 $\mu$L of anhydrous ether using a 1.0-mL syringe inserted through the septum.

In a dry, screw-capped vial, prepare a solution of 76 $\mu$L (113 mg, 0.72 mmol) of bromobenzene in 400 $\mu$L of anhydrous diethyl ether. *Automatic delivery pipets are used to transfer these reagents in the **hood**.* Recap the vial          HOOD
until just before injecting its contents into the reaction vial.

Draw the bromobenzene–ether solution into a 1.0-mL syringe, and insert the syringe through the rubber septum on the Claisen head.

Place an additional 300 $\mu$L of anhydrous diethyl ether rinse in the empty vial (which contains traces of bromobenzene), cap it, and set it aside for later use.

With slow stirring, add 6–8 drops of the bromobenzene solution to induce the initial formation of Grignard reagent. The evolution of tiny bubbles from the surface of the magnesium (the heat of reaction is vaporizing the low-boiling ether solvent) is evidence of successful reaction initiation.

Once the reaction gives evidence of initiation, add the remainder of the bromobenzene dropwise *slowly* over a 3–5-min period. (*In macroscale reactions it is extremely important to make sure that initiation of the reaction has occurred prior to adding large quantities of the organohalide, as Grignard reactions often go through an induction period before starting up. If significant quantities of the halide are present when the reaction commences, the sudden and rapid rate of reaction can produce a very rapid evolution of heat, and the reaction may well erupt out of control.) Warm the reactants gently to maintain gentle reflux.*

**Step I:**
BrC$_6$H$_5$, 76 $\mu$L
+ Mg, 18 mg
+ I$_2$, 1 crystal
+ (CH$_3$CH$_2$)$_2$O, 700 $\mu$L
**Step II:**
(C$_6$H$_5$)$_2$CO, 105 mg
+ (CH$_3$CH$_2$)$_2$O, 600 $\mu$L

Upon completion of the addition of the bromobenzene, draw the ether rinse from the capped vial into the syringe and also add it to the reaction vial through the septum in a single portion.

Heat the resulting heterogeneous reaction mixture gently with stirring for 15 minutes.

---

CAUTION:  DO NOT OVERHEAT! This overheating will cause loss of ether solvent and promote formation of byproducts. Small fragments of magnesium may remain at the end of the reaction. Maintain no more than a gentle reflux at all times. If solvent volume decreases rapidly, check for leaks around the Capseals and add additional anhydrous ether through the septum to make up for the lost volume of ether.

---

Cool the gray-brown mixture of the Grignard reagent, phenylmagnesium bromide, to room temperature.

### The Benzophenone Reagent

Prepare a solution of 105 mg (0.58 mmol) of benzophenone in 300 $\mu$L of anhydrous diethyl ether in a dry vial with a cap. *The ether is measured using a graduated 1-mL syringe and is dispensed in the **hood**.*

HOOD

Draw the solution immediately into a 1.0-mL syringe, and then insert the syringe needle through the rubber septum on the Claisen head.

Place an additional 300 $\mu$L of the anhydrous diethyl ether in the empty vial, cap it, and set it aside for later use.

### Reaction Conditions

*Carefully*, with stirring, add the benzophenone solution to the Grignard reagent over a period of approximately 30 s or at a rate that maintains the temperature of the ether solvent at a no more than gentle reflux.

Upon completion of this addition, add the rinse from the capped vial, in like manner, in a single portion.

Stir the reaction mixture for 2–3 min and then allow it to cool to room temperature. Remove the reaction vial from the Claisen head and cap it. During this cooling period the reaction mixture generally solidifies. *Once the reaction vial is detached from the Claisen head, it is recommended that the vial be placed in a 10-mL beaker to prevent loss of product by accidental tipping.*

**NOTE.**   *If the laboratory is done in two periods, one may either stop at this point, or following the hydrolysis sequence described in the next step.*

### Isolation of Product

Hydrolyze the magnesium alkoxide salt by the *careful*, dropwise addition of 3 $M$ HCl from a Pasteur pipet, while at the same time using a small stirring rod to break up the solid residue. Continue the addition until the aqueous phase tests acidic with litmus paper. A two-layer reaction mixture forms (ether–water) as the solid gradually dissolves.

---

CAUTION:  The addition of the acid may be accompanied by the evolution of heat and some frothing of the reaction mixture. An ice bath should be handy to cool the solution, if necessary. Additional ether may be added, if required, to maintain the volume of the organic phase. Check the acidity of the mixture periodically. The total reaction mixture must be acidic; both insufficient or excess amounts of hydrochloric acid will result in a decreased yield of product during the subsequent workup.

---

Now remove the magnetic spin vane with forceps and set it aside to be rinsed with an ether wash. Cap the vial tightly, shake, carefully vent, and allow the layers to separate.

Using a Pasteur filter pipet, transfer the lower aqueous layer to a clean 5.0-mL conical vial.

**IMPORTANT.**   *Save the ether layer as it should contain your product.*

Wash the acidic aqueous layer with three 0.5-mL portions of diethyl ether (calibrated Pasteur pipet). Rinse the spin vane with the first portion of ether as it is added to the vial. Cap the vial, shake (or use a Vortex mixer, if available), vent carefully and allow the layers to separate. After each extraction, combine the organic phase with the ether solution saved above. The bottom (aqueous) layer is set aside in a 10-mL Erlenmeyer flask until the experiment is completed.

Now extract the combined ether layers with 0.5 mL of cold water to remove any acidic residue. Combine the aqueous rinse with the previously extracted and stored aqueous layers in a 10-mL Erlenmeyer flask. Dry the ether solution (capped) over 250–300 mg of anhydrous granular sodium sulfate for approximately 10 min. Stir the drying agent intermittently with a glass rod or swirl the flask. If large clumps of sodium sulfate begin to develop and the solution remains cloudy, it may be necessary to transfer the ether extracts to another vial for a second treatment with the drying agent, or you may be able to simply add more $Na_2SO_4$ to the original vial. The ether solution should be *clear* following treatment with the anhydrous sodium sulfate. Make all these transfers with Pasteur filter pipets.

Transfer the dried ether solution to a *previously tared* Craig tube containing a boiling stone. The transfer is carried out in 0.5-mL portions, concentrating each ether aliquot by warming the vial in a sand bath **[HOOD]** between the transfers. Rinse the vial and drying agent with an     HOOD additional 0.5 mL of ether, add the rinse to the Craig tube, and finally, concentrate the solution to dryness to yield the product residue.

Determine the weight of the crude triphenylmethanol product.

## Purification and Characterization

The major impurity usually present in the triphenylmethanol is biphenyl, which is formed by a coupling reaction,

$$2 \text{ RX} + \text{Mg} \xrightarrow{\text{ether}} \text{R—R} + \text{MgX}_2$$

The purity of your crude product may be determined using *thin-layer chromatography.*

**TLC CONDITIONS.** *Use Eastman Kodak fluorescent silica gel sheets (1 × 4 cm). Develop the plates with methylene chloride and visualize the spots by UV light. Reference $R_f$ values for triphenylmethanol and biphenyl are about 0.6 and 0.9, respectively.*

The coupled byproduct can be separated from the desired alcohol by taking advantage of the difference in solubilities of the hydrocarbon and the alcohol in ligroin. Ligroin, a nonpolar alkyl solvent, readily dissolves the nonpolar biphenyl, whereas the polar tertiary alcohol is much less soluble.

Add 0.5 mL of cold ligroin to the crude product contained in a Craig tube and scrape and agitate the solid material into a suspension with a small stirring rod. Swirl and stir the solid product with the solvent for several minutes.

Recover the solid triphenylmethanol using the Craig tube in the usual manner. *Save* the ligroin solution that will contain any biphenyl, by transferring it using a Pasteur filter pipet, to a tared 10-mL Erlenmeyer flask.

Repeat the above extraction with a second 0.5-mL portion of ligroin, again stirring the solid suspension and combining the recovered ligroin solution with that saved above. Place the Erlenmeyer flask (cover the

HOOD mouth with filter paper held by a rubber band) in the **hood** overnight, or warm it in a sand bath to allow the ligroin to evaporate. Estimate the amount of biphenyl (and any other impurities) produced in the reaction.

Heat the Craig tube containing the solid triphenylmethanol in a 100 °C oven for 5 min, and then place the crystals on a clay plate to complete the drying process. A further check of the product purity may be carried out by TLC, as described above.

Weigh the purified triphenylmethanol product and calculate the percent yield. Determine the melting point of the material and compare your result to that recorded in the literature. If desired, the product may be purified further by recrystallization from isopropanol using the Craig tube.

Characterization of the triphenylmethanol is best done by obtaining the IR and NMR spectra and comparing the spectral data to that of an authentic sample or to the published spectra in *The Aldrich Library of IR Spectra* and *The Aldrich Library of NMR Spectra,* respectively.

Triphenylmethanol (0.1 g/L in methanol) and biphenyl (0.05 g/L in methanol) also have markedly different UV spectra. This electronic absorption data can further help to establish the identity of the products formed in this Grignard reaction.

*UV Spectral Data*

|  |  |
|---|---|
| Triphenylmethanol | $\lambda_{max}$ 240 nm (log $\varepsilon_{max}$ 3.16, dioxane) |
|  | $\lambda_{max}$ 253 nm (log $\varepsilon_{max}$ 3.26, dioxane) |
|  | $\lambda_{max}$ 260 nm (log $\varepsilon_{max}$ 3.28, dioxane) |
| Biphenyl | $\lambda_{max}$ 247 nm (log $\varepsilon_{max}$ 4.24, ethanol) |

**QUESTIONS**

**6-109.** Predict the product formed in each of the following reactions and give each reactant and product a suitable name.

a. $CH_3CH_2MgBr + CH_2O \xrightarrow[\text{2. H}^+]{\text{1. ether}}$

b. $p\text{-}CH_3C_6H_4MgBr + CH_3CH_2CHO \xrightarrow[\text{2. H}^+]{\text{1. ether}}$

c. $C_6H_5MgBr + D_2O \xrightarrow{\text{ether}}$

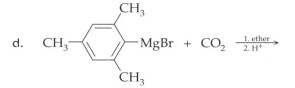

d.

**6-110.** Using the Grignard reaction, carry out the following transformations. Any necessary organic or inorganic reagents may be used. Name all reactants and products.

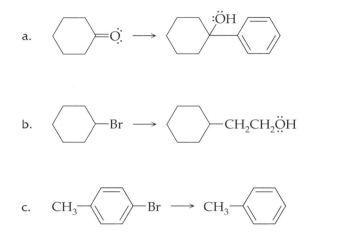

a.

b.

c.

**6-111.** Outline a synthetic reaction scheme for the preparation of triphenylmethanol from:
   a. Methyl benzoate
   b. Diethyl carbonate.

**6-112.** In the experiment, ligroin may be used as a solvent for the separation of the product from biphenyl.
   a. What is ligroin?
   b. Can you suggest an alternative solvent that might be used in this step?

**6-113.** Give the reaction scheme, showing the products formed (before hydrolysis), when one equivalent of ethylmagnesium bromide is treated with one equivalent of 5-hydroxy-2-pentanone. Does addition of two equivalents of the Grignard reagent to this ketone yield a different product(s)? If so, give the structure(s).

The alternate method of preparing the Grignard reagent is adapted from:

**BIBLIOGRAPHY**

Eckert, T. S. *J. Chem. Educ.* **1987,** *64,* 179.

General references on Grignard reagents:

Ashby, E. C. *Q. Rev.* **1967,** *21,* 259.

Coates, G. E.; Green, M. L. H.; Wade, K. *Organometallic Compounds;* 3rd ed.; Methuen: London; Vol. II, 1968.

Grignard, V. *Compt. Rend.* **1900,** *130,* 1322.

Jones, R. G.; Gilman, H. *Chem. Rev.* **1954,** *54,* 835.

Kharasch, M. S.; Reinmuth, O. *Grignard Reactions of Non-metallic Substances;* Prentice-Hall: New York; 1954.

Wakefield, B. J. *Organometal. Chem. Rev.* **1966,** *1,* 131.

Yoffe, S. T.; Nesmeyanov, A. N. *Handbook of Magnesium-Organic Compounds*; Pergamon: London; Vols. I–III, 1957.

A synthesis of triphenylmethanol (triphenylcarbinol) is reported in *Organic Syntheses*:

Bachmann, W. E.; Hetzner, H. P. *Organic Syntheses*; Wiley: New York, 1955; Collect. Vol. III, p. 839.

The preparation of a series of tertiary alcohols by the addition of Grignard reagents to diethyl carbonate is reported in *Organic Syntheses*:

Moyer, W. W.; Marvel, C. S. *Organic Syntheses*; Wiley: New York, 1943; Collect. Vol. II, pp. 602–603.

## EXPERIMENT [17]

### Grignard Reaction with an Aldehyde: 4-Methyl-3-heptanol

Common name: 4-Methyl-3-heptanol
CA number: [14979-39-6]
CA name as indexed: 3-Heptanol, 4-methyl-

### Purpose

To carry out a classic method for the synthesis of *secondary* alcohols: the addition of a Grignard reagent[16] to an aldehyde (other than formaldehyde).

#### Prior Reading

Technique 4:  Solvent Extraction
              Liquid–Liquid Extraction (see pp. 80–82).
              Drying of the Wet Organic Layer (see pp. 86–87).
Technique 6:  Chromatography
              Column Chromatography (see pp. 98–107).
Technique 1:  Gas Chromatography (see pp. 56–61).

## REACTION

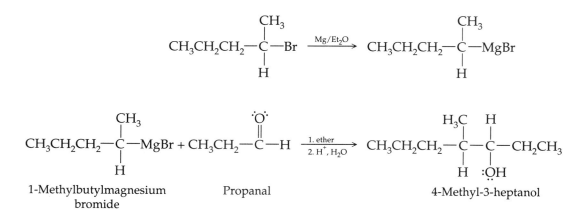

1-Methylbutylmagnesium bromide          Propanal          4-Methyl-3-heptanol

**NOTE.** *See Experiment [16] for a biography of François Auguste Victor Grignard, Nobel Laureate, who discovered and developed the Grignard reagents. This experiment also contains further details about the mechanism and use of these reagents that have had such a powerful influence on synthetic organic chemistry.*

[16] For references relating to the preparation of Grignard reagents, see Experiment [16].

In this experiment, the addition of a nucleophilic Grignard reagent (1-methylbutylmagnesium bromide), to the electrophilic carbonyl carbon of an aldehyde (propanal) is described. The product obtained is a 2° alcohol, 4-methyl-3-heptanol. Because it is possible to vary the structure of both the Grignard reagent and the aldehyde, a wide variety of 2° alcohols can be prepared by this route. Primary alcohols result when formaldehyde is used as the aldehyde. Secondary alcohols may also be obtained with these reagents when ethyl formate, an ester, acts as the electrophile. This latter reaction, however, requires two molar equivalents of the Grignard reagent. The mechanism for the reaction of an aldehyde with a Grignard reagent follows.

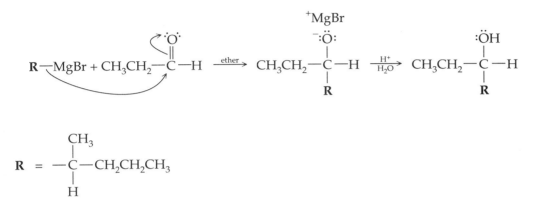

**EXPERIMENTAL PROCEDURE**

Estimated time to complete the experiment: 3.0 hours.

The chromatographic separation requires approximately an additional 15 min per student.

---

CAUTION: Ether is a flammable liquid. All flames must be extinguished during this experiment.

---

### Physical Properties of Reactants

| Compound | MW | Amount | mmol | bp(°C) | d | $n_D$ |
|---|---|---|---|---|---|---|
| 2-Bromopentane | 151.05 | 125 μL | 1.0 | 117 | 1.2 | 1.4413 |
| Diethyl ether | 74.12 | 700 μL | | 34.5 | | |
| Magnesium | 24.31 | 36 mg | 1.48 | | | |
| Iodine | 253.81 | 1 crystal | | | | |
| Propanal | 58.08 | 50 μL | 0.69 | 49 | 0.81 | 1.3636 |

## Reagents and Equipment

### Preparation of 1-Methylbutylmagnesium Bromide

**NOTE.** *All the glassware used in the preparation of the Grignard reagent should be cleaned and dried in an oven at 110 °C for at least 30 min. After removal from the drying oven, the hot glassware should be placed in a desiccator and cooled before*

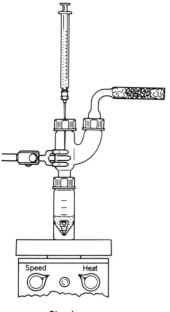

**Step I:**
Mg, 36 mg + I$_2$, 1 crystal
+ (CH$_3$CH$_2$)$_2$O, 500 μL
+ CH$_3$CHBr(CH$_2$)$_2$CH$_3$, 125 μL
**Step II:**
CH$_3$CH$_2$CHO, 50 μL
+ (CH$_3$CH$_2$)$_2$O, 200 μL

*being assembled. Flame drying of the apparatus with a microburner annealing flame (high gas mixture) is an alternative, if ovens are not available and if ether is not in use. This latter method preferably should be carried out prior to assembly, and as with the oven-dried equipment, the hot glassware should be placed in a desiccator to cool. Care must be taken when flaming the reaction vial, as thermal shock can easily crack this heavy-walled vessel. If the flame drying procedure is performed on the assembled apparatus, care also must be taken not to overheat the O-rings and plastic Capseals.*

**NOTE.**   *Scrape a 2–3 in. piece of magnesium ribbon clean of oxide coating, and cut it into sections 1 mm in length. This freshly cut material should be handled only with forceps.*

Prepare a 3.0-mL conical vial containing a magnetic spin vane and equipped with a Claisen head fitted with a calcium chloride drying tube and a rubber septum. Weigh and place 36 mg (1.5 mmol) of magnesium in the vial, and then add a small crystal of iodine, followed by 100 μL of anhydrous ether. (■)

**Alternate Procedure**
Place the magnesium metal and the iodine crystal in the vial and assemble the apparatus. *Gently* warm the mixture (microburner or hot plate) until evidence of purple vapor from the iodine is seen.

**NOTE.**   *If a microburner is used in this step, all flames must be extinguished before proceeding.*

Now add the 100 μL of anhydrous ether, using a 1.0-mL syringe inserted through the septum.
Prepare a solution of 125 μL (153 mg, 1.0 mmol) of 2-bromopentane in 300 μL of anhydrous diethyl ether in a dry, screw-capped vial. *Use an automatic delivery pipet to deliver these reagents.*
After the assembly has cooled to room temperature, draw the 2-bromopentane solution into a 1.0-mL syringe and then insert the syringe needle through the rubber septum on the Claisen head. Place an additional 100 μL of diethyl ether in the empty vial, cap it, and set it aside for later use.
While stirring the heterogeneous mixture, add 6–8 drops of the 2-bromopentane–ether solution to initiate the formation of the Grignard reagent. The evolution of tiny bubbles from the surface of the magnesium is evidence of reaction.
SLOWLY   When the reaction has started, add the remainder of the 2-bromopentane–ether solution dropwise [SLOWLY] over a 3–5 min period. Warm the reactants slightly . Upon completion of this addition, draw the rinse in the capped vial into the syringe and add it through the septum in a single portion to the reaction vial. Gently warm the resulting solution for 15 min.

---

**CAUTION: DO NOT OVERHEAT. This overheating will cause loss of ether solvent. Small fragments of magnesium may remain at the end of the addition of the alkyl halide.**

---

Cool the gray-colored solution of Grignard reagent to room temperature.

## The Propanal Reagent

Prepare a solution of the aldehyde by weighing 50 $\mu$L (40 mg, 0.7 mmol) of propanal into a tared, oven-dried, capped vial followed by the addition of 100 $\mu$L of anhydrous diethyl ether. The propanal is the limiting reagent and therefore an accurate weight should be recorded for the yield calculations. *Dispense the aldehyde and diethyl ether by automatic delivery pipets in the* **hood.**                                                                 HOOD

Immediately draw the aldehyde solution into a 1.0-mL syringe and insert the syringe needle through the rubber septum on the Claisen head.

Place an additional 100 $\mu$L of the anhydrous diethyl ether in the empty vial, cap it, and set it aside for later use.

## Reaction Conditions

Now add the propanal solution *carefully*, with stirring, to the Grignard reagent over a period of about 30 s at such a rate as to keep the ether solvent at a steady reflux.

Following this addition, add the rinse in the capped vial in one portion in a similar manner.

Stir the reaction mixture for 5 min and then allow it to cool to room temperature. Remove the conical vial and cap. It is recommended that the vial be placed in a beaker to prevent tipping and loss of product.

**NOTE.** *If the laboratory is done in two periods, one may stop at this point (recap the vial for storage), or after the hydrolysis sequence in the next step.*

## Isolation of Product

Hydrolyze the magnesium alkoxide salt by the *careful*, dropwise addition of 2–3 drops of water from a Pasteur pipet. Stir the resulting mixture for 5 min. A two-phase (ether–water) reaction mixture develops as the magnesium salt is hydrolyzed.

---

**CAUTION: The addition of water causes the evolution of heat. An ice bath should be handy to cool the solution if it begins rapid reflux.**

---

Now add 2–3 drops of 3 *M* HCl. Remove the vial, cap it, and allow it to stand at room temperature for 5 min. Test the aqueous layer with litmus paper. The solution should be slightly acidic. *Too much or too little aqueous HCl will cause problems in the subsequent workup.*

Remove the magnetic spin vane with forceps and set it aside to be rinsed with an ether wash. Cap the vial tightly, shake (or use a Vortex mixer), vent *carefully*, and allow the layers to separate.

Using a Pasteur filter pipet, transfer the aqueous (lower) layer to a clean 5.0-mL conical vial. *Save* the ether layer since it contains the crude reaction product.

Now wash the aqueous layer, previously transferred to the 5.0-mL vial, with three 0.5-mL portions of diethyl ether. Rinse the magnetic spin vane with the first portion as it is added to the vial. Upon addition of each portion of ether (calibrated Pasteur pipet), cap the vial, shake (or use a Vortex mixer), vent carefully, and allow the layers to separate. With the aid of a Pasteur filter pipet, remove each ether layer and combine it with the ether solution retained above. After the final extraction, save the aqueous (lower) layer until you have isolated and characterized the final product. Extract the combined ether fractions with 0.5 mL of cold water to remove any acidic material. Save the aqueous rinse until you have isolated and characterized the final product.

Dry the ether solution by transferring it, using a Pasteur filter pipet, to a shortened Pasteur filter pipet containing 500 mg of anhydrous sodium sulfate. Collect the eluate in a tared 10 × 75-mm test tube. Remove the ether solvent from the eluate **[HOOD]** by warming in a sand bath to concentrate the solution to a weight less than 90 mg.

HOOD

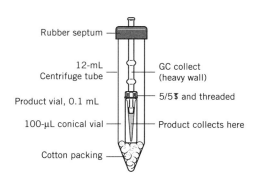

Rubber septum

12-mL
Centrifuge tube

Product vial, 0.1 mL

100-μL conical vial

Cotton packing

GC collect
(heavy wall)

5/5 $ and threaded

Product collects here

### Purification and Characterization

The product, 4-methyl-3-heptanol, is isolated and purified using gas chromatography.

Use a 100-μL syringe to inject the entire sample of crude material obtained above onto the GC column. Collect the components of interest as they elute from the column, employing the chromatography technique described in the Prior Reading section. (■)

*Gas Chromatographic Conditions:*

GOW-MAC Series-150, Thermal Conductivity Detector
20% Carbowax 20M column, $\frac{1}{4}$ in. × 8 ft
Temperature, 145 °C
Flow rate of 50 mL/min (He gas)

The retention time for 4-methyl-3-heptanol under these conditions is 7–8 min (2 min after any other peak).

Determine the weight of the 4-methyl-3-heptanol collected, and calculate the percent yield. Determine the boiling point and refractive index (optional) of the alcohol, and compare these with values in the literature.

Obtain an IR spectrum of the product as a thin film and compare it with that recorded in *The Aldrich Library of IR Spectra*. If enough of the hydrocarbon byproduct is collected, also obtain an IR spectrum to aid in its identification.

### Chemical Tests

The ignition test should indicate that this compound is an aliphatic species. Does your result confirm this fact? Perform the ceric nitrate test to demonstrate the presence of the —OH group and the Lucas test to demonstrate that a secondary alcohol has been prepared. If you were required to prepare a solid derivative of this alcohol, which one would you select? It may be of interest to determine the solubility of this product in water, ether, concentrated sulfuric and 85% phosphoric acids. Do your results agree with what you would predict? What test(s) would you perform to establish that one of the starting reagents was an aldehyde?

**QUESTIONS**

**6-114.** Show how one could carry out each of the following transformations using the Grignard reaction. Any necessary organic or inorganic reagents may be used. Name each reactant and product.

    **a.** $CH_3(CH_2)_2CH_2Br \rightarrow (CH_3CH_2CH_2CH_2)_2CH\overset{..}{O}H$

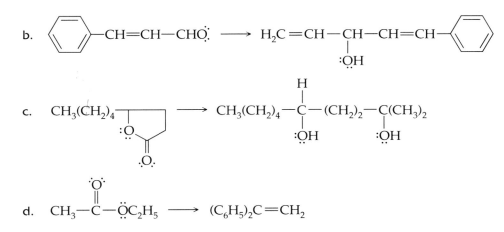

b.

c.

d.

**6-115.** Explain why Grignard reagents cannot be prepared from an organic halide that also contains a hydroxyl (—OH), a carboxyl (—CO$_2$H), a thiol (—SH), or an amino (—NH$_2$) group.

**6-116.** What would be the final product of the reaction between methyl benzoate and two equivalents of ethylmagnesium bromide?

**6-117.** Consider the same reaction as in Question 6-116 except that in this case it is carried out with ethyl benzoate. What product would be expected in this case?

**6-118.** Grignard reagents may be used to prepare other organometallic reagents, for example, ethylmagnesium bromide reacts with cadmium chloride to yield diethylcadmium.

$$2\ CH_3CH_2MgCl + CdCl_2 \rightarrow (CH_3CH_2)_2Cd + 2\ MgCl_2$$

Indicate the product of the following reactions and name the organometallic product.

$$4\ CH_3MgCl + SiCl_4 \rightarrow$$

$$2\ C_6H_5MgCl + HgCl_2 \rightarrow$$

---

A list of secondary alcohol preparations presented in *Organic Syntheses* is below:

**BIBLIOGRAPHY**

Boeckman, R. K., Jr.; Blum, D. M.; Ganer, B.; Halvey, N. *Organic Syntheses;* Wiley: New York, 1988; Collect. Vol. VI, p. 1033.

Coburn, E. R. *Organic Syntheses;* Wiley: New York, 1955; Collect. Vol. III, p 696.

Coleman, G. H.; Craig, D. *Organic Syntheses;* Wiley: New York, 1943; Collect. Vol. II, p. 179.

Drake, N. L. *Organic Syntheses;* Wiley: New York, 1943; Collect. Vol. II, p 406.

Overberger, C. G.; Saunders, J. H.; Allen, R. E.; Gander, R. *Organic Syntheses;* Wiley: New York, 1955; Collect. Vol. III, p. 200.

Skattebol, L.; Jones, E. R. H.; Whiting, M. C. *Organic Syntheses;* Wiley: New York, 1963; Collect. Vol. IV, p. 792.

Trust, R. I.; Ireland, R. E. *Organic Syntheses;* Wiley: New York, 1988; Collect. Vol. VI, p. 606.

**EXPERIMENT [18]**

## The Perkin Reaction: Condensation of Rhodanine with an Aromatic Aldehyde to Yield o-Chlorobenzylidene Rhodanine

Common name: o-Chlorobenzylidene rhodanine
CA number: [6318-36-1]
CA name as indexed: 4-Thiazolidinone, 5-[(2-chlorophenyl)methylene]-2-thioxo-

### Purpose

To explore the use of the interesting heterocyclic compound, rhodanine, as the source of an active (acidic) —CH₂— group. (The methylene group contained in the thiazolidinone ring system possesses the capacity to participate in base-catalyzed condensation reactions similar to those of the **aldol reaction**.) To carry out a base-catalyzed condensation reaction with an aromatic aldehyde. To examine the properties of this condensation product, which has the capacity to function as an intermediate in a number of synthetic pathways. Indeed, one of these routes yields the important class of aromatic amino acids, the phenylalanines.

*Prior Reading*
    *Technique 5:*   Crystallization
            Use of the Hirsch Funnel (see pp. 93–95).
            Craig Tube Crystallization (see pp. 95–96).

**REACTION**

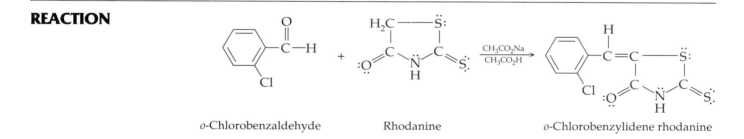

o-Chlorobenzaldehyde            Rhodanine            o-Chlorobenzylidene rhodanine

**Sir William Henry Perkin** (1838–1907, English, born in London). Perkin came under the influence of the German chemist, A. W. von Hofmann, at the Royal college of Chemistry (London), where Perkin was a student. In his second year at the school, he gained the title of Hofmann's *honorary assistant*, and following a publication the next year, when he was 17, he was advanced to the rank of assistant. At home, he set up his own laboratory where he worked evenings and on vacations. This laboratory is where he undertook, at Hofmann's suggestion, the synthesis of quinine.

Working with his friend Arthur Church in his home laboratory, he and Church prepared one of the first azo dyes derived from naphthalene (nitrosonaphthlene). This effort resulted in his first patent (with Church). In 1856 at the age of 18, Perkin discovered the first commercially significant synthetic coal-tar dye, *Mauve* or *Aniline Purple*. He also obtained a patent on the synthetic method for preparing this material. The method involved the treatment of an aniline salt with bichromate of potash (K₂Cr₂O₇) (this was an outgrowth of his quinine studies). Against Hofmann's wishes, Perkin withdrew from college, designed a factory, and started commercial production of this dye, which was marketed as *Perkin's Tyrian Purple*. The name *Tyrian Purple* was originally used for a prized purple dye from the Phoenician city of Tyre, which today is part of Lebanon. *Tyrian Purple* was produced in small quantities from a material isolated from a snail. The production of this dye became a lost art during the Dark Ages, but a species of mollusk containing this very rare pigment was rediscovered in Ireland in 1684. The natural dye is obtained by the air oxidation of a colorless fluid expressed from the glands of the snail. It required the contents of 10,000 snails to obtain a single gram of the fluid.

Perkin's dye synthesis was the beginning of the coal-tar dyestuffs industry. He later developed and manufactured magenta (violet dye) and alizarin (a red dye). The rapid acceptance of these dyes by fabric manufacturers was demonstrated by the fact that annual alizarin production reached 220 tons per year by 1871. At age 37 and a wealthy man, Perkin sold his commercial holdings and devoted the rest of his life to pure chemical research.

Perkin's later years were as productive as his earlier ventures. He developed methods for the preparation of aminoacetic acid. He established the structural relationships between tartaric, fumaric, and maleic acids, and *synthesized cinnamic acid. This last endeavor led to the development of what is now known as the* **Perkin reaction.** The reaction is widely used to prepare unsaturated acids from aromatic aldehydes. These studies led to his synthesis of coumarin (actually it was the first condensation product he obtained with the classic reaction, see also Experiment [3A$_{adv}$]). Other areas investigated by Perkin dealt with the relationship of physical properties and chemical structure.

Perkin was knighted in 1906 in recognition of his contributions to chemistry and to the practical application of many of his discoveries. In the United States, The Society of Chemical Industry awards a Perkin Medal each year to an outstanding industrial chemist.[17]

---

**DISCUSSION**

The classic Perkin reaction is the base-catalyzed condensation of an aromatic aldehyde with a carboxylic acid anhydride to yield and an $\alpha,\beta$-unsaturated carboxylic acid. The initial stages of the condensation can be viewed as an aldol-type reaction (see Experiments [20], [3A$_{adv}$], [3B$_{adv}$], and [A3$_a$]).

A variation of the Perkin reaction is the condensation of aromatic aldehydes with rhodanine, which plays a similar role to that of the anhydride in the original reaction. Rhodanine, a derivative of the thiazolidinone ring system was first synthesized in 1935 by Percy Julian (a future president of Howard University) and Bernard Sturgis (an undergraduate at DePauw University at the time). This heterocyclic molecule has an active (acidic) methylene group that can be deprotonated with a relatively mild base (in this case acetate, $CH_3COO^-$ or $AcO^-$, ion) to generate the nucleophile that attacks the carbonyl group of the aldehyde.

Under the conditions employed in the current experiment, dehydration–elimination rapidly follows the initial nucleophilic addition with formation of the benzylidene intermediate. The mechanism is shown here.

---

[17] See Edelstein, S. M. *American Dyestuff Reporter* **1956,** *45,* 598; Mendola, R. *J. Chem. Soc.* **1908,** *93,* 2214; Levinstein, H. *Chem. Ind.* **1938,** 1137; Rose, R. E. *Ind. Eng. Chem.* **1938,** 16, 608; *Dictionary of Scientific Biography* C.C. Gillespie, Ed., Scribner's: New York, 1974, Vol. X, p. 515.

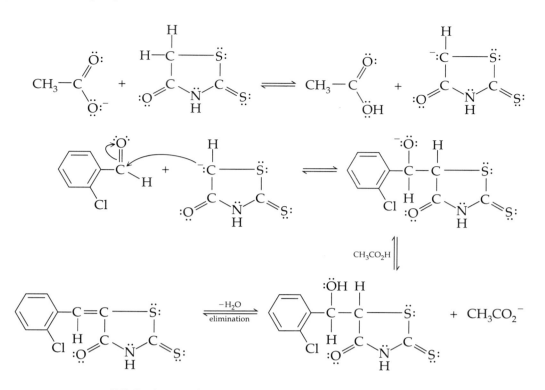

While the resulting rhodanine derivatives have been shown to exhibit antibacterial, antitubercular, antimalarial, antifungal, and antiparasitic activity, the principal focus of attention on these interesting compounds has been as reactive synthetic intermediates. For example, these particular compounds can be converted, in high yield, in three steps to nitriles in which the side chain of the original aldehyde has been extended by an additional carbon atom.

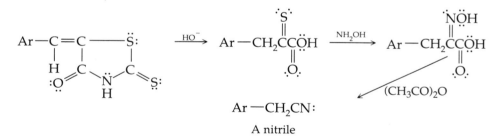

A nitrile

---

# EXPERIMENTAL PROCEDURE    Estimated time of the experiment: 1.5 hours.

## Physical Properties of Reactants and Product

| Compound | MW | Amount | mmol | mp(°C) | bp(°C) | d | $n_D$ |
|---|---|---|---|---|---|---|---|
| Rhodanine | 133.19 | 30 mg | 0.23 | 170 | | | |
| Sodium acetate | 82.03 | 52 mg | 0.63 | 324 | | | |
| Acetic acid, glacial | 60.05 | 1.0 mL | | | 118 | | |
| o-Chlorobenzaldehyde | 140.57 | 58 mg | 0.41 | | 212 | 1.25 | 1.5662 |
| o-Chlorobenzylidene rhodanine | 259.76 | | | 191 | | | |

## Reagents and Equipment

In a 3.0-mL conical vial containing a boiling stone and equipped with an air condenser, weigh out and place 30 mg (0.23 mmol) of rhodanine and 52 mg (0.63 mmol) of anhydrous sodium acetate. Now add 1.0 mL of glacial acetic acid **[HOOD]** dispensed from a graduated pipet. To this mixture, measure and add 58 mg (0.41 mmol) of *o*-chlorobenzaldehyde. (■)

HOOD

**NOTE.** *Dry the sodium acetate in the oven for 1 hour before use. The aldehyde* ***must*** *be free from the corresponding acid or lower yields of product will result. It is recommended that the purity of the aldehyde be checked by IR analysis. The reaction vial may be weighed before and after the addition of aldehyde to obtain an accurate weight.*

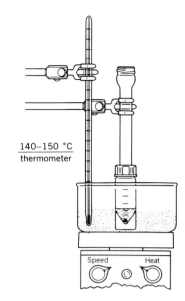

140–150 °C
thermometer

## Reaction Conditions

Heat the reaction mixture in a sand bath at 140–150 °C for 30 min.

**IMPORTANT.** *Immerse the vial in the sand up to the level of the top of the reaction mixture.*

As the reaction progresses, the mixture becomes homogeneous and turns yellow. At the end of the reaction period the resulting solution is cooled to room temperature. When the conical vial has reached ambient temperature, place it in an ice bath to complete crystallization of the condensation product.

Rhodanine, 30 mg
+ NaOAc, 52 mg
+ CH$_3$CO$_2$H, 1.0 mL
+ *o*-ClC$_6$H$_4$CHO, 58 mg
Note *It is important that the vial be immersed in sand to the level of the reaction mixture.*
Ac = acetyl group

## Isolation of Product

Collect the yellow crystals by vacuum filtration using a Hirsch funnel. Rinse the reaction vial with two 1.0-mL portions of cold glacial acetic acid transferred by a calibrated Pasteur pipet, and then use each rinse to wash the filter cake on the Hirsch funnel. (■) Complete the removal of the acid solvent by air drying the crystals on a porous clay plate.

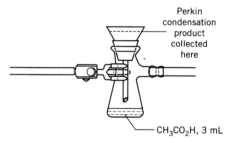

Perkin
condensation
product
collected
here

CH$_3$CO$_2$H, 3 mL

## Purification and Characterization

Weigh the dried product and calculate the percent yield of the crude material. Recrystallization (Craig tube) of a 5-mg portion of the benzylidene product from 0.5 mL of glacial acetic acid yields fine, bright yellow needles that can be used to complete the characterization of this interesting substance.

Determine the melting point and compare your results with the value given in the table at the beginning of the experiment.

## Chemical Tests

This compound is an interesting reaction product as it contains three different types of heteroatoms: chlorine, nitrogen, and sulfur. The sodium fusion test (see Chapter 10) may be used to substantiate the presence of these elements.

It would also be of interest to establish if a positive Beilstein test for chlorine is found when in the presence of sulfur and nitrogen, or whether the soda–lime test for nitrogen is observed with the elements of chlorine and sulfur also located within this molecule (see Chapter 10).

Does ignition of the material (see Chapter 10) indicate that the compound contains an aromatic ring ?

## OPTIONAL SEMIMICROSCALE PREPARATION

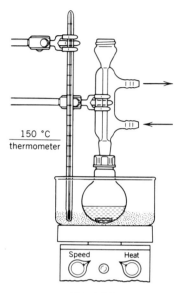

150 °C
thermometer

Speed    Heat

10-mL RB flask

This experiment may be scaled up and carried out at a level 5 or 10 times greater than the amounts used in the above microscale experiment. The data given below is for the 10-fold procedure. The procedure is identical to that given above in the microscale section, but with the following modifications:

**1.** A 10-mL round-bottom flask containing a spin bar and fitted with an air cooled condenser is used in place of the conical reaction vial. (■)

**2.** The reagent and solvent amounts are given in the following table.

**Physical Properties of Reactants and Products**

| Compound | MW | Amount | mmol | mp(°C) | bp(°C) | $d$ | $n_D$ |
|---|---|---|---|---|---|---|---|
| Rhodanine | 133.19 | 300 mg | 2.3 | 170 | | | |
| Sodium acetate | 82.03 | 520 mg | 6.3 | 324 | | | |
| Acetic acid, glacial | 60.05 | 5.0 mL | | | 118 | | |
| o-Chlorobenzaldehyde | 140.57 | 580 mg | 4.1 | | 212 | 1.25 | 1.5662 |
| o-Chlorobenzylidene rhodanine | 259.76 | | | 191 | | | |

**3.** After the product is air dried on a clay plate until no acetic acid (vinegar) odor remains, it is placed in a 100 °C oven to dry overnight. It may also be dried in a vacuum drying oven until no acetic acid (vinegar) odor remains.

## QUESTIONS

**6-119.** In which of the following two species, **A** or **B,** is the underlined hydrogen atom more acidic? Explain.

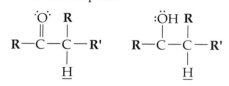

**6-120.** A number of compounds similar to rhodanine and possessing active methylene groups have been used in the Perkin condensation. Two are shown below.

Draw the structure of the product that would be formed if each underwent the Perkin condensation with p-chlorobenzaldehyde.

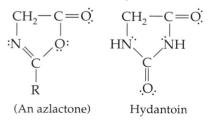

(An azlactone)          Hydantoin

**6-121.** para-Nitrobenzaldehyde reacts at a faster rate than benzaldehyde in the Perkin reaction while p-N,N-dimethylaminobenzaldehyde is much less reactive toward the same nucleophile. Explain.

**6-122.** Explain the fact that the C=O group in —C—CH— is effective in increasing the acidity of the $\alpha$-hydrogen atom.

**6-123.** There are a large number of condensations that are closely related to the Perkin reaction. Among these are the aldol (Experiments [20], [3B$_{adv}$], and [A3$_a$]), Knoevenagel, Claisen (Experiment [3A$_{adv}$]), and Dieckmann condensations. What general class of compounds can be prepared using each of these well-known reactions?

---

**BIBLIOGRAPHY**

For a review on the general Perkin reaction see:

Johnson, J. R. *Org. React.* **1942,** *1,* 210.

For examples of the general Perkin reaction see:

Corson, B. B. *Organic Syntheses;* Wiley: New York, 1943; Collect. Vol. II, p. 229.

Herbst, R. M.; Shemin, D. *Organic Syntheses;* Wiley: New York, 1943; Collect. Vol. II, p. 1.

Johnson, J. R. *Organic Syntheses;* Wiley: New York, 1955; Collect. Vol. III, p. 426.

Thayer, F. K. *Organic Syntheses;* Wiley: New York, 1941; Collect. Vol., I, p. 398.

Weiss, R. *Organic Syntheses;* Wiley: New York, 1943; Collect. Vol. II, p 61.

For references using rhodanine in the Perkin reaction see:

Andreasch, R. *Monatsh. Chem.* **1928,** *49,* 122.

Brown, F. C. *Chem. Rev.* **1961,** *61,* 463.

Campbell, N.; McKail, J. E. *J. Chem. Soc.* **1948,** 1215.

Foye, W. O.; Tovivich, P. *J. Pharm. Sci.* **1977,** *66,* 1607.

Julian, P. L.; Sturgis, B. M. *J. Am. Chem. Soc.* **1935,** *57,* 1126.

---

## Alkene Preparation by the Wittig Reaction: (*E*)-Stilbene; 1-Methylene-4-*tert*-butylcyclohexane; and *trans*-9-(2-Phenylethenyl)anthracene

Common names: (*E*)-Stilbene; *trans*-1,2-diphenylethene
CA number: [103-30-0]
CA name as indexed: Benzene, 1,1'-(1,2-ethenediyl)bis-, (*E*)-

Common name: 1-Methylene-4-*tert*-butylcyclohexane
CA number: [13294-73-0]
CA name as indexed: Cyclohexane, 1-(1,1-dimethylethyl)-4-methylene-

Common name: *trans*-9-(2-Phenylethenyl)anthracene
CA number: [42196-97-4]
CA name as indexed: Anthracene, 9-(2-phenylethenyl)-, (*E*)-

### Purpose

To investigate the conditions under which the Wittig reaction is carried out. The Wittig reaction involves the reaction of a *phosphorus ylide* (Experiments [19A] and [19C]) with an aldehyde or ketone, and is used extensively in organic synthesis to synthesize alkenes. To investigate (Experiments [19B] and [19D]) the use of the Horner–Wadsworth–Emmons modified Wittig reaction between an aldehyde and a phosphonate ester, using phase-transfer catalysis.

*Prior Reading*

*Technique 4:*  Solvent Extraction
Liquid–Liquid Extraction (see pp. 80–82).
Drying of a Wet Organic Layer (see pp. 86–87).
Concentration of Solutions (see pp. 106–108).
*Technique 5:*  Crystallization
Use of the Hirsch Funnel (see pp. 93–95).
Craig Tube Crystallization (see pp. 95–96).
*Technique 6:*  Chromatography
Packing the Column (see pp. 99–100).
Elution of the Column (see p. 100).
Thin-Layer Chromatography (see pp. 103–104).

**Georg Friedrich Karl Wittig** (1897–1987, German). Wittig obtained his Ph.D. in 1923 at the University of Marburg under von Auwers. He later became Professor of Chemistry at the Universities of Braunschweig, Freiburg, Tübingen, and Heidelberg. His initial research work was concerned with the concept of ring strain, diradical formation, and valance tautomerism. However, he soon became involved in carbanion chemistry. Wittig discovered halogen–metal exchange reactions and then moved extensively into research involving the chemistry of the ylides. In 1953, he discovered the reactive phosphonium ylides and subsequently their reaction with aldehydes and ketones to give alkenes under very mild conditions; this reaction is now known as the *Wittig reaction*. This discovery allowed the introduction of the C=C linkage at a specific location in the product. In recognition for this work, Wittig received the Nobel Prize in 1979 [with H. C. Brown (organoboron compounds)].

In other work, Wittig postulated the dehydrobenzene intermediate and proved its existence through trapping reactions of the Diels–Alder type. He discovered sodium tetraphenylborate, which is now used in the analytical determination of potassium and ammonium ions. His later work involved the chemistry of metalated Schiff bases. This work subsequently led to the development of the concept of directed aldol condensations.

Wittig coined a large number of technical terms–valence tautomerism, ylide, onium complexes, halogen–metal exchange, dehydrobenzene, and umpolung (polarity reversal). Wittig died in 1987 at the age of 90.[18]

---

## REACTION

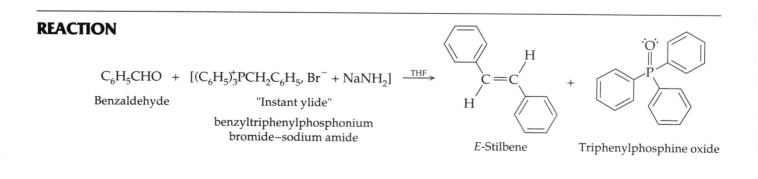

$C_6H_5CHO$  +  $[(C_6H_5)_3^+PCH_2C_6H_5,\ Br^-\ +\ NaNH_2]$  $\xrightarrow{\text{THF}}$

Benzaldehyde          "Instant ylide"

benzyltriphenylphosphonium
bromide–sodium amide

*E*-Stilbene          Triphenylphosphine oxide

---

[18] See Parker, S. P. Ed., *McGraw-Hill Modern Scientists and Engineers,* McGraw-Hill: NY, 1980, Vol. 3, p. 341; *The Annual Obituary* St. James Press: Chicago, 1987, p. 460; Oesper, R. E. *J. Chem. Ed.* **1954**, *31*, 357.

The Wittig reaction constitutes a valuable method for the preparation of alkenes. The major advantages of this approach are: (1) there is no ambiguity in the location of the C=C generated by the reaction, as there can be in many elimination reactions, and (2), there is no potential for rearrangements, as there can be in E1 elimination reactions.

An *ylide* is a neutral species whose Lewis structure contains opposite charges on adjacent atoms. The atoms involved are carbon, and an element from either Group 15 (VA) or 16 (VIA) of the periodic table, such as N, P, or S. The Wittig reaction uses phosphorus ylides, which are obtained by deprotonation of a phosphonium salt with a strong base. Phosphorus ylides are relatively stable, but reactive species, for which the following resonance structures may be written; the phosphorus atom can exceed an octet by accommodating electron donation into its *3d* orbitals.

$$[R_3\overset{+}{P}-\overset{-}{C}H_2: \leftrightarrow R_3P{=}CH_2]$$

The phosphonium salts are available through a nucleophilic displacement reaction of triphenylphosphine with various alkyl halides. Triphenylphosphine is a good nucleophile, and thus most phosphonium salts are easily prepared.

$$(C_6H_5)_3P: + RCH_2X \rightarrow (C_6H_5)_3\overset{+}{P}CH_2R, X^-$$
$$(X = I, Br, Cl)$$

The hydrogen atoms on the carbon attached to the resulting phosphorus cation are somewhat acidic because they are adjacent to a positive charge, a significant electron-withdrawing group. Thus, treatment of the phosphonium salt with a strong base, such as butyllithium in THF or sodium hydride in DMSO, removes one of these protons and produces the ylide.

$$(C_6H_5)_3\overset{+}{P}CH_2R, X^- + C_4H_9Li \xrightarrow{\text{THF}} (C_6H_5)_3\overset{+}{P}-\overset{\bar{}}{C}H-R + C_4H_{10} + LiX$$
$$\text{An ylide}$$

The ylides are generally not isolated, but rather generated in situ and reacted directly with the carbonyl compound.

The "instant ylides" are solid-phase mixtures of a phosphonium salt with sodium amide ($NaNH_2$, a strong base and the conjugate base of ammonia) and are now commercially available. In the solid phase, no reaction between the strong base and the phosphonium salt occurs. Thus, to generate the desired ylide, the "instant ylide" mixture need only be placed in a suitable solvent. This process is a marked advantage over the usual methods employed to obtain these species. Ylides, because of the significant negative charge density on carbon, are nucleophilic enough to attack electrophiles as reactive as the carbon of a carbonyl group. When the ylide is reacted with an aldehyde or a ketone, an intermediate *oxaphosphetane* (a four-membered ring containing an oxygen and a phosphorus atom) is formed, which decomposes to give the alkene product. The mechanistic sequence is outlined here.

*Generation of the Ylide*

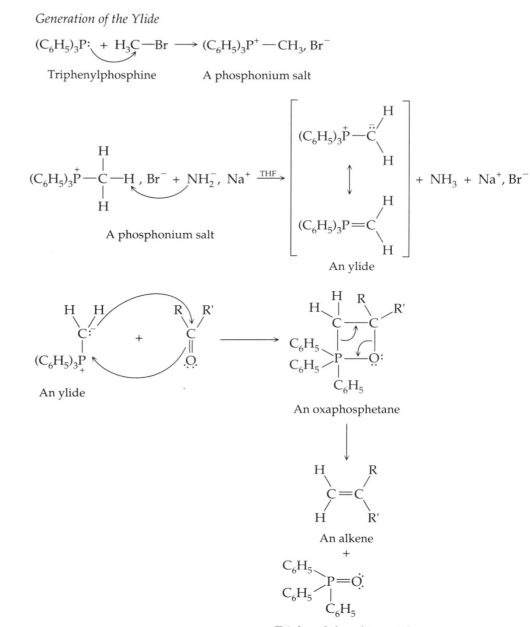

An oxaphosphetane

An alkene

Triphenylphosphine oxide

The Wittig reaction is very general. The significant limitation is that the carbonyl compound cannot contain acidic hydrogen atoms, such as in an alcohol or carboxyl group, or electrophiles more reactive than the aldehyde or ketone itself. The Wittig reaction can be quite stereoselective for the formation of either *Z* or *E* alkenes, but the factors involved in such stereoselectivity are sometimes difficult to predict, and often difficult to explain.

An important modification (often called the Horner–Wadsworth–Emmons reaction) of the Wittig reaction makes use of phosphonate esters, RPO(OR′)$_2$. It is highly stereoselective for the formation of *E*-alkenes. The reaction and mechanism are depicted below for the preparation of (*E*)-stilbene. Instead of using a phosphorus cation to stabilize the negative charge, as in the phosphonium ylide above, a phosphonate ester group is used to stabilize an adjacent carbanion.

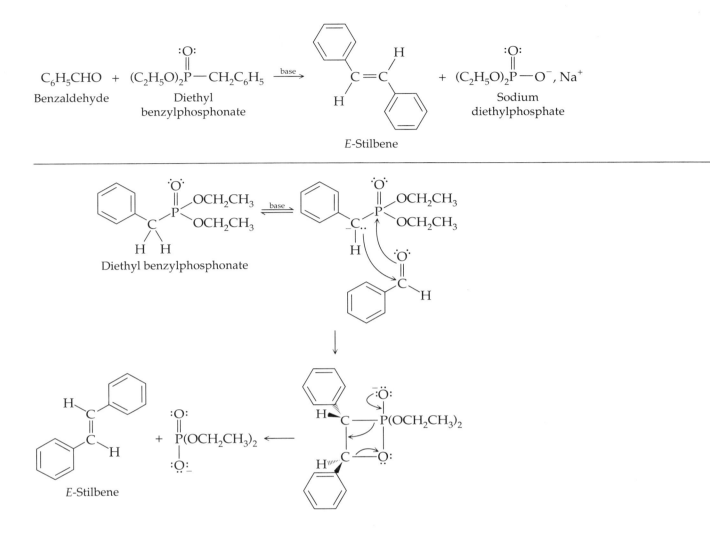

The use of the phosphonate ester (Horner–Wadsworth–Emmons re-action) allows much easier separation of the product alkene, since the sodium phosphate byproduct is water soluble; the byproduct of the Wittig reaction, triphenylphosphine oxide, is not water soluble. In the Horner–Wadsworth–Emmons modification, a conjugated, or electron-withdraw-ing, substituent (such as a phenyl or carbonyl group) on the nucleophilic carbon is used to assist in the stabilization of the carbanion. This modifica-tion (Experiment [19B]) may be used as an alternative to Experiment [19A] for the preparation of (E)-stilbene. The "instant-ylide" Wittig reaction yields predominantly the E isomer of stilbene (70%). The Horner–Wadsworth–Emmons reaction yields exclusively the E isomer. Both proce-dures are given below. The synthesis described in Experiment [19D] also employs the Horner–Wadsworth–Emmons modification.

Horner–Wadsworth–Emmons reactions lend themselves to the use of *phase-transfer catalysis* in Experiments [19B] and [19D]. Phase-transfer catalysis allows the use of an aqueous base (NaOH in H₂O) with the or-ganic compounds dissolved in an organic solvent (hexane in Experiment [19B], methylcyclohexane in Experiment [19D]) immiscible with water. The reaction system, as the name implies, involves two phases, an aqueous phase and an organic phase. The phase-transfer catalyst plays a very im-portant role, as without it, no reaction would occur as the initial reactants (hydroxide ion and diethyl benzylphosphonate) are dissolved in different, immiscible phases, and NaOH is insoluble in hexane (or methylcyclohexane), and diethyl benzylphosphonate is insoluble in water.

The key features of the phase-transfer catalyst, Aliquat® 336, are that it is a quaternary ammonium salt, with long-chain alkyl groups attached to the nitrogen.

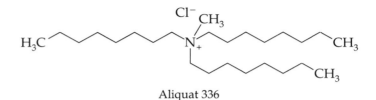

Aliquat 336

The phase-transfer catalyst is soluble in both water and the organic solvent. It is water soluble because it is an ion, and it is hexane soluble because of the three long-chain alkyl groups. Thus, the phase-transfer catalyst distributes itself in both phases, and freely shuttles back and forth through the phase boundary between solvent layers. In aqueous NaOH, the chloride anion exchanges with hydroxide anion, as the counterion to the ammonium cation. When it does this, the catalyst carries the hydroxide ion from the aqueous phase, as an ion-pair, across the phase boundary into the organic phase, where the base then reacts with the diethyl benzylphosphonate. The Horner–Wadsworth–Emmons reaction then occurs, producing the alkene and diethyl phosphate anion. This anion becomes associated with the ammonium cation of the phase-transfer catalyst, and is transported to the aqueous layer, where the catalyst picks up another hydroxide ion and repeats the entire process.

## Experiment [19A]   (E)-Stilbene by the "Instant Ylide" Method

> CAUTION: Tetrahydrofuran is a flammable liquid. All flames must be extinguished in the laboratory when this solvent is used.

# REACTION

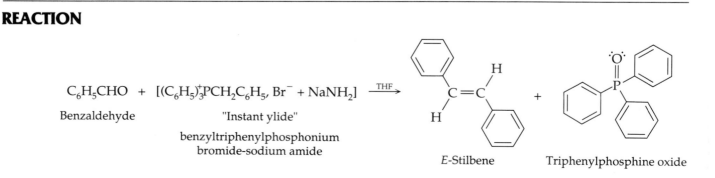

$C_6H_5CHO$ + [$(C_6H_5)_3^+PCH_2C_6H_5$, $Br^-$ + $NaNH_2$] $\xrightarrow{\text{THF}}$

Benzaldehyde           "Instant ylide"

benzyltriphenylphosphonium
bromide-sodium amide

E-Stilbene           Triphenylphosphine oxide

Estimated time to complete the reaction: Two 3-hour laboratory periods.

## EXPERIMENTAL PROCEDURE

### Physical Properties of Reactants

| Compound | MW | Amount | mmol | bp(°C) | d | $n_D$ |
|---|---|---|---|---|---|---|
| Benzaldehyde | 106.13 | 100 μL | 0.95 | 178 | 1.04 | 1.5463 |
| "Instant ylide" benzyltriphenyl phosphonium bromide–sodium amide | | 600 mg | ~1.2 | | | |
| Tetrahydrofuran | 72.12 | 1.0 mL | | 67 | | |

### Reagents and Equipment

Weigh and place 600 mg (~1.2 mmol) of benzyltriphenylphosphonium bromide–sodium amide *(instant ylide)* mixture in a dry 5.0-mL conical vial containing a magnetic spin vane. Add *freshly distilled* dry tetrahydrofuran (1.0 mL) using a calibrated Pasteur pipet, and immediately attach the vial to an air condenser protected by a calcium chloride drying tube. (■) Stir the mixture for 15 min at room temperature. During this time, the mixture turns orange. Following generation of the ylide, remove the air condenser for as short a time as possible and quickly add 100 μL (0.95 mmol) of benzaldehyde *(freshly distilled)* to the reaction flask, using an automatic delivery pipet, and immediately reattach the air condenser.

### Reaction Conditions

Stir the resulting heterogeneous mixture at room temperature for an additional 15 min. The system develops a light brown color during this time.

### Isolation of Product

Work up the reaction by adding 1.0 mL of a 25% aqueous NaOH solution (calibrated Pasteur pipet), and transfer the resulting mixture to a 15-mL centrifuge tube using a Pasteur pipet. Rinse the reaction vial with three 2.0-mL portions of diethyl ether, each of which is also transferred to the centrifuge tube (Pasteur filter pipet). Partially neutralize the resulting two-phase mixture by careful addition of 3.0 mL of 0.1 N HCl. Transfer the ether layer (top) by Pasteur filter pipet to a short microcolumn prepared from a Pasteur filter pipet containing 1.5 g of anhydrous sodium sulfate. (■)

Collect the dried eluate in a 25-mL Erlenmeyer flask containing a boiling stone. Extract the remaining aqueous layer with two additional 2.0-mL portions of ether and transfer these ether extracts, as before, to the microcolumn containing anhydrous sodium sulfate. Concentrate the eluate solution (~10 mL) to dryness using a gentle stream of nitrogen, or by warming in a sand bath **[HOOD]** to give a white solid.

HOOD

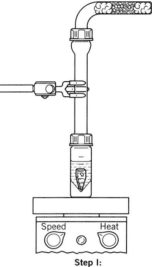

**Step I:**
$(C_6H_5)_3$-$\overset{+}{P}$-$CH_2$-$C_6H_5$, $\bar{B}r$ + $NaNH_2$
600 mg + THF, 1.0 mL
**Step II:**
$C_6H_5$ CHO, 100, μL

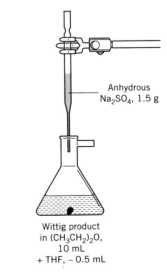

Anhydrous
$Na_2SO_4$, 1.5 g

Wittig product
in $(CH_3CH_2)_2O$,
10 mL
+ THF, ~ 0.5 mL

### Purification and Characterization

Purify the crude product by chromatography, using a silica gel column.

The triphenylphosphine oxide is relatively insoluble in the hexane solvent. First, separate the triphenylphosphine oxide byproduct from the

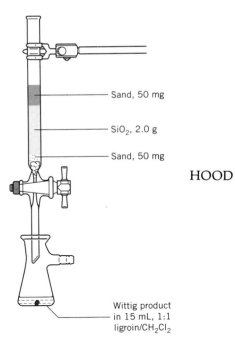

Sand, 50 mg

SiO₂, 2.0 g

Sand, 50 mg

HOOD

Wittig product in 15 mL, 1:1 ligroin/CH₂Cl₂

mixture of stilbenes by extracting the solid obtained above with hexane, using the following procedure.

To the 25-mL Erlenmeyer flask containing the crude product, add 2.0 mL of hexane and agitate the solution with swirling. Some breakup of the material with a microspatula may be necessary. Transfer the hexane solution by Pasteur filter pipet to a 10-mL Erlenmeyer flask containing a boiling stone. Extract the remaining crude solid with two additional 2.0-mL portions of hexane, and transfer these extracts to the 10-mL Erlenmeyer flask as before. Concentrate the hexane solution (~6 mL) to about 1.5 mL using a gentle stream of nitrogen gas, or by warming on a sand bath **[HOOD]**.

Pack a short (1 × 10 cm) chromatography column with 2.0 g of activated silica gel, and premoisten the column with ligroin (60–80 °C). (■) Add the above hexane solution directly to the column. Now elute the column with 15 mL of a 1:1 ligroin (60–80 °C)/methylene chloride solution. Collect the eluate, in a single fraction, in a tared 25-mL **filter flask** containing a boiling stone. Concentrate the solution to dryness under reduced pressure in a warm sand bath to yield the pure product mixture. (■)

**NOTE.** *A 25-mL side-arm filter flask, equipped with a Hirsch funnel and filter paper disks to control the pressure, is a convenient system for the removal of a small volume (5–20 mL) of solvent. A rotary evaporator, if available, is a nice alternative.*

Weigh the crude product residue and calculate the percent yield.

The isolated product mixture may be analyzed by gas chromatography and/or thin-layer chromatography. *This constitutes the second week of laboratory for this experiment.*

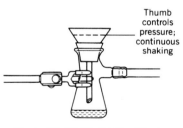

Thumb controls pressure; continuous shaking

Ligroin (60–80 °C)/CH₂Cl₂, 15 mL + Wittig product

### Gas Chromatographic Analysis

Dissolve the crude product mixture isolated above in the minimum amount of 1 : 1 ligroin(60–80 °C)/methylene chloride solution (~0.5–0.75 mL). Inject a 10-$\mu$L sample into a gas chromatograph, set up according to the following conditions:

Column: $\frac{1}{4}$-in. × 8-ft, 20% Carbowax 20M on Chromasorb P®(80/100 mesh)
Temperature: 220 °C
Flow rate: 30 mL/min (He gas)
Chart speed: 1 cm/min

The compounds elute in the order: *(Z)*-stilbene, followed by *(E)*-stilbene. Measure the ratio of peak heights and calculate the isomeric composition of your mixture.

### Thin-Layer Chromatographic Analysis

Spot a TLC plate with a sample from the product solution used for GC analysis, and also with a standard solution.

Use hexane as the elution solvent, silica gel (containing a fluorescent indicator) as the stationary phase, and UV light for visualization.

Typical $R_f$ values: *(E)*-stilbene 0.21; *(Z)*-stilbene, 0.27.

Concentrate the ligroin–methylene chloride product mixture as before. Separate, and purify, the *(E)*-stilbene by recrystallization from a minimum amount of 95% ethanol, using the Craig tube.

Weigh the dried and purified *(E)*-stilbene product and calculate the percent yield. The purity of this material may be determined by TLC, using the conditions outlined above.

Obtain a melting point and IR spectrum of the material and compare your results with those reported in the literature.

*(E)-* and *(Z)-*Stilbene also exhibit different absorptions in the ultraviolet region. The data are summarized below.

*(Z)-Stilbene: (1 mm cell)*

$\lambda_{max}$ 223 nm ($\varepsilon_{max}$ = 20,600, methanol, 0.05 g/L)
$\lambda_{max}$ 276 nm ($\varepsilon_{max}$ = 10,900, methanol, 0.1 g/L)

*(E)-Stilbene: (0.05 g/L, 1 mm cell)*

$\lambda_{max}$ 229 nm ($\varepsilon_{max}$ = 21,000, methanol)
$\lambda_{max}$ 294 nm ($\varepsilon_{max}$ = 33,200, methanol)
$\lambda_{max}$ 307 nm ($\varepsilon_{max}$ = 32,100, methanol)

These ultraviolet absorption data illustrate an interesting example of steric effects on the absorption pattern exhibited by geometrical isomers of an alkene. There is significant steric hindrance between the two phenyl groups in the *Z* isomer, which causes the phenyl groups to twist out of coplanarity with the alkene. Thus, conjugation is diminished. This result is reflected in the lower intensity of the 276-nm band as compared to the 294-nm band in the *E* isomer.

## Chemical Tests

Further characterization may be accomplished by performing the $Br_2$/$CH_2Cl_2$ test for unsaturation. Note that the dibromo compound is prepared in Experiment [A2$_b$] . It may be used here as a reference sample in the characterization of *(E)-*stilbene. The ignition test (Chapter 10) may be used to confirm the presence of the aromatic portion of the molecule.

## Experiment [19B] *(E)-*Stilbene by the Horner–Wadsworth–Emmons Reaction

**REACTION**

$$C_6H_5CHO \ + \ (C_2H_5O)_2\overset{\overset{\displaystyle \cdot\cdot O\cdot\cdot}{\|}}{P}\!-\!CH_2C_6H_5 \ \xrightarrow[\text{Aliquat 336}]{\overset{40\%}{\underset{\text{Hexane}}{\text{NaOH}}}} \ C_6H_5CH\!=\!CHC_6H_5 \ + \ (C_2H_5O)_2\overset{\overset{\displaystyle \cdot\cdot O\cdot\cdot}{\|}}{P}\!-\!O^-, Na^+$$

Benzaldehyde      Diethylbenzyl phosphonate      *E*-Stilbene      Sodium diethyl phosphate

**EXPERIMENTAL PROCEDURE**

Estimated time to complete the experiment: 2.5 hours.

**Physical Properties of Reactants**

| Compound | MW | Amount | mmol | bp(°C) | d | $n_D$ |
|---|---|---|---|---|---|---|
| Benzaldehyde | 106.12 | 100 μL | 0.98 | 178 | 1.04 | 1.5463 |
| Diethyl benzylphosphonate | 228.23 | 200 μL | 0.96 | 106–108 (@1 mm) | 1.095 | 1.4970 |
| Aliquat 336 (tricaprylmethylammonium chloride) | 404.17 | 88 mg | 0.22 | | | |
| Hexane | 86.18 | 2.0 mL | | | | |
| 40% Sodium hydroxide | | 2.0 mL | | | | |

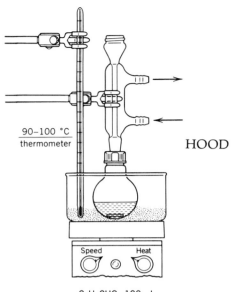

90–100 °C
thermometer

HOOD

C₆H₅CHO, 100 μL
diethyl benzylphosphonate, 200 μL,
+ Aliquat 336, 88 mg
40% NaOH, 2 mL, hexane, 2 mL
10–mL RB flask

## Reagents and Equipment

Weigh and place 88 mg (100 $\mu$L) of tricaprylmethylammonium chloride (Aliquat 336) in a 10-mL round-bottom flask containing a magnetic stirrer. Add 100 $\mu$L (0.98 mmol) of benzaldehyde, 200 $\mu$L (0.96 mmol) of diethyl benzylphosphonate, 2.0 mL of hexane, and 2 mL of 40 % sodium hydroxide solution. Attach the flask to a reflux condenser. (■)

**NOTE.** *The benzaldehyde, diethyl benzylphosphonate, hexane, and NaOH solution are dispensed in the* **hood** *using automatic delivery pipets. The Aliquat 336 is very viscous and is best measured by weighing. A medicine dropper is used to dispense this material. It is advisable to lightly grease the bottom joint of the condenser since strong base is being used.*

## Reaction Conditions

Heat the two-phase mixture at reflux on a sand bath (temperature ~90–100 °C) for 1 hour. Stir the reaction mixture vigorously during this period. Allow the resulting solution to cool to nearly room temperature. Crystals of product may appear as cooling occurs.

## Isolation of Product

Add methylene chloride (700 $\mu$L), which will dissolve any crystalline material that may have formed. Then, using a Pasteur pipet, transfer the contents of the round-bottom flask to a 15-mL centrifuge tube. Rinse the flask with an additional 300 $\mu$L of methylene chloride, and transfer the rinse to the same centrifuge tube. Remove the (upper) aqueous layer carefully, using a Pasteur filter pipet, and save it in a vial until you have successfully isolated and characterized the product. Wash the organic layer with two 1-mL portions of water. Stir the mixture with a small glass rod after each addition (or a Vortex mixer may be used) and then remove the water layer and add it to the one collected before. Dry the methylene chloride solution by addition of a small amount of anhydrous sodium sulfate. Use a Pasteur filter pipet to transfer the dried solution to a 10-mL Erlenmeyer flask. Wash the sodium sulfate remaining in the centrifuge tube with two 1-mL portions of methylene chloride. Remove these washings, using the Pasteur filter pipet, and transfer them to the same Erlenmeyer flask.

Concentrate the solution, which contains the desired product, to dryness on a warm sand bath under a stream of nitrogen.

## Purification and Characterization

The *(E)*-stilbene obtained is, in most cases, sufficiently pure for characterization. However, this should be confirmed by thin-layer chromatography as outlined earlier (see Purification and Characterization, Experiment [19A]). If only a trace of the *Z* isomer is detected, recrystallize the product directly, using absolute ethanol. Collect the recrystallized material by vacuum filtration using a Hirsch funnel. Maintain the vacuum for an additional 10 min to partially dry the crystalline product. Then place it on a clay plate, or on filter paper, and allow it to dry thoroughly. As an alternative, the product may be dried in a vacuum drying oven (or pistol) for 10–15 min at 30 °C (1–2 mm).

Weigh the dried *(E)*-stilbene product and calculate the percent yield. The *(E)*-stilbene may be characterized as outlined in Experiment [19A].

## Experiment [19C]   Methylene-4-*tert*-butylcyclohexane

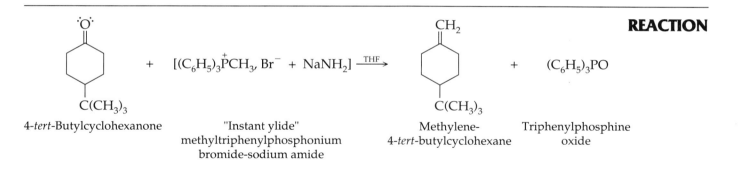

**REACTION**

4-*tert*-Butylcyclohexanone          "Instant ylide"          Methylene-          Triphenylphosphine
                          methyltriphenylphosphonium   4-*tert*-butylcyclohexane      oxide
                          bromide–sodium amide

---

Estimated time to complete the experiment: 4.0 hours.          **EXPERIMENTAL PROCEDURE**

**Physical Properties of Reactants and Product**

| Compound | MW | Amount | mmol | mp(°C) | bp(°C) | $n_D$ |
|---|---|---|---|---|---|---|
| 4-*tert*-Butylcyclohexanone | 154.26 | 100 mg | 0.64 | 50 | | |
| "Instant ylide" methyltriphenylphosphonium bromide–sodium amide | | 320 mg | ~0.72 | | | |
| Tetrahydrofuran | | 1.0 mL | | | 67 | |
| Methylene-4-*tert*-butylcyclohexane | 151.27 | | | | 185 | 1.4630 |

### Reagents and Equipment

In a dry 5.0-mL conical vial containing a magnetic spin vane, and equipped with an air condenser protected by a calcium chloride drying tube, weigh and place 320 mg (~0.72 mmol) of methyltriphenylphosphonium bromide–sodium amide (instant ylide) mixture. (■) Now add *freshly distilled* tetrahydrofuran (1.0 mL), using a calibrated Pasteur pipet, and stir the mixture for 15 min at room temperature. During this period it turns a bright yellow color.

Following generation of the ylide, weigh and add 100 mg (0.64 mmol) of 4-*tert*-butylcyclohexanone to the reaction flask.

### Reaction Conditions

Stir the resulting heterogeneous mixture, at room temperature, for an additional 90 min. The mixture develops a light-tan color over this period of time.

### Isolation of Product

Work up the reaction by adding 1.0 mL of a 25% aqueous NaOH solution (calibrated Pasteur pipet), and then transfer the resulting mixture to a 12-mL centrifuge tube using a Pasteur filter pipet. Rinse the reaction flask

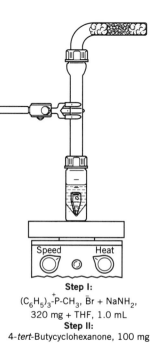

**Step I:**
$(C_6H_5)_3\overset{+}{-}P-CH_3$, $\bar{B}r + NaNH_2$,
320 mg + THF, 1.0 mL
**Step II:**
4-*tert*-Butycyclohexanone, 100 mg

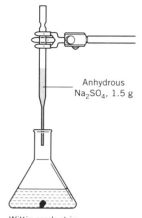

Anhydrous
Na$_2$SO$_4$, 1.5 g

Wittig product in
(CH$_3$CH$_2$)$_2$O, ~ 21 mL
+ THF, ~ 0.5 mL

HOOD

with three 2.0-mL portions of diethyl ether. Transfer each rinse to the same centrifuge tube. Partially neutralize the resulting two-phase system by the careful addition of 2.0 mL of 0.1 N HCl and mix briefly.

Separate the ether layer by Pasteur filter pipet, and place it in a 25-mL Erlenmeyer flask. Extract the remaining aqueous layer with three additional 5-mL portions of diethyl ether. Separate these extracts as before and combine them with the original ether layer.

Transfer the combined ether fractions, by Pasteur filter pipet, to a short microcolumn prepared from a Pasteur filter pipet containing 1.5 g of anhydrous sodium sulfate. (■) Collect the dried eluate in a 25-mL Erlenmeyer flask containing a boiling stone. Concentrate this solution to dryness, using a gentle stream of nitrogen, or by warming on a sand bath [HOOD], to yield a colorless liquid residue.

## Purification and Characterization

Purify the crude product isolated above by chromatography on an alumina column.

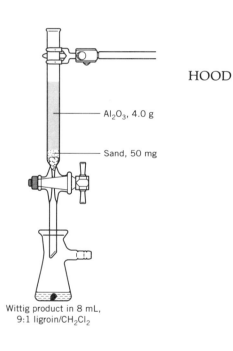

Al$_2$O$_3$, 4.0 g

Sand, 50 mg

Wittig product in 8 mL,
9:1 ligroin/CH$_2$Cl$_2$

HOOD

Pack a short (1 × 10 cm) buret column with 4.0 g of activated basic alumina, and premoisten the alumina with ligroin. (■) Dissolve the crude material in 0.5 mL of 9:1 ligroin (60–80 °C)/methylene chloride solvent and transfer the solution to the column using a Pasteur filter pipet. Elute the product in a single fraction using 8.0 mL of 9:1 ligroin (60–80 °C)/methylene chloride. Collect the eluate in a tared 25-mL filter flask containing a boiling stone. Evaporate the solvent [HOOD] under vacuum, with swirling, in a warm sand bath, leaving a liquid residue. *Product loss may occur if the concentrated residue is heated excessively.* (■)

The product is of sufficient purity for characterization. Weigh the methylene-4-*tert*-butylcyclohexane and calculate the percent yield. Determine the boiling point, and compare your result with that listed in the Reactants and Product table.

Obtain an IR spectrum of the alkene product, using the capillary film technique. Compare your result with Figure 6.32.

## Infrared Analysis

The spectrum (Fig. 6.8) of the ketone was discussed in Experiment [5B]. The major change observed in the spectrum of the product (Fig. 6.32), when compared with the starting material, is the loss of the carbonyl absorption band (1717 cm$^{-1}$), and its replacement by a less-intense new band at 1654 cm$^{-1}$. This latter absorption is associated with the stretching motion of the C=C system exocyclic to the six-membered ring. The key bands associated with group frequencies present in the reaction product are identified at: 3075, 3000–2850, 1783, 1654, 1394, 1368, and 883 cm$^{-1}$.

The presence of these bands confirms the projected structure of the Wittig reaction product. The data may be interpreted as follows: the bands at 3075 (sharp spike), 1783 (weak), 1654 (sharp–medium), and 883 (strong) cm$^{-1}$ form a frequency train that is defined as a "terminal alkene macrofrequency." The overall interpretation requires the presence of all four bands in the spectrum if the specific assignment of any one of them is to be correct. Thus, these four data points lead to a structural interpretation with very high confidence limits. The 3075-wavenumber band arises from the coupled antisymmetric stretch of the two C—H oscillators on the terminal alkenyl methylene group. The weak 1783-wavenumber peak is an overtone of the strong band observed near 883 cm$^{-1}$ (883 × 2 = 1766 cm$^{-1}$) and is unusually intense for an isolated harmonic. The fundamental can be assigned as the =CH$_2$ out-of-plane wag (C—H deformation). In this fairly rare example, the observed harmonic occurs at a higher wavenumber value than twice the fundamental frequency. A situation of this type is termed *negative anharmonicity* (see Infrared Discussions, Chapter 9). Finally, the C=C stretching mode is assigned to

Thumb controls pressure;
continuous shaking

9:1 Ligroin/CH$_2$Cl$_2$, 8 mL
+ THF ~ 0.5 mL
+ Wittig product

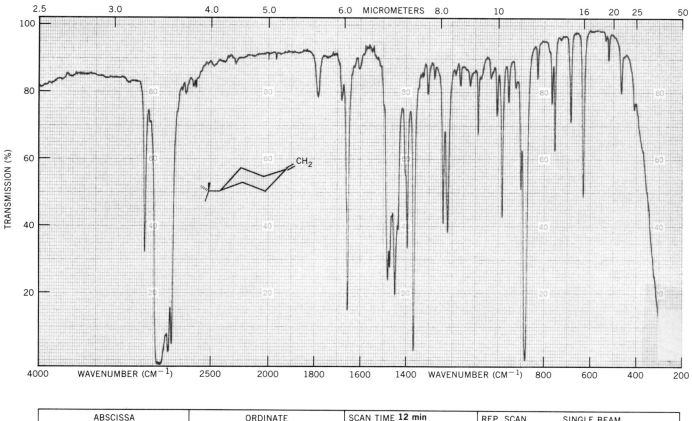

**FIGURE 6.32**    IR spectrum: Methylene-4-*tert*-butylcyclohexane.

the 1654-wavenumber band, which is consistent with the requirements of this frequency train as it is found below 1660 cm$^{-1}$ (see Chapter 9, infrared discussions of cis, terminal, or vinyl carbon–carbon double bonds). The bands identified at 1396 and 1368 cm$^{-1}$ indicate that the tertiary butyl group, as expected, has been preserved during the conversion of the ketone to a terminal alkene (see also Discussion, Experiment [5B])

# Experiment [19D]    *trans*-9-(2-Phenylethenyl)anthracene

**REACTION**

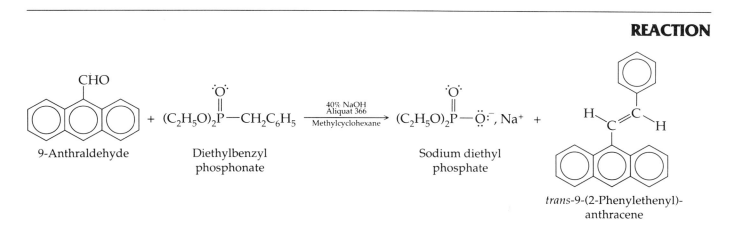

9-Anthraldehyde    Diethylbenzyl
phosphonate

$$+ (C_2H_5O)_2\overset{\overset{\displaystyle \ddot{O}\cdot}{\|}}{P}-CH_2C_6H_5 \xrightarrow[\text{Methylcyclohexane}]{\substack{40\% \text{ NaOH} \\ \text{Aliquat 366}}} (C_2H_5O)_2\overset{\overset{\displaystyle \ddot{O}\cdot}{\|}}{P}-\ddot{\underset{\displaystyle\cdot\cdot}{O}}{:}^-, Na^+ \;+$$

Sodium diethyl
phosphate

*trans*-9-(2-Phenylethenyl)-
anthracene

## EXPERIMENTAL PROCEDURE

Estimated time to complete the reaction: 2.5 hours.

### Physical Properties of Reactants and Product

| Compound | MW | Amount | mmol | mp(°C) | bp(°C) | d | $n_D$ |
|---|---|---|---|---|---|---|---|
| 9-Anthraldehyde | 206.24 | 106 mg | 0.51 | 104–105 | | | |
| Diethyl benzylphosphonate | 228.23 | 120 μL | 0.58 | | 106–8 @ 1 mm | 1.095 | 1.4970 |
| Aliquat 336 (tricaprylmethylammonium chloride) | 404.17 | 2 drops | | | | | |
| Methylcyclohexane | 98.2 | 2.0 mL | | | 101 | 0.77 | 1.4215 |
| 40% Aqueous sodium hydroxide | | 2.0 mL | | | | | |
| *trans*-(2-Phenylethenyl)anthracene | 280.4 | | | 130–132 | | | |

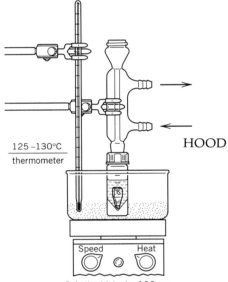

125–130°C thermometer

HOOD

9-Anthraldehyde, 106 mg,
+ diethylbenzyl phosphonate, 120 μL,
+ Aliquat 336, 2 drops, 40% NaOH, 2 mL
+ methylcyclohexane, 2 mL

Speed    Heat

### Reagents and Equipment

Weigh and place 106 mg (0.51 mmol) of 9-anthraldehyde in a 10-mL round-bottom flask containing a magnetic stirrer. Add 2 drops of Aliquat 336 (tricaprylmethylammonium chloride), 120 μL (0.58 mmol) of diethyl benzylphosphonate, 2.0 mL of methylcyclohexane and 2 mL of 40% sodium hydroxide solution. Attach the vial to a reflux condenser. (■)

**NOTE.** *The diethyl benzylphosphonate, methylcyclohexane, and NaOH solution are dispensed in the **hood** using automatic delivery pipets. The Aliquat 336 is very viscous. A medicine dropper is used to dispense this material.*

*It is advisable to lightly grease the bottom joint of the condenser, since strong base is being used.*

### Reaction Conditions

Heat the two-phase mixture, on a sand bath at a temperature of about 125–130 °C, for 45 min. Stir the reaction mixture vigorously during this period; the upper (organic) layer turns deep red. Allow the mixture to cool to room temperature.

### Isolation of Product

Transfer the two-phase solution to a 15-mL glass centrifuge tube using a Pasteur filter pipet. Rinse the flask with 1 mL of methylene chloride and transfer this rinse to the centrifuge tube. Carefully remove the aqueous layer, using a Pasteur filter pipet, and save it in a vial until you have successfully isolated and characterized the product. Wash the organic layer with two 2-mL portions of water. Stir the mixture with a small glass rod after each addition (or a Vortex mixer may be used) and then remove the water layer and save it in the same vial as before. Dry the methylene chloride solution by adding sodium sulfate. Using a Pasteur filter pipet, transfer the dried solution to a tared 10-mL Erlenmeyer flask containing a boiling stone. Wash the sodium sulfate remaining in the centrifuge tube with two 1-mL portions of methylene chloride. Also transfer these washings to the Erlenmeyer flask.

Concentrate the solution, which contains the desired product, almost to dryness on a sand bath under a slow stream of nitrogen gas. Now add approximately 1–2 mL of 2-propanol (isopropanol) to the flask and allow the resulting solution to stand at room temperature for 10–15 min, and then place it in an ice bath to complete the crystallization of the product.

## Purification and Characterization

Collect the yellow crystals by vacuum filtration, using a Hirsch funnel, and wash the filter cake with two 1-mL portions of *cold* methanol. Maintain the vacuum for an additional 10 min to partially dry the crystalline product. Then place the material on a clay plate, or on filter paper, and allow it to dry thoroughly. As an alternative, the product may be dried in a vacuum drying oven (or pistol) for 10–15 min at 30 °C (1–2 mm).

The purity of the product may be checked using *thin-layer chromatography* (see Technique 6A). Dissolve a small amount of the starting aldehyde and the product in ethanol and apply a sample to the TLC plate. Use toluene as the elution solvent, silica gel (containing a fluorescent indicator) as the stationary phase, and UV light for visualization. Typical $R_f$ values: *trans*-9-(2-phenylethenyl)anthracene, 0.92; 9-anthraldehyde, 0.50.

Weigh the product and calculate the percent yield. A portion of the material may be further purified by recrystallization from 2-propanol using the Craig tube.

Determine the melting point and compare it with the value given in the Reactants and Product table. Obtain the IR spectrum of the alkene product (KBr pellet technique). The material should have a strong absorption band at 962 cm$^{-1}$, which confirms the presence of a trans double bond.

This material has a characteristic UV spectrum in methanol. The following data were obtained at a concentration of $1.5 \times 10^{-5}$ (see Fig. 6.33).

$\lambda_{max}$ 255 nm ($\varepsilon_{max}$ = 35,207, methanol)
$\lambda_{max}$ 383 nm ($\varepsilon_{max}$ = 3618, methanol)

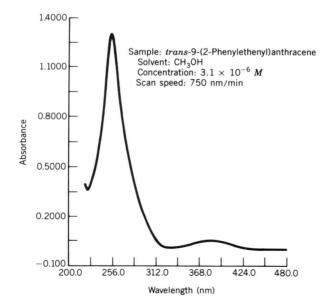

**FIGURE 6.33** UV-visible spectrum: *trans*-9-(2-phenylethenyl)anthracene.

**QUESTIONS**

**6-124.** Complete each of the following reactions by giving a suitable structure for the species represented by the letters. Give a suitable name for Compound **B** in each reaction.

a.  $(C_6H_5)_3\overset{+}{P}-CH_2C_6H_5, Cl^- \xrightarrow[C_2H_5OH]{NaOC_2H_5}$ **A**

$$A + C_6H_5CH=CH-\overset{\overset{\displaystyle H}{|}}{C}=O \longrightarrow B + (C_6H_5)_3PO$$

b.  $(C_6H_5)_3\overset{+}{P}-CH_3, Br^- \xrightarrow[DMSO]{NaH}$ **A**

$$A + \hspace{-0.2em}\left\langle \hspace{-0.3em}\bigcirc\hspace{-0.3em}\right\rangle \hspace{-0.6em}=\hspace{-0.3em}O \longrightarrow B + (C_6H_5)_3PO$$

**6-125.** Why is it important that any aldehyde used in the Wittig reaction be free of carboxylic acid impurities?

**6-126.** Reaction of triphenylphosphine with benzyl bromide produces the corresponding phosphonium salt. Suggest a suitable mechanism for this reaction.

**6-127.** Heteroatoms other than P are also capable of stabilizing the negative charge on C to yield ylides. For example, nitrogen is capable of forming such a system.

$$\left[\begin{array}{c}(C_6H_5)_3\overset{+}{P}-\overset{-}{\overset{..}{C}}H_2\\ \updownarrow\\ (C_6H_5)_3P=CH_2\end{array}\right] \qquad (CH_3)_3\overset{+}{N}-\overset{-}{\overset{..}{C}}H_2$$

A phosphorus ylide     A nitrogen ylide

Why are resonance structures not drawn for the nitrogen ylide as in the phosphorus system?

**6-128.** Would you expect that the sulfonium salt, $(C_6H_5)_2CH_3S^+, Br^-$, is capable of forming an ylide when reacted with a strong base? If so, would its structure be best represented in a manner resembling the P or N ylide? Explain.

**6-129.** Explain why the C=C stretching mode gives rise to a rather weak IR band in 1-methylcyclohexene, while in its isomer, methylenecyclohexene, the band is of medium to strong intensity.

**6-130.** Predict the C=C stretching frequencies of the alkenes formed when cyclopentanone, cyclobutanone, and cyclopropanone undergo the Wittig reaction with methyltriphenylphosphonium bromide–sodium amide reagent.

**6-131.** What concentration of a given compound (mol/L) in methanol, having a molecular weight of 165, would have been used if the solution gave an absorbance of 0.68 with a calculated $\varepsilon_{max} = 14,800$?

# BIBLIOGRAPHY

Of the many articles on the Wittig reaction, three are listed:

Maryanoff, B. E.; Reitz, A. B. *Chem. Rev.* **1989**, *89*, 863.

Silversmith, E. F. *J. Chem. Educ.* **1986,** *63*, 645.

Vedejs, E.; Marth, C. F. *J. Am. Chem. Soc.* **1990**, *112*, 3905.

Selected references pertaining to the Horner–Wadsworth–Emmons modification of the Wittig reaction:

Boutagy, J.; Thomas, R. *Chem. Rev.* **1974,** *74*, 87.

Horner, L.; Hoffmann, H.; Wippel, H. G.; Klahre, G. *Chem. Ber.* **1959**, *92*, 2499.

Wadsworth, W. S., Jr.; Emmons, W. D. *J. Am. Chem. Soc.* **1961,** *83*, 1733.

Below are selected examples of the Wittig reaction in *Organic Syntheses*:

Campbell, T. W.; McDonald, R. N. *Organic Syntheses*; Wiley: New York, 1973; Collect. Vol. V, p. 985.

Jorgenson, M. J.; Thacher, A. F. *Organic Syntheses*; Wiley: New York, 1973; Collect. Vol. V, p. 509.

McDonald, R. N.; Campbell, T. W. *Organic Syntheses*; Wiley: New York, 1973; Collect. Vol. V, p. 499.

Nagata, W; Wakabayashi, T.; Hayase, Y. *Organic Syntheses*; Wiley: New York, 1988; Collect. Vol. VI, p. 358.

Wadsworth, W. S., Jr.; Emmons, W. D. *Organic Syntheses*; Wiley: New York, 1973; Collect. Vol. V, p. 547.

Wittig, G.; Schöllkopf, U. *Organic Syntheses*; Wiley: New York, 1973; Collect. Vol. V, p. 751.

# Aldol Reaction: Dibenzalacetone

**EXPERIMENT [20]**

Common names: Dibenzalacetone, dibenzylideneacetone
CA number: [35225-79-7]
CA name as indexed: 1,4-Pentadien-3-one, 1,5-diphenyl-, *(E,E)-*

## Purpose

To investigate the synthetically useful **aldol reaction** as a method of forming carbon–carbon bonds. It is a general reaction of aldehydes that may also be extended to ketones. The specific case outlined in this experiment is known as the Claisen–Schmidt reaction. Experiments [A3$_a$] and [F$_1$] provide other examples of the aldol condensation.

*Prior Reading*

Technique 5:   Crystallization
                      Introduction (see pp. 90–92.)
                      Use of the Hirsch Funnel (see pp. 93–95.)
                      Craig Tube Crystallization (see pp. 95–96.)

**REACTION**

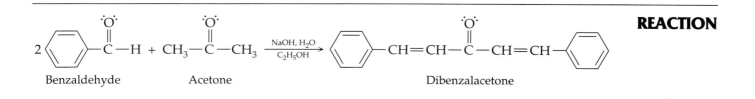

Benzaldehyde        Acetone                              Dibenzalacetone

**DISCUSSION**

The aldol reaction (aldol condensation) is one of the fundamental reactions of organic chemistry since it leads to the formation of a new carbon–carbon bond and is broadly applicable. A condensation reaction is one in which two molecules are joined with the concomitant expulsion of a small stable molecule, usually water or an alcohol. The aldol reaction may be used to condense various combinations of aldehydes and ketones. The mixed aldol condensation of an aldehyde having no $\alpha$-hydrogen atom with a ketone is specifically known as the Claisen–Schmidt reaction. This variation of the aldol condensation is illustrated here in the synthesis of dibenzalacetone.

The reaction conditions of this aldol condensation favor the formation of the product, dibenzalacetone. This product is insoluble in the aqueous ethanol solvent and precipitates from the reaction as it is formed, whereas the starting materials and the intermediate, benzalacetone, are all soluble in aqueous ethanol. These experimental conditions assist in driving the equilibrium reaction to completion.

The aldol condensation involves generation of an *enolate* by removal of an acidic proton from a carbon $\alpha$ to the carbonyl group of an aldehyde or ketone, and subsequent nucleophilic addition of this enolate to the carbonyl carbon of an aldehyde or ketone. The reaction is usually base catalyzed and involves several mechanistic steps.

$$R-\overset{\overset{\ddot{O}}{\|}}{C}-CH_3 + :\ddot{O}H^- \;\rightleftharpoons\; \left[ R-\overset{\overset{\ddot{O}}{\|}}{C}\overset{\frown}{-}\ddot{C}H_2: \;\longleftrightarrow\; R-\overset{\overset{:\ddot{O}:^-}{|}}{C}=CH_2 \right] + H_2O$$
$$\text{(an enolate)}$$

$$R'-\overset{\overset{\ddot{O}}{\|}}{C}-H + :\ddot{C}H_2-\overset{\overset{\ddot{O}}{\|}}{C}-R \;\rightleftharpoons\; R'-\overset{\overset{:\ddot{O}:^-}{|}}{\underset{H}{C}}-CH_2-\overset{\overset{\ddot{O}}{\|}}{C}-R \;\overset{HOH}{\rightleftharpoons}\; R'-\overset{\overset{:\ddot{O}H}{|}}{\underset{H}{C}}-CH_2-\overset{\overset{\ddot{O}}{\|}}{C}-R$$
$$\text{(nucleophilic attack)} \qquad\qquad\qquad\qquad \text{(protonation)}$$

The reaction involves several steps: (1) base-catalyzed generation of an enolate, (2) nucleophilic attack of this anion on a carbonyl carbon, and (3) protonation of the resulting anion to yield the initial aldol product, a $\beta$-hydroxy carbonyl compound. Note that each step in the sequence is in equilibrium and the entire reaction is, therefore, reversible. Treatment of the $\beta$-hydroxy carbonyl compound with base causes the reverse aldol (**retro-aldol**) reaction to occur.

$$:\ddot{O}H^- + R'-\overset{\overset{:O:}{|}}{\underset{R'}{C}}-CH_2-\overset{\overset{:O:}{\|}}{C}-R \;\longrightarrow\; R'-\overset{\overset{\ddot{O}}{\|}}{\underset{R'}{C}} + CH_2=\overset{\overset{:\ddot{O}:^-}{|}}{C}-R + HOH$$

$$R'-\overset{\overset{\ddot{O}}{\|}}{\underset{R'}{C}} + CH_3-\overset{\overset{\ddot{O}}{\|}}{C}-R + HO^-$$

The $\beta$-hydroxycarbonyl product may be isolated in most cases, if desired, as the subsequent dehydration is generally much slower than the addition reaction that precedes it. The final stage, as in the present reaction, is a hydroxide-catalyzed dehydration of this initial product by way of its enolate. Though hydroxide ion (HO⁻) is generally not a good leaving group, the hydrogen α to the ketone is quite acidic, the elimination produces a rather stable and conjugated α,β-unsaturated ketone, and under strongly basic conditions the hydroxide ion is an adequate leaving group.

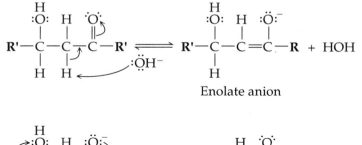

Enolate anion

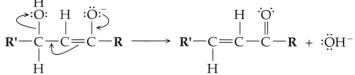

In the present reaction, a double aldol condensation occurs, which yields the dibenzalacetone product. An additional example of the aldol reaction is shown in Experiment [A3ₐ], where tetraphenylcyclopentadienone is prepared.

**NOTE.** *This reaction may be carried out in a 10 × 75-mm test tube. However, the reagents must be stirred efficiently with a glass rod at frequent intervals. If a larger test tube is used, a small magnetic stirring bar or vane is more efficient as an agitator.*

---

Estimated time to complete the experiment: 2.0 hours.

**EXPERIMENTAL PROCEDURE**

### Physical Properties of Reactants

| Compound | MW | Amount | mmol | bp (°C) | d | $n_D$ |
|---|---|---|---|---|---|---|
| Benzaldehyde | 106.13 | 80 μL | 0.79 | 178 | 1.04 | 1.5463 |
| Acetone | 58.08 | 29 μL | 0.40 | 56 | 0.79 | 1.3588 |
| NaOH catalyst solution | | 1 mL | | | | |

### Reagents and Equipment

**IMPORTANT.** *It is recommended that the purity of the benzaldehyde be checked by IR. The presence of benzoic acid in the benzaldehyde can substantially lower the yield of product. The benzaldehyde may be purified by distillation under reduced pressure (bp 178–179 °C; 57–59 °C @ 8 torr).*

*The benzaldehyde may be added to the vial by weight, or by volume (using an automatic delivery pipet). To prevent loss of acetone by evaporation, fit the reaction vial with a septum cap and cool it in an ice bath. Add the acetone (by volume) through the septum using a GC syringe.*

*Stoichiometric quantities of the reagents are used. An excess of benzaldehyde results in a more intractable product; excess acetone favors the formation of benzalacetone.*

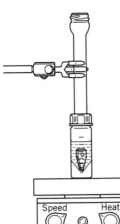

CAUSTIC

C₆H₅CHO, 80 μL, +
acetone, 29 μL +
ethanolic NaOH, 1.0 mL

Place in a 3.0-mL conical vial containing a magnetic spin vane and equipped with an air condenser, 80 μL (84 mg, 0.79 mmol) of benzaldehyde and 29.0 μL (23 mg, 0.40 mmol) of acetone. (■)

Now add to this reaction mixture 1.0 mL of the aqueous, ethanolic sodium hydroxide catalyst solution, delivered from a calibrated Pasteur pipet.

**INSTRUCTOR PREPARATION.**  *The catalyst solution is prepared by dissolving 0.4 g of sodium hydroxide [CAUSTIC] in 4.0 mL of water. To this solution add 3.0 mL of 95% ethanol.*

### Reaction Conditions

Stir the reaction mixture at room temperature for 30 min. During this time the solid yellow product precipitates from solution.

### Isolation of Product

Collect the crude yellow dibenzalacetone by vacuum filtration using a Hirsch funnel. (■) Remove the magnetic spin vane from the reaction vial with forceps. *Some of the product adheres to the magnetic spin vane. This material should be removed by carefully scraping the vane with a microspatula. The material is added to the product collected by filtration.*

Wash the filter cake with three 1.0-mL portions of water. The filtrate should be nearly neutral, as indicated by pH test paper. If not, repeat the washing until the test indicates that the filtrate is neutral.

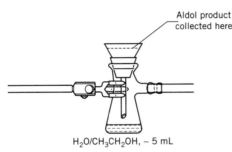

Aldol product
collected here

H₂O/CH₃CH₂OH, ~ 5 mL

**IMPORTANT.**  *It is essential to remove the NaOH completely. If it is not removed, the recrystallization step proves difficult.*

Air-dry the product by maintaining the suction on the Hirsch funnel for approximately 10 min. During this operation a piece of plastic food wrap may be placed over the mouth of the Hirsch funnel to aid in the drying process (see Prior Reading).

**NOTE.**  *An alternate procedure may be used to isolate and purify the dibenzalacetone. Transfer the reaction product directly to a large Craig tube. The washings and the ethanol recrystallization step (see Purification and Characterization section) may then be carried out with no transfer of the material. In this way, loss of product is minimized.*

### Purification and Characterization

The crude dibenzalacetone may be purified by recrystallization from 95% ethanol, using a Craig tube.

Weigh the dried product and calculate the percent yield. Determine the melting point and compare your result with those in the literature.

Obtain an infrared spectrum of the material using the KBr pellet technique and compare your spectrum with that shown in Figure 6.35.

A comparison of the infrared spectra of the starting reagents with that of the product is given below.

## Infrared Analysis

*Acetone:* This simple aliphatic ketone possesses the short macro group frequency train: 3415 (overtone of C=O stretch, $2 \times 1712 = 3424$ cm$^{-1}$), 3000–2850 ($sp^3$ C—H stretch), 1712 (C=O stretch), and 1360 cm$^{-1}$ (symmetric methyl bend $\alpha$ to a carbonyl, see discussion of methyl bends in acetates in Experiment [8B]).

*Benzaldehyde:* The infrared spectrum of this aromatic aldehyde (Fig. 6.34) is rich and interesting. The aromatic aldehyde macro group frequency consists of the following peaks: 3070, 2830 and 2750, 1706, 1602, 1589, and 1396 cm$^{-1}$.

**a.** 3070 cm$^{-1}$: C—H stretch on $sp^2$ carbon.

**b.** 2830 and 2750 cm$^{-1}$: This pair of bands is another example of Fermi coupling (see discussion in Experiment [7] and Chapter 9, Infrared Discussions).

**c.** 1706 cm$^{-1}$: The carbonyl stretch of the aldehyde group. The frequency observed in aliphatic aldehydes falls in the range 1735–1720 cm$^{-1}$, but when conjugated, the value drops (see Chapter 9, Infrared Discussions) and is found in the range 1720–1700 cm$^{-1}$.

**d.** 1602 and 1589 cm$^{-1}$: This pair of bands are related to the degenerate ring stretching vibrations $\nu_{8a}$ and $\nu_{8b}$ of benzene (see infrared discussions in Chapter 7, Experiment [1B$_{adv}$] and Chapter 9, Infrared Discussions).

**e.** 1396 cm$^{-1}$: The aldehyde C—H in-plane bending vibration. The first harmonic of this vibration is Fermi coupled to the aldehyde C—H stretching mode.

Benzaldehyde also possesses a second powerful macro group frequency train that reflects the substitution pattern of the aromatic ring system (see Chapter 9, Infrared Discussions). The monosubstituted benzene ring macro group frequency train contains peaks in the following regions: 1950, 1880, 1800, 1730, 750, and 690 cm$^{-1}$.

**a.** 1990, 1920, 1830, and 1770 cm$^{-1}$: this series of four weak bands with generally decreasing intensity from high-to-low wavenumber values (the third peak near 1800 cm$^{-1}$ may be intensified if the ring is conjugated as in this case) arise from combination bands (see Chapter 9, Infrared Discussions), which involve the out-of-plane bending frequencies of the ring C—H bonds (see below). The exact wavenumber positions are not very important, but the overall shape of the pattern can be used to determine the ring substitution pattern (see Chapter 9, Infrared Discussions).

**b.** 750 and 690 cm$^{-1}$: this pair of bands is very characteristic of monosubstituted phenyl groups. The 750-wavenumber peak arises from the all-in-phase out-of-plane bending vibration of the five C—H groups adjacent to each other on the ring (see Chapter 9, Infrared Discussions). The 690 cm$^{-1}$ companion band involves an out-of-plane displacement of the ring carbon atoms (see Chapter 9, Infrared Discussions).

*Dibenzalacetone:* The infrared spectrum of the product (Fig. 6.35) contains many features of the starting materials, plus new and shifted bands unique to the newly formed structure.

**a.** The monosubstituted aromatic macro group frequency remains: 1960, 1890, 1815, 1770, 760, and 690 cm$^{-1}$. In addition, the two pairs of degenerate ring stretching bands are present: 1598, 1580 and 1500, 1450 cm$^{-1}$. The 1500-wavenumber band is rather weak in benzaldehyde, but intensifies in the product.

**b.** The ketone carbonyl remains the most intense band in the spectrum but is shifted to 1653 cm$^{-1}$ because of the carbonyl's conjugation to two aromatic rings. The aldehyde's carbonyl stretching mode at 1706 cm$^{-1}$ has vanished.

**c.** The methyl C—H stretching bands between 3000–2850 cm$^{-1}$, and the coupled aldehyde C—H stretching peaks at 2830 and 2750 cm$^{-1}$ have disappeared.

**d.** New bands appear at 3058 and 3035 cm⁻¹ (alkene C—H, stretch) which are overlapped with the aromatic ring (C—H stretching) peaks (3075 cm⁻¹). In addition, bands appear at 1627 (conjugated C=C) and 985 cm⁻¹ (trans-substituted C=C, C—H in-phase out-of-plane bend). The latter band occurs slightly above its usual location near 965 cm⁻¹. This rise in frequency is the result of conjugation of the double bond to the carbonyl group. For an example of a much more dramatic rise in frequency, refer to the discussion of the cis C—H bending modes in the spectrum of maleic anhydride (Experiment [14]).

Examine the spectrum of your product in a potassium bromide matrix. Discuss the similarities and differences of the experimentally derived spectral data to the reference spectra (Figs. 6.34 and 6.35).

Dibenzalacetone is a compound that can be characterized by classical chemical tests.

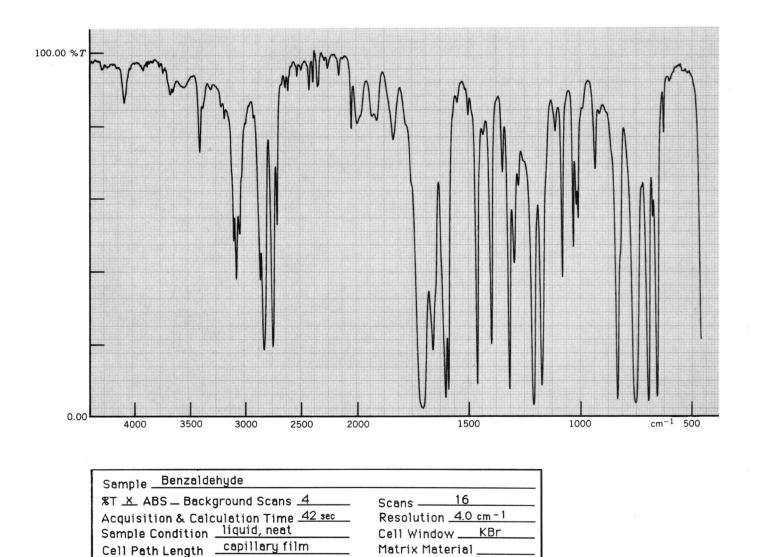

**FIGURE 6.34**   IR spectrum: Benzaldehyde.

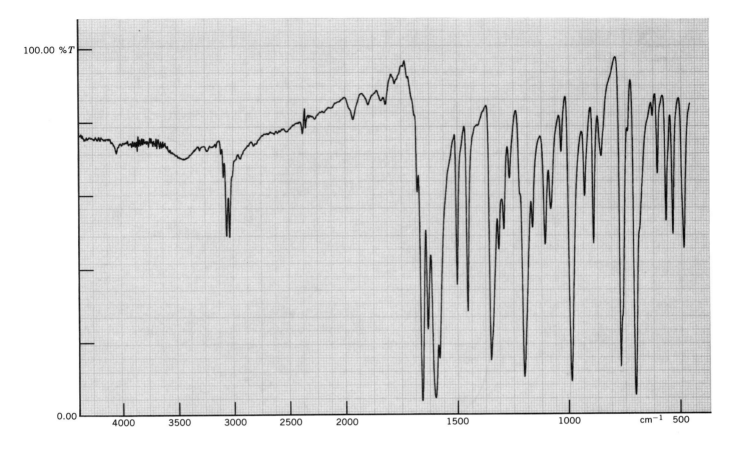

**FIGURE 6.35** IR spectrum: Dibenzalacetone.

**Chemical Tests**

An ignition test (See Table 10.1) should indicate that it contains an aromatic ring system. Perform the test to confirm this.

Several classification tests might also be of assistance in classifying this compound. Does the 2,4-dinitrophenylhydrazine test for an aldehyde or ketone give a positive result? Isolate the 2,4-dinitrophenylhydrazone derivative and determine its melting point. Does it correspond to the literature value of 180 °C? What further test should be run to determine whether the carbonyl is present as an aldehyde or ketone ?

A test for unsaturation should be enlightening. Would you perform the bromine–methylene chloride or the Baeyer test (Chapter 10, Classification Tests). Did the correct test give a positive result?

## OPTIONAL SEMIMICROSCALE PREPARATION

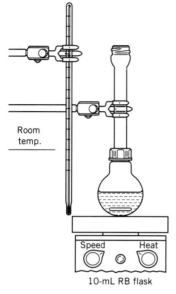

Room temp.

10-mL RB flask

This experiment may be scaled up to be carried out at 5 or 10 times the amounts used in the above micro preparation. The data summarized below is for the 10–fold procedure.

The procedure is identical to the above with the following exceptions.

**1.** Use a 10-mL round-bottom flask fitted with an air condenser. (■)
**2.** Increase the reagent and solvent amounts.

### Properties of Reactants

| Compound | MW | Amount | mmol | bp(°C) | d | $n_D$ |
|---|---|---|---|---|---|---|
| Benzaldehyde | 106.13 | 800 µL | 7.9 | 178 | 1.04 | 1.5463 |
| Acetone | 58.08 | 300 µL | 4.0 | 56 | 0.79 | 1.3588 |
| NaOH catalyst solution | | 5 mL | | | | |

**3.** The collected filter cake is transferred to a 10-mL beaker, stirred with 5.0 mL of water and then recollected by vacuum filtration. Repeat this process, usually about three times, until the filtrate is neutral to litmus paper.

## QUESTIONS

**6-132.** A key step in the total synthesis of the hydrocarbon azulene follows. Outline a suitable mechanism to account for the reaction.

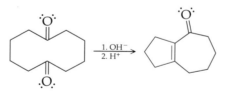

**6-133.** The aldol reaction has been utilized extensively for the generation of five- and six-membered rings. Suggest a suitable mechanism for the cyclization reactions shown below.

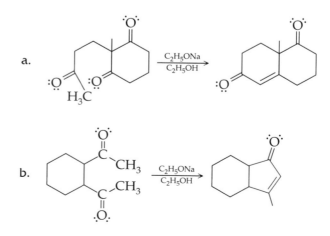

**6-134.** Predict the major organic product formed in each of the following reactions:

a. $CH_3CH_2NO_2 + CH_2O \xrightarrow{\text{NaOH}}$

b. $C_6H_5-CH=CH-CHO + CH_3-\overset{\overset{\displaystyle \cdot\ddot{O}\cdot}{\|}}{C}-C_6H_5 \xrightarrow[C_2H_5OH]{C_2H_5ONa}$

c. $C_6H_5CHO + C_6H_5CH_2CN \xrightarrow[C_2H_5OH]{C_2H_5ONa}$

**6-135.** "Crossed" or "mixed" aldol condensations are practical for synthesis, if one of the aldehydes (or ketones) has no $\alpha$-hydrogen atoms. Explain.

**6-136.** Give several examples of aldehydes or ketones that could be used in a "crossed" aldol condensation with propanal. Assign structures and names to the products that could be formed and point out any side reactions that might occur.

**6-137.** In the aldol condensation using the conditions of this experiment, why might it be essential that the benzaldehyde contain no benzoic acid?

---

**BIBLIOGRAPHY**

Review articles:

Mukaiyama, T. *Org. React.* **1982**, *28*, 203.

Neilsen, A. T.; Houlihan, W. *Org. React.* **1968**, *16*, 1.

Examples of the aldol and Claisen–Schmidt reactions:

Auerbach, R. A.; Crumrine, G. S.; Ellison, D. L.; House, H. O. *Organic Syntheses*; Wiley: New York, 1988; Collect. Vol. VI, p. 692.

Conrad, C. R.; Dolliver, M. A. *Organic Syntheses*; Wiley: New York, 1943; Collect. Vol. II, p. 167.

Hill, G. A.; Bramann, G. M. *Organic Syntheses*; Wiley: New York, 1941; Collect. Vol. I, p. 81.

Kohler, E. P.; Chadwell, H. M. *Organic Syntheses*; Wiley: New York, 1941; Collect. Vol. I, p. 78.

Leuck, G. J.; Cejka, L. *Organic Syntheses*; Wiley: New York, 1941; Collect. Vol. I, p. 283.

Russell, A.; Kenyon, R. L. *Organic Syntheses*; Wiley: New York, 1955; Collect. Vol. III, p. 747.

Wittig, G.; Hesse, A. *Organic Syntheses*; Wiley: New York, 1988; Collect. Vol. VI, p. 901.

---

# Quantitative Analysis of Grignard Reagents: 1-Methylbutylmagnesium Bromide and Phenylmagnesium Bromide

---

Common names: 1-Methylbutylmagnesium bromide, 2-pentylmagnesium bromide
CA number: [57325-22-1]
CA name as indexed: Magnesium, bromo(1-methylbutyl)-

Common name: phenylmagnesium bromide
CA number: [100-58-3]
CA name as indexed: Magnesium, bromophenyl-

## Purpose

To generate a Grignard reagent, a common and synthetically useful source of a nucleophilic carbanion, and to use an aqueous titration method to determine the amount prepared.

### Prior Reading

Moisture-Protected Reaction Apparatus (see pp. 29–31).

---

**REACTION**

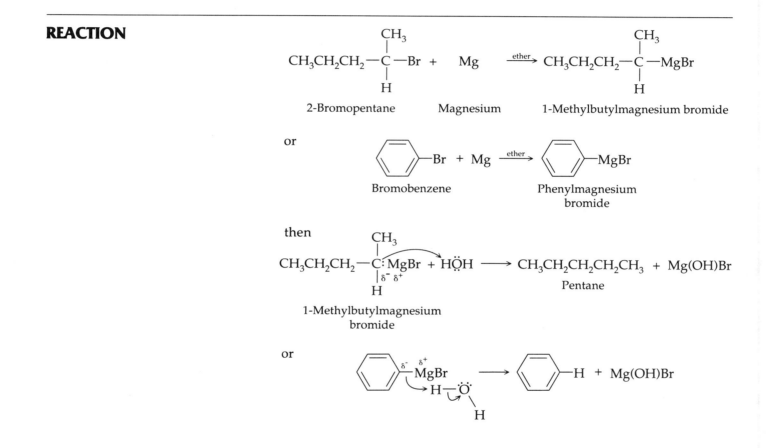

---

**DISCUSSION**

This experiment demonstrates the formation of a Grignard reagent (1-methylbutylmagnesium bromide or phenylmagnesium bromide) and a titration method by which the amount of the reagent prepared can be analyzed.

The discovery by Victor Grignard in 1900 that organic halides react with magnesium metal to give organomagnesium compounds was a landmark in organic chemistry. Grignard reagents are among the most useful and versatile reagents in organic synthesis.

The reaction of the Grignard reagent with water is the basis of the analytical method used in this experiment.

$$RMgX + HOH \rightarrow RH + Mg(OH)X$$

In the Grignard reagent, the carbon atom bound to the electropositive magnesium atom has a high negative charge density, which is responsible for the strong nucleophilic and basic character exhibited by this organometallic reagent. The carbon, acting as a base, can abstract even a weakly acidic proton from protic reagents, such as water, carboxylic acids, alcohols, and so on. In this process, the corresponding hydrocarbon (the conjugate acid of the R group carbanion) and the basic magnesium halide species are produced. This reaction sequence can be used in the laboratory as a synthetic method to convert organohalides to hydrocarbons. In the examples given in this experiment, 1-methylbutylmagnesium bromide would yield pentane, while phenylmagnesium bromide gives benzene upon protonation.

Titration of the Mg(OH)X species with standardized acid solution allows one to determine the amount of Grignard reagent originally formed in the solution.

$$2 \text{ Mg(OH)X} + \text{H}_2\text{SO}_4 \rightarrow 2 \text{ HOH} + \text{MgSO}_4 + \text{MgX}_2$$

An excess of the sulfuric acid is generally added to ensure that the Mg(OH)X is completely reacted. The excess acid is then neutralized with standard sodium hydroxide solution. The difference between the total amount of sulfuric acid used and the amount of sodium hydroxide required corresponds to the number of equivalents of the acid actually used to neutralize the Mg(OH)X species. This value is then directly related to the equivalents of Grignard reagent by the reactions given above.

## Preparation of the Grignard Reagent[19]

CAUTION:   Ether is a flammable liquid. Extinguish all flames during this experiment.

NOTE.   *All the glassware used in the experiment should be cleaned, dried in an oven at 110 °C for at least 30 min, and then cooled in a desiccator before use.*

Estimated time for the experiment: 1.5 hours.

**EXPERIMENTAL PROCEDURE**

## Part 1   1-Methylbutylmagnesium Bromide

### Physical Properties of Reactants

| Compound | MW | Amount | mmol | bp(°C) | $d$ | $n_D$ |
|----------|-----|--------|------|--------|-----|-------|
| 2-Bromopentane | 151.05 | 125 $\mu$L | 1.0 | 117 | 1.21 | 1.4413 |
| Magnesium | 24.31 | 36 mg | 1.5 | | | |
| Iodine | 253.81 | 1 crystal | | | | |
| Diethyl ether | 74.12 | 400 $\mu$L | | 34.5 | | |

[19] For references relating to the preparation of Grignard reagents, see Experiment [16].

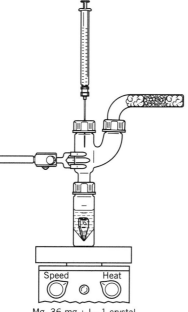

Mg, 36 mg + I₂, 1 crystal
+ (CH₃CH₂)₂O, 400 μL +
CH₃CHBr(CH₂)₂CH₃, 125 μL

Prepare the Grignard reagent exactly as described in Experiment [17]. The reagents, amounts of reagents, order of addition, work-up manipulations, and precautions are the same. The equipment is also identical. (■) Cool the gray-colored Grignard reagent mixture to room temperature, and then assay it by the titration method outlined in Part 2.

## Part 2    Phenylmagnesium Bromide[20]

### Physical Properties of Reactants

| Compound | MW | Amount | mmol | mp(°C) | bp(°C) | $d$ | $n_D$ |
|---|---|---|---|---|---|---|---|
| Bromobenzene | 157.02 | 76 μL | 0.72 | | 156 | 1.50 | 1.5597 |
| Diethyl ether | 74.12 | 1.3 mL | | | 34.5 | 0.73 | |
| Magnesium | 24.3 | 17.5 mg | 0.73 | | | | |
| Iodine | | 1 crystal | | | | | |

**NOTE.**  *If not used for analysis, the solution of Grignard reagent may be treated with various reagents to prepare a wide variety of compounds. For example, see Experiments [16], [17], and [4ₐdᵥ].*

### Analysis of the Grignard Reagent

Place 10 mL of freshly boiled distilled water and one drop of phenolphthalein indicator in a 50-mL Erlenmeyer flask. Using a syringe, transfer the cool Grignard reagent solution to the Erlenmeyer flask. Rinse the reaction vial with 0.5 mL of diethyl ether and add the rinse to the Erlenmeyer flask.

**NOTE.**  *The addition of water to the Grignard reagent results in the hydrolysis of this reagent to form the corresponding hydrocarbon and a basic magnesium halide, Mg(OH)X. The water is initially boiled to remove any dissolved carbon dioxide that might interfere with the titration.*

*It is important to make sure that in the transfer of the Grignard reagent solution all small pieces of unreacted magnesium are excluded.*

### Analysis by Titration

Add, from a 10-mL buret, 5 or 6 mL of standard 0.2 N H₂SO₄ solution to the ethereal Grignard solution. *The resulting solution should be acidic and colorless. If not, add an additional portion of the acid.*

HOOD

Add a boiling stone to the flask and heat the mixture at a sand bath temperature of 90–95 °C in the **hood** for 5 min.

While the solution is still warm, add a drop of phenolphthalein indicator solution, and neutralize the excess acid by back titration with 0.1 N NaOH solution. Back titration produces a very light colored, pink end point. *It may be necessary to add an additional drop of acid, and then more base, to get the best possible end point.*

### Data and Calculations

The difference between the initial and final buret readings is the volume of standard acid and base used in the titration of the Grignard reagent.

---

[20] This Grignard reagent is prepared exactly as described in Experiment [16]. The reagents, amount of reagents, order of addition, workup manipulations, and precautions are the same. The equipment is also identical. (■)

From the data, calculate the equivalents of Grignard reagent formed. Also, as a percentage, determine the amount of Grignard reagent analyzed compared to its theoretical yield of formation.

**6-138.** Technical grade ether often contains ethanol. Would you recommend this material as a suitable solvent for the preparation of Grignard reagents? If not, why not?

**6-139.** It is likely that the amount of Grignard reagent your analysis indicates was formed is greater than the amount of Grignard reagent actually present just before you added water. Explain.

**6-140.** What hydrocarbon would you expect to obtain by the action of water on each of the Grignard reagents listed below?
   a. Butylmagnesium bromide
   b. *sec*-Butylmagnesium bromide
   c. *iso*-Butylmagnesium bromide
   d. *tert*-Butylmagnesium bromide

**6-141.** What product would each of the Grignard reagents in question 6-140 yield when treated with $D_2O$?

**6-142.** What product would each of the Grignard reagents listed in Question 6-140 yield when treated with ethanol? With isopropyl alcohol?

**6-143.** The solubility of Grignard reagents in ether plays a crucial role in their formation. The reagents are soluble because the magnesium is coordinated to the ether oxygen in a Lewis acid–base interaction. Each ether molecule donates an electron pair to the magnesium to complete an octet.

$$(CH_3CH_2)_2\ddot{O}: \rightarrow \underset{\underset{Br}{|}}{\overset{\overset{R}{|}}{Mg}} \leftarrow :\ddot{O}(CH_2CH_3)_2$$

Grignard reagents are normally insoluble in hydrocarbon solvents. However, they can be rendered soluble by the addition of a tertiary amine to the hydrocarbon–Grignard reagent mixture. Explain.

For the many references related to the preparation and use of the Grignard reagent cited in *Organic Syntheses*, see the Reaction Indexes in Collected Volumes I–VI under Grignard Reactions.

# Williamson Synthesis of Ethers: Propyl *p*-Tolyl Ether and Methyl *p*-Ethylphenyl Ether

Common names: Propyl *p*-tolyl ether, 4-propoxytoluene
CA number: [5349-18-8]
CA name as indexed: Benzene, 1-methyl-4-propoxy-

Common names: Methyl *p*-ethylphenyl ether, *p*-ethylanisole
CA number: [1515-95-3]
CA name as indexed: Benzene, 1-ethyl-4-methoxy-

## Purpose

To explore the conditions under which ethers are prepared by the well-known Williamson' ether synthesis. To prepare alkyl aryl ethers by $S_N2$ reactions of alkyl halides with substituted phenoxide anions. To demonstrate the use of phase-transfer catalysis.

### Prior Reading

Technique 4: Solvent Extraction
Liquid-Liquid Extraction (see pp. 80–82).
Technique 6: Chromatography
Column Chromatography (see pp. 98–101).
For Optional Scaleup: Separatory Funnel Extraction (see pp. 82–83).

## REACTION

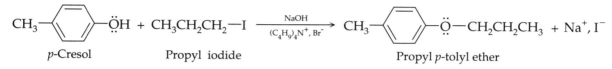

| *p*-Cresol | Propyl iodide | Propyl *p*-tolyl ether |

## DISCUSSION

The two compounds whose preparations are described in Experiments [22A] and [22B] are alkyl aryl ethers. The general method of preparation is the Williamson synthesis, an $S_N2$ reaction specifically between a phenoxide ion ($ArO^-$) nucleophile and an alkyl halide. This reaction is often used for the synthesis of symmetrical and unsymmetrical ethers where at least one of the ether carbon atoms is primary or methyl, and thus amenable to an $S_N2$ reaction. Elimination (E2) is generally observed if secondary or tertiary halides are used, as phenoxide ions are also bases.

The conditions under which these reactions are conducted lend themselves to the use of phase-transfer catalysis. The reaction system involves two phases, the aqueous phase and the organic phase. In the present case, the alkyl halide reactant acts as the organic solvent, as does the product formed. The phase-transfer catalyst plays a very important role. In effect, it carries the phenoxide ion, as an ion-pair, from the aqueous phase, across the phase boundary into the organic phase, where the $S_N2$ reaction then occurs. The ether product and the corresponding halide salt of the catalyst are produced in this reaction. The halide salt then migrates back into the aqueous phase, where the halide ion is exchanged for another phenoxide ion, and the process repeats itself. The catalyst can play this role since the large organic groups (the four butyl groups) allow the solubility of the ion-pair in the organic phase, while the charged ionic center of the salt renders it soluble in the aqueous phase. For further discussions of phase-transfer catalysis, see Experiments [19B] and [19D].

In the reactions described below, the mechanism is a classic $S_N2$ process, and involves a backside nucleophilic attack of the phenoxide anion on the alkyl halide.

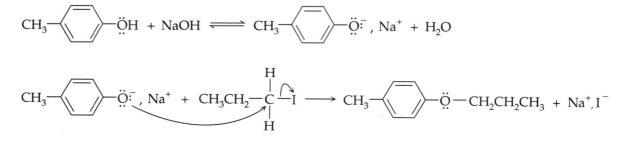

It is of interest to contrast the acidity of phenols with that of simple alcohols. A phenol is more acidic than an alcohol. In a typical aliphatic alcohol (e.g., ethanol) loss of the proton forms a strong anionic base, alkoxide ion (ethoxide ion).

$$R-CH_2-OH \rightleftharpoons H^+ + R-CH_2-O^-$$
$$\text{Alkoxide ion}$$

The strongly basic characteristics of the alkoxide species is due to the fact that the negative charge is localized on the oxygen atom. Ethanol has a $pK_a = 16$. In contrast, the conjugate base of a phenol can delocalize its negative charge.

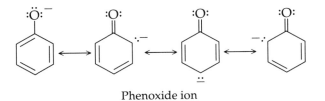

Phenoxide ion

Thus, the phenoxide ion is stabilized by this resonance delocalization; therefore it is a weaker base than the alkoxide ion. Conversely, the phenol is a stronger acid than a typical aliphatic alcohol. Phenol has a $pK_a = 10$ and is thus 1 million times more acidic than ethanol.

## Experiment [22A]  Propyl p-Tolyl Ether

The reaction for Experiment [22A] is shown above.

Estimated time of the experiment: 2.5 hours.

**EXPERIMENTAL  PROCEDURE**

### Physical Properties of Reactants

| Compound | MW | Amount | mmol | mp(°C) | bp(°C) | d | $n_D$ |
|---|---|---|---|---|---|---|---|
| p-Cresol | 108.15 | 160 μL | 1.56 | 32–34 | 202 | 1.02 | 1.5312 |
| 25% NaOH solution | | 260 μL | | | | | |
| Tetrabutyl-ammonium bromide | 322.38 | 18 mg | 0.056 | 103–104 | | | |
| Propyl iodide | 169.99 | 150 μL | 1.54 | | 102 | 1.75 | 1.5058 |

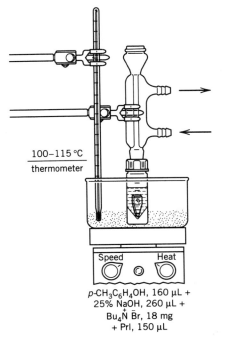

100–115 °C
thermometer

### Reagents and Equipment

Weigh and place 160 μL (168 mg, 1.56 mmol) of p-cresol in a 5.0-mL conical vial containing a magnetic spin vane. Now add 260 μL of 25% aqueous sodium hydroxide, and thoroughly mix the resulting solution. (■) To this solution weigh and add the tetrabutylammonium bromide (Bu₄N⁺,Br⁻) catalyst (18 mg), followed by 150 μL (262 mg, 1.54 mmol) of propyl iodide. Immediately attach the vial to a reflux condenser.

p-CH₃C₆H₄OH, 160 μL +
25% NaOH, 260 μL +
Bu₄N⁺ Br, 18 mg
+ PrI, 150 μL

HOOD

**NOTE.**  *Warm the cresol in a hot water bath to melt it. Dispense this reagent and the propyl iodide in the **hood** using an automatic delivery pipet.*

---

**CAUTION:  Propyl iodide is a cancer suspect agent.**

---

## Reaction Conditions

Place the reaction vessel in a sand bath, and stir vigorously at 110–115 °C for 45–60 min.

## Isolation of Product

Cool the resulting two-phase mixture to room temperature, and remove the spin vane with forceps. Rinse the spin vane with 1.0-mL of diethyl ether, adding the rinse to the two-phase mixture. Cap the vial, agitate, vent, and transfer the bottom aqueous layer, using a Pasteur filter pipet, to a 3.0-mL conical vial. A Vortex mixer, if available, can be used to good advantage in this extraction step. Wash this aqueous fraction with 1.0 mL of diethyl ether. Save this, and all subsequent aqueous fractions together in a small Erlenmeyer flask until your final product has been isolated and characterized. Now transfer this diethyl ether wash to the 5-mL conical vial containing the ether solution of the product. Extract the resulting ether solution with a 400-μL portion of 5% aqueous sodium hydroxide solution. Cap the vial, agitate, vent, and remove and save the bottom aqueous layer, using a Pasteur filter pipet. Wash the product–ether solution with 200 μL of water. Remove, and save, the aqueous phase to obtain the crude, wet ether solution of the product. Add a boiling stone to the vial, and concentrate the solution on a warm sand bath under a gentle stream of nitrogen to isolate the crude product.

## Purification and Characterization

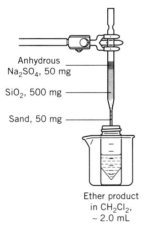

Anhydrous
Na$_2$SO$_4$, 50 mg

SiO$_2$, 500 mg

Sand, 50 mg

Ether product
in CH$_2$Cl$_2$,
~ 2.0 mL

The crude product is purified by chromatography on silica gel. Prepare a microchromatographic column by placing 500 mg of activated silica gel in a Pasteur filter pipet, followed by 50 mg of anhydrous sodium sulfate. (■) Dissolve the crude product in 250 μL of methylene chloride, and transfer the resulting solution to the dry column by use of a Pasteur pipet. Elute the material with 2.0 mL of methylene chloride, and collect the eluate in a tared 3.0-mL conical vial containing a boiling stone. Fit the vial with an air condenser, and evaporate the solvent by placing the vial in a sand bath maintained at a temperature of 60–65 °C.

Weigh the pure propyl *p*-tolyl ether and calculate the percent yield. Determine the boiling point and density (optional) and compare the experimental values with those in the literature.

Obtain an IR spectrum of the compound and compare it to that shown in Figure 6.36 for 4-propoxytoluene (propyl *p*-tolyl ether).

## Nuclear Magnetic Resonance Analysis

If facilities permit, you can obtain both $^1$H and $^{13}$C NMR spectra of your propyl *p*-tolyl ether in CDCl$_3$, and compare your spectra with those in Figures 6.37 and 6.38. There are two extraneous peaks in the $^1$H spectrum: the small singlet at 7.24 ppm is due to residual CHCl$_3$ in the CDCl$_3$, and the small singlet at 1.55 ppm is probably due to a trace amount of H$_2$O in either the sample or the NMR solvent. The 1:1:1 triplet at 77 ppm in the $^{13}$C spectrum is from the CDCl$_3$ solvent.

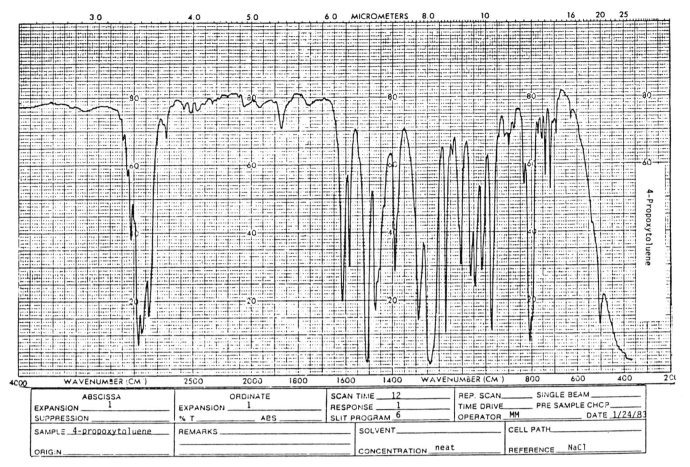

**FIGURE 6.36**  IR spectrum: Propyl *p*-tolyl ether.

300 MHz NMR SPECTRUM OF PROPYL *p*-TOLYL ETHER IN CDCl$_3$

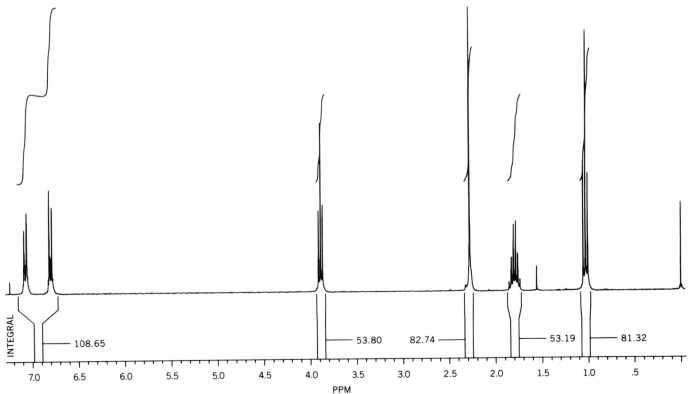

**FIGURE 6.37**  $^1$H-NMR spectrum: Propyl *p*-tolyl ether.

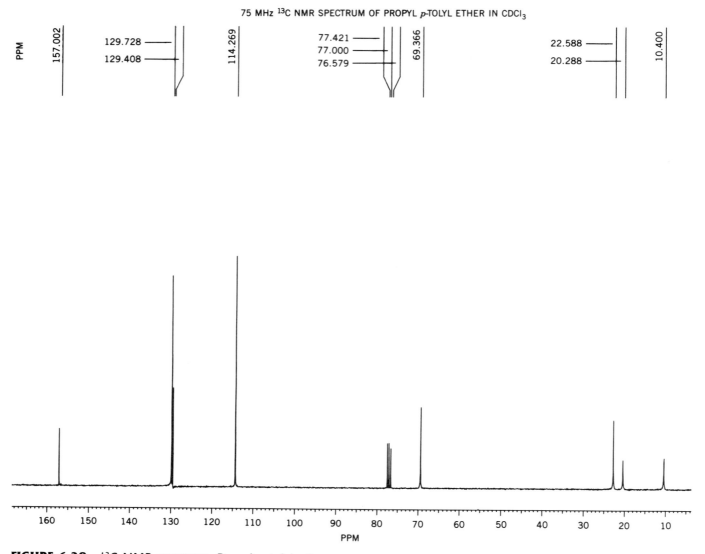

**FIGURE 6.38**   ¹³C-NMR spectrum: Propyl p-tolyl ether.

Since the ¹H spectrum is entirely first order, it can be readily interpreted. You should be able to use the splitting patterns to assign peaks to each of the different groups of protons in the molecule. The integration can assist you.

### Chemical Tests

Qualitative chemical tests can also be used to assist in characterizing this compound as an ether. Perform the ignition test (Table 10.1) to determine whether the material contains an aromatic ring.

A key factor to investigate is the solubility characteristics of this material (see Chapter 10). Determine its solubility in water, 5% sodium hydroxide, 5% hydrochloric acid, concentrated sulfuric acid, and 85% phosphoric acid. Do the results place this compound in the solubility class of an ether containing less than 8, or more than 8, carbon atoms? Does the ferrox test (Chapter 10) confirm the presence of oxygen in the compound?

This ether may be prepared on a larger scale (~100-fold increase) using a procedure similar to that just outlined, with the following modifications.

**1.** Use a 100-mL round-bottom flask containing a magnetic stirrer and equipped with a reflux condenser. (■)
**2.** The reagent and solvent amounts are summarized in the following table. Note that propyl bromide is used in place of propyl iodide at this scale.

**Physical Properties of Reactants**

| Compound | MW | Amount | mol | mp(°C) | bp(°C) | d | $n_D$ |
|---|---|---|---|---|---|---|---|
| p-Cresol | 108.15 | 16.3 g | 0.15 | 32–34 | 202 | 1.02 | 1.5312 |
| NaOH | 40.0 | 6.05 g | 0.15 | | | | |
| Tetrabutyl-ammonium bromide | 322.38 | 0.41 g | 0.0012 | 103–104 | | | |
| Propyl bromide | 122.99 | 12.71 g | 0.1 | | 71 | 1.35 | 1.4341 |
| Water | | 25 mL | | | | | |

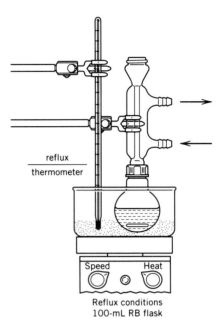

reflux
thermometer

Speed    Heat

Reflux conditions
100-mL RB flask

**3.** Stir the reaction mixture at reflux temperature for 90 min.

## Isolation of Product

Allow the solution to cool to room temperature, transfer it to a 125-mL separatory funnel, and remove the aqueous layer. Store the aqueous layer in a 125-mL Erlenmeyer flask. Wash the organic layer successively with 5% NaOH solution (20 mL) and distilled H₂O (20 mL). After each washing, remove the aqueous phase and add it to the 125-mL Erlenmeyer flask, which should be kept until the final product has been purified and characterized. Collect the remaining deep-red organic layer in a 125-mL Erlenmeyer flask, and dry it over anhydrous sodium sulfate.

Remove the drying agent by filtration through a glass wool plug and collect the product ether in a tared container. Weight, calculate the percent yield, and then purify and characterize a small amount of the material as described in the microscale procedure.

## Experiment [22B]   Methyl *p*-Ethylphenyl Ether

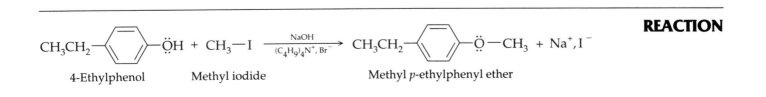

4-Ethylphenol          Methyl iodide                    Methyl *p*-ethylphenyl ether

## EXPERIMENTAL PROCEDURE    Estimated time to complete the experiment: 2.5 hours.

### Physical Properties of Reactants

| Compound | MW | Amount | mmol | mp(°C) | bp(°C) | d | $n_D$ |
|---|---|---|---|---|---|---|---|
| 4-Ethylphenol | 122.17 | 150 mg | 1.2 | 42–45 | | | |
| 25% NaOH solution | | 250 μL | | | | | |
| Tetrabutylammonium bromide | 322.38 | 15 mg | 0.047 | 103–104 | | | |
| Methyl iodide | 141.94 | 90 μL | 1.45 | | 41–43 | 2.28 | 1.5304 |

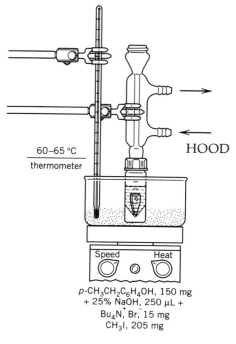

60–65 °C
thermometer

HOOD

$p$-CH₃CH₂C₆H₄OH, 150 mg
+ 25% NaOH, 250 μL +
Bu₄N⁺ Br⁻, 15 mg
CH₃I, 205 mg

### Reagents and Equipment

Use the same apparatus as in Experiment [22A] for this synthesis. Weigh and add 150 mg (1.2 mmol) of 4-ethylphenol to the reaction vial followed by 250 μL of 25% aqueous sodium hydroxide solution. (■) Stir the mixture at room temperature until dissolution occurs. The phase-transfer catalyst [tetrabutylammonium bromide ($Bu_4N^+$, $Br^-$), 15 mg, 0.05 mmol] is now added, followed by 90 μL (205 mg, 1.45 mmol) of methyl iodide.

**NOTE.** *Methyl iodide is toxic and must be dispensed in the **hood**. Dispense both the alkaline solution and the methyl iodide using an automatic delivery pipet. Because of its volatility, methyl iodide is used in slight excess.*

### Reaction Conditions

Place the reaction assembly in a sand bath maintained at 60–65 °C and stir the mixture for 1 hour.

### Isolation of Product

Work up the resulting product mixture as described in Experiment [22A].

### Purification and Characterization

Purify the crude product by chromatography on silica gel as described in Experiment [22A], Purification and Characterization. (■)

Weigh the pure methyl $p$-ethylphenyl ether and calculate the percent yield. Determine the boiling point and density (optional). Compare your results with the values reported in the literature.

Obtain the IR spectrum of the compound and compare it with that in Figure 6.39.

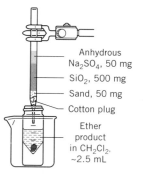

Anhydrous
Na₂SO₄, 50 mg
SiO₂, 500 mg
Sand, 50 mg
Cotton plug
Ether
product
in CH₂Cl₂,
~2.5 mL

### Nuclear Magnetic Resonance Analysis

If facilities permit, you can obtain both ¹H and ¹³C NMR spectra of your methyl $p$-ethylphenyl ether in CDCl₃, and compare your spectra with those in Figures 6.40 and 6.41. There are two extraneous peaks in the ¹H spectrum: The small singlet at 7.24 ppm is due to residual CHCl₃ in the

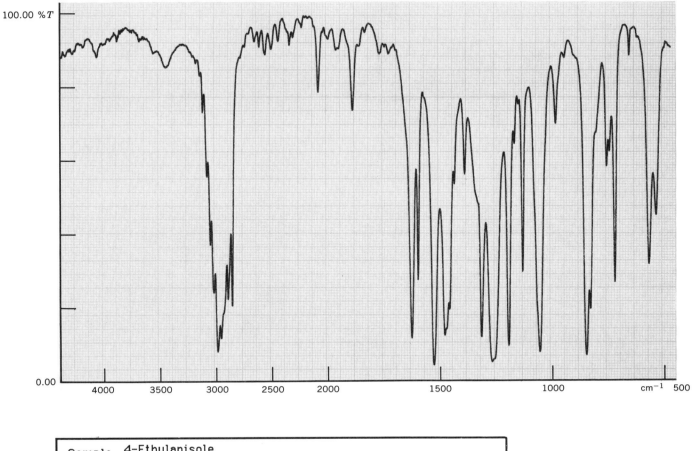

| | |
|---|---|
| Sample __4-Ethylanisole__ | |
| %T _X_ ABS __ Background Scans _4_ | Scans _____4_____ |
| Acquisition & Calculation Time _42 sec_ | Resolution _4.0 cm⁻¹_ |
| Sample Condition _liquid, neat_ | Cell Window ____KBr____ |
| Cell Path Length _capillary film_ | Matrix Material _____ |

**FIGURE 6.39**   IR spectrum: Methyl *p*-ethylphenyl ether.

CDCl$_3$ and the small singlet at 1.55 ppm is probably due to a trace amount of water in either the sample or the NMR solvent. The 1:1:1 triplet at 77 ppm in the $^{13}$C spectrum is from the CDCl$_3$ solvent.

Since the $^1$H spectrum is entirely first order, it can be readily interpreted. You should be able to use the splitting patterns to assign peaks to each of the different groups of protons in the molecule. The integration can assist you.

## Chemical Tests

Qualitative chemical tests can also be used to assist in characterizing this compound as an ether. Perform the ignition test (Table 10.1) to determine whether the material contains an aromatic ring.

A key factor to investigate is the solubility characteristics of this material (see Chapter 10). Determine its solubility in water, 5% sodium hydroxide, 5% hydrochloric acid, concentrated sulfuric acid, and 85% phosphoric

300 MHz $^1$H NMR SPECTRUM OF METHYL $p$-ETHYLPHENYL ETHER IN CDCl$_3$

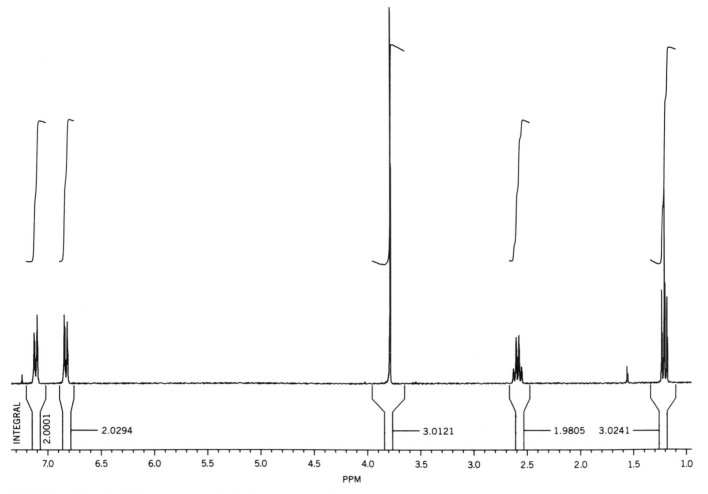

**FIGURE 6.40**    $^1$H-NMR spectrum: Methyl $p$-ethylphenyl ether.

acid. Do the results place this compound in the solubility class of an ether containing fewer than 8, or more than 8, carbon atoms?

Does the ferrox test (Chapter 10) confirm the presence of oxygen in the compound?

## OPTIONAL SEMIMICROSCALE AND MACROSCALE PREPARATIONS

If desired, this experiment can be scaled up by a factor of 10 or more.

### Tenfold Scaleup

The reagent and solvent amounts are given in the following table.

75 MHz $^{13}$C NMR OF METHYL $p$-ETHYLPHENYL ETHER IN CDCl$_3$

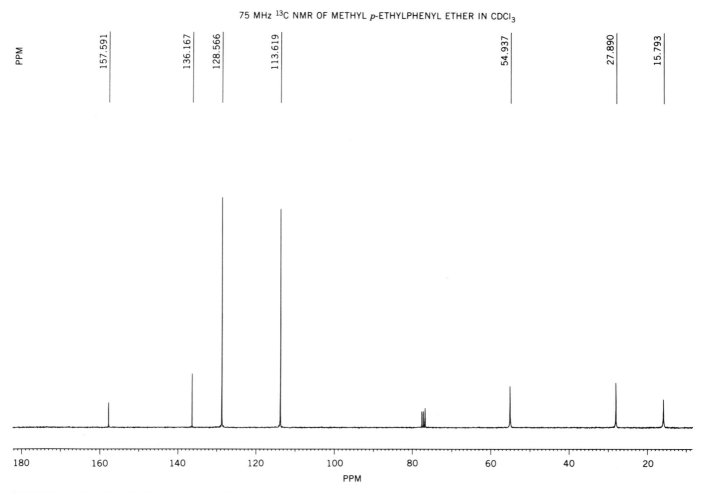

**FIGURE 6.41** $^{13}$C-NMR spectrum: Methyl $p$-ethylphenyl ether.

## Physical Properties of Reactants

| Compound | MW | Amount | mmol | mp(°C) | bp(°C) | $d$ | $n_D$ |
|---|---|---|---|---|---|---|---|
| 4-Ethylphenol | 122.17 | 1.5 g | 12.3 | 42–45 | | | |
| NaOH | 40.0 | 625 mg | 15.6 | | | | |
| Tetrabutylammonium bromide | 322.38 | 150 mg | 0.47 | 103–104 | | | |
| Methyl iodide | 141.94 | 2.05 g | 14.5 | | 41–43 | 2.28 | 1.5304 |
| Water | | 2.5 mL | | | | | |

## Reagents and Equipment

In a 25-mL round-bottom flask equipped with a stirring bar and a reflux condenser, place 2.5 mL of water and 1.5 g (12 mmol) of $p$-ethylphenol (the phenol will not be water soluble). Cool the mixture by immersing the flask

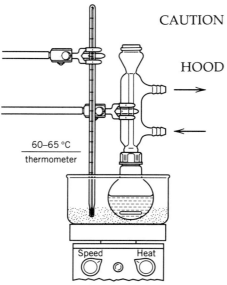

CAUTION

HOOD

60–65 °C
thermometer

Speed    Heat

25-mL RB flask

in a beaker of cold water, and then, with stirring, add 625 mg of sodium hydroxide **[CAUTION]**.

After dissolution of the sodium hydroxide, weigh and add 150 mg of the phase-transfer catalyst, tetrabutylammonium bromide, followed by 2.05 g (900 μL) of methyl iodide **[HOOD]** using an automatic delivery pipet. (■)

## Reaction Conditions

Place the reaction vessel in a sand bath, and heat the reaction mixture, with stirring, at 60–65 °C for 1 hour.

## Isolation of Products

Cool the resulting two-phase mixture to room temperature, and then transfer it by Pasteur pipet to a 15-mL centrifuge tube. Wash the reaction flask with three 1-mL portions of diethyl ether and transfer the washings to the centrifuge tube. Cap the tube, shake, and vent (this operation may be done using a Vortex mixer if available), and allow the layers to separate. Remove the lower aqueous layer using a Pasteur filter pipet, and save it in a 10-mL Erlenmeyer flask until the final product has been purified and characterized.

**NOTE.**  *Do not remove any precipitated material that settles between the two layers.*

Extract the ether layer with one 2-mL portion of 5% sodium hydroxide solution and then with 1 mL of water. Add these washings to the 10-mL Erlenmeyer flask.

Purify the crude product by chromatography on silica gel. Prepare a column by placing 5 g of activated silica gel in a 25-mL buret, followed by 0.5 g of anhydrous sodium sulfate. Dissolve the wet ether extract obtained above in 2.5 mL of methylene chloride, and transfer the resulting solution by Pasteur pipet to the dry column. Elute the sample with an additional 20 mL of methylene chloride. Collect the eluate in a tared 50-mL Erlenmeyer flask containing a boiling stone. Evaporate the solvent by placing the flask in a sand bath maintained at 60–65 °C **[HOOD]**. Use a gentle stream of nitrogen or dry air to hasten the process.

HOOD

Weigh the resulting product and calculate the percent yield. Characterize the product as outlined above in the microscale procedure.

## Eightyfold Scaleup

The reagent and solvent amounts are given in the following table.

### Physical Properties of Reactants

| Compound | MW | Amount | mol | mp(°C) | bp(°C) | d | $n_D$ |
|---|---|---|---|---|---|---|---|
| 4-Ethylphenol | 122.17 | 12.5 g | 0.102 | 42–45 | | | |
| NaOH | 40.0 | 4.0 g | 0.10 | | | | |
| Tetrabutylammonium bromide | 322.38 | 0.40 g | 0.0012 | 103–104 | | | |
| Methyl iodide | 141.94 | 14.5 g | 0.102 | | 41–43 | 2.28 | 1.5304 |
| Water | | 25 mL | | | | | |

## Reagents and Equipment

Use a 100-mL round-bottom flask containing a magnetic stirring bar and equipped with a reflux condenser. (■)

## Reaction Conditions

Heat the reaction mixture at reflux, with stirring, for 90 min.

## Isolation of Product

Allow the two-phase mixture to cool to room temperature. Transfer this mixture to a 125-mL separatory funnel, and remove the aqueous layer. Store this in a 125-mL Erlenmeyer flask until the final product has been purified and characterized. Wash the organic layer successively with 5% sodium hydroxide solution (20 mL) and distilled water (20 mL). After each washing remove the aqueous layer and add it to the 125-mL Erlenmeyer flask. Finally, collect the reddish-brown organic layer in another 125-mL Erlenmeyer flask, and dry it over anhydrous sodium sulfate.

Remove the drying agent by filtration through a glass wool plug and collect the product ether in a tared container. Weight, calculate the percent yield, and then purify and characterize a small amount of the material as described in the microscale procedure.

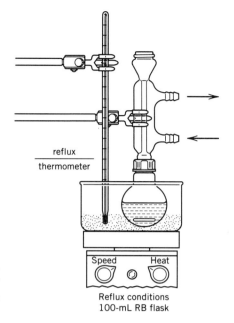

reflux
thermometer

Speed    Heat

Reflux conditions
100-mL RB flask

**QUESTIONS**

**6-144.** Sulfides are often prepared using an $S_N2$ reaction. For example,

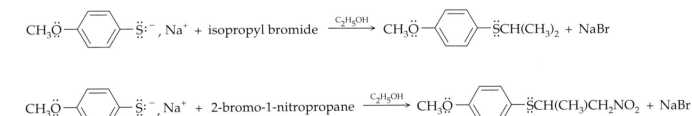

The reaction with isopropyl bromide is 16 times faster than the reaction with 2-bromo-1-nitropropane. Explain.

**6-145.** If 3-bromo-1-propanol is treated with NaOH, a compound of molecular formula $C_3H_6O$ is formed. Suggest a structure for this product.

**6-146.** Arrange the substituted phenols given below in order of increasing reactivity toward ethyl iodide in the Williamson reaction. Explain your order.

a. $CH_3$⟨⟩$\ddot{O}H$     b. $O_2N$⟨⟩$\ddot{O}H$     c. $CH_3\ddot{O}$⟨⟩$\ddot{O}H$

**6-147.** *trans*-2-Chlorocyclohexanol reacts readily with NaOH to form cyclohexene oxide, but the cis isomer will not undergo this reaction. Explain.

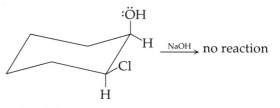

*cis*-2-Chlorocyclohexanol

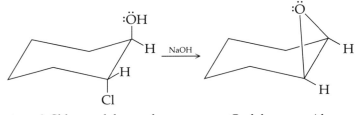

*trans*-2-Chlorocyclohexanol          Cyclohexene oxide

**6-148.** *tert*-Butyl ethyl ether might be prepared two ways using different starting materials.

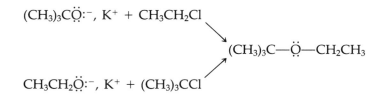

Which route would you choose to prepare the above ether, and why?

**6-149.** What product(s) would you expect to form when tetrahydrofuran is treated with excess hydroiodic acid (HI)?

**6-150.** Write a suitable mechanism for the cleavage of butyl isopropyl ether with HI at 100 °C to form exclusively isopropyl alcohol and 1-iodobutane. Explain why butyl alcohol and isopropyl iodide are not formed in the reaction.

**6-151.** There are only four lines in the aromatic region of the fully [1]H decoupled [13]C NMR spectrum of propyl *p*-tolyl ether (110–160 ppm), yet there are six aromatic carbon atoms. Explain.

---

**BIBLIOGRAPHY**

References selected from the large number of examples of the Williamson reaction in *Organic Syntheses* include:

Allen, C. F. H.; Gates, J. W., Jr. *Organic Syntheses;* Wiley: New York, 1955; Collect. Vol. III, p. 140. *ibid.,* p. 418.

Boehme, W. R. *Organic Syntheses;* Wiley: New York, 1963; Collect. Vol. IV, p. 590.

Fuson, R. C.; Wojcik, B. H. *Organic Syntheses;* Wiley: New York, 1943; Collect. Vol. II, p. 260.

Gassman, P. G.; Marshall, J. L. *Organic Syntheses;* Wiley: New York, 1973; Collect. Vol. V, p. 424.

Gokel, G. W.; Cram, D. J.; Liotta, C. L.; Harris, H. P.; Cook, F. L. *Organic Syntheses;* Wiley: New York, 1988; Collect. Vol. VI, p. 301.

Kuryla, W. C.; Hyve, J. E. *Organic Syntheses;* Wiley: New York, 1973; Collect. Vol. V, p 684.

Marvel, C. S.; Tanenbaum, A. L. *Organic Syntheses;* Wiley: New York, 1941; Collect. Vol. I, p. 435.

Mirrington, R. N.; Feutrill, G. I. *Organic Syntheses;* Wiley: New York, 1988; Collect. Vol. VI, p. 859.

Pedersen, C. J. *Organic Syntheses;* Wiley: New York, 1988; Collect. Vol. VI, p. 395.

Vyas, G. N.; Shah, N. M. *Organic Syntheses;* Wiley: New York, 1963; Collect. Vol. IV, p. 836.

Wheeler, T. S.; Willson, F. G. *Organic Syntheses;* Wiley: New York, 1941; Collect. Vol. I, p. 296.

Review articles on phase transfer catalysis:

Gokel, G. W.; Weber, W. P. *J. Chem. Educ.* **1978,** *55,* 350. *ibid.,* 429.

Jones, R. A. *Aldrichimica Acta* **1976,** *9,* 35.

Varughese, P. J. *J. Chem. Educ.* **1977,** *54,* 666.

The procedures used in these experiments for the preparation of the ethers were adapted from the work of:

McKillop, A.; Fiaud, J. C.; Hug, R. P. *Tetrahedron* **1974,** *30,* 1379.

Rowe, J. E. *J. Chem. Educ.* **1980,** *57,* 162.

## Amide Synthesis: Acetanilide and N,N'-Diacetyl-1,4-phenylenediamine

**EXPERIMENT [23]**

Common name: Acetanilide
CA number: [103-84-4]
CA name as indexed: Acetamide, *N*-phenyl-

Common name: *N,N'*-Diacetyl-1,4-phenylenediamine
CA number: [140-50-1]
CA name as indexed: Acetamide, *N,N'*-1,4-phenylenebis-

### Purpose

To carry out one of the major synthetic routes used in the preparation of amides; the method involves the reaction of ammonia, or a primary or secondary amine, with an active acylating reagent. To explore the use of acetic anhydride as an acylating agent. The acetanilide product (Experiment [23A]) may be used in Experiment [28].

### Prior Reading

Technique 5:   Crystallization
             Use of the Hirsch Funnel (see pp. 93–95).
             Craig Tube Crystallizations (see pp. 95–96).

**REACTION**

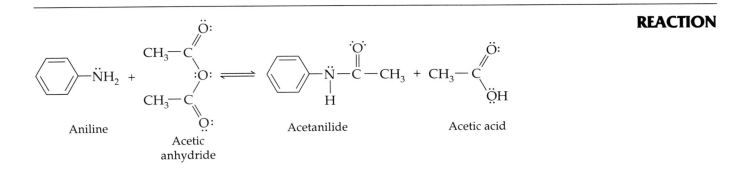

## DISCUSSION

A number of important chemical and biochemical synthetic sequences are initiated by the addition of a nitrogen nucleophile to a carbonyl carbon atom to yield **carboxylic amides**. Amides are classified as **primary (1°)**, **secondary (2°)**, or **tertiary (3°)** based on the number of carbon atoms attached to the nitrogen.

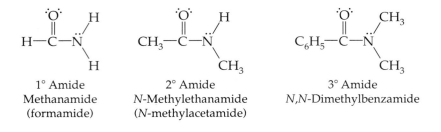

| 1° Amide | 2° Amide | 3° Amide |
|---|---|---|
| Methanamide | N-Methylethanamide | N,N-Dimethylbenzamide |
| (formamide) | (N-methylacetamide) | |

Cyclic amides are called **lactams** and are classified by ring size. Imides contain a nitrogen bonded to two carbonyl groups and are nitrogen analogues of anhydrides.

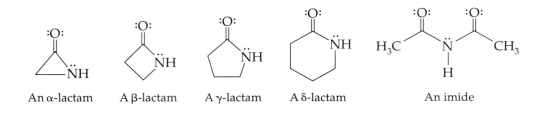

An α-lactam    A β-lactam    A γ-lactam    A δ-lactam    An imide

Amides appear in such diverse compounds as penicillin V (a β-lactam and an amide), polypeptides (α-amino acids linked by amide bonds); and an imide, 1,2-benzenecarboxylic imide, is used in the Gabriel synthesis of amines. An imide is prepared in Experiment [24] and anhydrides are prepared in Experiments [25A] and [25B]. The polyamide polymer, nylon, is prepared in Chapter 8, Sequence B.

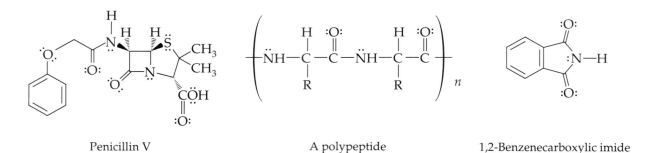

Penicillin V                A polypeptide                1,2-Benzenecarboxylic imide

The experiments outlined here illustrate the preparation of 2° amides. The process involves the attack of a primary amine on the acetyl group of acetic anhydride. Ammonia or secondary amines also react readily with this reagent to yield 1° and 3° amides, respectively. The mechanism shown here is an example of the attack of a nucleophilic reagent on a carbonyl carbon of the anhydride.

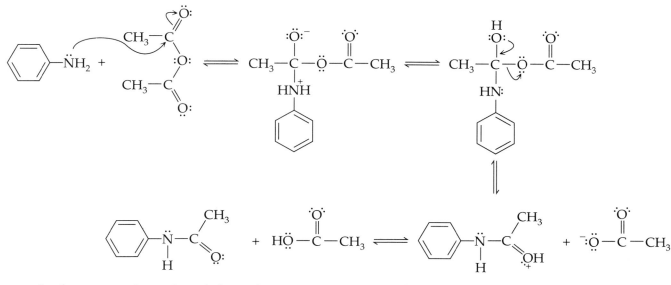

In the preparations given below, the amine reagents (aniline and *p*-phenylenediamine) are purified as their hydrochloride salts. Arylamines are relatively weak bases ($K_b$ values in the order of $10^{-10}$) but when treated with a strong acid, such as HCl, they are completely protonated, yielding the corresponding water-soluble hydrochloride salt.

$$C_6H_5—NH_2 + HCl \rightarrow C_6H_5—NH_3^+, Cl^-$$

Aniline                Anilinium
hydrochloride salt
(water soluble)

As directed in the experiment, decolorizing charcoal is added to the aqueous solution of the arylamine salt. The charcoal absorbs impurities and subsequent filtration of the mixture, which removes the charcoal, yields an aqueous solution of the purified arylamine salt.

The second stage of the reaction sequence requires that a solution of sodium acetate be added to the reaction mixture after initial addition of acetic anhydride to the purified anilinium hydrochloride salt solution.

$$C_6H_5—NH_3^+, Cl^- + CH_3COO^-, Na^+ \rightleftharpoons C_6H_5—NH_2 + CH_3COOH + Na^+, Cl^-$$

Addition of the sodium acetate solution serves to liberate the arylamine so that the desired nucleophilic substitution reaction may occur; ammonium cations are not nucleophilic as they are positively charged and do not even possess a lone pair of electrons.

Sodium acetate is the conjugate base of acetic acid, which is a weak acid. Furthermore, the anilinium ion ($pK_a = 4.6$) is a slightly stronger acid than acetic acid ($pK_a = 4.8$). Thus, the equilibrium reaction is shifted to the right, producing the arylamine.

In Experiment [23B], the *p*-phenylenediamine forms the corresponding dihydrochloride salt, $Cl^- \bullet H_3N^{\pm}\!\!-\!C_6H_4\!\!-^{\pm}\!NH_3 \bullet Cl^-$. As in Experiment [23A], the aqueous solution of the salt is purified with charcoal, and upon addition of the sodium acetate solution, the free amine is regenerated.

This overall process illustrates an important transformation for most amines. These amines can be converted to water-soluble ionic salts by reaction with acids and can be recovered from these acid salts by treatment with base. This technique was used in Experiment [4C] to extract ethyl 4-aminobenzoate, as its water-soluble salt, from a mixture.

## Experiment [23A]   Acetanilide

The reaction is shown above.

---

**EXPERIMENTAL PROCEDURE**    Estimated time to complete the experiment: 1.5 hours.

### Physical Properties of Reactants

| Compound | MW | Amount | mmol | mp(°C) | bp(°C) | d | $n_D$ |
|---|---|---|---|---|---|---|---|
| Aniline | 93.13 | 100 μL | 1.09 | | 184 | 1.02 | 1.5863 |
| Concd HCl | | 3 drops | | | | | |
| Sodium acetate trihydrate | 136.08 | 150 mg | 1.10 | 58 | | | |
| Acetic anhydride | 102.09 | 150 μL | 1.59 | | 140 | 1.08 | 1.3901 |

### Reagents and Equipment

HOOD

Place 100 μL of aniline in a tared 10 × 75-mm test tube (standing in a small beaker or Erlenmeyer flask) **[HOOD]**. Fit the tube with a cork stopper.

---

**WARNING:   Aniline is a toxic material and a cancer suspect agent.**

---

**NOTE.**   *Dispense the aniline using an automatic delivery pipet. Again weigh the test tube and container to determine the exact amount of aniline delivered.*

HOOD

Now using a 1.0-mL graduated pipet add, with swirling, 0.5 mL of water followed by 3 drops of concentrated hydrochloric acid **[HOOD]** (Pasteur pipet). Add 10 mg of powdered decolorizing charcoal, or the pelletized form (Norit™), to the resulting solution.

Transfer the well-mixed suspension (Pasteur pipet) to a 25-mm funnel fitted with fast-grade filter paper to remove the charcoal by gravity filtration. *Wet the filter paper in advance with distilled water and blot the excess water from the stem of the funnel.*

Collect the filtrate in a 3.0-mL conical vial. Use an additional 0.5 mL of water to rinse the test tube and the collected charcoal. Combine the rinse with the original filtrate. Place a magnetic spin vane in the vial and attach it to an air condenser. (■)

**NOTE.**   *If the pelletized form of charcoal is used, transfer through a Pasteur filter pipet directly to the 3.0-mL conical vial should be sufficient.*

**NOTE.**   *Tap all of the filtrate from the funnel stem into the collecting vial. As a result of this purification step, a clear, colorless solution of aniline hydrochloride should be obtained.*

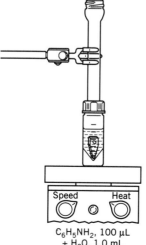

$C_6H_5NH_2$, 100 μL
+ $H_2O$, 1.0 mL
+ concd HCl, 3 drops
+ NaOAc·3$H_2O$, 150 mg
+ $(CH_3CO)_2O$, 150 μL

Dissolve 150 mg (1.10 mmol) of sodium acetate trihydrate in 0.5 mL of distilled water in a 10 × 75-mm test tube. Cap the tube and set the solution aside for use in the next step.

Remove the air condenser, and then use an automatic delivery pipet to add [**HOOD**], with stirring, 150 µL of acetic anhydride to the solution of     HOOD aniline hydrochloride, followed quickly by addition (Pasteur pipet) of the previously prepared solution of sodium acetate. Reattach the air condenser.

## Reaction Conditions

The reaction is very rapid and the product begins to precipitate immediately upon mixing of the reagents. Stir to thoroughly mix the reagents. Allow the reaction mixture to stand at room temperature for approximately 5 min, and then place it in an ice bath for an additional 5–10 min to complete the crystallization process.

## Isolation of Product

Collect the acetanilide product by filtration under reduced pressure using a Hirsch funnel. (■) Rinse the conical vial with two 0.5-mL portions of water (calibrated Pasteur pipet) and use the rinse to wash the collected filter cake. Place a piece of plastic food wrap over the mouth of the funnel and continue the suction for 5–8 min (see Prior Reading). The snow-white crystals are further dried on a porous clay plate or on filter paper in a desiccator.

Acetanilide
collected here

H₂O, ~ 2 mL + water-soluble
reaction products

## Purification and Characterization

Further purification of the product is generally not required. However, the acetanilide may be recrystallized from hot water or from ethanol–water using the Craig tube.

**NOTE.** *Acetanilide (150 mg) can be recrystallized from approximately 3 mL of water, or from 2 mL of ethanol–water (1 : 10 v/v), with better than 80% recovery.*

Weigh the dried crystals and calculate the percent yield. Determine the melting point of the material and compare your result to that reported in the literature.

Obtain an IR and/or NMR spectrum of the product and compare it with that of an authentic sample or to that recorded in *The Aldrich Library of IR Spectra* and/or *The Aldrich Library of NMR Spectra*, respectively.

## Chemical Tests

Characterization of the product may be enhanced by performing several chemical tests given in Chapter 10.

Check the solubility of acetanilide in water. Is the aqueous solution acidic, basic, or does it remain neutral as indicated by pH paper? Does the ignition test indicate that an aromatic group is present? Does the soda lime or sodium fusion test indicate the presence of nitrogen? Does the hydroxamate test for amides give a positive result?

## OPTIONAL SEMIMICROSCALE PREPARATION

This experiment can be scaled up to be run at five times the amounts used in the above microscale preparation.

The procedure is identical to the above with the following exceptions.

**1.** Use two 15 × 100-mm test tubes.
**2.** Carry out the reaction in a 10-mL round-bottom flask containing a magnetic spin bar and fitted with an air condenser. (■)
**3.** Increase the amounts of all reagents and the solvent by a factor of 5.

**Physical Properties of Reactants**

| Compound | MW | Amount | mmol | mp(°C) | bp(°C) | d | $n_D$ |
|---|---|---|---|---|---|---|---|
| Aniline | 93.13 | 500 µL | 5.45 | | 184 | 1.02 | 1.5863 |
| Concd HCl | | 0.75 mL | | | | | |
| Sodium acetate trihydrate | 136.08 | 750 mg | 5.50 | 58 | | | |
| Acetic anhydride | 102.09 | 750 µL | 7.93 | | 140 | 1.08 | 1.3901 |

Room temp thermometer

Speed   Heat

A 10-mL RB flask

## Experiment [23B]   N,N'-Diacetyl-1,4-phenylenediamine

## REACTION

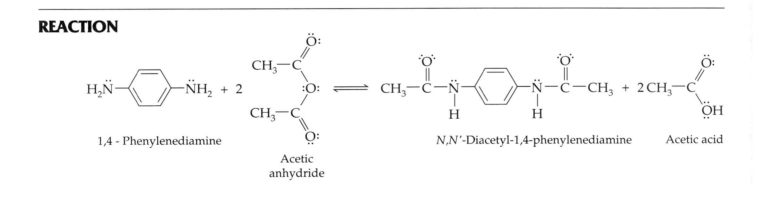

1,4 - Phenylenediamine · · · Acetic anhydride · · · N,N'-Diacetyl-1,4-phenylenediamine · · · Acetic acid

## EXPERIMENTAL PROCEDURE

Estimated time to complete the experiment: 1.5 hours.

**Physical Properties of Reactants**

| Compound | MW | Amount | mmol | mp(°C) | bp(°C) | d | $n_D$ |
|---|---|---|---|---|---|---|---|
| 1,4-Phenylenediamine | 108.14 | 117 mg | 1.08 | 138 | | | |
| Concd HCl | | 6 drops | | | | | |
| Sodium acetate trihydrate | 136.08 | 300 mg | 2.20 | 58 | | | |
| Acetic anhydride | 102.09 | 350 µL | 3.71 | | 140 | 1.08 | 1.3901 |

## Reagents and Equipment

Weigh and place 117 mg (1.08 mmol) of 1,4-phenylenediamine in a 10 × 75-mm test tube (standing in a small beaker or Erlenmeyer flask).

---

**CAUTION:   This reagent is toxic and a cancer suspect agent.**

---

With gentle swirling, add 1.0 mL of distilled water and 6 drops (Pasteur pipet) of concentrated hydrochloric acid. After dissolution, add 30 mg of either powdered decolorizing charcoal (Norit) or the pelletized form. Transfer the well-mixed suspension, by use of a Pasteur pipet, to a 25-mm funnel fitted with fast-grade filter paper previously wet with water. The charcoal is removed by gravity filtration. Collect the filtrate, which is clear to slightly yellow, in a 5-mL conical vial containing a magnetic spin vane. Use three 0.5-mL portions of water (calibrated Pasteur pipet) to rinse the test tube, and in turn use the rinse to wash the collected charcoal. The rinse is combined with the original filtrate. *Blot the excess water from the stem of the funnel.* Attach the vial to an air condenser. (■)

**NOTE.**   *If the pelletized form of charcoal is used, transfer through a Pasteur filter pipet directly to the 5.0-mL conical vial should be sufficient.*

Dissolve 300 mg (2.20 mmol) of sodium acetate trihydrate in 0.5 mL of distilled water in a 10 × 75-mm test tube. Stir the mixture with a spatula to aid the dissolution process. Cap the tube and set it aside for use in the next step.

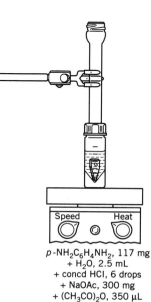

p-NH₂C₆H₄NH₂, 117 mg
+ H₂O, 2.5 mL
+ concd HCl, 6 drops
+ NaOAc, 300 mg
+ (CH₃CO)₂O, 350 μL

## Reaction Conditions

Remove the air condenser, and use an automatic delivery pipet to add 350 μL of acetic anhydride to the solution of 1,4-phenylenediamine dihydrochloride, and stir the mixture briefly using a magnetic stirrer.

**NOTE.**   *Dispense the acetic anhydride in the **hood** using an automatic delivery pipet. A slight amount of white precipitate may be observed at this stage.*

HOOD

Now add the previously prepared sodium acetate solution by Pasteur pipet to the reaction mixture with stirring. Reattach the air condenser.

## Isolation of Product

The reaction is very rapid and the desired product begins to precipitate almost immediately. After stirring briefly, allow the mixture to stand at room temperature for a few min and then place it in an ice bath for an additional 5–10 min.

## Purification and Characterization

Collect the crude N,N'-diacetyl-1,4-phenylenediamine by vacuum filtration using a Hirsch funnel. (■) Rinse the vial with two 0.5-mL portions of water and use the rinse to wash the filter cake. Place a piece of plastic food wrap over the mouth of the funnel and continue the suction for an additional 5–8 min. Place the collected material on a porous clay plate or filter paper to dry further.

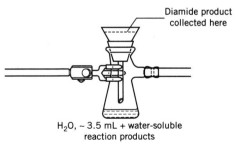

Diamide product
collected here

H₂O, ~ 3.5 mL + water-soluble
reaction products

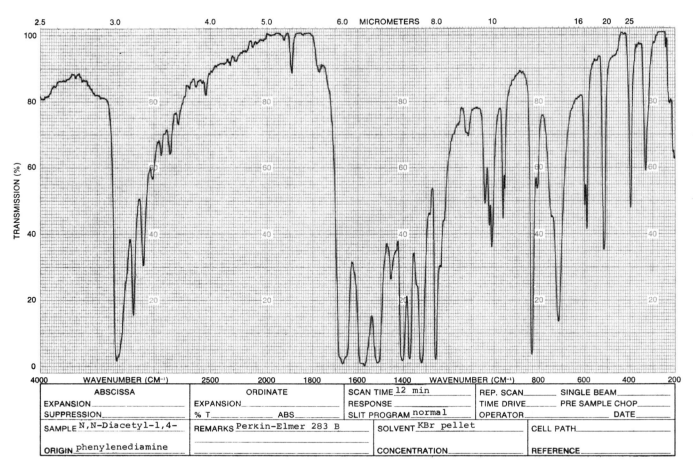

**FIGURE 6.42** IR spectrum: *N,N'*-Diacetyl-1,4-diphenylenediamine.

**NOTE.** *If this material is to be used in Experiment [29B], recrystallization from methanol is suggested.*

Weigh the dried crystals and calculate the percent yield. Determine the melting point and compare it with the value in the literature. Obtain an IR and/or the NMR spectrum of the product. The infrared spectrum is shown in Figure 6.42.

Chemical characterization tests might also be run as outlined in Experiment [23A].

**QUESTIONS**

**6-152.** What is the function of the sodium acetate in the reactions outlined in this experiment?

**6-153.** Which is the stronger base; aniline or cyclohexylamine? Explain.

**6-154.** Arrange the following substituted anilines in increasing order of reactivity toward acetic anhydride.

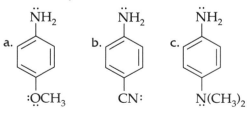

**6-155.** Suggest a mechanism for the preparation of acetic anhydride from acetic acid and acetyl chloride in the presence of pyridine (an amine base).

**6-156.** Anhydrides generally react more slowly with an amine than acid chlorides, though both reactions produce amides. Explain this observation.

**6-157.**

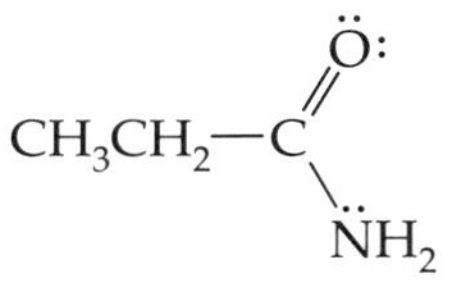

$$CH_3CH_2-\overset{\overset{\ddot{O}:}{\|}}{C}\underset{\ddot{N}H_2}{} \quad \text{is less basic than} \quad CH_3CH_2-\ddot{N}H_2$$

Explain.

**BIBLIOGRAPHY**

Review articles:

Beckwith, A. L. J. In *The Chemistry of the Amides*; J. Zabicky, Ed.; Wiley: New York, 1966; p. 86.

Satchell, D. P. N. *Q. Rev.* **1963**, *17*, 160.

Selected acylation reactions in *Organic Syntheses* between anhydrides and amines:

Cava, M. P.; Deana, A. A.; Muth, K.; Mitchell, M. J. *Organic Syntheses;* Wiley: New York, 1973; Collect. Vol. V, p. 944.

Fanta, P. E.; Tarbell, D. S. *Organic Syntheses;* Wiley: New York, 1955; Collect. Vol. III, p. 661.

Herbst, R. M.; Shemin, D. *Organic Syntheses;* Wiley: New York, 1943; Collect. Vol. II, p. 11.

Jacobs, T. L.; Winstein, S.; Linden, G. B.; Robson, J. H.; Levy, E. F.; Seymour, D. *Organic Syntheses;* Wiley: New York, 1955; Collect. Vol. III, p. 456.

Noyes, W. A.; Porter, P. K. *Organic Syntheses;* Wiley: New York, 1941; Collect. Vol. I, p. 457.

Wiley, R. H.; Borum, O. H. *Organic Syntheses;* Wiley: New York, 1963; Collect. Vol. IV, p. 5.

# Imide Synthesis: *N*-Phenylmaleimide

Common name: *N*-phenylmaleimide
CA number: [941-69-5]
CA name as indexed: 1*H*-Pyrrole-2,5-dione, 1-phenyl-

**Purpose**

To extend the amide synthesis (Experiment [23]) to the preparation of imides. In this experiment, the condensation of a cyclic anhydride with aniline to form an imide is described. The initial reaction to give the carboxylic amide is followed by an intramolecular condensation to produce the desired imide derivative.

*Prior Reading*

Technique 5: Crystallization
Use of the Hirsch Funnel (see pp. 93–95).
Craig Tube Crystallizations (see pp. 95–96).

**REACTION**

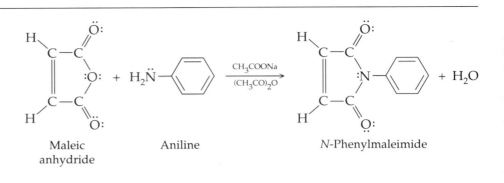

Maleic anhydride      Aniline      N-Phenylmaleimide

**DISCUSSION**

Imides are diacyl derivatives of ammonia or primary amines. The reaction is similar in its scope and mechanism to the acetylation of aniline or 1,4-phenylenediamine presented in Experiments [23A] and [23B]. As illustrated in the present experiment, cyclic anhydrides produce cyclic imides. Cyclic anhydrides are prepared in Experiments [25A] and [25B]. Derivatives of imides have been suggested for use in the treatment of arthritis, tuberculosis, and epilepsy. Several also have been found to be growth stimulants. Imide-based polymers are used in many applications, including fire-resistant woven fabrics. The N-phenylmaleimide prepared in this experiment is also a good dienophile in the Diels–Alder reaction (see Experiments [14] and [15] ), and in fact has been used as a reagent to characterize 1,3 dienes.

The first step in the reaction between the primary amine and the cyclic anhydride is an addition–elimination reaction, which involves a nucleophilic attack by the amine on a carbonyl carbon of the maleic anhydride. This results in the formation of an amide and a carboxylic acid, which are linked together to constitute maleanilic acid. Since anhydrides are considerably more reactive toward nucleophiles, in order to promote ring closure to the imide, the carboxylic acid is then converted to another (mixed) anhydride by reaction with acetic anhydride. This anhydride then undergoes an intramolecular nucleophilic addition–elimination by the amide nitrogen, which gives the desired imide, N-phenylmaleimide. The second acylation of a nitrogen nucleophile is much slower than the first. That is, attack of an amide nitrogen on the carbonyl carbon of the anhydride is slower than the attack of the amine nucleophile on the anhydride carbonyl carbon. The mechanistic sequence is given below (R = phenyl).

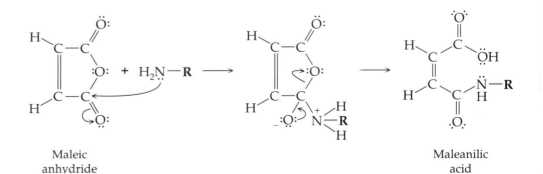

Maleic anhydride                                    Maleanilic acid

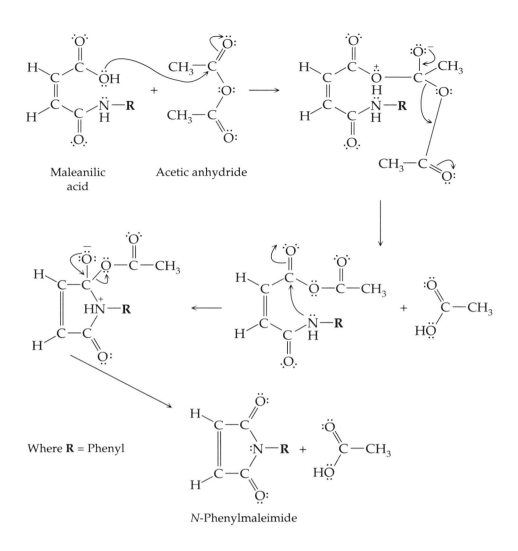

Maleanilic acid

Acetic anhydride

Where **R** = Phenyl

*N*-Phenylmaleimide

---

Estimated time to complete the experiment: 1.5 hours.

**EXPERIMENTAL PROCEDURE**

## Experiment [24A]   Maleanilic Acid

---

**REACTION**

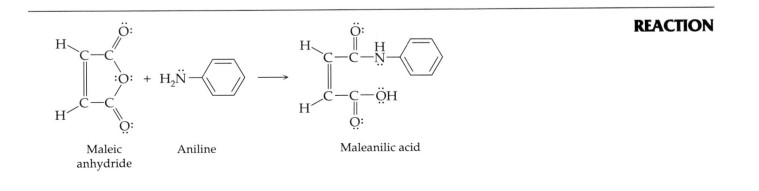

Maleic anhydride        Aniline        Maleanilic acid

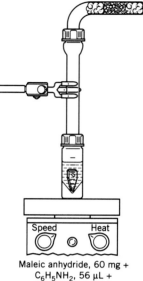

Maleic anhydride, 60 mg +
C₆H₅NH₂, 56 µL +
(CH₃CO)₂O, 1.2 mL

### Physical Properties of Reactants

| Compound | MW | Amount | mmol | mp(°C) | bp(°C) | d | $n_D$ |
|----------|-----|--------|------|--------|--------|-----|--------|
| Maleic anhydride | 98.06 | 60 mg | 0.61 | 60 | | | |
| Diethyl ether | 74.12 | 1.2 mL | | | 34.5 | | |
| Aniline | 93.13 | 56 µL | 0.62 | | 184 | 1.02 | 1.5863 |

## Reagents and Equipment

In a 3.0-mL conical vial containing a magnetic spin vane, and equipped with an air condenser protected with a drying tube, place 60 mg (0.61 mmol) of maleic anhydride and 1.0 mL of anhydrous diethyl ether. (■) Stir the mixture at room temperature until all the maleic anhydride has dissolved.

HOOD

> **WARNING:** Diethyl ether is highly flammable. All flames in the laboratory should be extinguished. Dispense this reagent in the *hood* using a calibrated Pasteur pipet.

In a separate, dry ½-dram vial prepare a solution of 56 µL (57 mg, 0.62 mmol) of aniline in 100 µL of anhydrous diethyl ether.

HOOD

> **WARNING:** Aniline is highly toxic and is a cancer suspect agent. It should be dispensed in the *hood* using an automatic delivery pipet.

Using a Pasteur pipet, add the aniline–ether solution in one portion to the stirred maleic anhydride–ether solution. Rinse the ½-dram vial with 100 µL of anhydrous ether, and also transfer this rinse to the reaction solution.

## Reaction Conditions

Stir the reaction mixture at room temperature for 15 min, and then cool it in an ice bath for 5–10 min.

## Isolation of Product

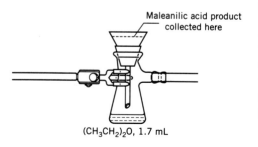

Maleanilic acid product collected here

(CH₃CH₂)₂O, 1.7 mL

Collect the deposit of fine, cream-colored powder by vacuum filtration using a Hirsch funnel. (■) Wash the maleanilic acid crystals with 0.5 mL of cold diethyl ether (calibrated Pasteur pipet), and air-dry them in the funnel for 5 min while maintaining the suction.

## Purification and Characterization

Weigh the maleanilic acid and calculate the percent yield. Determine the melting point and compare your value to that given in Cava et al. (Bibliography section). Obtain an IR spectrum using the KBr pellet technique and compare it with an authentic sample. The air-dried product is suitable for use in the next step without further purification.

## Experiment [24B]   *N*-Phenylmaleimide

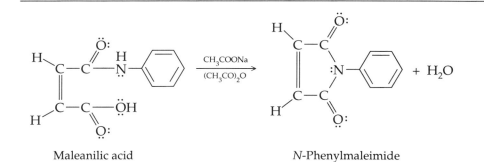

**REACTION**

Maleanilic acid                              *N*-Phenylmaleimide

**EXPERIMENTAL PROCEDURE**

Estimated time to complete the experiment: 1.5 hours.

### Physical Properties of Reactants

| Compound | MW | Amount | mmol | mp(°C) | bp(°C) | *d* | $n_D$ |
|---|---|---|---|---|---|---|---|
| Maleanilic acid | 191.18 | 100 mg | 0.52 | 201–202 | | | |
| Sodium acetate | 82.03 | 25 mg | 0.30 | 324 | | | |
| Acetic anhydride | 102.09 | 200 µL | 2.12 | | 140 | 1.08 | 1.3901 |

### Reagents and Equipment

In a 3.0-mL conical vial containing a magnetic spin vane, and equipped with an air condenser protected by a drying tube, place 25 mg (0.30 mmol) of anhydrous sodium acetate and 200 µL (216 mg, 2.12 mmol) of acetic anhydride. (■)

---

**CAUTION.**  Acetic anhydride is corrosive and a lachrymator. It should be dispensed in the *hood* by use of an automatic delivery pipet.

---

HOOD

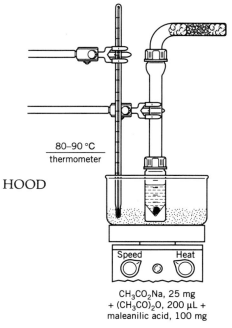

80–90 °C
thermometer

Speed        Heat

CH$_3$CO$_2$Na, 25 mg
+ (CH$_3$CO)$_2$O, 200 µL +
maleanilic acid, 100 mg

Now add 100 mg (0.52 mmol) of maleanilic acid (prepared in Experiment [24A]) to the reaction vial.

### Reaction Conditions

Heat the reaction mixture, with stirring, at a sand bath temperature of 80–90 °C for 30 min. Then cool the resulting mixture to room temperature, add 1.0 mL of cold water (calibrated Pasteur pipet), stir for a few min and then place the vial in an ice bath for 5–10 min to complete crystallization.

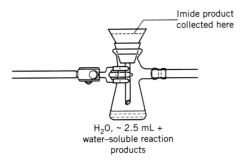

Imide product collected here

H₂O, ~ 2.5 mL + water–soluble reaction products

## Isolation of Product

Collect the solid product by vacuum filtration using a Hirsch funnel, and then wash the filter cake with three 0.5-mL portions of cold water (calibrated Pasteur pipet). (■) Cover the mouth of the funnel with plastic food wrap and continue the suction for an additional 5–10 min.

## Purification and Characterization

Recrystallize the crude N-phenylmaleimide from cyclohexane using the Craig tube, to yield canary-yellow needles. After drying the product on filter paper, or on a porous clay plate, weigh the crystals and calculate the percent yield, Determine the melting point and compare your result with the value given by Cava et al. (Bibliography section). Obtain an IR spectrum and compare it with that of an authentic sample or with that shown in *The Aldrich Library of IR Spectra*.

## QUESTIONS

**6-158.** As stated in the discussion section, the second step in the reaction to form the imide is much slower than that of the first stage (formation of the acid amide). Explain.

**6-159.** Phthalimide has a $K_a = 5 \times 10^{-9}$. Write an equation for the reaction of phthalimide with potassium amide (a strong base) in N,N-dimethyl formamide (DMF) solvent. Name the product.

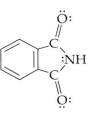

**6-160.** Predict which of the following species is the most acidic. Explain.

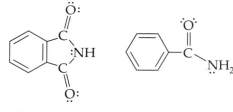

Phthalimide          Benzamide

**6-161.** The phthalimide anion is a strong nucleophile. It can react easily with primary alkyl halides to form substituted phthalimides.

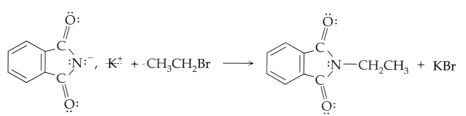

Suggest a suitable mechanism for this reaction.

**6-162.** *N*-Phenylmaleimide, the product prepared in Experiment [24B], can act as a dienophile in the Diels–Alder reaction (see Experiments [14] and [15]). Draw the structure of the product that would be formed by the treatment of *N*-phenylmaleimide with (a) 3-sulfolene under the conditions given in Experiment [14 ] and (b) furan.

---

**BIBLIOGRAPHY**

Review article on cyclic imides:

Hargreaves, M. K.; Pritchard, J. G.; Dave, H. R. *Chem. Rev.* **1970**, *70,* 439.

Selected imide preparations in *Organic Syntheses* include:

Cava, M. P.; Deana, A. A.; Muth, K.; Mitchell, M. J. *Organic Syntheses;* Wiley: New York, 1973; Collect. Vol. V, p. 944.

Noyes, W. A.; Porter, P. K. *Organic Syntheses;* Wiley: New York, 1941; Collect. Vol. I, p. 457.

Smith, L. I.; Emerson, O. H. *Organic Syntheses;* Wiley: New York, 1955; Collect. Vol. III, p. 151.

Soine, T. O.; Buchdahl, M. R. *Organic Syntheses;* Wiley: New York, 1963; Collect. Vol. IV, p. 106.

---

## Synthesis of Cyclic Carboxylic Acid Anhydrides: Succinic Anhydride and Phthalic Anhydride

**EXPERIMENT [25]**

Common name: Succinic anhydride
CA number: [108-30-5]
CA name as indexed: 2,5-Furandione, dihydro-

Common names: Phthalic anhydride, Benzene-1,2-dicarboxylic anhydride
CA number: [85-44-9]
CA name as indexed: 1,3-Isobenzofurandione

### Purpose
To carry out one of the important methods for preparing cyclic carboxylic acid anhydrides. The reaction demonstrates the use of acetic anhydride, an important industrial and research chemical, as a dehydrating agent.

### *Prior Reading*
    *Technique 5:*  Crystallization
                   Use of the Hirsch Funnel (see pp. 93–95).
                   Craig Tube Crystallizations (see pp. 95–96).
    *Technique 9:*  Sublimation (see pp. 116–117).

---

**REACTION**

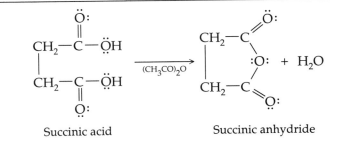

Succinic acid                 Succinic anhydride

## DISCUSSION

Five- and six-membered cyclic anhydrides can be easily formed when the corresponding dicarboxylic acid is heated in the presence of a dehydrating agent. One of the most commonly used dehydrating agents is acetic anhydride. The formation of an anhydride from its corresponding acid by reaction with another anhydride is referred to as *anhydride exchange*.

It is possible to prepare five- and six-membered cyclic anhydrides in the absence of acetic anhydride by direct dehydration at elevated temperatures. Maleic anhydride, for example, is easily obtained by this method in greater than 85% yield. Heating of succinic acid at 300 °C yields succinic anhydride in 95% yield.

The similarity of this reaction for the preparation of anhydrides to that for the synthesis of imides (Experiment [24]) should be noted. The ring closure to form the imide is mechanistically related to that of anhydride formation, in that in one case an amide nitrogen makes a nucleophilic attack on a carbonyl, while in the other, an acid oxygen acts as the nucleophile.

The mechanistic sequence for anhydride exchange follows:

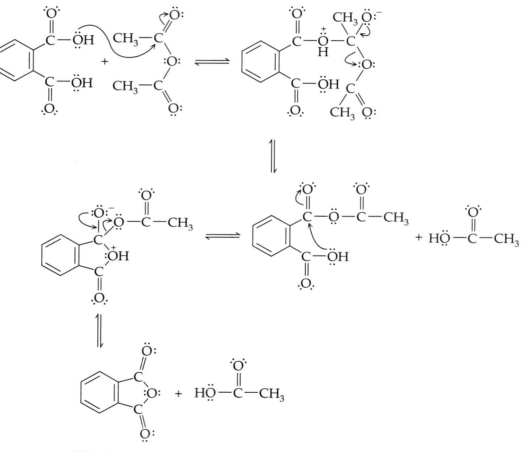

Phthalic anhydride

The equilibrium above is driven toward the products since formation of three molecules (two molecules of acetic acid and one molecule of anhydride) is entropically favored over the two molecules of reactants. The equilibrium could be further driven towards the products by distilling off the more-volatile acetic acid as it is formed.

Acetic anhydride is an important industrial reagent. Over one-half its annual production of approximately 750,000 tons is used for the manufacture of cellulose acetate. Cellulose acetate is a widely used textile fiber and is the chief component of cigarette filters. Acetic anhydride is the acetylation reagent used for the production of aspirin (acetylsalicylic acid). Succinic anhydride finds use in the succinylation of gelatin used as a blood plasma substitute, as a food preservative in chicken against *Salmonella,* and as a dog food preservative. Phthalic anhydride finds extensive use in plasticizer formulations for many resins, and in the manufacture of dyes.

## Experiment [25A]   Succinic Anhydride

The reaction is shown on p. 353.

Estimated time to complete the experiment: 1.5 hours.

### **EXPERIMENTAL PROCEDURE**

### Physical Properties of Reactants

| Compound | MW | Amount | mmol | mp(°C) | bp(°C) | d | $n_D$ |
|---|---|---|---|---|---|---|---|
| Succinic acid | 118.09 | 150 mg | 1.27 | 188 | | | |
| Acetic anhydride | 102.09 | 200 μL | 2.12 | | 140 | 1.08 | 1.3901 |

### Reagents and Equipment

**NOTE.**   *All equipment must be dried in an oven (110 °C) for 30 min before use.*

Weigh and place 150 mg (1.3 mmol) of succinic acid in a 1.0-mL conical vial containing a magnetic spin vane. Add 200 μL (2.12 mmol) of acetic anhydride, and then attach the vial to a reflux condenser protected by a calcium chloride drying tube. (■)

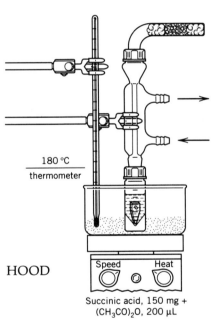

180 °C thermometer

---

**CAUTION:   Acetic anhydride is moisture sensitive and an irritant. Dispense it in the *hood* using an automatic delivery pipet.**

HOOD

Succinic acid, 150 mg + (CH₃CO)₂O, 200 μL

### Reaction Conditions

Heat the reaction mixture, with stirring, in a sand bath at a temperature of 180 °C for 45 min, timing the reaction from the point at which the succinic acid is completely dissolved.

### Isolation of Product

Cool the mixture to room temperature. A voluminous precipitate of succinic anhydride deposits. Further cool the vial in an ice bath for 5 min, and collect the solid material by vacuum filtration using a Hirsch funnel. (■) Wash the white needles with three 0.5-mL portions of diethyl ether (calibrated Pasteur filter pipet) and then place them on a porous clay plate or filter paper to dry.

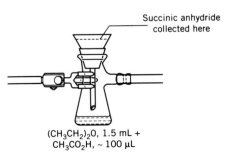

Succinic anhydride collected here

(CH₃CH₂)₂O, 1.5 mL + CH₃CO₂H, ~ 100 μL

## Purification and Characterization

The succinic anhydride crystals should be sufficiently pure for characterization. These crystals may, however, be recrystallized from absolute ethanol using a Craig tube.

Weigh the product and calculate the percent yield. Determine the melting point and obtain an infrared spectrum using the KBr pellet technique. Compare your results to those listed in the literature and *The Aldrich Library of IR Spectra*. What characteristic absorptions do you observe for the anhydride group in the carbonyl region of the spectrum?

## Experiment [25B]    Phthalic Anhydride

**REACTION**

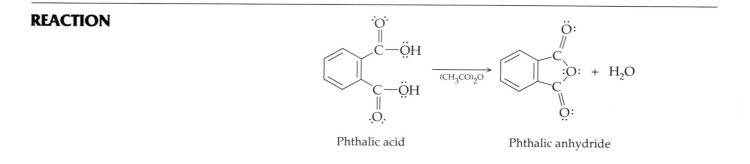

Phthalic acid          Phthalic anhydride

**EXPERIMENTAL PROCEDURE**

Estimated time to complete the experiment: 1.5 hours.

### Physical Properties of Reactants

| Compound | MW | Amount | mmol | mp(°C) | bp(°C) | d | $n_D$ |
|---|---|---|---|---|---|---|---|
| Phthalic acid | 166.14 | 100 mg | 0.60 | 210 | | | |
| Acetic anhydride | 102.09 | 200 μL | 2.12 | | 140 | 1.08 | 1.3901 |

### Reagents and Equipment

Weigh and add 100 mg (0.60 mmol) of phthalic acid to a 3.0-mL conical vial containing a magnetic spin vane. Add 200 μL (2.1 mmol) of acetic anhydride and then attach the vial to a reflux condenser protected by a calcium drying tube. (■)

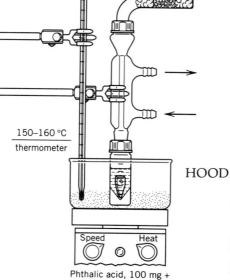

150–160 °C
thermometer

HOOD

Speed    Heat

Phthalic acid, 100 mg +
(CH₃CO)₂O, 200 μL

---

**CAUTION:** Acetic anhydride is moisture sensitive and an irritant. Dispense it in the *hood* using an automatic delivery pipet.

---

### Reaction Conditions

Heat the reaction solution, with stirring, at a sand bath temperature of 150–160 °C for 30 min.

**NOTE.** *Position the vial firmly on the bottom of the sand bath vessel to maintain this reaction temperature.*

## Isolation of Product

Cool the mixture to room temperature, whereupon the product crystallizes from solution. Cool the vial and contents in an ice bath for 10 min and collect the solid by vacuum filtration using a Hirsch funnel. (■) Rinse the filter cake carefully by dropwise addition of 0.5 mL of cold hexane (Pasteur pipet) and continue the suction for several min. Complete the drying of the solid product by placing the crystals on a porous clay plate or on filter paper.

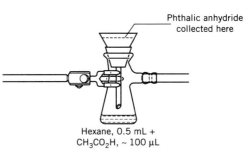

Phthalic anhydride collected here

Hexane, 0.5 mL + $CH_3CO_2H$, ~ 100 μL

## Purification and Characterization

The phthalic anhydride should be sufficiently pure for characterization. It may be purified further by sublimation, or by recrystallization from absolute ethanol using the Craig tube.

Weigh the product and calculate the percent yield. Determine the melting point and obtain an infrared spectrum using the KBr pellet technique. Compare your results with those reported in the literature and *The Aldrich Library of IR Spectra*. What characteristic absorptions do you observe for the anhydride group in the carbonyl region of the spectrum?

## QUESTIONS

**6-163.** As stated in the discussion, direct dehydration can be used as a method for the preparation of five- and six-membered cyclic anhydrides. Propose a suitable mechanism for the reaction below.

$$\text{maleic acid} \xrightarrow{\Delta} \text{maleic anhydride} + H_2O$$

**6-164.** Propose a suitable mechanism for the formation of the mixed anhydride obtained in the following reaction.

$$C_6H_5CH_2CO_2H + (CF_3CO)_2O \longrightarrow C_6H_5CH_2\overset{\overset{\displaystyle\cdot\cdot}{O}}{\overset{\|}{C}}-\overset{\cdot\cdot}{\underset{\cdot\cdot}{O}}-\overset{\overset{\displaystyle\cdot\cdot}{O}}{\overset{\|}{C}}CF_3 + CF_3CO_2H$$

**6-165.** There are two stereoisomeric 1,3-cyclobutane dicarboxylic acids. One can form a cyclic anhydride, the other cannot. Draw the structures of these compounds and indicate which one can be converted to a cyclic anhydride. Explain.

**6-166.** When maleic acid is heated to about 100 °C it forms maleic anhydride. However, fumaric acid requires a much higher temperature (250–300 °C) before it dehydrates. In addition, it forms only maleic anhydride. Explain.

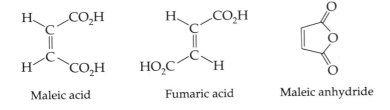

Maleic acid          Fumaric acid          Maleic anhydride

**6-167.** What product would you expect to obtain from reaction of one equivalent of propanol with phthalic anhydride?

## BIBLIOGRAPHY

Selected references from *Organic Syntheses* in which anhydrides are prepared, using acetic anhydride as the dehydrating agent, include:

Cason, J. *Organic Syntheses;* Wiley: New York, 1963; Collect. Vol. IV, p. 630.

Clarke, H. T.; Rahrs, E. J. *Organic Syntheses;* Wiley: New York, 1944; Collect. Vol. I, p. 91.

Grummitt, O.; Egan, R.; Buck, A. *Organic Syntheses;* Wiley: New York, 1955; Collect. Vol. III, p. 449.

Horning, E. C.; Finelli, A. F. *Organic Syntheses;* Wiley: New York, 1963; Collect. Vol. IV, p. 790.

Nicolet, B. H.; Bender, J. A. *Organic Syntheses;* Wiley: New York, 1944; Collect. Vol. I, p. 410.

Shriner, R. L.; Furrow, C. L. Jr. *Organic Syntheses;* Wiley: New York, 1963; Collect. Vol. IV, p. 242.

The synthesis of succinic anhydride is described in:

Fieser, L. F.; Martin, E. L. *Organic Syntheses;* Wiley: New York, 1943; Collect. Vol. II, p. 560.

## EXPERIMENT [26]

## Diazonium Coupling Reaction: Methyl Red

Common names: Methyl Red
CA number: [493-52-7]
CA name as indexed: Benzoic acid, 2-[[4-(dimethylamino)phenyl]azo]-

### Purpose

To learn the process of generating arenediazonium salts in solution. The arenediazonium salt generated will be used in an electrophilic aromatic substitution reaction (*diazo coupling*) to prepare an azobenzene derivative. Many azobenzene derivatives, including the one prepared here, have extensively conjugated $\pi$-electron systems. As these are highly colored compounds, they are generally referred to as *azo dyes*.

### *Prior Reading*

*Technique 5:*   Crystallization
Use of the Hirsch Funnel (see pp. 93–95).
Craig Tube Crystallization (see pp. 95–96).

## REACTION

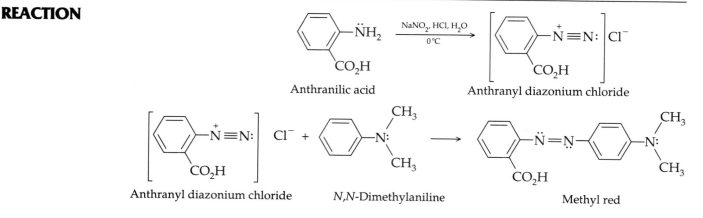

Anthranilic acid → Anthranyl diazonium chloride

Anthranyl diazonium chloride + N,N-Dimethylaniline → Methyl red

The coupling of a diazonium salt to a suitable aromatic substrate is an example of an aromatic electrophilic substitution reaction. When primary aromatic (and also aliphatic) amines ($ArNH_2$) are treated with nitrous acid ($NaNO_2 + HCl \rightarrow HONO$), they are converted into diazonium cations, $ArN_2^+$. In solution, nitrous acid (HONO) is in equilibrium with its anhydride, dinitrogen trioxide ($N_2O_3$), which is the actual diazotizing agent. The primary amine reacts with the dinitrogen trioxide to form a nitrosamine.

$$Ar-NH_2 + N_2O_3 \longrightarrow Ar-NH-N=\ddot{O} + HONO$$
(a nitrosamine)

The nitrosamine is in equilibrium with its tautomer, a diazoic acid. The diazoic acid then undergoes dehydration to form the diazonium salt. Diazonium salts are explosive when dry, and therefore are generally not isolated.

$$Ar-\ddot{N}H-\ddot{N}=\ddot{O}: + H_3O^{:+} \rightleftharpoons Ar-\ddot{N}=\ddot{N}-\ddot{O}H + H_3O^{:+}$$
(a nitrosamine)                    (a diazoic acid)

$$Ar-\ddot{N}=\ddot{N}-\ddot{O}H + H_3O^{:+} \longrightarrow Ar-\overset{+}{N}\equiv N: + 2 H_2\ddot{O}:$$
Diazonium ion

Reaction of the diazonium salt with various aromatic compounds leads to the formation of azo derivatives by what is generally called a "coupling reaction," but is mechanistically simply an ordinary electrophilic aromatic substitution reaction. The mechanism of the reaction is given here.

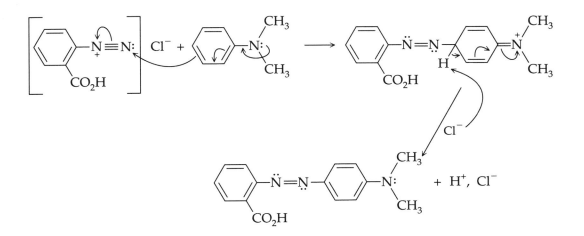

Azo dyes find use as acid–base indicators. For example, Methyl Red prepared in this experiment, Methyl Orange and Congo Red are well-known acid–base indicators. Azo dyes are commonly used in the textile, food and cosmetic industries; FD & C Yellow No. 6, a yellow azo dye is used to color candy, ice cream, beverages, and so on. Several azo dyes (including Butter Yellow and FD & C Red No. 2) have been banned by the FDA from use in foods, drugs, and cosmetics in the United States because of suspected carcinogenic properties.

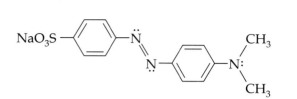

Methyl Orange

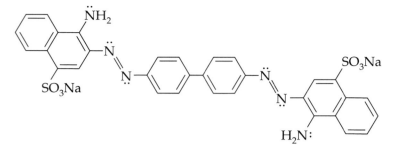

Congo Red

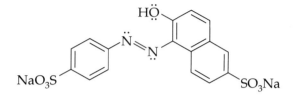

FD & C Yellow No. 6

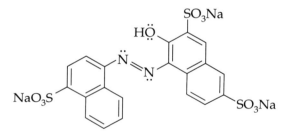

FD & C Red No. 2

Butter Yellow

---

**EXPERIMENTAL PROCEDURE**    Estimated time for the completion of the experiment: 3.0 hours. The reaction is shown on p. 358.

## Physical Properties of Reactants

| Compound | MW | Amount | mmol | mp(°C) | bp(°C) | d | $n_D$ |
|---|---|---|---|---|---|---|---|
| Anthranilic acid | 137.14 | 65 mg | 0.47 | 146–147 | | | |
| Concd HCl | | 150 μL | | | | | |
| Water | | 800 μL | | | | | |
| Sodium nitrite | 69.0 | 36 mg | 0.52 | 271 | | | |
| N,N-dimethylaniline | 121.18 | 89 μL | 0.71 | | 194 | 0.96 | 1.5582 |
| Sodium acetate | 82.03 | 68 mg | 0.83 | 324 | | | |
| 10% aq. NaOH | | 100 μL | | | | | |

CAUTION: When dry, benzenediazonium 2-carboxylate detonates violently upon being scraped or heated. It must, therefore, be kept in solution at all times.

## Reagents and Equipment

Equip a 3.0-mL conical vial with a magnetic spin vane and an air condenser. (■) Weigh and add 65 mg (0.48 mmol) of anthranilic acid to the vial. Now add a solution of 150 μL of concentrated hydrochloric acid dissolved in 400 μL of water to the vial, using a Pasteur pipet.

CAUTION: When preparing the acid solution, the acid must be added *to* the water. Dispense these reagents using automatic delivery pipets.

If necessary, warm the mixture, with stirring, on a hot plate magnetic stirrer to obtain a homogeneous solution. Cool the solution in an ice bath, with stirring, for 10 min.

In a 10 × 75-mm test tube, or a small vial, prepare a solution of 36 mg (0.52 mmol) of sodium nitrite dissolved in 200 μL of water. Cool this solution in an ice bath.

## Reaction Conditions

When both solutions in the ice bath are cooled to a temperature below 5 °C, slowly add (dropwise) the nitrite solution to the stirred anthranilic acid solution, while maintaining the temperature below 5 °C. This transfer is accomplished using a Pasteur pipet. *The solution must be kept cool so that the diazonium salt will not hydrolyze to the corresponding phenol.*

After a period of 4–5 min, check the clear solution of anthranyldiazonium chloride for the presence of excess nitrous acid by placing a drop of the solution on a piece of potassium iodide–starch test paper. If an excess is present, the test paper gives an immediate blue color. If no color is obtained, prepare additional nitrite solution and add as before until a positive test is observed.

Remove the air condenser from the reaction vial containing the solution of anthranyldiazonium chloride. Fairly rapidly, add 89 μL (85 mg, 0.71 mmol) of *N,N*-dimethylaniline (automatic delivery pipet). Reattach the air condenser.

WARNING. This aniline derivative is toxic and should be dispensed in the *hood.*

HOOD

Stir the solution for an additional 15 min, keeping the temperature below 5 °C.

Prepare a solution of 68 mg (0.83 mmol) of sodium acetate dissolved in 200 μL of water in a 10 × 75-mm test tube. Transfer this solution (Pasteur pipet) to the reaction mixture. Make this addition *without* removing the air condenser. Maintain the resulting solution at 5 °C, with stirring, for an additional 20 min.

< 5 °C
thermometer

Speed   Heat

**Addition 1:**
Anthranilic acid, 65 mg
+ concd HCl, 150 μL +
H$_2$O, 600 μL +
NaNO$_2$, 36 mg
**Addition 2:**
C$_6$H$_5$N(CH$_3$)$_2$, 85 mg
**Addition 3:**
CH$_3$CO$_2$Na, 68 mg +
H$_2$O, 200 μL

Remove the reaction vial from the ice bath and allow it to stand for 15 min in order to warm to ambient temperature.

Now add 100 $\mu$L of 10% aqueous NaOH solution (automatic delivery pipet) to the solution. Allow the reaction mixture to stand at room temperature for about 30 min. *The formation of the azo compound is a very slow reaction, but the rate of formation is increased by raising the pH of the solution.*

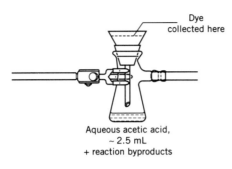

Dye collected here

Aqueous acetic acid,
~ 2.5 mL
+ reaction byproducts

## Isolation of Product

Collect the precipitate of crude Methyl Red dye by vacuum filtration using a Hirsch funnel. (■) Rinse the reaction flask with 0.5 mL of water and use this rinse to wash the crystals. Then wash the crystals with 0.5 mL of 3 $M$ acetic acid, to remove unreacted $N,N$-dimethylaniline from the product, followed by another wash with 0.5 mL of water. This last wash is usually pale pink in color.

**NOTE.**  *Dispense the small amounts of water and acetic acid using a calibrated Pasteur pipet.*

## Purification and Characterization

Dissolve the crude product in 500 $\mu$L of methanol. If necessary, warm the mixture in a beaker of hot water to aid in the dissolution. Cool the solution in an ice bath and collect the resulting crystals of Methyl Red by vacuum filtration using a Hirsch funnel. Dry the material on filter paper, or under vacuum at room temperature.

Weigh the product and calculate the percent yield. Determine the melting point and compare it to the value given in the literature. If further purification is desired, recrystallize the material from toluene using a Craig tube.

## QUESTIONS

**6-168.**  In the experiment, a point is made that the formation of the azo compound is a slow reaction, but that the rate is increased by raising the pH of the solution. Why is this necessary? In other words, how does the pH of the solution affect the reactivity of the $N,N$-dimethylaniline reagent?

**6-169.**  In relation to Question 6-168, diazonium salts couple with phenols in slightly alkaline solution. What effect does the pH of the solution have on the reactivity of the phenol?

**6-170.**  Starting with the appropriate aromatic amine and using any other organic or inorganic reagent, outline a synthetic sequence for the preparation of the following azo dyes.

a.

$$\text{Chrysoidine}$$

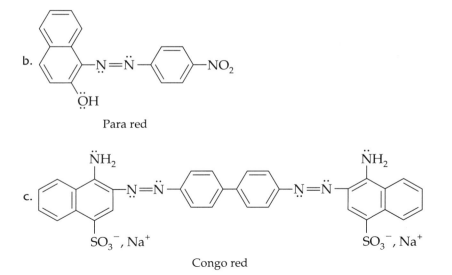

Para red

Congo red

**6-171.** What is the main structural feature of the azo dyes that causes them to be colored compounds?

**6-172.** Methyl Orange is an acid–base indicator. In dilute solution at pH > 4.4, it is yellow.

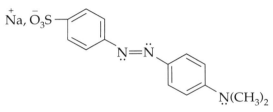

At pH = 3.2 the solution appears red. Draw a structure of the species that is formed at the lower pH if the acid proton adds to the azo nitrogen atom adjacent to the aromatic ring containing the —$SO_3^-$ group. Why does the proton add to this particular nitrogen when two other nitrogen atoms are available in the molecule?

---

**BIBLIOGRAPHY**

Selected coupling reactions with diazonium salts from *Organic Syntheses*:

Cleland, G. H. *Organic Syntheses*; Wiley: New York, 1988; Collect. Vol. VI, p. 21.

Conant, J. B.; Lutz, R. E.; Corson, B. B. *Organic Syntheses*; Wiley: New York, 1941; Collect. Vol. I, p. 49.

Fieser, L. F. *Organic Syntheses*; Wiley: New York, 1943; Collect. Vol. II, p. 35. ibid., p. 39.

Hartwell, J. L.; Fieser, L. F. *Organic Syntheses*; Wiley: New York, 1943; Collect. Vol. II, p. 145.

Santurri, P.; Robbins, F.; Stubbins, R. *Organic Syntheses*; Wiley: New York, 1973; Collect. Vol. V, p. 341.

The synthesis of Methyl Red is also given in *Organic Syntheses*:

Clarke, H. T.; Kirrer, W. R. *Organic Syntheses*; Wiley: New York, 1941; Collect. Vol. I, p. 374.

The present experiment is an adaptation of that given in:

Vogel, A. I. *A Textbook of Practical Organic Chemistry*, 4th ed.; Longmans: London, 1978, p. 716.

## EXPERIMENT [27]

# Friedel–Crafts Acylation: Acetylferrocene and Diacetylferrocene

Common name: Acetylferrocene
CA number: [1271-55-2]
CA name as indexed: Ferrocene, acetyl-

Common names: Diacetylferrocene, 1,1'-Diacetylferrocene
CA number: [1273-94-5]
CA name as indexed: Ferrocene, 1,1'-diacetyl-

## Purpose

To investigate the conditions under which the synthetically important Friedel–Crafts acylation (alkanoylation) reaction is carried out. The reaction described here illustrates electrophilic aromatic substitution on an aromatic ring contained in an organometallic compound. The highly colored products are easily separated by both thin-layer and dry-column chromatography.

### Prior Reading

Technique 4:   Solvent Extraction
                       Liquid–Liquid Extraction (see pp. 80–82).
                       Drying of the Wet Organic Layer (see pp. 86–87).
Technique 6:   Chromatography
                       Column Chromatography (see pp. 98–101).
                       Thin-Layer Chromatography (see pp. 103–104).
                       Concentration of Solutions (see p. 107).

## REACTION

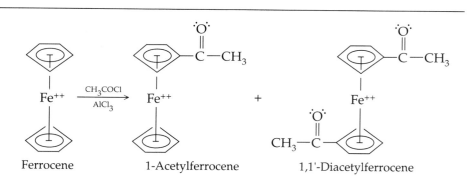

Ferrocene                    1-Acetylferrocene              1,1'-Diacetylferrocene

## DISCUSSION

The generation of the appropriate electrophile (carbocation, carbocation complex, or acylium ion) in the presence of an aromatic ring system (nucleophile) can lead to alkylation or acylation of the aromatic ring. This set of reactions, discovered by Charles Friedel and James Crafts in 1877, originally used aluminum chloride as the catalyst. The reaction is now known to be catalyzed by a wide range of Lewis acids, including ferric chloride, zinc chloride, boron trifluoride, and strong acids, such as sulfuric, phosphoric, and hydrofluoric acids.

**Alkylation** is accomplished by use of haloalkanes, alcohols, or alkenes; any species that can function as a carbocation precursor. The alkylation reaction is accompanied by two significant and limiting side reactions:

*polyalkylation*, due to ring activation by the added alkyl groups, and *rearrangement* of the intermediate carbocation. These lead to diminished yields, and mixtures of products that can be difficult to separate as shown here.

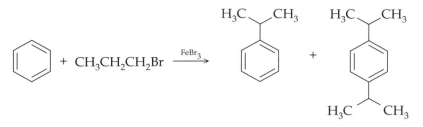

**Acylation** reactions generally do not suffer from these limitations, and can be conducted using acid chlorides or anhydrides as the electrophilic reagents. Since the introduction of a carbonyl group onto the aromatic ring in an acylation reaction deactivates the ring, the problem of multiple substitution is avoided. The acylium cation, since it is resonance stabilized, is unlikely to rearrange.

The mechanism involves three steps: (1) formation of a cationic electrophile; (2) nucleophilic attack on this electrophile by an aromatic ring; and (3) loss of a proton from the resulting cation to regenerate the aromatic ring system. The mechanism shown here represents the $AlCl_3$ catalyzed generation of the acylium ion electrophile from acetyl chloride (ethanoyl chloride), followed by subsequent nucleophilic attack by the ferrocene ring system.

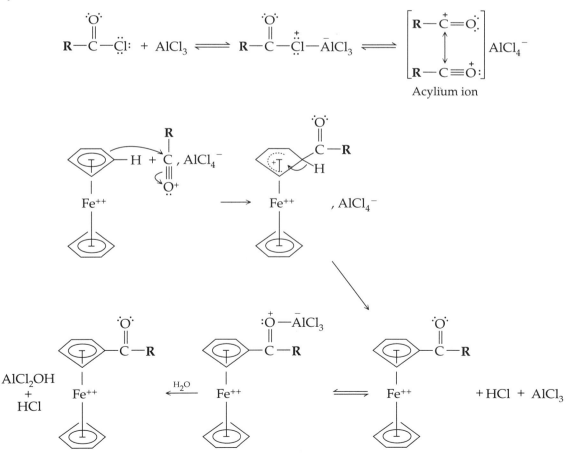

The present experiment also demonstrates the practical value of monitoring reaction progress by TLC analysis.

**Charles Friedel** (1832–1899, French). Professor of Chemistry at the Sorbonne. Friedel did extensive work on ketones, lactic acid, and glycerol and he discovered isopropyl alcohol. He is best known for his studies of the use of aluminum chloride in the synthesis of aromatic products (**Friedel–Crafts reaction**, 1877). Friedel prepared a series of esters of silicic acid and demonstrated the analogy between the compounds of carbon and silicon, meanwhile confirming the atomic weight of silicon. He determined the vapor densities and molecular weights of the chlorides of aluminum, iron and gallium.[21]

**James Mason Crafts** (1839–1917, American). Professor of Chemistry at Cornell University and later at Massachusetts Institute of Technology, where he eventually became President. Crafts studied with Bunsen (Germany) and Wurtz (France) and also worked on the organic compounds of silicon. Crafts was, of course, the codiscoverer of the **Friedel–Crafts reaction**. He also carried out investigations in the area of thermochemistry, catalytic effects in concentrated solutions, and determination of the densities of the halogens at high temperatures.[22]

## EXPERIMENTAL PROCEDURE

Estimated time to complete the experiment: two 3.0-hour laboratory periods.

### Physical Properties of Reactants

| Compound | MW | Amount | mmol | mp(°C) | bp(°C) | d | $n_D$ |
|---|---|---|---|---|---|---|---|
| Aluminum chloride | 133.34 | 150 mg | 1.12 | 190 | | | |
| Acetyl chloride | 78.50 | 80 μL | 1.12 | | 51 | 1.11 | 1.3898 |
| Ferrocene | 186.04 | 100 mg | 0.54 | 173 | | | |
| Methylene chloride | | 4.0 mL | | | 40 | | |

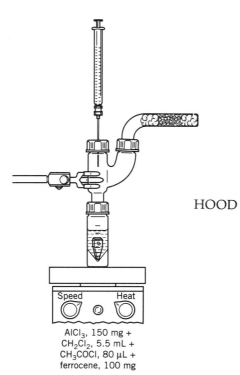

HOOD

AlCl₃, 150 mg +
CH₂Cl₂, 5.5 mL +
CH₃COCl, 80 μL +
ferrocene, 100 mg

### Reagents and Equipment

**NOTE.** *Dry the glassware in an oven at 110 °C for 30 min and allow it to cool in a desiccator before starting the experiment.*

Equip a tared 5.0-mL conical vial containing a magnetic spin vane with a Claisen head protected by a calcium chloride drying tube and a septum cap. Weigh and add 150 mg (1.12 mmol) of fresh, anhydrous aluminum chloride. (■)

Using a calibrated Pasteur pipet, add 2.5 mL of methylene chloride to the reaction vial. With swirling, add 80 μL (1.12 mmol) of acetyl chloride from an automatic delivery pipet [HOOD]. Use a syringe to add a solution of 100 mg (0.54 mmol) of ferrocene dissolved in 1.5 mL of methylene chloride to the resulting mixture.

[21] See *Berichte.* **1899,** *32,* 372; Crafts, J. M. *J. Chem. Soc.* **1900,** *77,* 993; *Bull. Soc. Chim. Fr.* **1900,** *23,* 1; Béhal, A. *ibid.,* **1932,** *51,* 1423; Willemart, A. *J. Chem. Educ.* **1949,** *26,* 3.

[22] See Ashdown, A. A. *J. Chem. Educ.* **1928,** *5,* 911; Talbot, H. P. *J. Am. Chem. Soc.* **1917,** *39,* 171; Richards, T. W. *Proc. Am. Acad.* **1917–1919,** *53,* 801; Cross, C. R. *J. Natl. Acad. Sci.* **1914,** *9,* 159.

**NOTE.**    *Use a capped vial and recap it between the addition of each reagent. After addition of the acetyl chloride, attach the vial to the Claisen head and add the ferrocene solution through the septum as shown in the figure. Do this in one or two portions depending on whether a 1- or 2-mL syringe is used.*

At this stage, the reaction mixture turns a deep-violet color.

---

WARNING:   It is important to minimize the exposure to moist air during these transfers. Both the aluminum chloride and the acetyl chloride are highly moisture sensitive so that rapid, yet accurate, manipulations are necessary to minimize deactivation of these reagents, and hence poor results. In addition, both chemicals are irritants. Avoid breathing the vapors or allowing the reagents to come in contact with skin. These reagents are dispensed in the *hood.*                                   HOOD

---

## TLC Sample Instructions

Obtain an aliquot for TLC analysis by removing a small amount of the reaction mixture by touching the open end of a Pasteur pipet to the surface of the solution. First, remove the cap from the straight neck of the Claisen head, and then insert the pipet down the neck so as to touch the surface of the solution. Dissolve this aliquot in about 10 drops of cold methylene chloride in a small capped vial. Mark the vial and *save* it for TLC analysis.

## Reaction Conditions

Following the addition of the ferrocene solution, note the time, and begin stirring. Allow the reaction to proceed at room temperature for 15 min.

## Isolation of Product

Quench the reaction by transferring the mixture by Pasteur pipet to a 15-mL capped centrifuge tube (or a 15-mL screw-capped vial) containing 5.0 mL of ice water. Cool the tube in an ice bath and neutralize the resulting solution by dropwise addition (calibrated Pasteur pipet) of about 0.5 mL of 25% aqueous sodium hydroxide.

**IMPORTANT.**    *Avoid an excess of base. Use litmus or pH paper to confirm the neutralization.*

Now extract the mixture with three 3-mL portions of methylene chloride. Cap the tube, shake, vent, and allow the layers to separate (a Vortex mixer may be used in this step). Remove the lower (methylene chloride) layer using a Pasteur filter pipet. Combine the methylene chloride extracts in a 25-mL Erlenmeyer flask, and dry the wet solution over about 200 mg of granular anhydrous sodium sulfate for 20 min. Transfer the dried solution to a tared 10-mL Erlenmeyer flask, using a Pasteur filter pipet, in aliquots of 4 mL each. After each transfer, concentrate the solution **[HOOD]** under     HOOD
a stream of dry nitrogen gas in a warm sand bath to a volume of about 0.5 mL. Rinse the drying agent with an additional 2.0 mL of methylene chloride and combine this rinse with the concentrate. Remove several drops of this solution by Pasteur pipet and place them in a capped vial

containing 10 drops of *cold* methylene chloride. Mark the vial and *save* it for
TLC analysis. Remove the remaining solvent by warming in a sand bath
HOOD   **[HOOD]** to yield the crude, solid product (~130 mg). Weigh the residue.

HOOD   **OPTION.**   *As an option, the combined extracts and washes may be left to evaporate* **[HOOD]** *in a 25-mL Erlenmeyer flask, with the mouth covered by filter paper,
until the following week.*

*If the reaction is performed over a 2-week period, this is a convenient point at
which to stop. However, if time permits, perform the TLC analysis now.*

### Thin-Layer Chromatographic Analysis

Use TLC to analyze the two samples saved above. Also analyze a standard
mixture of the substituted ferrocenes (supplied by the instructor) at the
same time. Use the developed TLC plates as a guide to determine the
product mixture obtained in the reaction, and as an aid in determining the
appropriate elution solvent required for separation of the mixture by dry-
column chromatography.

**INFORMATION.**   *Good results have been achieved by conducting the TLC analysis with Eastman Kodak silica gel–polyethylene terephthalate plates (#13179).
Activate the plates at an oven temperature of 100 °C for 30 min. Place them in a
desiccator for cooling and storing until used. Elute the plates using pure methylene
chloride as the elution solvent. Visualization of unreacted ferrocene can be enhanced
with iodine vapor. See Prior Reading for methods of TLC analysis and determination of $R_f$ values.*

### Purification and Characterization

Now purify the reaction products formed in the reaction by dry-column
chromatography. The term *dry-column chromatography* refers to the fact that
the column is packed with dry alumina, rather than with a slurry (see Prior
Reading). Dissolve the solid product residue isolated above in 0.5 mL
(calibrated Pasteur pipet) of methylene chloride in a small vial. Mix this
solution with 300 mg of alumina (activity III, see Glossary) in a tared vial,
HOOD   and evaporate the solvent under a stream of dry nitrogen **[HOOD]** to give
a product–alumina mixture. Assemble a chromatographic buret column in
the following order (bottom to top): prewashed cotton plug, 5 mm of sand,
60–80 mm of alumina (~5.0 g, activity III), *one-half* of the product–alumina
mixture, and 10 mm of alumina. (■)

**INFORMATION.**   *This procedure prevents overloading of the chromatographic
column during the separation of the reaction products. If the yield of crude reaction
products exceeds 75 mg (the usual case), introduce one-half of the alumina–product
mixture to the column. If the crude products, however, are obtained in quantities of
less than 75 mg, add the entire alumina–product mixture to the top of the column.*

*If only one-half of the alumina–crude ferrocene acylation product mixture is
placed on the column, it is important to reweigh the tared vial in order to establish a
reasonably accurate estimate of the overall yields obtained in the reaction.*

Given the polar nature of the alumina, the products will elute in order
of increasing polarity: ferrocene followed by acetylferrocene, followed by
diacetylferrocene. Begin elution of the column with pure hexane if TLC
analysis indicates that unreacted ferrocene is present in the product mixture. Be sure to add the initial solvent down the side of the column so as
not to disturb the alumina bed. During elution, the ferrocenes will separate

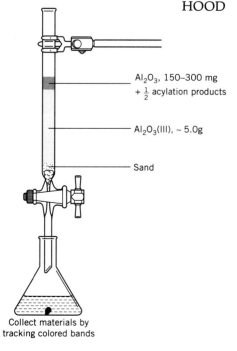

Al$_2$O$_3$, 150–300 mg
+ $\frac{1}{2}$ acylation products

Al$_2$O$_3$(III), ~ 5.0g

Sand

Collect materials by
tracking colored bands

into two or three bands of different colors on the column. The volume of each eluted fraction should be in the range of 2–5 mL if the band is carefully tracked down the column. Once the ferrocene band has been collected, continue the elution with a 1:1 mixture of $CH_2Cl_2$/hexane to obtain the monosubstituted product. Further elution with a 9:1 mixture of $CH_2Cl_2$/$CH_3OH$ will elute the disubstituted material. Collect and save each chromatographic band separately. Store the solvent that elutes without color, in an Erlenmeyer flask or beaker, until you have isolated all your product. Remove the solvent **[HOOD]** under a stream of dry nitrogen gas,    HOOD using a warm sand bath. During concentration of the solvent, spot each fraction on a TLC plate to verify the separation and purity of its contents.

Determine the melting point of each of the isolated products and compare your results to those reported in the literature.

Obtain an IR spectrum of each material and compare the results with an authentic sample or to those spectra given in *The Aldrich Library of IR Spectra.* Interpretation of the spectra allows an unambiguous determination of substitution based on the presence or absence of absorption in the 1100–900-wavenumber region of the spectrum.

*Characterization of the Fractions (Total Sample) Isolated from the Reaction Workup*

Acetylferrocene: mp _____ °C; _____ mg, _____ mmol
Diacetylferrocene: mp _____ °C; _____ mg, _____ mmol
**Total:** _____ mmol, _____% yield

**QUESTIONS**

**6-173.** In the formation of diacetylferrocene, the product is always the one in which each ring is monoacetylated. Why is no diacetylferrocene produced in which both acetyl groups are on the same aromatic ring?

**6-174.** Ferrocene cannot be nitrated using the conventional $HNO_3$–$H_2SO_4$ mixed acid conditions, even though nitration is an electrophilic aromatic substitution reaction. Explain.

**6-175.** In contrast to nitration (Question 6-174), ferrocene undergoes the acetylation and sulfonation reaction. Explain.

**6-176.** The bonding in ferrocene involves sharing of the 6 electrons from each cyclopentadienyl ring with the iron atom. Based on the electronic configuration of the iron species in the compound, show that a favorable 18-electron inert gas configuration is established at the iron atom.

**6-177.** In a manner similar to that in Question 6-176, predict whether ruthenocene and osmocene (the ferrocene analogues of ruthenium and osmium) would be stable compounds? Explain.

**6-178.** Would you predict that bis(benzene)chromium (0) would be a stable compound? Explain.

**6-179.** Predict the major product(s) in each of the following Friedel–Crafts reactions. Name each product. If the reaction does not occur, offer a reasonable explanation for that fact.

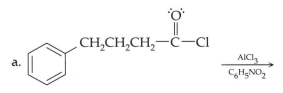

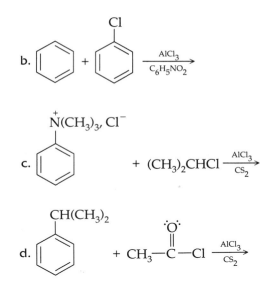

## BIBLIOGRAPHY

The acetylation of ferrocene has been monitored using chromatographic techniques. Several references are listed:

Bohen, J. M.; Joullié, M. M.; Kaplan, F. A. *J. Chem. Educ.* **1973,** *50,* 367.

Bozak, R. E. J. *J. Chem. Educ.* **1966,** *43,* 73.

Herz, J. E. *J. Chem. Educ.* **1966,** *43,* 599.

Several reviews on the Friedel–Crafts reaction, selected from many cited in the literature:

Berliner, E. *Org. React.* **1949,** *5,* 229.

Gore, P. H. *Chem. Rev.* **1955,** *55,* 229.

Johnson, W. S. *Org. React.* **1944,** *2,* 114.

Olah, G. A., Ed.; *Friedel–Crafts and Related Reactions*; Interscience: New York, 1963–1965; Vols. I–IV.

Several of the large number of Friedel–Crafts acetylation reactions are given in *Organic Syntheses:*

Adams, R.; Noller, C. R. *Organic Syntheses*; Wiley: New York, 1941; Collect. Vol. I, p. 109.

Arsenijivic, L; Arsenijivic, V.; Horeau, A.; Jacques, J. *Organic Syntheses*; Wiley: New York, 1988; Collect. Vol. VI, p. 34.

Fieser, L. F. *Organic Syntheses*; Wiley: New York, 1941; Collect. Vol. I, p. 517.

Fieser, L. F. *Organic Syntheses*; Wiley: New York, 1955; Collect. Vol. III, p. 6.

Lutz, R. E. *Organic Syntheses*; Wiley: New York, 1955; Collect. Vol. III, p. 248.

Marvel, C. S.; Sperry, W. M. *Organic Syntheses*; Wiley: New York, 1941; Collect. Vol. I, p. 95.

Sims, J. J.; Selman, L. H.; Cadogan, M. *Organic Syntheses*; Wiley: New York, 1988; Collect. Vol. VI, p. 744.

## EXPERIMENT [28]

## Halogenation: Electrophilic Aromatic Substitution to Yield 4-Bromoacetanilide

Common names: 4-Bromoacetanilide, *p*-bromoacetanilide
CA number: [103-88-8]
CA name as indexed: Acetamide, *N*-(4-bromophenyl)-

## Purpose

To extend our understanding of the experimental conditions under which electrophilic aromatic substitution reactions are carried out (also see Experiments [27] and [29A]–[29D]). This experiment deals with electrophilic aromatic halogenation. The directive influence of the acetamido, —NHCOCH₃, group on the bromination of acetanilide is explored.

### Prior Reading

Technique 5:   Crystallization
                Use of the Hirsch funnel (see pp. 93–95).
                Craig Tube Crystallization (see pp. 95–96).

---

**REACTION**

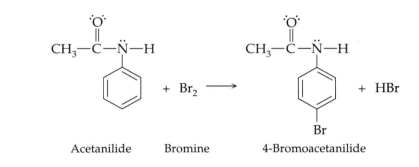

Acetanilide      Bromine      4-Bromoacetanilide

---

**DISCUSSION**

Aromatic compounds may be brominated by treatment with bromine in the presence of a Lewis acid catalyst, such as ferric chloride. For very electron-rich aromatic rings, such as arylamines, the reaction may proceed in the absence of a catalyst. With amines or phenols, in many cases, it is difficult to stop the bromination at monosubstitution, and all open ortho and para positions are brominated. For this reason, primary aromatic amines are often converted to a corresponding amide derivative, if a mono-brominated product is desired. This strategy is demonstrated in the present experiment. The —NHCOCH₃ group is a less powerful o,p directing group than —NH₂ due to the presence of the electron-withdrawing carbonyl group, which renders the ring less nucleophilic. Electrophilic substitution by bromine is still, however, effectively directed electronically to the ortho and para positions on the ring. The acetamido group, —NHCOCH₃, effectively blocks the ortho positions by steric hindrance. For these reasons, only para substitution is observed. The acetanilide used in this experiment may be prepared using the procedure described in Experiment [23A].

The mechanism of the bromination reaction is a classic illustration of an electrophilic substitution on an aromatic ring. The mechanism shown below is presented as proceeding without the aid of a catalyst.

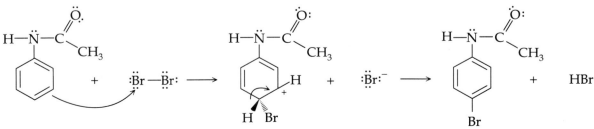

## EXPERIMENTAL PROCEDURE

Estimated time for completion of the experiment: 1.5 hours.

**Physical Properties of Reactants**

| Compound | MW | Amount | mmol | mp(°C) | bp(°C) |
|---|---|---|---|---|---|
| Acetanilide | 135.17 | 25 mg | 0.19 | 114 | |
| Glacial acetic acid | 60.05 | 4 drops | | | 118 |
| Bromine–acetic acid solution | | 3 drops | | | |

### Reagents and Equipment

*[handwritten margin note: use hood]*

Weigh and place 25 mg (0.19 mmol) of acetanilide in a 3.0-mL conical vial fitted with a cap. Add 4 drops of glacial acetic acid using a medicine dropper. Stirring with a glass rod may be necessary to help dissolve the acetanilide. Now add to the clear solution three drops of the bromine–acetic acid solution [**HOOD**]. Cap the vial immediately.

HOOD

*[handwritten margin note: – leave glassware under hood in beaker.]*

HOOD

> **WARNING:** Bromine is a severe irritant. It is suggested that plastic gloves be worn since bromine burns can be severe and require extended periods of time to heal. Dispense bromine only in the *hood*. Prepare the bromine–acetic acid reagent by mixing 2.5 mL of bromine with 5.0 mL of glacial acetic acid.

### Reaction Conditions

*[handwritten margin note: recrystalize in craig tube. + use centrifuge tube.]*

Allow the reddish-brown solution to stand at room temperature for 10 min with intermittent shaking. During this period, yellow-orange colored crystals precipitate from the solution.

*[handwritten margin note: Be careful of Br.]*

### Isolation of Product

*[handwritten note: count drops of 2ml/4 = 0.5ml.]*

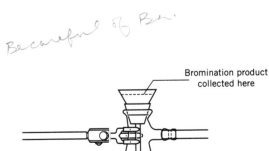

Bromination product collected here

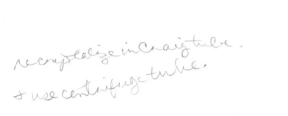

Aqueous acetic acid, ~2.0 mL and other reaction byproducts

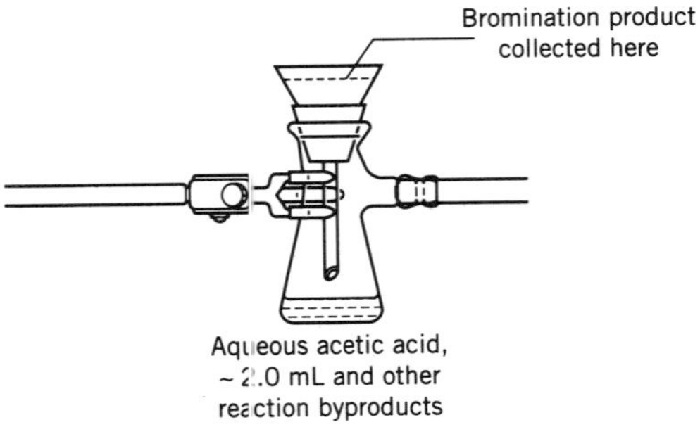

Add 0.5 mL of water (calibrated Pasteur pipet) to the reaction mixture with swirling, followed by 5 drops of aqueous sodium bisulfite solution (33%). This treatment destroys the unreacted bromine (and its residual color), and results in white crystals. Cool the reaction mixture in an ice bath for 10 min to maximize the yield of product.

Collect the white crystals of 4-bromoacetanilide by vacuum filtration using a Hirsch funnel. (■) Wash the filter cake with three 0.25-mL portions of cold water (calibrated Pasteur pipet) and partially dry by drawing air through the crystals under reduced pressure for approximately 5 min. A sheet of plastic food wrap over the funnel mouth aids this process (see Prior Reading).

### Purification and Characterization

*[handwritten margin note: weigh + do % yield + turn in product]*

Purify the crude 4-bromoacetanilide by recrystallization from 95% ethanol using the Craig tube. Weigh the dried product and calculate the percent yield. Determine the melting point and compare your result to the value given in the literature.

Obtain an IR spectrum of the material (KBr disk) and compare it with that recorded in *The Aldrich Library of IR Spectra*. If possible, obtain $^1$H and/or $^{13}$C NMR spectra of your material in DMSO-$d_6$.

## Nuclear Magnetic Resonance Analysis

Figures 6.43 and 6.44 are, respectively, the $^1$H and $^{13}$C NMR spectra of *p*-bromoacetanilide in DMSO-$d_6$. These can be used to compare with the NMR spectra you may obtain of your product.

In the $^{13}$C NMR spectrum, the DMSO-$d_6$ appears as a septet at 39.7 ppm. The resonance from the methyl group of the *p*-bromoacetanilide occurs at 24 ppm and the amide carbonyl carbon resonates at 169 ppm. The carbon atoms of the benzene ring are observed between 110 and 140 ppm.

In the $^1$H spectrum, the peak from trace amounts of DMSO-$d_5$ is seen at about 2.6 ppm. The peak at 3.4 ppm is probably due to water or another impurity in the sample. Note the two small peaks located equidistant to the tall singlet near 2.0 ppm. The small "satellite" peaks are the result of the

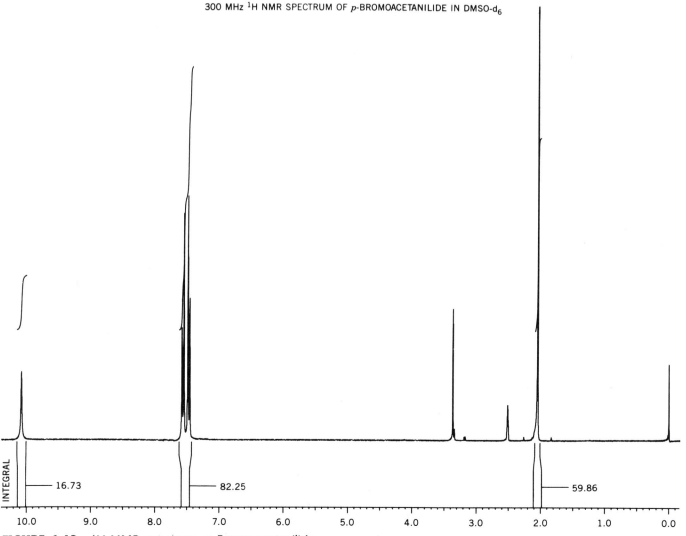

300 MHz $^1$H NMR SPECTRUM OF *p*-BROMOACETANILIDE IN DMSO-$d_6$

16.73      82.25      59.86

INTEGRAL

**FIGURE 6.43**   $^1$H-NMR spectrum: *p*-Bromoacetanilide.

75 MHz ¹³C NMR SPECTRUM OF *p*-BROMOACETANILIDE IN DMSO-d₆

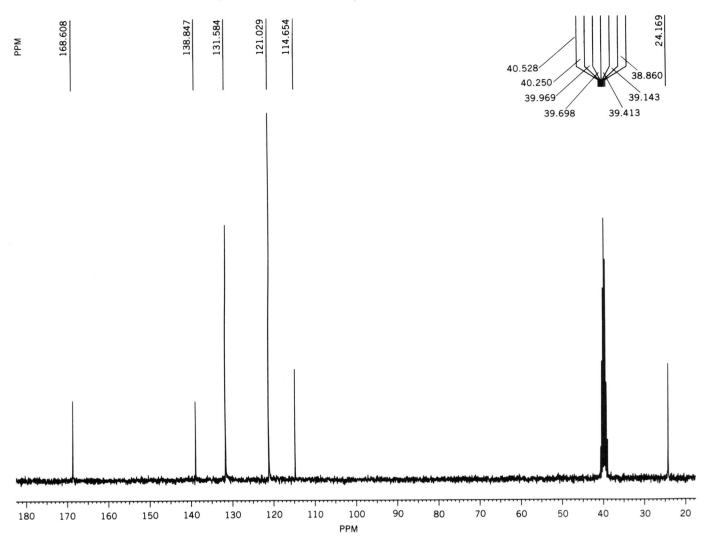

**FIGURE 6.44** ¹³C-NMR spectrum: *p*-Bromoacetanilide.

1.1% of the methyl groups that have ¹³C instead of ¹²C, and thus here coupling between the carbon and the protons is observed. The two doublets for the aromatic protons are observed near 7.5 ppm. The amide NH proton, which is probably hydrogen bonded to the basic (Lewis) sulfoxide functional group in DMSO-d₆, occurs rather downfield, near 10.1 ppm. This chemical shift may vary in your sample due to subtle differences in concentration, temperature, and moisture content of your DMSO-d₆.

## Chemical Tests

Chemical classification tests may also be performed on the amide product. The ignition and the Beilstein test (Chapter 10) are used to confirm the presence of the aromatic ring and the halogen group, respectively. Does the hydroxamate test for amides (Chapter 10) give a positive result?

**6-180.** Use resonance structures to show why the group shown is a less powerful ortho-para directing group than the —NH$_2$ group.

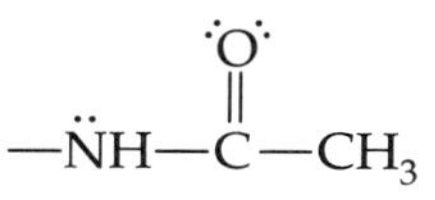

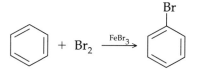

**6-181.** Benzene is brominated in the presence of FeBr$_3$ catalyst.

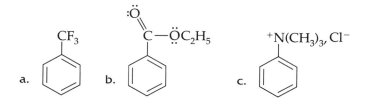

Suggest an appropriate mechanism for this reaction.

**6-182.** Draw the structure of the major monobrominated product(s) formed when each of the following compounds is reacted with Br$_2$ in the presence of FeBr$_3$.

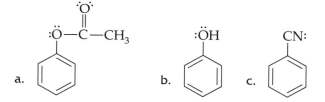

**6-183.** Arrange the following compounds in order of increasing reactivity toward electrophilic aromatic substitution. Explain the reason(s) for your decisions.

**6-184.** In the experiment, sodium bisulfite solution is added at the end in order to destroy the unreacted bromine. What reaction is occurring here? Is HSO$_3^-$ acting as an oxidizing or reducing agent? Write a balanced equation as part of your answer.

**6-185.** Both the $^1$H and $^{13}$C NMR spectra (Figs. 6.43 and 6.44) provide unambiguous evidence that the bromination of acetanilide gave exclusively para substitution. Explain.

The following references are selected from a large number of examples given in *Organic Syntheses* that illustrate electrophilic aromatic substitution using bromine:

**BIBLIOGRAPHY**

Adams, R.; Marvel, C. S. *Organic Syntheses*; Wiley: New York, 1941; Collect. Vol. I, p. 128.

Coleman, C. H.; Talbot, W. F. *Organic Syntheses*; Wiley: New York, 1943; Collect. Vol. II, p. 592.

Hartman, W. W.; Dickey, J. B. *Organic Syntheses*; Wiley: New York, 1943; Collect. Vol. II, p. 173.

Johnson, J. R.; Sandborn, L. T. *Organic Syntheses*; Wiley: New York, 1941; Collect. Vol. I, p. 111.

Langley, W. D. *Organic Syntheses*; Wiley: New York, 1941; Collect. Vol. I, p. 127.

Sandin, R. B.; McKee, R. A. *Organic Syntheses*; Wiley: New York, 1943; Collect. Vol. II, p. 100.

Smith, L. I. *Organic Syntheses*; Wiley: New York, 1943; Collect. Vol. II, p. 95.

Wilson, C. V. *Ibid.*, p. 575.

**EXPERIMENT [29]**

## Nitration: 2,5-Dichloronitrobenzene; *N,N'*-Diacetyl-2,3-dinitro-1,4-phenylenediamine; 5-Nitrosalicylic Acid; and 2- and 4-Nitrophenol

Common name: 2,5-Dichloronitrobenzene
CA number: [89-61-2]
CA name as indexed: Benzene, 1,4-dichloro-2-nitro-

Common name: *N,N'*-Diacetyl-2,3-dinitro-1,4-phenylenediamine
CA number: [7756-00-5]
CA name as indexed: Acetamide, *N,N'*-(2,3-dinitro-1,4-phenylene)bis-

Common names: 5-Nitrosalicylic acid, anilotic acid
CA number: [96-97-9]
CA name as indexed: Benzoic acid, 2-hydroxy-5-nitro-

Common names: 2-Nitrophenol, *o*-nitrophenol
CA number: [88-75-5]
CA name as indexed: Phenol, 2-nitro-

Common names: 4-Nitrophenol, *p*-nitrophenol
CA number: [100-02-7]
CA name as indexed: Phenol, 4-nitro-

### Purpose

Aromatic nitration is an important synthetic reaction. This experiment explores two experimental methods used for placing a nitro group on an aromatic ring system via an electrophilic aromatic substitution reaction. In Experiments [29A], [29B], and [29C] anhydrous nitric acid is used as the nitrating agent. In Experiment [29D], nitration is accomplished using a $SiO_2 \bullet HNO_3$ reagent.

### *Prior Reading*

*Technique 2:* Simple Distillation at the Semimicroscale Level (see pp. 65–68).
*Technique 5:* Crystallization
  Use of the Hirsch Funnel (see pp. 93–95).
  Craig Tube Crystallization (see pp. 95–96).
*Technique 6:* Chromatography
  Column Chromatography (see pp. 98–101).
  Thin-Layer Chromatography (see pp. 103–104).
  Concentration of Solutions (see pp. 106–108).
*Chapter 4:* Mixture Melting Points (see p. 53).

**GENERAL REACTION**

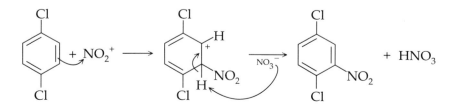

The nitration reactions described in this experiment all demonstrate one of
the classic electrophilic aromatic substitution reactions. Nitration has been
used extensively in organic synthesis since a nitro group on an aromatic
ring may be readily reduced to an amino group.

Once introduced onto the aromatic ring, the electron-withdrawing
nitro group deactivates the ring toward further reactions with electrophiles. For example, bromination of nitrobenzene leads only to *m*-nitro-
bromobenzene; no dibromonitrobenzene is readily formed. However,
when activating groups ($\pi$-electron donors) are present on the ring, it is
possible to nitrate the ring twice. This phenomenon can be illustrated by
comparing the results of the nitration of 1,4-dichlorobenzene (Experiment
[29A]) with that of *N,N'*-diacetyl-1,4-phenylenediamine (Experiment
[29B]). Because of the presence of the activating acetamido (CH$_3$CONH—)
groups, the dinitro derivative forms readily. In Experiment [29C] (the preparation of 5-nitrosalicylic acid), the directing influences of the 1-CO$_2$H and
2-OH substituents on the entering NO$_2$ group are illustrated. In this example, these two groups compliment each other since they both direct the
entering nitro group to the 5 position. The 5 position and the 3 position are
both electronically favored since the CO$_2$H group is meta directing; the OH
group is ortho–para directing. The nitro group ends up at the 5 position,
and not at the 3 position, due to steric effects.

The use of a silica gel based reagent to accomplish nitration under
fairly mild conditions is illustrated in Experiment [29D]. The nitrating reagent, SiO$_2$•HNO$_3$, is prepared by treatment of silica gel with nitric acid. In
the experiment, phenol is nitrated to produce a mixture of products. *Thin-
layer chromatography* is used to analyze the mixture, and the ortho and para
nitrated phenols are separated by *column chromatography* using a silica gel
column. If unreacted phenol is detected in the TLC analysis, an extraction
technique is used to separate it from the para isomer. This separation
technique is based on the fact that a nitrated phenol is more acidic than
phenol itself.

It is generally accepted that the nitronium ion (NO$_2^+$) is the electrophile that adds to the aromatic ring. The overall mechanism for nitration
follows:

$$HONO_2 + HONO_2 \rightleftharpoons H_2\overset{+}{O}-NO_2 + NO_3^-$$

$$H_2\overset{+}{O}-NO_2 + HONO_2 \rightleftharpoons H_3O^+ + NO_2^+ + NO_3^-$$

The above mechanism illustrates the reaction of two HNO$_3$ molecules reacting to generate the nitronium ion as when using the anhydrous nitric

**DISCUSSION**

acid reagent. Sulfuric acid is often used to enhance the production of $NO_2^+$, as shown here.

$$HONO_2 + HOSO_3H \rightleftharpoons H_2\overset{+}{O}-NO_2 + HSO_4^-$$

$$H_2\overset{+}{O}-NO_2 \rightleftharpoons H_2O + \overset{+}{N}O_2$$

Thus a commonly used nitrating reagent is a mixture of concentrated sulfuric and nitric acids.

---

## SEMIMICROSCALE PREPARATION OF ANHYDROUS NITRIC ACID

HOOD

Anhydrous nitric acid ($HNO_3$) is prepared by the following procedure.

---

**CAUTION:**  The reagents and the product of this preparation are highly corrosive. The distillation must be conducted in a *hood*. Appropriate gloves are strongly suggested. Prevent contact with eyes, skin, and clothing. Any spill should be neutralized using solid sodium carbonate or bicarbonate.

---

## EXPERIMENTAL PROCEDURE

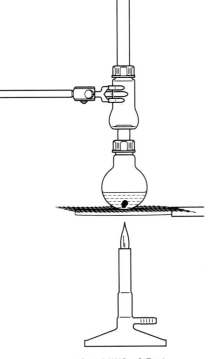

Concd HNO₃, 0.7 mL
+ concd H₂SO₄, 1.0 mL

Estimated time of preparation: 0.5 hour.

**IMPORTANT.**  *Use this anhydrous nitric acid immediately for the nitration experiments given below. The amount obtained at the scale used here is sufficient for the preparation of two of the nitro compounds described in this experiment.*

### Physical Properties of Reactants and Product

| Compound | MW | Amount | bp(°C) | d |
|---|---|---|---|---|
| Concd nitric acid (68%) | | 0.7 mL | 120.5 | 1.41 |
| Concd sulfuric acid (96–98%) | | 1.0 mL | 338 | 1.84 |
| Anhydrous nitric acid | 63.01 | | 83 | 1.40 |

### Reagents and Equipment

Using two clean, dry 1.0-mL graduated pipets, add 0.7 mL of concentrated nitric acid, followed by 1.0 mL of concentrated sulfuric acid, to a 10-mL round-bottom flask containing a boiling stone. Swirl the flask gently to mix the reagents. Attach the flask to a Hickman still fitted with an air condenser. (■)

---

**CAUTION:**  Sulfuric acid can cause severe burns. Nitric acid is a strong oxidizing agent. Prevent contact with eyes, skin, and clothing. A spill can be neutralized using sodium carbonate or bicarbonate.

## Reaction Conditions

Heat the acid solution very gently with a microburner, keeping the micro-burner in constant motion, until approximately 0.2 mL of anhydrous nitric acid has been collected as distillate in the collar of the still.

## Purification and Characterization

Use the anhydrous nitric acid as collected. No further purification is required.

**NOTE.** *Anhydrous nitric acid (white fuming nitric acid) is a colorless liquid, bp 83 °C. It is estimated that the nitric acid obtained in this preparation is at least 99.5–100% pure. If it is necessary to store the distillate, remove the acid from the collar of the still (Pasteur pipet) and place it in a 1.0-mL conical vial fitted with a glass stopper. It may be necessary to slightly bend the end of the pipet in a flame so that it can reach the collar of a still that does not have a side port. The anhydrous nitric acid is colorless or faintly yellow.*

## Experiment [29A]   2,5-Dichloronitrobenzene

**REACTION**

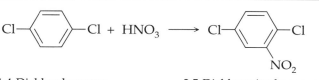

1,4-Dichlorobenzene          2,5-Dichloronitrobenzene

**EXPERIMENTAL PROCEDURE**

Estimated time to complete the experiment: 0.5 hour.

### Physical Properties of Reactants

| Compound | MW | Amount | mmol | mp(°C) | bp(°C) | d |
|---|---|---|---|---|---|---|
| 1,4-Dichlorobenzene | 147.01 | 38 mg | 0.26 | 53 | | |
| Anhydrous nitric acid | 63.01 | 100 μL | 2.4 | | 83 | 1.50 |

## Reagents and Equipment

Equip a 3.0-mL conical vial with an air condenser. (■) Weigh and add 38 mg (0.26 mmol) of 1,4-dichlorobenzene, followed by 100 μL of anhydrous nitric acid delivered from a calibrated Pasteur pipet (9 in.).

> **CAUTION:  The nitric acid reagent is highly corrosive. Prevent contact with eyes, skin, and clothing. A spill is neutralized using solid sodium carbonate or bicarbonate.**

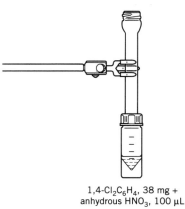

1,4-Cl$_2$C$_6$H$_4$, 38 mg + anhydrous HNO$_3$, 100 μL

## Reaction Conditions

Allow the resulting solution to stand at room temperature for a period of 15 min. Next add 1.0 mL of water (calibrated Pasteur pipet) dropwise, while stirring with a thin glass rod, and then place the vial in an ice bath to cool.

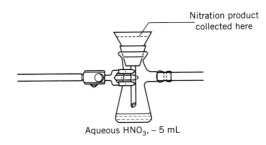

Nitration product collected here

Aqueous HNO₃, ~ 5 mL

## Isolation of Product

Collect the crystalline precipitate by vacuum filtration using a Hirsch funnel. (■) Wash the filter cake with four 1.0-mL portions of water (calibrated Pasteur pipet), and then place it on a porous clay plate or on filter paper to dry.

## Purification and Characterization

The product, consisting of fine, white needles, is sufficiently pure for characterization. It may be recrystallized from ethanol–water, using a Craig tube, if desired.

Weigh the 2,5-dichloronitrobenzene and calculate the percent yield. Determine the melting point and compare your result to that reported in the literature. Notice that the starting material and the nitrated product have very close melting points. It is recommended that a mixed melting point be carried out to establish that the desired product has been isolated (see Prior Reading).

A further or alternative check on the purity of the material may be made using thin-layer chromatography (TLC).

**INFORMATION.**  *Carry out the TLC analysis with Eastman Kodak silica gel–polyethylene terephthalate plates (#13179). Activate the plates at an oven temperature of 100 °C for 30 min. Place them in a desiccator for cooling and storing until used. Elute the plates using hexane solvent. Visualization is accomplished with UV light. See Prior Reading for the methods of TLC analysis and determination of $R_f$ values.*

## Chemical Tests

Additional chemical tests (Chapter 10) may also be performed to further characterize the product. Does the ignition test confirm the presence of the aromatic ring? Does the Beilstein test detect the presence of chlorine? Can the sodium fusion test detect the presence of nitrogen? Can the specific presence of the nitro group be detected by reaction with ferrous hydroxide solution?

## Experiment [29B]   *N,N′*-Diacetyl-2,3-dinitro-1,4-phenylenediamine

**REACTION**

$N,N'$-Diacetyl-1,4-phenylenediamine    +    HNO₃ ⟶    $N,N'$-Diacetyl-2,3-dinitro-1,4-phenylenediamine

Estimated time to complete the experiment: 0.5 hour.        **EXPERIMENTAL PROCEDURE**

### Physical Properties of Reactants and Product

| Compound | MW | Amount | mmol | mp(°C) | bp(°C) | d |
|---|---|---|---|---|---|---|
| N,N'-Diacetyl-1,4-phenylenediamine | 192 | 48 mg | 0.25 | 312–315 | | |
| Anhydrous nitric acid | 63.01 | 100 μL | 2.4 | | 83 | 1.50 |
| N,N'-Diacetyl-2,3-dinitro-1,4-phenylenediamine | 282 | | | 257 | | |

### Reagents and Equipment

To a 3.0-mL conical vial equipped with an air condenser, weigh and add 48 mg (0.25 mmol) of N,N'-diacetyl-1,4-phenylenediamine. Now add drop-wise **[CAUTION]** 100 μL of anhydrous nitric acid delivered from a calibrated Pasteur pipet (9 in.). (■) *The N,N'-diacetyl-1,4-phenylenediamine is prepared by the procedure outlined in Experiment [23B].*

CAUTION

EXOTHERMIC

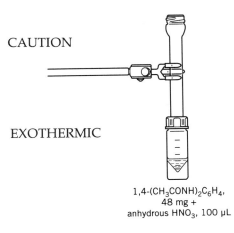

1,4-(CH₃CONH)₂C₆H₄,
48 mg +
anhydrous HNO₃, 100 μL

> CAUTION: The reaction is highly *EXOTHERMIC.* A vigorous reaction occurs if the acid is added too rapidly. The nitric acid reagent is highly corrosive; prevent contact with eyes, skin, and clothing. A spill is neutralized using sodium carbonate or bicarbonate.

### Reaction Conditions

Allow the resulting solution to stand at room temperature for a period of 10 min. Add 1.0 mL of water (calibrated Pasteur pipet) dropwise and then place the vial in an ice bath to cool.

### Isolation of Product

Collect the resulting yellow precipitate by vacuum filtration using a Hirsch funnel. (■) Wash the filter cake with four 1.0-mL portions of water (calibrated Pasteur pipet). Maintain suction to aid in drying of the product. A piece of plastic food wrap over the mouth of the funnel can aid this process (see Prior Reading). Dry the material further on a porous clay plate, on filter paper, or under vacuum at room temperature.

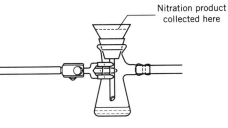

Nitration product
collected here

Aqueous HNO₃, ~ 5 mL

### Purification and Characterization

The product needs no further purification. Weigh the dried material and calculate the percent yield. Determine the melting point and compare your result with the melting point listed in the above Reactant Product table. Obtain an IR spectrum of your product using the KBr pellet technique and compare it with that shown in Figure 6.45.

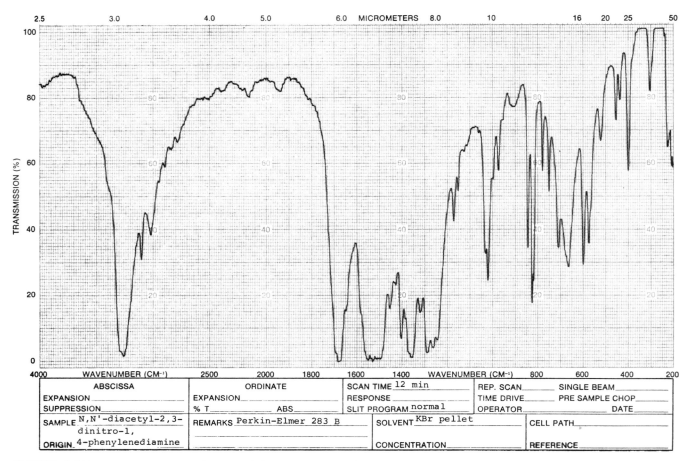

| ABSCISSA | | ORDINATE | | SCAN TIME 12 min | REP. SCAN_____ SINGLE BEAM_____ |
|---|---|---|---|---|---|
| EXPANSION_____ | | EXPANSION_____ | | RESPONSE_____ | TIME DRIVE_____ PRE SAMPLE CHOP_____ |
| SUPPRESSION_____ | | % T_____ ABS_____ | | SLIT PROGRAM normal | OPERATOR_____ DATE_____ |
| SAMPLE N,N'-diacetyl-2,3-dinitro-1, | | REMARKS Perkin-Elmer 283 B | | SOLVENT KBr pellet | CELL PATH_____ |
| ORIGIN 4-phenylenediamine | | | | CONCENTRATION_____ | REFERENCE_____ |

**FIGURE 6.45** IR spectrum: *N,N'*-Diacetyl-2,3-dinitro-1,4-phenylenediamine.

## Experiment [29C]   5-Nitrosalicylic Acid

*This material may be used to prepare the caffeine 5-nitrosalicylate derivative (see Experiment [11B]).*

**REACTION**

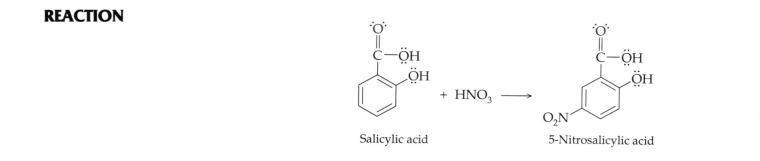

Salicylic acid          + HNO₃ ⟶          5-Nitrosalicylic acid

Estimated time to complete the experiment: 0.5 hour.

**EXPERIMENTAL PROCEDURE**

**Physical Properties of Reactants**

| Compound | MW | Amount | mmol | mp(°C) | bp(°C) | d |
|---|---|---|---|---|---|---|
| Salicylic acid | 138.12 | 50 mg | 0.36 | 159 | | |
| Anhydrous nitric acid | 63.01 | 100 μL | 2.4 | | 83 | 1.50 |

CAUTION:   This reaction should be conducted in a *hood*.

HOOD

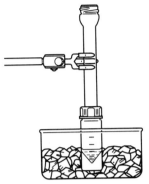

CAUTION

### Reagents and Equipment

Weigh and add 50 mg (0.36 mmol) of salicylic acid to a 3.0-mL conical vial equipped with an air condenser. Place the vial in an ice bath to cool. Also place a stoppered conical vial containing 100 μL of anhydrous nitric acid [CAUTION] (Pasteur pipet) in the ice bath. (■)

Salicylic acid, 50 mg +
anhydrous HNO₃, 100 μL

CAUTION:   The nitric acid reagent is highly corrosive; prevent contact with eyes, skin, and clothing. A spill is neutralized using solid sodium carbonate or bicarbonate.

### Reaction Conditions

Add the cold nitric acid dropwise from a calibrated Pasteur pipet (9 in.) [CAUTION] to the salicylic acid. Keep the vial containing the salicylic acid in the ice bath during the addition. The evolution of a red-brown gas ($NO_2$, [WARNING: TOXIC]) is observed during the addition, if the acid is not pure.

CAUTION

WARNING: TOXIC

CAUTION:   The reaction is highly *EXOTHERMIC*. A very
vigorous reaction occurs if the acid is added too rapidly.

EXOTHERMIC

Allow the vial to stand in the ice bath for an additional 20 min and then add 1.0 mL of distilled water dropwise (calibrated Pasteur pipet) to the reaction mixture.

### Isolation of Product

Collect the orange-pink solid by vacuum filtration using a Hirsch funnel. (■) Wash the filter cake with four 1.0-mL portions of cold water (calibrated Pasteur pipet). Maintain suction to aid in drying of the product. A piece of plastic food wrap over the mouth of the funnel can aid this process (see Prior Reading). Dry the material further on a porous clay plate, on filter paper, or under vacuum at room temperature.

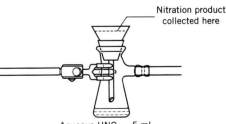

Nitration product
collected here

Aqueous HNO₃,~ 5 mL

### Purification and Characterization

Recrystallize the product using a Craig tube by dissolving the material in the *minimum* amount of absolute ethanol, followed by the dropwise addition of water until precipitation occurs. Cool the mixture in an ice bath and separate the light-yellow crystals. Dry them on a porous clay plate or under vacuum as noted above.

Weigh the crystals and calculate the percent yield. Determine the melting point and compare your value to that listed in the literature. Obtain an IR spectrum of the material (KBr disk) and compare it with that recorded in the literature.

## Experiment [29D]   2- and 4-Nitrophenol

### Instructor Preparation

*In a 250-mL Erlenmeyer flask containing a magnetic stirring bar, weigh and place 20.0 g of silica gel (70–230 mesh; the removal of fines is not necessary). Now add 50 mL of 7.5 M nitric acid and stir the mixture for 3 hours at room temperature. Remove the nitrated silica gel by gravity filtration (do not rinse), place it on a clay* HOOD *plate and allow it to air dry in a **hood** overnight. Store the product in an airtight container.*

*Determine the nitric acid content of the silica gel by titration of a water suspension of the gel with a 0.1 N NaOH solution. The acid content of the gel should be in the range of 16–20% by weight.*

## Nitration of Phenol

### REACTION

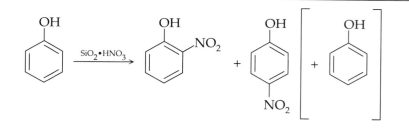

### EXPERIMENTAL PROCEDURE

Estimated time to complete the experiment: 2.0 hours.

#### Physical Properties of Reactants

| Compound | MW | Amount | mmol | mp(°C) | bp(°C) |
|---|---|---|---|---|---|
| Phenol | 94.11 | 240 mg | 2.55 | 40.5–41.5 | 182 |
| SiO$_2$•HNO$_3$ | | 1.0 g | | | |
| Methylene chloride | | 5 mL | | | 40 |

## Reagents and Equipment

Weigh and add 240 mg (2.55 mmol) of phenol to a 10-mL round-bottom flask containing a stir bar. Now add 5 mL of methylene chloride. To this solution, weigh and add 1.0 g of nitrated silica gel (~16% $HNO_3$). Attach the flask to an air condenser. (■)

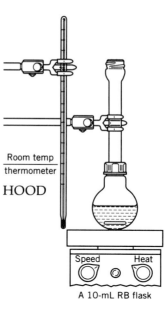

Room temp
thermometer

HOOD

Speed   Heat

A 10-mL RB flask

---

> **CAUTION:** Phenol is highly toxic and corrosive. Prevent contact with eyes and skin. It is best dispensed by warming the container of phenol in a warm water bath and then using an automatic delivery pipet. This should be done in the *hood*.

---

## Reaction Conditions

Stir the resulting mixture at room temperature for 5 min.

## Isolation of Product

Separate the silica gel from the reaction mixture by gravity filtration through a filter funnel containing a small plug of glass wool. Collect the filtrate in a 10-mL Erlenmeyer flask containing a boiling stone. Wash the collected silica gel with two 0.5-mL portions of methylene chloride, and collect these washings in the same Erlenmeyer flask.

Concentrate the filtrate to a volume of about 1.0 mL using a warm sand bath under a slow stream of nitrogen **[HOOD]**.

HOOD

Use thin-layer chromatography to obtain an analysis of the product mixture (see Prior Reading). Use methylene chloride as the elution solvent, silica gel (with a fluorescent indicator) as the stationary phase, and UV light for visualization. Typical $R_f$ values: 4-nitrophenol, 0.04; phenol, 0.15; 2-nitrophenol, 0.58.

**NOTE.**   *2,4-Dinitrophenol has a typical $R_f$ value of 0.33 under the chromatography conditions above; it is not usually formed under these reaction conditions.*

## Characterization and Purification

Column chromatography is now used to separate the mixture of products. Pack (dry) a 1.0-cm diameter buret column with 7.5 g of activated silica gel. Place the above product solution on the column using a Pasteur pipet and then elute the column with 25 mL of 60:40 methylene chloride/pentane solvent. Collect the first 20 mL of eluate in a tared 25-mL Erlenmeyer flask. Concentrate this fraction to dryness in a warm sand bath under a slow steam of nitrogen to isolate the 2-nitrophenol. Weigh the product.

Now elute the column with about 30 mL of 1:1 ethyl acetate/methylene chloride solvent. Collect the first 20 mL of eluate in a tared 25-mL Erlenmeyer flask containing a boiling stone. Concentrate this fraction as above to a volume of about 2 mL. Use thin-layer chromatography to determine the purity of the product (see conditions outlined above). The main constituent of this fraction is 4-nitrophenol. However, the presence of unreacted phenol or possibly quinone is often detected. Now concentrate the product solution to dryness, and weigh the 4-nitrophenol isolated.

If the TLC analysis indicates the 4-nitrophenol to be impure, it may be purified as follows.

Dissolve the solid residue in 7 mL of saturated sodium bicarbonate solution, and transfer the resulting solution to a 15-mL centrifuge tube. Extract this solution twice with 2-mL portions of methylene chloride (a Vortex mixer may be used to good advantage here). Remove the methylene chloride layers using a Pasteur pipet, and save them (together) in a small Erlenmeyer flask until you have isolated and characterized the product.

Cool the resulting aqueous solution in an ice bath, and add 6 *M* HCl dropwise, with mixing (glass rod or Vortex mixer), until it becomes neutral or slightly acidic toward litmus or pH paper. *Do this step carefully. Too vigorous a reaction may result in loss of product.* Now extract the resulting solution with three 2-mL portions of methylene chloride. Following each extraction, remove the methylene chloride using a Pasteur filter pipet and transfer it to a 10-mL Erlenmeyer flask. Dry the wet solution over anhydrous sodium sulfate, and then transfer it by Pasteur filter pipet to a tared 10-mL Erlenmeyer flask containing a boiling stone. Concentrate this solution using a warm sand bath under a slow stream of nitrogen gas to yield the 4-nitrophenol product. Weigh the product. The purity of the material may be checked again, using TLC as outlined above. It may also be recrystallized from water using the Craig tube, if necessary.

To characterize the 2- and 4-nitrophenols, determine their melting points and obtain their infrared spectra (KBr disk). Compare your results to those reported in the literature and in *The Aldrich Library of IR Spectra.*

### Chemical Tests

Chemical classification tests (see Chapter 10) are also of value to establish the identity of these compounds. Perform the ferric chloride test for phenols. Is a positive result obtained for each compound? Does the sodium fusion test detect the presence of nitrogen? The test for nitro groups might also be performed. The phenyl- or α-naphthylurethane derivatives of the phenols may be prepared to further establish their identity (see Chapter 10, Preparation of Derivatives).

## QUESTIONS

**6-186.**  Predict the most likely mononitration product from each of the following compounds. Explain the reasons for your choice.

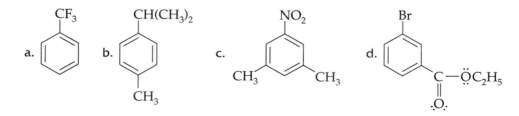

**6-187.**  Write equations to show how nitronium ions might be formed using a mixture of nitric and sulfuric acids.

**6-188.**  Which ring of phenyl benzoate would you expect to undergo nitration more readily? Explain.

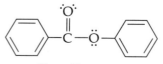

Phenyl benzoate

**6-189.** Arrange the following compounds in order of increasing reactivity toward nitration. Explain.

Acetanilide    Acetophenone    Bromobenzene    Toluene

**6-190.** Offer a reasonable explanation of why nitration of 1,4-dichlorobenzene yields the mononitro derivative while N,N'-diacetyl-1,4-phenylenediamine forms the dinitro compound.

**6-191.** Explain why p-nitrophenol is a stronger acid than phenol itself. Would p-methoxyphenol be a stronger or weaker acid than phenol? Explain.

---

**BIBLIOGRAPHY**

Related to anhydrous nitric acid:

Stern, S. A.; Mullhaupt, J. T.; Kay, W. B. *Chem. Rev.* **1960,** *60,* 185.

The preparation used in this experiment is an adaptation of that reported by:

Cheronis, N. D.; Entrikin, J. B. *Semimicro Qualitative Organic Analysis*; Crowell: New York, 1947, p. 258.

The preparation of the nitrated silica gel was adapted from that reported by:

Tapia, R.; Torres, G.; Valderrama, J. A. *Synth. Commun.* **1986,** *16,* 681.

These references are selected from a large number of examples of nitration given in *Organic Syntheses*. None of these examples use anhydrous nitric acid or the nitrated silica gel as the nitrating agent.

Braum, C. E.; Cook, C. D. *Organic Syntheses*; Wiley: New York, 1963; Collect. Vol. IV, p. 711.

Corson, B. B.; Hasen, R. K. *Organic Syntheses*; Wiley: New York, 1943; Collect. Vol. II, p. 434.

Fanta, P. E.; Tarbell, D. S. *Organic Syntheses*; Wiley: New York, 1955; Collect. Vol. III, p. 661.

Fetscher, C. A. *Organic Syntheses*; Wiley: New York, 1963; Collect. Vol. IV, p. 735.

Fitch, H. M. *Organic Syntheses*; Wiley: New York, 1955; Collect. Vol. III, p. 658.

Hartman, W. W.; Smith, L. A. *Organic Syntheses*; Wiley: New York, 1943; Collect. Vol. II, p. 438.

Howard, J. C. *Organic Syntheses*; Wiley: New York, 1963; Collect. Vol. IV, p. 42.

Huntress, E. H.; Shriner, R. L. *Organic Syntheses*; Wiley: New York, 1943; Collect. Vol. II, p. 459.

Kamm, O.; Segur, J. B. *Organic Syntheses*; Wiley: New York, 1941; Collect. Vol. I, p. 372.

Kobe, K. A.; Doumani, T. F. *Organic Syntheses*; Wiley: New York, 1955; Collect. Vol. III, p. 653.

Mendenhall, G. D.; Smith, P. A. S. *Organic Syntheses*; Wiley: New York, 1973; Collect. Vol. V, p. 829.

Newman, M. S.; Boden, H. *Organic Syntheses*; Wiley: New York, 1973; Collect. Vol. V, p. 1029.

Robertson, G. R. *Organic Syntheses*; Wiley: New York, 1941; Collect. Vol. I, p. 376.

## EXPERIMENT [30]

# Nucleophilic Aromatic Substitution: 2,4-Dinitrophenylthiocyanate

Common name: 2,4-Dinitrophenylthiocyanate
CA number: [1594-56-5]
CA name as indexed: Thiocyanic acid, 2,4-dinitrophenyl ester

### Purpose

To investigate conditions under which a specific type of **nucleophilic aromatic substitution reaction** can occur. To carry out the conversion in a two-phase solvent system through the use of a phase-transfer catalyst. The experiment is an example of nucleophilic aromatic substitution on an aromatic ring having two strongly electron-withdrawing substituents. In this experiment, a bromine substituent (leaving group) is replaced by a thiocyanate, $-S-C\equiv N$, group (nucleophile).

### Prior Reading

Technique 4:   Solvent Extraction
   Liquid–Liquid Extraction (see pp. 80–82).
   Drying of the Wet Organic Layer (see pp. 86–87).
Technique 5:   Crystallization
   Craig Tube Crystallization (see pp. 95–96).
Technique 6:   Chromatography
   Column Chromatography (see pp. 98–101).
   Concentration of Solutions (see pp. 106–108).

## REACTION

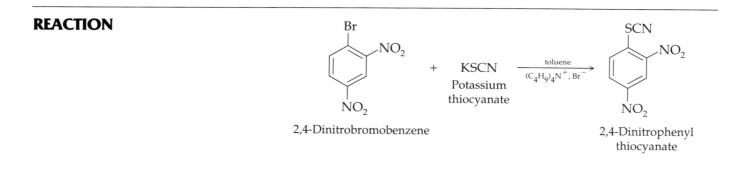

2,4-Dinitrobromobenzene                          2,4-Dinitrophenyl thiocyanate

## DISCUSSION

Nucleophilic aromatic substitution reactions generally take place only if activating groups are present at the positions ortho and/or para to a good leaving group. Activating groups, in this case, are groups that are electron withdrawing and render the aromatic ring more electron-deficient, and thus more susceptible to nucleophilic attack. On the other hand, $\pi$-electron donating groups hinder the reaction. Groups such as $-NO_2$, $-SO_2CH_3$, $^+N(CH_3)_3$, $-CF_3$, $-CN$, $-SO_3^-$, $-Br$, $-Cl$, $-I$, $-CO_2^-$, especially if one $-NO_2$ group is already present, are strongly activating. These groups not only make the ring more susceptible to nucleophilic attack, they assist in the resonance stabilization of the intermediate pentadienyl anion. In the reaction illustrated in this experiment, two nitro groups are present on the ring, both ortho and para to the departing bromine substituent. While functional groups containing a carbonyl group *do*, in principle, activate the aromatic ring toward nucleophilic attack, the carbonyl groups themselves are usually more likely to react with a nucleophile than is the aromatic ring.

The conditions under which this reaction is conducted lend themselves to the use of phase-transfer catalysis. The system involves two phases, the aqueous phase and the organic phase (toluene). The phase-transfer catalyst plays a very important role. As an ammonium cation, it carries the ¯SCN ion from the aqueous phase into the organic phase where the substitution reaction then occurs. The product and the corresponding bromide salt of the phase-transfer catalyst are produced in this conversion. The bromide salt of the phase-transfer catalyst then migrates back into the aqueous phase, and the process repeats itself. The catalyst can play this role since the large organic groups (the four butyl groups) increase the solubility of the phase-transfer catalyst in the organic phase, while the charged ionic center of the catalyst renders it water soluble. Phase-transfer catalysts can also be used in many other reactions, including the preparation of ethers by the Williamson method (see Experiments [22A] and [22B], and the Horner–Wadsworth–Emmons modification of the Wittig Reaction (see Experiments [19B] and [19D]).

The reaction in this experiment proceeds by an *addition–elimination two-step* sequence, of which the first step is generally rate determining. A tetrahedral intermediate (pentadienyl anion) is formed by the attack of the nucleophile on the carbon atom to which the leaving group is attached. In the subsequent step, the leaving group then departs with its bonding electrons, regenerating the aromatic system. It is important to note that, in contrast to nucleophilic substitution reactions on alkyl carbon atoms, the reaction does not proceed by an $S_N2$ mechanism, since in this case an intermediate is formed. The mechanism is outlined here.

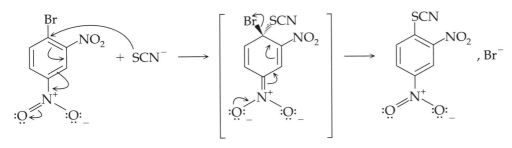

Estimated time to complete the reaction: 2.0 hours.

## EXPERIMENTAL PROCEDURE

### Physical Properties of Reactants and Product

| Compound | MW | Amount | mmol | mp(°C) | bp(°C) |
|---|---|---|---|---|---|
| 2,4-Dinitrobromobenzene | 247.01 | 50 mg | 0.20 | 75 | |
| Tetrabutylammonium bromide | 322.38 | 5 mg | 0.02 | 103–104 | |
| Aqueous potassium thiocyanate (50%) | | 150 μL | | | |
| Toluene | 92.15 | 350 μL | | | 111 |
| 2,4-Dinitrophenylthiocyanate | 244 | | | 138–139 | |

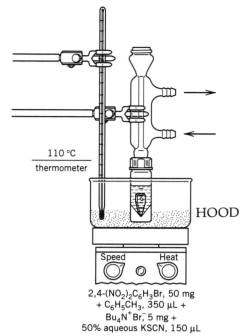

110 °C
thermometer

HOOD

Speed    Heat

2,4-(NO$_2$)$_2$C$_6$H$_3$Br, 50 mg
+ C$_6$H$_5$CH$_3$, 350 μL +
Bu$_4$N$^+$Br$^-$, 5 mg +
50% aqueous KSCN, 150 μL

## Reagents and Equipment

Assemble a 3.0-mL conical vial containing a magnetic spin vane and equipped with a reflux condenser. Weigh and place in the vial, 50 mg (0.20 mmol) of 2,4-dinitrobromobenzene followed by 350 μL of toluene. (■) To this solution add 5.0 mg (0.02 mmol) of tetrabutylammonium bromide and 150 μL of a 50% aqueous potassium thiocyanate (wt/wt) solution.

---

**CAUTION: Tetrabutylammonium bromide is an irritant. HANDLE WITH CARE! It is also hygroscopic and should be protected from moisture prior to its use. Dispense the toluene and the thiocyanate solution in the *hood* using automatic delivery pipets.**

---

## Reaction Conditions

Heat the resulting mixture at a sand bath temperature of 110° C with stirring for 1 hour. Allow the solution to cool to room temperature.

## Isolation of Product

SAVE    Separate the toluene layer [**SAVE**] from the aqueous layer [**SAVE**] by use of a Pasteur pipet and place the toluene layer in a 3.0-mL capped vial. Extract the toluene layer with two 1.0-mL portions of water and combine the water extracts with the original water layer you saved. Now extract the combined aqueous layers with two 0.5-mL portions of toluene, and combine these toluene extracts with the original toluene layer.

**NOTE.** *The volumes of liquid used in the extraction are measured using calibrated Pasteur pipets. For each extraction, shake the vial, vent it carefully, and allow the layers to separate. The transfers are made using Pasteur filter pipets.*

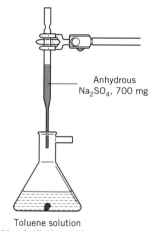

Anhydrous
Na$_2$SO$_4$, 700 mg

Toluene solution
with substitution product,
~ 1.0 mL

Transfer the toluene solution by Pasteur pipet to a microcolumn (Pasteur filter pipet) containing 700 mg of anhydrous sodium sulfate previously wetted with toluene. (■) Allow the solution to elute through the column, and collect the dried toluene eluate in a 10-mL Erlenmeyer flask containing a boiling stone. Concentrate this solution by warming in a sand bath under HOOD a gentle stream of nitrogen gas [**HOOD**]. A yellow, solid product is obtained.

## Purification and Characterization

The product is generally nearly pure. Recrystallize the material from chloroform, using a Craig tube, if desired.

---

HOOD    **CAUTION:   Chloroform is toxic! Dispense and use only in the *hood*. Do not breathe the vapors.**

---

Weigh the dried product and calculate the percent yield. Determine the melting point and compare it with that listed in the Reagent and Product table. Obtain an IR spectrum (KBr disk) and compare your result to that of an authentic sample, or to the spectrum shown in *The Aldrich Library of IR Spectra*.

**6-192.** Explain why the first reaction below requires substantially more vigorous reaction conditions than the second reaction.

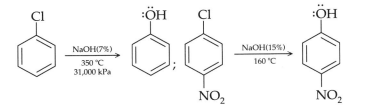

**6-193.** Explain why the intermediate pentadienyl anion in the nucleophilic aromatic substitution reaction (see Discussion section) is not aromatic, even though it has the same number of $\pi$ electrons (6) as the starting benzene derivative.

**6-194.** Complete each of the following reactions. Name the expected product.

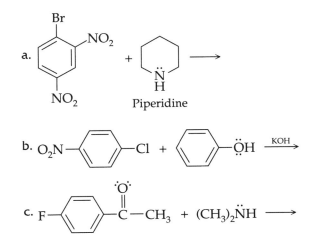

**6-195.** Compare the potential energy diagram of an $S_N2$ substitution reaction on an aliphatic halide to that of the nucleophilic substitution reaction carried out in this experiment. Discuss the main differences in the diagrams in terms of the mechanisms.

**6-196.** Would you expect the rate of the reaction in this experiment to depend on the concentrations of *both* the thiocyanate ion *and* 2,4-dinitrobromobenzene? Explain.

**6-197.** Show the intermediates, including all the resonance hybrid structures for the pentadienyl anion, that are formed in the following aromatic nucleophilic substitution reactions.

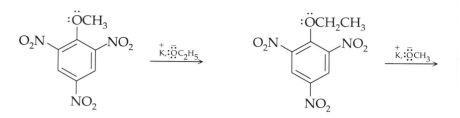

**6-198.**   On workup of each of the reactions given in Question 6-197, what product(s) would you expect to form? If more than one, indicate their relative amounts.

---

**BIBLIOGRAPHY**

Selected examples of nucleophilic aromatic substitution reactions in *Organic Syntheses* include:

Brewster, R. Q.; Groening, T. *Organic Syntheses*; Wiley: New York, 1943; Collect. Vol. II, p. 445.

Bunnett, J. F.; Conner, R. M. *Organic Syntheses*; Wiley: New York, 1973; Collect. Vol. V, p. 478.

Hartman, W. W. *Organic Syntheses*; Wiley: New York, 1941; Collect. Vol. I, p. 175.

Hartman, W. W.; Byers, J. R.; Dickey, J. B. *Organic Syntheses*; Wiley: New York, 1943; Collect. Vol. II, p. 451.

Kharasch, N.; Langford, R. B. *Organic Syntheses*; Wiley: New York, 1973; Collect. Vol. V, p. 474.

Reverdin, F. *Organic Syntheses*; Wiley: New York, 1941; Collect. Vol. I, p. 219.

Sahyun, M. R. V.; Cram, D. J. *Organic Syntheses*; Wiley: New York, 1973; Collect. Vol. V, p. 926.

Skorcz, J. A.; Kuminski, F. E. *Organic Syntheses*; Wiley: New York, 1973; Collect. Vol. V, p. 263.

---

**EXPERIMENT [31]**

## Halogenation Using *N*-Bromosuccinimide: 9-Bromoanthracene

Common name: 9-Bromoanthracene
CA number: [1564-64-3]
CA name as indexed: Anthracene, 9-bromo-

### Purpose

To run a free radical halogenation reaction using *N*-bromosuccinimide (NBS), a highly specific brominating agent. This material has the advantage that it is a solid and is therefore easier to handle than bromine, which is a toxic liquid.

### *Prior Reading*

*Technique 5:*   Crystallization
Use of the Hirsch Funnel (see pp. 93–95).
Craig Tube Crystallization (see pp. 95–96).

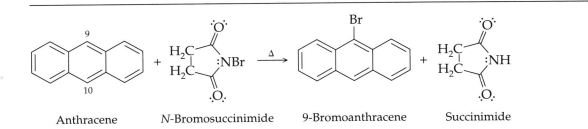

Anthracene     N-Bromosuccinimide     9-Bromoanthracene     Succinimide

**REACTION**

**DISCUSSION**

$N$-Bromosuccinimide (NBS) is a highly specific brominating agent. Using this reagent, anthracene is brominated in the 9-position. $N$-Bromosuccinimide may also be used to brominate positions $\alpha$ to (1) a carbonyl group, (2) a triple bond, (3) an alkene (allylic position), and (4) aromatic rings (benzylic position). Other polynuclear hydrocarbons that have been brominated using NBS include naphthalene, phenanthrene, and acenaphthene.

In the preparation of 9-bromoanthracene, the reaction progress is easily followed, since NBS (a reactant) and succinimide (a product) are both nearly insoluble in carbon tetrachloride. The NBS is more dense than the carbon tetrachloride solvent, and as the reaction proceeds this solid disappears from the bottom of the reaction flask, and the less dense succinimide forms and floats on the surface of the reaction solution.

Free radicals have been identified in the mechanism of bromination using $N$-bromosuccinimide. In fact, the reaction proceeds only under conditions likely to produce radicals: photochemical conditions, by heating, or in the presence of a free radical initiator. The NBS reagent provides a steady source of small amounts of $Br_2$ via the rapid reaction of NBS with traces of hydrogen bromide.

The initiation step in bromination with NBS is the formation of a bromine radical by the homolytic dissociation of the $Br_2$ molecule.

### Initiation Step

$$Br_2 \xrightarrow[\text{free radical initiators}]{h\nu \text{ or}} 2\ Br\cdot$$

The bromine radical then abstracts a hydrogen atom from the 9 position of the anthracene.

### Propagation Step

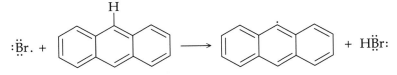

The HBr so formed then reacts with NBS to produce a bromine molecule and succinimide.

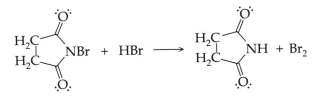

The Br$_2$ molecule then reacts with the anthracenyl radical formed above to yield the product and a bromine radical.

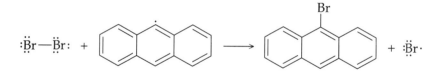

This bromine radical starts the sequence over again. That is, a chain reaction is initiated. In the present reaction, a trace amount of an iodine–carbon tetrachloride solution is added to the reaction mixture. The iodine acts as a moderator or a retarder in the reaction (see Dauben et al. in the Bibliography section), and thus only the monobrominated product is formed; 9,10-dibromoanthracene is not generated under these conditions.

## EXPERIMENTAL PROCEDURE

Estimated time to complete the experiment: 2.0 hours.

### Physical Properties of Reactants

| Compound | MW | Amount | mmol | mp(°C) | bp(°C) |
|---|---|---|---|---|---|
| Anthracene | 178.24 | 50 mg | 0.28 | 216 | |
| N-Bromosuccinimide | 177.99 | 50 mg | 0.28 | 180–183 | |
| Carbon tetrachloride | 153.82 | 0.4 mL | | | 77 |
| Iodine–CCl$_4$ solution | | 1 drop | | | |

### Reagents and Equipment

Weigh and add 50.0 mg (0.28 mmol) of anthracene and 50 mg (0.28 mmol) of N-bromosuccinimide to a 3.0-mL conical vial containing a magnetic spin vane. To this mixture add 0.4 mL of carbon tetrachloride followed by one drop of I$_2$–CCl$_4$ solution delivered from a Pasteur pipet.

---

**WARNING:** Carbon tetrachloride is a Cancer Suspect Agent. Dispense it in the *hood* using an automatic delivery pipet.

---

Attach the vial to a reflux condenser fitted with a calcium chloride drying tube. (■)

### Instructor Preparation

**I$_2$–CCl$_4$ SOLUTION.** *Iodine (0.2 g, 0.01 mol) is dissolved in 10 mL of carbon tetrachloride. Place the solution in a **hood** for student use.*

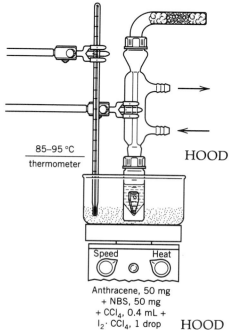

85–95 °C thermometer

HOOD

Speed    Heat

Anthracene, 50 mg
+ NBS, 50 mg
+ CCl$_4$, 0.4 mL +
I$_2$· CCl$_4$, 1 drop    HOOD

## Reaction Conditions

Heat the reaction mixture, with stirring, to reflux in a sand bath at 85–95 °C for 1 hour. During this time the solution turns brown and crystals of succinimide appear at the surface of the reaction solution.

## Isolation of Product

Collect the succinimide product by vacuum filtration of the warm solution using a Hirsch funnel. (■) Wash the filter cake of succinimide with three or four 0.5-mL portions of cyclohexane (calibrated Pasteur pipet). Combine the washings with the original filtrate.

Concentrate the filtrate to dryness **[HOOD]** under reduced pressure to obtain yellow-green crystals of 9-bromoanthracene. *Accelerate evaporation of the solvent by immersing the flask in warm water.*

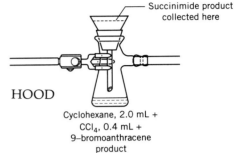

Succinimide product collected here

HOOD

Cyclohexane, 2.0 mL +
CCl₄, 0.4 mL +
9–bromoanthracene
product

## Purification and Characterization

Weigh the air-dried succinimide and calculate the percent yield. Determine the melting point and compare your value to that in the literature. Obtain an IR spectrum (KBr disk) and compare your spectrum with that shown in *The Aldrich Library of IR Spectra.*

The crude 9-bromoanthracene is purified by recrystallization from 95% ethanol using the Craig tube. Weigh the dried product and calculate the percent yield. Determine the melting point and compare your result with that listed in the literature.

## Chemical Tests

The ignition test should establish the presence of the aromatic nature of the substituted anthracene compound. Does the Beilstein test indicate the presence of a halogen?

**QUESTIONS**

**6-199.** Predict and give a suitable name for the product(s) formed in the following reactions with NBS.

a. 1-Propene + NBS $\xrightarrow[h\nu]{CCl_4}$

b. (cyclohexene) + NBS $\xrightarrow[h\nu]{CCl_4}$

c. $CH_3CH_2CH_2-CH=CH-CH_3$ + NBS $\xrightarrow[h\nu]{CCl_4}$

d. (3-methylthiophene) + NBS $\xrightarrow[h\nu]{CCl_4}$

e. (cyclohexenone) + NBS $\xrightarrow[h\nu]{CCl_4}$

**6-200.** When 1-octene is treated with NBS, three monobromo straight-chain alkenes having molecular formula $C_9H_{17}Br$ are isolated from the reaction mixture. Identify these compounds and give each a suitable name.

**6-201.** Benzyl bromide ($C_6H_5CH_2Br$) can be prepared by treating toluene with NBS in the presence of a peroxide initiator. Suggest a suitable mechanism to account for this reaction.

**6-202.**   The benzyl radical

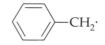

has unusual stability. Account for this fact by drawing appropriate resonance structures.

**6-203.**   Suggest a suitable mechanism for the following reaction (* = ¹³C)

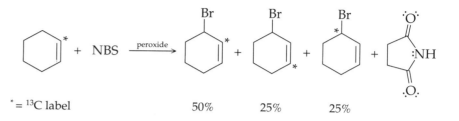

* = ¹³C label            50%        25%        25%

---

**BIBLIOGRAPHY**

Dauben, H. J. Jr.; McCoy, L. L. *J. Am. Chem. Soc.* **1959,** *81,* 4863.
Djerassi, C. *Chem. Rev.* **1948,** *43,* 271.
Horner, L.; Winkelmann, E. H. *Angew. Chem.* **1959,** *71,* 349.
Horner, L.; Winkelmann, E. H. *Newer Methods Prep. Org. Chem.* **1964,** *3,* 151.

Brominations in *Organic Syntheses* using NBS:

Campaigne, E.; Tullar, B. F. *Organic Syntheses;* Wiley: New York, 1963; Collect. Vol. IV, p. 921.

Corbin, T. F.; Hahn, R. C.; Schechter, H. *Organic Syntheses;* Wiley: New York, 1973; Collect. Vol. V, p. 328.

Greenwood, F. L.; Kellert, M. D.; Sedlak, J. *Organic Syntheses;* Wiley: New York, 1963; Collect. Vol. IV, p. 108.

Kalir, A. *Organic Syntheses;* Wiley: New York, 1973; Collect. Vol. V, p. 825.

Nakagawa, N.; Saegusa, J.; Tomozuka, M.; Ohi, M.; Kiuchi, M.; Hino, T.; Ban, Y. *Organic Syntheses;* Wiley: New York, 1988; Collect. Vol. VI, p. 462.

---

**EXPERIMENT [32]**

## Hypochlorite Oxidation of an Alcohol: Cyclohexanone

Common name: Cyclohexanone
CA number: [108-94-1]
CA name as indexed: Cyclohexanone

### Purpose

To explore the oxidation of a secondary alcohol to a ketone using the hypochlorite reagent. To use the technique of steam distillation to separate the product from the reaction mixture. To use chromatographic techniques to purify and isolate the cyclohexanone product.

*Prior Reading*
   *Technique 1:*   Gas Chromatography (see pp. 56–61).
   *Technique 2:*   Semimicroscale Distillation
                          Simple Distillation at the Semimicroscale Level (see pp. 65–68).

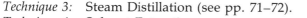

*Technique 3:* Steam Distillation (see pp. 71–72).
*Technique 4:* Solvent Extraction
Liquid–Liquid Extraction (see pp. 80–82).
Drying of the Wet Organic Layer (see pp. 86–87).
Salting Out (see p. 88).

**REACTION**

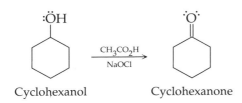

Cyclohexanol       Cyclohexanone

**DISCUSSION**

Sodium hypochlorite solutions (1.8–2.0 $M$) sold as liquid bleach are often described as having 11.5–12.5% available chlorine. The term "available chlorine" compares the oxidizing capacity of the solution relative to that of the same weight of chlorine, $Cl_2$. Sodium hypochlorite solutions are used extensively in swimming pool sanitation, and as a bleach in the pulp and textile industries. A less-concentrated product (5% available chlorine) is used in laundries and as household bleach. Consumption statistics for 1982 indicate that $210 \times 10^3$ tons of sodium hypochlorite were consumed in the United States alone. The reaction described in this experiment illustrates the use of liquid bleach (11.5–12.5% available chlorine) as an oxidizing agent in the organic laboratory.

Sodium hypochlorite is prepared commercially by passing chlorine gas through a solution of aqueous sodium hydroxide.

$$Cl_2 + NaOH \rightleftharpoons NaOCl + NaCl$$

The actual oxidizing agent in the present experiment is the chloronium ion ($Cl^+$) which is reduced in the reaction to chloride ion ($Cl^-$). The cyclohexanol acts as a reducing agent, and thus becomes oxidized to cyclohexanone.

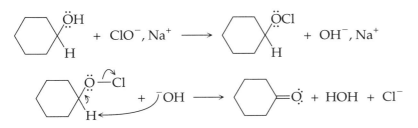

Chromium-based oxidants are commonly used to accomplish these transformations in the research laboratory (see Experiments [33A] and [E1]); the hypochlorite oxidation presents an opportunity to perform an oxidation using the much less costly and much less toxic hypochlorite solution as the oxidant.

In this experiment, a Hickman still is used to isolate the crude cyclohexanone product from the reaction mixture in an example of the steam distillation technique (see Prior Reading). The crude mixture collected in the collar of the still consists of cyclohexanone, water, and acetic acid. If any unreacted cyclohexanol is present in this mixture, it is removed in the subsequent chromatographic purification sequence using alumina. Gas chromatographic analysis may be used to determine the purity of the cyclohexanone product.

## EXPERIMENTAL PROCEDURE

Estimated time to complete the experiment: 3.0 hours.

### Physical Properties of Reactants

| Compound | MW | Amount | mmol | bp(°C) | $d$ | $n_D$ |
|---|---|---|---|---|---|---|
| Cyclohexanol | 100.16 | 100 mg | 1.0 | 161 | 0.96 | 1.4641 |
| Glacial acetic acid | 60.05 | 250 μL | | 118 | | |
| Sodium hypochlorite soln (~12.5% available chlorine) | | 2 mL | | | | |

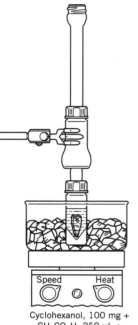

Cyclohexanol, 100 mg +
CH$_3$CO$_2$H, 250 μL +
11.5–12.5% NaOCl, 2 mL

### Reagents and Equipment

Weigh and place 100 mg (1.0 mmol) of cyclohexanol in a 5.0-mL conical vial containing a magnetic spin vane. Add 250 μL of glacial acetic acid and then attach the vial to a Hickman still. (■) *Dispense the glacial acetic acid in the hood by use of an automatic delivery pipet.*

HOOD

Cool the resulting solution in an ice bath and add dropwise with stirring, 2.0 mL of aqueous sodium hypochlorite (NaOCl) solution (~12.5% available chlorine, 1.8–2.0 M) by use of a graduated pipet. Remove the ice bath following the addition. *Add the NaOCl solution by inserting the pipet down the neck of the still just into the throat of the vial.*

**NOTE.** *Solid calcium hypochlorite (65% available chlorine) may be used as a source of oxidant in place of the sodium hypochlorite solution.*

### Reaction Conditions

Stir the resulting solution at room temperature for 1 hour. *Maintain an excess of hypochlorite oxidizing agent throughout the reaction period; overoxidation is not likely. Monitor the aqueous layer periodically using KI–starch paper. If a positive test is not obtained (the paper should turn blue if an oxidant is present), add additional sodium hypochlorite solution (1–3 drops) to ensure that an excess of the oxidizing agent is present. Use a Pasteur pipet inserted down the neck of the still to add the reagent and a clean pipet to remove a drop of solution for testing with KI–starch paper.*

### Isolation of Product

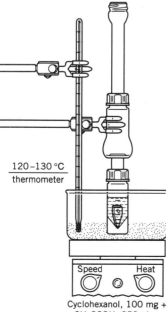

120–130 °C
thermometer

Cyclohexanol, 100 mg +
CH$_3$COOH, 250 μL +
11.5–12.5% NaOCl, 2 ml +
NaHSO$_3$(sat sol),
~8–10 drops

Using a Pasteur pipet, add a saturated, aqueous sodium bisulfite solution dropwise to the reaction mixture until the solution gives a negative KI–starch test.

Separate the crude product from the reaction mixture by steam distillation. With stirring, heat the mixture in a sand bath at a temperature of 120–130 °C. (■) Collect the first 0.5–1.0 mL of distillate in the ring collar of the condenser. Transfer this material to a 3.0-mL conical vial containing a spin vane, using a Pasteur pipet and remove the heat from the still pot. After the distillation apparatus has cooled, rinse the condenser collar with 0.5 mL of diethyl ether, and also transfer this rinse to the conical vial.

**NOTE.** *The distillate collected in the condenser collar consists of cyclohexanone, water, and acetic acid.*

To neutralize the acetic acid present in the separated product mixture, add anhydrous sodium carbonate (~100 mg) in small portions, with stirring, to the solution until evolution of $CO_2$ gas ceases. Now add 50 mg of NaCl to the mixture. The resulting two-phase system is stirred until all the solid material dissolves.

Separate the ether layer containing the cyclohexanone product from the aqueous phase using a Pasteur filter pipet, and transfer it to a microscale drying and chromatography column. Assemble the column using a Pasteur filter pipet packed first with about 50 mg of sand, then with 300 mg of alumina (activity I, see Glossary), followed by 200 mg of anhydrous magnesium sulfate. (■) Wet the column with diethyl ether before the transfer.

Collect the eluate in a tared 3.0-mL conical vial containing a boiling stone. Extract the aqueous layer remaining in the conical vial with three 0.5-mL portions of ether, and also pass each extract through the column and combine these fractions with the original eluate.

Fit the 3.0-mL conical vial with an air condenser and remove the ether by gentle warming on a sand bath in the **hood**.

*Anhydrous MgSO₄, 200 mg*

*Al₂O₃(I), 300 mg*

*Sand, 50 mg*

*Oxidation product in (CH₃CH₂)₂O, ~ 1.0 mL*

## Purification and Characterization

The liquid residue of cyclohexanone isolated upon evaporation of the ether solvent is sufficiently pure for characterization. It may be further purified by prep-GC (see below). Weigh the product and calculate the percent yield.

The purity of the cyclohexanone product may be determined using *gas chromatographic analysis (see Prior Reading and Experiment [5A] to review this technique and the experimental conditions for its use)*. Determine the boiling point and density of your product; compare your results to those given in the literature.

Obtain the IR spectrum of your product, and compare it with that recorded in *The Aldrich Library of IR Spectra*.

## Chemical Tests

Chemical classification tests may also be run to assist in the characterization of this material. The 2,4-dinitrophenylhydrazine test (Chapter 10) should give a positive result. Isolation of this derivative and the determination of its melting point would further establish the identity of the product as cyclohexanone.

**6-204.** In the experiment, why is a solution of sodium bisulfite added to the reaction product mixture (see Isolation of Product)? Write a reaction to account for what is happening. Is the bisulfite ion acting as an oxidizing or reducing agent?

**6-205.** In the isolation of the cyclohexanone product, 50 mg of sodium chloride is added to the water–cyclohexanone–diethyl ether mixture. Explain how the addition of sodium chloride aids in isolation of the cyclohexanone product.

**6-206.** Predict the product(s) for each of the following oxidation reactions. Give a suitable name for each reactant and product.

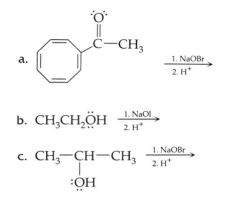

a.

$\xrightarrow[\text{2. H}^+]{\text{1. NaOBr}}$

b. $CH_3CH_2\ddot{O}H$  $\xrightarrow[\text{2. H}^+]{\text{1. NaOI}}$

c. $CH_3$—$\underset{\underset{:\!\ddot{O}H}{|}}{CH}$—$CH_3$  $\xrightarrow[\text{2. H}^+]{\text{1. NaOBr}}$

**6-207.** 2,3-Dimethyl-2,3-butanediol (*pinacol*), upon heating in aqueous acid, rearranges to form 3,3-dimethyl-2-butanone (*pinacolone*). Suggest a mechanism for this reaction.

**6-208.** What chemical tests might be used to distinguish between pentanal and 2-pentanone; between benzyl alcohol and diphenylmethanol?

## BIBLIOGRAPHY

As a general reference for the use of sodium hypochlorite as an oxidant, see:

Fieser, L. F; Fieser, M. *Reagents for Organic Synthesis;* Wiley: New York, 1967; Vol. I, p. 1084.

Experiment [32] is based on work reported by:

Stevens, R. V.; Chapman, K. T.; Weller, H. N. *J. Org. Chem.* **1980,** *45,* 2030. See also Zuczek, N. M.; Furth, P. S. *J. Chem. Educ.* **1981,** *58,* 824.

## EXPERIMENT [33]

## Chromium Trioxide—Resin Oxidation of an Alcohol: 9-Fluorenone; Conversion to the 2,4-Dinitrophenylhydrazone

Common name: 9-Fluorenone
CA number: [486-25-9]
CA name as indexed: 9*H*-Fluoren-9-one

Common name: 9-Fluorenone, 2,4-dinitrophenylhydrazone
CA number: [15884-61-4]
CA name as indexed: 9*H*-Fluoren-9-one, (2,4-dinitrophenyl)hydrazone

### Purpose

To investigate the use of a polymer-bound chromium trioxide oxidizing agent in the oxidation of a secondary (2°) alcohol to a ketone. The progress of the reaction is monitored by thin-layer chromatography (TLC). The product ketone is characterized by formation of its 2,4-dinitrophenylhydrazone (2,4-DNP) derivative.

*Prior Reading*

    *Technique 5:*  Crystallization

                   Use of the Hirsch Funnel (see pp. 93–95).

                   Craig Tube Crystallizations (see pp. 95–96).

    *Technique 6:*  Chromatography

                   Thin-Layer Chromatography (see pp. 103–104).

                   High-Performance Liquid Chromatography (see

                        pp. 105–106).

                   Concentration of Solutions (see pp. 106–108).

**REACTION**

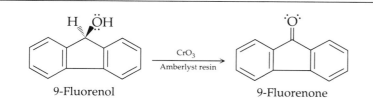

9-Fluorenol                            9-Fluorenone

**DISCUSSION**

This experiment illustrates the oxidation of a 2° alcohol to a ketone. The oxidizing agents commonly used for this purpose are sodium dichromate or chromic oxide in sulfuric acid. In the present case, a convenient and advantageous polymer-bound chromium trioxide reagent is used. It is not only easy to prepare, but is also easy to separate from the product mixture and can be recycled. Today, the use of polymer-bound reagents in organic synthesis is developing at a rapid pace. The mechanism of the oxidation is outlined here.

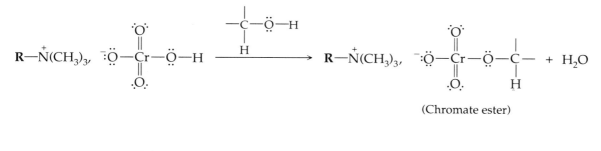

(Chromate ester)

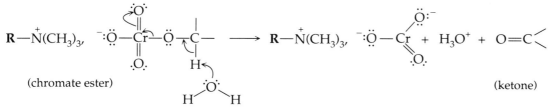

(chromate ester)                                         (ketone)

As seen, the oxidation proceeds through the formation of a chromate ester. Note that the oxidation state of chromium, Cr(VI), at this stage remains the same as in the starting reagent. The second stage is equivalent to an E2 elimination reaction, with the water molecule acting as a base. The donation of an electron pair to the metal atom changes its oxidation state to Cr(IV).

It should be noted that a solution of chromic oxide in aqueous sulfuric acid (the Jones reagent) is used as a test reagent for 1° and 2° alcohols. A positive test is observed when the clear orange test reagent gives a greenish opaque solution upon addition of the alcohol (see Chapter 10).

### Experiment [33A]   9-Fluorenone

## EXPERIMENTAL PROCEDURE

Estimated time to complete the experiment: 3–4 hours.

**Physical Properties of Reactants**

| Compound | MW | Amount | mmol | mp(°C) | bp(°C) |
|----------|-----|--------|------|--------|--------|
| 9-Fluorenol | 182.23 | 100 mg | 0.55 | 154 | |
| Chromic oxide–resin | | 500 mg | | | |
| Toluene | 92.15 | 3.5 mL | | | 111 |

### Reagents and Equipment

### Instructor Preparation

*Prepare the CrO$_3$ resin by adding 35 g of Amberlyst A-26 resin to a solution of 15 g of CrO$_3$ in 100 mL of water. Stir the mixture for 30 min at room temperature. Then collect the resin by vacuum filtration using a Büchner funnel and successively rinse it with water and acetone. Partially dry the material on the Büchner funnel by drawing air through the resin under vacuum for 1 hour. Allow it to air-dry overnight.*

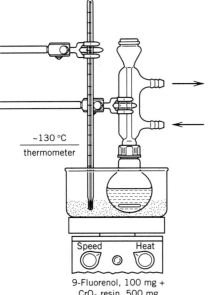

~130 °C
thermometer

Speed    Heat

9-Fluorenol, 100 mg +
CrO$_3$ resin, 500 mg
CH$_3$C$_6$H$_5$, 3.5 mL,
10-mL RB flask

Weigh and add 100 mg (0.55 mmol) of 9-fluorenol to a 10-mL round-bottom flask containing a magnetic stirring bar. Add 3.5 mL of toluene, and sample the resulting solution for TLC analysis (see Prior Reading). Now add 500 mg of the oxidizing resin and attach the flask to a reflux condenser. (■)

### Reaction Conditions

Heat the contents of the flask to reflux, with stirring, using a sand bath temperature of approximately 130 °C.

Sample the solution for TLC analysis after a period of 5 min and every 15–20 min thereafter. In this manner, the progress of the reaction may be monitored until the conversion is complete (~35–40 min). This point is reached when the TLC analysis shows that the 9-fluorenol has been completely consumed.

**SUGGESTED TLC CONDITIONS.**   *Use Eastman Kodak fluorescent silica gel sheets (1 x 6 cm). Elute them with methylene chloride solvent, and visualize the spots by UV light. Determine reference R$_f$ values for 9-fluorenol and 9-fluorenone using known reference samples under the same conditions.*

### Isolation of Product

Cool the reaction mixture and remove the resin by gravity filtration through a cotton plug placed in a small funnel. Transfer the solution to the filter funnel using a Pasteur pipet, and collect the filtrate in a tared 10-mL

Erlenmeyer flask containing a boiling stone. Rinse the reaction flask and resin with two 1.0-mL portions of methylene chloride (calibrated Pasteur pipet). Combine the rinse with the original filtrate.

Remove the solvent **[HOOD]** from the filtrate under a stream of nitrogen by warming in a sand bath to yield a yellow colored residue of crude 9-fluorenone product.

HOOD

## Purification and Characterization

Obtain the weight of the crude product and calculate the percent yield.

Recrystallize a 50-mg portion of the 9-fluorenone from hexane (~1.0 mL of hexane/50 mg of ketone) using a Craig tube. Determine the melting point of this purified material and compare your data to that reported in the literature. Calculate the percent recovery on recrystallization.

**NOTE.**   *High-performance liquid chromatography (see Prior Reading) is effective in separating the pure ketone. Use a $C_{18}$ reversed-phase column and methanol–water as the elution solvent.*

## Chemical Tests

The 2,4-dinitrophenylhydrazine test for aldehydes and ketones (see Chapter 10) may be done to confirm the presence of the carbonyl group. This hydrazone derivative is prepared on a 38-mg scale in Experiment [33B]. Does the ignition test indicate that this compound is aromatic?

This material has a characteristic UV spectrum. The following data were obtained at a concentration of $8.66 \times 10^{-6} \, M$ (see Chapter 9, Ultraviolet–Visible Discussions).

$\lambda_{max}$ 248 nm ($\varepsilon_{max} = 5.27 \times 10^4$, methanol)
$\lambda_{max}$ 255 nm ($\varepsilon_{max} = 7.83 \times 10^4$, methanol)
$\lambda_{max}$ 290 nm ($\varepsilon_{max} = 3.93 \times 10^3$, methanol)

## Experiment [33B]   9-Fluorenone 2,4-Dinitrophenylhydrazone

### Purpose

To prepare a known derivative of the material obtained in Experiment [33A] as an aid in establishing that the desired compound was prepared.

**REACTION**

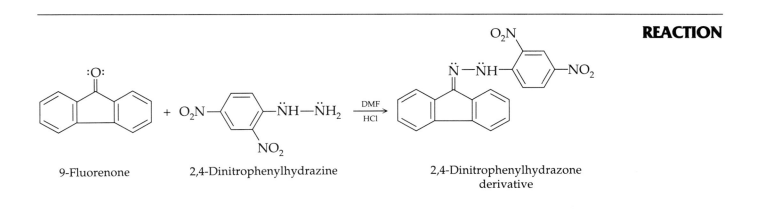

9-Fluorenone          2,4-Dinitrophenylhydrazine          2,4-Dinitrophenylhydrazone
                                                              derivative

**EXPERIMENTAL PROCEDURE**    Estimated time to complete the experiment: 30 min.

**Physical Properties of Reactants and Product**

| Compound | MW | Amount | mmol | mp(°C) | bp(°C) |
|---|---|---|---|---|---|
| 9-Fluorenone | 180.22 | 38 mg | 0.21 | 84 | |
| 2,4-Dinitrophenylhydrazine | 198.14 | 50 mg | 0.25 | 194 | |
| N,N-Dimethylformamide | 73.09 | 1.0 mL | | | 149–150 |
| HCl, concd | | 2 drops | | | |
| 2,4-Dinitrophenylhydrazone derivative of 9-fluorenone | 360.56 | | | 283 | |

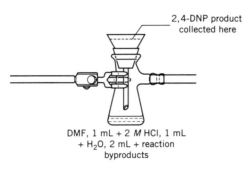

2,4-DNP product collected here

DMF, 1 mL + 2 M HCl, 1 mL + H₂O, 2 mL + reaction byproducts

Prepare a solution of 50 mg (0.25 mmol) of 2,4-dinitrophenylhydrazine in 1.0 mL of N,N-dimethylformamide (DMF) in a 10 × 75-mm test tube. To this solution add 38 mg (0.21 mmol) of the 9-fluorenone prepared in Experiment [33A]. After shaking the test tube so that all solids are in solution, add 2 drops (Pasteur pipet) of concentrated hydrochloric acid. Allow the resulting mixture to stand at room temperature for 5 min, and then place it in an ice bath for an additional 5 min to complete precipitation of the solid derivative. Collect the material by vacuum filtration using a Hirsch funnel. (■) Wash the crystals with 1.0 mL of 2 M HCl (dropwise) to remove unreacted 2,4-dinitrophenylhydrazine and DMF, followed by two 1.0-mL portions of water to remove any residual acid. Dry the product in a desiccator or in a vacuum drying oven.

Weigh the derivative and calculate the percent yield. Determine the melting point and compare it to the literature value given in the Reactants and Product table.

**QUESTIONS**

**6-209.** Suggest a suitable mechanism for the reaction of 9-fluorenone with 2,4-dinitrophenylhydrazine to form the corresponding 2,4-dinitrophenylhydrazone.

**6-210.** It is also possible to characterize 9-fluorenone by preparation of an oxime or semicarbazone. Formulate equations showing clearly the formation of these two derivatives and name each reagent used in the preparation.

**6-211.** As indicated in the Discussion, a solution of chromic oxide in aqueous sulfuric acid is used as a test reagent for 1° and 2° alcohols.
   a. What is this test (consult Chapter 10)?
   b. Predict which of the following alcohols will give a positive test with the chromic oxide reagent. Give the structure for each of the alcohols: 1-heptanol, 2,2,3-trimethyl-3-pentanol, cholesterol, 3-methyl-2-butanol, and 4-tert-butylcyclohexanol

**6-212.** There are actually **two** isomeric 2,4-dinitrophenylhydrazones of 2-pentanone. Draw the structures of these isomers.

**6-213.** 2-Pentanone, in reference to Question 6-212 also forms a derivative on treatment with semicarbazide.

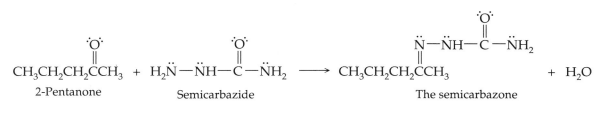

2-Pentanone     Semicarbazide     The semicarbazone

Note that semicarbazide has **two** —NH$_2$ groups that might react with the carbonyl of the ketone to form the semicarbazone. Explain why it reacts as depicted above.

**BIBLIOGRAPHY**

Experiment [33] is an adaptation of that reported by:

Wade, L. G., Jr.; Stell, L. M. *J. Chem. Educ.* **1980**, *57*, 438.

Introduction to the use of polymer-bound reagents:

Hodge, P. *Chem. in Br.* **1978**, *14*, 237.

Leznoff, C. C. *Acc. Chem. Res.* **1978**, *1*, 327.

Selected chromate oxidations reported in *Organic Syntheses* include:

Boeckman, R. K., Jr.; Blum, D. M.; Ganem, B.; Halvey, N. *Organic Syntheses;* Wiley: New York, 1988; Collect. Vol. VI, p. 1033.

Bruce, W. F. *Organic Syntheses;* Wiley: New York, 1943; Collect. Vol. II, p. 139.

Conant, J. B.; Quayle, O. R. *Organic Syntheses;* Wiley: New York, 1941; Collect. Vol. I, p. 211.

Eisenbraun, E. J. *Organic Syntheses;* Wiley: New York, 1973; Collect. Vol. V, p. 310.

Fieser, L. F. *Organic Syntheses;* Wiley: New York, 1963; Collect. Vol. IV, p. 189. ibid., p. 195.

Sandborn, L. T. *Organic Syntheses;* Wiley: New York, 1941; Collect. Vol. I, p. 340.

# Hypochlorite Oxidation of Methyl Ketones by the Haloform Reaction: Benzoic Acid and *p*-Methoxybenzoic Acid

**EXPERIMENT [34]**

Common name: Benzoic acid
CA number: [65-86-0]
CA name as indexed: Benzoic acid

Common names: *p*-Methoxybenzoic acid, 4-methoxybenzoic acid, *p*-anisic acid
CA number: [100-09-4]
CA name as indexed: Benzoic acid, 4-methoxy-

## Purpose

To explore the well-known haloform reaction as a synthetic route to the preparation of organic acids. To investigate the use of a basic aqueous solution of hypochlorite ion as a source of molecular chlorine (Cl$_2$).

*Prior Reading*
Technique 4:   Solvent Extraction
                Liquid–Liquid Extraction (see pp. 80–82).
                Separatory Funnel Extraction (Scaleup) (see
                pp. 82–83).
Technique 5:   Crystallization
                Use of the Hirsch Funnel (see pp. 93–95).
                Craig Tube Crystallizations (see pp. 95–96).

## REACTION

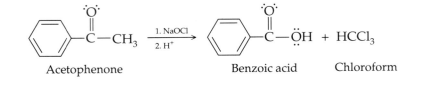

Acetophenone                              Benzoic acid          Chloroform

## DISCUSSION

The reaction of methyl ketones with a halogen in an alkaline medium is known as the haloform reaction. In this experiment, halogen and base are present because of the equilibrium reaction shown here.

$$H_2O + NaOCl + NaCl \rightleftharpoons 2\,NaOH + Cl_2$$

In the haloform reaction, two products are formed: (1) a haloform ($CHCl_3$, $CHBr_3$, or $CHI_3$, depending on the halogen used); and (2) the carboxylic acid having one less carbon atom than the starting ketone. It is the formation of the carboxylic acid that gives the reaction its synthetic utility. An advantage of this oxidation is that it does not affect carbon–carbon double bonds, as illustrated here.

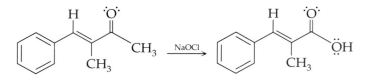

If $I_2$ in a basic solution is used, iodoform ($CHI_3$) is generated in the reaction. This compound is a yellow solid that precipitates from the reaction medium. This observation has been used extensively as a chemical test (**iodoform test**) for methyl ketones and the $RCH(OH)CH_3$ structural group (see Chapter 10).

In the present experiment, a solution of sodium hypochlorite (NaOCl) is used as the source of the chlorine. Sodium hypochlorite solutions are marketed with the hypochlorite concentration described in terms of the "available chlorine" content, which is a term comparing the oxidizing potential of the solution with that of the equivalent mass of chlorine. For the purposes of this experiment, a solution of NaOCl that has 5% available chlorine is needed, and common household bleach will suffice. For another example of an oxidation with an aqueous hypochlorite solution, see Experiment [32].

The reaction takes place in two stages. In the first stage, the methyl group is trihalogenated in a stepwise fashion. In the second stage, hydroxide ion attacks the carbonyl carbon of the trihaloketone to generate the haloform along with the alkali metal salt of the acid; acidification yields the carboxylic acid. The mechanistic sequence follows:

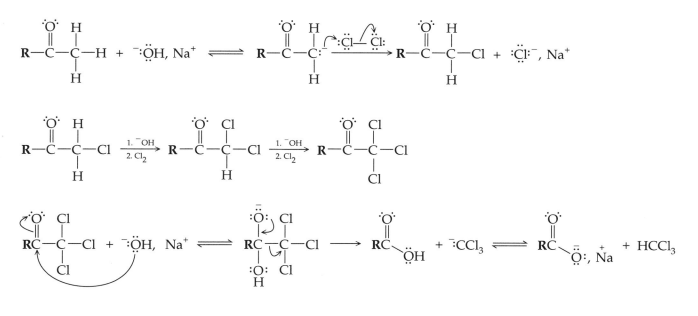

## Experiment [34A]   Benzoic Acid

The reaction is shown above.

---

Estimated time to complete the experiment: 1.5 hours.     **EXPERIMENTAL PROCEDURE**

### Physical Properties of Reactants

| Compound | MW | Amount | mmol | mp(°C) | bp(°C) | d | $n_D$ |
|----------|-----|--------|------|--------|--------|---|-------|
| Acetophenone | 120.16 | 20 μL | 0.17 | 20.7 | 202.6 | 1.03 | 1.5372 |
| Aqueous NaOCl (household bleach, 5% available chlorine) | | 700 μL | | | | | |
| Sodium sulfite | 120.6 | 75 mg | | | | | |

### Reagents and Equipment

To a 3.0-mL conical vial containing a magnetic spin vane and equipped with an air condenser, add 20 μL (21 mg, 0.17 mmol) of acetophenone and 700 μL of household bleach (NaOCl, 5% available chlorine). (■)

---

**CAUTION:** Both reagents are irritants and should be dispensed in the *hood* using automatic delivery pipets.

---

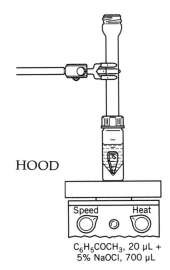

HOOD

$C_6H_5COCH_3$, 20 μL +
5% NaOCl, 700 μL

### Reaction Conditions

Stir the mixture at room temperature for 30 min.

## Isolation of Product

Add approximately 5 mg of sodium sulfite to destroy any unreacted bleach, and stir the reaction mixture briefly. Extract the resulting mixture with two 0.5-mL portions of ether (calibrated Pasteur filter pipet). Separate each portion of ether extract using a Pasteur filter pipet.

---

> **BE CAREFUL:** The ether layer is the top layer; the lower, alkaline, layer contains the acid product. The ether extraction removes the chloroform generated in the reaction and any unreacted acetophenone. Save the ether extracts in a small vial until you have successfully isolated and characterized the final product.

---

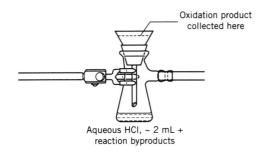

Oxidation product collected here

Aqueous HCl, ~ 2 mL + reaction byproducts

Acidify the aqueous layer (check with pH paper) by adding dropwise 3 $M$ HCl from a Pasteur pipet. A thick, white precipitate of benzoic acid then appears. Collect the solid by vacuum filtration using a Hirsch funnel, and wash the filter cake with three 0.5-mL portions of water (calibrated Pasteur pipet). (■) Maintain the vacuum for approximately 5 min by covering the funnel with plastic food wrap (see Prior Reading) to partially dry the product. Transfer the material to a porous clay plate or filter paper to complete the drying process.

## Purification and Characterization

The product is of sufficient purity for characterization. Weigh the material and calculate the percent yield.

Determine the melting point and compare your result with the value listed in the literature. If desired, obtain an IR spectrum using the KBr pellet technique and compare it with an authentic sample of benzoic acid, or with that given in *The Aldrich Library of IR Spectra*.

### Chemical Tests

Chemical classification tests may also be used to assist in product analysis. You might wish to perform the following:

**1.** The ignition test should indicate the presence of the aromatic ring (Chapter 10).

**2.** The solubility of the compound in water, 5% NaOH, and in 5% $NaHCO_3$ should be checked (see Chapter 10). Does the water solution turn blue litmus paper red? Is there evidence of $CO_2$ evolution when the benzoic acid is added to the bicarbonate solution? If positive results are obtained in these tests, how do they confirm that a carboxylic acid is present?

**3.** The preparation of an amide derivative (see Chapter 10) would also aid in establishing the identity of the acid product.

## Experiment [34B]  *p*-Methoxybenzoic Acid

**REACTION**

$$CH_3\ddot{O}-\underset{p\text{-Methoxyacetophenone}}{\underbrace{\phantom{xxxxx}}}-\overset{\overset{\displaystyle ..}{\overset{\displaystyle O}{\|}}}{C}-CH_3 \xrightarrow[\text{2. H}^+]{\text{1. NaOCl}} CH_3\ddot{O}-\underset{p\text{-Methoxybenzoic acid}}{\underbrace{\phantom{xxxxx}}}-\overset{\overset{\displaystyle ..}{\overset{\displaystyle O}{\|}}}{C}-\ddot{O}H + \underset{\text{Chloroform}}{HCCl_3}$$

Estimated time to complete the experiment: 1.5 hours.

**EXPERIMENTAL PROCEDURE**

### Physical Properties of Reactants

| Compound | MW | Amount | mmol | mp(°C) | bp(°C) |
|---|---|---|---|---|---|
| p-Methoxyacetophenone | 150.8 | 29 mg | 0.19 | 38–39 | 258 |
| Aqueous NaOCl (household bleach, 5% available chlorine) | | 700 μL | | | |
| Sodium sulfite | 120.6 | 75 mg | | | |

### Reagents and Equipment

Weigh and add 29 mg (0.19 mmol) of p-methoxyacetophenone to a 3.0-mL conical vial containing a magnetic spin vane. Now add 700 μL of household bleach (5% available chlorine), and then attach the vial to an air condenser. (■)

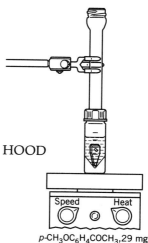

HOOD

p-CH₃OC₆H₄COCH₃, 29 mg + 5% NaOCl, 700 μL

---

CAUTION: The bleach is an irritant to skin and eyes. It is dispensed using an automatic delivery pipet in the *hood.*

---

### Reaction Conditions

Stir the mixture for 30 min with very gentle heating. *Use the lowest possible setting on a hot plate magnetic stirrer.*

### Isolation of Product

Weigh and add approximately 5 mg of sodium sulfite. Stir the reaction medium briefly and then cool it to room temperature. Extract the resulting mixture with two 0.5-mL portions of ether (calibrated Pasteur pipet). Separate each portion of ether extract using a Pasteur filter pipet.

---

BE CAREFUL: The ether layer is the top layer; the bottom aqueous layer contains the product. The ether extraction removes the chloroform generated in the reaction and any unreacted p-methoxyacetophenone. Save the ether extracts in a small vial until you have successfully isolated and characterized the final product.

---

Acidify the aqueous layer by the dropwise addition (check with pH paper) of 3 $M$ HCl from a Pasteur pipet. A thick, white precipitate of p-methoxybenzoic acid is formed. Cool the mixture in an ice bath for 5 min and collect the solid by vacuum filtration using a Hirsch funnel. (■) Wash the filter cake with three 0.5-mL portions of cold water (calibrated Pasteur pipet). Maintain the vacuum for approximately 5 min by covering the funnel with plastic food wrap (see Prior Reading) to partially dry the product. Transfer the material to a porous clay plate or filter paper to complete the drying process.

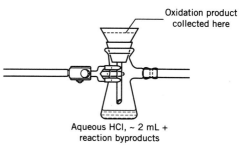

Oxidation product collected here

Aqueous HCl, ~ 2 mL + reaction byproducts

## Purification and Characterization

The product is reasonably pure but may be recrystallized from ethanol–water, if desired, using a Craig tube. Weigh the product and calculate the percent yield. Determine the melting point and obtain an IR spectrum using the KBr pellet technique. Compare your results with those reported in the literature and *The Aldrich Library of IR Spectra*, respectively. Chemical characterization tests can be used to assist in the classification of this product, as outlined in Experiment [34A].

---

## OPTIONAL SEMIMICROSCALE PREPARATION

This preparation may be scaled up by a 15-fold increase in reagent amounts. The procedure is similar to that given above for the micropreparation with the exceptions noted below. *A major difference is that a separatory funnel is used in the extraction step.*

**1.** Use a 25-mL round-bottom (RB) flask containing a magnetic stirring bar and equipped with an air condenser. The reaction conditions are as given in the microscale experiment. (■)

**2.** The reagent and solvent amounts are increased 15-fold.

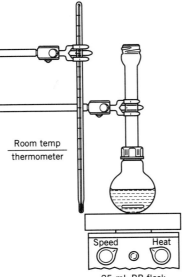

Room temp thermometer

Speed    Heat

25-mL RB flask

### Physical Properties of Reactants

| Compound | MW | Amount | mmol | mp(°C) | bp(°C) |
|---|---|---|---|---|---|
| *p*-Methoxyacetophenone | 150.8 | 430 mg | 2.85 | 38–39 | 258 |
| Aqueous NaOCl (household bleach, 5% available chlorine) | | 10.5 mL | | | |
| Sodium sulfite | 120.6 | 75 mg | | | |

**3.** After addition of the sodium sulfite, stirring and cooling, transfer the product mixture to a 125-mL separatory funnel. Extract this mixture with two 7.5-mL portions of diethyl ether. Save the ether extracts in a small Erlenmeyer flask until you have successfully isolated and characterized the final product. The product is in the aqueous layer.

**NOTE.** *Carry out the extraction by first returning the lower aqueous layer to the reaction flask and then removing the ether layer. Then return the aqueous layer to the separatory funnel and rinse the reaction flask with the second portion of ether. This ether rinse is then added to the separatory funnel to complete the extraction.*

**4.** After collecting the product by vacuum filtration, wash the filter cake with three 7-mL portions of cold water.
**5.** The pale-yellow product is characterized as described in the microprocedure.

## QUESTIONS

**6-214.** In the haloform reaction, once the first α-hydrogen atom is replaced by a halogen atom, each successive hydrogen is more easily substituted until the trihalo species is obtained. Explain.

**6-215.** The haloform reaction using $I_2$ and NaOH is referred to as the "iodoform" test for methyl ketones (see Chapter 10). The test also gives positive results for compounds containing the $-CH(OH)CH_3$ group. This results from the oxidation of the alcohol to the methyl ketone in the first stage. Write a balanced equation for the conversion of $C_6H_5CHOHCH_3$ to the methyl ketone in the presence of $I_2$ and NaOH. Identify which species is being oxidized and which is being reduced.

**6-216.** If you were carrying out an industrial scale synthesis in which one step involved a haloform reaction to convert a methyl ketone into the corresponding acid having one less carbon atom, would you use NaOH and $Cl_2$, NaOH and $Br_2$, or NaOH and $I_2$ as the reagent? Give reasons for your choice.

**6-217.** Can you explain the fact that even though dibenzoylmethane ($C_6H_5COCH_2COC_6H_5$) is not a methyl ketone, it gives a positive iodoform test when treated with the NaOH and $I_2$.

**6-218.** Do you think that acetaldehyde would give a positive iodoform test? Explain your reasoning.

**6-219.** A water-soluble phenol is also acidic toward litmus paper as is a water-soluble carboxylic acid. How would you distinguish the difference between an aromatic acid and a phenol using a chemical test?

---

**BIBLIOGRAPHY**

Selected examples of the haloform reaction from *Organic Syntheses*:

Newman, M. S.; Holmes, H. L. *Organic Syntheses*; Wiley: New York, 1943; Collect. Vol. II, p. 428.

Sanborn, L. T.; Bousquet, E. W. *Organic Syntheses*; Wiley: New York, 1941; Collect. Vol. I, p. 526.

Smith, L. I.; Prichard, W. W.; Spillane, L. J. *Organic Syntheses*; Wiley: New York, 1955; Collect. Vol. III, p. 302.

Smith, W. T.; McLeod, G. L. *Organic Syntheses*; Wiley: New York, 1963; Collect. Vol. IV, p. 345.

Staunton, J.; Eisenbraun, E. J. *Organic Syntheses*; Wiley: New York, 1973; Collect. Vol. V, p. 8.

---

# Photochemical Isomerization: *cis*-Azobenzene

**EXPERIMENT [35]**

Common name: *cis*-Azobenzene
CA number: [1080-16-6]
CA name as indexed: Diazene, diphenyl-, (Z)-

## Purpose
To demonstrate the photochemically promoted isomerization of *trans*-azobenzene to *cis*-azobenzene. The course of the reaction is followed using thin-layer chromatography.

## *Prior Reading*
Technique 6:    Chromatography
                Thin-Layer Chromatography (see pp. 103–104).

# REACTION

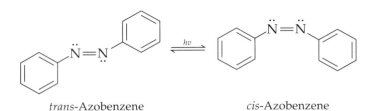

*trans*-Azobenzene          *cis*-Azobenzene

# DISCUSSION

The photochemical conversion of *trans*- to *cis*-azobenzene proceeds by way of an excited electronic state of the trans isomer. Since both compounds are colored and have different $R_f$ values, the conversion is conveniently followed using TLC. We should note that photochemical isomerization of an alkene was carried out in Experiment [6], and a photochemical isomerization of a diazabicyclohexene system is presented in Experiment [F4].

A $\pi$ bond is created by overlap of the $sp^2$ hybridized nitrogen $p$ orbitals. Rotation about the axis of the double bond requires a good deal of energy because it destroys the $p$ orbital overlap, and thus the $\pi$ bond. Unless the material is irradiated with light of the appropriate wavelength, absorption of the radiation does not take place, and the isomers do not interconvert unless sufficient thermal energy (typically > 150 °C ) is supplied to break the $\pi$ bond. If one of the electrons in the $\pi$ bond ($\pi$ molecular orbital) is photochemically excited into an antibonding ($\pi^*$) molecular orbital, as occurs in this experiment, the $\pi$ bond is weakened significantly. Rotation about this bond can then occur rapidly at room temperature.

Azo compounds, because of their intense color, are used commercially as synthetic dyes. Methyl Red (prepared in Experiment [26]) is an example of this class of materials.

There has been reported a novel, reversible chemical method for information storage that has potential use in optical memory devices.[23] The method involves the ability of a film of an azobenzene derivative to be "switched" between three chemical states, photochemically or electrochemically, as shown here.

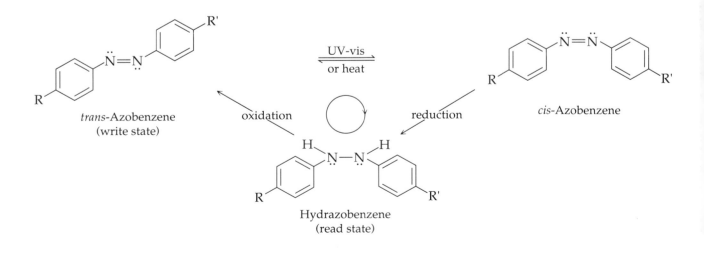

*trans*-Azobenzene
(write state)

oxidation          UV-vis
or heat

reduction

*cis*-Azobenzene

Hydrazobenzene
(read state)

[23] Fujishima, A.; Liu, Z. F.; Hashimoto, K. *Nature (London)* **1990**, *347*, 658; see also *Chem. Eng. News* Oct. 22, 1990, 6.

As seen, the trans isomer of azobenzene can be isomerized to the cis isomer photochemically (as in this experiment) with the reverse reaction occurring either thermally or by exposure to visible light. The less-stable cis isomer may be reduced to hydrazobenzene (read state) electrochemically. To complete the cycle, electrochemical oxidation of the hydrazobenzene produces the *trans*-azobenzene isomer (write state). This sequence of reactions offers an exciting possibility for the development of an erasable, photochromic molecular digital memory system.

## EXPERIMENTAL PROCEDURE

Estimated time to complete the experiment: 1.5 hours.

### Physical Properties of Reactants

| Compound | MW | Amount | mmol | mp(°C) | bp(°C) |
|---|---|---|---|---|---|
| *trans*-Azobenzene | 182.23 | 25 mg | 0.14 | 68.5 | |
| Methylene chloride | 84.94 | 1.0 mL | | | 40 |

### Reagents and Equipment

Weigh and place 25 mg (0.14 mmol) of azobenzene in a 3-mL conical vial. Now add 1.0 mL of methylene chloride.

> **CAUTION:** Azobenzene is a cancer suspect agent. Dispense the methylene chloride in the *hood* using a graduated, or automatic delivery, pipet.

HOOD

### Determination of $R_f$ Values

Using a capillary TLC spotter prepared from a melting point tube, spot an activated silica thin-layer chromatographic plate (1.5 x 7 cm) with the above solution.

**NOTE.**   *Refer to Prior Reading for instructions in setting up and carrying out a TLC analysis.*

Elute the chromatogram in a 4 × 2 in. screw-cap bottle, using about 5 mL of hexane as the solvent. *Two spots are generally observed on the plate:*

Typical $R_f$ values:      *trans*-Azobenzene ($R_f = 0.6$)
                            *cis*-Azobenzene ($R_f = 0.1$)

### The Isomerization Reaction

Spot a second thin-layer plate with the same solution used above, but *do not* elute the plate yet. Place the plate under a sunlamp (1 ft. away) for 1 hour. Then elute this plate as above, using hexane as the elution solvent.

### Characterization

After elution of the TLC plate, only one spot, near the origin, should be observed.
Determine the $R_f$ value for this material and compare it to the control

sample developed previously. The results should show that all the trans isomer present in the starting sample has been converted to the cis isomer.

**QUESTIONS**

**6-220.** Which of the two isomers of azobenzene would you expect to be the most thermodynamically stable? Explain. In this isomerization experiment, was an equilibrium established between the two? How do you know?

**6-221.** Over the years the two isomers of azobenzene have been designated by various terms: cis–trans, syn–anti, and E–Z. Using each set of terms, assign them to the isomers of azobenzene.

**6-222.** Listed below are the activation energies (in kcal/mol) for the thermal decomposition of a series of azo compounds.

| Compound | Activation Energy (kJ/mol) |
|---|---|
| $CH_3-\ddot{N}=\dot{N}-CH_3$ | 210 |
| $(CH_3)_2CH-\ddot{N}=\dot{N}-CH(CH_3)_2$ | 170 |
| $(C_6H_5)_2CH-\ddot{N}=\dot{N}-CH(C_6H_5)_2$ | 110 |

Explain the trend.

**6-223.** You have found that two isomers of a given compound can be separated by TLC. When the solvent front had moved 60 mm above the original line where the spotting of samples of each of the isomeric species had been done, the spot of one species (X) was at 40 mm and that of the other (Y) at 25 mm. Calculate the $R_f$ values for the isomers X and Y.

**6-224.** Discuss what you would observe after elution and visualization of a TLC plate having made the following mistakes in carrying out the analysis.

   a. The solvent level in the developing jar was higher than the original line on which the samples were spotted.
   b. Too much sample was applied to the TLC plate.

**BIBLIOGRAPHY**

Experiment [35] is adapted from the work of:

Helmkamp, G. K.; Johnson, H. W., Jr. *Selected Experiments In Organic Chemistry*, 2nd ed.; W. H. Freeman: San Francisco, 1968; p. 32.

# 7

# Advanced Microscale Organic Laboratory Experiments

$C_7H_6$, [1,2]Spirene
Simmons and Fukunaga (1967).

**INTRODUCTION**

From a theoretical perspective, the chemistry described in this chapter is more demanding of the student–investigator than that described in Chapter 6.

1. The organic reactions performed are less familiar and are not as likely to be found in most introductory texts.
2. The mechanisms proposed for these systems are more involved and not generally developed at the introductory level.
3. Many of the reagents employed are rarely used in the introductory organic laboratory.

   Thus, the experiments contained in Chapter 7 are specifically tailored to challenge the more advanced undergraduate students, those who already are able to access the chemical literature. This chapter can also offer a few beginning students who are particularly interested in the subject, and wish to spend extra preparation time, a special laboratory experience. The techniques involved are not, in most instances, any different from those employed in the introductory microscale laboratory reactions described in Chapter 6, and therefore the manipulations involved should not be considered a barrier to undertaking any of the advanced experiments. The reaction conditions, however, are less forgiving to slight deviations from the suggested ones and the ultimate success of the transformations is less secure.

   Thus, the experimentation contained in Chapter 7 can be adapted to several levels of undergraduate laboratory programs. For example, the study of these reactions can potentially make significant contributions to advanced undergraduate programs where microscale techniques are being introduced to research oriented students for the first time.

The formats for the discussions and experimental procedures are similar to those employed in Chapter 6. The reactions selected for study in this chapter include:

- An unusual borane reduction of a carbonyl directly to a methylene group.
- The trapping of an $\alpha,\beta$-unsaturated ketone as its enol acetate, by acylium ion formation with chlorotrimethylsilane and acetic anhydride.
- The synthesis of a heterocyclic ring employing diethyl carbonate and sodium hydride. The discovery of the medicinal properties of this class of heterocycles led directly to modern anticoagulation therapy.
- The synthesis of an isotopically labeled molecule through the use of a unique Grignard cross-coupling reaction in the presence of dichloro[1,3-bis(diphenylphosphino)propane]nickel(II).
- An oxidative coupling of a naphthol by ferric chloride in a reaction that mimics nature's method of coupling phenolic substances into important pigments, such as Hypericin (**I**).

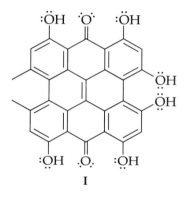

I

- An important molecular rearrangement of oximes to amides discovered by Ernst Otto Beckmann in 1886 and so named in his honor. The modern version utilizes one of the most powerful acid reagents employed in organic chemistry, triflic acid.

The study of the reactions outlined in Chapter 7 should help to facilitate the student's smooth transition into the organic research laboratory. We hope you find the collection as exciting as we did.

## EXPERIMENT [1$_{adv}$]

## Diborane Reductions: Thioxanthene and Xanthene

Common name: Thioxanthene
CA number: [261-31-4]
CA name as indexed: 9*H*-Thioxanthene

Common name: Xanthene
CA number: [92-83-1]
CA name as indexed: 9*H*-Xanthene

## Purpose

To investigate an unusual example of a hydroboration in which a carbonyl group is directly and fully reduced to a methylene group by borane (BH$_3$). To recognize that this example is an atypical reduction of an aldehyde or ketone, as the use of this reagent usually leads to the formation of the corresponding alcohol. To explore the mechanistic rationale for this unexpected product.

### *Prior Reading*

    *Standard Experimental*
        *Apparatus:*   Moisture Protected Reaction Apparatus (see pp. 29–31).
       *Technique 6:*   Chromatography
               Packing the Column (see pp. 99–100).
               Elution of the Column (see p. 100).
       *Technique 9:*   Concentration of Solutions
               Removal of Solvent Under Reduced Pressure (see pp. 107–108).

**NOTE.** *See Experiment [13] for a biography of Herbert C. Brown, Nobel Laureate, who discovered and developed the boron hydride reagents. This experiment also contains further details about the use of these powerful reagents, which have revolutionized reduction reactions in organic chemistry.*

---

**DISCUSSION**

Borane is a useful and selective reducing agent. It is prepared by the reaction of boron trifluoride etherate with sodium borohydride. The borane produced, as the etherate, may be distilled as the dimer, which is a colorless, toxic gas (B$_2$H$_6$). Collection of the dimer distillate in tetrahydrofuran (THF) again forms the monomer, in this case as the BH$_3$•THF complex. The latter is commercially available as a 1.0 *M* solution.

$$3\ NaBH_4 + 4\ BF_3\text{•}O(CH_2CH_3)_2 \rightarrow 3\ NaBF_4 + 2\ B_2H_6 \uparrow \text{(gas)}$$

Borane complexes can also be formed with other ethers, such as diethyl ether (as just discussed) or diglyme (diethylene glycol dimethyl ether). These complexes form readily because the ether, acting as a Lewis base (electron donor), can satisfy the electron-deficient boron atom, which acts as a Lewis acid (electron acceptor). Borane reacts rapidly with water, and therefore procedures using the BH$_3$•THF complex must be conducted under anhydrous conditions.

Borane is a Lewis acid that is attacked by electron-rich centers. Thus, when aldehydes or ketones are treated with the BH$_3$•THF complex, the borate ester (H$_2$B—OR) is rapidly formed, which, upon hydrolysis, gives the corresponding alcohol. The reduction of the carbonyl group is believed to take place by addition of the oxygen atom to the electron-deficient boron atom, followed by irreversible transfer of hydride ion from the now anionic boron to the carbon atom of the (former) carbonyl.

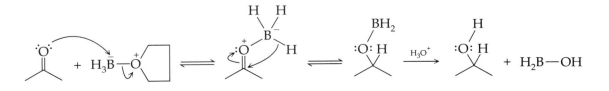

In the case of the xanthone and thioxanthone ring systems, the corresponding alcohol is not formed. The intermediate borate ester undergoes an elimination reaction, forming a borate anion and a resonance-stabilized carbocation. The second stage of the reaction is initiated by displacement of THF from a second BH₃•THF complex by a lone pair from either the oxygen atom of the xanthene carbocation or the sulfur atom of the thioxanthene carbocation. This *new complex* then undergoes an internal hydride (:H⁻) transfer from boron to the C-9 ring position (this is the second hydride attack on this position in the overall reaction) to form a stable oxonium (or sulfonium) ion intermediate. On aqueous–alcohol workup of the reaction mixture the oxonium (sulfonium) product is quickly hydrolyzed to yield xanthene (or thioxanthene) possessing a fully reduced methylene group, —CH₂—, at the 9 position.

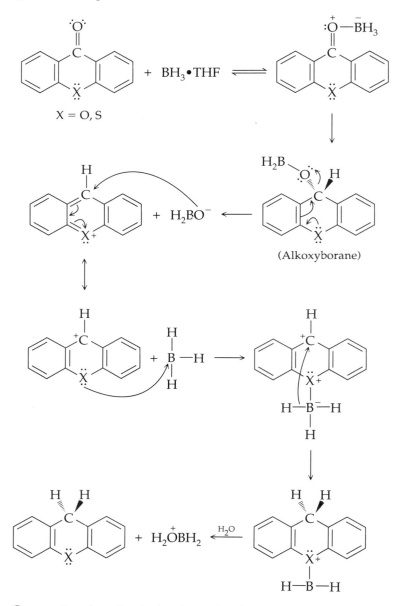

Conventional methods for the reduction of a carbonyl to a methylene group that do not require the conjugative assistance of a heteroatom are the well-known Clemmensen [Zn(Hg), HCl], and Wolff–Kishner (H₂NNH₂/KOH) reductions, and the desulfurization of the corresponding thioacetal with Raney nickel.

# Experiment [1A$_{adv}$]    Thioxanthene

## REACTION

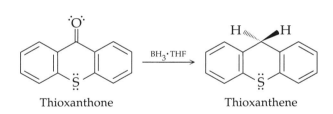

Thioxanthone    $\xrightarrow{BH_3 \cdot THF}$    Thioxanthene

## EXPERIMENTAL PROCEDURE

Estimated time to complete the experiment: 1.5 hours.

**Physical Properties of Reactants**

| Compound | MW | Amount | mmol | mp(°C) | bp(°C) |
|---|---|---|---|---|---|
| Thioxanthone | 212.28 | 50 mg | 0.24 | 209 | |
| Tetrahydrofuran | 72.12 | 1.7 mL | | | 67 |
| Borane•THF, 1 $M$ | | 1.0 mL | | | |

## Reagents and Equipment

Attach a 5.0-mL conical vial containing a magnetic spin vane to a Claisen head. Then fit the Claisen head with a nitrogen inlet tube (prepared from a syringe), and a rubber septum. Weigh and add 50 mg (0.24 mmol) of thioxanthone to the conical vial.

Flush the reaction vial with a gentle stream of nitrogen gas for several minutes, add 1.7 mL of *dry* (see Note) tetrahydrofuran (THF) through the septum (syringe), and then place a small balloon over the Claisen-head outlet so as to maintain a dry atmosphere in the system. (■)

**NOTE.** *The THF must be **absolutely** dry. It is recommended that HPLC grade reagent be used and that if not a fresh bottle, it be distilled once from calcium hydride and stored over molecular sieves. It may then be stored and used safely for up to a week without adversely affecting the yield of product.*

Heat the mixture with stirring in a sand bath at 55–60 °C until the thioxanthone dissolves, yielding a yellow solution. Then, with continued stirring, add 1.0 mL of a 1.0 $M$ solution of BH$_3$•THF in one portion through the rubber septum with a 1.0-mL syringe.

> **CAUTION: The BH$_3$•THF solution reacts *violently* with water.**

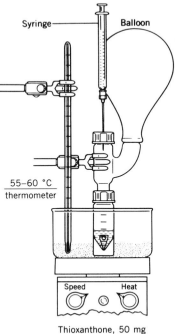

Thioxanthone, 50 mg
+ THF, 1.7 mL
+ 1$M$ BH$_3$•THF, 1.0 mL

## Reaction Conditions

Heat the solution with stirring in a sand bath at 55–60 °C for 5 min. The solution should become colorless during this time.

## Isolation of Product

Quench the reaction by *carefully* adding dropwise approximately 10 drops of 95% ethanol (Pasteur pipet), with stirring, **or** until the observed foaming subsides. *The aqueous alcohol is added to decompose any unreacted BH₃•THF reagent and to hydrolyze the sulfonium ion intermediate.*

After the solution has cooled to room temperature, transfer the mixture by Pasteur pipet to a 25-mL filter flask containing a boiling stone. Carefully remove roughly one-half of the tetrahydrofuran solvent under reduced pressure with continuous swirling of the flask (see Prior Reading). (■) Then, use a calibrated Pasteur pipet to add two 1.0-mL portions of water to the reaction mixture. Carefully remove the remaining tetrahydrofuran and ethanol solvent under reduced pressure with continuous swirling of the flask. As the tetrahydrofuran and ethanol evaporate, white crystals of thioxanthene appear, and a slurry of these crystals in water will remain in the flask after the tetrahydrofuran and ethanol are removed. Collect the product crystals under reduced pressure by use of a Hirsch funnel, and wash the filter cake on the Hirsch filter bed with two 1.0-mL portions of water. Dry the crystals in air on a porous clay plate or on filter paper.

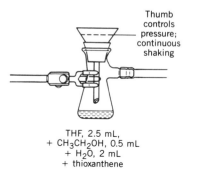

Thumb controls pressure; continuous shaking

THF, 2.5 mL,
+ CH₃CH₂OH, 0.5 mL
+ H₂O, 2 mL
+ thioxanthene

## Purification and Characterization

The thioxanthene product is essentially pure as isolated. It may be recrystallized from an ethanol–chloroform mixture, if necessary. Weigh the product and calculate the percent yield. Determine the melting point of the material and compare it with the value reported in the literature. Obtain IR spectra of thioxanthone and thioxanthene and compare them to each other as well as to those given in *The Aldrich Library of IR Spectra*.

## Experiment [1B$_{adv}$]   Xanthene

**REACTION**

Xanthone  →(BH₃•THF)→  Xanthene

**EXPERIMENTAL PROCEDURE**

Estimated time to complete the experiment: 3.0 hours.

### Physical Properties of Reactants

| Compound | MW | Amount | mmol | mp(°C) | bp(°C) |
|---|---|---|---|---|---|
| Xanthone | 196.22 | 50 mg | 0.26 | 174 | |
| Tetrahydrofuran | 72.12 | 0.7 mL | | | 67 |
| Borane•THF, 1 *M* | | 0.75 mL | | | |

## Reagents and Equipment

Using the experimental apparatus described in Experiment [1A<sub>adv</sub>], weigh and place in the reaction flask, 50 mg (0.26 mmol) of xanthone followed by 0.7 mL of dry THF (see Experiment [1A<sub>adv</sub>], Reagents and Equipment). Maintain a dry nitrogen atmosphere in the system by the same procedure as used in Experiment [1A<sub>adv</sub>]. Heat the reaction mixture, with stirring, to 55–60 °C using a sand bath. After the xanthone dissolves, use a 1.0-mL syringe to add, in one portion, 0.75 mL of a 1.0 $M$ solution of BH$_3$•THF through the rubber septum on the screw-cap port of the Claisen head.

---

**CAUTION:**   The BH$_3$•THF solution reacts *violently* with water.

---

## Reaction Conditions

Stir the reaction solution in a sand bath at 55–60 °C for a period of 1 hour.

## Isolation of Product

While stirring the warm reaction mixture, quench the reaction by the careful addition (Pasteur pipet) of approximately 10 drops of 95% ethanol or until the observed foaming subsides. *The alcohol is added to decompose any unreacted BH$_3$•THF reagent and to hydrolyze the oxonium ion intermediate.*

Transfer the solution by Pasteur pipet to a 25-mL filter flask containing a boiling stone. Remove roughly one-half of the tetrahydrofuran solvent under reduced pressure with continuous swirling of the flask. (■) Now add two 1.0-mL portions of water (calibrated Pasteur pipet) to the solution. Carefully remove the **remaining** tetrahydrofuran–ethanol solvent under reduced pressure with continuous swirling of the flask. As the tetrahydrofuran and ethanol are removed and the solution becomes more concentrated, white crystals of xanthene appear as a slurry in the remaining water. Collect the crystals under reduced pressure using a Hirsch funnel.

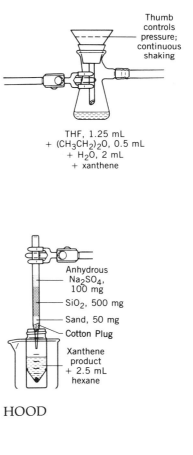

Thumb controls pressure; continuous shaking

THF, 1.25 mL + (CH$_3$CH$_2$)$_2$O, 0.5 mL + H$_2$O, 2 mL + xanthene

Anhydrous Na$_2$SO$_4$, 100 mg

SiO$_2$, 500 mg

Sand, 50 mg

Cotton Plug

Xanthene product + 2.5 mL hexane

## Purification and Characterization

Purify the crude xanthene by column chromatography. Place 0.5 g of activated silica gel followed by 0.1 g of anhydrous sodium sulfate in a Pasteur filter pipet. (■)

Wet the column first with a small amount of hexane, and then place a solution of the crude xanthene, dissolved in 0.25 mL of methylene chloride, on the column using a Pasteur filter pipet. Elute the xanthene by adding additional hexane (~2.5 mL). Collect the eluate in a tared 5-mL conical vial containing a boiling stone. Remove the hexane solvent by evaporation **[HOOD]** while warming in a sand bath to yield pure xanthene. This compound may be recrystallized from ethanol if further purification is found to be necessary after characterization of the product.

HOOD

After air-drying, weigh the solid and calculate the percent yield of xanthene. Determine the melting point and compare it with the value found in the literature.

Obtain an IR spectrum of your xanthene, using the KBr pellet technique, and compare the spectrum to that shown in Figure 7.1.

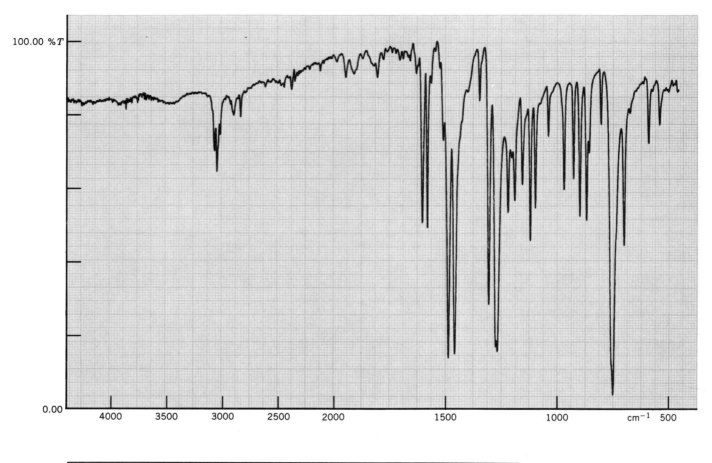

| Sample | Xanthene |
| --- | --- |

%T _X_ ABS __ Background Scans _4_     Scans _16_

Acquisition & Calculation Time _42 sec_     Resolution _4.0 cm⁻1_

Sample Condition ____solid____     Cell Window _____

Cell Path Length _____     Matrix Material _KBr_

**FIGURE 7.1**    IR spectrum: Xanthene.

In this experiment, the carbonyl group of an aromatic ketone was reduced to a methylene group. Examine the infrared spectra of the starting material and of the reduction product to see what evidence is present to indicate that the desired reaction has occurred.

### Infrared Analysis

We will first consider the spectrum of xanthone (Fig. 7.2). The macro group frequency associated with six-membered carbocyclic aromatic ring systems applies in this instance (peaks at 3100–3000, 1600, 1585, 1500, and 1450 cm⁻¹). This frequency train involves the bands located at 3085–3020, 1610–1570, 1484, and 1460 cm⁻¹. These peaks are assigned as follows:

**a.** 3085–3020 cm⁻¹: C—H stretch on $sp^2$ hybridized carbon. The breadth and complexity of this set of absorption bands indicates the presence of a fairly complex aromatic system.

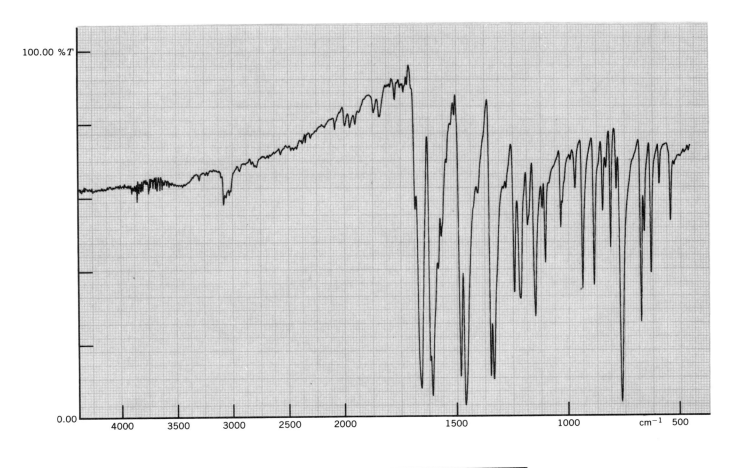

| Sample | Xanthone | | |
| --- | --- | --- | --- |
| %T _X_ ABS __ Background Scans _4_ | | Scans _16_ | |
| Acquisition & Calculation Time _42 sec_ | | Resolution _4.0 cm⁻¹_ | |
| Sample Condition _solid_ | | Cell Window _____ | |
| Cell Path Length _____ | | Matrix Material _KBr_ | |

**FIGURE 7.2**   IR spectrum: Xanthone.

**b.** $1610-1570$ cm$^{-1}$: The peaks observed in this region are related to the two degenerate fundamental stretching motions, $\nu_{8a}$ and $\nu_{8b}$, of the simple aromatic ring system. Normally, $\nu_{8a}$ is found near 1600 cm$^{-1}$ and is considerably more intense than $\nu_{8b}$, which is located near 1580 cm$^{-1}$. In xanthone the situation is more complicated, as the central $\gamma$-pyrone ring introduces a pseudoaromatic six-membered ring system. Thus, this ring system might be expected to possess somewhat shifted fundamental frequencies for the carbon rings. It is not unexpected then, that we observe a band system of four major components in this region.

**c.** 1484 and 1460 cm$^{-1}$: Aromatic ring stretch related to $\nu_{19a}$ and $\nu_{19b}$. These frequencies are less disturbed by the presence of the pyrone system and occur near their expected locations of 1500 and 1450 cm$^{-1}$.

The very strong band at 756 cm$^{-1}$, in the absence of strong absorption near 700 cm$^{-1}$, is supporting evidence for the presence of four adjacent ring C—H groups, which implies ortho substitution.

The intense band observed at 1656 cm$^{-1}$ is strong evidence for the presence of a highly conjugated carbonyl group which, of course, is consistent with the structure of the starting material.

The IR spectrum of the product (Fig. 7.1) supports the complete reduction to the methylene system. The macro group frequency train defined for six-membered carbocyclic aromatic systems still applies. The expected frequencies are very close to the observed values:

**a.** 3075–3025, 1603, 1589, 1490, and 1457 cm$^{-1}$.
**b.** The carbonyl band has vanished, and in its place two weak bands have arisen near 2902 and 2840 cm$^{-1}$. These latter peaks are assigned to the antisymmetric and symmetric stretching modes of the newly formed methylene group.
**c.** The most intense band in the spectrum occurs at 750 cm$^{-1}$ (four in a row, C—H all-in-phase, out-of-plane bend) and indicates that the basic substitution pattern has not changed on the ring system during the reaction.

## QUESTIONS

**7-1.** In the reaction performed in this experiment, assume that the first stage of the reaction is the rate-determining step. Would you then predict that the relative rate of reduction of the carbonyl group in Compound **A** to the methylene group to be faster or slower than that of xanthone under the conditions of this experiment? Explain.

Compound A:
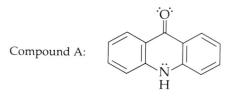

**7-2.** The reduction of aldehydes or ketones to the methylene group occurs with hydride reagents only when some special feature of the substrate promotes cleavage of the C—OH linkage. Suggest a suitable mechanism by which the reduction given below might occur.

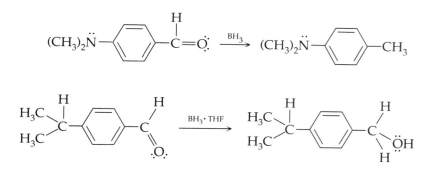

**7-3.** Borane also forms complexes with sulfides and amines. Draw a suitable structure to represent the complex formed between BH$_3$ and dimethyl sulfide, and also that formed between BH$_3$ and triethylamine.

**7-4.** Using your lecture textbook as a reference, find three different methods for the conversion of 4-methylcyclohexanone to methylcyclohexane.

**7-5.** The infrared spectrum of the xanthene reduction product contains evidence that demonstrates that conjugation of the rings is still maintained following removal of the carbonyl group. What is this evidence?

The reactions outlined in Experiments [1A$_{adv}$] and [1B$_{adv}$] were adapted from the work of:

Wechter, W. J. *J. Org. Chem.* **1963**, *28*, 2935.

Borane as a reducing agent:

Carruthers, W. *Some Modern Methods of Organic Synthesis*, 3rd ed.; Cambridge University Press: New York, 1986, Chapter 5.

Fieser, L. F.; Fieser, M. *Reagents For Organic Synthesis*; Wiley: New York, 1967, Vol. I, p. 199. Subsequent volumes of this series have further examples of diborane as a reducing agent.

Lane, C. F. *Aldrichimica Acta* **1973,** *6*, 36.

Pelter, A. *Chem. Ind. (London)* **1976,** 888.

Borane as a hydroborating agent:

Brown, H. C. *Hydroboration*; Benjamin: New York, 1962.

Brown, H. C. *Boranes in Organic Chemistry*; Cornell University Press: New York, 1972.

Brown, H. C. *Organic Synthesis Via Boranes*; Wiley: New York, 1975.

Zweifel, G.; Brown, H. C. *Org. React.* **1963**, *13*, 1.

**BIBLIOGRAPHY**

## Heterocyclic Ring Synthesis: Benzimidazole

**EXPERIMENT [2$_{adv}$]**

Common name: Benzimidazole
CA number: [51-17-2]
CA name as indexed: 1*H*-Benzimidazole

### Purpose

To investigate conditions under which one of the important heterocyclic ring systems, the benzimidazole ring, may be formed. The method used involves the condensation of a 1,2-diaminobenzene with formic acid. The simplest possible benzimidazole ring system, benzimidazole itself, is prepared in this experiment.

### *Prior Reading*

Technique 5:   Crystallization
              Use of the Hirsch Funnel (see pp. 93–95).
              Craig Tube Crystallizations (see pp. 95–96).

**REACTION**

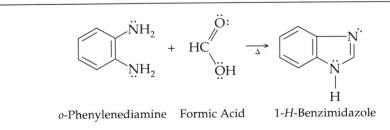

*o*-Phenylenediamine   Formic Acid     1-*H*-Benzimidazole

**DISCUSSION**

This experiment illustrates the classic method of forming the benzimidazole ring system. This heterocycle is generally prepared from 1,2-diaminobenzene (*o*-phenylenediamine) derivatives by reaction with carboxylic acids, or their derivatives, under acidic conditions. The ring system is aromatic; thus it is difficult to oxidize or reduce, and it is stable to both acids and bases. It is an important heterocyclic ring system that occurs in vitamin $B_{12}$ and in many other biologically important compounds. Benzimidazole itself inhibits the growth of certain yeasts and bacteria.

The parent compound of this class of heterocyclic compounds is imidazole. This ring system exhibits basic properties and is protonated to give a conjugate acid with $pK_a = 6.95$. Once the imidazole ring is protonated, the two nitrogen atoms are indistinguishable, since the resonance forms of the protonated species are equivalent.

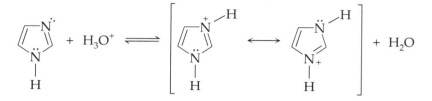

The reaction to form the ring system used in this experiment proceeds in two stages. The first involves the in situ formation of an N-substituted formamide, via the usual nucleophilic addition–elimination reaction. The second involves an intramolecular nucleophilic addition to a carbonyl group, and subsequent elimination of water to form the unsaturated heterocyclic ring. The sequence is outlined below.

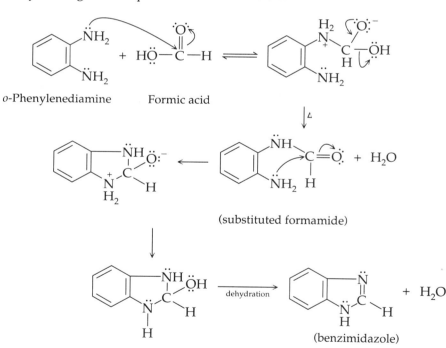

## EXPERIMENTAL PROCEDURE

Estimated time to complete the experiment: 2.0 hours.

### Physical Properties of Reactants

| Compound | MW | Amount | mmol | mp(°C) | bp(°C) | d | $n_D$ |
|---|---|---|---|---|---|---|---|
| o-Phenylenediamine | 108.1 | 108 mg | 1.0 | 102 | | | |
| 90% Formic acid | 46.03 | 64 μL | 1.7 | | 101 | 1.22 | 1.3714 |

## Reagents and Equipment

Weigh and add 108 mg (1.0 mmol) of *o*-phenylenediamine to a 3.0-mL conical vial containing a magnetic spin vane. Now add 64 $\mu$L (79 mg, 1.7 mmol) of 90% formic acid, and attach the vial to a reflux condenser protected by a calcium chloride drying tube. (■)

---

WARNING:  *o*-Phenylenediamine is toxic and is a cancer suspect agent. Formic acid is very corrosive to the skin and should be dispensed in the *hood* using an automatic delivery pipet.

---

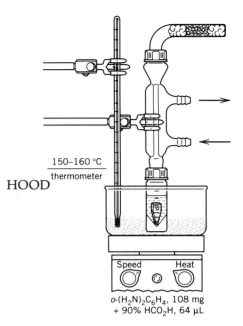

150–160 °C
thermometer
HOOD

Speed    Heat

*o*-$(H_2N)_2C_6H_4$, 108 mg
+ 90% $HCO_2H$, 64 $\mu$L

## Reaction Conditions

Heat the reaction mixture, with stirring, at a sand bath temperature of 150–160 °C for 1 hour.

## Isolation of Product

Allow the reaction mixture to cool to room temperature. Add 630 $\mu$L of 10% aqueous sodium hydroxide solution (automatic delivery pipet). Crude benzimidazole precipitates at this point. Collect the product by vacuum filtration using a Hirsch funnel, and wash the filter cake with three 0.5-mL portions of cold water (calibrated Pasteur pipet). (■)

## Purification and Characterization

Recrystallize the crude material from water, using a Craig tube. Dry the product in a desiccator or in a vacuum drying apparatus (see Prior Reading). Weigh the crystals and calculate the percent yield. Determine the melting point and compare your result with that listed in the literature.

The UV spectrum of benzimidazole in 95% ethanol has been reported.[1]

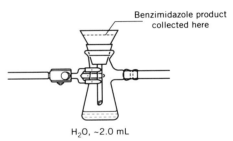

Benzimidazole product
collected here

$H_2O$, ~2.0 mL

## Chemical Tests

Several tests may be run to assist in the identification of this material. Does the ignition test confirm the presence of the aromatic ring system? Does the soda lime or sodium fusion test indicate that nitrogen is present? Is the material soluble in 10% hydrochloric solution?

**7-6.**  The parent compound of the imidazole series, imidazole (**I**) itself, was first prepared in 1858.

**QUESTIONS**

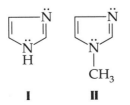

I          II

[1] Steck, E. A.; Nackod, F. C.; Ewing, G. W.; Gorman, N. H. *J. Am. Chem. Soc.* **1948,** 70, 2406.

Can you account for the fact that it has a very high boiling point (256 °C) whereas 1-methyl imidazole (**II**) has the somewhat lower boiling point of 199 °C?

**7-7.**  The imidazole ring system has a great deal of aromatic character. Can you formulate two resonance structures that account for this characteristic?

**7-8.**  Imidazole is a weak acid, and thus reacts with strong bases to form the corresponding anion. Show this reaction, and draw resonance structures that account for the stability of the conjugate base.

**7-9.**  Suggest a mechanism for the dehydration involved in the last step in the synthesis of benzimidazole.

**7-10.**  Imidazole, acting as a nucleophile, catalyzes the hydrolysis of phenyl acetate by attack on the carbonyl carbon atom of the ester. The imidazole displaces the phenoxide anion and forms acetyl imidazole. In turn, the acetyl imidazole is quite unstable in water and hydrolyzes to form acetic acid, and regenerates the imidazole molecule. Write a suitable mechanism outlining these steps.

## BIBLIOGRAPHY

The conditions of this reaction were adapted from those reported by:

Wagner, E. C.; Millett, W. H. *Organic Syntheses*; Wiley: New York, 1943; Collect. Vol. II, p. 65.

## EXPERIMENT [3_adv]

## Heterocyclic Ring Synthesis: 4-Hydroxycoumarin and Dicoumarol

Common name: 4-Hydroxycoumarin
CA number: [1076-38-6]
CA name as indexed: 2*H*-1-Benzopyran-2-one, 4-hydroxy-

Common names: Dicoumarol, dicoumarin
CA number: [66-76-2]
CA name as indexed: 2*H*-1-Benzopyran-2-one, 3,3'-methylenebis[4-hydroxy-

### Purpose
To synthesize a material, dicoumarol, that was the prototype for the oral anticoagulants widely used in medicine to lower blood coagulation rates. To illustrate the addition of a carbon nucleophile to a carbonyl carbon to form a C—C bond. To utilize a Claisen condensation reaction to prepare a $\beta$-ketoester which, upon cyclization, forms a lactone, 4-hydroxycoumarin. Further condensation of two mole equivalents of 4-hydroxycoumarin with formaldehyde yields dicoumarol.

### Prior Reading
*Technique 2:*   Simple Distillation at the Semimicroscale Level (see pp. 65–68).
*Technique 4:*   Solvent Extraction
                 Liquid–Liquid Extraction (see pp. 80–82).
*Technique 5:*   Crystallization
                 Use of the Hirsch Funnel (see pp. 93–95).
                 Craig Tube Crystallizations (see pp. 95–96).

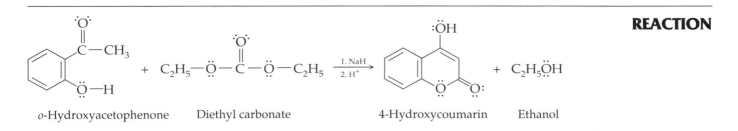

**REACTION**

o-Hydroxyacetophenone      Diethyl carbonate                4-Hydroxycoumarin       Ethanol

**DISCUSSION**

This reaction sequence illustrates the well-known Claisen condensation, which is widely used to form C—C bonds. The bond formation is brought about by the nucleophilic attack of an enolate anion on the carbonyl carbon of an ester. The enolate is generated by removal of a slightly acidic hydrogen from the $\alpha$-carbon atom of a ketone, nitrile, or ester, using a relatively strong base. The reaction mechanism is shown here. The methyl ketone is deprotonated by the base, sodium hydride. Sodium hydride (NaH) provides a basic and nonnucleophilic source of hydride ion (H$^-$). The resulting enolate then attacks the ester, diethyl carbonate. The $\beta$-ketoester product thus formed is esterified in an *intramolecular* reaction with the phenolic —OH group to form the lactone product **(I)**. Note that the methylene hydrogen atoms in this lactone **(I)** are quite acidic because they are adjacent to two carbonyl groups. In the basic medium of this reaction, the sodium enolate of the lactone **(II)** will be formed. This enolate is water soluble, since it is a salt, which explains why the aqueous solution must be acidified to precipitate the neutral 4-hydroxycoumarin when isolating the product.

**1.** 4-Hydroxycoumarin

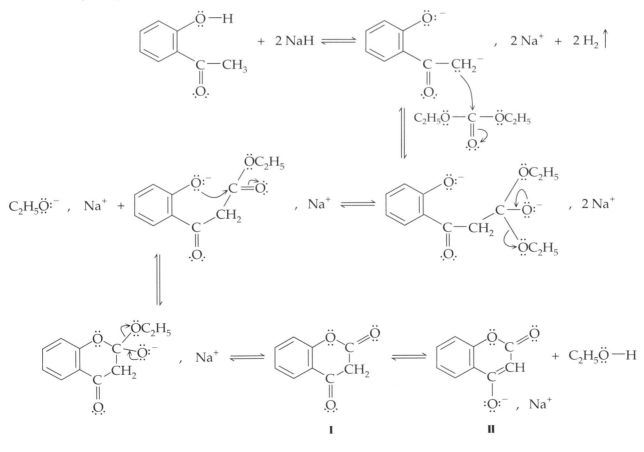

I                                       II

**2.** Dicoumarol

The aldol condensation of 4-hydroxycoumarin with formaldehyde provides an $\alpha,\beta$-unsaturated carbonyl compound, which then undergoes a conjugate (1,4) addition of a second molecule of 4-hydroxycoumarin. This reaction could be catalyzed by either trace base or by trace acid; the acid-catalyzed reaction is shown and discussed here. The enol portion of 4-hydroxycoumarin is the nucleophile in an aldol reaction with protonated formaldehyde. The resulting product dehydrates to provide the $\alpha,\beta$-unsaturated carbonyl compound, which, after protonation renders it a more reactive electrophile, then reacts with another nucleophilic molecule of 4-hydroxycoumarin in a conjugate addition reaction. This product, upon loss of a proton to the aqueous solvent, leads to dicoumarol. This substance is present in moldy sweet clover. It is a blood anticoagulant, and its ingestion leads to the haemorrhagic sweet clover disease that kills cattle.

## Experiment [3A$_{adv}$]   4-Hydroxycoumarin

The reaction is shown on page 429.

---

Estimated time to complete the reaction: 4.0 hours.

**EXPERIMENTAL PROCEDURE**

### Physical Properties of Reactants

| Compound | MW | Amount | mmol | bp(°C) | d | $n_D$ |
|---|---|---|---|---|---|---|
| Sodium hydride (60% in oil dispersion) | 24.0 | 85 mg | 2.13 | | | |
| Toluene | 92.15 | 6.0 mL | | 111 | | |
| o-Hydroxyacetophenone | 136.16 | 133 μL | 1.1 | 218 | 1.13 | 1.5584 |
| Diethyl carbonate | 118.13 | 333 μL | 2.75 | 126 | 0.96 | 1.3845 |

### Reagents and Equipment

**NOTE.** *It is **very important** that all equipment used in this reaction be thoroughly dried in an oven at 110 °C for 30 min just prior to use. Upon removal from the drying oven, it should be allowed to cool to ambient temperature in a desiccator.*

Weigh and place 85 mg (2.13 mmol) of sodium hydride (60% dispersion in oil) in a 10-mL round-bottom flask containing a magnetic stirring bar. Now add 3.0 mL of dry toluene. Attach the flask to a Hickman still fitted with an air condenser protected by a calcium chloride drying tube. Wrap the Hickman still 14/10 ꙇ male joint with Teflon tape to prevent joint freeze-up. (■)

---

CAUTION:   Sodium hydride (NaH) is a flammable solid. Dispense in the *hood*. Toluene is distilled and stored over molecular sieves. Also dispense this solvent in the *hood*.

---

In rapid order, place 133 μL (150 mg, 1.1 mmol) of o-hydroxyacetophenone, 3.0 mL of dry toluene, and 333 μL (324 mg, 2.75 mmol) of diethyl carbonate in a stoppered 10-mL Erlenmeyer flask.

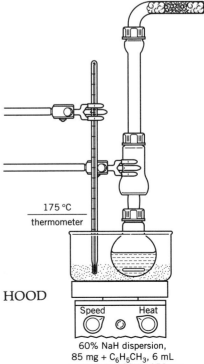

175 °C
thermometer

HOOD

Speed    Heat

60% NaH dispersion,
85 mg + C$_6$H$_5$CH$_3$, 6 mL
+ o-HOC$_6$H$_4$COCH$_3$, 150 mg
+ (CH$_3$CH$_2$O)$_2$CO, 342 mg

**NOTE.** *Dispense the small volumes of the liquid reagents using an automatic delivery pipet. Be sure to dry the removable plastic tips in the oven before use. Use a 10-mL graduated cylinder to measure the toluene.*

Remove the air condenser from the distilling head and add the o-hydroxyacetophenone solution, with stirring, to the reaction flask as rapidly as possible using a Pasteur pipet. The resulting solution turns yellow. Immediately reattach the air condenser.

## Reaction Conditions

Place the reassembled apparatus in a sand bath and rapidly raise the temperature of the bath to about 175 °C.

---

**CAUTION:** **To avoid excessively high temperatures, calibrate the hot plate temperature control prior to the experiment.**

---

Collect the ethanol and toluene distillate (1.0 mL in 2–3 fractions) in the collar of the still, and then remove the apparatus from the heat source. Allow the reaction solution to cool to room temperature, and then add 3.0 mL of water with stirring.

## Isolation of Product

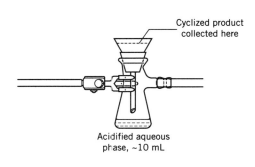

Cyclized product collected here

Acidified aqueous phase, ~10 mL

Transfer the resulting two-phase solution, using a Pasteur pipet, to a 15-mL centrifuge tube. Rinse the reaction flask with an additional 2.0 mL of water and add this to the centrifuge tube. Separate the toluene layer using a Pasteur filter pipet and transfer it to a second 15-mL centrifuge tube. Then extract this organic phase with 3.0 mL of water, and add the water extract to the original water phase. *This aqueous solution contains the water-soluble sodium enolate* (**II**). Cool the combined aqueous layers in an ice bath and acidify them by dropwise addition of concentrated hydrochloric acid delivered from a Pasteur pipet. The solid product precipitates from the aqueous phase. Add acid until the yellow color of the solution disappears (~10 drops). Collect the product by vacuum filtration using a Hirsch funnel. (■)

## Purification and Characterization

Recrystallize the crude 4-hydroxycoumarin from 50% ethanol using a Craig tube. Dry the material on a porous clay plate or in a vacuum apparatus (see Prior Reading).

Weigh the product and calculate the percent yield. Determine the melting point and obtain an IR spectrum using the KBr pellet technique. Compare your results with those reported in the literature and in *The Aldrich Library of IR Spectra.*

## Chemical Tests

You may wish to perform the ignition test to establish that the compound is aromatic. Does the ferric chloride test (Chapter 10) for phenols give a positive result?

## Experiment [3B$_{adv}$]   Dicoumarol

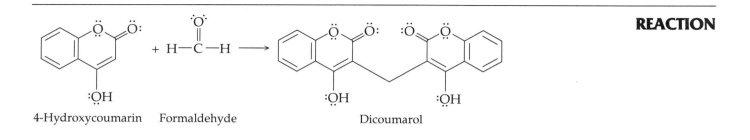

4-Hydroxycoumarin    Formaldehyde                Dicoumarol

---

Estimated time to complete the reaction: 0.5 hour.

### Physical Properties of Reactants

| Compound | MW | Amount | mmol | mp(°C) | bp(°C) |
|---|---|---|---|---|---|
| 4-Hydroxycoumarin | 162.15 | 50 mg | 0.31 | 213–214 | |
| Water | | 15 mL | | | 100 |
| Formaldehyde (37% in H$_2$O) | | 0.5 mL | 6 | | |

### Reagents and Equipment

Weigh and place 50 mg (0.31 mmol) of 4-hydroxycoumarin, followed by 15 mL of water, in a 50-mL Erlenmeyer flask containing a boiling stone. Heat the mixture to boiling on a hot plate. Now add 0.5 mL (~500 mg, ~6 mmol) of formaldehyde (37% aqueous solution) to the resulting solution.

> **WARNING:** Formaldehyde is a cancer suspect agent. Dispense in the *hood*.

HOOD

### Reaction Conditions

White crystals of the product form immediately on addition of the formaldehyde solution. Cool the flask in an ice bath.

### Isolation of Product

Collect the solid product by vacuum filtration using a Hirsch funnel. (■) Wash the filter cake with three 1-mL portions of ice water. Remove the crystals and dry them on filter paper or on a porous clay plate.

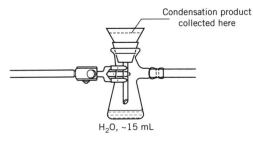

Condensation product collected here

H$_2$O, ~15 mL

### Purification and Characterization

Recrystallize the material from a mixture of toluene/cyclohexanone (~2:1) using a Craig tube. After drying the isolated material, determine its melting point and compare your value with that in the literature. Obtain an IR spectrum using the KBr disk technique, and compare your spectrum with that in *The Aldrich Library of IR Spectra*.

## QUESTIONS

**7-11.** Why must water be excluded from the reaction that generates 4-hydroxycoumarol?

**7-12.** In the first step of the above reaction, formation of 4-hydroxycoumarin, what is the purpose of removing the ethanol generated in the reaction?

**7-13.** The Dieckmann condensation is simply an intramolecular Claisen condensation. For example,

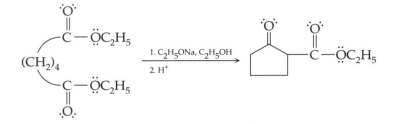

Suggest a suitable mechanism to account for the formation of the cyclic β-ketoester.

**7-14.** Predict the product formed in each of the following reactions.

**a.** Acetophenone + diethyl carbonate $\xrightarrow[\text{2. H}^+]{\text{1. NaH, toluene}}$

**b.** Acetophenone + ethyl formate $\xrightarrow[\text{2. H}^+]{\text{1. C}_2\text{H}_5\text{ONa, ethanol}}$

**7-15.** Suggest a synthesis for each of the following compounds utilizing the Claisen condensation. Any necessary organic or inorganic reagents may be used.

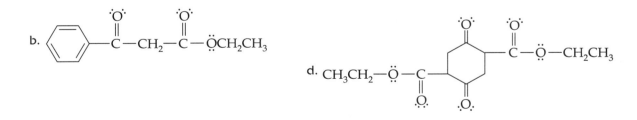

**7-16.** Draw structures for the two possible mono enols of 1,3-cyclohexanedione. Explain which one is more stable.

---

## BIBLIOGRAPHY

A review of the Claisen condensation is given in:

Hauser, C. R.; Swamer, F. W.; Adams, J. T. *Org. React.* **1954,** *8*, 59.

Selected references of Claisen condensations from *Organic Syntheses* include:

Ainsworth, C. *Organic Syntheses*; Wiley: New York, 1963; Collect. Vol. IV, p. 536.

John, J. P.; Swaminathan, S.; Venkataraman, P. S. *Organic Syntheses*; Wiley: New York, 1973; Collect. Vol. V, p. 747.

Magnani, A.; McElvain, S. M. *Organic Syntheses*; Wiley: New York, 1955; Collect. Vol. III, p. 251.

Marvel, C. S.; Dreger, E. E. *Organic Syntheses*; Wiley: New York, 1941; Collect. Vol. I, p. 238.

Riegel, E. R.; Zwilgmeyer, F. *Organic Syntheses*; Wiley: New York, 1943; Collect. Vol. II, p. 126.

Snyder, H. R.; Brooks, L. A.; Shapiro, S. H. *Organic Syntheses*; Wiley: New York, 1943; Collect. Vol. II, p. 531.

---

## Grignard and Aryl Halide Cross-Coupling Reaction: 1-Methyl-2-(methyl-$d_3$)-benzene

**EXPERIMENT [4_adv]**

---

Common names: 1-Methyl-2-(methyl-$d_3$)-benzene, $\alpha,\alpha,\alpha$-$d_3$-$o$-xylene
CA number: [25319-53-3]
CA name as indexed: Benzene, 1-methyl-2-(methyl-$d_3$)-

### Purpose
To illustrate a selective cross-coupling reaction between an alkyl Grignard reagent[2] and an aryl halide in the presence of a nickel phosphine catalyst to form a C—C bond. The reaction is also used to demonstrate the technique of "labeling" specific hydrogen atoms with an isotope for identification purposes.

### Prior Reading
Technique 4:   Solvent Extraction
                Liquid–Liquid Extraction (see pp. 80–82).
                Drying of the Wet Organic Layer (see pp. 86–87).
Technique 6:   Chromatography
                Column Chromatography (see pp. 98–101).
                Concentration of Solutions (see pp. 106–108).

---

The Grignard reagent is prepared in Step 1.

**REACTIONS**

### Step 1

$$CD_3I \quad + \; Mg \xrightarrow{\text{ether}} \quad D_3CMgI$$
Methyl-$d_3$ iodide          Methyl-$d_3$-magnesium iodide

The coupling reaction occurs in Step 2.

### Step 2

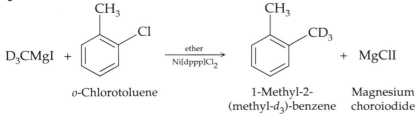

D_3CMgI  +          $\xrightarrow[\text{Ni[dppp]Cl}_2]{\text{ether}}$          + MgClI

$o$-Chlorotoluene          1-Methyl-2-          Magnesium
                         (methyl-$d_3$)-benzene   choroiodide

Alternatively, the use of methyl iodide in the formation of the Grignard reagent produces $o$-xylene in Step 2.

---

[2] For general reference on Grignard reagents refer to Experiment [16].

**DISCUSSION**

Before the discovery of the nickel-phosphine catalyst, the cross-coupling of Grignard reagents with organic halides was seldom employed in synthetic practice. This fact was mainly due to the formation of homocoupling products along with significant elimination side reactions. With the use of this new catalyst, the reaction now has wide application in the synthesis of unsymmetrical alkanes and alkenes.

In this experiment we will synthesize an isotopically labeled *o*-xylene. The *o*-xylene product will be labeled with a trideuteromethyl group. Deuterium, one of the hydrogen **isotopes**, can be used as a **label** in organic compounds. Its incorporation into a molecule can be detected by IR, NMR, or MS. Deuterium labeling is a particularly powerful way of investigating infrared spectra because the resultant frequency shifts (which are inversely proportional to the square root of the mass ratio; see Infrared Discussions, Chapter 9) are the largest obtainable with stable isotopes. Isotopic labeling is often used to follow hydrogen-deuterium exchange reactions, such as enolizations, to study biological reactions, and for gaininginsight into organic reaction mechanisms. The nickel catalyst used in this experiment is [1,3-bis(diphenylphosphino)propane]nickel (II) chloride, mercifully abbreviated as Ni[dppp]Cl$_2$.

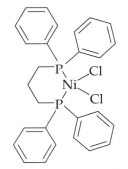

[1,3-Bis(diphenylphosphino)propane]nickel(II) chloride
(Ni[dppp]Cl$_2$)

It has been proposed that the catalytic ability of the dihalodiphosphinenickel reagent stems from its ability to react with a Grignard reagent to form a diorganonickel complex (**A**).[3] This complex is then converted to the halo-organonickel complex **B**, by reacting it with an *organic halide*.

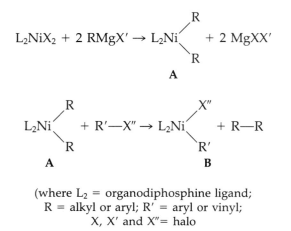

(where L$_2$ = organodiphosphine ligand;
R = alkyl or aryl; R′ = aryl or vinyl;
X, X′ and X″= halo)

[3] Tamao, K.; Sumitani, K.; Kumada, M. *J. Am. Chem. Soc.* **1972**, *94*, 4374.

Reaction of Complex **B** with the Grignard reagent forms a new diorganonickel Complex **C**, from which the cross-coupling product is formed by attack of the organic halide. In the reaction, Complex **B** is regenerated and thus the catalytic cycle is completed.

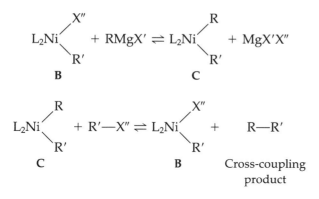

---

Estimated time to complete the experiment: 5.0 hours.

## EXPERIMENTAL PROCEDURE

### Physical Properties of Reactants and Product

| Compound | MW | Amount | mmol | mp(°C) | bp(°C) | d | $n_D$ |
|---|---|---|---|---|---|---|---|
| Magnesium | 24.31 | 65 mg | 2.7 | 649 | | | |
| Diethyl ether | 74.12 | 700 μL | | | 34.5 | 0.71 | 1.3526 |
| Methyl-$d_3$-iodide | 144.96 | 250 μL | 3.93 | | 42 | 2.28 | 1.5262 |
| Ni[dppp]Cl$_2$ | 542.1 | 10 mg | 0.02 | | | | |
| o-Chlorotoluene | 126.59 | 150 μL | 1.28 | | 159 | 1.08 | 1.5268 |
| 1-Methyl-2-(methyl-$d_3$)-benzene | 109.17 | | | | 144 | | 1.5055 |

**NOTE.** *All the glassware used in the experiment should be cleaned, dried in an oven at 110 °C for at least 30 min, and then cooled in a desiccator before use.*

### Reagents and Equipment

Equip a 5.0-mL conical vial containing a magnetic spin vane with a reflux condenser protected by a calcium chloride drying tube.

**Step 1.** Scrape a 4–5-in. piece of magnesium ribbon to remove the oxide coating. Then cut it into 1-mm-long sections. *Using forceps,* weigh and add 65 mg (2.7 mmol) of the magnesium sections to the conical vial. In the **hood,** use an automatic delivery pipet to dispense 300 μL of anhydrous diethyl ether into the conical vial containing the magnesium. Reassemble the apparatus.

HOOD

*Using automatic delivery pipets,* prepare a solution of 250 μL of methyl-$d_3$ iodide in 200 μL of anhydrous diethyl ether in a capped vial.

---

**CAUTION:** Methyl-$d_3$ iodide is a suspected carcinogen.

---

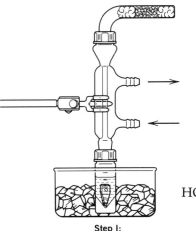

**Step I:**
Mg, 65 mg +
(CH₃CH₂)₂O, 0.5 mL +
CD₃I, 250 μL

Draw this solution into a Pasteur pipet. Remove the drying tube, insert the pipet down the length of the condenser, allowing the pipet bulb to rest on the condenser lip.

### Reaction Conditions

Place the reaction vial in an ice bath, and add the methyl-$d_3$ iodide dropwise with stirring. Withdraw the pipet, reinstall the drying tube, and stir the mixture for 20 min. (■)

HOOD    **STEP 2.   [This step must be carried out in the *hood*]**   In this step, all liquids should be dispensed in the **hood** using automatic delivery pipets. Weigh and add to a capped 0.5 dram vial, 10 mg (0.02 mmol) of Ni[dppp]Cl₂.

---

**CAUTION:   Ni[dppp]Cl₂ is a cancer suspect agent.**

---

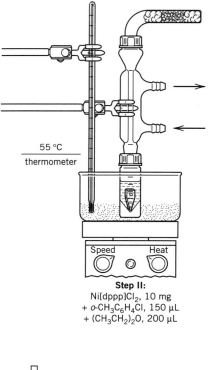

55 °C
thermometer

**Step II:**
Ni[dppp]Cl₂, 10 mg
+ o-CH₃C₆H₄Cl, 150 μL
+ (CH₃CH₂)₂O, 200 μL

Now add 150 μL of o-chlorotoluene and 200 μL of anhydrous diethyl ether. Draw this solution into a Pasteur pipet and, in one portion, add it to the reaction vial through the condenser as described above. Reinsert the drying tube. Place this mixture in a preheated sand bath at a temperature of 55 °C. Heat with stirring for 2 hours. (■)

On addition of the catalyst, the solution turns green. After approximately 30 min of heating, the mixture becomes dark brown.

### Isolation of Product

Cool the reaction mixture in an ice bath and quench the reaction by the dropwise addition (calibrated Pasteur pipet) of 1.5 mL of 1.0 $M$ HCl solution. *The acid is added slowly since frothing occurs.*

**NOTE.**   *The following extracting solutions are measured using calibrated Pasteur pipets. For each extraction operation, the vial is capped, shaken, and vented, and the layers are allowed to separate. The aqueous (lower) phase is then removed.*

Add 1.0 mL of ether to the solution. Using a Pasteur filter pipet, remove the aqueous layer. Extract the remaining ether phase with 1.0 mL of water followed by 1.0 mL of saturated sodium bicarbonate solution and then 1.0 mL of deionized water. Then extract with two 1.0-mL portions of 1.0 $N$ sodium thiosulfate (Na₂SO₃) solution and 1.0 mL of water.

### Purification and Characterization

Isolate and purify the reaction product using column chromatography. In a modified Pasteur filter pipet, place 200 mg of activated silica gel (100 mesh) followed by 1.0 g of anhydrous sodium sulfate. Wet the column with 0.5 mL of ether (calibrated Pasteur pipet). (■)

Transfer the ether solution of product to the column by Pasteur pipet, and elute the material from the column with two 0.5-mL portions of ether. Collect the eluate in a tared 10-mL Erlenmeyer flask and then transfer the

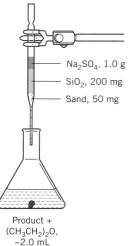

Na₂SO₄, 1.0 g
SiO₂, 200 mg
Sand, 50 mg

Product +
(CH₃CH₂)₂O,
~2.0 mL

eluate to a 5-mL conical vial for concentration. Evaporate the solvent by warming the vial in a sand bath **[HOOD]** to yield the "labeled" *o*-xylene.    HOOD
   Weigh the product and calculate the percent yield. Determine the boiling point, density, and refractive index (optional) and compare your values with those listed for *o*-xylene in the literature. Obtain an IR spectrum of the crude (dry) reaction product using the capillary film sampling technique and analyze it following the analysis given below.

## Infrared Analysis

The spectrum of *o*-chlorotoluene (Fig. 7.3) nicely mimics the large macro group frequency train for an ortho-substituted dialkylbenzene system, which has the following frequency train: 3100–3000, 3000–2850, 1950, 1920, 1880, 1840, 1790, 1690, 1600, 1580, 1500, 1450, 1380, and 750 cm$^{-1}$.

   **a.** 3072 and 3026 cm$^{-1}$: C—H stretch on aromatic ring.
   **b.** 3000–2850 cm$^{-1}$: C—H stretch on alkyl substituents.
   **c.** 1955, 1915, 1880, 1835, 1795, and 1690 cm$^{-1}$: Combination band pattern for ortho-disubstituted rings. The fit in this case is extremely good.

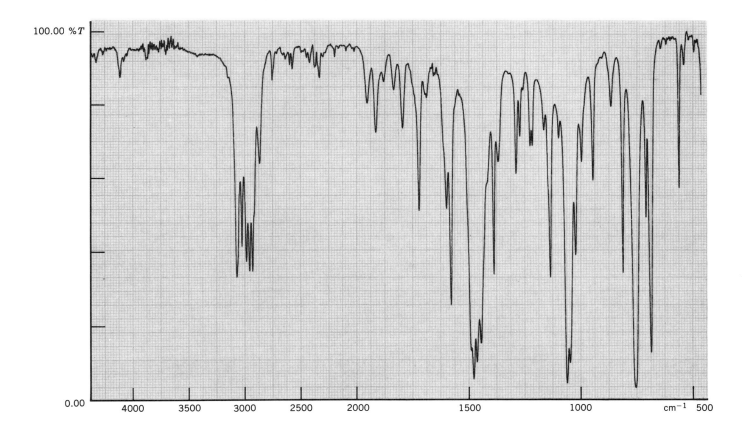

**FIGURE 7.3**   IR spectrum: 2-Chlorotoluene.

**d.** 1596 and 1577, 1475 and 1450 cm⁻¹: Two degenerate pairs of ring stretching modes. The 1450 wavenumber band is overlapped by the antisymmetric methyl deformation vibrations.

**e.** 1382 cm⁻¹: Symmetric methyl deformation mode.

**f.** 749 cm⁻¹: All in-phase out-of-plane bend of four adjacent C—H groups on the aromatic ring.

The C—Cl stretch falls below the region of measurement in these aromatic systems.

The IR spectrum of 1-methyl-2-(methyl-$d_3$)-benzene is shown in Figure 7.4. The reaction has replaced the —Cl group with a —CD₃ group. Thus, the only spectral changes expected to be observed will involve the vibrations of the labeled group. To a first approximation, the frequency should shift by a factor of $1/\sqrt{2} = 0.707$, but in practice the full shift is not observed as most vibrations are not pure modes. A factor of 0.75–0.72 is generally observed. In 1-methyl-2-(methyl-$d_3$)-benzene, we have two sets of doublets, with the doublet centers at 2220 and 2090 cm⁻¹. Since we would expect the C—D stretching modes to give rise to only two band systems, coupling with overtones of lower lying fundamental vibrations is likely occurring. The corresponding C—H stretching modes also evidence some coupling. If we use the major band centers observed at 2965 and 2870 cm⁻¹, however, as the C—H stretching values, the observed ratios are 2220/2965 = 0.749 and 2090/2870 = 0.728, results that are quite reasonable to expect for this type of isotopic shift. The bending modes are moved into the cluttered fingerprint region and do not easily lend themselves to analysis.

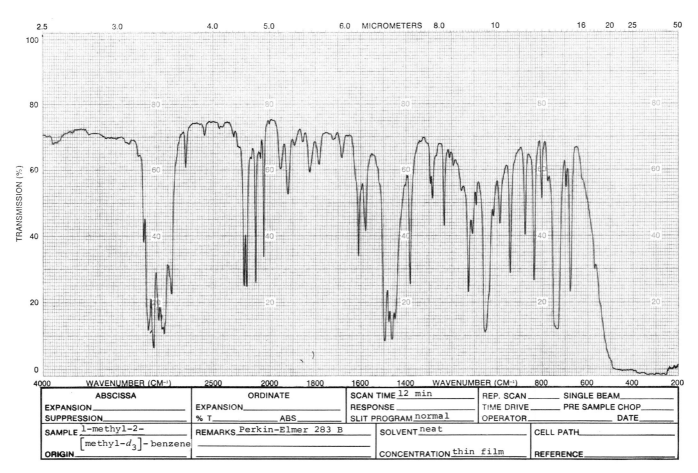

**FIGURE 7.4** IR spectrum: 1-Methyl-2-(methyl-$d_3$)-benzene.

Examine the spectrum of the reaction product you have obtained as a capillary film. Discuss the similarities and differences of the experimentally derived spectral data to the reference spectra (Figs 7.3 and 7.4 )

**7-17.** Predict the reagents that could be used to prepare each of the compounds by the Grignard coupling reaction. Give a suitable name to each reactant.

a. 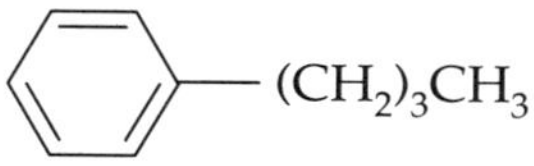 ⬡—$(CH_2)_3CH_3$

c. 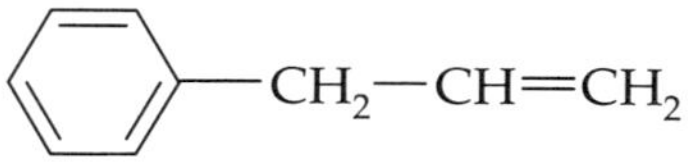 ⬡—$CH_2$—$CH$=$CH_2$

b. 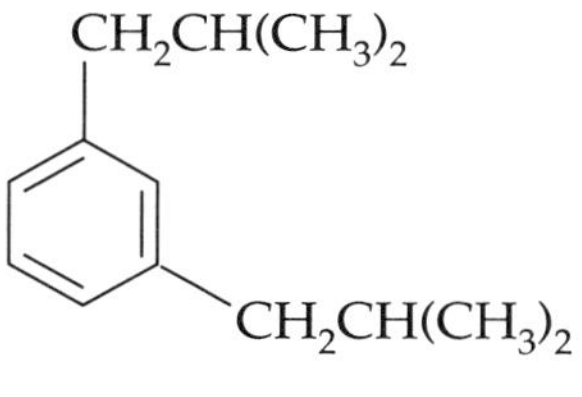 $CH_2CH(CH_3)_2$ / ⬡ / $CH_2CH(CH_3)_2$

d. 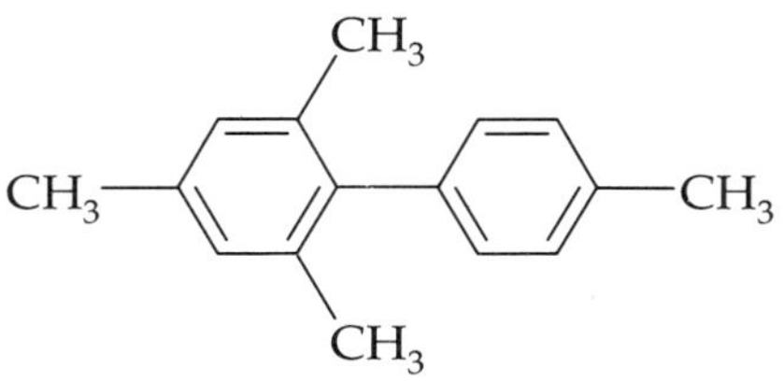 $CH_3$— (with $CH_3$ substituents) —biphenyl— $CH_3$

**7-18.** The Wurtz coupling reaction involves the treatment of haloalkanes with an active metal, such as sodium.
   a. Butyl bromide + Na
   b. 1-Bromo-3-chlorocyclobutane + Na
What products would be formed in the above reactions? Name the products.

**7-19.** In reference to Question 7-18, explain why the coupling reaction of *n*-butyl bromide and *n*-propyl bromide under conditions of the Wurtz reaction is synthetically inefficient for the preparation of heptane.

**7-20.** Describe what you would expect to see in the $^1H$ and $^{13}C$ NMR spectra of 1-methyl-2-(methyl-$d_3$)-benzene.

**7-21.** The spectrum of *o*-chlorotoluene given in Figure 7.3 has two additional peaks that are not present in the library reference standard spectrum. Both these bands are weak. One occurs near 1714 cm$^{-1}$ and the other at 1362 cm$^{-1}$. Can you offer an explanation for the presence of the extra absorption bands? How successful do you think the coupling reaction would be if carried out utilizing this sample as the starting material?

**7-22.** In the spectrum of *o*-chlorotoluene, the lower wavenumber band of the 1596, 1577 pair is the more intense member. In the labeled product, the 1610, 1580 pair has the higher wavenumber peak more intense. Explain this reversal of intensities.

---

The present coupling reaction is based on the work reported by:

Kumada, M.; Tamao, K.; Sumitani, K. *Organic Syntheses*; Wiley: New York, 1988; Collect. Vol. VI, p. 407, and references cited therein.

Selected Grignard coupling reactions presented in *Organic Syntheses*:

Gilman, H.; Catlin, W. E. *Organic Syntheses*; Wiley: New York, 1941; Collect. Vol. I, p. 471.

Gilman, H.; Robinson, J. *Organic Syntheses*; Wiley: New York, 1943; Collect. Vol. II, p. 47.

Lespieau, R.; Bourguel, M. *Organic Syntheses;* Wiley: New York, 1941; Collect. Vol. I, p. 186.

Smith, L. I. *Organic Syntheses;* Wiley: New York, 1943; Collect. Vol. II, p. 360.

Turk, A.; Chanan, H. *Organic Syntheses;* Wiley: New York, 1955; Collect. Vol. III, p. 121.

The reaction was first described by:

Mayo, D. W.; Bellamy, L. J.; Merklin, G. T.; Hannah, R. W. *Spectrochim. Acta* **1985**, *41A*, 355.

## EXPERIMENT [5ₐdᵥ]    Oxidative Coupling of 2-Naphthol: 1,1′-Bi-2-Naphthol

Common name: 1,1′-Bi-2-naphthol
CA number: [602-0-5]
CA name as indexed: [1,1′-Binaphthalene]-2,2′-diol

### Purpose
To investigate the coupling reaction that aromatic phenols undergo in the presence of transition metal oxidants. This reaction mimics the biogenetic process that occurs in nature.

### Prior Reading
Technique 5:  Crystallization
             Use of the Hirsch Funnel (see pp. 93–95).
             Craig Tube Crystallizations (see pp. 95–96).

## REACTION

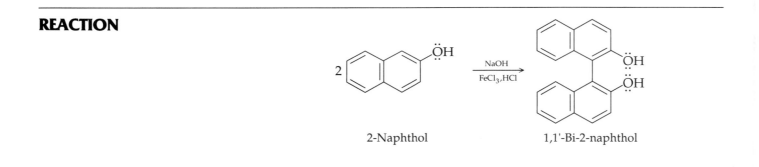

2-Naphthol                    1,1′-Bi-2-naphthol

## DISCUSSION

This reaction is important since it illustrates oxidative coupling of phenols, which is an important biogenetic pathway in nature, leading to the formation of many different natural products.

The coupling reaction involves oxidation of 2-naphthol by electron transfer to give an aryloxy radical, which then dimerizes to yield the product. The mechanism is shown at top of page 443.

This binaphthol compound, by virtue of restricted rotation about the single bond joining the two naphthalene groups, is a chiral compound. That is, the molecule cannot readily exist in a planar form because of the steric interference of the bulky —OH substituents, and in fact, the two

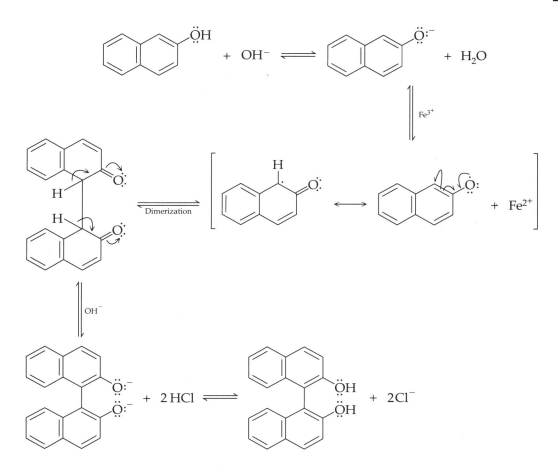

enantiomers have been separated. Such enantiomeric compounds contain no stereocenter, but rather have an axis of chirality, or a *chiral axis*, which in this case contains the bond joining the two napthalene rings (see Fig. 7.5).[4]

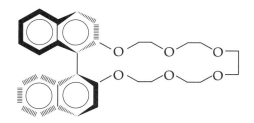

An investigation of the complexation of this class of molecules with various ionic species may lead to important insights into the catalytic nature of enzymes. For example, the crown ether similar to that above containing *two* 2,2'-substituted 1,1'-binaphthyl groups as chiral barriers, complexes preferentially with one enantiomer of some primary amine salts.

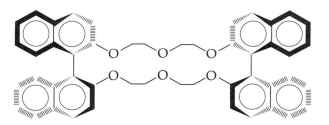

[4] Donald J. Cram (Nobel Laureate, 1987) and coworkers have incorporated this binaphthyl group into cyclic crown ethers.

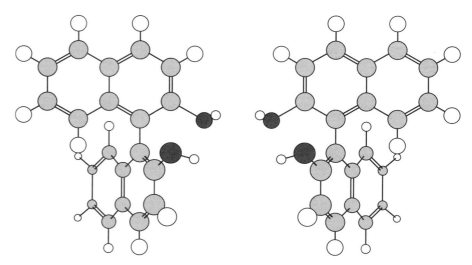

**FIGURE 7.5** Enantiomers of 1,1′-bi-2-naphthol.

Complexation of enantiomerically pure binaphthols with metal hydride reducing agents allow enantiospecific reductions of some carbonyl compounds.

## EXPERIMENTAL PROCEDURE

Estimated time to complete the experiment: 2.0 hours.

### Physical Properties of Reactants

| Compound | MW | Amount | mmol | mp(°C) |
|---|---|---|---|---|
| 2-Naphthol | 144.19 | 100 mg | 0.69 | 123–124 |
| Sodium hydroxide | 40.00 | 30 mg | 0.75 | 318.4 |
| Ferric chloride•6 H₂O | 270.30 | 297 mg | 1.1 | 37 |
| HCl (concd) | 36.46 | 100 μL | | |
| Water | 18.06 | 3 mL | | |

### Reagents and Equipment

Weigh and add 100 mg (0.69 mmol) of 2-naphthol and 30 mg (0.75 mmol) of sodium hydroxide to a 10-mL round-bottom flask containing a stirring bar.

---

**CAUTION: Sodium hydroxide is a corrosive and toxic chemical. Do not allow it to touch the skin or to come in contact with the eyes.**

---

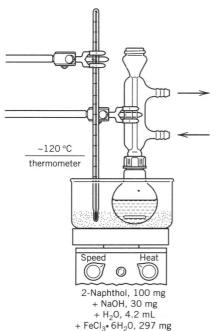

~120 °C
thermometer

Speed    Heat

2-Naphthol, 100 mg
+ NaOH, 30 mg
+ H₂O, 4.2 mL
+ FeCl₃• 6H₂O, 297 mg
+ concd HCl, 100 μL

Add 3.0 mL of water. Attach the flask to a reflux condenser and place the assembly in a sand bath on a magnetic stirring hot plate. (■)

Heat the reaction mixture to reflux, with stirring, using a sand bath temperature of 120 °C.

In a 10-mL Erlenmeyer flask prepare a solution of: 180 mg (1.1 mmol) of *anhydrous* ferric chloride (MW 162) *or* 297 mg (1.1 mmol) of ferric chloride *hexahydrate* (MW 270), 1.0 mL of water (calibrated Pasteur pipet), and 100 μL of concentrated hydrochloric acid.

> **WARNING:   BE CAREFUL when mixing acid with water. Add the acid TO the water. Avoid contact with the skin. Dispense the acid with an automatic delivery pipet.**

Using a Pasteur pipet, transfer the ferric chloride solution, through the top of the condenser, to the reaction flask. Rinse the Erlenmeyer flask with 200 μL of water, and add this rinse to the reaction flask as before.

### Reaction Conditions

Heat the resulting mixture at reflux, with stirring, for 45–60 min using a sand bath temperature of 120 °C. Allow the mixture to cool to room temperature, and then place the flask in an ice bath to complete crystallization of the product.

### Isolation of Product

Collect the solid product by vacuum filtration using a Hirsch funnel, and wash the filter cake with two 1.0-mL portions of cold water (calibrated Pasteur pipet). (■)

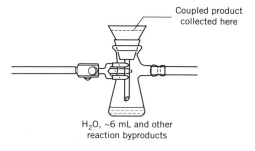

Coupled product collected here

H$_2$O, ~6 mL and other reaction byproducts

### Purification and Characterization

Recrystallize the crude product from 95% ethanol using a Craig tube, and dry the resulting crystals on a porous clay plate, or on filter paper.

Weigh the crystals and calculate the percent yield. Determine the melting point and compare it with the literature value. Obtain the IR spectrum (KBr disk) and compare it to that shown in *The Aldrich Library of IR Spectra.*

### Chemical Tests
Chemical classification tests (Chapter 10) may be used to assist in characterization of this material. Perform the ignition test and the ferric chloride test.

## QUESTIONS

**7-23.**   In the oxidative coupling reaction of 1-naphthol, it is *possible* to obtain three products.

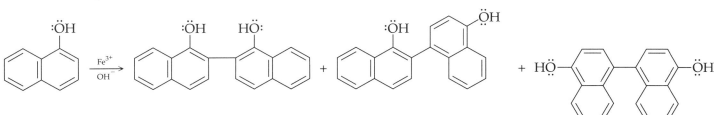

Account for the formation of this mixture by suggesting a mechanistic sequence similar to that presented in the discussion section of this experiment.

**7-24.**  Predict the structure of the diphenylquinone product, $C_{28}H_{40}O_2$, formed by the oxidative coupling of 2,6-di-*tert*-butylphenol with oxygen in the presence of base.

**7-25.**  Substituted phenols, such as BHT, are used as antioxidants in processed foods.

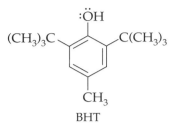

BHT

The role of the antioxidant is to stop spoilage caused by the free radical reactions brought about by reaction of oxygen with compounds containing C=C bonds.
   a. Give a suitable chemical name for BHT.
   b. Can you suggest why this compound is an effective antioxidant?

**7-26.**  Naphthalene is the simplest of the polycyclic aromatic hydrocarbons and can be represented by three resonance structures. Draw them. Indicate which of the structures are equivalent.

**7-27.**  Some relatively simple chiral compounds contain a chiral axis. Use molecular models to convince yourself that the two molecules below are enantiomers, even though they contain no stereocenters.

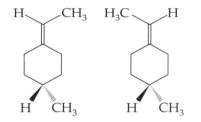

**BIBLIOGRAPHY**

For references related to the oxidative coupling of phenolic derivatives, see:

Dewar, M. J. S.; Nakaya, T. *J. Am. Chem. Soc.* **1968**, *90*, 7134.
Scott, A. I. *Q. Rev.* **1965**, *19*, 1.

For an overview of Cram's work on the binaphthyl crown ethers, see:

Cram, D. J.; Cram, J. M. *Science* **1974**, *183*, 803.

**EXPERIMENT [6_adv]**

## Beckmann Rearrangement: Benzanilide

Common name: Benzanilide
CA number: [93-98-1]
CA name as indexed: Benzamide, *N*-phenyl-

## Purpose

To carry out the *Beckmann rearrangement* in which a ketone is converted, via an oxime, to an amide. The reaction is an example of a group of reactions in which migration to an electron-deficient nitrogen occurs.

### Prior Reading

Technique 4:   Solvent Extraction
                    Liquid–Liquid Extraction (see pp. 80–82).
Technique 5:   Crystallization
                    Craig Tube Crystallizations (see pp. 95–96).

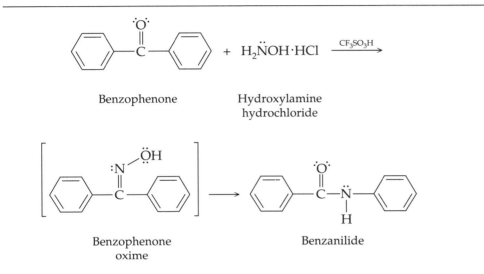

Benzophenone            Hydroxylamine
                            hydrochloride

Benzophenone                Benzanilide
    oxime

The Beckmann rearrangement was discovered in 1886 by E. Beckmann and is the reaction of an oxime of a ketone, in the presence of acid, to yield the corresponding amide or amides. It is a very general reaction and a wide variety of reagents have been used to cause the rearrangement to take place. These include sulfuric, hydrochloric and polyphosphoric acids, phosphorus pentachloride, and aromatic sulfonyl chlorides. The Beckmann rearrangement is the nitrogen analogue, both functionally and mechanistically, of the Baeyer–Villiger oxidation, which converts a ketone to an ester.

In the present case, only one amide is formed, since benzophenone is a symmetrical ketone. Because the oximes are prepared from ketones, the reaction was often used, before the advent of modern spectroscopic techniques, to determine the structure of the starting ketone. This structure determination was accomplished by the subsequent identification of the acid and amine obtained upon hydrolysis of the amide product of the rearrangement.

As depicted in the following mechanism, the acid catalyst converts the oxime's hydroxyl group to a good leaving group (water, a small neutral molecule). The acid used in this experiment is a very strong acid, trifluoromethanesulfonic acid, usually called *triflic acid*. The first part of the overall reaction in this experiment constitutes the formation of the oxime, which is followed by the actual migration of an aryl (or alkyl) group from the former carbonyl carbon to the nitrogen atom, the Beckmann rearrangement.

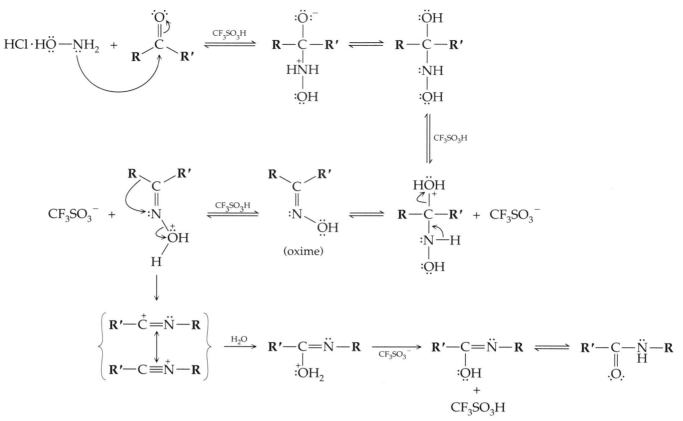

There are two significant points concerning the stereo- and regio-chemistry of the reaction: (1) the group that migrates is the one *anti* to the hydroxyl group on the C=N bond and (2), the stereochemistry, if any, of the migrating group is retained.

## EXPERIMENTAL PROCEDURE

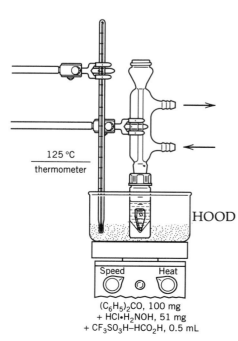

125 °C
thermometer

HOOD

Speed    Heat

$(C_6H_5)_2CO$, 100 mg
+ HCl•$H_2$NOH, 51 mg
+ $CF_3SO_3H$–$HCO_2H$, 0.5 mL

Estimated time to complete the experiment: 2.5 hours.

### Physical Properties of Reactants

| Compound | MW | Amount | mmol | mp(°C) |
|---|---|---|---|---|
| Benzophenone | 182.21 | 100 mg | 0.55 | 48.1 |
| Hydroxylamine hydrochloride | 69.49 | 51 mg | 0.73 | 155–57 |
| Triflic acid–formic acid solution | | 500 µL | | |

### Reagents and Equipment

Weigh and place 100 mg (0.55 mmol) of benzophenone and 51 mg (0.73 mmol) of hydroxylamine hydrochloride in a 5.0-mL conical vial containing a magnetic spin vane. Using an automatic delivery pipet, now add 500 µL of triflic acid–formic acid solution [HOOD] to this mixture. Attach the vial to a reflux condenser and mount the assembly in a sand bath on a magnetic stirring hot plate. (■)

---

**CAUTION:** Triflic acid (trifluoromethanesulfonic acid) is one of the strongest acids known. It is very corrosive and toxic!

## Instructor Preparation

*The acid solution is prepared by adding 2 drops of triflic acid to 5.0 mL of 90% formic acid.*

## Reaction Conditions

With stirring, heat the reaction mixture at reflux for 1 hour in a sand bath at 125 °C. Then cool the resulting solution to room temperature.

## Isolation of Product

To the cooled reaction solution add 1.0 mL of water (calibrated Pasteur pipet), and extract the resulting mixture with three 1.0-mL portions of methylene chloride. Separate the organic phase using a Pasteur filter pipet and transfer it to a 10-mL Erlenmeyer flask. For each extraction, after addition of the methylene chloride, cap the vial, shake, vent, and then allow the layers to separate.

Dry the combined methylene chloride extracts over granular, anhydrous sodium sulfate. Using a Pasteur filter pipet, transfer the dried solution to a clean, dry 10-mL Erlenmeyer flask containing a boiling stone. Evaporate the solvent **[HOOD]** under a gentle stream of nitrogen on a warm sand bath to yield the crude reaction product.          HOOD

## Characterization and Purification

Transfer the crude benzanilide from the Erlenmeyer flask to a Craig tube, and recrystallize the material from 95% ethanol.

Weigh the product and calculate the percent yield. Determine the melting point and compare your result to that listed in the *CRC Handbook*. Obtain an infrared spectrum using the KBr pellet technique.

## Infrared Analysis

In this experiment, an aromatic ketone has been rearranged to a secondary amide. By examining the infrared spectra of starting material and product, we can confirm this molecular transformation.

The spectrum of benzophenone (Fig. 7.6) possesses two macro group frequency trains:

**1.** *Conjugated aromatic ketone:* This ketone is defined by the bands at 3325 (overtone of ketone carbonyl stretch), 1663 (C=O stretch, doubly conjugated), 1601 and 1580 ($\nu_{8a}$ and $\nu_{8b}$ degenerate ring stretch), 1492 and 1450 cm$^{-1}$ ($\nu_{19a}$ and $\nu_{19b}$ degenerate ring stretch). The intensification of the 1580-wavenumber peak confirms the conjugation of the carbonyl to the ring. The 1500-wavenumber ring stretch, which is generally a bit variable in intensity, is quite weak in this case (2 different benzene rings).

**2.** *Monosubstituted phenyl group:* This group is defined by weak bands located at 1969, 1913, 1823, 1724 cm$^{-1}$, and strong bands recorded at 701 and 640 cm$^{-1}$. For discussions of these modes see Chapter 9, and Experiment [20]. In the case of phenyl rings conjugated to carbonyl groups, the 750- and 690-wavenumber bands often appear on the low side of their usual range, and can be down as much as 40–50 cm$^{-1}$, as occurs here.

The rearrangement product has incorporated a heteroatom into its structure, but the carbonyl group and the ring systems have survived. The infrared spectrum of benzanilide (Fig 7.7) must be consistent with the formation of a secondary amide. We expect to observe a macro group frequency for the presence of monosubstituted phenyl groups, plus a second macro group frequency for a secondary aromatic amide.

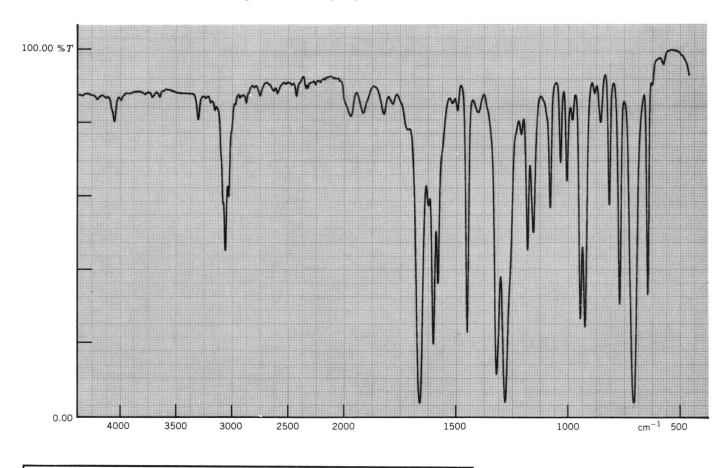

**FIGURE 7.6**   IR spectrum: Benzophenone.

The first macro group frequency is similar to that of the starting material with bands occurring at:

**a.** 1950, 1912, 1840, and 1725 cm⁻¹: All four combination bands are doubled in this case. The values given are for the centers of the doublets.

**b.** 752 and 718 cm⁻¹: C—H out-of-plane bend. The lower value likely corresponds to the ring conjugated directly with the carbonyl group.

**c.** 693 cm⁻¹: Ring out-of-plane bend (puckering) vibration.

The second macro group frequency for the aromatic secondary amide utilizes the following bands: 3350, 3060, 1659, 1602, 1587, 1539, 1322, and 680 (broad) cm⁻¹.

**a.** 3350 cm⁻¹: This strong band corresponds to the highly hydrogen-bonded amide N—H stretch.

**b.** 3060 cm⁻¹: C—H stretch on $sp^2$ carbon, aromatic C—H.

**c.** 1659 cm⁻¹: C=O stretch of a conjugated and hydrogen-bonded amide. It is the most intense band in the spectrum.

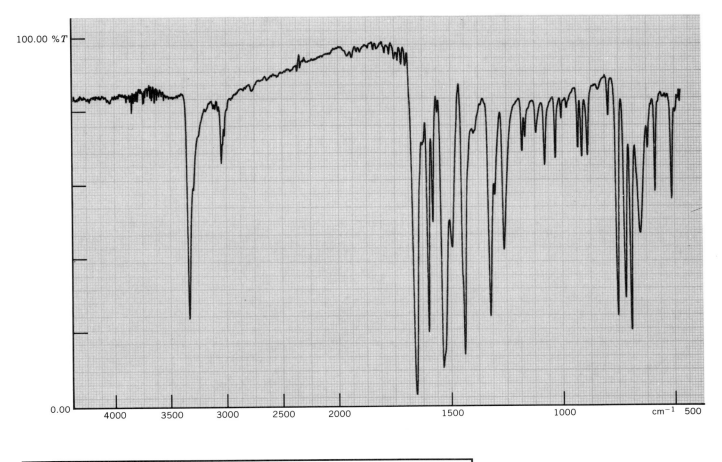

| Sample | Benzanilide | | |
|---|---|---|---|
| %T X ABS Background Scans 4 | | Scans 16 | |
| Acquisition & Calculation Time 42 sec | | Resolution 4.0 cm$^{-1}$ | |
| Sample Condition solid | | Cell Window | |
| Cell Path Length | | Matrix Material KBr | |

**FIGURE 7.7**   IR spectrum: Benzanilide.

**d.** 1602 and 1587 cm$^{-1}$: Ring stretch degenerate pair, $\nu_{8a}$ and $\nu_{8b}$.

**e.** 1539 and 1322 cm$^{-1}$: These two bands involve both N—H bending (in-plane) and C—N stretch. The two fundamentals fall somewhere between 1450–1400 cm$^{-1}$. Thus, they can mechanically interact and split apart with one component falling at 1539 cm$^{-1}$, and the other near 1322 cm$^{-1}$. As the fundamental frequencies will vary somewhat from molecule to molecule, the interaction term that is sensitive to the frequency match (see Infrared discussions, Chapter 9) will also vary in magnitude, and thus the wavenumber separation will be affected.

**f.** 680 cm$^{-1}$: Broad and weak. This ill-defined band arises from the out-of-plane bend of the N—H group. It is similar to the O—H bend found in alcohols in this spectral region (see infrared discussion in Experiments [5A], [5B], and [8B]).

Examine the spectrum of your reaction product in a potassium bromide matrix. Discuss the similarities and differences of the experimentally derived spectral data to the reference spectra (Figs. 7.6 and 7.7).

### Chemical Tests

Chemical classification tests (Chapter 10), such as the ignition test and the soda lime test (or sodium fusion test), may also be conducted to further establish the identity of the product. The hydroxamate test may be used to establish the presence of the amide functional group.

**QUESTIONS**

**7-28.** Draw the structure of the product expected in each of the Beckmann rearrangements presented below. Give a suitable name for each product.

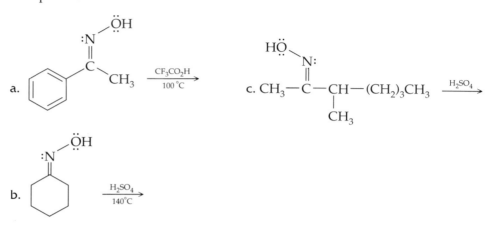

**7-29.** The picryl iminoethers (an iminoether is C=N—O—C) of oximes undergo the Beckmann rearrangement without an acid catalyst.

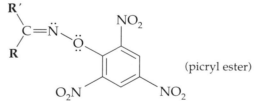

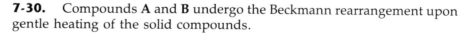

Explain why a catalyst is not required in this rearrangement.

**7-30.** Compounds **A** and **B** undergo the Beckmann rearrangement upon gentle heating of the solid compounds.

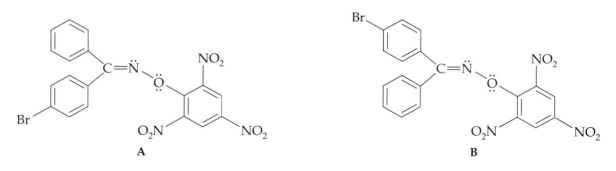

a. Write the structure of the products expected.
b. If a mixture of the two iminoethers, **A** and **B**, are heated in the same reaction flask, what products would be formed? Explain.

**7-31.** When acetophenone oxime was allowed to rearrange in $^{18}$O-enriched solvent, the amide product contained the same percent of $^{18}$O as the solvent. Explain this observation.

**7-32.** Oximes are usually crystalline materials and have been prepared as a means of identifying liquid ketones or aldehydes. It has been found in the preparation of these derivatives that if the acid concentration is too high (low pH), that the oxime does not form. Explain.

**7-33.** The mechanical coupling between the N—H bending vibration and the C—N stretching vibration depends on a close frequency match. Normally the C—C, C—N, and C—O stretching vibrations are found in the 1200–800-wavenumber region. Why, in the case of the amides, do we find the C—N stretch approaching 1400 cm$^{-1}$?

**BIBLIOGRAPHY**

Reviews on the Beckmann rearrangement:

Blatt, A. H. *Chem. Rev.* **1933**, *12*, 215.

Donaruma, L. G.; Heldt, W. Z. *Org. React.* **1960,** *11*, 1.

Jones, B. *Chem. Rev.* **1944,** *35*, 335.

Smith, P. A. In *Molecular Rearrangements*; De Mayo, P., Ed.; Interscience: New York, Vol. I, 1973, pp. 483–507.

An example of the Beckmann rearrangement using different reagents is given in:

Ohno, M.; Naruse, N.; Terasawa, I. *Organic Syntheses*; Wiley: New York, 1973; Collect. Vol. V, p. 266.

The present reaction was adapted from the work of:

Ganboa, I.; Palomo, C. *Synth. Commun.* **1983,** *13*, 941.

## Preparation of an Enol Acetate: Cholesta-3,5-dien-3-ol Acetate

Common name: Cholesta-3,5-dien-3-ol acetate
CA number: [2309-32-2]
CA name as indexed: Cholesta-3,5-dien-3-ol, acetate

### Purpose
To investigate the conditions under which the enol acetate of an $\alpha,\beta$-unsaturated ketone is prepared. The combined use of acetic anhydride with chlorotrimethylsilane is an illustration of a technique for the generation of *acylium ions*, a reactive electrophile for acylation reactions.

*Prior Reading*
    *Technique 5:*  Crystallization
                    Craig Tube Crystallizations (see pp. 95–96).
    *Technique 6:*  Chromatography
                    Column Chromatography (see pp. 98–101).
                    Thin-Layer Chromatography (see pp. 103–104).
                    Removal of Solvent Under Reduced Pressure (see pp. 107–108).

## REACTION

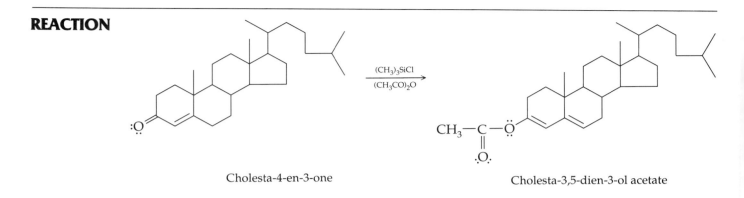

Cholesta-4-en-3-one                    Cholesta-3,5-dien-3-ol acetate

## DISCUSSION

Acylium ions are useful reactive intermediates in organic synthesis because their high reactivity allows reactions with relatively weak nucleophiles, in this case the carbonyl oxygen of an $\alpha,\beta$-unsaturated enone. The product in this experiment, a dienol acetate of a cholesterol derivative, is itself a useful intermediate in the synthesis of steroids such as cortisone.

Acetic anhydride and chlorotrimethylsilane react, as shown here, to generate a small equilibrium concentration of acylium ion ($CH_3C\equiv O^+$), along with trimethylsilyl acetate, $CH_3CO_2Si(CH_3)_3$. The acylium ion is a much more reactive electrophilic acyl group than neutral sources of an electrophilic acyl group, such as acetic anhydride or acetyl chloride, $CH_3COCl$.

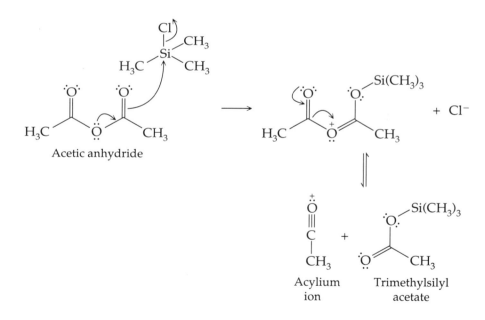

The acylium ion thus generated is a very reactive electrophile, capable of reacting with the relatively weakly nucleophilic lone pairs on the carbonyl oxygen of the cholesta-4-en-3-one. The resulting oxonium ion renders the $\gamma$-protons of the enone quite acidic, and thus they are readily removed by a very weak base, such as chloride ion, to generate the product, cholesta-3,5-dien-3-ol acetate. The proposed mechanism for this transformation is shown here.

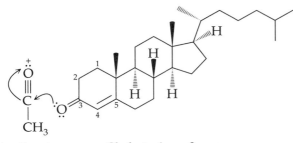

Acylium ion     Cholesta-4-ene-3-one

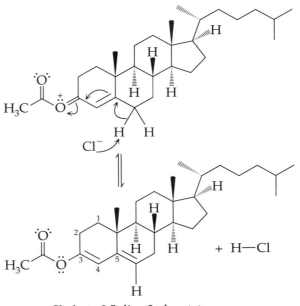

Cholesta-3,5-dien-3-ol acetate

Estimated time to complete the experiment: 4.0 hours.

## EXPERIMENTAL PROCEDURE

### Physical Properties of Reactants and Product

| Compound | MW | Amount | mmol | mp(°C) | bp(°C) | d | $n_D$ |
|---|---|---|---|---|---|---|---|
| Cholesta-4-en-3-one | 384.65 | 96 mg | 0.25 | 81–82 | | | |
| Acetic anhydride | 102.09 | 1.0 mL | 10.6 | | 140 | 1.08 | 1.3901 |
| Chlorotrimethyl-silane | 108.64 | 200 μL | 1.58 | | 58 | 0.86 | |
| Cholesta-3,5-dien-3-ol acetate | 425.68 | | | 79–80 | | | |

### Reagents and Equipment

**NOTE.** *The glass equipment should be dried in an oven (110 °C) and cooled to room temperature in a desiccator before use.*

Weigh and place 96 mg (0.25 mmol) of cholesta-4-ene-3-one in a dry 5.0-mL conical vial containing a magnetic spin vane. Now add 1.0 mL of acetic anhydride and 200 μL of chlorotrimethylsilane **[HOOD]**. Immediately attach the vial to a reflux condenser protected by a calcium chloride drying tube. (■)    HOOD

**NOTE.** *The quality of the acetic anhydride has a significant influence on the reaction. For best results, the anhydride should be distilled and stored over molecular sieves before use. It is convenient to dispense [HOOD] the anhydride and the chlorotrimethylsilane (which hydrolyzes rapidly in moist air) using automatic deliv-*    HOOD

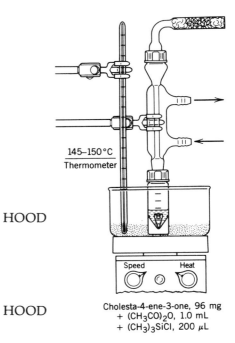

145–150 °C
Thermometer

Cholesta-4-ene-3-one, 96 mg
+ (CH₃CO)₂O, 1.0 mL
+ (CH₃)₃SiCl, 200 μL

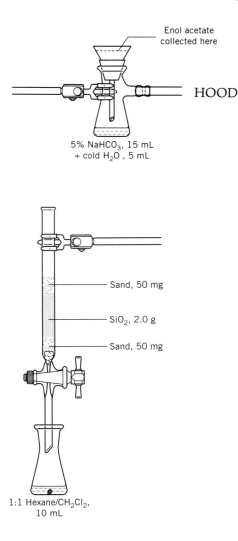

Enol acetate collected here

HOOD

5% NaHCO₃, 15 mL + cold H₂O, 5 mL

Sand, 50 mg

SiO₂, 2.0 g

Sand, 50 mg

1:1 Hexane/CH₂Cl₂, 10 mL

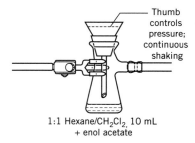

Thumb controls pressure; continuous shaking

1:1 Hexane/CH₂Cl₂, 10 mL + enol acetate

*ery pipets. An alternative is to place each reagent in a small bottle sealed with a septum cap. The reagents are then removed through the septum using a 1-mL syringe with a Teflon-tipped plunger while the bottle is connected to another needle providing dry nitrogen. The chlorotrimethylsilane (bp 57 °C) may also be purified by distillation [HOOD].*

## Reaction Conditions

Heat the reaction mixture with stirring at reflux for 1–2 hours in a sand bath (145–150 °C). Follow the course of the reaction using TLC.

**TLC Directions.** *Use activated silica gel plates with 1:1 methylene chloride/ hexane as the elution solvent. Visualization of the separated components is achieved by placing the plate in a closed jar containing a few crystals of iodine (see Prior Reading for further details).*

## Isolation of Product

Allow the reaction mixture to cool to room temperature and then place it in an ice bath for 15–30 min. A solid product forms during this time. Collect this solid by vacuum filtration using a Hirsch funnel, and wash the filter cake with 15 mL of 5% aqueous sodium bicarbonate, followed by 5 mL of cold water. (■)

## Purification and Characterization

Purify the crude product by column chromatography. In a 1 × 10-cm buret, place 2.0 g of activated silica gel (100 mesh) packed wet (slurry) with methylene chloride. (■) Dissolve the product in about 1.0 mL of hexane and transfer the solution by Pasteur pipet to the column. Elute the material from the column using approximately 10 mL of 1:1 methylene chloride/ hexane solvent. Collect the eluate in a tared 25-mL filter flask.

Remove the solvent by warming in a sand bath under reduced pressure to give the solid product, cholesta-3,5-dien-3-ol acetate. (■) A rotary evaporator, if available, is a more rapid alternative.

Recrystallize this material from methanol using a Craig tube, and dry the resulting crystals on a clay plate. Weigh the product and calculate the percent yield.

Determine the melting point and compare it with the literature value shown in the Reactant and Product table. Obtain an IR spectrum of the material and compare it with that of an authentic sample, as well as with the spectrum of the starting material.

**QUESTIONS**

**7-34.** Reaction of $(CH_3)_3SiCl$ with alcohols produces a trimethylsilyl ether. For example,

$$CH_3CH_2CH_2\ddot{O}H + (CH_3)_3SiCl \xrightarrow[\text{ether}]{(C_2H_5)_3\ddot{N}:} CH_3CH_2CH_2\ddot{O}Si(CH_3)_3 + (C_2H_5)_3\overset{+}{N}H, Cl^-$$

The trimethylsilyl ether is more volatile than the corresponding alcohol, and is often used to facilitate GC analysis. Explain.

**7-35.** Chlorotrimethylsilane reacts with enolate anions to form stable silyl enol ethers. For example,

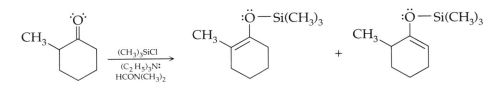

Assuming that the reaction is run under equilibrium conditions, predict which of the above silyl enol ethers is formed in the largest amount and why.

**7-36.** In this experiment, the protons at *both* the $\gamma$ and $\alpha'$ positions relative to the intermediate oxonium ion are quite acidic. Removal of a proton from the $\gamma$-position results in the formation (pathway *a* below) of the actual product, cholesta-3,5-dien-3-ol acetate (see Discussion section). Removal of a proton from the $\alpha'$ position would result in the formation (pathway *b* below) of an isomeric product, cholesta-2,4-dien-3-ol acetate. Since the reaction is conducted under equilibrium conditions, the product obtained must be the thermodynamically more stable of the two possibilities. Why is cholesta-2,4-dien-3-ol acetate less stable than cholesta-3,5-dien-3-ol acetate?

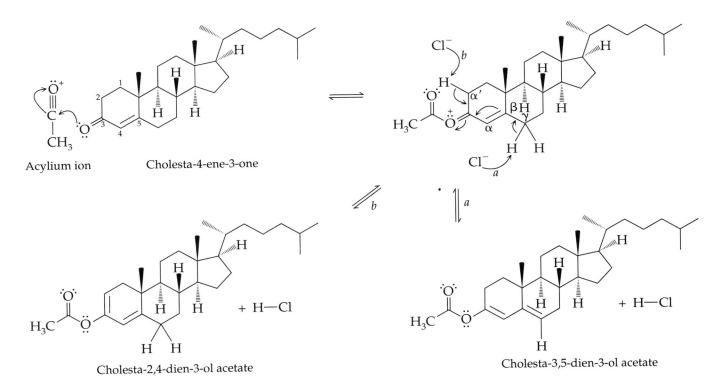

Acylium ion          Cholesta-4-ene-3-one

Cholesta-2,4-dien-3-ol acetate

Cholesta-3,5-dien-3-ol acetate

**7-37.** There are two reasons why the enols of $\beta$-dicarbonyl compounds, such as pentane-2,4-dione are relatively stable. One is that they are conjugated and the $\pi$-electron overlap due to the conjugation provides additional stability. What is the other reason that these species are relatively stable? Would you expect the enol of cyclohexane-1,3-dione to be as stable as the enol of pentane-2,4-dione?

**BIBLIOGRAPHY**

The present experiment is adapted from the work reported by:

Chowdhury, P. K.; Sharma, R. P.; Barua, J. N. *Tetrahedron Lett.* **1983**, *24*, 3383.

For information on the formation and reactivity of enols, see:

House, H. O. *Modern Synthetic Reactions*, 2nd ed.; Benjamin: Reading, MA, 1972; Chapter 9, p. 492.

# 8

# Sequential Syntheses: The Transition from Macro to Micro

**C₈H₈, Cubane**
Eaton and Cole (1964).

The synthesis of a vast array, now numbering in the millions, of new organic molecules in academic and industrial laboratories over the past 100 years is one of the great achievements of modern science. Many of these new compounds have had profound effects on our way of life, both good and bad. It is now becoming apparent that a great challenge in the next century will be how society applies these powerful materials, and the new molecules yet to be born, to the common good.

**INTRODUCTION**

Our ability to synthesize highly complex organic substances has taken a number of dramatic jumps during this century, and has resulted in a bewildering collection of substances that have been devised, synthesized, and applied to practically every facet of our lives. Many of these materials are now vital to our daily life (e.g., consider penicillin) and we all too often take them for granted. In just the last 30–40 years, new advances in pharmaceutical compounds have saved, extended, and made more comfortable the lives of literally hundreds of millions of people. The list could go on and on, including textiles, surfactants, plastics, and synthetic oils to name only a few.

Historically, the synthesis of complex organic substances was primarily driven by the need to obtain large quantities of biologically active material that occurred as the product of plant or animal metabolism, but that could only be obtained in very small quantities from nature. For example, the synthesis of the adrenal cortex hormone, cortisone, was a major breakthrough for hormone therapy. The synthesis of this material initially required 33 steps. That is, the research chemist carried out a sequence of 33 reactions in which stable isolable intermediates were formed sequentially, leading ultimately to the desired cortisone molecule. Industrial sequences of this length are now rare, but those requiring three to six steps are common.

**459**

The strategy of the synthetic chemist is to devise a route whereby the desired compound can be prepared efficiently, using inexpensive, readily available starting materials in the fewest steps. For each individual step, the yield of intermediate should be as high as possible with a minimum of side reactions. In industry the overall cost of the proposed synthetic sequence must be considered, including the time involved, type of equipment required, and safety factors. Today, with the worldwide demand for organic materials in vast quantities (e.g., petroleum products) it is becoming increasingly important to assess the impact on our environment of these synthetic materials prior to large scale production.

As organic transformations most always take place with some loss of material (similar bond energies lead to alternative reaction pathways and easy byproduct formation) the yield of intermediate from each individual step can have a significant impact on the overall yield of the final product. *In a multistep synthesis, the overall yield is the mathematical product of the individual steps.* For example, if we assume that in a five-step sequence for the preparation of a new dye, each step takes place in 85% yield, the overall yield would be $(0.85)^5$ x 100 or 44% (in a 33-step synthesis with an 85% yield for every step, the overall yield would be ~0.5%). This property of organic synthesis emphasizes why a synthetic route devised to produce a particular molecular structure must be carefully planned to minimize losses at each stage of the chosen pathway. This problem also illustrates why the initial steps of a sequence usually employ larger quantities of reactants (macro or semimicro quantities), and why in the last stages, experience at running reactions at the microscale level can be invaluable.

In this chapter we describe a set of six sequential experiments. These experiments vary in the number of intermediates that are required from seven to three, and they vary in the complexity of the chemistry from straightforward extensions of Chapter 6, to relatively demanding experimentation similar to that described in Chapter 7.

The target molecules include:

- The drug **sulfanilamide** (the first of the antibiotics), which is obtained in a novel three-step sequence not usually found in the introductory laboratory:

$$H_2N - \langle \bigcirc \rangle - SO_2NH_2$$

*p*-Aminobenzenesulfonamide
(sulfanilamide)

- The industrially important polymer, **nylon-6,6** (the first of the commercial synthetic textile fibers), formed in three steps closely paralleling the original synthesis.

$$\left[ -\overset{\overset{\displaystyle \cdot\cdot O \cdot\cdot}{\|}}{C}(CH_2)_4\overset{\overset{\displaystyle \cdot\cdot O \cdot\cdot}{\|}}{C} - \overset{\cdot\cdot}{N}H(CH_2)_6\overset{\overset{\displaystyle H}{|}}{\underset{\cdot\cdot}{N}} - \right]_n$$

Nylon-6,6

- The synthesis of **2'-bromostyrene**, which requires three steps. This compound is an interesting substance because of its commercial use as a fragrance in soap products.

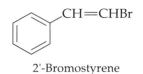

2'-Bromostyrene

- The synthesis of **piperonylonitrile** is an example of a novel conversion, in three steps, of an aromatic aldehyde to an aromatic nitrile.

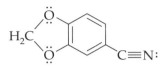

Piperonylonitrile

- The synthesis of **hexaphenylbenzene** requires two parallel three-step sequences to obtain two intermediates that then react with each other to give the target compound after a seven-step synthesis. The final product is a most unusual organic material with one of the highest melting points observed for an organic substance.

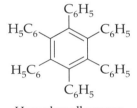

Hexaphenylbenzene

- The synthesis of a **photochromic imine** in four steps yields perhaps the most intriguing substance in the entire chapter. The ability of this material to turn a deep-blue color when exposed to light, and then to lose its color when placed back in the dark makes this structure one of the most interesting of the sequential products. It also involves the most challenging chemistry of the multistep syntheses.

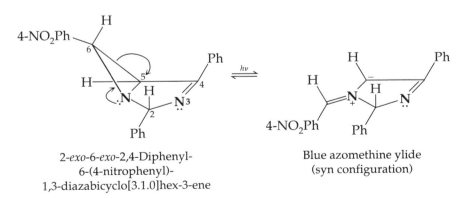

2-*exo*-6-*exo*-2,4-Diphenyl-
6-(4-nitrophenyl)-
1,3-diazabicyclo[3.1.0]hex-3-ene

Blue azomethine ylide
(syn configuration)

In a number of the sequences, the stereochemistry of the reactions is vitally important and controlled by the mechanisms that are operating under the prescribed conditions. This aspect of the transformations is discussed in detail. By performing one or all six of these multistep sequences you will have a chance to challenge your experimental technique under conditions essentially identical to those found in the modern synthetic organic research laboratory. You will quickly recognize the reason why laboratory technique is so vital to the success of multistep syntheses.

## SEQUENCE A                    The Synthesis of Hexaphenylbenzene

### INTRODUCTION

The molecule to be prepared in the Sequence A synthesis is hexaphenyl-benzene (**I**).

**I**

Hexaphenylbenzene

This system, which contains seven aromatic rings, was first made by Dilthey at the University of Bonn in 1933 by employing a classic Diels–Alder reaction with exactly the same two reactants as you will generate in Sequence A. The Bonn group showed that an earlier claim, by Durand, to have prepared this compound via a massive Grignard attack by phenyl-magnesium bromide on hexachlorobenzene, had not actually yielded hexaphenylbenzene. The compound isolated by Durand melted at 266 °C, while Dilthey's material melted at 421–422 °C. Hexaphenylbenzene was later synthesized photochemically by Büchi at the Massachusetts Institute of Technology (MIT) in 1962. The MIT group improved the purity of the isolated material, and reported a melting point of 439–441 °C. Fieser, at Harvard University, refined Dilthey's synthetic route, and published the definitive preparation in *Organic Syntheses* in 1966. Fieser obtained melting points *without decomposition* in evacuated melting point capillaries (see Chapter 4) in the range 454–456 °C.

This hydrocarbon system possesses a number of interesting structural and physical properties. First, we should note that it has a relatively high molecular weight, near 534, and a molecular formula of $C_{42}H_{30}$. Hexaphenylbenzene exhibits an extremely high melting point for a nonionic organic material. For example, of the 15,000 plus substances listed in the Table of Physical Constants for Organic Compounds in the *CRC Handbook of Physics and Chemistry,* only two materials melt above 450 °C (and both of these compounds decompose at their melting point), and only 10 melt above 400 °C. Indeed, hexaphenylbenzene melts above hexabenzobenzene (**II**) the completely delocalized seven-ring fused system, mp = 438–440 °C.

**II**
Hexabenzobenzene
(coronene)

From molecular modeling studies of hexaphenylbenzene, it appears that the six substituent rings surrounding the central system will be sterically restricted from lying in the plane of that ring by ortho-position interaction (**III**)

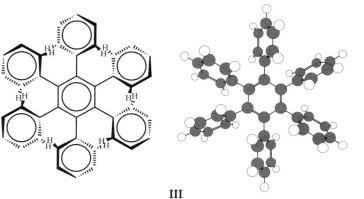

**III**
Hexaphenylbenzene with steric interactions
between ortho positions

and Figure 8.1 shows a molecular model of one of the chiral rotamers of hexaphenylbenzene.

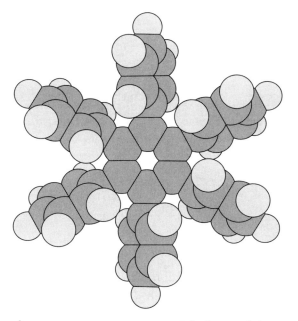

**FIGURE 8.1**  Molecular model of one of the chiral rotamers of hexaphenylbenzene.

The twisted structure presents a particularly interesting problem in stereoisomerism. It is clear that when all the rings are coplanar (dihedral angle of 0°), we have maximum overlap of the π system and delocalization energy, but we also have a maximum of steric repulsion energies. On the other hand, when the dihedral angle approaches 90°, all delocalization is blocked, though steric repulsion between rings is at a minimum. It would seen reasonable to expect the system to reach some energetic compromise between these two extreme orientations. If this is the case, is it possible to establish the angle at which the external rings are positioned? An X-ray crystallographic study carried out by Bart in 1968 on solid crystalline hexaphenylbenzene did, in fact, show that in the crystal lattice the phenyl groups are twisted 65° out of the plane of the central ring. In the crystal lattice, the molecule adopts a conformation, similar to a six-bladed propeller, which is chiral. That is, in the solid state hexaphenylbenzene can exist in two enantiomeric forms. Indeed, if this molecule happened to undergo resolution of the optical isomers during crystallization, in much the same fashion as Pasteur's tartaric acid salts, it should be possible to mechanically separate the racemate into crystals in which all the rings are tilted only in one possible conformation, or in the other. These enantiomers, however, should possess a relatively low barrier to rotation, so that when dissolved in solution rapid racemization, via rotation of the rings (propeller blades) to the opposite pitch, would be expected.

This does seem to be the case. Gust at Arizona State University showed in 1977 that if a derivative of hexaphenylbenzene were synthesized in which two different groups were substituted on adjacent rings in the ortho positions (e.g., a methyl group and a methoxy group, **IV**), two possible sets of diastereomers would result.

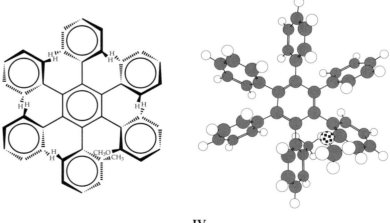

**IV**
Substituted hexaphenylbenzene with steric
interactions between ortho position

It is presumed that it would be impossible for the rings to rotate the two ortho substituents past one another; but that other rotations may or may not be facile. If a large rotational barrier were present, the external rings would remain tilted at 65° with the same pitch. If this were the case, we would expect four diastereomers, and thus four different C—CH₃ resonances in the ¹H NMR spectrum. If rapid interconversion of the tilted conformers occurred, we would expect that the two bulky ortho groups would restrict full rotation of those two rings, even though the barrier to pitch inversion is low. Thus, in this latter case we would expect two diastereomeric pairs of enantiomers (one with the two ortho groups up and one with one up and one down), and two different C—CH₃ resonances in the NMR. When the compound was synthesized, and its ¹H NMR spectrum obtained, two resonances for methyl groups attached to aromatic rings, and two O—CH₃ resonances were observed. These two diastereomers, **V**, were separated by column chromatography, and it was found that they slowly interconverted upon being heated to 215 °C. Thus, on the NMR time scale, it would appear that in hexaphenylbenzene, a reasonably rapid interconversion of the propeller conformations is taking place in solution at room temperature.

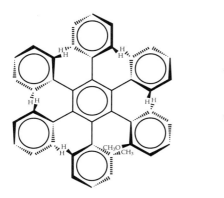
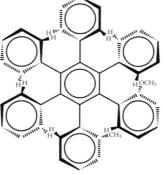

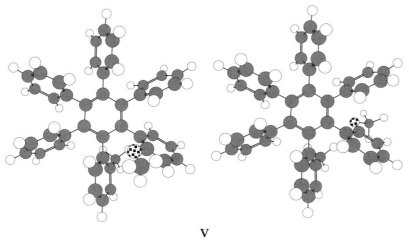

**V**
*ortho*-Disubstituted hexaphenylbenzene
exists as two diastereomers

**(EXPERIMENTS [A1ₐ], [A2ₐ], [A3ₐ], [A1_b], [A2_b], [A3_b], and [A4_ab])**

# The Synthesis of Hexaphenylbenzene from Benzaldehyde:

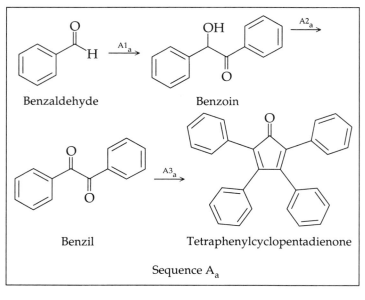

Benzaldehyde    Benzoin

Benzil    Tetraphenylcyclopentadienone

Sequence Aₐ

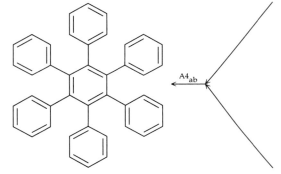

Hexaphenylbenzene

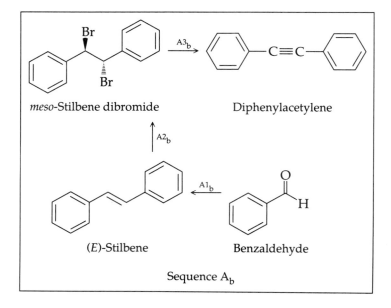

*meso*-Stilbene dibromide    Diphenylacetylene

(E)-Stilbene    Benzaldehyde

Sequence A_b

Benzaldehyde

Benzoin [A1$_a$]

Benzil [A2$_a$]

Tetraphenylcyclopentadienone [A3$_a$]

$\longrightarrow$ Hexaphenylbenzene [A4$_{ab}$]

Diphenylacetylene [A3$_b$]

*meso*-Stilbene dibromide [A2$_b$]

(*E*)-Stilbene [A1$_b$]

Benzaldehyde

As the flow charts illustrate, the total synthesis of hexaphenylbenzene involves two parallel sets of three reactions each, the **a** series and the **b** series. The two series culminate in synthesizing tetraphenylcyclopentadienone (**a** series) and diphenylacetylene (**b** series), which are then reacted together (Experiment [A4$_{ab}$]) to produce the final product, hexaphenylbenzene.

The **a** series uses benzaldehyde as a starting material, which is first converted to the $\alpha$-hydroxyketone, benzoin in Experiment [A1$_a$]. Benzoin is then oxidized to benzil (Experiment [A2$_a$]). Benzil (along with diphenylacetone), is utilized in the construction of tetraphenylcyclopentadienone in Experiment [A3$_a$], the *third* and last of the Sequence A$_a$ intermediates.

The **b** series of synthetic experiments also begins with benzaldehyde, which is converted in Experiment [A1$_b$] into (*E*)-stilbene. (*E*)-Stilbene is the precursor to *meso*-1,2-dibromo-1,2-diphenylethane (*meso*-stilbene dibromide) prepared in Experiment [A2$_b$]. This dibromide is in turn converted by the double dehydrohalogenation reaction in Experiment [A3$_b$] into diphenylacetylene.

The diphenylacetylene is then reacted with the tetraphenylcyclopentadiene, synthesized in the last step of the **a** series, in a Diels–Alder reaction to produce the final product of the sequences, hexaphenylbenzene, in Experiment [A4$_{ab}$].

This preparation of hexaphenylbenzene demonstrates the manner in which a variety of basic organic reactions can be integrated to prepare a desired end product.

**BIBLIOGRAPHY**

Bart, J. C. J. *Acta Crystallogr., Sect. B* **1968**, *24*, 1277.

Büchi, G.; Perry, C. W.; Robb, E. W. *J. Org. Chem.* **1962**, *27*, 4106.

Dilthy, W.; Schommer, W.; Trosken, O. *Berichte* **1933**, *66B*, 1627.

Durand, J. F.; Hsun, L. W. *C. R. Hebd. Seances Acad. Sci.* **1931**, *191*, 1460.

Fieser, L. F. *Organic Syntheses*; Wiley: New York, 1973; Collect. Vol. V, p. 604.

Gust, D. *J. Am. Chem. Soc.* **1977**, *99*, 6980.

## Experiment [A1$_a$]  The Benzoin Condensation of Benzaldehyde: Benzoin

Common name: Benzoin
CA number: [579-44-2]
CA name as indexed: Ethanone, 2-hydroxy-1,2-diphenyl-

### Purpose

To carry out one of the classic reactions of organic chemistry, the *benzoin condensation*. To examine the properties of the $\alpha$-hydroxyketone product of this well-known reaction. The particular $\alpha$-hydroxyketone generated in this experiment is the compound from which the reaction gains its name, *benzoin*.

To provide quantities of benzoin for use in the multistep synthesis of hexaphenylbenzene (see Experiment [A4$_{ab}$]). Benzoin is synthesized in this first step of the **a** series of the **Sequential Experiments**. In this sequence of reactions, benzoin is converted by oxidation (Experiment [A2$_a$]) to benzil and then to tetraphenylcyclopentadienone (Experiment [A3$_a$]). The latter compound undergoes a Diels–Alder addition with diphenylacetylene (Experiment [A3$_b$]) to give hexaphenylbenzene (Experiment [A4$_{ab}$]).

**NOTE.**  *If the benzoin product is to be employed in the reaction sequence, it is recommended that one of the semimicro procedures be followed so that sufficient material will be available for the subsequent steps. The conditions for a one-step microscale experiment are listed following the two semimicroscale procedures.*

### Prior Reading

Technique 5:  Crystallization
Use of the Hirsch Funnel (see pp. 93–95).
Craig Tube Crystallizations (see pp. 95–96).

**REACTION**

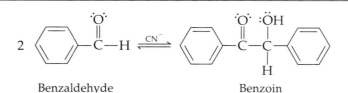

Benzaldehyde                     Benzoin

**DISCUSSION**

Aromatic aldehydes, in the presence of catalytic cyanide ion, dimerize to form the corresponding $\alpha$-hydroxyketone (acyloin). This reaction, which is reversible, is known as the benzoin condensation, even though we now know that it is not actually a condensation reaction, since no water or alcohol is produced. Cyanide ion is a *specific* catalyst for the reaction with *aromatic* aldehydes, and can function in this capacity because: it is a good nucleophile, it stabilizes the intermediate carbanion, and it is a good leaving group. In the mechanism outlined below, it is observed that the nucleophilic cyanide ion attacks a molecule of the aromatic aldehyde to form the conjugate base of a cyanohydrin. The effect of the —CN group is to increase the acidity of the $\alpha$-hydrogen atom, thus allowing the formation of the anion **(I)**

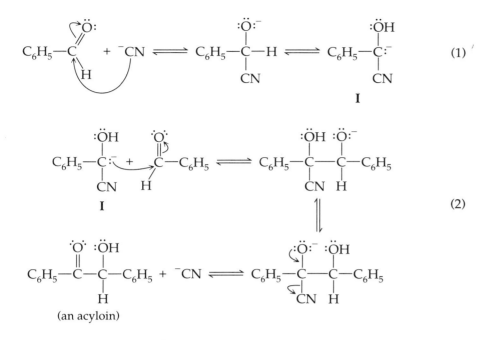

(1)

(2)

(an acyloin)

Once generated, the nucleophilic carbanion (**I**) attacks a second molecule of the aromatic aldehyde to yield a substituted cyanohydrin. This species can then be stabilized by loss of cyanide ion to form the $\alpha$-hydroxyketone product.

The electronic effects of various substituents on the aromatic ring have been investigated. Since, in this reaction, the same aldehyde functions as both the nucleophile and the electrophile, electronic effects that enhance one of these functions are likely to inhibit the other. If a strongly electron-donating group is in the para position of the ring, the reaction fails due to the increase in electron density at the carbonyl carbon, brought about by the presence of the electron donor, which renders the carbonyl carbon less electrophilic. The benzoin condensation is also hindered by strong electron-withdrawing groups on the ring. The presence of a para-nitro group decreases the electron density on the carbonyl carbon atom in the cyanohydrin anion, making its carbanion less nucleophilic, which drastically retards the rate of addition of the anion to the second molecule of aldehyde.

The cyanide-ion catalysis works only for aromatic aldehydes. Aliphatic aldehydes can, however, be condensed to $\alpha$-hydroxyketones in the presence of thiazolium salts, such as *N*-dodecylthiazolium bromide (see below).

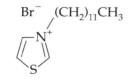

*N*-Dodecylthiazolium bromide

(The microscale reaction quantities are increased by a factor of 2)

Estimated time to complete the experiment: 2.0 hours.

**SEMIMICROSCALE
EXPERIMENTAL PROCEDURE**

**Physical Properties of Reactants**

| Compound | MW | Amount | mmol | bp(°C) | d | $n_D$ |
|---|---|---|---|---|---|---|
| Benzaldehyde | 106.13 | 400 μL | 3.99 | 178 | 1.04 | 1.5450 |
| Sodium cyanide (0.54 *M*) in ethanol (95%) | | 2 mL | | | | |

## Reagents and Equipment

HOOD

HOOD

Place, using an automatic delivery pipet, 400 μL (416 mg, 3.99 mmol) of fresh, *acid-free*, benzaldehyde **[HOOD]** in a weighed 10-mL round-bottom flask containing a magnetic spin bar. Reweigh the flask and record the weight. Now add 2 mL of a 0.54 *M* solution of sodium cyanide in ethanol **[HOOD]**. Remember to use a fresh tip on the automatic delivery pipet.

**NOTE.**   *The 0.54 M NaCN solution should be prepared by the instructor.*

---

**WARNING!   Sodium cyanide (NaCN) is extremely toxic.**

---

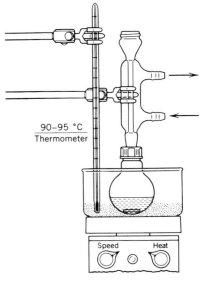

90–95 °C
Thermometer

Attach the flask to a reflux condenser and mount the assembly in a sand bath on a magnetic stirring hot plate. (■)

Speed        Heat

10–mL RB flask
C₆H₅CHO, 400 μL
+ NaCN solution, 2mL

## Reaction Conditions

Heat the mixture with stirring in a sand bath at 90–95 °C. Maintain this temperature for 30 min. The reaction solution turns yellow, and may then become cloudy within approximately 5 min.

---

**CAUTION:   Do not overheat the reaction mixture. If the solution begins to darken, immediately remove the vial from the sand bath.**

---

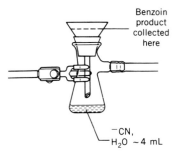

Benzoin
product
collected
here

⁻CN,
H₂O ~ 4 mL

## Isolation of Product

At the end of the reflux period, cool the solution to room temperature and then place it in an ice bath for 10 min. Collect the benzoin product by filtration under reduced pressure using a Hirsch funnel. (■) Wash the filter cake on the Hirsch filter bed with two 1-mL portions of cold water, and air-dry the material under suction using plastic food wrap (see Prior Reading) for 5 min. The crude material is further dried on a porous clay plate or on filter paper.

## Purification and Characterization

This crude material may be purified by recrystallization from methanol or ethanol (95%) in a Craig tube.

Weigh the benzoin product and calculate the percent yield. Determine the melting point and compare your value to that found in the literature.

Obtain the IR spectrum of the compound as a KBr pressed disk. Compare your spectrum to that recorded in *The Aldrich Library of IR Spectra*.

This compound also has a characteristic ultraviolet spectrum showing a peak of maximum absorption ($\lambda_{max}$) at 247 nm ($\varepsilon_{max} = 13,200$, at a concentration of $6.0 \times 10^{-5}$ $M$ in ethanol) characteristic of the benzoyl group, $C_6H_5C{=}O$.

## Chemical Tests

Benzoin contains an aromatic ring. Confirm this fact by performing the ignition test (Chapter 10). To confirm the presence of the alcohol and ketone functions in benzoin carry out the chromic anhydride test for the OH group and the 2,4-dinitrophenylhydrazine test for the C=O group. Isolate the solid 2,4-dinitrophenylhydrazone derivative and compare its melting point to the literature value.

There is a specific test for the presence of benzoin. Place a few crystals of your material in 800 $\mu$L of 95% ethanol. The addition of a few drops of 10% sodium hydroxide solution produces a purple coloration. The color fades when shaken in air but reappears if the solution is allowed to stand.

---

These procedures are identical to that given above with the following exceptions:

*Fivefold Scaleup*

**1.** Increase the scale of the reaction by a factor of 5 compared to the microscale procedure.

## Physical Properties of Reactants

| Compound | MW | Amount | mmol | bp(°C) | $d$ | $n_D$ |
|---|---|---|---|---|---|---|
| Benzaldehyde | 106.13 | 1.0 mL | 9.8 | 178 | 1.04 | 1.5450 |
| Sodium cyanide (0.54 $M$) in ethanol (95%) | | 5.0 mL | | | | |

**2.** Use a 10-mL round-bottom flask. (■)
**3.** The product is washed with two 2-mL portions of cold water.

**OPTIONAL SCALES**

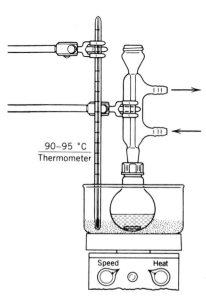

90–95 °C
Thermometer

Speed    Heat

10–mL RB flask
$C_6H_5CHO$, 1.0 mL
+ NaCN solution, 5 mL

# MICROSCALE REACTION PROCEDURE

### Physical Properties of Reactants

| Compound | MW | Amount | mmol | bp(°C) | d | $n_D$ |
|---|---|---|---|---|---|---|
| Benzaldehyde | 106.13 | 200 μL | 1.96 | 178 | 1.04 | 1.5450 |
| Sodium cyanide (0.54 $M$) in ethanol (95%) | | 1 mL | | | | |

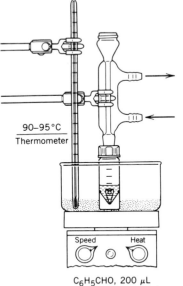

90–95 °C
Thermometer

Speed    Heat

$C_6H_5CHO$, 200 μL
+ NaCN solution, 1 mL

**1.** The product is washed with two 0.5-mL portions of cold water.

**2.** Use a 5-mL conical vial. (■)

# QUESTIONS

**8-1.** The benzoin produced in this experiment contains a chiral carbon atom (a stereocenter), but the product itself is not optically active. Explain.

**8-2.** The cyanide ion is a highly specific catalyst for the benzoin condensation. Can you list three functions this ion performs in this catalytic role?

**8-3.** Can you suggest a reason why $p$-cyanobenzaldehyde does not undergo the benzoin condensation to yield a symmetrical benzoin product?

**8-4.** Tollens' reagent is used as a qualitative test for the presence of the aldehyde functional group (see Chapter 10, Classification Tests). However, benzoin, which does not contain an aldehyde, gives a positive test with this reagent. Explain.

**8-5.** Show how one might accomplish the conversion of $p$-methylbenzaldehyde to each of the following compounds.

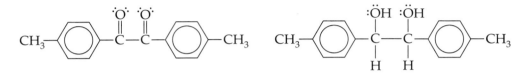

# BIBLIOGRAPHY

Adams, R.; Marvel, C. S. *Organic Syntheses* ; Wiley: New York, 1941; Collect. Vol. I, p. 94.

Ide, W. S.; Buck, J. S. *Org. React.* **1948**, *4*, 269.

Kuebrich, I. P.; Schowen, R. L.; Wang, M.; Lupes, M. E. *J. Am. Chem. Soc.* **1971**, *93*, 1214.

## Experiment [A2ₐ]   Copper(II) Ion Oxidation of Benzoin: Benzil

Common name: Benzil
CA number: [134-81-6]
CA name as indexed: Ethanedione, diphenyl-

### Purpose

Benzil is the *second* of three synthetic intermediates in the **a** series of Sequential Reactions, which lead to the synthesis of hexaphenylbenzene. Benzoin, the starting material for this step, is prepared in Experiment [A1ₐ]. Benzil, the product formed in the present reaction, is used in the synthesis of tetraphenylcyclopentadienone in Experiment [A3ₐ]. Tetraphenylcyclopentadienone is then converted to hexaphenylbenzene in Experiment [A4ₐᵦ].

This experiment also affords an excellent opportunity for you to investigate the use of a soluble, metal–ion catalyst as an oxidizing agent. In this case, a secondary alcohol is oxidized to a ketone. Since nitrogen gas is formed as a byproduct, the progress and rate of the reaction can be followed by measuring the evolution of nitrogen.

**NOTE.** *If it is planned to continue the synthetic sequence to hexaphenylbenzene, the semimicroscale procedure described below should be used. If you wish to study this reaction as an individual microscale experiment, those conditions follow the semimicro discussion.*

### Prior Reading

Technique 5:   Crystallization
               Use of the Hirsch Funnel (see pp. 93–95).
               Craig Tube Crystallizations (see pp. 95–96).
Technique 6:   Chromatography
               Column Chromatography (see pp. 98–101).
               Thin–Layer Chromatography (see pp. 103–104).
               Concentration of Solutions (see pp. 106–108).
Technique 7:   Collection or Control of Gaseous Products (see pp. 109–111).

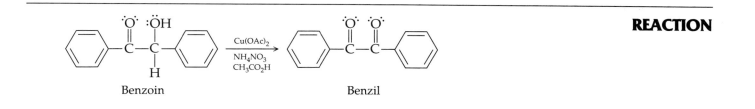

Benzoin                                    Benzil

**REACTION**

**DISCUSSION**

Benzil, a diketone, is obtained by the catalytic oxidation of benzoin using $Cu^{2+}$ ion as the catalytic oxidant. The reaction is general for $\alpha$-hydroxyketones, and is the basis of Fehling's qualitative test for certain sugars. The mechanism of the oxidation shows the catalytic effect of the $Cu^{2+}$ ion as it is continuously reduced and reoxidized in the sequence outlined below. A key ingredient is the nitrate ion, which oxidizes the $Cu^+$ to the $Cu^{2+}$ oxidation state, and is in turn reduced to nitrite ion. The nitrite ion in the presence of acid and ammonium ion, decomposes to yield nitrogen gas and water.

$$2\,Cu^+ + 2\,H^+ + NO_3^- \rightarrow 2\,Cu^{2+} + NO_2^- + H_2O$$

$$NH_4NO_2 \xrightarrow{H^+} N_2 + 2\,H_2O$$

## SEMIMICROSCALE EXPERIMENTAL PROCEDURE

(The microscale reaction is increased fourfold.)

Estimated time of the experiment: 2.5 hours.

### Physical Properties of Reactants

| Compound | MW | Amount | mmol | mp(°C) |
|---|---|---|---|---|
| Benzoin | 212.25 | 400 mg | 1.88 | 137 |
| Cupric acetate solution | | 1.4 mL | | |

**INSTRUCTOR PREPARATION.**  *Prepare the catalyst solution by dissolving 0.1 g of cupric acetate and 5 g of ammonium nitrate in 7.0 mL of deionized water (may require warming), followed by addition of 28 mL of glacial acetic acid. Place the container in the **hood** and dispense the solution by use of an automatic delivery pipet.*

### Reagents and Equipment

Equip a 5.0-mL conical vial, containing a magnetic spin vane, with a reflux condenser, to which is attached a gas exit delivery tube. (■) Weigh and add to the vial 400 mg (1.88 mmol) of benzoin, followed by 1.4 mL of cupric acetate catalyst solution (calibrated Pasteur pipet).

Arrange the gas delivery tube so that it fits into an inverted 100-mL graduated cylinder that is filled with, and immersed in, a beaker of water.

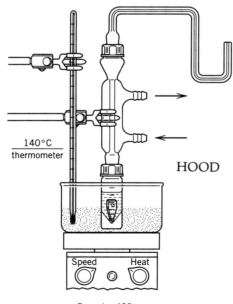

140°C thermometer

HOOD

Speed    Heat

Benzoin, 400 mg
+ cupric acetate catalyst, 1.4 mL,
5–mL conical vial

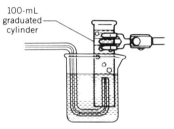

100-mL graduated cylinder

(■) This arrangement facilitates the measurement of the nitrogen gas evolved during the course of the reaction.

**NOTE.** *It is absolutely necessary that all connections be tight to prevent leakage of the gas evolved. Lightly grease the ground-glass joint on the gas delivery tube.*

## Reaction Conditions

Heat the reaction mixture with stirring at a sand bath temperature of 140–145 °C for about 1 hour or until the collected gas volume remains constant. *As the benzoin dissolves, the reaction mixture turns green and evolution of nitrogen gas commences. The theoretical volume of gas from the oxidation of 400 mg of benzoin is 42.4 mL at standard temperature and pressure (STP).*

## Isolation of Product

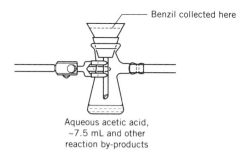

Benzil collected here

Aqueous acetic acid, ~7.5 mL and other reaction by-products

If the gas delivery tube is used, *disconnect* it from the top of the condenser *before* removing the reaction vial from the heat source.

Cool the reaction mixture to room temperature, add 2 mL of cold water (calibrated Pasteur pipet), and then place the reaction vial in an ice bath for 10 min. Collect the yellow crystals of benzil by vacuum filtration using a Hirsch funnel. (■) Rinse the reaction vial and crystals with two additional 2-mL portions of cold water.

## Purification and Characterization

Purify the crude product by recrystallization from methanol, or from 95% ethanol, using the Craig tube. Dry the recrystallized yellow benzil on a porous clay plate, or on filter paper. *The benzil obtained after recrystallization often contains a small amount of benzoin impurity. It may be purified by chromatography on silica gel using the procedure outlined below.* Before carrying out the column chromatography purification of benzil, the purity of the initial product may be assessed using thin-layer chromatography (see Prior Reading). Use methylene chloride as the elution solvent, silica gel as the stationary phase, and UV light for visualization. Typical $R_f$ values: benzil, 0.62; benzoin, 0.33.

Pack a 1.0-cm diameter column to a height of 5 cm with a slurry of activated silica gel in methylene chloride. (■) Introduce the sample of benzil to the column, followed by 100 mg of sand. Use approximately 10–15 mL of methylene chloride to elute the benzil, which is easily identified on the column because of its yellow color. Concentrate the eluate collected in a 25-mL filter flask to obtain the pure benzil (see Prior Reading). A rotary evaporator is an effective alternative.

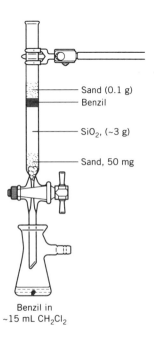

Sand (0.1 g)
Benzil
SiO$_2$, (~3 g)
Sand, 50 mg

Benzil in ~15 mL CH$_2$Cl$_2$

Weigh the dried product and calculate the percent yield. Determine the melting point and compare your value to that reported in the literature.

Obtain an IR spectrum of the product and compare it with that of the starting material and to that reported in *The Aldrich Library of IR Spectra.*

Benzil also has a characteristic UV spectrum (see Fig 8.2). It exhibits a wavelength maximum ($\lambda_{max}$) at 259 nm ($\varepsilon_{max}$ = 16,329, methanol). It is of interest to compare this absorption spectrum of benzil with that of benzoin (Experiment [A1$_a$]). If the melting point and infrared spectrum compare reasonably closely to the literature values, this material may be used in the synthesis of tetraphenylcyclopentadienone (Experiment [A3$_a$]). If the melting point is low, check the product's purity by thin-layer chromatography.

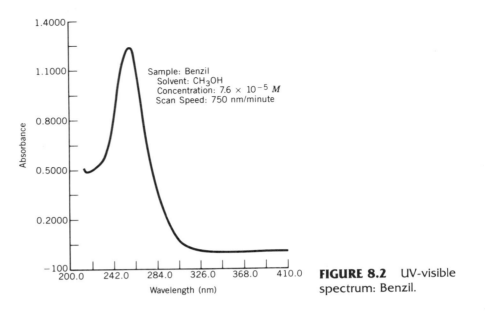

**FIGURE 8.2**  UV-visible spectrum: Benzil.

## Chemical Tests

Ketones and aldehydes are often characterized by the preparation of a solid derivative. To assist in the characterization of benzil, prepare its 2,4-dinitrophenylhydrazone or semicarbazone (Chapter 10). The melting points of these derivatives are listed in Appendix A, Table A.5.

## OPTIONAL MICROSCALE PREPARATION

The microscale procedure is similar to that outlined above for the semi-microscale preparation, with the following exceptions:

**1.** Use a 3.0-mL conical vial containing a magnetic spin vane fitted with a reflux condenser to which is attached a gas delivery tube. (■)

**2.** Collect the gas in an inverted calibrated collection tube. (■)

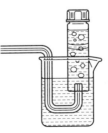

**3.** Decrease the amount of the reagents and solvents fourfold.

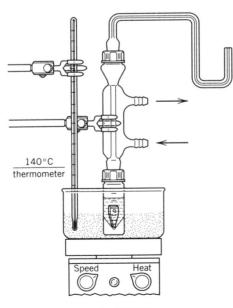

140°C thermometer

Speed    Heat

Benzoin, 100 mg + oxidation catalyst, 350 μL, 3–mL conical vial

### Physical Properties of Reactants

| Compound | MW | Amount | mmol | mp(°C) |
|---|---|---|---|---|
| Benzoin | 212.25 | 100 mg | 0.47 | 137 |
| Cupric acetate solution | | 350 μL | | |

**4.** Heat the reaction mixture at about 140 °C until the evolution of gas has ceased.

**5.** Add 0.5 mL of cold water to the cooled reaction product and, after filtration, wash the material with two 0.5-mL portions of cold water.

**6.** Purify and characterize the benzil product as described in the semimicroscale procedure.

**8-6.** In the directions given for the experiment, it is emphasized that the gas delivery tube must be disconnected from the top of the condenser before removing the reaction vial from the heat source. Why is this necessary?

**8-7.** What qualitative chemical tests would you perform to distinguish between benzoin and benzil? (See Chapter 10.)

**8-8.** 1,2-Dicarbonyl compounds, such as benzil, can be characterized by reaction with *o*-phenylenediamine to form a substituted quinoxaline.

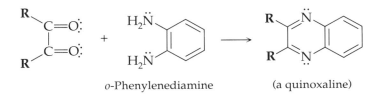

o-Phenylenediamine          (a quinoxaline)

   a. Write the structure for the quinoxaline derivative obtained when benzil is the reactant. Do you think this compound would be colored? If so, why?
   b. Suggest a suitable mechanism for the formation of the quinoxaline compounds based on the reaction scheme shown above.
   c. What reagent would you react with *o*-phenylenediamine to prepare the unsubstituted compound quinoxaline? Show a reaction scheme giving the structure of reactants and products.

**8-9.** Make a sketch of the $^1$H NMR spectrum that would be observed for benzil.

**8-10.** Suggest a method for the synthesis of $C_6H_5CH(OH)CH(OH)C_6H_5$ from benzil. Discuss the stereochemical aspects of this 1,2 diol.

**8-11.** Based on the ultraviolet data given for benzil in the Purification and Characterization section of this experiment, what concentration of the benzil must have been used if a 1-cm cell was employed, and a maximum absorption of 0.5 was observed?

**BIBLIOGRAPHY**

The synthesis of benzil is reported in *Organic Syntheses*:

Adams, R.; Marvel, C. S. *Org. Synth.* **1921**, *1*, 25.

Clarke, H. T.; Dreger, E. E. *Organic Syntheses*; Wiley: New York, 1941; Collect. Vol. I, p. 87.

For further information on the oxidation see:

Weiss, M.; Appel, M. *J. Am. Chem. Soc.* **1948**, *70*, 3666; Depreux, P.; Bethegines, G.; Marcinal-Lefebure, A. *J. Chem. Educ.* **1988**, *65*, 553.

## Experiment [A3$_a$] Tetraphenylcyclopentadienone

Common name: Tetraphenylcyclopentadienone
CA number: [479-33-4]
CA name as indexed: 2,4-Cyclopentadien-1-one, 2,3,4,5-tetraphenyl-

### Purpose

The cyclic dienone product of this aldol condensation is the *third* intermediate in the **a** series of Sequential Reactions which lead to the target molecule, hexaphenylbenzene. It is the last intermediate in the **a** series, and when reacted with the *last* intermediate in the **b** series (Experiment [A4$_{ab}$]), will give the target molecule.

In addition to supplying a key intermediate in the **a** series synthetic sequence, the experiment illustrates the use of the aldol condensation for the synthesis of a five-membered carbocyclic ring. The product also is a good demonstration of the impact of extended conjugation on the absorption of visible light. Starting with bright-yellow benzil, we form an even more extended $\pi$ system in this experiment, and as a result, the tetraphenyldienone product absorbs strongly over a significant portion of visible spectrum, and is thus deeply colored.

**NOTE.**  *If it is planned to continue the synthetic sequence to hexaphenylbenzene, the first microscale procedure described on page 479 should be used. If you wish to study this reaction as an individual microscale experiment, follow the second set of conditions.*

### Prior Reading

> *Experiment [20]*  (see pp. 313–321).
> *Technique 5:*  Crystallization
> Use of the Hirsch Funnel (see pp. 93–95).
> Craig Tube Crystallizations (see pp. 95–96).

**REACTION**

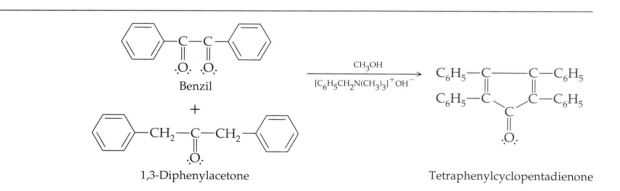

Benzil

+

1,3-Diphenylacetone

Tetraphenylcyclopentadienone

**DISCUSSION**

This experiment is a further example of the aldol condensation reaction (see Experiment [20] for discussion and Experiment [F1] for another example).[1] The reaction carried out here differs from the earlier example in that in this case two ketones, one of which has no $\alpha$-hydrogen atoms, are the reactants. It is also different because the reagents selected lead to the formation of a carbocyclic ring system. The initially formed aldol product undergoes an elimination reaction to yield a material that has a highly

---

[1] For references to the aldol condensation reaction refer to Experiment [20].

conjugated system of double bonds. In general, the more extended the conjugation in a molecule, the less energy is required to promote the $\pi$ electrons to a higher energy level. In this case, energy in the visible region of the spectrum is absorbed, which results in a product possessing a deep purple color.

The mechanism involves a sequence of two aldol condensations. The first is intermolecular; the second is intramolecular. The mechanism is similar to that outlined in Experiment [20].

The product of this reaction is the dienone intermediate used in the final step of the synthesis of hexaphenylbenzene (see Experiment [A4ab]).

---

(The first microscale reaction is increased by a factor of 2.)

Estimated time to complete the experiment: 1.5 hours.

## MICROSCALE REACTION PROCEDURE (1)

### Physical Properties of Reactants

| Compound | MW | Amount | mmol | mp(°C) | bp(°C) |
|---|---|---|---|---|---|
| 1,3-Diphenylacetone | 210.28 | 100 mg | 0.48 | 35 | |
| Benzil | 210.23 | 100 mg | 0.48 | 95 | |
| Triethylene glycol | 150.18 | 0.5 mL | | | 278 |
| Benzyltrimethylammonium hydroxide (40% solution CH$_3$OH) | | 100 $\mu$L | | | |

### Reagents and Equipment

Measure and place 100 mg (0.48 mmol) of 1,3-diphenylacetone and 100 mg (0.48 mmol) of benzil, followed by 0.5 mL of triethylene glycol, in a 3.0-mL conical vial containing a magnetic spin vane. Equip the vial with an air condenser. (■)

**NOTE.** *The benzil used in this reaction must be free of benzoin impurity. If benzil is prepared according to Experiment [A2$_a$], it should be purified by the chromatographic procedure cited therein if benzoin is detected by thin-layer chromatography.*

### Reaction Conditions

Heat the mixture with stirring in a sand bath at 150–160 °C for 5–10 min. The benzil should dissolve during this time.

Next, the apparatus containing the homogeneous reaction solution is removed from the heat source, and immediately 100 $\mu$L of a 40% benzyltrimethylammonium hydroxide–methanol solution is added to the hot reactants, with gentle shaking. As cooling occurs, the appearance of dark purple-colored crystals of tetraphenylcyclopentadienone are evident. *Cooling may be accelerated by placing the vial under a stream of cold water.*

### Isolation of Product

Add 1.4 mL of cold methanol with stirring (glass rod), and then cool the mixture in an ice bath for 5–10 min. The dark crystals are collected by filtration under reduced pressure by use of a Hirsch funnel. The reaction

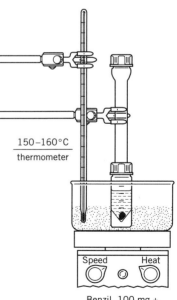

150–160°C thermometer

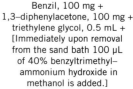

Benzil, 100 mg +
1,3–diphenylacetone, 100 mg +
triethylene glycol, 0.5 mL +
[Immediately upon removal
from the sand bath 100 µL
of 40% benzyltrimethyl–
ammonium hydroxide in
methanol is added.]

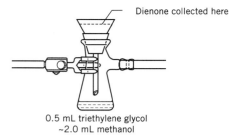

Dienone collected here

0.5 mL triethylene glycol
~2.0 mL methanol

vial and crystals are rinsed with a few drops of cold methanol. Continue dropwise addition of cold methanol to the crystals on the filter bed until the product appears purple and is no longer brownish in color. Finally, the crystalline product is allowed to air-dry on a porous clay plate (or in an oven at about 80 °C for 1 hour). (■)

## Purification and Characterization

The product is often of sufficient purity for direct use in the preparation of hexaphenylbenzene (Experiment [A4ab]).

If further purification is required, the intermediate dienone may be recrystallized from triethylene glycol using a Craig tube.

Weigh the tetraphenylcyclopentadienone product and calculate the percent yield. Determine the melting point and compare it with the literature value. Obtain an IR spectrum of the material and compare it with that of an authentic sample.

The comparision of the UV-visible spectra of benzil (see data given in Experiment [A2a]) and the product may be used to demonstrate the shift of absorption bands with increased conjugation in the molecule.

The UV-visible data for tetraphenylcyclopentadienone are summarized as follows and as shown is Figure 8.3.

$$\lambda_{max} \ 510 \ nm \ (\varepsilon_{max} = 1080, \ chloroform)$$

$$\lambda_{max} \ 345 \ nm \ (\varepsilon_{max} = 6380, \ chloroform)$$

**NOTE.**   *If you have synthesized the tetraphenylcyclopentadienone from benzaldehyde, calculate the overall yield to this point in the synthesis of hexaphenylbenzene. Base these calculations on the amount of benzaldehyde you started with.*

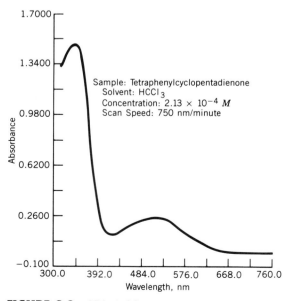

**FIGURE 8.3**   UV-visible spectrum: Tetraphenylcyclopentadienone.

When the microscale procedure is employed as a single-step experiment, and not as part of the synthesis of hexaphenylbenzene, the scale is conveniently reduced to one half that outlined above, with the following experimental modifications.

**Physical Properties of Reactants**

| Compound | MW | Amount | mmol | mp(°C) | bp(°C) |
|---|---|---|---|---|---|
| 1,3-Diphenylacetone | 210.28 | 50 mg | 0.24 | 35 | |
| Benzil | 210.23 | 50 mg | 0.24 | 95 | |
| Triethylene glycol | 150.18 | 0.25 mL | | | 278 |
| Benzyltrimethylammonium hydroxide (40% solution MeOH) | | 50 μL | | | |

    **1.** The reaction is carried out in a 3-mL conical vial.
    **2.** The reaction mixture is maintained at 155–165 °C for 10 min.
    **3.** Following addition of the base, the vial is reheated to 150–160 °C (2–3 min) and then allowed to cool.

**QUESTIONS**

**8-12.** The above reaction carried out to construct the intermediate dienone parallels an earlier example of an aldol condensation in Experiment [20] in which another dienone was synthesized. In both cases all the reactants possessed carbonyl groups. What further structural similarities between the key reactants were required so that both pathways would lead to aldol condensations?

**8-13.** Outline a complete mechanistic sequence to account for the formation of the tetraphenylcyclopentadienone compound.

**8-14.** Cyclopentadienone is unstable and rapidly undergoes the Diels–Alder reaction with itself. Write the structure for this Diels–Alder addition product.

**8-15.** The Diels–Alder addition product of Question 8-14 undergoes a fragmentation reaction on heating to produce a bicyclotrienone compound plus carbon monoxide. Suggest a structure for this product.

**8-16.** Using the Hückel [4n + 2] rule for aromaticity, predict which of the following species might be expected to show aromatic properties.

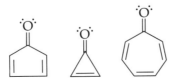

**8-17.** Based on Questions 8-14 and 8-15, why is tetraphenylcyclopentadienone such a stable compound?

**BIBLIOGRAPHY**

An *Organic Syntheses* preparation of tetraphenylcyclopentadienone is available:

Johnson, J. R.; Grummitt, O. *Organic Syntheses*; Wiley, New York, 1955; Collect. Vol. III, p. 80.

### Experiment [A1ᵦ]    *(E)*-Stilbene

Common names: *(E)*-Stilbene; *trans*-1,2-diphenylethene
CA number: [103-30-0]
CA name as indexed: Benzene, 1,1'-(1,2-ethenediyl)bis-, *(E)*-

**Purpose**

The purpose of this experiment is to prepare a sufficient quantity of *(E)*-stilbene to complete a multistep sequence of synthetic reactions to obtain the target compound, hexaphenylbenzene (Experiment [A4ₐᵦ]).

A further purpose of experiment [A1ᵦ] is to investigate the use of the Horner–Wadsworth–Emmons modified Wittig reaction to complete the synthetic objective. To study a reaction that involves the condensation of an aldehyde and a phosphonate ester, using a highly effective *phase-transfer catalyst*.

**NOTE.**  *(E)-Stilbene is the first intermediate to be synthesized in the **b** series of Sequential Reactions en route to hexaphenylbenzene. If it is planned, therefore, to carry out the entire sequence to hexaphenylbenzene, the semimicroscale procedure presented here should be used. Conditions to run the reaction as an individual microscale experiment are given in detail in Experiment [19B]*

*Prior Reading*
    *Technique 4:*  Solvent Extraction
                 Liquid–Liquid Extraction (see pp. 80–82).
                 Drying of a Wet Organic Layer (see pp. 86–87).
    *Technique 5:*  Crystallization
                 Use of the Hirsch Funnel (see pp. 93–95).
                 Craig Tube Crystallization (see pp. 95–96).
    *Technique 6:*  Chromatography
                 Packing the Column (see pp. 99–100).
                 Elution of the Column (see p. 100).
                 Thin-Layer Chromatography (see pp. 103–104).
                 Concentration of Solutions (see pp. 106–108).

**REACTION**

$$C_6H_5CHO + (C_6H_5O)_2\overset{\overset{\displaystyle ::O:}{\|}}{P}-CH_2C_6H_5 \xrightarrow[\substack{Hexane \\ Aliquat\ 336}]{40\%\ NaOH} C_6H_5CH{=}CHC_6H_5 + (C_2H_5O)_2\overset{\overset{\displaystyle ::O:}{\|}}{P}-O^-, Na^+$$

     Benzaldehyde       Diethylbenzyl                    *E*-Stilbene     Sodium diethyl phosphate
                      phosphonate

**DISCUSSION**

For a discussion of the Wittig reaction and a list of references, including the mechanism and modifications, see Experiment [19]. The role of the phase-transfer catalyst in the Horner–Wadsworth–Emmons modification of the Wittig reaction is also discussed in some detail in that experiment.

Estimated time to complete the experiment: 2.5 hours.

## EXPERIMENTAL PROCEDURE

**Physical Properties of Reactants**

| Compound | MW | Amount | mmol | bp(°C) | d | $n_D$ |
|---|---|---|---|---|---|---|
| Benzaldehyde | 106.12 | 600 μL | 5.88 | 178 | 1.04 | 1.5463 |
| Diethyl benzylphosphonate | 228.23 | 1.20 mL | 5.57 | 106–108 (@1 mm) | 1.095 | 1.4970 |
| Aliquat 336 (tricaprylmethylammonium chloride) | 404.17 | 528 mg | 1.30 | | | |
| Hexane | 86.18 | 12.0 mL | | | | |
| 40% Sodium hydroxide | | 12.0 mL | | | | |

### Reagents and Equipment

Weigh and place 528 mg (600 μL) of tricaprylmethylammonium chloride (Aliquat 336) in a 50-mL round-bottom flask containing a magnetic stirrer. Add 600 μL (5.88 mmol) of *freshly distilled* benzaldehyde, 1.20 mL (5.57 mmol) of diethyl benzylphosphonate, 12.0 mL of hexane, and 12 mL of 40% sodium hydroxide solution. Attach the flask to a reflux condenser. (■)

**NOTE.** *The benzaldehyde, (automatic delivery pipet), diethyl benzylphosphonate (2-mL glass pipet), hexane, and NaOH solution are dispensed in the* **hood.**
*The Aliquat 336 is very viscous and is best measured by weighing. A medicine dropper is used to dispense this material.*
*It is advisable to lightly grease the bottom joint of the condenser since strong base is being used in the reacting medium. At the end of the reaction, it is also important to loosen the Cap-seal and twist the joint to make sure it is free to rotate as the apparatus is cooling.*

### Reaction Conditions

Heat the two-phase mixture at reflux on a sand bath at about 90–100 °C for 1 hour. Stir the reaction mixture vigorously during this period. Allow the resulting orange solution to cool to nearly room temperature. Crystals of product may appear as cooling occurs.

### Isolation of Product

Add 4.0 mL of methylene chloride, which will dissolve any crystalline material that may have formed. Now transfer the contents of the round-bottom flask to a 125-mL separatory funnel. Rinse the flask with an additional 2.0–3.0 mL of methylene chloride, and transfer this rinse to the separatory funnel using a Pasteur filter pipet. Remove the aqueous layer carefully (*as the densities of the aqueous and organic layers are rather close, it is wise to test the solubility of a few drops of the bottom layer in water to ascertain which phase is the aqueous one in the separatory funnel*), and then wash the remaining organic layer with two 6.0-mL portions of water. Save the combined aqueous extracts in a 50-mL Erlenmeyer flask until you have successfully isolated and characterized the product. Now transfer the remaining wet methylene chloride solution to a 50-mL Erlenmeyer flask. Dry the solution by addition of anhydrous sodium sulfate (~1–2 g). Use a 50-mL

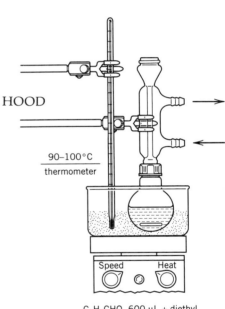

HOOD

90–100 °C
thermometer

Speed    Heat

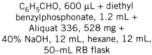

$C_6H_5CHO$, 600 μL + diethyl benzylphosphonate, 1.2 mL + Aliquat 336, 528 mg + 40% NaOH, 12 mL, hexane, 12 mL, 50–mL RB flask

glass pipet to transfer the dried solution to a second 50-mL Erlenmeyer flask. Wash the sodium sulfate remaining in the first Erlenmeyer flask with two 3-mL portions of methylene chloride. Remove these washings, using a Pasteur filter pipet, and transfer them to the same Erlenmeyer flask.

You may concentrate the solution, which contains the desired product, to dryness on a warm sand bath under a gentle stream of nitrogen gas, but a more efficient and rapid procedure is to remove the solvent by rotary evaporation (see Concentration of Solutions, Technique 6B).

## Purification and Characterization

Recrystallize the crude (E)-stilbene from 95% ethanol in a small Erlenmeyer flask or a test tube. Collect the recrystallized material by vacuum filtration using a Hirsch funnel. Maintain the vacuum for an additional 10 min to partially dry the crystalline product. Now place the material on a clay plate, or on filter paper, and allow it to air-dry thoroughly. As an alternative procedure, the final traces of water may be removed by placing the sample (use an open test tube with the mouth covered by filter paper retained by a rubber band) in a vacuum drying oven (or pistol) for 10–15 min at 30 °C (1–2 mm).

The recrystallized compound, generally, is sufficiently pure to use in the next step of the **b** series of Sequential Reactions, Experiment [A2$_b$].

Weigh the (E)-stilbene and calculate the percent yield. Obtain a melting point and IR spectrum of the material, and compare your results with those reported in the literature.

Further characterization of the (E)-stilbene, including thin-layer chromatography, gas chromatography, and ultraviolet–visible spectroscopy may be carried out as outlined in Experiment [19A]

## Chemical Tests

Further characterization may be accomplished by performing the Br$_2$/CH$_2$Cl$_2$ test for unsaturation. Note that the dibromo compound is prepared in Experiment [A2$_b$]. It may be used here as a derivative to characterize the (E)-stilbene. The ignition test (see Chapter 10) may be used to confirm the presence of the aromatic portion of the molecule.

## QUESTIONS

**8-18.**   Give the structure of the phosphorus ylide and carbonyl compound you might use to prepare the following alkenes.

   Methylenecyclohexane      2-Methyl-2-hexene      $C_6H_5CH$=$C(CH_3)_2$

**8-19.**   Trimethylphosphine is less expensive than triphenylphosphine. However, it cannot be used in preparation of most phosphorus ylides. Explain.

**8-20.**   How, starting from triphenylphosphine, $(C_6H_5)_3P$:, can you prepare the following ylide: $(C_6H_5)_3P$=$C(CH_3)CH_2CH_2CH_3$.

**8-21.**   Consult *The Aldrich Library of IR Spectra* for IR spectra of (E)- and (Z)-Stilbene. Which absorption bands are most useful in determining the difference between the two compounds?

**8-22.**   Compare the mechanisms of the aldol condensation (Experiment [20]) with that of the Wittig reaction. Point out any similarities and/or differences.

# Experiment [A2_b]  Bromination of E-Stilbene: *meso*-Stilbene Dibromide

Common names: *meso*-Stilbene dibromide, *meso*-1,2-dibromo-1,2-di-
    phenylethane
CA number: [13440-24-9]
CA name as indexed: Benzene, 1,1'-(1,2-dibromo-1,2-ethanediyl)-
    bis-, (R*,S*)-

## Purpose

To synthesize the *second* intermediate in the **b** series of Sequential Reactions by carrying out the bromination of (E)-stilbene to obtain *meso*-stilbene
dibromide. This product is the precursor to diphenylacetylene, the next
synthetic intermediate in the **b** series. A further purpose of this experiment
is to demonstrate the *stereospecific* addition of bromine to alkenes.

NOTE.  *If it is planned to continue the synthetic sequence to hexaphenylbenzene,
the semimicroscale procedure described below should be used. If you wish to study
this reaction as an individual microscale experiment, those conditions and other
scaleup options follow the semimicro discussion.*

## Prior Reading
    Technique 5:   Crystallization
                Use of the Hirsch Funnel (see pp. 93–95).

**REACTION**

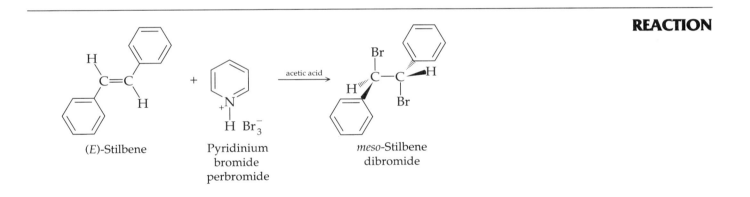

(E)-Stilbene | Pyridinium bromide perbromide | *meso*-Stilbene dibromide

**DISCUSSION**

The bromination of alkenes is an example of an electrophilic addition reaction (also see Experiments [D2] and [F2]).

In the present reaction, bromination of (E)-stilbene yields *meso*-
stilbene dibromide. Thus, this reaction is classed as *stereospecific* since the
other possible diastereomers are not formed.

The reaction proceeds in two stages. The first step involves the formation of an intermediate cyclic *bromonium ion*. The concept of a three-membered cyclic intermediate was first proposed as early as 1937. Subsequent
studies have provided solid evidence that cyclic halonium ions do, indeed,
exist. For example, stable solutions of cyclic bromonium ions in liquid $SO_2$
($-60\ °C$) have been prepared as $SbF_6^-$ salts. Two examples are given here.

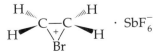

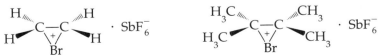

Ethylene bromonium ion salt    Tetramethylethylene bromonium ion salt

Nuclear magnetic resonance spectroscopic measurements have provided powerful evidence that these and other selected alkenes form stable *bridged* bromonium ion salts. A solid bromonium ion tribromide salt of adamantylidene adamantane has been isolated, and its structure determined by X-ray crystallography.

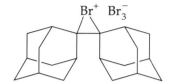

Tribromide salt of adamantylidene adamantane

The bromine molecule ($Br_2$) is normally symmetrical. However, as it approaches the nucleophilic and electron-rich $\pi$ bond of the alkene, it becomes polarized by induction and can then function as the electrophile in an addition reaction. The result is the generation of a cyclic bromonium ion.

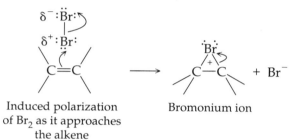

Induced polarization          Bromonium ion
of $Br_2$ as it approaches
the alkene

In the present reaction, both the bromine and the *(E)*-stilbene are achiral. However, the bromonium ion that is produced is chiral. In this ion, the bromine atom bridges both carbon atoms of the original carbon–carbon double bond to form a three-membered ring intermediate. The generation of a cyclic species has a profound effect on the *stereochemistry* of the second step of the bromine addition.

The second stage of the bromination involves nucleophilic attack by bromide ion on the intermediate bromonium ion. As the nucleophile must approach from the face opposite the leaving group, bond formation involves inversion of configuration at the carbon center under attack in the second stage of the bromination reaction.

Note that *either* carbon can be approached by the nucleophile (one attack is shown). This second step is a classic backside $S_N2$ sequence. The bromination of cyclic alkenes provides further evidence that this type of halogenation is an anti addition, with the bromine atoms introduced trans to one another.

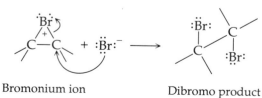

Bromonium ion          Dibromo product

It is important to realize that if two different groups are present on one or both of the $sp^2$ carbon atoms of the alkene linkage, chiral carbon centers are generated on addition of bromine, though if a chiral product were formed from achiral reagents, one would expect it to be racemic. In the case with *(E)*-stilbene, two chiral centers are generated. However, due to the symmetry of the reactants and the stereoselectivity of the reaction, only the meso diastereomer is formed.

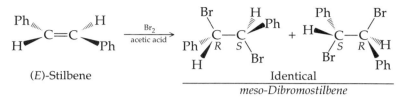

(E)-Stilbene

Identical

meso-Dibromostilbene

Refer to the Discussion section of Experiment [D2] for further information on the stereochemistry of bromination reactions.

Bromination of alkenes using a $Br_2$–$CCl_4$ solution (a red-brown color) is frequently used as a qualitative test for the presence of unsaturation in a compound. Rapid loss of color from the reagent solution is a positive test (see Chapter 10).

Pyridinium bromide perbromide, a solid brominating agent, is used as a source of bromine in this experiment. The material is more convenient to handle than liquid bromine (see Experiment [D2]).

(The microscale reaction is increased by a factor of 2.6.)

Estimated time to complete the reaction: 1.0 hour.

## SEMIMICROSCALE EXPERIMENTAL PROCEDURE

**Physical Properties of Reactants**

| Compound | MW | Amount | mmol | mp(°C) | bp(°C) |
|---|---|---|---|---|---|
| (E)-Stilbene | 180.25 | 600 mg | 3.3 | 122–124 | |
| Glacial acetic acid | | 12 mL | | | 118 |
| Pyridinium bromide perbromide | 319.83 | 1.2 g | 3.7 | 205 | |

## Reagents and Equipment

In a 50.0-mL round-bottom flask containing a magnetic spin bar and equipped with an air condenser, weigh and place 600 mg (3.3 mmol) of (E)-stilbene. Next add 6 mL of glacial acetic acid (using a graduated cylinder), and warm the resulting mixture in a sand bath at 130–140 °C with stirring until the solid dissolves (~5 min). (■)

---

**WARNING:** Glacial acetic acid is corrosive and toxic. It is dispensed in the *hood* using an automatic delivery pipet.

---

Remove the condenser from the flask, and to the warm solution **[HOOD]** add 1.2 g (3.7 mmol) of pyridinium bromide perbromide in one portion. Wash down any perbromide adhering to the sides of the flask with an additional 6 mL of acetic acid using a Pasteur pipet. Reattach the air condenser.

---

**WARNING:** The brominating agent is a mild lachrymator. It should be dispensed in the hood. An alternative solid brominating agent is tetra-N-butylammonium tribromide.

---

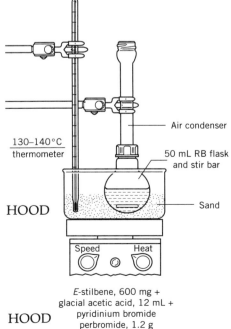

HOOD

HOOD

E-stilbene, 600 mg + glacial acetic acid, 12 mL + pyridinium bromide perbromide, 1.2 g

### Reaction Conditions

With stirring, heat the reaction mixture at a sand bath temperature of 130–140 °C for an additional 5–6 min. (The product often begins to precipitate during this period.)

### Isolation of Product

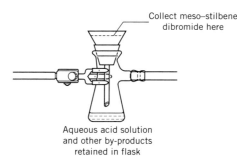

Collect meso–stilbene dibromide here

Aqueous acid solution and other by-products retained in flask

Remove the reaction flask from the heat source and allow it to cool to approximately 40–50 °C (water bath). Add 12 mL of water, with swirling, and then place the flask in an ice bath for 5–8 min. Collect the crystalline solid by vacuum filtration using a Hirsch funnel. (■)

### Purification and Characterization

Wash the material with three 2-mL portions of cold water to obtain white crystals, and then with two 2-mL portions of acetone. Air-dry the product on a clay plate or on filter paper.

Weigh the *meso*-stilbene dibromide and calculate the percent yield. Determine the evacuated melting point, and compare your result with the literature value. Obtain IR and NMR spectra and compare them with those reported in *The Aldrich Library of IR Spectra* or *The Aldrich Library of NMR Spectra.*

Generally, the material is sufficiently pure to be used in the next stage of the **b** series of Sequential Reactions, the preparation of diphenylacetylene. If desired, a small portion (~10–20 mg) may be recrystallized from hot xylene using the Craig tube.

### Chemical Tests

You may wish to perform several classification tests on the product (see Chapter 10). Carry out the ignition test to confirm the presence of an aromatic group. The Beilstein test can be used to detect the presence of bromine. The silver nitrate test for alkyl halides should also give a positive result.

## OPTIONAL MACROSCALE AND MICROSCALE PREPARATIONS

### Macroscale Reaction Procedure

The procedure is similar to that for the 2.6 fold scaleup preparation with the following exceptions.

**1.** The reagent and solvent amounts are increased approximately 4.3-fold over the microscale preparation.

#### Physical Properties of Reactants

| Compound | MW | Amount | mmol | mp(°C) | bp(°C) |
|---|---|---|---|---|---|
| (E)-Stilbene | 180.25 | 1.0 g | 5.5 | 122–124 | |
| Glacial acetic acid | | 12 mL | | | 118 |
| Pyridinium bromide perbromide | 319.83 | 2.0 g | 6.2 | 205 | |

**2.** After cooling the reaction mixture, add 20 mL of water to assist in precipitating the product. Wash the collected crystals with three 3-mL portions of cold water followed by two 3-mL portions of acetone.

## Microscale Reaction Procedure

The procedure is similar to that for the 2.6-fold scaleup preparation with the following modifications:

1. Use a 10-mL round-bottom flask
2. The reagent and solvent amounts are *decreased* by a factor of approximately 2.6.

### Physical Properties of Reactants

| Compound | MW | Amount | mmol | mp(°C) | bp(°C) |
|----------|-----|--------|------|--------|--------|
| (E)-Stilbene | 180.25 | 230 mg | 1.28 | 122–124 | |
| Glacial acetic acid | | 4.2 mL | | | 118 |
| Pyridinium bromide perbromide | 319.83 | 450 mg | 1.4 | 205 | |

3. Add 2.2 mL of glacial acetic acid at the same time as the addition of the (E)-Stilbene
4. An additional 2 mL of glacial acetic acid is added with the brominating reagent.
5. The reaction mixture is diluted with 4.5 mL of water, swirled, and placed in an ice bath for 5–8 min.
6. The filter cake is washed with three 2-mL portions of cold water, followed by two 2-mL portions of acetone.

**QUESTIONS**

**8-23.** Using suitable structures, draw the sequence for the addition of bromine to (Z)-stilbene.

**8-24.** Are the results different for the answer in Question 8-23 than for the result in this experiment? If so, how? What is the stereochemical relationship between the products formed in the two reactions?

**8-25.** Bromine undergoes addition to ethylene in the presence of a high concentration of $Cl^-$ ion to give 1-bromo-2-chloroethane, as well as 1,2-dibromoethane. Chloride ion does not add to the C=C unless bromine is present. Suggest a suitable mechanism to explain these results. Is the rate of bromination significantly affected by the presence of the $Cl^-$ ion?

**8-26.** Offer an explanation for the fact that bromine adds to 2,3-dimethyl-2-butene 920,000 times faster than to ethylene, to produce the respective dibromides.

**8-27.** A student adds a few drops of $Br_2$–$CCl_4$ solution to an unknown organic compound. The color of the bromine solution disappears. The student reports that the unknown contains a C=C. Would you arrive at the same conclusion? If not, why not?

**BIBLIOGRAPHY**

A large number of examples of the bromination of alkenes appear in *Organic Syntheses*. Selected references are given below:

Allen, C. F. H.; Abell, R. D.; Normington, J. B. *Organic Syntheses*; Wiley: New York, 1941; Collect. Vol. I, p. 205.

Cromwell, N. H.; Benson, R. *Organic Syntheses*; Wiley: New York, 1955; Collect. Vol. III, p. 105.

Fieser, L. F. *Organic Syntheses*; Wiley: New York, 1963; Collect. Vol. IV, p. 195.

Khan, N. A. *Organic Syntheses*; Wiley: New York, 1963; Collect. Vol. IV, p. 969.

McElvain, S. M.; Kundiger, D. *Organic Syntheses*; Wiley: New York, 1955; Collect. Vol. III, p. 123.

Paquette, L. A.; Barrett, J. H. *Organic Syntheses*; Wiley: New York, 1973; Collect. Vol. V, p. 467.

Rhinesmith, H. S. *Organic Syntheses*; Wiley: New York, 1943; Collect. Vol. II, p. 177.

Snyder, H. R.; Brooks, L. A. *Organic Syntheses*; Wiley: New York, 1943; Collect. Vol. II, p. 171.

## Experiment [A3$_b$]   Dehydrohalogenation of *meso*-Stilbene Dibromide: Diphenylacetylene

Common names: Diphenylacetylene, diphenylethyne
CA number: [501-65-5]
CA name as indexed: Benzene, 1,1'-(1,2-ethynediyl)bis-

### Purpose

The product formed in this multiple elimination reaction is the *third* intermediate in the **b** series of Sequence A, and is one of the immediate precursors to our target molecule, hexaphenylbenzene.

To investigate the synthesis and properties of alkynes, and to become familiar with E2 elimination reactions.

**NOTE.**   *If it is planned to continue the synthetic sequence to hexaphenylbenzene, the semimicroscale procedure described below should be used. If you wish to study this reaction as an individual microscale experiment, those conditions, and other scaleup options, follow the semimicro discussion.*

### Prior Reading

Technique 5:   Crystallization
Use of the Hirsch Funnel (see pp. 93–95).

**REACTION**

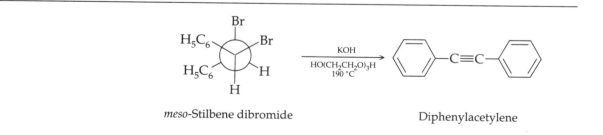

*meso*-Stilbene dibromide            Diphenylacetylene

**DISCUSSION**

This reaction illustrates the dehydrohalogenation of a *vicinal* dibromo compound to form an alkyne. It is a useful reaction for the synthesis of alkynes, as the starting dibromides are readily available from alkenes (see, e.g., Experiment [A2$_b$]).

The double dehydrohalogenation reaction is usually run in the presence of a strong base, such as KOH or NaNH$_2$, and proceeds in two stages.

In the first, an intermediate bromoalkene is formed, which can be isolated under more mildly basic conditions. In fact, this reaction is a valuable route to vinyl halides. The mechanism of elimination involves the abstraction of the proton on the carbon atom $\beta$ to the halogen. The E2 mechanism, which operates under these strongly basic conditions, is fastest when it involves removal of a proton, $H^+$, antiperiplanar to the leaving group, $Br^-$. The E2 sequence of bond breakage and formation involves a smooth transition from reactant to product without the formation of an intermediate (concerted mechanism). The general mechanism is shown here.

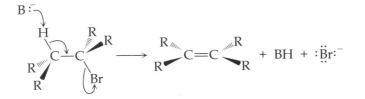

This type of elimination reaction is stereospecific since the geometry of the transition state requires that the H, both Cs, and the Br, all lie in the same plane.

If *meso*-stilbene dibromide is treated with KOH in ethanol solvent, it *is* possible to isolate the monodehydrohalogenation product, the bromoalkene.

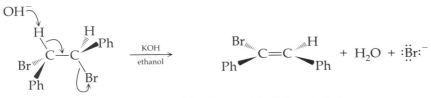

*meso*-Stilbene dibromide          (E)-1-Bromo-1,2-diphenylethylene

The second stage of the reaction involves a higher activation energy, and therefore it requires higher temperatures to proceed. In the presence of a strong base near 200 °C, the bromoalkene undergoes an E2 elimination to form the triple bond. Part of the reluctance to eliminate, in this particular case, results from the fact that the elimination proceeds by a syn pathway.

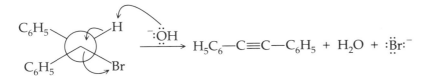

Thus, the stereochemistry of the reactant employed necessitates somewhat higher temperatures for the second elimination reaction.

---

(The microscale reaction is increased by a factor of 5.)

Estimated time to complete the reaction: 1.0 hour.

**SEMIMICROSCALE
EXPERIMENTAL PROCEDURE**

**Physical Properties of Reactants**

| Compound | MW | Amount | mmol | mp(°C) | bp(°C) |
|---|---|---|---|---|---|
| *meso*-Stilbene dibromide | 340.07 | 400 mg | 1.2 | 241 dec | |
| Potassium hydroxide | 56.11 | 387 mg | 6.9 | 360 | |
| Triethylene glycol | 150.18 | 2 mL | | | 278 |

## Reagents and Equipment

Weigh and place 400 mg (1.2 mmol) of *meso*-stilbene dibromide and 387 mg (6.9 mmol) of KOH flakes in a 10-mL Erlenmeyer flask containing a magnetic stir bar. Using a graduated cylinder, measure and add 2 mL of triethylene glycol to the flask.

## Reaction Conditions

Place the reaction flask in a **preheated** sand bath set at a temperature of 190–195 °C, and stir the reaction for 7–8 min.

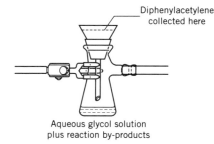

Diphenylacetylene collected here

Aqueous glycol solution plus reaction by-products

## Isolation of Product

Allow the resulting dark-colored reaction mixture to cool to approximately 40–50 °C (water bath), and then add 5.0 mL of water. Now place the flask in an ice bath for 15 min. Collect the solid product by filtration under reduced pressure using a Hirsch funnel. (■)

## Purification and Characterization

Rinse the product crystals with two 1-mL portions of cold 70% ethanol, and air-dry them on a porous clay plate or on filter paper. These crystals can be recrystallized from 95% ethanol (~2.0 mL). If desired, a small portion may be recrystallized from 95% ethanol using the Craig tube.

Weigh the recrystallized product and calculate the percent yield. Determine the melting point and compare your result with the literature value. Obtain IR and NMR spectra of the material and compare them with those recorded in *The Aldrich Library of IR Spectra* or *The Aldrich Library of NMR Spectra*.

**NOTE.**   *If you have synthesized the diphenylacetylene from benzaldehyde, calculate the overall yield to this point in the synthesis of hexaphenylbenzene. Base these calculations on the starting amount of benzaldehyde.*

### Chemical Test

The ignition test for aromatic groups indicates the presence of the phenyl groups. Decolorization of a $Br_2$–$CH_2Cl_2$ solution should give a positive test for unsaturation (see Chapter 10).

---

## OPTIONAL MACROSCALE AND MICROSCALE PREPARATIONS

## Macroscale Reaction Procedure

(This reaction is scaled up by a factor of 10 over the microscale procedure.)
The procedure is similar to that outlined above with the following exceptions:

**1.** Carry out the reaction in a 25-mL Erlenmeyer flask containing a boiling stone. Run the reaction in the **hood.**

HOOD

**2.** Increase the reagent and solvent amounts approximately twofold over the semimicroscale procedure, as indicated here.

**Physical Properties of Reactants**

| Compound | MW | Amount | mmol | mp(°C) | bp(°C) |
|---|---|---|---|---|---|
| *meso*-Stilbene dibromide | 340.07 | 800 mg | 2.4 | 241 dec | |
| Potassium hydroxide | 56.11 | 756 mg | 13 | 360 | |
| Triethylene glycol | 150.18 | 4 mL | | | 278 |

**3.** After cooling the reaction mixture, add 10 mL of water.

**4.** Rinse the product crystals with two 1-mL portions of cold 70% ethanol. They can be recrystallized from 95% ethanol (~5.0 mL).

## Microscale Reaction Procedure

The procedure is similar to that outlined above with the following exceptions:

**1.** Carry out the reaction in a 3-mL conical vial containing a boiling stone.

**2.** The reaction is heated in a sand bath at 190 °C for 5 min.

**3.** Decrease the amounts of reagents and solvents as given here.

**Physical Properties of Reactants**

| Compound | MW | Amount | mmol | mp(°C) | bp(°C) |
|---|---|---|---|---|---|
| *meso*-Stilbene dibromide | 340.07 | 80 mg | 0.24 | 241 dec | |
| Potassium hydroxide | 56.11 | 75 mg | 1.3 | 360 | |
| Triethylene glycol | 150.18 | 0.4 mL | | | 278 |

**4.** After cooling of the reaction mixture to 40–50 °C (water bath), add 1.0 mL of water, and place the reaction vessel in an ice bath for 15 min.

**5.** Rinse the product crystals with one 0.25-mL portion of cold 70% ethanol. The alkyne can be recrystallized from 95% ethanol (~0.5 mL).

**8-28.** Both *(E)*- and *(Z)*-2-chlorobutenedioic acids dehydrochlorinate to give acetylene dicarboxylic acid.

QUESTIONS

$$HO_2C—C(Cl)=CH—CO_2H \rightarrow HO_2C—C\equiv C—CO_2H$$

The *Z* acid reacts about 50 times faster than the *E* acid. Explain.

**8-29.** Compounds containing a carbon–carbon triple bond undergo the Diels–Alder reaction. Formulate the product formed by the reaction of *(E,E)*-1,4-diphenyl-1,3-butadiene with diethyl acetylenedicarboxylate.

**8-30.** Alkynes can be hydrated in the presence of acid and $HgSO_4$, by electrophilic addition of a molecule of water to the triple bond. The reaction proceeds by way of a carbocation intermediate. Hydration of acetylene (ethyne) produces acetaldehyde (ethanal). Outline the steps that occur in this transformation.

**8-31.** Use the IR tables to locate the absorption bands of the stretching frequencies of the alkyne C—H bond, the alkyne C≡C bond, and the alkene C—H bond. Using these data, explain how you would distinguish between 1-butyne, 2-butyne, and 2-butene.

---

**BIBLIOGRAPHY**

For a review on the preparation of alkynes see:

Jacobs, T. L. *Org. React.*, **1949**, *5*, 1.

A large number of elimination reactions leading to the formation of acetylenes appear in *Organic Syntheses*. Selected references are given below:

Abbott, W. T. *Organic Syntheses*; Wiley: New York, 1943; Collect. Vol. II, p. 515.

Allen, C. F. H.; Abell, R. D.; Normington, J. B. *Organic Syntheses*; Wiley: New York, 1941; Collect. Vol. I, p. 205.

Campbell, K. N.; Campbell, B. K. *Organic Syntheses*; Wiley: New York, 1963; Collect. Vol. IV, p. 763.

Guha, P. C.; Sankaren, D.K. *Organic Syntheses*; Wiley: New York, 1955; Collect. Vol. III, p. 623.

Hessler, J. C. *Organic Syntheses*; Wiley: New York, 1941; Collect. Vol. I, p. 438.

Khan, N. A. *Organic Syntheses*; Wiley: New York, 1963; Collect. Vol. IV, p. 967.

The synthesis of diphenylacetylene has been reported:

Smith, L. I.; Falkof, M. M. *Organic Syntheses*; Wiley: New York, 1955; Collect. Vol. III, p. 350.

### Experiment [A4$_{ab}$]   Hexaphenylbenzene

Common name: Hexaphenylbenzene
CA number: [992-04-1]
CA name as indexed: 1,1':2',1"-Terphenyl, 3',4',5',6'-tetraphenyl-

#### Purpose

To complete the Sequence A experiments. To study the use of the Diels–Alder reaction as a method to form six-membered aromatic rings. To carry out the decarbonylation and aromatization of an intermediate bicyclic Diels–Alder adduct. To examine the properties of our synthetic target molecule, hexaphenylbenzene.

**NOTE.** *If it was planned to continue the synthetic sequence to hexaphenylbenzene, you should have enough of the two starting reactants to carry out the first true microscale reaction used in Sequence A. The details of this interesting Diels–Alder addition, first carried out in 1933, are described below. You may, of course, wish to study this reaction as an individual microscale experiment, in which case employ the same conditions given here.*

#### Prior Reading

Technique 5: Crystallization
Use of the Hirsch Funnel (see pp. 93–95).
Craig Tube Crystallizations (see pp. 95–96).
Experiment [14]: (see pp. 262–272).
Experiment [15]: (see pp. 272–277).

**REACTION**

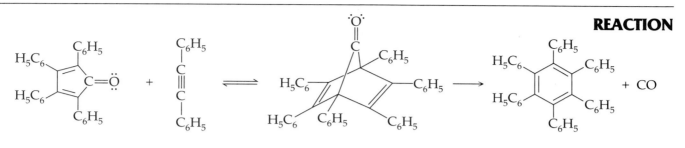

Tetraphenylcyclopentadienone    Diphenylacetylene                                      Hexaphenylbenzene

**DISCUSSION**

This experiment (Experiment [A4$_{ab}$]) completes the Sequence A set of seven experiments that lead to the synthesis of hexaphenylbenzene. As we have noted earlier (see introduction to Sequence A), this compound is a rather unique and interesting organic system possessing a number of unusual properties. For example, it has one of the highest known melting points for a nonionic organic molecule, 465 °C, and it is perhaps even more intriguing that it melts without decomposition. Indeed, its melting point exceeds that of all 15,000 organic compounds listed in the *CRC Handbook* for 1991–1992. Hexaphenylbenzene also contains particularly novel stereochemistry as discussed in the introduction.

The Diels–Alder reaction is one of the most useful synthetic tools in organic chemistry. It is an example of a cycloaddition reaction between a conjugated diene and a dienophile, which leads to the formation of six-membered rings. Here, the initial bicyclic Diels–Alder adduct can undergo a reaction that is the reverse of a concerted cycloaddition reaction between a benzene ring and the lone electron pair on the carbon of carbon monoxide. This retro cycloaddition is thermodynamically favored here because the retro reaction generates an aromatic system, along with the quite stable carbon monoxide molecule. Under the high-temperature conditions used in this experiment, the initial bicyclic Diels–Alder adduct is quickly decarbonylated to yield hexaphenylbenzene, and is not itself isolated.

By varying the nature of the diene and dienophile, a very large number of structures can be prepared using the Diels–Alder reaction. In the majority of cases, carbocyclic rings are generated, but ring closure can also occur with reactants containing heteroatoms. This leads to the synthesis of compounds containing heterocyclic rings. For further, and more detailed, discussion of the Diels–Alder reaction see Experiments [14] and [15].

In the present reaction, an excess of diphenylacetylene is used to ensure that all the tetraphenylcyclopentadienone is consumed in the reaction, as diphenylacetylene is far easier to separate from hexaphenylbenzene in the purification steps.

**EXPERIMENTAL PROCEDURE**

Estimated time to complete the experiment: 1.0 hour.

**Physical Properties of Reactants and Product**

| Compound | MW | Amount | mmol | mp(°C) |
| --- | --- | --- | --- | --- |
| Tetraphenylcyclopentadienone | 384.48 | 100 mg | 0.26 | 220–221 |
| Diphenylacetylene | 178.23 | 100 mg | 0.56 | 61 |
| Hexaphenylbenzene | 534.66 | | | 465 |

### Reagents and Equipment

In a $13 \times 100$-mm Pyrex test tube, place 100 mg (0.26 mmol) of tetraphenylcyclopentadienone and 100 mg (0.56 mmol) of diphenylacetylene. Then transfer to the test tube about 1 mL of high-boiling silicone oil (calibrated Pasteur pipet). Clamp the test tube at a slight angle, facing it away from both yourself and your laboratory neighbors.

### Reaction Conditions

Bring the mixture gently to a boil over a 3–5 min period, by heating the test tube with the moving flame of a microburner. On melting, the reagents dissolve in the hot silicone oil to yield a dark red-purple solution. Continue to gently boil the solution for an additional 10 min. During this latter period, the deeply colored solution fades and hexaphenylbenzene begins to separate from solution as tan crystals.

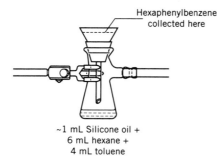

Hexaphenylbenzene collected here

~1 mL Silicone oil + 6 mL hexane + 4 mL toluene

### Isolation of Product

After heating for 15 min, cool the test tube to room temperature and add 4 mL of hexane, with stirring, to dilute the silicone oil and any unreacted starting material. The crude, precipitated hexaphenylbenzene is then collected by filtration on a Hirsch funnel. (■)

### Purification and Characterization

Wash the filter cake with 2 mL of hexane to yield tan crystals of the addition product. Then wash it twice with 2-mL portions of cold toluene, to yield white crystalline hexaphenylbenzene. Air-dry the product on filter paper or a porous clay plate. Weigh the Diels–Alder adduct and calculate the percent yield. Obtain an IR spectrum and compare it to that of an authentic sample. The melting point of this material is well over 400 °C, therefore, melting point determinations with apparatus that employ oil baths should **NOT** be attempted. The best approach, if a melting point is required, is to carry out an evacuated melting point determination with one of the metal heating block systems that accept the normal capillaries, *but remember to first check the maximum temperature reading on the thermometer used in the apparatus* (see evacuated melting points, Chapter 4, p. 51).

CAUTION

CAUTION

If necessary, the product can be recrystallized in a Craig tube from diphenyl ether (5–10-mg maximum, as this very high-melting material is very insoluble even in this high-boiling aromatic ether; recrystallization is rather difficult).

This experiment completes the seven-step Sequence A synthesis of hexaphenylbenzene from benzaldehyde. Calculate the overall yield based on both the earlier calculations for each pathway (for the diene and the dienophile used in the final Diels–Alder reaction), see Experiments [A3$_a$] and [A3$_b$], and the actual yield for this step in the synthesis.

## QUESTIONS

**8-32.**  What starting materials would you use to prepare each of the following compounds by the Diels–Alder reaction?

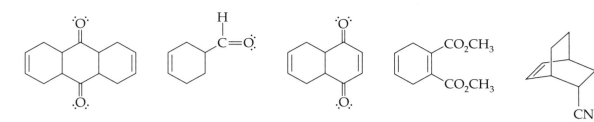

**8-33.** Diels–Alder reactions with benzene are rare, and require a very reactive (electron-deficient) dienophile, since benzene is a poor dienophile. Two are shown below. Give the structures of the product produced in each reaction.

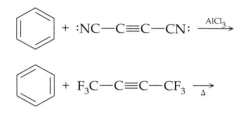

**8-34.** Below are shown two heteroatom compounds that undergo the Diels–Alder reaction. Formulate the product obtained in each reaction.

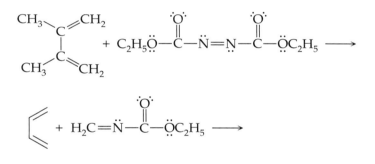

---

Review articles:

Butz, L. W.; Rytina, A. W. *Org. React.* **1949,** *5,* 136.

Holm, H. L. *Org. React.* **1948,** *4,* 60.

Kloetzel, M. C. *Org. React.* **1948,** *4,* 1.

Norton, J. A. *Chem. Rev.* **1942,** *31,* 319.

Sauer, J. *Angew. Chem. Int. Ed. Engl.* **1966,** *5,* 211; *Ibid.,* **1967,** *6,* 16.

An *Organic Syntheses* preparation using tetraphenylcyclopentadienone in a Diels–Alder reaction to obtain tetraphenylphthalic anhydride has been recorded:

Grummitt, O. *Organic Syntheses*; Wiley: New York, 1955; Collect. Vol. III, p. 807.

**BIBLIOGRAPHY**

---

# The Stepwise Synthesis of Nylon-6,6

**SEQUENCE B**

## Purpose

To prepare the important industrial polymer, nylon-6,6, by the technique of step-growth polymerization. To examine the physical properties of the polymer. To synthesize the two monomers used in the polymerization.

## INTRODUCTION

### Background of an Industrial Polymer

The type of polymerization used in the nylon preparation described in this series of experiments is termed "step-growth" polymerization. The technique employs two different difunctional monomers that undergo ordinary organic reactions. In the present case an acid chloride is treated with an amine to produce an amide linkage.

Nylon is a polyamide. In industry it is produced by reaction of two difunctional monomers (or **co**monomers), a dicarboxylic acid and a diamine. The polymer that you are going to study is of great historical significance in polymer chemistry, as it was the first of the polyamides to be recognized as possessing excellent physical properties for forming very strong fibers. Nylon-6,6 was, in fact, the first commercially produced synthetic polyamide. The "6,6" nomenclature refer to the number of carbon atoms in each of the two comonomers. Industrially, Nylon-6,6 is prepared from 1,6-hexanediamine (hexamethylenediamine) and hexanedioic acid (adipic acid).

$$n \; \overset{\ddot{O}}{\underset{\text{Adipic acid}}{\text{HÖCCH}_2\text{CH}_2\text{CH}_2\text{CH}_2\text{CÖH}}} + n \; \underset{\text{1,6-Hexanediamine}}{\text{H}_2\ddot{\text{N}}(\text{CH}_2)_6\ddot{\text{N}}\text{H}_2} \xrightarrow{\text{heat}} \underset{\text{Nylon-6,6}}{\left[\overset{\ddot{O}}{\text{C}}(\text{CH}_2)_4\overset{\ddot{O}}{\text{C}}-\ddot{\text{N}}\text{H}(\text{CH}_2)_6\overset{\text{H}}{\text{N}}\right]_n}$$

In the industrial process, the diacid and diamine are mixed to form the corresponding amine salt (hexamethylene diammonium adipate), which is then heated under steam pressure (250 psig) at 275 °C to form the amide bonds. The resulting polymer has an average molecular weight of about 10,000, with an average of over 400 repeating monomer units in each molecule of polymer, and a melting point of about 150 °C. Fibers can be drawn from the melted polymer by a "cold-drawing" technique. This method of drawing fibers physically orients the polymer molecules into linear chains that are stabilized by the presence of hydrogen bonds between C=O and the N—H groups of adjacent chains, and the strength of the fiber is thereby increased. The synthetic polyamide linkages in the various forms of nylon are very similar (identical in some cases) to those found in proteins. For example, silk fibers (an example of a biopolymer) gain their great strength from this type of interaction.

Numerous combinations of diacids and diamines have been evaluated as fiber materials. However, only a few have reached commercial production, which depends on low-cost, easy-to-access intermediates, and satisfactory general and physical properties of the polymer. One such group of materials are the "Aramid" class of fibers, which are prepared from aromatic monomers (see Experiment [B3], Question 8-37). One trade name for a material prepared from these types of fibers is Nomex®. It has a high degree of heat and flame resistance. Race car driving suits are made from it, and it is also used as an insulator in the space shuttles.

Nylon-6,6 was first synthesized in 1899 by Gabriel and Maas in Germany. It was not until 1929, however, that the substance was shown to possess practical commercial properties. It was Carother's research program on polyamides at DuPont that made the major discoveries that initiated the world's polymer industry. DuPont began production of nylon in October of 1939, and the first nylon stockings were manufactured in May of 1940. By 1950, 14 chemical plants in 10 countries produced 55,000 metric tons of polyamide fiber. By 1980 worldwide production had expanded to $3.05 \times 10^6$ tons, with about one third of the polymer synthesized in the United States.

Thus, you should appreciate that the chemical industry carries out organic reactions on massive quantities of material for use in today's highly technological society. The discovery and characterization of these materials all starts in the research laboratory, with many of them initially prepared in microscale quantities. One of the great triumphs of our technology has been the successful scaleup of synthetic organic reactions, but that is a story for another day.

## Experiment [B1] Oxidation of Cyclohexanol: Adipic Acid

Common name: Adipic acid
CA number: [124-04-9]
CA name as indexed: Hexanedioic acid

### Purpose
To carry out the nitric acid oxidation of cyclohexanol to obtain adipic acid, an intermediate in the route to prepare nylon, a polyamide.

### Prior Reading

Technique 5: Crystallization
Use of the Hirsch Funnel (see pp. 93–95).
Standard Experimental
Apparatus: Reflux Apparatus (see pp. 24–26).

---

**REACTION**

---

**DISCUSSION**

Industrially, the production of adipic acid is a two-step sequence. The main route involves the oxidation of cyclohexane with air to form a mixture of cyclohexanol and cyclohexanone. This mixture is then further oxidized using nitric acid, oxygen, and a Cu–V catalyst to yield the acid.

Ninety percent of all the synthetically produced adipic acid is used in the manufacture of Nylon-6,6. In the United States, 688,000 tons/year of adipic acid were manufactured in 1974. In the early years of nylon production, adipic acid was also used to prepare the 1,6-hexanediamine (more commonly known as hexamethylene diamine) comonomer. Treatment of the adipic acid with ammonia gave hexanedinitrile which, on catalytic hydrogenation, produced the diamine. This monomer is now generally obtained from 1,3-butadiene. A recent DuPont industrial process involves direct *regioselective* addition of two molecules of HCN to the diene, in the presence of a transition metal catalyst, to produce the dinitrile intermediate.

The oxidation of cyclohexanol by concentrated nitric acid is mechanistically complex. A reasonable mechanistic route to the dicarboxylic acid is given here. The first stage of the oxidation is considered to proceed by a mechanism similar to that found in chromic acid oxidations of alcohols (see Experiment [33]). The reaction here involves the initial formation of a nitrate ester intermediate which, under the reaction conditions, cleaves by proton abstraction to form the ketone. This reaction is accompanied by reduction of the nitrate to nitrite. The proton transfer may involve a cyclic intramolecular rearrangement during the oxidation–reduction cleavage step. A likely mechanism is outlined below:

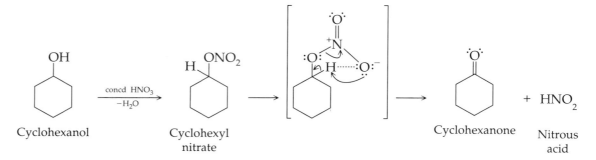

Cyclohexanol      Cyclohexyl nitrate      Cyclohexanone    Nitrous acid

The next stage of the reaction can be viewed as a further oxidation to yield a diketone. This stage is initiated by nucleophilic attack on a nitronium ion ($NO^+$) derived from either the nitric or nitrous acid. The nucleophile is the enol tautomer of the ketone, and the reaction forms an $\alpha$-nitrosoketone, which is in tautomeric equilibrium with a mono-oxime. This species rapidly hydrolyzes under acidic conditions to yield an $\alpha$-diketone intermediate. This sequence is shown here.

$$H^+ + HNO_2 \rightleftharpoons NO^+ + H_2O$$

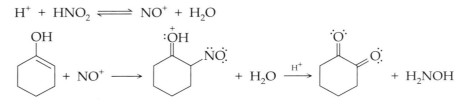

Under strongly acidic conditions, the diketone (these highly electrophilic systems are reactive toward weak nucleophiles) likely undergoes nitrate addition, which is followed by attack of water, ring opening, and reduction of nitrate to nitrite. All this activity ultimately leads to the formation of the desired compound, the open-chain dicarboxylic acid, adipic acid.

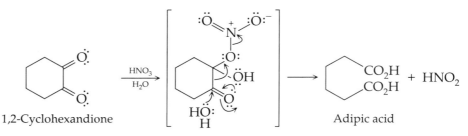

1,2-Cyclohexandione                        Adipic acid

---

**EXPERIMENTAL PROCEDURE**      Estimated time to complete the experiment: 2.0 hours.

**Physical Properties of Reactants**

| Compound | MW | Amount | mmol | mp(°C) | bp(°C) | d |
|---|---|---|---|---|---|---|
| Cyclohexanol | 100.16 | 1.0 mL | 9.6 | 25.1 | 161.1 | 0.96 |
| Concd HNO₃ | 31 | 2.0 mL | | | | |

## Reagents and Equipment

Using a graduated glass pipet, measure and add 2.0 mL of concentrated nitric acid to a 10-mL round-bottom flask containing a boiling stone.

---

> **WARNING:**   Nitric acid is very corrosive. Dispense the material in the hood.

---

Attach the flask to a water-cooled reflux condenser, place the assembly in a sand bath and heat the acid solution to 55–60 °C. Now add dropwise, using a calibrated 9-in. Pasteur pipet inserted down the throat of the reflux condenser, 1.0 mL of cyclohexanol at a rate of one drop every 30 seconds. (*Gently swirl the reaction mixture in the bath after each addition.*) The slow addition is necessary to control the reaction temperature. (■)

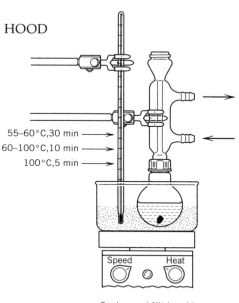

HOOD

55–60°C,30 min →
60–100°C,10 min →
100°C,5 min →

2 mL concd Nitric acid,
1 mL cyclohexanol

## Reaction Conditions

Maintain the sand bath temperature at 55–60 °C for 30 min after the addition of cyclohexanol is complete. Now gradually raise the sand bath temperature to 100 °C over 10 min, and then maintain this temperature for an additional 5 min.

## Isolation of Product

Remove the assembly from the hot sand bath and allow the solution to cool to room temperature. Detach the round-bottom flask, and clamp it in an ice bath for 5–10 min. Collect the light yellow crystals by vacuum filtration using a Hirsch funnel.

---

> **CAUTION:**   This solution is still strongly acidic.

---

Wash the product crystals with 200 $\mu$L portions of *ice cold* water until the crystals turn white.

## Purification and Characterization

Dry the adipic acid crystals in an oven at 110–125 °C for 10 min. Weigh the product and calculate the percent yield. Determine the melting point, and obtain an IR spectrum using the KBr disk technique. Compare your results to those recorded in the literature.

## Chemical Test

Add several crystals (~5 mg) of the adipic acid to 1 mL of a 10% aqueous solution of sodium bicarbonate. Does evolution of $CO_2$ indicate the presence of a carboxylic acid?

## QUESTIONS

**8-35.** Which of the following compounds is the stronger acid: $CF_3CH_2CO_2H$ or $CH_3CH_2CO_2H$? Explain.

**8-36.** What spectroscopic method would you use to unambiguously distinguish between the following isomeric acids? Give an explanation of how the spectra are interpreted to give you an assignment for each compound.

$$CH_3(CH_2)_3CO_2H \qquad (CH_3)_2CHCH_2CO_2H \qquad (CH_3)_3CCO_2H$$

**8-37.** Indicate how you would use both $^1H$ and $^{13}C$ NMR spectroscopy to tell the difference between the following isomeric carboxylic acids.

$$HO_2CCH_2CH_2CO_2H \qquad CH_3CH(CO_2H)_2$$

**8-38.** The two carboxyl groups in 3-chlorohexanedioic acid are not equivalent and thus have different dissociation constants. Which carboxylic group is more acidic? Explain.

**8-39.** Write an equation for the formation of the salt that could be formed by reaction of one molecule of adipic acid with two molecules of ethylamine.

---

**BIBLIOGRAPHY**

For detailed information on the production and use of adipic acid see:

*Kirk–Othmer Encyclopedia of Chemical Technology*, 3rd ed., Vol. 1, Wiley: New York, 1978, p. 511.

The synthesis of adipic acid is given in *Organic Syntheses*:

Ellis, B. A. *Organic Syntheses*; Wiley: New York, 1941; Collect. Vol. I, p. 18.

### Experiment [B2]   Preparation of an Acid Chloride: Adipoyl Chloride

Common name: Adipoyl chloride
CA number: [111-50-2]
CA name as indexed: Hexanedioyl dichloride

#### Purpose
To convert adipic acid to its corresponding acid chloride by reaction with thionyl chloride. To further understand the nucleophilic substitution reaction pathway by which carbonyl-containing compounds undergo reaction.

#### Prior Reading
*Standard Experimental Apparatus:*   Reflux Apparatus (see pp. 24–26).
*Collection or Control of Gaseous Products:*   (see pp. 109–111).

---

**REACTION**

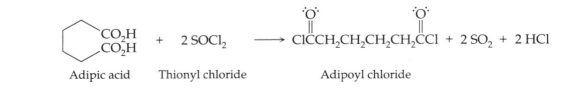

Adipic acid      Thionyl chloride          Adipoyl chloride

---

**DISCUSSION**

Carboxylic acids react with thionyl chloride ($SOCl_2$) to produce the corresponding acid chlorides, as shown in the above reaction. Thionyl chloride is an attractive reagent due to its low cost, and the fact that both byproducts produced in the reaction are gases. Thus, the reaction is driven to completion by the evolution of HCl and SO$_2$, and a nearly pure acid chloride is obtained. The major drawback to the reaction is that it produces a strong acid (HCl) and thus cannot be used with compounds that are acid sensitive. Oxalyl chloride is often used as an alternative reagent.

The reaction proceeds by a nucleophilic acyl substitution pathway. With thionyl chloride, a chlorosulfite intermediate is generated. Thus, the —OH group is converted into a relatively good leaving group. The chlorosulfite intermediate then undergoes attack by the chloride ion at the carbonyl carbon to yield the final product. The sequence is shown here.

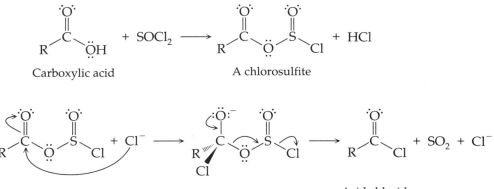

Acid halides are important intermediates and they are used extensively for the conversion of carboxylic acids into other derivatives. For example, acid halides can be used to prepare (in addition to amides): anhydrides, esters, aldehydes, and ketones. Acid halides readily undergo reaction with water (hydrolysis) to form the corresponding carboxylic acid. For this reason the reaction system must be protected from atmospheric moisture when acid halides are formed and/or used.

In the present sequence leading to the formation of nylon, the adipoyl chloride provides a reactive species which, when treated with a diamine, forms the desired amide linkage inherent to nylon.

Estimated time to complete the experiment: 2.0 hours.

## EXPERIMENTAL PROCEDURE

### Physical Properties of Reactants

| Compound | MW | Amount | mmol | mp(°C) | bp(°C) | d |
|---|---|---|---|---|---|---|
| Adipic acid | 146.14 | 500 mg | 3.4 | 153 | 265 | 1.35 |
| Thionyl chloride | 118.97 | 0.75 mL | 10.4 | −105 | 78.8 | 1.66 |

### Reagents and Equipment

Place and dry in an oven at about 125 °C for 30 min, a 10-mL round-bottom flask, a water-cooled reflux condenser, and a drying tube packed with glass wool and alumina. After removing them from the oven, place these items in a desiccator and allow them to cool to room temperature. Measure and

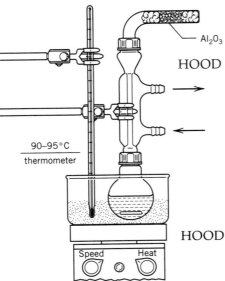

90–95 °C
thermometer

Speed    Heat

Adipic acid, 500 mg +
thionyl chloride, 0.75 mL
reflux conditions,
10–mL RB flask

place 500 mg (3.4 mmol) of adipic acid and 0.75 mL (10.4 mmol) of thionyl chloride into the 10-mL flask. (■)

**NOTE.**  *Store the dry adipic acid and the container of thionyl chloride in a desiccator. Dispense the thionyl chloride in the* **hood** *using an automatic delivery pipet.*

---

**CAUTION:   Thionyl chloride is both corrosive and a lachrymator.**

---

Add a magnetic stir bar to the flask, quickly attach the flask to the reflux condenser protected by the drying tube, and mount the assembly in a sand bath.

**NOTE.**  *If a larger scale reaction is to be run outside of the* **hood**, *it will be necessary to construct a gas trap to control the evolution of the SO$_2$ and HCl gases (see Prior Reading).*

## Reaction Conditions

With stirring, heat the reaction mixture to 90–95 °C within 5 min. Continue to heat the system within this temperature range for an additional 1 hour.

**NOTE.**  *Heating the solution above 95 °C causes decomposition of the product. The presence of unreacted adipic acid in the flask indicates incomplete reaction.*

## Isolation and Characterization

The progress of the reaction may be followed by IR analysis. With a glass capillary, remove a small sample from the flask and obtain the spectrum of the material using the capillary film technique. Remove the sample from the instrument sampling compartment immediately following the spectral scan in order to prevent the HCl gas-buildup from damaging the instrument. The reaction is considered incomplete if the IR spectrum displays a weak band on the low-wavenumber side of the acid halide carbonyl peak.

If the reaction is determined to be incomplete after 1 hour, continue to heat the mixture until IR analysis (the disappearance of the weak side band) indicates that completion has occurred. The adipoyl chloride is quite labile, and therefore, it is not purified further, but it is used directly in the preparation of nylon as described in Experiment [B3].

## QUESTIONS

**8-40.**  Diagram a complete mechanistic sequence showing the hydrolysis reaction of ethanoyl chloride with water to form ethanoic acid.

**8-41.**  Give an explanation of why acid chlorides are more reactive toward nucleophilic substitution than are the corresponding ethyl esters. *Hint:* Consider the nature of the leaving group and the rate-determining step in an addition–elimination sequence.

**8-42.**  What major organic product would you expect to be formed when acetyl chloride reacts with each of the following reagents?
a. H$_2$O
b. NH$_3$ (excess)

    c. 1-Butanol and pyridine
    d. CH$_3$COO$^-$, Na$^+$

**8-43.** Acid chlorides are used extensively as electrophiles in the Friedel–Crafts reaction to prepare aromatic ketones. The reaction involves the treatment of an aromatic hydrocarbon with an acyl chloride in the presence of a Lewis acid, such as aluminum chloride (see Experiment [27]).

    Using this reaction, outline the reaction sequence you would use to prepare
        a. Ethyl phenyl ketone      b. Benzophenone

**8-44.** Formulate a suitable mechanism to account for the reaction of thionyl chloride with carboxylic acids to yield the corresponding chlorosulfite (see the reaction diagrammed in the discussion section). *Hint:* The first step is the nucleophilic attack of the oxygen of the —OH group of the acid on the sulfur atom of the S=O group of the thionyl chloride.

---

Adipoyl chloride is prepared as an intermediate in several preparations reported in *Organic Syntheses:*

Fuson, R. C.; Walker, J. T. *Organic Syntheses*; Wiley: New York, 1943; Collect. Vol. II, p. 169.

Guha, P.C.; Sankaran, O. K. *Organic Syntheses*; Wiley: New York, 1955; Collect. Vol. III, p. 623.

**BIBLIOGRAPHY**

## Experiment [B3]   Preparation of a Polyamide: Nylon-6,6

Common names: Nylon-6,6, polyhexamethylene adipamide
CA number: [32131-17-2]
CA name as indexed: Poly[imino(1,6-dioxo-1,6-hexanediyl)imino-1,6-hexanediyl]

### Purpose
To prepare the polyamide, nylon, by the step-growth condensation polymerization of adipoyl chloride with 1,6-hexanediamine. To use an interfacial (emulsion) polymerization technique to generate nylon fibers.

---

**REACTION**

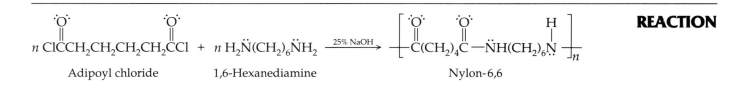

Adipoyl chloride      1,6-Hexanediamine          Nylon-6,6

---

**DISCUSSION**

The preparation of nylon outlined in this experiment is not the industrial method (see initial discussion). The use of the reactive diacid chloride reagent allows one to carry out the step-growth polymerization reaction under very mild conditions more convenient to the instructional laboratory. The interfacial (emulsion) polymerization technique employed consists of dissolving the adipoyl chloride reagent in a water-immiscible solvent (cyclohexane) and bringing this solution into contact with an aqueous solution of the diamine. A thin film forms at the *interface* of the two solvents as the condensation reaction proceeds. A "rope" of nylon polymer can be pulled from the interface of the two solvents since the film is continuously generated as the reaction occurs. This polymer has an average molecular weight of ~10,000! In this particular experiment, about 5–7 meter lengths of nylon polymer can be obtained. This particular synthesis of nylon is often used in lecture demonstrations and chemical magic shows.

## EXPERIMENTAL PROCEDURE

Estimated time to complete the experiment: 0.5 hours.

### Physical Properties of Reactants

| Compound | MW | Amount | mmol | mp(°C) |
|----------|-----|--------|------|--------|
| Adipoyl chloride | 183.05 | ~622 mg | 3.4 | |
| 1,6-Hexanediamine (5% aq) | 116.21 | 8 mL | | 41–42 |
| Cyclohexane | 84.16 | 8 mL | | |

### Reagents and Equipment

Transfer the clear solution of adipoyl chloride prepared in Experiment [B2] to a 50-mL beaker using a Pasteur pipet. Rinse the flask with 2 mL of cyclohexane and transfer this rinse to the same beaker. Add an additional 8.0 mL of cyclohexane. Now slowly add 8.0 mL of a 5% aqueous solution of 1,6-hexanediamine containing eight drops of 25% NaOH solution.

**NOTE.**  *Add the solution using a Pasteur pipet, taking care to run it down the side of the beaker.*

### Isolation and Characterization

Using a copper wire bent into a small hook, hook the film in the center of the beaker and draw up the nylon fiber from the solution interface. A slow, steady pull will result in long strands of the polymer. Wash the fibers in a beaker of water before handling them.

## QUESTIONS

**8-45.**  Explain why the 25% NaOH solution is added to the reaction mixture.

**8-46.**  Draw a structure to illustrate the hydrogen bonding that may occur when two polymer molecules of this polyamide are cold-drawn together.

**8-47.**  Predict the structure of the polymer that would result in the condensation of the following reactants. These monomers are used to produce the polyamide Nomex, a high-melting material used as an insulator in space shuttles, and as the fire-resistant fabric in clothing worn in race cars.

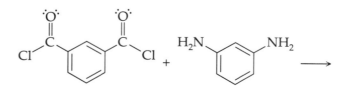

**8-48.** Amides undergo hydrolysis to carboxylic acids on treatment with alkali. Diagram a suitable mechanism for the conversion of benzamide to benzoic acid using sodium hydroxide as the base.

---

For detailed information on the production and use of nylon see:

Heckert, W. W. *J. Chem. Educ.* **1953,** *30,* 166.

*Kirk–Othmer Encyclopedia of Chemical Technology,* 3rd ed., Vol. 18, Wiley: New York, 1982, p. 328 (general); p. 372 (fibers); p. 406 (plastics).

For information on the interfacial polymerization technique see:

Sprague, B. S.; Singleton, R. W. *Text. Res. J.* **1965,** *35,* 999.

Morgan, P. W.; Kwolek, S. L. *J. Chem. Educ.* **1959,** *36,* 182.

---

# The Synthesis of Sulfanilamide

### Purpose

To carry out the multistep synthesis of the epoch-making antibacterial drug, sulfanilamide. To learn the techniques for handling moisture sensitive materials. To investigate the strategies involved in solving a synthetic organic problem.

---

The sulfanilamides were the first effective systemic bactericides. The discovery, in the early 1930s, of these substances transformed medical care. Indeed, a major fraction of the world's population is alive today as a result of these compounds and their microbiological successors. Early recognition of the importance of these materials was underlined when their discoverer was honored by receiving the 1938 Nobel Prize in Medicine. Even 60 years later, "sulfa" drugs are still some of the most effective antibacterial substances available to physicians today.

These compounds, to which human existence owes a considerable debt of gratitude, are all derivatives of the parent material, *p*-aminobenzenesulfonamide (sulfanilamide).

$$H_2N-\langle\ \rangle-SO_2NH_2$$

*p*-Aminobenzenesulfonamide
(sulfanilamide)

This weakly basic compound contains two —$NH_2$ groups located in quite different environments, so that each one possesses quite different chemical properties and reactivities. One is substituted directly on the aromatic ring; the other is contained in a sulfonamide group. A wide variety of these derivatives can be prepared by introducing different substituents on the nitrogen atom of the sulfonamide functional group. Any attempt at modifying the ring —$NH_2$ group, however, was found to completely destroy the biological activity.

Prontosil, *p*-[(2,4-diaminophenyl)azo]benzenesulfonamide, was first synthesized in Germany in 1932 as a product of azo-dye research at I. G.

Farbenindustrie. It was soon tested for chemotherapeutic activity, as it bound very strongly to protein fibers. In 1933, it became the first drug to effectively cure blood stream infections. The early investigations of the biological activity of Prontosil required direct testing on animal infections, as the drug appeared to be inactive in vitro.

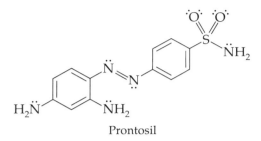

Prontosil

In 1935, Trefouels (Pasteur Institute) established that Prontosil breaks down in the body to yield, as one of the metabolites, sulfanilamide. It was shown later that year, by Fourneau, that the biological activity of Prontosil was entirely contained in the sulfanilamide section of the molecule, and that no biological activity was lost during the cleavage. This later discovery, and the fact that sulfanilamide itself had been earlier synthesized by Gelmo in 1908 (unfortunately by not exploring the biological activity of the substance, he missed the opportunity at a Nobel Prize), voided any patent protection and led to a worldwide effort to synthesize a wide variety of these materials. By 1944 over 5000 sulfanilamide derivatives had been prepared and tested.

Currently the sulfa drugs are still very important therapeutic agents, but have to a large extent been replaced by antibiotics, such as the penicillins. Their low cost and general effectiveness for urinary tract infections, however, still make sulfanilamide derivatives attractive alternatives. These compounds have also found a valuable role in veterinary medicine.

Production of sulfanilamide reached a peak of 9000 metric tons/year in 1943; present production is about one half that amount.

The following experiments in this sequence outline a novel and efficient procedure for the preparation of this important substance. The chemistry involves the trifluoroacetylation of aniline (Experiment [C1]), the chlorosulfonation of the resulting acetanilide (Experiment [C2]), which is followed by concomitant aminolysis and hydrolysis of the sulfonyl chloride intermediate (Experiment [C3]) to produce the final product, sulfanilamide.

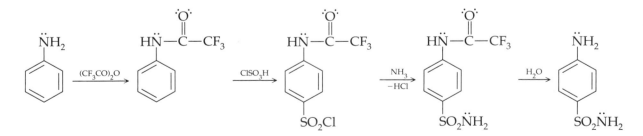

## Experiment [C1]   Acetylation of Aniline: 2,2,2-Trifluoroacetanilide

Common name: 2,2,2-Trifluoroacetanilide
CA number: [404-24-0]
CA name as indexed: Acetamide, 2,2,2-trifluoro-$N$-phenyl-

## Purpose

To protect the easily oxidized amine group of aniline from electrophilic attack during the sulfonation of the benzene ring in the next step of the synthesis. This deactivation is accomplished by reducing the electron density on the amine nitrogen (and thus its nucleophilicity) by acetylating aniline with the strongly electron-withdrawing trifluoroacetyl group. To become familiar with techniques for handling moisture sensitive materials.

*Prior Reading*

| | |
|---|---|
| *Standard Experimental Apparatus:* | Moisture-Protected Claisen Head with 3- or 5-mL Conical Vial, Arranged for Syringe Addition (see pp. 29–31). |
| *Technique 6B:* | Concentration of Solutions: Evaporation with Nitrogen Gas (see p. 107). |

**REACTION**

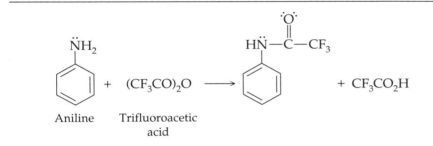

Aniline     Trifluoroacetic acid

**DISCUSSION**

Acid anhydrides react with amines, or ammonia, to yield amides. The reaction of anhydrides is similar to that of acid chlorides (see Experiment [B3]), although the anhydrides normally react at slower rate.

Acetic anhydride, $CH_3C(O)OC(O)CH_3$, is an important industrial reagent used to prepare acetamides, $CH_3C(O)NR_2$ from a variety of amines. For example, acetaminophen, an over-the-counter analgesic, is synthesized by the reaction of *p*-hydroxyaniline with acetic anhydride.

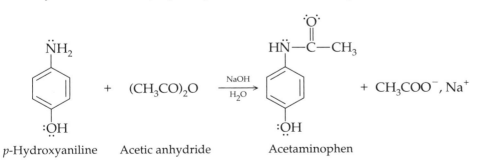

*p*-Hydroxyaniline    Acetic anhydride      Acetaminophen

In the present reaction, *trifluoro*acetic anhydride is used in place of acetic anhydride. The halogenated anhydride is considerably more reactive because of the presence of the three fluorine atoms on the α-carbon atom of the anhydride. The electron-withdrawing power of the fluorine substituents helps to enhance nucleophilic attack on the anhydride by the weak aromatic nucleophile. Once the amide is formed, the trifluoromethyl group makes the amide a better protecting group than the acetyl group, as the trifluoroacetyl group renders the amine lone electron pair less nucleophilic.

Furthermore, the byproduct of the reaction, trifluoroacetic acid, is fairly volatile (bp 72 °C) and is therefore easily removed from the reaction mixture by evaporation. This enhanced reactivity helps avoid the use of an added base, as is required in the preparation of acetaminophen. Chemically, the procedure is simplified and the intermediate trifluoroacetanilide is usually obtained in a relatively pure condition. *This highly reactive anhydride, however, requires considerable care in its use.*

---

**WARNING:   No moisture can remain on the surface of the clean glassware or hydrolysis may occur with explosive force!**

---

## EXPERIMENTAL PROCEDURE

Estimated time to complete the experiment: 0.75 hour.

### Physical Properties of Reactants

| Compound | MW | Amount | mmol | mp(°C) | bp(°C) | d |
|---|---|---|---|---|---|---|
| Aniline | 93.13 | 230 μL (235 mg) | 2.5 | −6.2 | 184.3 | 1.02 |
| Trifluoroacetic anhydride | 201.04 | 500 μL (744 mg) | 3.54 | −65 | 39.5 | 1.49 |
| Methylene chloride | 84.93 | 500 μL | | | | |

### Reagents and Equipment

Place and dry in an oven at about 125 °C for 30 min, a 5-mL conical vial, a Claisen head, a drying tube packed with glass wool and calcium chloride, a 1-mL glass syringe, and a 1/2-dram screw-cap vial. After removal from the oven, place these items in a desiccator and allow them to cool to room temperature. Measure and place 230 μL (2.5 mmol) of aniline and 500 μL of methylene chloride in the 5-mL conical vial. Immediately attach the vial to the Claisen head equipped with a septum cap and protected by the drying tube (see Prior Reading). Cool the vial in a cold water bath.

HOOD    **NOTE.**   *Dispense the aniline and methylene chloride in the* **hood** *using automatic delivery pipets.*

---

**CAUTION:   Aniline is a highly toxic substance and a cancer suspect agent.**

---

HOOD    In the **hood**, place 500 μL of trifluoroacetic anhydride and 500 μL of methylene chloride in the dried 1/2-dram vial.

---

**CAUTION:   The anhydride hydrolyzes rapidly in moist air, so make the transfer quickly.**

---

Now draw this solution into the dried syringe, and insert the needle through the septum cap of the Claisen head. Dropwise **slowly** add this solution to the cold aniline–methylene chloride solution.

SLOWLY

**NOTE.** *Heat is produced in the reaction. If the addition is too rapid, the methylene chloride starts to reflux and fumes are emitted through the drying tube.*

## Reaction Conditions

After the addition is complete, allow the reaction mixture to stand at room temperature for 10 min.

## Isolation of Product

Remove the reaction vial and, in the **hood**, concentrate the solution on a warm sand bath under a slow stream of nitrogen gas. The crude 2,2,2-trifluoroacetanilide intermediate is obtained as a white powder. Cap the vial.

HOOD

## Purification and Characterization

The crude material is not purified further, but rather used directly in the next reaction of the sequence without further characterization.

**8-49.** Outline a complete mechanistic sequence for the reaction to prepare acetaminophen as presented in the discussion section of this experiment.

**QUESTIONS**

**8-50.** In the preparation of amides, acid chlorides or anhydrides may be used to react with the selected amine. It is known that acid chlorides are more reactive than the corresponding anhydrides in this type of nucleophilic acyl substitution reaction. Offer a reasonable explanation for this observation.

**8-51.** Offer an explanation of why 2,2,2-trifluoroacetic anhydride is more reactive toward aniline than acetic anhydride in the nucleophilic acyl substitution reaction presented in this experiment.

**8-52.** Offer a reasonable explanation of why acetamide has a higher melting and boiling point then *N,N*-dimethylacetamide even though it has a lower molecular weight.

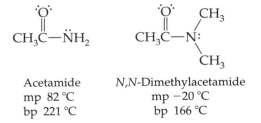

Acetamide
mp 82 °C
bp 221 °C

*N,N*-Dimethylacetamide
mp −20 °C
bp 166 °C

**8-53.** Outline a simple chemical test that would distinguish between ethyl benzoate and benzamide (Hint: See Chapter 10).

## Experiment [C2] Chlorosulfonation of 2,2,2-Trifluoroacetanilide: *p*-(Trifluoroacetamido)benzenesulfonyl Chloride

Common name: *p*-(Trifluoroacetamido)benzenesulfonyl chloride
CA number: [31143-71-2]
CA name as indexed: Benzenesulfonyl chloride, 4-[(trifluoroacetyl)amino]-

### Purpose
To carry out the substitution of the aniline ring by introduction of the sulfonyl group. This substitution reaction is the second stage on the route to preparing sulfanilamide. To prepare *p*-(trifluoroacetamido)benzene-sulfonyl chloride by treatment of the intermediate, 2,2,2-trifluoroacetani-lide, with chlorosulfonic acid. To gain experience at handling highly reac-tive moisture-sensitive reagents.

### Prior Reading
Technique 5: Crystallization.
Use of the Hirsch Funnel (see pp. 93–95).

**REACTION**

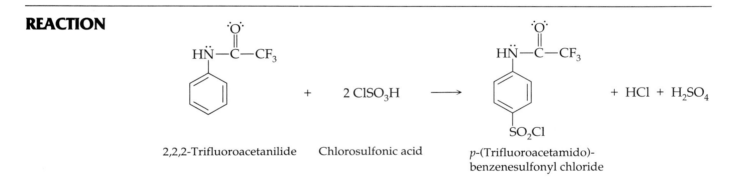

2,2,2-Trifluoroacetanilide    Chlorosulfonic acid    *p*-(Trifluoroacetamido)-
benzenesulfonyl chloride

**DISCUSSION**

As shown here, the sulfonyl chloride group (—SO$_2$Cl) can be conveniently introduced to an aromatic ring via an electrophilic aromatic substitution reaction employing chlorosulfonic acid. The reaction is usually referred to as *chlorosulfonation*. It has been determined that two equivalents of the acid are required per equivalent of the aromatic compound. It is also known that in the initial attack the system first forms the corresponding sulfonic acid, which in turn is converted to the sulfonyl chloride. It is believed that the initial stage of the reaction involves SO$_3$ as the electrophile. It is likely that this reagent results from the establishment of the equilibrium reaction shown here.

$$ClSO_3H \rightleftharpoons SO_3 + HCl$$

The intermediate sulfonic acid is converted to its sulfonyl chloride derivative by reaction with a second equivalent of the chlorosulfonic acid.

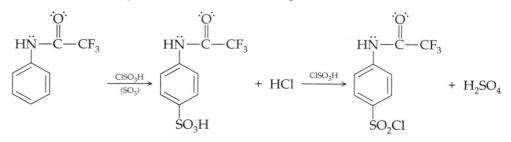

Product isolation is achieved by transferring the solution of the sulfonyl chloride onto crushed ice to precipitate the insoluble acid chloride. This isolation procedure can be used because sulfonyl chlorides are much less susceptible to hydrolysis than their carboxylic acid chloride counterparts (see Experiment [B2]). Sulfonyl chlorides *do* undergo this hydrolysis reaction, albeit rather slowly, resulting in the formation of the corresponding sulfonic acid.

Notice that only the para substituted product is formed in the reaction. Why is this? (See Question 8-57.)

---

Estimated time to complete the experiment: 0.75 hour.

## EXPERIMENTAL PROCEDURE

**Physical Properties of Reactants**

| Compound | MW | Amount | mmol | mp(°C) | bp(°C) | d |
|---|---|---|---|---|---|---|
| 2,2,2-Trifluoroacetanilide | 189.06 | 478 mg (theoretical value) | 2.5 | 87.6 | | |
| Chlorosulfonic acid | 116.52 | 900 μL | 13.7 | −80 | 158 | 1.77 |

**NOTE.** *All reagent transfers must be made and the reaction must be conducted in the hood.*                          HOOD

## Reagents and Equipment

Place and dry in an oven at about 125 °C for 30 min, an air condenser and a 9-in. Pasteur pipet calibrated to deliver 0.9 mL. After removal from the oven, place these items in a desiccator and allow them to cool to room temperature.

Attach the vial containing the 2,2,2-trifluoroacetanilide (prepared in Experiment [C1]) to the *dry* air condenser. Measure, using the 9-in. Pasteur pipet, 0.9 mL of chlorosulfonic acid, and **SLOWLY** add this reagent to the     SLOWLY
flask by inserting the pipet down the neck of the air condenser.

---

**CAUTION:** **Chlorosulfonic acid is a very corrosive substance. It reacts violently with water and causes serious burns on contact with the skin. Dispense the reagent in the *hood*. Leave all     HOOD
equipment used to transfer this material in the *hood* until it is obvious that the residual reagent has reacted with the moist air (white fumes subside), and leave these materials in contact only with *glass* surfaces.**

---

**NOTE.** *An alternative procedure for making the transfer is to measure the amount of this reagent using an oven-dried 1-mL glass pipet, and then transfer it to an oven-dried screw-cap vial. The material is then transferred to the reaction vial as described above, using the dry 9-in. pipet.*

## Reaction Conditions

Place the reaction assembly on a sand bath [**HOOD**] and heat the mixture     HOOD
at 60–70 °C for 10 min.

## Isolation and Characterization

SLOWLY

Allow the reaction mixture to cool to room temperature, and then place the flask in an ice bath. Using a second, dry 9-in. Pasteur pipet, transfer the cold reaction solution **SLOWLY** to a 10-mL beaker containing ice (~1/2 full) **[HOOD]**.

CAUTION
HOOD

---

**CAUTION:** This is a very vigorous reaction. Use **CAUTION**. Leave all equipment in the *hood* until it is obvious the residual reagent has reacted with the moist air (white fumes subside). Make sure that only glass surfaces are in contact with the reagent residues.

---

A precipitate of the product, *p*-(trifluoroacetamido)benzenesulfonyl chloride, forms at this stage. Collect the tannish-white precipitate by vacuum filtration and wash the filter cake with three 1-mL portions of ice cold water.

Allow the material to air-dry. Determine the weight of the crude product and calculate the percent yield from aniline. Determine its melting point, and compare it to the reported value of 142–145 °C.

## Chemical Test

Carry out a sodium fusion test to confirm the presence of nitrogen, sulfur, and chlorine in the product (see Chapter 10).

## QUESTIONS

**8-54.** Outline a suitable mechanism to illustrate the reaction of benzenesulfonyl chloride with excess ethylamine to form the corresponding sulfonamide.

**8-55.** In reference to Question 8-54, why is an excess of ethylamine used?

**8-56.** Account for the fact that the amide group ($-NHCOCF_3$) is ortho and para directing. Use resonance structures to illustrate your written explanation.

**8-57.** As mentioned in the discussion, offer an explanation of why only the *para*-isomer is obtained in the chlorosulfonation reaction carried out in this experiment.

## BIBLIOGRAPHY

Selected acylation reactions in *Organic Syntheses* between anhydrides and amines:

Herbst, R. M.; Shemin, D. *Organic Syntheses*; Wiley: New York, 1943; Collect. Vol. II, p. 11.

Noyes, W. A.; Porter, P. K. *Organic Syntheses*; Wiley: New York, 1941; Collect. Vol. I, p. 457.

Wiley, R. H.; Borum, O. H. *Organic Syntheses*; Wiley: New York, 1963; Collect. Vol. IV, p. 5.

# Experiment [C3] Preparation of an Arene Sulfonamide: Sulfanilamide

Common names: Sulfanilamide, *p*-aminobenzenesulfonamide
CA number: [63-74-1]
CA name as indexed: Benzenesulfonamide, 4-amino-

## Purpose
To complete the sequential synthesis and obtain the sulfa drug, *sulfanil-amide*, by treatment of the sulfonyl chloride intermediate prepared in Experiment [C2], with an aqueous ammonia solution.[2] To fully characterize this interesting product.

### Prior Reading
> Technique 5: Crystallization
> Use of the Hirsch Funnel (see pp. 93–95).
> Craig Tube Crystallization (see pp. 95–96).

---

The final step in the synthesis of the target molecule, sulfanilamide, is the conversion of the sulfonyl chloride intermediate to a sulfonamide and the removal of the 2,2,2-trifluoroacetyl protecting group. These transformations are accomplished in one step since both reactions take place upon heating the protected sulfonyl chloride with aqueous ammonia. The reaction sequence is shown here.

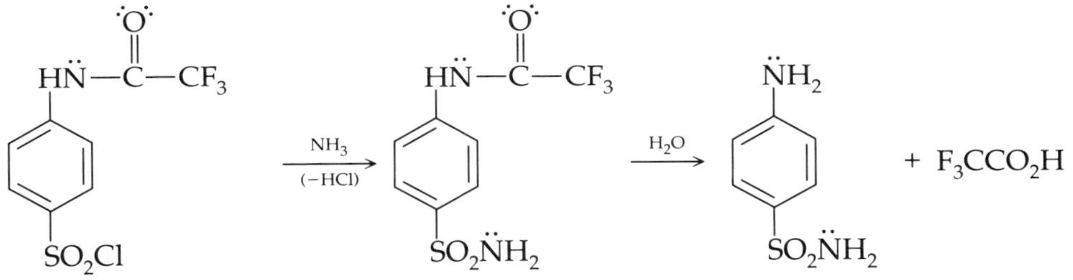

**DISCUSSION**

*p*-(Trifluoroacetamido)-
benzenesulfonyl chloride

Sulfanilamide

We can now see several reasons for first protecting the reactive amino group on the aniline molecule. First, to attempt the introduction of the sulfonyl group directly on aniline would very likely lead to a species (*p*-aminobenzenesulfonyl chloride) that would react with itself (act as a difunctional monomer) to produce a sulfonamide polymer.

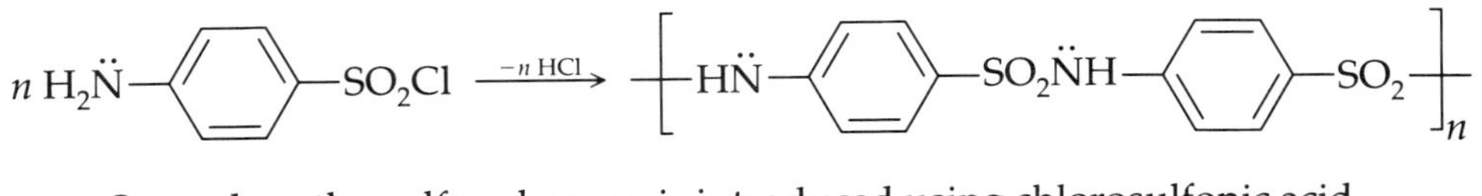

Second, as the sulfonyl group is introduced using chlorosulfonic acid, an acidic medium is present. Under these conditions, since free HCl is available (see discussion in Experiment [C2]) the amino group would be protonated and become a meta directing group. This situation would lead to the formation of the wrong isomer, or at best, a mixture of isomers.

---

[2] This experiment is adapted from the work of S. Danishefsky, Yale University (personal communication).

Third, as mentioned above, the amino group is highly prone to oxidation and the unavoidable presence of traces of the strong oxidant, $SO_3$, would be expected to lead to considerable oxidation of the aromatic amine.

The 2,2,2-trifluoroacetyl group meets all the requirements of a good protecting group. First, the acetylation reaction goes rapidly in nonaqueous media to give a clean, dry product. Second, the protecting group is easily and rapidly removed in the final stage without disturbing the other functional groups in the molecule.

## EXPERIMENTAL PROCEDURE

Estimated time to complete the experiment: 0.75 hour.

**Physical Properties of Reactants**

| Compound | MW | Amount | mmol | mp(°C) |
|---|---|---|---|---|
| (2,2,2-Trifluoroacetamido) benzenesulfonyl chloride | 287 | 400 mg | 1.39 | 142–145 |
| Aqueous ammonia | | 600 μL | | |

HOOD

**CAUTION:** All reagent transfers, and the *reaction*, must be conducted in the *hood*.

### Reagents and Equipment

Weigh and place in a 10-mL Erlenmeyer flask 400 mg (1.39 mmol) of the *p*-(2,2,2-trifluoroacetamido)benzenesulfonyl chloride, which was prepared in Experiment [C2]. In a 10 × 75-mm test tube, prepare a solution of 0.6 mL of fresh, concentrated aqueous ammonia (ammonium hydroxide) and 0.4 mL of deionized water. Add this solution to the solid sulfonyl chloride in the Erlenmeyer flask. Now add a boiling stone. Use a glass rod to break up any lumps of the solid that may form.

HOOD

**CAUTION:** Dispense the ammonia solution in the *hood*.

**CAUTION:** If the sulfonyl chloride reagent contains acidic impurities, a vigorous reaction may occur when the reagents are mixed.

### Reaction Conditions

HOOD

Place the Erlenmeyer flask on a hot (100–110 °C) sand bath [**HOOD**] and heat the mixture until the solid material dissolves. Agitate with a glass rod, if necessary, to assist the dissolution process. Now heat the solution to boiling for an additional minute.

## Isolation and Characterization

Remove the flask from the sand bath and allow the reaction mixture to cool to room temperature. Place the flask in an ice bath for 15–20 min. During this time light-yellow crystals of product precipitate.

Collect the crystalline product by vacuum filtration, and wash the crystals with three 0.5-mL portions of ice-cold water. Air-dry the sulfanilamide, weigh and calculate a crude yield, both for this step and for the overall sequence based on the amount of aniline used in the first step. Determine the melting point of this material.

Recrystallize the crude material from water using the Craig tube. Air-dry the white crystals overnight, or in an oven (110 °C) for 10–15 min. Determine the melting point and compare your value to that reported in the literature. To further characterize the material obtain an IR and NMR spectrum and compare them to those found in *The Aldrich Library of Infrared Spectra* and/or in *The Aldrich Library of NMR Spectra*.

## Chemical Test

Carry out a sodium fusion test to confirm the presence of nitrogen and sulfur in the product (see Chapter 10).

**QUESTIONS**

**8-58.** Based on the results of this experiment, which is more nucleophilic: hydroxide ion (HO⁻) or ammonia ($NH_3$)? Explain.

**8-59.** Outline a suitable mechanism to account for the conversion of the sulfonyl chloride group to the sulfonamide group by reaction with ammonia.

**8-60.** Outline a synthesis for the sulfa drug, sulfathiazole, starting from benzene. The primary aminothiazole ring system is available.

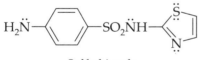

Sulfathiazole

**8-61.** Sulfonyl chlorides also react readily with alcohols to yield sulfonate esters.

$$ArSO_2Cl + HOR \xrightarrow[\substack{or \\ 3° \ amine}]{pyridine} ArSO_2OR + HCl$$

Propose a suitable mechanism to account for this reaction.

**8-62.** In reference to the Hinsberg test used to distinguish between primary, secondary, and tertiary amines (see Chapter 10), primary amines yield a sulfonamide that is soluble in aqueous sodium hydroxide, whereas secondary amines give a sulfonamide that is not soluble in this reagent. Offer an explanation for this observation.

**BIBLIOGRAPHY**

The use of the perfluoroacetic anhydride reagent in this synthesis is reported in:

Hurdis, E. C.; Yang, J. W. *J. Chem. Educ.* **1969**, *46*, 697.

For an overall view of the sulfa drug story see:

*Kirk–Othmer Encyclopedia of Chemical Technology*, 3rd ed., Vol. 2, Wiley: New York, 1978, p. 795.

---

## SEQUENCE D

## The Synthesis of 2'-Bromostyrene

### Purpose

To carry out a three-step synthesis of 2'-bromostyrene starting with benzaldehyde. To gain experience working with semimicroscale quantities of organic materials. (See Sequence A, where benzaldehyde is also used as the starting material, in that case for the formation of two different intermediates in the seven-step synthesis of hexaphenylbenzene.) To become familiar with the stereochemistry involved in the addition of molecular halogen to an alkene (see also Experiment [A2$_b$]). To observe the elimination of a hydrogen halide promoted by a concerted decarboxylation reaction (see also Experiment [A3$_b$]). As an option, to use $^1$H NMR spectroscopy to determine the cis/trans ratio of isomers in the product (see also Experiment [5B]).

---

## THE SYNTHESIS OF A FRAGRANCE

Numerous classes of organic compounds have characteristic odors. For example, volatile esters have pleasant odors (see Experiment [8B]) and are often used in perfumes and artificial flavorings where they contribute to the fragrance of the material; acid chlorides have sharp penetrating odors (see Experiment [B2]); the alkyl amines have a "fishy" smell (see Experiment [B3]), benzaldehyde has the odor of bitter almond oil (see Experiment [20]), and cinnamaldehyde, the odor of cinnamon (see Experiment [11C]). Experiment [11C] also contains an expanded discussion of naturally occurring fragrant materials, termed *essential oils*. As chain length increases, the odors of the short-chain alkyl carboxylic acids progress from the sharp, irritating odors of formic and acetic acids to the very rank, disagreeable odors of butyric (rancid butter), valeric and caproic (dirty socks and goats) acids. Low molecular weight thiols (mercaptans), sulfides, and disulfides have intensely disagreeable odors. Examples of these materials are the active principals in the chemical defense spray used by the skunk. Via evolution, this animal has developed a highly effective mixture of 3-methylbutane-1-thiol, *trans*-2-butene-1-thiol, and *trans*-2-buten-1-yl methyl disulfide, which is able to deter most skunk predators, not just humans. Our sense of smell can detect one part of ethanethiol in 50 billion parts of air. This sounds spectacular, but humans have developed other highly sensitive detectors (the retinas of our eyes) to a band of electromagnetic radiation in the region 400–750 nm (the visible region of the spectrum). We have, as a consequence, become less dependent on our sense of chemical communication (our sense of smell). When we compare our detection limit of mercaptans to the male silkworm's detection of the female sex *pheromone*, we find that the *Bombyx mori* respond to threshold levels close to 1 million times lower (in the concentration range of 100 molecules per milliliter)!

Clearly, chemical communication (odor) plays a critical role in mediating animal and plant behavior within and between species. These metabolites are often stored and released when the animal (or plant) responds to certain stimuli. Historically, these substances served as the major communication link between living systems during the evolutionary development of single cell organisms. Thus, they predated hormones which function in a similar fashion, but within a host collection of cells. While these substances may be used by a given insect species as a sex attractant, pheromones can also mediate other intraspecies functions, such as acting as trail guides and alarm signals. Thus, pheromones are substances that are used to communicate information between individuals of the same species to their adaptive advantage. The activity of a pheromone frequently depends on the configuration around a double bond (*E* or *Z*) as well as the absolute configuration around an *R* or *S* chiral center, just as humans select for these structural features (see discussion in Experiment [11D]). In a number of cases, the response has been shown to depend on the ratio of the isomers. There are several other classes of chemical communication substances employed by plants and animals. The *allomones* are interspecies materials which, on release, are used to the adaptive advantage of the host but to the disadvantage of the receiver. A typical function of an allomone is as a chemical defense agent; the thiols and disulfides used by the skunk are excellent examples. Three typical chemical communication substances are shown here.

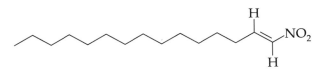

Termite defense allomone

and

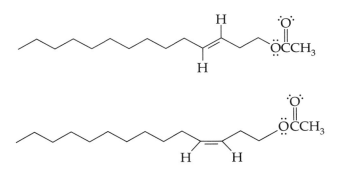

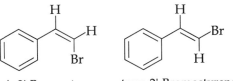

Sex pheromones of the European corn borer (a mixture of cis + trans acetates)

The synthetic route undertaken in the Sequence D experiments leads ultimately to a mixture of the diastereomeric cis–trans isomers of 2′-bromostyrene.

*cis*-2′-Bromostyrene    *trans*-2′-Bromostyrene

This material is often used as an additive to impart a pleasant fragrance to soaps, as it has a very pleasant aroma, similar to that of hyacinth. It is not, however, a naturally occurring material. As a nice illustration of the fact that stereoisomeric compounds can have markedly different physiological effects, it has been demonstrated that a single diastereomer, *trans*-2'-bromostyrene, is responsible for the hyacinth-like odor.

The chemistry involved in this particular synthesis of 2'-bromostyrene contains a number of fundamental organic reactions and is rather interesting to study. The first stage (Experiment [D1]), involves the condensation of benzaldehyde with malonic acid to produce the intermediate, *trans*-cinnamic acid. This reaction is mechanistically similar to the well-known *aldol* reaction, and is often referred to as the *Knoevenagel* condensation.

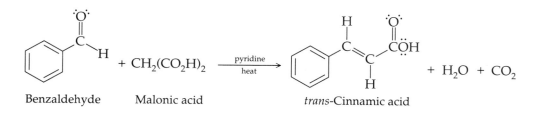

| Benzaldehyde | Malonic acid | | *trans*-Cinnamic acid |

The second stage (Experiment [D2]), involves the addition of bromine to the intermediate formed in the first step, *trans*-cinnamic acid. The product of this halogenation is *erythro*-2,3-dibromo-3-phenylpropanoic acid. The stereochemistry involved in the formation of this second intermediate is a result of the nature of the anti addition of molecular bromine to a trans alkene. The details are given in the discussion in Experiment [D2].

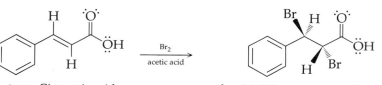

trans-Cinnamic acid                    *erythro*-2,3-Dibromo-3-phenylpropanoic acid

In Experiment [D3], the third step in Sequence D, the target molecule, 2'-bromostyrene, is generated by a novel elimination reaction that involves both the loss of $CO_2$ and HBr from the dibromophenylpropanoic acid. This stage of the synthesis offers an excellent opportunity to study the effect of solvent on the stereochemical course of a reaction by spectral analysis. In this case NMR (or infrared) data can be used to establish whether a cis–trans mixture is obtained, or just one pure isomer (cis).

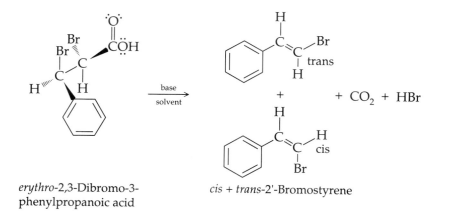

*erythro*-2,3-Dibromo-3-
phenylpropanoic acid                    *cis + trans*-2'-Bromostyrene

## Experiment [D1]  The Verley–Doebner Modification of the Knoevenagel Reaction: *trans*-Cinnamic Acid

Common name: *trans*-Cinnamic acid
CA number: [140-10-3]
CA name as indexed: 2-Propenoic acid, 3-phenyl-, *(E)-*

### Purpose

To prepare *trans*-cinnamic acid as the first intermediate in the Sequence D set of reactions that ultimately lead to 2′-bromostyrene. To review the chemistry associated with an important variety of aldol-type condensation, the modified Knoevenagel reaction.

### Prior Reading

Technique 5:  Crystallization
Use of the Hirsch Funnel (see pp. 93–95).

---

**REACTION**

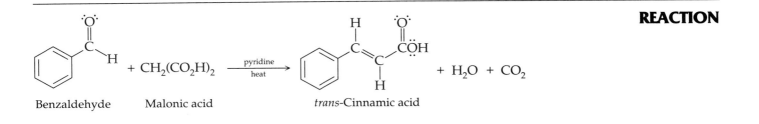

Benzaldehyde      Malonic acid      *trans*-Cinnamic acid

---

**DISCUSSION**

Those reactions that are called aldols derive their name from the early nineteenth-century organic literature. The term was first applied to a self-addition product of acetaldehyde that forms under basic conditions. The product, 3-hydroxybutanal, can form in yields as high as 50% in the presence of sodium hydroxide. This substance eventually became referred to as *aldol* as it was both an <u>ald</u>ehyde and an alco<u>hol</u>.

The aldol reaction can be defined in the broad sense as a reaction in which a nucleophile generated alpha to an electron-withdrawing functional group (in the large majority of cases, carbonyl groups are responsible for the $\alpha$-hydrogen acidity although they are not required) adds to a carbonyl group (it may be self-condensation, as in the example of acetaldehyde, but it may also involve attack on another carbonyl-containing species, if it is present). Several experiments in this text illustrate this type of reactivity. Experiments [20] and [A3$_a$] are classic aldol reactions in which conditions are vigorous enough to result in elimination of the $\beta$-hydroxyl group to yield $\alpha,\beta$-unsaturated ketones as products. In the first step of Sequence D, we now add the *Knoevenagel* reaction as another condensation possessing a mechanism similar to that of the aldol.

**Emil Knoevenagel** (1865–1921) was born in Hanover, Germany. He was the son of a chemist and started his studies at the Technical Institute at Hanover. Later he studied with both Victor Meyer and Gattermann at Göttingen where he received the Ph.D. in 1889. When Victor Meyer moved to Heidelberg, Knoevenagel accompanied him. In 1896 he was appointed assistant professor of organic chemistry at Heidelberg and, in the same year, published his studies on the reaction that now bears his name. He eventually became professor of organic chemistry in 1900.

Knoevenagel had a variety of interests, including stereoisomerism. He worked extensively with aldol-type reactions, and pioneered the use of amine bases to promote these condensations. He was particularly interested in pyridine chemistry, and was the first one to synthesize the pyridine ring by heating hydroxylamine with 1,5-diketones.

Following World War I, in which he saw active service, he continued his studies only to succumb suddenly, at the age of 56, during an appendectomy.

Interestingly, although Knoevenagel demonstrated the effectiveness of amine bases in promoting aldol type reactions and though, as noted above, he was particularly interested in pyridine, he overlooked this material's potential application to aldol condensations. It was left to Verley (1899) and Doebner (1900) to introduce successful modifications of the condensation in which pyridine appears to play a number of roles: as a solvent, as a base, and assists in the decarboxylation.

The Knoevenagel reaction in its simplest form is the condensation of malonic esters (or their analogues) with aldehydes or ketones in the presence of an amine base catalyst plus a small amount of a carboxylic acid (or amino acid) cocatalyst. The condensation products are often $\alpha,\beta$-unsaturated carbonyl compounds. For example,

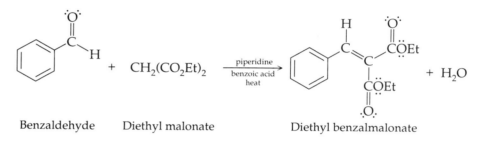

Benzaldehyde     Diethyl malonate                    Diethyl benzalmalonate

Other substances possessing an acidic methylene group have been incorporated in aldol-type reactions. These materials include ethyl cyanoacetate and ethyl acetoacetate, as well as phenylacetonitrile, benzyl ketones, and aliphatic nitro compounds. The reaction is often run in benzene or toluene solvent, so that the water formed can be continuously removed. Other cocatalysts, besides carboxylic acids, are various ammonium salts, such as ammonium acetate and piperidinium acetate. The Knoevenagel reaction gives highest yields when aldehydes are used as the electrophile, although selected ketones can sometimes give acceptable yields. It should be noted that one of the more important properties of these reactions from a synthetic perspective is that they offer a route to the formation of C—C bonds.

In the formation of the first synthetic intermediate in Sequence D, the very effective Verley–Doebner modification of the fundamental Knoevenagel condensation is used. This modification employs malonic acid in place of the conventional ester to promote enolization. In addition, the heterocyclic amine, pyridine, functions as both the base catalyst and the solvent. A cocatalyst, $\beta$-alanine (an amino acid), is also introduced. Mechanistically, the reaction closely resembles the aldol condensation in that in both cases a carbanion is generated by abstraction, by base, of a proton *alpha* to a carbonyl group. The resulting carbanion is stabilized as an enolate anion (see top of page 523).

The enolate anion then acts as a nucleophile and attacks the carbonyl carbon of a second carbonyl-containing molecule in the reaction. (An example would be the aldol reaction in Experiment [20].) The intermediate aldol product, the $\beta$-hydroxyacid, undergoes rapid dehydration under

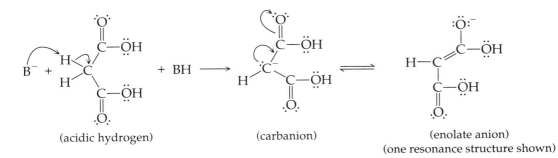

(acidic hydrogen)          (carbanion)          (enolate anion)
(one resonance structure shown)

these reaction conditions, to give the $\alpha,\beta$-unsaturated diacid shown here. Conjugate addition (1,4 addition) of pyridine and ionization of a carboxylic acid are followed by decarboxylation and concomitant elimination of pyridine to yield the $\alpha,\beta$-unsaturated carboxylic acid as shown here.

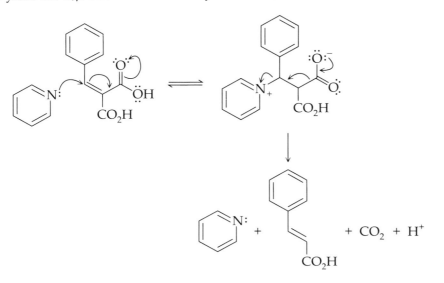

---

Estimated time to complete the experiment: 2.5 hours

**EXPERIMENTAL PROCEDURE**

### Physical Properties of Reactants

| Compound | MW | Amount | mmol | mp(°C) | bp(°C) | d |
|---|---|---|---|---|---|---|
| Benzaldehyde | 106.12 | 580 $\mu$L (603 mg) | 5.7 | −26 | 179 | 1.04 |
| Malonic acid | 104.06 | 1.5 g | 14.4 | 135 dec | | |
| Pyridine | 79.10 | 3 mL | | | 115 | |
| $\beta$-Alanine | 89.09 | 120 mg | 1.35 | 205 dec | | |

### Reagents and Equipment

In a 10-mL round-bottom flask containing a magnetic stir bar, weigh and place 1.5 g, (14.4 mmol) of malonic acid followed by 120 mg of $\beta$-alanine. Now add 3 mL of pyridine and 580 $\mu$L of freshly distilled benzaldehyde [**HOOD**].

HOOD

**NOTE.** *Dispense the benzaldehyde using an automatic delivery pipet. Measure the pyridine using a 10-mL graduated cylinder.*

---

HOOD

**CAUTION:** **Pyridine is a toxic amine with a strong unpleasant odor. Both reagents should be stored and dispensed in the hood.**

---

Attach the flask to a water-jacketed reflux condenser, and place the assembly in a sand bath on a magnetic stirring hot plate.

## Reaction Conditions

Heat the mixture with stirring at a sand bath temperature of about 130 °C for 1.5 hours.

## Isolation of Product

HOOD

Allow the reaction mixture to cool to room temperature and remove the flask [**HOOD**]. Transfer the reaction solution using a Pasteur pipet to a 50-mL Erlenmeyer flask containing 12 mL of ice-cold water. Rinse the round-bottom flask with an additional 3 mL of ice-cold water, and transfer the rinse to the same Erlenmeyer flask in like manner. Now add, in small portions, 6 $M$ HCl (~8 mL) until a white precipitate forms and the solution tests weakly acidic to pH paper.

Collect the precipitate of *trans*-cinnamic acid by vacuum filtration using a Hirsch funnel. Wash the white crystals with three 2-mL portions of cold water. To partially dry the material, continue the suction for an additional 15 min (remember to cover the Hirsch funnel with filter paper to protect the filter cake from contamination with dust from the laboratory atmosphere during the extended air-drying period).

## Purification and Characterization

Oven-dry the partially dried product at 100 °C overnight. Weigh the crude acid and calculate the percent yield. Determine the melting point. Compare your result with the literature value. This material is likely to be of sufficient purity to be used in the next step of Sequence [D], Experiment [D2].

If your material melts below the literature value, for characterization recrystallize approximately 30 mg from hot water using the Craig tube. Dry as before and redetermine the melting point. To further characterize the material, obtain IR (KBr disk) and $^1$H NMR spectra. Compare your spectra with those recorded in *The Aldrich Library of IR Spectra* and/or *The Aldrich Library of NMR Spectra*.

**NOTE.** *Approximately 500 mg of purified product with a melting point within 2–3 °C of the literature value is suggested for continuing the sequence on to Experiment [D2].*

## Chemical Tests

Add several crystals (~5 mg) of the *trans*-cinnamic acid to 1 mL of 5% sodium bicarbonate on a watch glass. Does evolution of $CO_2$ indicate the presence of a carboxylic acid? Does the material give a positive bromine test for unsaturation (see Chapter 10)?

**8-63.** Write a complete mechanism for the addition of diethyl malonate to ethanal in the presence of base to form a β-hydroxy ester.

**8-64.** Outline a synthesis that forms at least one C—C bond for each of the following compounds.
   a. $CH_3CH_2CH_2CH_2CH_2CH=CHCO_2H$
   b. $CH_3CH_2CH_2CH=C(CN)(CO_2CH_2CH_3)$

**8-65.** As mentioned in the discussion, ketones generally give poor yields in the Knoevenagel reaction with diethyl malonate. However, the reaction with ketones gives good yields with ethyl cyanoacetate and malononitrile. Explain.

**8-66.** Give the structure of the products of the Knoevenagel reaction for the following pairs of reactants.
   a. Cyclopentanone + malononitrile (dicyanomethane)
   b. Benzaldehyde + ethyl acetoacetate
   c. Propanal + nitromethane

---

Descriptions of the Knoevenagel reaction:

Cope, A. C. *J. Am. Chem. Soc.* **1937**, *59*, 2327.

Cope, A. C.; Hofmann, C. M.; Wyckoff, C.; Hardenberg, E. *J. Am. Chem. Soc.* **1941,** *63*, 3452.

Doebner, O. *Chem. Ber.* **1900**, *33*, 2141.

House, H. O. *Modern Synthetic Reactions*; Benjamin, Menlo Park, CA, 1972, pp. 649–653.

Johnson, J. R. *Org. React.* **1942,** *1*, 210.

Jones, G. *Org. React.* **1967,** *15*, 204.

Verley, A. *Bull. Soc. Chim. Fr.* **1899**, *21*, 414.

Selected references of the use of the Knoevenagel reaction recorded in *Organic Syntheses*:

Allen, C. F. H.; Spanger, F. W. *Organic Syntheses*; Wiley: New York, 1955; Collect. Vol. III, p. 377.

Cope, A. C.; Hancock, E. M. *Organic Syntheses*; Wiley: New York, 1955; Collect. Vol. III, p. 399.

Horning, E. C.; Koo. J.; Fish, M. S.; Walker, G. N. *Organic Syntheses*; Wiley: New York, 1963; Collect. Vol. IV, p. 408.

McElvain, S. M.; Clemens, D. H. *Organic Syntheses*; Wiley: New York, 1963; Collect. Vol. IV, p. 463.

This step in the sequence was adapted from the work of:

Kolb, K. E.; Field, K. W.; Schatz, P. F. *J. Chem. Educ.* **1990,** *67*, A304.

## Experiment [D2]   Bromination of *trans*-Cinnamic Acid: *erythro*-2,3-Dibromo-3-phenylpropanoic Acid

Common name: *erythro*-2,3-Dibromo-3-phenylpropanoic acid
CA number: [31357-31-0]
CA name as indexed: Benzenepropanoic acid, α,β-dibromo-, *(R\*,S\*)*-

### Purpose
To carry out the bromination of *trans*-cinnamic acid in order to obtain *erythro*-2,3-dibromo-3-phenylpropanoic acid, the direct precursor to 2'-bromostyrene, which is the synthetic target of Sequence D. To review the stereospecificity of the addition of molecular bromine to an alkene.

*Prior Reading*
   *Technique 5:*   Crystallization
                    Use of the Hirsch Funnel (see pp. 93–95).

## REACTION

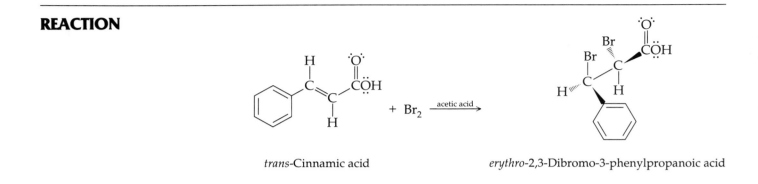

*trans*-Cinnamic acid

*erythro*-2,3-Dibromo-3-phenylpropanoic acid

## DISCUSSION

The bromination of alkenes is known to be a **stereospecific** reaction. An example and detailed discussion of this addition to *trans*-stilbene is given in Experiment [A2$_b$], and another example is illustrated in Experiment [F2]. In the present case, bromine undergoes a similar addition, and the halogenated product obtained is the *erythro* diastereomer of 2,3-dibromo-3-phenylpropanoic acid, which is produced as a racemic mixture of the two enantiomers.

The term erythro is derived from carbohydrate chemistry, and is used to describe the *relative* configurations of the two adjacent chiral centers (stereocenters). Specifically, the term erythro describes structures whose Fischer projections place identical substituents (on adjacent carbon atoms) on the *same side* of the (otherwise identical) Fischer projection. The opposite of erythro is threo, which would describe structures with identical substituents (on adjacent carbons) on opposite sides of the (otherwise identical) Fischer projection. Since erythro and threo describe only the relative configurations of the two chiral centers with respect to one another, there are two enantiomers of each erythro diastereomer, and of each threo diastereomer, as shown here both in Fischer projections and more familiar illustrations.

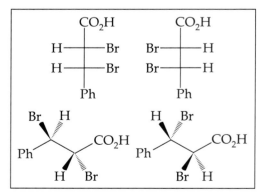

Enantiomers of the erythro diastereomer
of 2,3-dibromo-3-phenylpropanoic acid

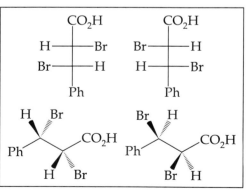

Enantiomers of the threo diastereomer
of 2,3-dibromo-3-phenylpropanoic acid

The result of the halogenation of *trans*-cinnamic acid, as in the case of *(E)*-stilbene in Experiment [A2$_b$], is an anti addition of molecular bromine. The enantiomeric products are shown here.

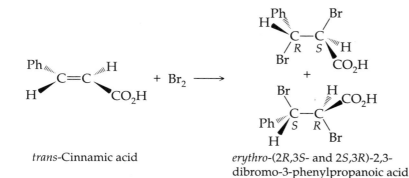

*trans*-Cinnamic acid

*erythro*-(2R,3S- and 2S,3R)-2,3-
dibromo-3-phenylpropanoic acid

What mechanism best explains this observed stereochemistry? The starting unsaturated acid and bromine are both *achiral*, and thus the product, if a chiral molecule, *must* be racemic. The outer-shell electrons of molecular bromine, however, are highly polarizable and molecular bromine reacts as an electrophile with the nucleophilic $\pi$ electrons of the alkene group of cinnamic acid, in a reaction that is effectively a nucleophilic substitution reaction on a bromine atom. The product is a carbocation, which is, however, stabilized by the ability of the bromine atom to donate a lone pair to the carbocation. This interaction results in the formation of a *chiral*, cyclic *bromonium ion* intermediate. This intermediate ion is a rigid structure in which the ring section is only open to further attack by a nucleophile ($S_N2$ attack by the $Br^-$ ion generated in the first stage) from the side opposite the bromine atom.

As both cinnamic acid and molecular bromine are achiral, two enantiomeric bromonium ions are formed at equal rates. Ring opening of each bromonium ion may preferentially occur as shown here at the carbon bearing the phenyl group, as this carbon bears more fractional positive charge than the carbon adjacent to the carboxyl group. As the two reaction pathways shown here are enantiomeric, they proceed at equal rates to produce racemic *erythro*-2,3-dibromo-3-phenylpropanoic acid.

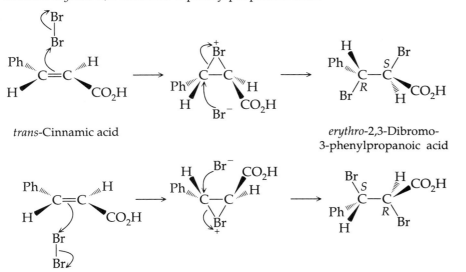

*trans*-Cinnamic acid

*erythro*-2,3-Dibromo-
3-phenylpropanoic acid

Recent investigations have shed further light on the nature of the cyclic intermediate. Stable solutions of cyclic bromonium ions in liquid $SO_2$ have been prepared as $SbF_6^-$ salts. For example,

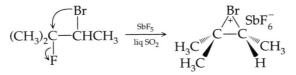

See the discussion in Experiment [A2ᵦ] for further evidence relating to these intermediate species.

The *erythro*-2,3-dibromo-3-phenylpropanoic acid product prepared in this experiment has a melting point of 203–204 °C. The corresponding threo diastereomer has been synthesized and its racemate melts at 91–93 °C. Thus, the experimental results support the stereospecific nature of the bromine addition that is rationalized by the proposed mechanism.

**NOTE.** *Pyridinium bromide perbromide, a solid reagent, is used as a source of bromine in this experiment. This material is much easier to handle than liquid bromine (also see Experiment [A2ᵦ]).*

## EXPERIMENTAL PROCEDURE

Estimated time to complete the experiment: 2.0 hours

### Physical Properties of Reactants and Product

| Compound | MW | Amount | mmol | mp(°C) | bp(°C) | d |
|---|---|---|---|---|---|---|
| *trans*-Cinnamic acid | 148.2 | 500 mg | 3.37 | 133 | | |
| Pyridinium bromide perbromide (tech) | 319.84 | 1.2 g | | | | |
| Acetic acid | 60.05 | 12 mL | 16.2 | | 116–118 | 1.05 |
| *erythro*-2,3-Dibromo-3-phenylpropanoic acid | 308.0 | | | 203–204 | | |

### Reagents and Equipment

HOOD

Weigh and place 500 mg (3.37 mmol) of *trans*-cinnamic acid in a 25-mL round-bottom flask containing a magnetic stir bar. Now weigh and place [**HOOD**] 1.2 g of pyridinium bromide perbromide in the same flask.

---

HOOD

**CAUTION: Pyridinium bromide perbromide is a corrosive reagent and a lachrymator. Dispense this material in the hood.**

---

Using a graduated cylinder, measure 12 mL of glacial acetic acid and add this to the solid reagents. Attach the flask to a water-cooled reflux condenser and place the assembly in a sand bath.

### Reaction Conditions

Heat the mixture with stirring in a sand bath at 120–125 °C for 45 min.

### Isolation of Product

Cool the orange solution to room temperature. With a Pasteur pipet, transfer, in small portions, the solution to a 50-mL Erlenmeyer flask. Rinse the reaction flask with an additional 2 mL of acetic acid, and transfer this rinse to the same Erlenmeyer flask.

Cool the solution in an ice bath for 5 min, and then slowly add 15 mL of cold water in three 5-mL portions. Swirl the contents of the flask between additions. A pale yellow precipitate should form. Collect the crude product by vacuum filtration using a Hirsch funnel. Wash the filter cake with three 2-mL portions of cold water, during which time the solid product should become white.

## Purification and Characterization

Dry the product in an oven at 110 °C overnight. Weigh the product and calculate the crude yield. Determine the melting point and compare it to that listed in the Reactant and Product table.

A 20–30-mg sample of the *erythro*-2,3-dibromo-3-phenylpropanoic acid may be recrystallized (Craig tube) from chloroform [**HOOD**], if desired.

HOOD

## Chemical Test

Perform the Beilstein test to detect the presence of a halogen (see Chapter 10).

**NOTE.** *Approximately 300 mg of purified (washed) product, with a melting point within 3–5 °C of the literature value, is suggested for continuing the sequence on to Experiment [D3].*

**8-67.** The product prepared in this experiment has 2 chiral centers (stereocenters) which give rise to 4 stereoisomers. Draw each of these stereoisomers, and indicate the relationships between each of the stereoisomers.

**8-68.** The addition of bromine to ethylene in the presence of high concentrations of chloride ion in an inert solvent results in the formation of 1,2-dibromoethane and 1-bromo-2-chloroethane. No 1,2-dichloroethane is obtained. Account for these results using a suitable mechanistic sequence.

**8-69.** 2,3-Dibromobutane contains 2 chiral centers. Therefore, the possibility of 4 stereoisomers exists. However, the addition of bromine to an equimolar mixture of *cis*- and *trans*-2-butene generates only three stereoisomers. Explain.

**8-70.** Cyclohexene undergoes bromination in methanol solvent to give *trans*-1-bromo-2-methoxycyclohexane. Draw a suitable mechanism to account for this.

**QUESTIONS**

**BIBLIOGRAPHY**

*erythro*-2,3-Dibromo-3-phenylpropanoic acid has been previously prepared:

Corvari, L.; McKee, J. R.; Zanger, M. *J. Chem. Educ.* **1991,** *68,* 161.

Murahashi, S.; Naota, T.; Tanigawa, Y. *Organic Syntheses;* Wiley: New York, 1990; Collect. Vol. VII, p. 172.

Sudborough, J. J.; Thompson, K. J. *J. Chem. Soc.* **1902,** *83,* 666.

The bromination of the ethyl ester of *trans*-cinnamic acid has been reported:

Abbott, T. W.; Althousen, D. *Organic Syntheses;* Wiley: New York, 1943; Collect. Vol. II, p. 270.

McElvain, S. M.; Kundiger, D. *Organic Syntheses;* Wiley: New York, 1955; Collect. Vol. III, p. 123.

Additional selected references of bromination of alkenes from *Organic Syntheses*:

Allen, C. F. H.; Edens, C. O., Jr. *Organic Syntheses*; Wiley: New York, 1955; Collect. Vol. III, p. 731.

Cromwell, N. H.; Benson, R. *Organic Syntheses*; Wiley: New York, 1955; Collect. Vol. III, p. 105.

Fieser, L. F. *Organic Syntheses*; Wiley: New York, 1963; Collect. Vol. IV, p. 195.

Snyder, H. R.; Brooks, L. A. *Organic Syntheses*; Wiley: New York, 1943; Collect. Vol. II, p. 171.

## Experiment [D3]   An Elimination Reaction with *erythro*-2,3-Dibromo-3-phenylpropanoic Acid: 2'-Bromostyrene

Common names: *(Z)-β*-Bromostyrene, *cis*-2'-bromostyrene
CA number: [588-73-8]
CA name as indexed: Benzene, (2-bromoethenyl)-, *(Z)-*

Common names: *(E)-β*-Bromostyrene, *trans*-2'-bromostyrene
CA number: [588-72-7]
CA name as indexed: Benzene, (2-bromoethenyl)-, *(E)-*

### Purpose

To carry out an elimination reaction using *erythro*-2,3-dibromo-3-phenylpropanoic acid, the direct precursor to 2'-bromostyrene. To illustrate the influence of solvents and base on the course of the elimination reaction. To consider factors that control the stereospecificity of a reaction. To explore the option of employing NMR spectroscopy to establish the cis/trans isomer ratio in the 2'-bromostyrene product.

### Prior Reading

Technique 4:    Solvent Extraction
                     Liquid–Liquid Extraction (see pp. 80–82).
Technique 6B:   Concentration of Solutions (see pp. 106–108).

**REACTION**

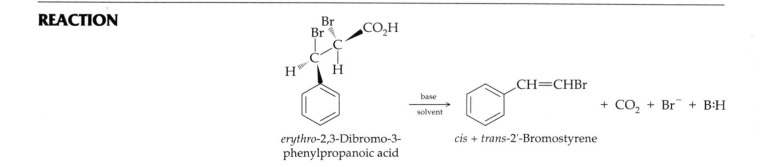

*erythro*-2,3-Dibromo-3-
phenylpropanoic acid          *cis* + *trans*-2'-Bromostyrene

**DISCUSSION**

It has been observed that the course of the elimination depicted in the above reaction is both base and solvent dependent. In this experiment you will have the opportunity to investigate this solvent dependence. In Part

A, you will study the effect of carbonate base in aqueous solvent. Under these conditions, the reaction sequence proceeds via an E1 mechanism in which the first step involves the ionization of one of the C—Br bonds. Clearly, of the two possibilities, halide formation that results in a resonance stabilized benzylic carbocation will be highly favored over development of a positive charge on a carbon alpha to a carboxyl group (see Question 8-71). Formation of the carbocation intermediate is then rapidly followed by decarboxylation. Although the conjugated section of the positively charged intermediate is planar, free rotation or partially hindered rotation remains possible about the C—C bond leading to the remaining tetrahedral carbon. Thus, both *cis*- and *trans*-2'-bromostyrene may be obtained during the subsequent decarboxylation. The proposed sequence is shown here.

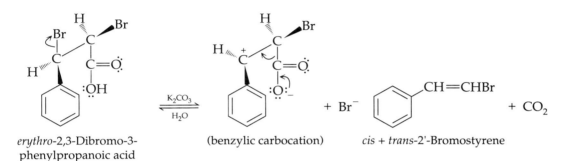

*erythro*-2,3-Dibromo-3-     (benzylic carbocation)     *cis* + *trans*-2'-Bromostyrene
phenylpropanoic acid

In Part B of the experiment, the solvent is changed from water (used in Part A) to a much less polar solvent, acetone. Under these nonaqueous conditions, elimination also occurs, but in this case the experimental data demand a different mechanism, one involving a smooth concerted electron flow without short-lived intermediate formation. Furthermore, when carried out in acetone, the reaction yields exclusively *cis*-2'-bromostyrene. Details of the mechanism follow:

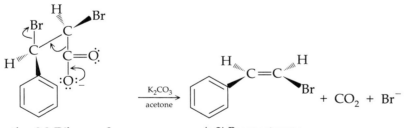

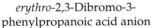
*erythro*-2,3-Dibromo-3-          *cis*-2'-Bromostyrene
phenylpropanoic acid anion

Under these conditions, therefore, the reaction becomes stereospecific. The mechanism is classified as an E2 elimination. Note that when the two leaving groups are anti to one another, as is thermodynamically preferred, elimination yields the cis alkene.

In the two sets of reaction conditions to be studied in Parts A and B of this experiment, potassium carbonate is used as the base in both cases and it is the solvent that is varied, as mentioned above. In another experimental procedure, sodium azide is used as the base and N,N-dimethylformamide (DMF) is used as the solvent, and higher selectivity for the formation of the cis isomer is reported (see Bibliography, Corvari et al.).

## EXPERIMENTAL PROCEDURE

Estimated time to complete the reaction: 1.5 hours Part A; 2.0 hours Part B.

### Physical Properties of Reactants

| Compound | MW | Amount | mmol | mp(°C) |
|---|---|---|---|---|
| *erythro*-2,3-Dibromo-3-phenylpropanoic acid | 308.0 | 300 mg | 0.97 | 203–204 |
| Potassium carbonate | 138.21 | 300 mg | 2.1 | 891 |

### PART A   Conditions: Water Solvent

#### Reagents and Equipment

Weigh and place 300 mg (0.97 mmol) of *erythro*-2,3-dibromo-3-phenylpro-panoic acid (prepared in Experiment [D2]) and 300 mg (2.1 mmol) of potassium carbonate in a 25-mL Erlenmeyer flask. Add 5 mL of distilled water.

#### Reaction Conditions

Warm the mixture in a 120 °C sand bath until the solids dissolve (5–10 min) and then heat for an additional 10 min. A yellow oil should separate from the aqueous phase during this period.

#### Isolation of Product

Cool the two-phase system to room temperature and using a Pasteur pipet, transfer the contents of the flask to a 12-mL centrifuge tube. Rinse the Erlenmeyer flask with a 2-mL portion of methylene chloride, which is then transferred to the centrifuge tube. Extract the aqueous phase with the $CH_2Cl_2$ rinse and follow this extraction with two more extractions of the aqueous phase with 2-mL portions of methylene chloride. Transfer the methylene chloride extracts (the organic phase should be the lower phase, but check solubility in water to be sure) to a second 25-mL Erlenmeyer flask containing a sufficient amount of anhydrous sodium sulfate to dry the solution. Allow the extracts to dry for about 10 min.

Transfer a portion of the dried solution, using a Pasteur filter pipet, to a *tared* 5-mL conical vial containing a boiling stone. Concentrate this por-

HOOD    tion on a warm sand bath under a gentle stream of nitrogen gas [**HOOD**].

Now transfer the remaining dried solution to the same tared vial and concentrate as before. Finally, rinse the sodium sulfate drying agent with an additional 1-mL portion of methylene chloride, transfer it to the vial and concentrate again to yield the *cis/trans*-2'-bromostyrene as a yellow oil.

**NOTE.** *Exercise care during the concentration steps. Product yield will be **greatly** reduced by overheating or leaving the solution on the sand bath for too long a period.*

#### Purification and Characterization

No further purification of the product is usually necessary. Note the hya-cinth-like odor of the liquid. Determine the refractive index and compare it to the literature value (lit. value = 1.6070 at 20 °C, 78% trans isomer). Obtain IR and $^1H$ NMR spectra of the oil. In the infrared, the cis isomer shows a phenyl C—H bend at 770 cm$^{-1}$, while the trans isomer shows this bending mode at 731 and the alkene C—H out-of-plane mode is found at 941 cm$^{-1}$ (see Bibliography, Strom et al.). Compare your results with the spectrum recorded in *The Aldrich Library of IR Spectra*.

## NMR Analysis

The $^1$H NMR spectrum of a mixture of *cis*- and *trans*-2'-bromostyrene is shown in Figure 8.4. The doublets corresponding to the two olefinic protons of the major isomer can be clearly seen, centered at 6.75 and 7.1 ppm. One of the corresponding doublets from the minor isomer is evident at 6.42 ppm. This doublet has a coupling constant of 8.0 Hz. The two larger doublets from the major isomer have a coupling constant of 14.0 Hz, and thus, in this sample, the major isomer must be *trans*-2'-bromostyrene.

We can readily locate the second doublet from the minor, *cis*-isomer by examining the $^1$H–$^1$H COSY two-dimensional NMR spectrum (see Chapter 9, NMR discussions) (Figure 8.5). The spectrum here is shown at a large vertical scale in order to discern the minor cis isomer. By looking below the doublet at 6.42 ppm, a cross-peak can be seen at about 7.1 ppm.

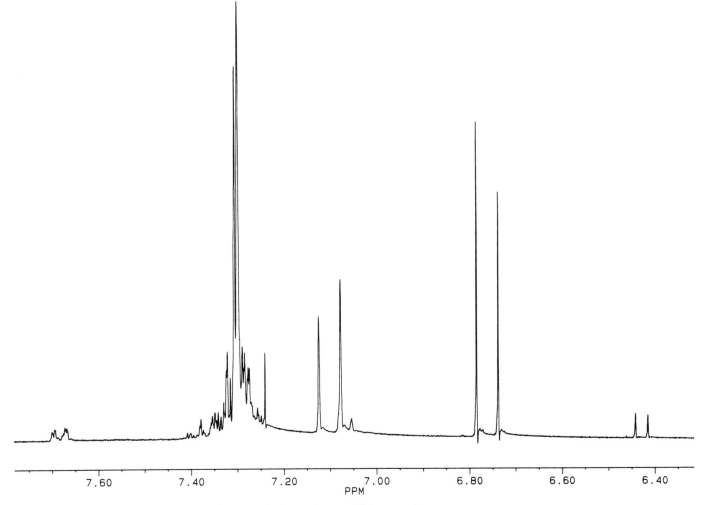

**FIGURE 8.4**  $^1$H NMR spectrum: Mixture of *cis*- and *trans*-2'-bromostyrene.

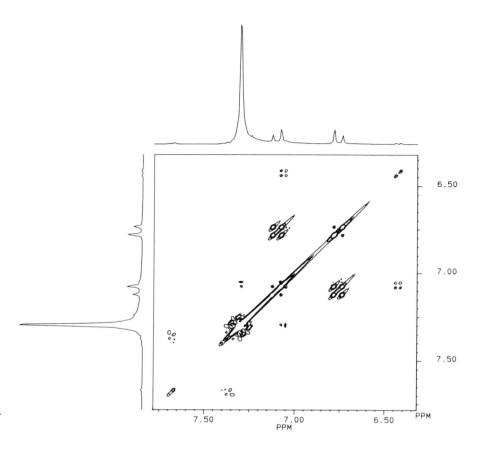

**FIGURE 8.5** ¹H−¹H COSY NMR spectrum: Mixture of *cis-* and *trans-2′-* bromostyrene.

Careful inspection reveals that these signals must be slightly upfield from the doublet at 7.1 ppm from the major (trans) isomer. Referring back to the ¹H NMR spectrum (Figure 8.4) one can see one peak of the hidden doublet just emerging from the upfield end of the doublet from the major isomer. Thus, the second doublet from the minor (cis) isomer must coincidentally be obscured by the NMR signal from the major isomer.

## PART B  Conditions: Acetone Solvent

### Reagents and Equipment

Weigh and place 300 mg (0.97 mmol) of *erythro*-2,3-dibromo-3-phenylpropanoic acid (prepared in Experiment [D2]) and 300 mg (2.1 mmol) of potassium carbonate in a 10-mL round-bottom flask containing a magnetic stir bar. Add 5 mL of acetone that has been dried over sodium sulfate. Attach the flask to a reflux condenser and assemble the apparatus on an 80 °C sand bath.

### Reaction Conditions

Heat (bath temperature ~80 °C) the mixture, with stirring, at reflux for 1 hour.

### Isolation of Product

HOOD

Allow the solution to cool to room temperature and then concentrate it to dryness [**HOOD**] on a warm sand bath under a gentle stream of nitrogen gas. (**DO NOT OVERHEAT!**)

Now add 5 mL of distilled water to dissolve the solid cake. A yellow oil appears at this point, forming a two-phase system. Using a Pasteur pipet, transfer this aqueous-oily mixture to a 12-mL centrifuge tube. Rinse the round-bottom flask with a 2-mL portion of methylene chloride, which is then transferred to the centrifuge tube. Extract the aqueous phase with the $CH_2Cl_2$ rinse and follow this extraction with two more extractions of the aqueous phase with 2-mL portions of methylene chloride. Transfer the organic extracts to a second 25-mL Erlenmeyer flask using a Pasteur filter pipet and dry the combined extracts over anhydrous sodium sulfate for at least 10 min.

Transfer the dried solution in 2–3-mL portions to a *tared* 5-mL conical vial containing a boiling stone. Concentrate the solution [**HOOD**] on a *warm* sand bath under a gentle stream of nitrogen gas. Rinse the sodium sulfate with an additional 1 mL of methylene chloride and combine this rinse with the dried solution. Concentrate to yield a yellow oil, *cis*-2'-bromostyrene.

HOOD

**NOTE.** *Exercise care during the concentration steps. Product yield will be* **greatly** *reduced by overheating or leaving the solution on the sand bath for too long a period.*

## Purification and Characterization

No further purification of the product is usually necessary. Weigh the product and calculate the crude percent yield. Obtain IR and ${}^1H$ NMR spectra of the oil. In the NMR, the peaks due to both isomers, if present, can be discerned (see Discussion, Part A). Compare your IR and NMR results with the spectra obtained for the cis–trans mixture formed in Part A and to those recorded in the Corvari et al. reference (see Bibliography) for pure *cis*-2'-bromostyrene. This reference also gives conditions under which the purity of the material may be more carefully determined by gas chromatography.

## Chemical Tests

Several tests can be run to determine the presence of unsaturation, aromatic character, and the presence of bromine. Select the appropriate tests from Chapter 10. Do your results confirm the presence of these groups?

**8-71.** 1-Chloro-1-phenylethane ionizes easily under E1 conditions to form a benzylic carbocation intermediate. The ion is stabilized due to the delocalization of the positive charge to the aromatic ring. Draw resonance structures that indicate the stability of this ion.

**QUESTIONS**

**8-72.** Comment on the fact that *erythro*-2,3-dibromo-3-phenylpropanoic acid undergoes elimination by an E1 pathway in water solvent (Part A conditions), but by an E2 pathway (Part B conditions) when acetone is used as the solvent.

**8-73.** Explain in terms of the mechanism why conditions in Part A lead to a mixture of cis–trans isomers but that those employed in Part B give only the cis isomer.

**8-74.** A benzylic carbocation is generated under the conditions of Part A. Would the presence of a para $CH_3O$— group on the benzene ring increase or decrease the stability of the benzylic carbocation? Explain.

**8-75.** Consider the stereochemistry of the carbocation intermediate formed under the conditions of Part A. As the starting material used in this experiment, *erythro*-2,3-dibromo-3-phenylpropanoic acid, is racemic, the enantiomer of the carbocation shown must also be generated. Does the other enantiomer lead to the same diastereomer (cis) of the product or does it lead to the trans diastereomer?

---

## BIBLIOGRAPHY

The synthesis of 2'-bromostyrene has been reported:

Corvari, L.; McKee, J. R.; Zanger, M. *J. Chem. Educ.* **1991,** *68,* 161.

Cristol, S. J.; Norris, W. P. *J. Am. Chem. Soc.* **1953,** *75,* 2645.

Grovenstein, E., Jr.; Lee, D. E. *J. Am. Chem. Soc.* **1953,** *75,* 2639.

L'abbé, G.; Miller, M. J.; Hassner, A. *Chem. Ind. (London)* **1970,** 1321.

Mestdagh, H.; Puechberty, A. *J. Chem. Educ.* **1991,** *68,* 515.

Murahashi, S.; Naota, T.; Tanigawa, Y. *Organic Syntheses*; Wiley: New York, 1990; Collect. Vol. VII , p. 172.

Strom, L. A.; Anderson, J. R.; Gandler, J. R. *J. Chem. Educ.* **1992,** *69,* 588.

---

## SEQUENCE E

# The Synthesis of Piperonylonitrile from Piperonyl Alcohol

### Purpose

To obtain the aromatic nitrile, piperonylonitrile, via a multistep synthesis that does not depend on potentially dangerous diazonium intermediates. To investigate the selective oxidation of a primary alcohol to an aldehyde. To explore, at the same time, the use of resin-bound reagents that simplify the isolation of products. To carry out a novel elimination reaction using an oxime to obtain the target nitrile.

## INTRODUCTION

The first step in this sequence of reactions is the oxidation of piperonyl alcohol to obtain piperonal, an aromatic aldehyde. The reaction is selective, and only the aldehyde is obtained; no oxidation to the carboxylic acid is observed.

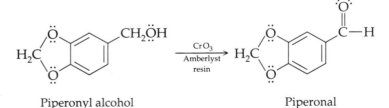

Piperonyl alcohol                                   Piperonal

Piperonal has an agreeable odor and is marketed to the perfume industry as *heliotropine*. This commercial term comes from *heliotrope*, which was a name given by early herbalists to any plant that turned to the sun (from the Greek: *helios* the sun + *trepein* to turn), such as the sunflower. The name now is more narrowly defined as applying to those plants of the genus *Heliotropium* and specifically to *H. peruvianum*, a common species that is widely cultivated. The fragrance of this particular species also is referred to as *heliotrope*. As piperonal possesses a very similar fragrance, the origin of the commercial terminology is clear. The commercial source of piperonal, however, is safrole, the formaldehyde ketal (acetal) of 3,4-dihy-droxyallylbenzene. Safrole is the chief constituent of the oil of sassafras and is itself obtained commercially from oil of camphor. When safrole is heated in the presence of a strong base, isomerization of the side-chain double bond yields isosafrole. Subsequent oxidation of the isomerized double bond yields piperonal. For more detailed discussions of odor–fragrance molecules and their role in living systems see Sequence D and Experiment [11C].

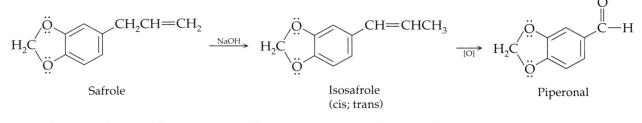

| Safrole | Isosafrole (cis; trans) | Piperonal |

The second step of the sequence is the conversion of the intermediate aldehyde to a substituted oxime. By design, the oxime intermediate is constructed with a good leaving group as the oxygen substituent on the nitrogen atom. The second stage, therefore, creates a molecular system prone to undergo the subsequent desired elimination reaction.

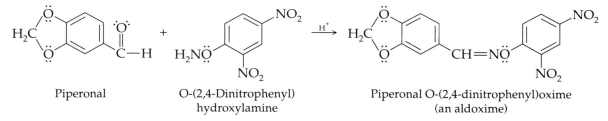

| Piperonal | O-(2,4-Dinitrophenyl) hydroxylamine | Piperonal O-(2,4-dinitrophenyl)oxime (an aldoxime) |

Treatment of the oxime intermediate with alcoholic KOH causes an elimination reaction that yields the target nitrile product, piperonylonitrile.

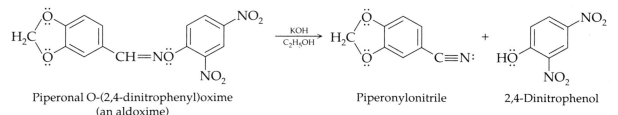

| Piperonal O-(2,4-dinitrophenyl)oxime (an aldoxime) | Piperonylonitrile | 2,4-Dinitrophenol |

The present synthetic sequence offers an alternative approach to the preparation of aromatic nitriles. The classic route to this class of compounds is to employ the Sandmeyer reaction. The key steps in this latter pathway involve the diazotization of an aromatic amine followed by reaction of the diazonium salt, which is not isolated (explosive!), with CuCN to give the aromatic nitrile directly. Aromatic halides (Br or Cl) may also be prepared in this manner using the corresponding copper(I) halides, as shown here.

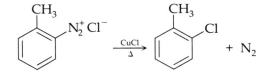

### Experiment [E1]   Piperonal

Common names: Piperonal, 3,4-(methylenedioxy)benzaldehyde
CA number: [120-57-0]
CA name as indexed: 1,3-Benzodioxole-5-carboxaldehyde

#### Purpose

To investigate the selective oxidation of a primary (1°) alcohol to an aldehyde. To explore the use of polymer-bound chromium trioxide as an oxidizing agent. To prepare the intermediate aldehyde, piperonal, as the first step in the sequence to synthesize piperonylonitrile.

#### Prior Reading

Technique 5:   Crystallization
Use of the Hirsch Funnel (see pp. 93–95).

Technique 6:   Chromatography
Column Chromatography (see pp. 98–101).
Concentration of Solutions (see pp. 106–108).

---

**REACTION**

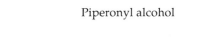

Piperonyl alcohol                              Piperonal

---

**DISCUSSION**

Many of the preferred reagents for the oxidation of primary alcohols to aldehydes (secondary alcohols to ketones) contain the transition metal chromium in its highest oxidation state, VI. Upon reaction with an alcohol, the yellow-orange chromium(VI) species is reduced to the blue-green chromium(III) state. Normally the reaction is carried out in aqueous acid solution using the sodium dichromate salt, $Na_2Cr_2O_7$, or the oxide, $CrO_3$. A typical reaction is shown here.

$$3\ CH_3CH_2CH_2OH + 4\ H_2SO_4 + Na_2Cr_2O_7 \rightarrow 3\ CH_3CH_2CHO + Cr_2(SO_4)_3 + Na_2SO_4 + 7\ H_2O$$

In this case, the aldehyde, propanal, can be isolated in moderate yield since it is relatively volatile and can be removed from the reaction mixture by distillation as it is formed. If not removed, the aldehyde, which is in equilibrium with the corresponding hydrate (geminal diol), is oxidized further to the carboxylic acid.

$$CH_3CH_2CH_2OH \xrightarrow[H_2SO_4]{Na_2Cr_2O_7} CH_3CH_2CHO \underset{}{\overset{H_2O}{\rightleftharpoons}} CH_3CH_2CH(OH)_2 \xrightarrow[H_2SO_4]{Na_2Cr_2O_7} CH_3CH_2COOH$$

1-Propanol                    Propanal            1,1-Dihydroxypropane          Propanoic acid

Since the hydrate is the species oxidized to the acid, oxidation of the aldehyde can largely be prevented by running the oxidation in a nonaqueous medium.

There are a number of selective oxidizing agents, such as pyridinium chlorochromate, frequently used for this purpose. With this reagent, the oxidation stops at the aldehyde stage, since the oxidation, as pointed out earlier, is conducted in a nonaqueous solution. Thus, decanal can be obtained from 1-decanol using this reagent (methylene chloride solvent) in 92% yield. Pyridinium chlorochromate (PCC) is a solid, yellow salt prepared from chromium trioxide as shown here.

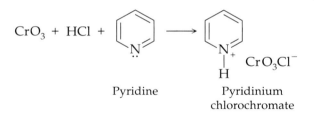

Pyridine            Pyridinium
                    chlorochromate

An alternate reagent is the chromium trioxide(pyridine)₂ complex, CrO₃•(py)₂ (where py = pyridine), often referred to as the Collins reagent.

$$RCH_2CH_2\ddot{O}H + CrO_3(py)_2 \xrightarrow[\underset{Na^+, \;:\ddot{O}CCH_3}{:\ddot{O}:}]{HCl, CH_2Cl_2} RCH_2\overset{\overset{\displaystyle :\ddot{O}:}{\|}}{C}H \quad (80\text{–}90\% \text{ average yield})$$

The chief difficulty with this reagent is that the complex is highly hygroscopic. However, it can be prepared in situ, thus avoiding this major drawback. Pyridinium dichromate and chromium trioxide•3,5-dimethylpyrazole are also effective as selective oxidizing agents for these reactions.

In the present experiment, the selective oxidation of a 1° alcohol to an aldehyde is carried out using a chromium trioxide polymer-bound oxidizing agent. This reagent is also used in Experiment [33A] to oxidize a 2° alcohol to a ketone. As noted there, the reagent is easy to prepare and has the advantage that the toxic reduced chromium species is bound to a polymer resin and can thus be easily separated from the reaction mixture by simple filtration.

As shown here, the oxidation proceeds through the formation of a chromate ester. Note that the oxidation state of chromium, Cr(VI), at this stage remains the same as in the starting reagent. The second stage is equivalent to an E2 elimination reaction, with the water molecule acting as a base. The donation of an electron pair to the metal atom changes its oxidation state to Cr(IV).

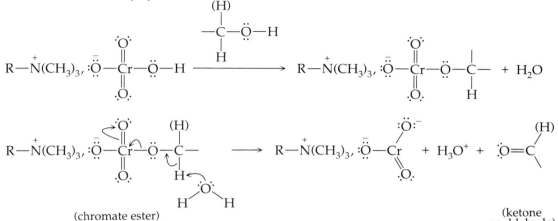

(chromate ester)                                          (ketone
                                                          or aldehyde)

**NOTE.**  *If you plan to carry out the total sequence of reactions, the scaled-up procedure should be followed. If an individual oxidation reaction is to be studied, the microscale procedure may be used.*

## EXPERIMENTAL PROCEDURE

### Semimicroscale Preparation

(This procedure is scaled up to 10 times the amounts of the microscale preparation.)

Estimated time to complete the experiment: 3.5 hours.

**Physical Properties of Reactants**

| Compound | MW | Amount | mmol | mp(°C) | bp(°C) |
|---|---|---|---|---|---|
| Piperonyl alcohol | 152.16 | 1.0 g | 6.57 | 58 | |
| Chromic oxide resin | | 7.5 g | | | |
| Dioxane | 88.12 | 10 mL | | | 101 |

### Reagents and Equipment

Weigh and place 1.0 g (6.57 mmol) of piperonyl alcohol and 7.5 g of freshly prepared chromic oxide resin in a 25-mL round-bottom flask containing a magnetic stirring bar. Using a graduated cylinder, measure [**HOOD**] and add to the flask 10 mL of dioxane. Attach the flask to a reflux condenser. (■)

HOOD

HOOD

---

**CAUTION:** Dioxane is a cancer suspect agent. It should be dispensed and handled in the *hood*. That is, the entire experiment should be set up and run in the *hood*, at least until the solvent is removed.

---

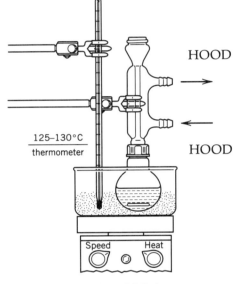

125–130°C
thermometer

25–mL RB flask
piperonyl alcohol, 1.0 g
+ chromic oxide oxidizing resin, 7.5 g
+ dioxane, 10 mL

Speed  Heat

### Instructor Preparation

*The CrO₃ resin is prepared by adding 35 g of Amberlyst A-26 resin to a solution of 15 g of $CrO_3$ in 100 mL of water. Stir the mixture for 30 min at room temperature. Collect the resin by filtration under reduced pressure and successively rinse the material with water and acetone. Partially dry the resin on a Büchner funnel by drawing air through it for 1 hour, then air-dry the material overnight.*

### Reaction Conditions

HOOD

With stirring, heat the reaction mixture at reflux in the **hood** for a period of 1.0 hour using a sand bath temperature of 125–130 °C.

### Isolation of Product

Cool the reaction product to room temperature. Transfer the solution by Pasteur filter pipet to a funnel fitted with fast grade filter paper. Collect the filtrate in a 25-mL Erlenmeyer flask containing approximately 1 g of anhydrous sodium sulfate. Rinse the reaction flask and resin (include the walls of the condenser) with three 1.0-mL portions of methylene chloride solvent (calibrated Pasteur pipet). Transfer each rinse by Pasteur filter pipet to the funnel. Combine each rinse filtrate with the original filtrate.

Now transfer the solution by Pasteur filter pipet to a 10-mL round-bottom flask and concentrate the solution using a rotary evaporator. If a rotary evaporator is not available, transfer the solution to a 25-mL Erlenmeyer flask containing a boiling stone and remove the solvent [**HOOD**] by

HOOD

warming the flask on a sand bath at 125 °C. A gentle stream of nitrogen aids this concentration process.

The crude piperonal is obtained as an oil and is purified by column chromatography.

## Purification and Characterization

Preweigh (and number 1–10) ten 25-mL Erlenmeyer flasks containing a boiling stone. Wet-pack a 25-mL buret using hexane and silica gel in the following manner. Place a cotton plug in the buret followed by a 1-in. layer of sand. Add 15 mL of hexane to the column, and then slowly add 8 g of silica gel while tapping the sides of the buret to release any air bubbles. Finally, add an additional 1 in. of sand to the top of the column.

Allow the hexane to drain from the buret until it reaches the top of the sand level in the column. Collect the eluted hexane in a waste container.

Dilute the crude piperonal oil with five drops of methylene chloride and then using a Pasteur pipet, add this solution to the top of the column. Again, drain a portion of the elution solvent until the solvent head is level with the sand at the top of the column. Rinse the sides of the column with 0.5 mL of 4:1 methylene chloride/hexane solution, and again drain the column to the same sand level.

Now add, in order, the following amounts of elution solvents: three 10-mL portions of 4:1 methylene chloride/hexane; three 10-mL portions of methylene chloride; and three 10-mL portions of 9:1 methylene chloride/ethyl acetate.

Collect the first 20 mL of eluate in a waste container. Now collect 10-mL fractions in each of the nine preweighed Erlenmeyer flasks. After collecting fraction No. 9, use flask No. 10 to collect a final fraction as the column completely drains.

Concentrate each of the collected fractions to dryness on a warm sand bath under a gentle stream of nitrogen gas [**HOOD**]. Weigh the flasks to determine the total amount of piperonal product. Now preweigh an additional 25-mL Erlenmeyer flask containing a boiling stone. To each of the flasks containing white or light yellow fanlike crystals, add sufficient methylene chloride to just dissolve them and then transfer the resulting solutions to the tared Erlenmeyer flask using a Pasteur filter pipet. Concentrate this final solution as before, and weigh the flask to obtain the total amount of piperonal isolated.

HOOD

**NOTE.** *If a rotary evaporator is available, the eluate can be collected in 10-mL round-bottom flasks. Transfer of each fraction to a single evaporation flask shortens the above procedure.*

Determine the melting point of your product and compare your value to that found in the literature. Obtain IR (KBr disk method) and/or ¹H NMR spectra and compare your results to those reported in *The Aldrich Library of IR Spectra* and/or *The Aldrich Library of NMR Spectra*, respectively.

## Chemical Tests

Chemical classification data may also be useful. The ignition test, the Tollens test, and the 2,4-dinitrophenylhydrazine test should all give a positive result for the piperonal compound.

**NOTE.** *Approximately 300 mg of purified product with a melting point within 2–4 °C of the literature value is suggested for continuing the sequence on to Experiment [E2].*

## OPTIONAL MICROSCALE PREPARATION

The microscale preparation is similar to that of the scaled-up synthesis outlined above, with the following exceptions.

HOOD

**NOTE.** *It should be emphasized that even in the case of the microscale preparation this reaction should be **entirely** carried out in the **hood** at all times until the dioxane solvent has been removed.*

**1.** The reagent and solvent amounts are one tenth of those used above.

### Physical Properties of Reactants

| Compound | MW | Amount | mmol | mp(°C) | bp(°C) |
|---|---|---|---|---|---|
| Piperonyl alcohol | 152.16 | 100 mg | 0.66 | 58 | |
| Chromic oxide resin | | 750 mg | | | |
| Dioxane | 88.12 | 1.0 mL | | | 101 |

**2.** A 5-mL round-bottom flask is used.

**3.** The reaction mixture is heated for 1 hour.

**4.** After the reaction product is cooled to room temperature, transfer the solution by Pasteur filter pipet to a funnel containing a loose cotton plug covered with 500 mg of anhydrous sodium sulfate. Collect the filtrate in a 10-mL Erlenmeyer flask. Rinse the reaction flask and resin (include the walls of the condenser) with three 0.5-mL portions of methylene chloride solvent (calibrated Pasteur pipet). Transfer each rinse by Pasteur filter pipet to the funnel and, finally, rinse the sodium sulfate with an additional 0.1 mL of methylene chloride. Combine each rinse filtrate with the original filtrate. Concentrate the solution as described.

**5.** Purify the crude oil using chromatography as follows:

Pack a short buret column with approximately 1.75 g of silica gel, and transfer the crude product by Pasteur pipet to the column. Rinse the flask with 1.0 mL of methylene chloride (calibrated Pasteur pipet) and also transfer the rinse to the column.

Now add methylene chloride to the column, 2.0 mL at a time (calibrated Pasteur pipet), and collect the eluate in three tared 10-mL Erlenmeyer flasks. Set aside the first 4.0 mL of eluate collected in the first flask. Collect the next 6.0 mL in one flask; this fraction contains the bulk of the product. Also collect one additional fraction of 3.0 mL in the third flask.

HOOD    Concentrate the second fraction (6.0 mL) **[HOOD]** using a warm sand bath with a slow stream of nitrogen to assist solvent removal. *Do not forget to add a boiling stone to the flask.*

Weigh the flask and contents. Reheat for 1 min, cool, and weigh again. If the two weights are within 2.0 mg, the product is quite pure. If not, repeat the evaporation process until a constant weight is obtained.

Cool the product. If it does not solidify, place it in an ice bath and scratch the sides and bottom with a glass rod to induce crystallization.

A small amount of additional product can be isolated from the third fraction in a like manner.

**8-76.** Primary alcohols can be oxidized to aldehydes and carboxylic acids. Often it is difficult to stop at the aldehyde oxidation state. One method frequently used to accomplish this, for those that boil below about 100 °C, is to remove the product from the reaction mixture as it is formed. This method is based on the fact that aldehydes have a lower boiling point than their corresponding alcohols. Explain this difference in boiling point.

**8-77.** A specific oxidizing agent for the conversion of primary alcohols to aldehydes is pyridinium chlorochromate, abbreviated as $\overset{+}{py} \cdot CrO_3Cl^-$. Generally, the oxidation is run in methylene chloride solvent. For example,

$$CH_3-\overset{\overset{\displaystyle CH_3}{|}}{C}=CH-CH_2CH_2-CH_2\ddot{O}H \xrightarrow[\substack{CH_2Cl_2 \\ RT}]{py^+ \cdot CrO_3Cl^-} CH_3-\overset{\overset{\displaystyle CH_3}{|}}{C}=CH-CH_2CH_2\overset{\overset{\displaystyle \cdot\ddot{O}\cdot}{||}}{CH}$$

Oxidation of this alcohol with the conventional $Na_2Cr_2O_7$–$H_2SO_4$–water system produces the carboxylic acid. Offer an explanation for the difference in these results.

**8-78.** In reference to Question 8-77, can you see another advantage of the pyridine chlorochromate reagent over that of the conventional conditions?

**8-79.** An excellent way to identify the presence of an aldehyde group is by $^1H$ NMR. The chemical shift of the aldehydic proton occurs in the 9–10-ppm region, where few other proton signals occur. Why is this chemical shift so far downfield?

**8-80.** A series of compounds with increasing boiling point is listed below. Offer an explanation for this trend.

Butane (8 °C)     Propanal (56 °C)     1-Propanol (97 °C)

**BIBLIOGRAPHY**

The present procedure is based on the work reported by:

Cainelli, G.; Cardillo, G.; Orena, M.; Sandri, S. *J. Am. Chem. Soc.* **1978**, *98*, 6737.

Several preparations involving the oxidation of a 1° alcohol to an aldehyde are given in *Organic Syntheses*:

Collins, J. C.; Hess, W. W. *Organic Syntheses*; Wiley: New York, 1988; Collect. Vol. VI, p. 644.

Hurd, C. D.; Meinert, R. N. *Organic Syntheses*; Wiley: New York, 1943; Collect. Vol. II, p. 541.

Ratcliffe, R. W. *Organic Syntheses*; Wiley: New York, 1988; Collect. Vol. VI, p. 373.

## Experiment [E2]  Piperonal *O*-(2,4-dinitrophenyl)oxime

Common name: Piperonal *O*-(2,4-dinitrophenyl)oxime
CA number: [17188-61-3]
CA name as indexed: 1,3-Benzodioxole-5-carboxaldehyde, *O*-(2,4-dinitrophenyl)oxime

### Purpose

To carry out the second step in the sequence of synthetic reactions leading to an aromatic nitrile, piperonylonitrile. To prepare a specific oxime derivative of the aldehyde obtained in Experiment [E1]. This oxime derivative is derivatized on oxygen so as to allow the oxygen to function as a good leaving group, which will allow an elimination reaction to form a nitrile in the final step of the synthetic sequence leading to piperonylonitrile.

### *Prior Reading*

*Technique 5:* Crystallization
Use of the Hirsch Funnel (see pp. 93–95).

---

**REACTION**

| Piperonal | O-(2,4-Dinitrophenyl) hydroxylamine | Piperonal O-(2,4-dinitrophenyl)oxime |

---

**DISCUSSION**

The preparation of this oxime is the second step in this sequence to obtain the aromatic nitrile target molecule. In Experiment [E2] you are going to convert the aromatic aldehyde formed in Experiment [E1] into an *O-phenylated oxime,* which on treatment with base in Experiment [E3] yields the desired nitrile via an elimination reaction. Oxime formation is also involved in the well-known Beckmann rearrangement (see Experiment [$6_{adv}$]), which is used for the synthesis of amides.

The generation of the oxime intermediate involves a nucleophilic addition of the amine group of the hydroxylamine to the carbonyl carbon, followed by a dehydration to form the carbon–nitrogen double bond (and the oxime group). A general mechanism for the reaction is given here.

The first step is a nucleophilic addition to the carbonyl group. Rapid proton transfer gives an intermediate that generally is not isolated.

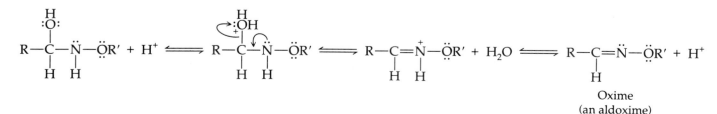

The next stage is an acid-catalyzed dehydration reaction in which water is eliminated to produce the oxime. This stage has been shown to be the rate-determining step in oxime formation.

$$R-\overset{\overset{\ddot{O}:}{|}}{\underset{\underset{H}{|}}{C}}-\overset{\overset{H}{|}}{\underset{\underset{H}{|}}{\ddot{N}}}-\ddot{O}R' + H^+ \rightleftharpoons R-\overset{\overset{H}{\underset{+}{\ddot{O}H}}}{\underset{\underset{H}{|}}{C}}-\overset{\overset{}{\underset{\underset{H}{|}}{N}}}-\ddot{O}R' \rightleftharpoons R-\overset{}{\underset{\underset{H}{|}}{C}}=\overset{+}{\underset{\underset{H}{|}}{N}}-\ddot{O}R' + H_2O \rightleftharpoons R-\overset{}{\underset{\underset{}{}}{C}}=\ddot{N}-\ddot{O}R' + H^+$$

Oxime
(an aldoxime)

**NOTE.** *If you are conducting the total sequence of reactions to obtain pipero-nylonitrile, follow the scaled up procedure. If the individual reaction is to be stud-ied, follow the microscale procedure.*

## Semimicroscale Preparation

**EXPERIMENTAL PROCEDURE**

(This procedure is scaled up to 10 times the amounts of the microscale preparation.)

Estimated time to complete the experiment: 1 hour.

### Physical Properties of Reactants and Product

| Compound | MW | Amount | mmol | mp(°C) | bp(°C) |
|---|---|---|---|---|---|
| Piperonal | 150.14 | 300 mg | 2.0 | 37 | 263 |
| O-(2,4-Dinitrophenyl)hydroxylamine | 199.12 | 400 mg | 2.0 | 111–112 | |
| Ethanol | 46.07 | 30 mL | | | 78 |
| HCL (12 *M*) | | 20 mL | | | |
| Piperonal O-(2,4-dinitrophenyl)oxime | 330.24 | | | 194–195 | |

### Reagents and Equipment

To a 100-mL three-necked round-bottom flask containing a magnetic stir bar and equipped with a reflux condenser protected by a calcium chloride drying tube, and two caps or glass stoppers, weigh and place 400 mg (2.0 mmol) of O-(2,4-dinitrophenyl)hydroxylamine, followed by 30 mL of abso-lute ethanol. (■)

**NOTE.** *The three-necked flask may be replaced with a conventional round-bottom 100-mL flask. In this case, addition of reagents can be carried out in the first instance by removing the condenser and in the second case by addition directly down the condenser with a 9-in. Pasteur pipet.*

**NOTE.** *The preparation of O-(2,4-dinitrophenyl)hydroxylamine is described in the Instructor's Manual, which is available from the publisher.*

Attach the flask to the condenser and warm (60–65 °C) the contents, with stirring, in a sand bath until a homogeneous solution is obtained. Now add 300 mg (0.2 mmol) of piperonal via one of the unused flask necks. After the aldehyde has dissolved, add slowly, with stirring, 20 mL of 12 *M* HCl through an unused flask neck using a Pasteur pipet. Stir the reaction mixture for an additional 1 min after the addition is complete.

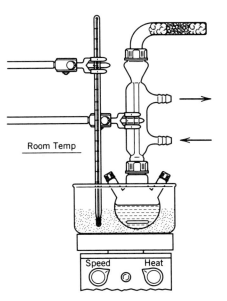

Room Temp

Speed    Heat

100–mL RB three necked flask containing
*O*–(2,4–dinitrophenyl)hydroxylamine, 400 mg +
absolute ethanol, 30 mL + piperonal, 300 mg +
12 *M* HCl, 20 mL

### Reaction Conditions

The oxime forms immediately. Complete the precipitation of the product by cooling the reaction flask in an ice bath for 10 min.

### Isolation of Product

Collect the solid product by filtration under reduced pressure using a Hirsch funnel. Rinse the flask with two 3-mL portions of cold absolute ethanol, using the rinse to wash the crystals on the filter funnel. Now rinse

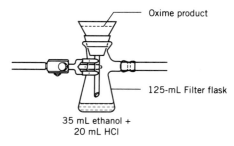

Oxime product

125-mL Filter flask

35 mL ethanol +
20 mL HCl

the crystals with three additional 3-mL portions of cold absolute ethanol. (■)

**NOTE.** *Approximately 50–70 mg of purified product with a melting point within 2–4 °C of the literature value is the minimum quantity and quality suggested for continuing the sequence in Experiment [E3].*

Air-dry the product on a clay plate or on filter paper. Collect and refrigerate the filtrate for at least 24 hours. This procedure generally produces another crop of oxime crystals. This second crop, collected by the same technique, may be combined with the initial product, if its melting point is above 180 °C.

## Purification and Characterization

Determine the melting point of the oxime. It is of sufficient purity to proceed with the next step of the sequence if the mp is 187 °C or greater. If necessary, recrystallize the material from acetic acid or hot acetone.

Obtain an IR spectrum of the oxime using the KBr disk technique and compare it with that of the reference standard shown in Figure 8.6.

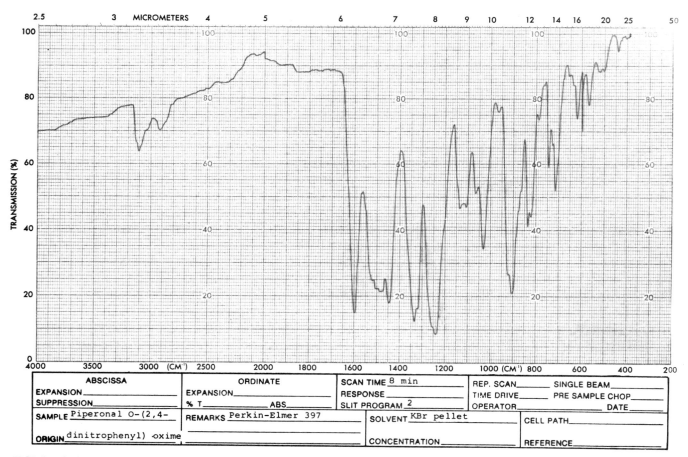

**FIGURE 8.6** IR spectrum: Piperonal *O*-(2,4-dinitrophenyl)oxime.

The microscale preparation is similar to that of the scaled up synthesis outlined above, with the following exceptions.

**1.** The reagent and solvent amounts are one tenth of those used above.

### Physical Properties of Reactants and Product

| Compound | MW | Amount | mmol | mp(°C) | bp(°C) |
|---|---|---|---|---|---|
| Piperonal | 150.14 | 30 mg | 0.20 | 37 | 263 |
| O-(2,4-Dinitrophenyl)hydroxylamine | 199.12 | 40 mg | 0.20 | 111–112 | |
| Ethanol | 46.07 | 3.0 mL | | | 78 |
| HCL (12 M) | | 2 drops | | | |
| Piperonal O-(2,4-dinitrophenyl)oxime | 330.24 | | | 194–195 | |

**2.** A 5-mL conical vial is used.

**3.** The hydrochloric acid is added dropwise through the top of the condenser using a 9-in. Pasteur pipet, after removing the drying tube. As the acid is delivered, the tip of the pipet should be held just above the surface of the solution. The drying tube is then reattached, and the contents are mixed by swirling the flask.

**8-81.** Referring to the first step in the mechanism of oxime formation outlined in the discussion section, offer an explanation of why a high acid concentration would hinder the formation of the intermediate generated in this first stage.

**8-82.** Oximes prepared from aldehydes or ketones by reaction with hydroxylamine can be reduced to primary amines in high yields. One can use various reagents for the reduction step including Ni–H$_2$ in methanol or LiAlH$_4$ in ether. Using a reduction reaction, carry out the following transformations.

   3-Pentanone to 3-aminopentane
   Propanal to 1-aminopropane
   Benzaldehyde to benzylamine

**8-83.** Oximes prepared from unsymmetrical ketones are likely to exist as mixtures of geometrical isomers. For example,

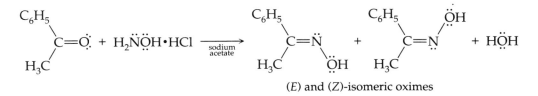

(E) and (Z)-isomeric oximes

Which isomer is designated as E? How do you account for both of these isomers being formed?

## BIBLIOGRAPHY

For the preparation of oximes using hydroxylamine hydrochloride as the reagent see:

Bousquet, E. W. *Organic Syntheses;* Wiley: New York, 1943, Collect. Vol. II, p. 313.

Hach, C. C.; Banks, C. V.; Diehl, H. *Organic Syntheses;* Wiley: New York, 1963, Collect. Vol. IV, p. 229.

Lachman, A. *Organic Syntheses;* Wiley: New York, 1943, Collect. Vol. II, p. 70.

Pasto, D. J.; Johnson, C. R.; Miller, M. J. *Experiments and Techniques in Organic Chemistry;* Prentice Hall: Englewood Cliffs, NJ, 1992, p. 332.

Semon, W. W.; Damerell, V. R. *Organic Syntheses;* Wiley: New York, 1943, Collect. Vol. II, p. 204.

Shriner, R. L.; Fuson, R. C.; Curtin, D. Y.; Morrill, T. C. *The Systematic Identification of Organic Compounds,* 6th ed., Wiley: New York, 1980, p. 181.

## Experiment [E3]   Piperonylonitrile

Common names: Piperonylonitrile, 3,4-methylenedioxybenzonitrile
CA number: [4421-09-4]
CA name as indexed: 1,3-Benzodioxole-5-carbonitrile

### Purpose

To convert the piperonal *O*-(2,4-dinitrophenyl)oxime intermediate, prepared in the previous experiment (Experiment [E2]), into the target molecule, piperonylonitrile. This completes the set of Sequence E reactions for the conversion of a substituted benzyl alcohol into an aromatic nitrile. To investigate the use of a novel elimination reaction to convert an oxime derivative to a nitrile.

### Prior Reading

*Technique 4:*   Solvent Extraction
Liquid–Liquid Extraction (see pp. 80–82).
*Technique 6:*   Chromatography
Column Chromatography (see pp. 98–101).
Concentration of Solutions (see pp. 106–108).

## REACTION

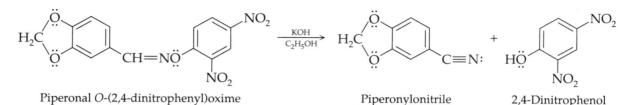

Piperonal *O*-(2,4-dinitrophenyl)oxime          Piperonylonitrile          2,4-Dinitrophenol

## DISCUSSION

The Sequence E reactions illustrate the oxidation of a benzylic alcohol to its corresponding aldehyde and the subsequent two-step conversion of this aldehyde via an intermediate aldoxime to an aromatic nitrile. This synthetic route is an attractive alternative to the preparation of aromatic nitriles via the Sandmeyer reaction.

Aromatic nitriles are usually prepared by the diazotization of the corresponding aromatic amine followed by treatment of the diazonium salt with copper(I) cyanide. This sequence is an example of the Sandmeyer reaction, a specific example of which is shown here.

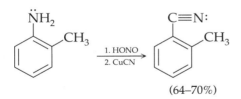

(64–70%)

This particular variation of the Sandmeyer reaction is useful since it allows the conversion of an aromatic amine, readily available by reduction of the corresponding nitro compound, to a reactive carbon substituent. This substitution involves replacement of the C—N bond with a C—C bond

In the present reaction, treatment of an *O*-phenylated oxime with base yields the nitrile by an elimination reaction. The role of the dinitrophenyl group is to enhance the oxime oxygen to function as a good leaving group. Thus, the phenoxy system departs as the conjugate base of the relatively acidic 2,4-dinitrophenol. The mechanistic sequence is outlined below.

$$^-:\ddot{O}H + R—C \!\!=\!\! \overset{\overset{\displaystyle \ddot{O}R'}{|}}{\underset{\overset{\displaystyle |}{H}}{N}} \longrightarrow H\ddot{O}H + R—C \!\!\equiv\!\! N: + ^-:\ddot{O}R' \rightleftharpoons R—C \!\!\equiv\!\! N: + H\ddot{O}R' + ^-:\ddot{O}H$$

Oximes derived from aldehydes (aldoximes) can be dehydrated to nitriles by many different dehydrating reagents; acetic anhydride is one of the most common reagents used.

$$\underset{\overset{\displaystyle |}{H}}{\overset{\displaystyle R}{\diagdown}} C \!\!=\!\! \overset{\overset{\displaystyle \ddot{O}H}{/}}{\underset{\displaystyle \ddot{N}}{}} \xrightarrow{\text{Ac}_2\text{O}} R—C \!\!\equiv\!\! N:$$

The reaction proceeds more rapidly when the hydrogen and the hydroxyl group are trans to one another (see also Experiments [10], [D3], and [A3$_b$]). Various derivatives of the hydroxylamine other than the ethers, RCH=NOR, illustrated in the present reaction, also undergo the conversion to nitriles. Among these are the RCH=NOCOR and RCH=NOSO$_2$Ar compounds. In some cases it is also possible to convert aldehydes to nitriles in one step by refluxing the reagents in concentrated hydrochloric acid (or by reaction with sodium formate in formic acid or sodium acetate in acetic acid) as follows:

$$R—\overset{\overset{\displaystyle H}{|}}{C} \!\!=\!\! \ddot{O}: + H_2\ddot{N}\ddot{O}H \bullet HCl \xrightarrow[\Delta]{\text{concd HCl}} R—C \!\!\equiv\!\! N:$$

When oximes are treated with strong acid they are converted to amides by a rearrangement sequence known as the Beckmann rearrangement. This reaction is illustrated in Experiment [6$_{adv}$].

Nitriles are synthetically versatile functional groups since they are readily converted to carboxylic acids by hydrolysis under acidic or basic conditions, reduced with LiAlH$_4$ to form primary amines, and reaction with Grignard reagents leads to the formation of ketones. These reactions are illustrated here.

$$R-C\equiv N: \xrightarrow[H_2O]{H^+ \text{ or NaOH}} R-\overset{\overset{\displaystyle \ddot{O}:}{\|}}{C}-\ddot{O}H$$

$$R-C\equiv N: \xrightarrow[2.\ H_2O]{1.\ LiAlH_4,\ ether} R-CH_2\ddot{N}H_2$$

$$R-C\equiv N: + R-MgX \xrightarrow[2.\ H_3O^+]{1.\ ether} R-\overset{\overset{\displaystyle \ddot{O}:}{\|}}{C}-R$$

Estimated time to complete the experiment: 2.5 hours.

## EXPERIMENTAL PROCEDURE

### Physical Properties of Reactants and Product

| Compound | MW | Amount | mmol | mp(°C) | bp(°C) |
|---|---|---|---|---|---|
| Piperonal *O*-(2,4-dinitrophenyl)oxime | 330.24 | 50 mg | 0.15 | 194–195 | |
| Ethanol | | 5 mL | | | 78.5 |
| Potassium hydroxide (0.2 *M*) | | 2 mL | | | |
| Piperonylonitrile | 147.13 | | | 92–93 | |

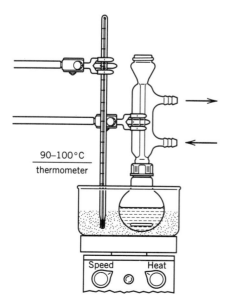

90–100°C
thermometer

Speed    Heat

Piperonal *O*-(2,4–dinitrophenyl)oxime, 50 mg +
95% ethanol, 5.0 mL +
0.2 *N* ethanolic KOH, 2.0 mL
10 mL RB flask

### Reagents and Equipment

Weigh and place 50 mg (0.15 mmol) of piperonal *O*-(2,4-dinitrophenyl)oxime in a 10-mL round-bottom flask containing a magnetic stir bar. Now add 5.0 mL of 95% ethanol and 2.0 mL of 0.2 *M* ethanolic KOH. Attach the flask to a reflux condenser. (■)

**NOTE.**  *The oxime is prepared in Experiment [E2]. The 0.2 M ethanolic KOH is prepared using 95% ethanol.*

### Reaction Conditions

Slowly heat the reaction mixture, while stirring, to reflux by use of a sand bath (100–110 °C) and maintain the mixture at this temperature (gentle reflux) for 1 hour. During the initial warming period, the solution turns a deep yellow.

**NOTE.**  *If the laboratory time is not sufficient to continue the isolation and purification of the product after the heating is terminated, cool the solution and remove the reaction vial. Cap the vial and store it until the next laboratory period.*

## Isolation of Product

Remove the reaction flask and concentrate the reaction mixture to a volume of 0.5 mL or less with a gentle stream of nitrogen gas and/or warming in a sand bath [**HOOD**]. *This concentration process takes a considerable length of time.*    HOOD

Now prepare an alkaline solution by diluting 1 mL of 5% aqueous NaOH with 5 mL of distilled water. Use this solution to transfer the concentrated reaction mixture to a 12-mL centrifuge tube in the following manner.

Add a 2-mL portion of the alkaline solution to the reaction flask, mix by swirling, and then transfer the resulting suspension to the centrifuge tube using a Pasteur pipet. Repeat this operation twice, using 2 mL of the alkaline solution each time.

Extract the resulting suspension with four 2-mL portions of methylene chloride (calibrated Pasteur pipet). Remove the methylene chloride extract (bottom layer) using a Pasteur filter pipet, and place the combined fractions in a 10-mL Erlenmeyer flask. Dry the solution over granular anhydrous sodium sulfate (0.5 g).

By use of a Pasteur filter pipet, transfer the dried solution to a 25-mL Erlenmeyer flask containing a boiling stone. Rinse the drying agent with two 1-mL portions of methylene chloride. Combine the rinses with the original solution. Remove the solvent [**HOOD**] under a gentle stream of    HOOD nitrogen and/or by warming in a sand bath to obtain the crude piperonylonitrile.

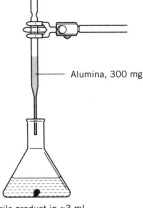

Alumina, 300 mg

Nitrile product in ~3 mL of CH$_2$Cl$_2$/hexane solvent

## Purification and Characterization

Purify the crude product by column chromatography using a Pasteur filter pipet filled with 300 mg of alumina (neutral, activity 1, see Glossary). Wet the column with 1.0 mL of 1:1 methylene chloride/hexane solution.

Dissolve the residue of crude nitrile in a minimum amount of 1:1 methylene chloride/hexane solvent, and transfer the resulting solution by Pasteur pipet to the column. Elute the nitrile from the column with 2.0 mL of the 1:1 CH$_2$Cl$_2$/hexane solvent and collect the eluate in a 10-mL Erlenmeyer flask containing a boiling stone. (■)

Evaporate the solvent under a gentle stream of nitrogen while warming in a sand bath [**HOOD**]. Dry the white needles of piperonylonitrile on a    HOOD porous clay plate or on filter paper.

Weigh the product and calculate the percent yield. Determine the melting point and compare it with the literature value.

Obtain an IR spectrum and $^1$H NMR spectra of the sample and compare your results with those in *The Aldrich Library of IR Spectra* and/or *The Aldrich Library of NMR Spectra*.

**8-84.** The Sandmeyer reaction is based on the replacement of the diazonium group in aryldiazonium salts by chloro, bromo, or cyano groups. Copper salt reagents are used.    **QUESTIONS**

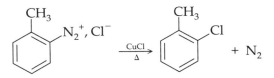

Show how one could carry out the following transformations using the Sandmeyer reaction.

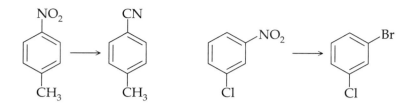

**8-85.**   When CuCN is used in the Sandmeyer reaction, the preparation is generally carried out in a neutral medium. Can you offer an explanation of why this is done?

**8-86.**   Outline a synthetic route for the preparation of nitriles using a carboxylic acid as the starting material.

**8-87.**   A yellow azo dye once used to color margarine has been outlawed because it is carcinogenic. Outline a synthesis of this dye, butter yellow, starting from benzene and *N,N*-dimethylaniline.

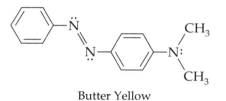

Butter Yellow

## BIBLIOGRAPHY

The procedure outlined above for the preparation of piperonylonitrile is based on the work of:

Miller, M. J.; Loudon, G. M. *J. Org. Chem.* **1975**, *40*, 126.

Several examples of the conversion of an aromatic aldehyde to the corresponding nitrile are given in *Organic Syntheses:*

Back, J. S.; Ide, W. S. *Organic Syntheses;* Wiley: New York, 1943; Collect. Vol. II, p. 622.

Clarke, H. T.; Nagy, S. M. *Organic Syntheses;* Wiley: New York, 1955; Collect. Vol. III, p. 690.

There are numerous references to the use of the Sandmeyer reaction in *Organic Syntheses.* Several are cited here:

Bigelow, L. A. *Organic Syntheses;* Wiley: New York, 1941; Collect. Vol. I, pp. 135, 136.

Cleland, G. H. *Organic Syntheses;* Wiley: New York, 1988; Collect. Vol. VI, p. 21.

Grundstone, F. D.; Tucker, S. H. *Organic Syntheses;* Wiley: New York, 1963; Collect. Vol. IV, p. 160.

Hartwell, J. L. *Organic Syntheses;* Wiley: New York, 1955; Collect. Vol. III, p. 185.

Marvel, C. S. *Organic Syntheses;* Wiley: New York, 1941; Collect. Vol. I, p. 170.

Rutherford, K. G.; Redmond, W. *Organic Syntheses;* Wiley: New York, 1973; Collect. Vol. IV, p. 133.

## Introduction to Photochromism: The Synthesis of a Photochromic Imine

The photochromic effect is a property of relative rarity in both organic and inorganic molecular structures. When it is present, a material is found to exhibit reversible color change upon exposure to radiation.

$$\text{Photochromic Substance A} \underset{\Delta \text{ or } h\nu_B}{\overset{h\nu_A}{\rightleftharpoons}} \text{Photochromic Substance B}$$

The photoproduct generally reverts to the initial system via thermal pathways, but there are examples known where the reverse reaction is induced by radiation of a different wavelength from that driving the forward reaction, or by both thermal and photochemical processes. Generally, sensitivity to thermal effects controls the concentrations obtained from the forward reactions and therefore, their effectiveness in producing the product.

## Cis–Trans Isomerizations

Experiment [6] studies the photochromic properties of *trans*-dibenzoyl-ethylene. In this case, the highly conjugated bright-yellow trans diastereomer is rapidly isomerized under intense sun-lamp visible radiation, via excited electronic states, to the colorless cis alkene. A $\pi$ electron is promoted to an antibonding $\pi^*$ molecular orbital, which destroys the $\pi$ bond, and thus permits facile rotation about the remaining $\sigma$ bond and formation of the cis alkene

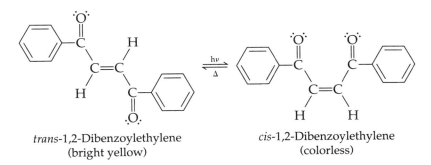

*trans*-1,2-Dibenzoylethylene
(bright yellow)

*cis*-1,2-Dibenzoylethylene
(colorless)

This is an endothermic reaction that does not reach equilibrium, but goes to completion because the cis isomer's electronic transitions are shifted to the ultraviolet and do not absorb visible radiation. The cis alkene is a structurally shorter chromophore than its trans isomer, and it is also likely to experience steric crowding with resultant distortion of the $\pi$ system. The cis isomer, therefore, absorbs at shorter wavelengths (higher energies) and has a lower molecular extinction coefficient (weaker) than the trans isomer. Thus once formed, the cis isomer is trapped. Upon heating, however, the cis isomer undergoes exothermic isomerization back to the original, more stable, trans alkene under equilibrium conditions.

Experiment [35] examines the photochromic behavior of the highly colored *cis*- and *trans*-azobenzene. In this case the $\pi \rightarrow \pi^*$ transition is promoted by ultraviolet light so that nonequilibrium isomerization to the cis isomer requires UV irradiation. The cis isomer is considerably less stable, however, and it undergoes relatively easy reversion back to the trans isomer by other mechanisms (the thermal conversion of visible radiation absorbed by the colored cis compound is an additional isomerization route apparently open to the cis isomer).

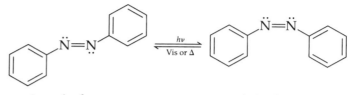

*trans*-Azobenzene                    *cis*-Azobenzene

## Tautomerism

A number of proton and valence tautomers are subject to photochemical induction. One example is 2-methylbenzophenone (**I**), a colorless compound that can be photochemically induced to tautomerize to a system with extended conjugation. The tautomer (**II**) is a yellow material that will revert to the colorless form under thermal conditions.

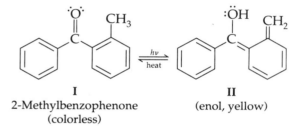

**I**                                         **II**
2-Methylbenzophenone                (enol, yellow)
(colorless)

## Homolysis and Heterolysis of Bonds

Photochromic homolysis has been observed with materials such as 2,3,4,4-tetrachloro-1-(4*H*)-naphthalenone (**III**).

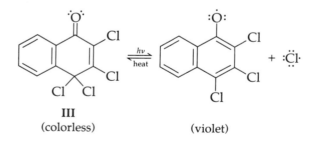

**III**
(colorless)                              (violet)

Excitation leading to heterolysis and zwitterion formation has been observed in numerous spiropyrans, as shown here for **IV**. These compounds undergo ring opening to yield a zwitterion (**V**).

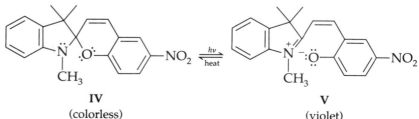

**IV**                                        **V**
(colorless)                              (violet)

The spiropyran example in which the photochromism develops following the transformation of a colorless isomer to a violet system is closely related to the structural isomerism observed in the target photochromic substance, a diazabicyclo[3.1.0]hex-3-ene (**VI**) synthesized in Sequence F. In both instances, the photoisomerization involves heterocyclic ring opening with formation of zwitterion **VII**.

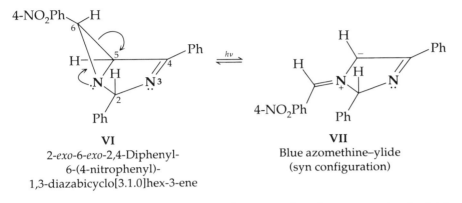

| VI | VII |
|---|---|
| 2-*exo*-6-*exo*-2,4-Diphenyl-<br>6-(4-nitrophenyl)-<br>1,3-diazabicyclo[3.1.0]hex-3-ene | Blue azomethine–ylide<br>(syn configuration) |

The diazabicyclo[3.1.0]hexenes form a series of compounds of which many exhibit photochromic properties.

## APPLICATIONS OF PHOTOCHROMISM

One successful application of inorganic photochromic systems has been in the manufacture of sunglasses. When a particular silver salt is incorporated in the lenses, the glass will darken on exposure to sunlight in order to protect the eyes, but then bleach quickly when the light intensity drops, so that the same glasses can be used at night or indoors.

As noted in Experiment [35], the cis–trans isomerism of azobenzene has been incorporated in a novel chemical method for information storage at ultrahigh densities with nondestructive readout. A device based on this chemical information storage system would have, it is estimated, the capacity of 100 million bits per square centimeter.

There also have been some investigations by the military directed toward developing camouflage "photochromic paints."

In most applications the ability of the system to undergo essentially endless recycling is an important factor. Thus, the degradation response time impacts heavily on the effectiveness of the system. In this regard, the valence bond tautomeric isomerizations would appear to possess the most promising properties, while those mechanisms that involve fragmentation open the system up to the possibility of irreversible byproduct formation. The heterolytic processes described in the photochromism of the target compound (internal ring opening) would appear to fall in between the two boundaries.

## THE REACTIONS OF SEQUENCE F[3]

The synthesis begins with an aldol reaction between 4-nitrobenzaldehyde and acetophenone in Experiment [F1] to yield 4-nitrochalcone.

[3] This synthetic sequence is based on a set of experiments first developed for the undergraduate instructional laboratory by Professor R. Marshall Wilson and Laboratory Director D.L. Lieberman of the University of Cincinnati. We are grateful to Paulette M. Fickett and Joanne M. Holland of Bowdoin College for further development and optimization work.

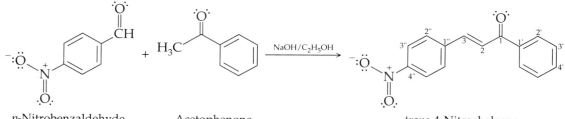

p-Nitrobenzaldehyde          Acetophenone                    trans-4-Nitrochalcone

The chalcone is brominated in the second step (Experiment [F2]) to yield *erythro*-2,3-dibromo-3-(4-nitrophenyl)propiophenone.

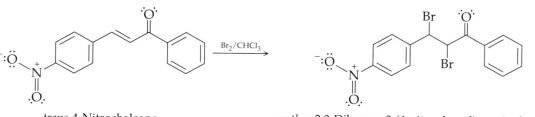

*trans*-4-Nitrochalcone          *erythro*-2,3-Dibromo-3-(4-nitrophenyl)propiophenone

When this halogenated compound is treated with ethanolic ammonium hydroxide for several days, as in Experiment [F3], the system undergoes several reactions (dehydrohalogenation, amination, and ring closure) to ultimately yield a trans substituted aziridine product.

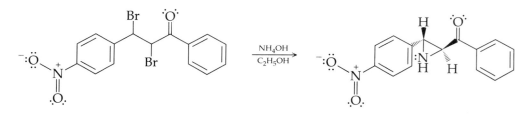

*erythro*-2,3-Dibromo-3-(4-nitrophenyl)propiophenone       *trans*-2-(4-Nitrophenyl)-3-benzoylaziridine

The photochromic substance, a diazabicyclo[3.1.0]hex-3-ene, is obtained in the fourth step (Experiment [F4]) when the aziridine is treated with benzaldehyde, anhydrous ammonia, and ammonium bromide. The reaction requires several days to go to completion.

*trans*-2-(4-Nitrophenyl)-3-benzoylaziridine          2-*exo*-6-*exo*-2,4-Diphenyl-6-(4-nitrophenyl)-1,3-diazabicyclo[3.1.0]-hex-3-ene

The product is a colorless crystalline substance that possesses the property of turning a deep blue when exposed to indoor light.

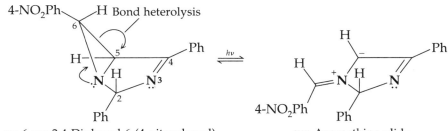

2-*exo*-6-*exo*-2,4-Diphenyl-6-(4-nitrophenyl)-
    1,3-diazabicyclo[3.1.0.]hex-3-ene

*syn*-Azomethine-ylide

The mechanism involved in the photochromic isomerization reaction is relatively rare, which makes these substances all the more interesting to study.

---

Coyle, J. D.; Hill, R. R.; Roberts, D. R. *Light, Chemical Change and Life: A Source Book in Photochemistry;* The Open University: Milton Keynes, England, 1982, pp. 306–309.

de la Mare, P. D. H.; Suzuki, H. *J. Chem. Soc.* **1968,** 648.

**BIBLIOGRAPHY**

## Experiment [F1]    An Aldol Reaction: *trans*-4-Nitrochalcone

Common name: 4-Nitrochalcone
CA number: [2960-55-6]
CA name as indexed: 2-Propen-1-one, 3-(4-nitrophenyl)-1-phenyl-, *(E)*-

### Purpose

To prepare the first of three intermediates on the synthetic pathway to our target molecule, a photochromic imine. To carry out a base-catalyzed aldol reaction in which an aromatic aldehyde is condensed with an aryl alkyl ketone. This addition reaction is followed by dehydration to form an $\alpha,\beta$-unsaturated ketone; this particular product is commonly called a chalcone. To isolate and purify this intermediate for use as the starting material in the next stage of the synthesis. To carry out a semimicroscale reaction to gain experience at conducting larger scale organic reactions.

### Prior Reading

    *Technique 4:*   Solvent Extraction
                     Liquid–Liquid Extraction (see pp. 80–82).
    *Technique 5:*   Crystallization
                     Use of the Hirsch Funnel (see pp. 93–95).

---

**REACTION**

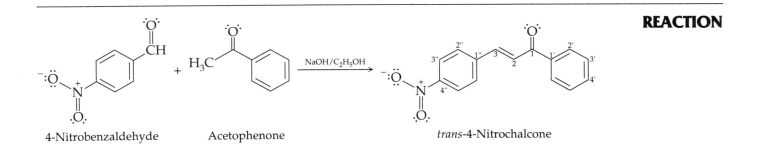

4-Nitrobenzaldehyde      Acetophenone      *trans*-4-Nitrochalcone

## DISCUSSION

The aldol reaction (aldol condensation) is one of the fundamental reactions of organic chemistry since it leads to the formation of a new carbon–carbon bond (see Experiment [20] for a very similar example of the Claisen–Schmidt type of aldol reaction). In this version, the condensation of 4-nitrobenzaldehyde (an aldehyde without an $\alpha$-hydrogen atom) with acetophenone (a ketone) gives *trans*-4-nitrochalcone. The aldol condensation of the unsubstituted aromatic aldehyde, benzaldehyde with acetophenone, yields *trans*-1,3-diphenyl-2-propenone (PhCH=CHCOPh), which has the common name, chalcone. Thus, the substituted derivatives of this system are known collectively as chalcones.

The extended conjugation in the product favors the formation of the chalcone product. Furthermore, this product is insoluble in the aqueous ethanol solvent and rapidly precipitates from the reaction medium as it is formed, whereas the starting materials are all soluble in aqueous ethanol. Thus, the experimental conditions assist in driving this equilibrium reaction to completion.

The classic aldol condensation involves generation of an *enolate* by removal of an acidic proton from a carbon alpha to the carbonyl group of an aldehyde or ketone, and subsequent nucleophilic addition of this enolate to the carbonyl carbon of an aldehyde or ketone. This reaction is base catalyzed and involves the following mechanistic steps.

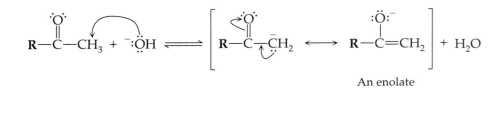

An enolate

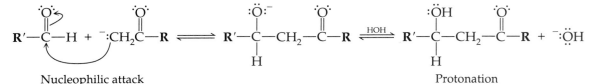

Nucleophilic attack    Protonation

The reaction involves: (a) base-catalyzed generation of the enolate, (b) nucleophilic attack of this anion on a carbonyl carbon, and (c) proton transfer to the resulting anion to yield the initial aldol product, a $\beta$-hydroxycarbonyl compound. The $\beta$-hydroxycarbonyl product may be isolated in many cases, if desired, as the subsequent dehydration is generally much slower than the addition reaction that precedes it. The final stage of the aldol reaction, as in the present reaction, is a hydroxide-catalyzed dehydration of the initial product by way of the enolate. Though hydroxide ion ($HO^-$) is generally not a good leaving group, the H alpha to the carbonyl in the $\beta$-hydroxyketone is quite acidic. In addition, the elimination produces a highly conjugated $\alpha,\beta$-unsaturated ketone. Under these strongly basic conditions, the hydroxide ion becomes an adequate leaving group. In these systems, both during the loss of the proton in the formation of the enolate anion and during the loss of hydroxide to yield the $\alpha,\beta$-unsaturated ketone, the molecular conformations involved favor development of the more-stable trans product.

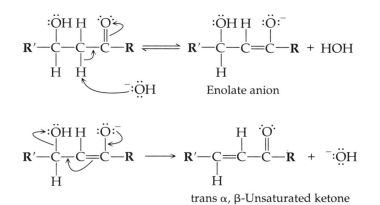

Enolate anion

trans α, β-Unsaturated ketone

In the present experiment an aldol condensation yields a benzalace-tophenone (chalcone) product. In Experiment [20], a nearly identical dou-ble aldol reaction yields dibenzalacetone. A further example of a double aldol reaction is found in Experiment [A3_a], where tetraphenylcyclopenta-dienone is the product of the reaction of benzil and 1,3-diphenylacetone.

Estimated time to complete the experiment: 3.0 hours.

**EXPERIMENTAL PROCEDURE**

### Physical Properties of Reactants

| Compound | MW | Amount | mmol | mp(°C) | bp(°C) |
|---|---|---|---|---|---|
| Acetophenone | 120.16 | 488 μL | 4.16 | 20.5 | 202.6 |
| Ethanol (95%) | | 20 mL | | | 78.5 |
| 4-Nitrobenzaldehyde | 151.12 | 500 mg | 3.31 | 106 | |
| Sodium hydroxide (10%) | | 2 mL | | | |

### Reagents and Equipment

Weigh and place 500 mg (3.31 mmol) of 4-nitrobenzaldehyde in a 50-mL Erlenmeyer flask containing a magnetic stir bar. Now add 20 mL of 95% ethanol and 488 μL of acetophenone.

### Reaction Conditions

Warm the reaction mixture, while stirring in a sand bath at 65–70 °C, until the aldehyde dissolves to yield a clear light yellow solution. At this point, cool the Erlenmeyer flask in an ice bath for at least 5 min and then, while continuing to cool the system, *carefully* add dropwise with stirring, 2 mL of 10% (aq) sodium hydroxide over a second 5-min period. During this time, the reaction mixture often turns a dark orange color and considerable pre-cipitation may occur. Cool the Erlenmeyer flask for an additional 15 min following the last addition of base.

### Isolation of Product

Collect the solid tan precipitate, which formed during the reaction, on the filter bed of a Hirsch funnel under reduced pressure. (■)

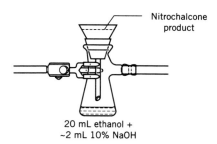

Nitrochalcone product

20 mL ethanol + ~2 mL 10% NaOH

If solid continues to form in the filtrate, refilter the reaction solution and combine the second collection of crystals with the first batch. You want to maximize your total yield of the aldol product, as this is the first step of a four-step synthesis. Thus, you will need efficient recovery of product at each intermediate stage of the synthesis in order to successfully obtain a reasonable quantity of the photochromic target molecule. Rinse the Erlenmeyer flask once or twice with ice cold water to effect as closely as possible a quantitative transfer of the chalcone to the Hirsch funnel.

### Purification and Characterization

Wash the tan filter cake containing the reaction product, dropwise with an ice cold 80:20 ethanol/water solution until the product appears as pale yellow crystals. The wash dissolves and removes a red-brown amorphous material that contaminates the crude product in many instances. Transfer the purified and partially dried chalcone to a watch glass, and then place it in a desiccator for final drying. Characterize the anhydrous intermediate accurately by an evacuated melting point and infrared spectrum; the latter can be compared to that recorded in *The Aldrich Library of IR Spectra*. The chalcone normally is of sufficient purity to be carried directly on to Experiment [F2].

**NOTE.** *Approximately 400 mg of purified product with a melting point within 2–4 °C of the literature value is the minimum quantity of intermediate suggested for continuing the sequence on to Experiment [F2].*

**QUESTIONS**

**8-88.** In a number of cases it is possible to successfully isolate the β-hydroxyketone intermediate prior to the dehydration that forms the α,β-unsaturated ketone.
   a. Does the isolation of the β-hydroxyketone suggest which step in the aldol condensation is the rate-determining step in this case?
   b. If so, which one *is* the rate-determining step?
   c. Suggest what reaction conditions may have a significant impact on determining which step becomes rate determining?
   d. What structural changes might lead to a change in the rate-determining step?

**8-89.** Ketones also undergo the aldol condensation, although a successful reaction often requires "enhanced" conditions, as the addition involves an unfavorable equilibrium constant. This is the situation in the reaction in which 4-nitrochalcone is synthesized. With the odds against it, why is the reaction successful in this case?

**8-90.** If you had obtained both the *cis*- and *trans*-chalcone products, and had purified them by recrystallization, how could you instantly know which one was cis and which one was trans without any further characterization?

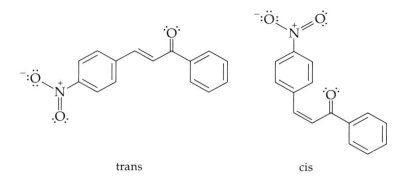

trans                                           cis

**8-91.** The mixed aldol reaction between propionaldehyde and acetone gives an 85% yield of 4-hydroxy-2-hexanone when run in THF at −78 °C with lithium diisopropylamide (LDA, a powerful base). The reaction is carried out by first adding the ketone to the base, cooling the solution, and then adding the aldehyde.

    a. Why does this mixed system give essentially a single product?
    b. Why is there no self-condensation of the acetone?
    c. Why does the system not rapidly go on to dehydrate?
    d. Why is the ketone added *to* the base rather than vice versa?

**8-92.** Give the aldol product or products from the following reaction:

$$CH_3CH_2CH_2CH_2CH_2CHO + CH_3CH_2CH_2CH_2CHO \xrightarrow{\text{base}}$$

**8-93.** The chalcone structure is particularly interesting. It was known in nature long before it was synthesized in the laboratory. This structure is incorporated biosynthetically into a large class of over 300 natural pigments called flavonoids. These substances heavily contribute to the spectacular New England autumn colors and many flower pigments. Flavonoids arise from chain extension of shikimic acid-derived cinnamic acids (see Experiment [10C] for a more detailed discussion of biological origin of these materials). A typical example of a flavone would be luteolin (5,7,3′,4′-tetrahydroxyflavone, **I**) the orange-yellow pigment of the snapdragon.

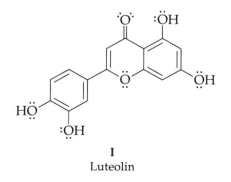

**I**
Luteolin

Flavonoids may be synthesized by employing reactions similar to those used in the chalcone synthesis. For example, the basic flavone structure (**II**) can be simply derived from a Claisen condensation between ethyl benzoate and 2-methoxyacetophenone, followed by treatment with HI.

**II**

Flavone (skeleton)

Show the mechanistic route leading to this flavone from the ester and the ketone.

---

**BIBLIOGRAPHY**

Selected examples of the Claisen–Schmidt reaction from *Organic Syntheses* are given here:

Conrod, C. R.; Dolliver, M. A. *Organic Syntheses*; Wiley: New York, 1943, Collect. Vol. II, p. 413.

Kohler, E. P.; Chadwell, H. M. *Organic Syntheses*; Wiley, New York, 1941, Collect. Vol. I, p. 78.

Leuck, G. J.; Cejka, L. *Organic Syntheses*; Wiley: New York, 1941, Collect. Vol. I, p. 283.

Wawzonek, S.; Smolin, E. M. *Organic Syntheses*; Wiley: New York, 1955, Collect. Vol. III, p. 715.

## Experiment [F2]
### *erythro*-2,3-Dibromo-3-(4-nitrophenyl)propiophenone

Common name: *erythro*-2,3-Dibromo-3-(4-nitrophenyl)propiophenone
CA number: [24213-17-0]
CA name as indexed: 1-Propanone, 2,3-dibromo-3-(4-nitrophenyl)-1-
  phenyl-, *(R\*,S\*)-*

### Purpose

To prepare the appropriate dibromide, an intermediate in your synthetic sequence, to act as the precursor to the aziridine ring system. To carry out a semimicroscale halogenation with bromine as the active reagent. A further purpose of this experiment is to demonstrate the *stereospecific* addition of bromine to alkenes.

### *Prior Reading*

*Technique 4:*   Solvent Extraction
        Liquid–Liquid Extraction (see pp. 80–82).
        Concentration of Solutions (see pp. 106–108).
*Technique 5:*   Crystallization
        Use of the Hirsch Funnel (see pp. 93–95).

**REACTION**

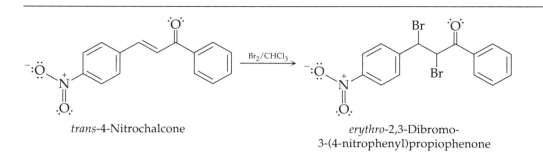

*trans*-4-Nitrochalcone        *erythro*-2,3-Dibromo-
3-(4-nitrophenyl)propiophenone

**REACTION**

**DISCUSSION**

The bromination of alkenes is an example of an electrophilic addition reaction. (See Experiments [A2$_b$] and [D2] for detailed discussions of the mechanism involved in this reaction. In particular, refer to Experiment [D2], which very closely resembles this reaction, for a discussion of the erythro and threo nomenclature used in this experiment). In the present reaction, bromination of 4-nitrochalcone yields *erythro*-2,3-dibromo-3-(4-nitrophenyl)propiophenone. This reaction is *stereospecific* since the other possible diastereomer [threo or (R*, R*)] is not formed.

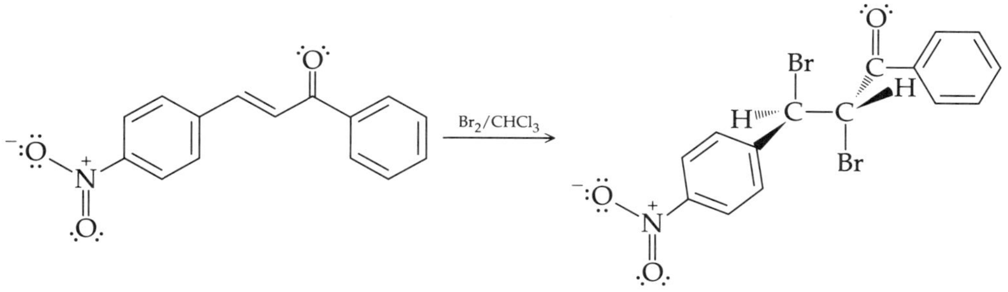

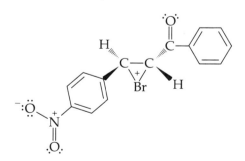

*trans*-4-Nitrochalcone        *erythro*-2,3-Dibromo-
3-(4-nitrophenyl)propiophenone

The reaction proceeds in two steps. The first involves the formation (from either side of the plane of the double bond; attack from below is shown here) of an intermediate cyclic *bromonium* ion.

The bromine molecule (Br$_2$) is normally symmetrical. However, as it approaches the electron-rich, and nucleophilic, $\pi$ bond of the alkene, it becomes polarized by induction and then functions as the electrophile in an addition reaction. The result is the generation of a cyclic bromonium ion.

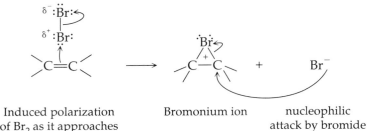

Induced polarization        Bromonium ion        nucleophilic
of Br$_2$ as it approaches                         attack by bromide
the alkene

In the present reaction, both the bromine and the 4-nitrochalcone are achiral, though the bromonium ion that is produced is chiral. Because it results from the reaction of achiral molecules in an achiral environment, the bromonium ion must be racemic. In this ion, the bromine atom bridges both carbon atoms of the original carbon–carbon double bond to form a three-membered ring intermediate. The generation of this high-energy cyclic species has a profound effect on the *stereochemistry* of the second step of the bromine addition, as the ring restricts rotation about the C—C single bond in the carbocation.

The second stage of the bromination involves nucleophilic attack by bromide ion on the intermediate bromonium ion. As the nucleophile must approach from the face opposite the leaving group, bond formation involves inversion of configuration at the carbon center under attack in the second stage of the bromination reaction.

Note that *either* carbon can be approached by the nucleophile (one attack is shown). This second step is a classic backside S$_N$2 type displacement.

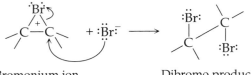

Bromonium ion                    Dibromo product

In the case with 4-nitrochalcone, two chiral centers are generated in the bromonium ion and we might, therefore, expect that two diastereomeric pairs would be formed. However, due to the stereoselectivity of the reaction, only a single diastereomer is generated, as a racemic pair of enantiomers (refer to Experiment [D2] for a further discussion of the stereochemistry of this halogenation).

## EXPERIMENTAL PROCEDURE        Estimated time to complete the experiment: 2.0 hours.

**Physical Properties of Reactants and Product**

| Compound | MW | Amount | mmol | mp(°C) | bp(°C) |
|---|---|---|---|---|---|
| *trans*-4-Nitrochalcone | 253 | 400 mg | 1.58 | 161–164 | |
| Chloroform | | ~8 mL | | | 61.7 |
| Bromine/chloroform (2.5%) | 179.8 | (200 μL Br$_2$) | 3.91 | | 58.8 |
| *erythro*-2,3-Dibromo-3-(4-nitrophenyl)propiophenone | 413 | | | 151–153 | |

## Reagents and Equipment

Weigh and place 400 mg (1.58 mmol) of 4-nitrochalcone in a tared 50-mL, round-bottom flask containing a magnetic stir bar. Now add about 8 mL of chloroform (dispensed in the **hood**). Connect the flask to a water-jacketed    HOOD reflux condenser fitted with a drying tube.

## Instructor Preparation

*The active brominating reagent in this reaction is liquid Br$_2$ dissolved in chloroform. Prepare a solution of 200 μL (624 mg) of bromine dissolved in 8 mL of chloroform multiplied by the number of students carrying out the experiment. The reagent should be prepared, dispensed and added to the reaction in the* **hood**.    HOOD

---

CAUTION:   Bromine is a highly reactive substance. Even in chloroform solution you must handle it with care. Be very careful not to get this reagent on your skin. All transfers of the    HOOD reagent should be made in the *hood*.

Chloroform itself is highly toxic and a cancer suspect agent. Handle it with respect.

---

## Reaction Conditions

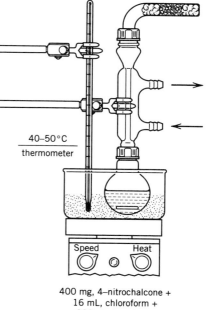

40–50°C
thermometer

Speed    Heat

400 mg, 4–nitrochalcone +
16 mL, chloroform +
200 μL, bromine +
50–mL RB flask

Warm the round-bottom flask in a sand bath between 60 and 70 °C, with stirring, until the 4-nitrochalcone dissolves to yield a clear light yellow solution. Once dissolution has occurred, continue to maintain the bath temperature at 60–70 °C and add the bromine reagent. The addition is carried out dropwise using a 9-in. Pasteur pipet inserted down the condenser (briefly remove the drying tube during this operation), with stirring, over a 10-min period. By the end of the Br$_2$ addition the solution turns a dark orange. Continue to heat the stirred reaction mixture for an additional 20 min. (■)

After cooling the reaction mixture to room temperature, remove a small aliquot and spot it on a silica gel TLC plate next to a reference spot of the 4-nitrochalcone starting material. Elute the plate with 50 : 50 methylene chloride/hexane and visualize the spots by UV. (There are a number of unidentified byproducts formed in small quantities during the reaction that are often observed on the TLC plates.) There also will probably be a trace of unreacted 4-nitrochalcone left in the reaction solution. The major (and highest $R_f$) spot ($R_f = 0.3$) is the brominated product. As an excess of bromine was used in the reaction, however, only small quantities of 4-nitrochalcone are normally detected at this stage. If significant unreacted substrate remains (i.e., more than a faint or weak spot on TLC), add an additional small amount of the bromine reagent and then reanalyze the reaction mixture again by TLC.

## Isolation of Product

Once it is established that the reaction is largely complete, remove the solvent and excess reagent from the reaction mixture by rotary evaporation. Weigh the crude residue.

**NOTE.**   *At this point the crystalline residue that remains following rotary evaporation may be stored until the next laboratory period by first flushing the round-bottom flask with dry nitrogen (or argon) and then quickly sealing the flask with a*

*ground-glass standard-taper stopper that is sealed with Parafilm. The flask should be labeled and given to your instructor for storage in the freezer. While it is possible to safely interrupt the workup at this stage, it should be pointed out that most organic materials are much more stable when they are stored in as pure a state as possible, thus, **if time permits, you are urged to finish the workup of the bromination.***

## Purification and Characterization

The crude material, a yellow-orange solid is now partially purified by column chromatography.

The chromatographic column (short buret) is packed with silica gel (10 g) after first positioning a plug of cotton and about 1 cm of sand at the bottom. Then a portion of 50:50 methylene chloride/ hexane is added to the column (~15 mL), which is followed by 10 g of silica gel. The solid substrate is slowly added while tapping the column to promote even settling of the packing material. During this process the column stopcock is slightly opened to create a slow drip rate of the packing solvent out of the column. As a result of this drainage more solvent may be required to keep the solvent level above that of the silica gel during the settling operation. Finally, carefully drain any excess solvent to the top of the column and close the stopcock after the packing procedure is complete.

The crude *erythro*-2,3-dibromo-3-(4-nitrophenyl)propiophenone is then dissolved in a minimum amount of methylene chloride (~10 mL) and applied to the top of the column by slowly pipetting the solution down the side of the column without disturbing the silica gel (as this solution is added it is also slowly drained onto the column by cracking open the stopcock). As the final quantity of crude product drains to the top of the column, elution is started with 50:50 methylene chloride/hexane solution (again by careful addition so as not to disturb the upper layers of silica gel containing the adsorbed reaction products). Collect 3 × 30-mL fractions in 50-mL Erlenmeyer flasks (labeling each flask with the fraction number). You may observe some yellow zones of material slowly moving down the column during the elution. This colored material usually does not begin to elute with this chromatographic scheme. Once the fractions have been collected, use TLC analysis (silica gel plates, 50:50 methylene chloride/hexane) to determine the composition of the fractions. The purest fractions are combined to give sufficient material to continue on to the next step (Experiment [F3]) in the sequence. Separate and remove the solvent ($N_2$ and warm sand bath). Determine the weight and melting point (evacuated) of your brominated product. The *erythro*-2,3-dibromo-3-(4-nitrophenyl)-propiophenone should appear as white to light yellow needles. It may be further recrystallized from 95% ethanol if desired.

If the melting point is only a few degrees low (~142–147 °C), increased purity often is quickly obtained by simply adding a few milliliters of ice cold chloroform to the product residue that is cooled in an ice bath. Triturate the residue for a few seconds with the cold solvent. Withdraw the solvent by Pasteur filter pipet leaving the washed crystals behind (this treatment may be repeated if necessary). Remove traces of the solvent remaining on the residue by a short rotary evaporation. Recheck the melting point (evacuated) to determine if product purity is improved enough to continue on to the third step. (■)

Weigh your purified *erythro*-2,3-dibromo-3-(4-nitrophenyl)propiophenone intermediate and calculate the percent yield based on both the starting 4-nitrochalcone and the 4-nitrobenzaldehyde (the starting material used in Experiment [F1]). Compare your spectrum to that of a reference standard shown in Figure 8.7.

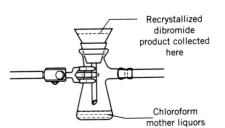

Recrystallized dibromide product collected here

Chloroform mother liquors

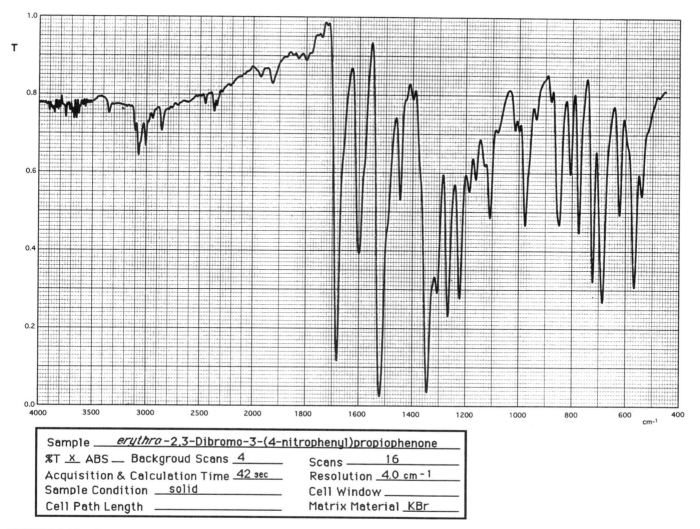

Sample ___erythro-2,3-Dibromo-3-(4-nitrophenyl)propiophenone___

%T _X_ ABS __ Backgroud Scans _4_    Scans ___16___

Acquisition & Calculation Time _42 sec_    Resolution _4.0 cm⁻¹_

Sample Condition __solid__    Cell Window _____

Cell Path Length _____    Matrix Material _KBr_

**FIGURE 8.7**    IR spectrum: *erythro*-2,3-Dibromo-3-(4-nitrophenyl)propiophenone.

**Optional.**    *Obtain the ¹H NMR spectrum of the product in CDCl₃. Identify the resonances for the two protons attached to the halogenated carbon atoms, and ascertain that the product is a single diastereomer. This information will be useful in Experiment [F3].*

**NOTE.**    *The erythro-2,3-dibromo-3-(4-nitrophenyl)propiophenone is a particularly sensitive substance and will decompose in contact with air at room temperature over several days. If you do not have time to continue on to Experiment [F3] during this laboratory period you should store your purified material as described above for the crude product.*

*If you have time to start the reaction, you are urged to continue on to the next step. Once you have the pure dibromide in hand, it takes only a relatively short time to set up and get the next reaction running. As this latter reaction will be left to run for a week, you have a lot to gain by getting the third step started during the period when the second intermediate is worked up.*

**NOTE.**    *Approximately 450 mg of purified product with a melting point within 2–4 °C of the literature value is the minimum quantity of intermediate suggested for continuing the sequence on to Experiment [F3].*

## QUESTIONS

**8-94.** A considerable excess of $Br_2$ in chloroform is required to successfully drive the halogenation of 4-nitrochalcone to completion. Offer a suggestion as to the role of the excess reagent.

**8-95.** What product(s) would you expect to obtain from the bromination ($Br_2$ in $CCl_4$) of cyclobutene?

**8-96.** In the bromination of 4-nitrochalcone a racemic dibromide is formed. A second diastereomeric dibromide is structurally possible, however, it is not formed in the reaction.
   a. Give stereochemically detailed drawings of the stereoisomer(s) isolated from the reaction mixture.
   b. Give stereochemically detailed drawings of the stereoisomer(s) that is/are not formed.
   c. Why is this reaction stereoselective?
   d. How could the second diastereomer be synthesized if its preparation was required?
   e. Assign, by the R and S convention, the stereocenters in each of the diastereomers.

**8-97.** Show the stereoisomer(s) generated by bromination of each of the enantiomers of *cis-* and *trans-*4-bromo-2-pentene. Show the relationship (enantiomer, diastereomer, etc.) and assign R and S centers in all product(s) and starting materials.

**8-98.** Show, with the correct absolute configuration, the stereoisomer(s) formed on bromination of *(S)*-4-*tert*-butyl-1-cyclohexene. Would you expect them to be formed in equal amounts?

## BIBLIOGRAPHY

A large number of examples of the bromination of alkenes appear in *Organic Syntheses*. Selected references are given below:

Allen, C. F. H.; Abell, R. D.; Normington, J. B. *Organic Syntheses;* Wiley: New York, 1941; Collect. Vol. I, p. 205.

Cromwell, N. H.; Benson, R. *Organic Syntheses;* Wiley: New York, 1943; Collect. Vol. II, p. 105.

Fieser, L. F. *Organic Syntheses;* Wiley: New York, 1963; Collect. Vol. IV, p. 195.

Khan, N. A. *Organic Syntheses;* Wiley: New York, 1963; Collect. Vol. IV, p. 969.

McElvain, S. M.; Kundiger, D. *Organic Syntheses;* Wiley: New York, 1955; Collect. Vol. III, p. 123.

Paquette, L. A.; Barrett, J. H. *Organic Syntheses;* Wiley: New York, 1973; Collect. Vol. V, p. 467.

Rhinesmith, H. S. *Organic Syntheses;* Wiley: New York, 1943; Collect. Vol. II, p. 177.

Snyder, H. R.; Brooks, L. A. *Organic Syntheses;* Wiley: New York, 1943; Collect. Vol. II, p. 171.

## Experiment [F3]
### *trans-*2-(4-Nitrophenyl)-3-benzoylaziridine

Common names: *trans-*2-(4-Nitrophenyl)-3-benzoylaziridine
CA number: [76336-95-3]
CA name as indexed: Methanone, [3-(4-nitrophenyl)-2-aziridinyl]phenyl-, *trans-*

## Purpose

To synthesize the third intermediate on the pathway to the target photochromic imine. To form a heterocyclic three-membered ring, an aziridine derivative. This is the first ring of the diazabicyclohexene system that you will ultimately convert into the photochromic imine. To study a process that involves three reactions and the formation of two intermediates en route to the final product. To study a number of interesting stereoselective reactions. To work with organic reactions that require several days to come to completion.

## Prior Reading

Technique 4: Solvent Extraction
Liquid–Liquid Extraction (see pp. 80–82).

Liquid–Liquid Extraction (see pp. 80–82).

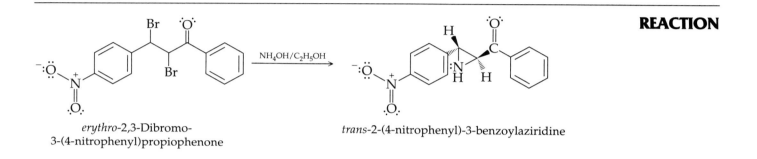

**REACTION**

erythro-2,3-Dibromo-
3-(4-nitrophenyl)propiophenone

trans-2-(4-nitrophenyl)-3-benzoylaziridine

**DISCUSSION**

The conversion of erythro-2,3-dibromo-3-(4-nitrophenyl)propiophenone to a substituted aziridine involves a number of interesting steps and intermediates. The first stage of the reaction involves attack on the halogenated intermediate by base (concd ammonium hydroxide, $NH_4OH \rightleftharpoons NH_3 + H_2O$). Under the highly polar conditions, the reaction likely proceeds via an E1cb mechanism involving the initial attack by ammonia, acting as a base, on the 2 proton to yield the resonance stabilized anion.

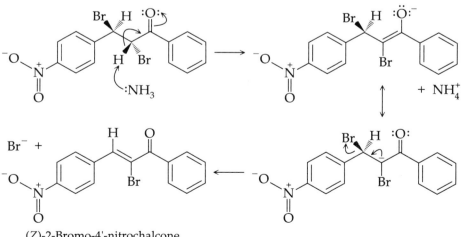

(Z)-2-Bromo-4'-nitrochalcone

You should refer to the detailed discussions in Experiments [9] and [10], which describe in detail the chemistry associated with the E1 and E2 elimination mechanisms. The E1cb mechanism, like the E1 mechanism, involves a two-step process, but in this case the order of charge development is the reverse of the E1 mechanism. Here proton abstraction precedes loss of the leaving group. A generalized scheme is shown here:

E1cb Mechanism

$$RCH_2CHR' + B^- \longrightarrow R\overset{-}{C}HCHR' + BH \longrightarrow RCH{=}CHR' + X^-$$
$$\underset{X}{|} \qquad\qquad\qquad \underset{X}{|}$$

It should be pointed out that the mechanisms of a large majority of elimination reactions can be explained by invoking various positions along the continuum between the three elimination mechanisms mentioned here; E1cb and E1 are at the extremes of the continuum, and the E2 mechanism lies exactly halfway between the two.

In the present case, carbanion (or near-carbanion) formation following α-proton abstraction appears to be favored, and formation of the anionic intermediate has the attractive feature that it then allows rotation about the incipient π bond. Stereoelectronic requirements of the elimination mechanism require the carbon–bromine bond to be parallel to the *p* orbital of the adjacent enolate (or α-keto carbanion, depending on which resonance structure is being discussed), just as the proton being removed and the leaving group prefer to be anti in an E2 elimination. There are two possible conformations (or rotamers) that meet this requirement. The most stable one will be that with the carbonyl and nitrophenyl groups anti to one another, and elimination of bromide ion from this conformation leads to the alkene with the carbonyl and nitrophenyl groups trans, which is the Z alkene, as illustrated here.

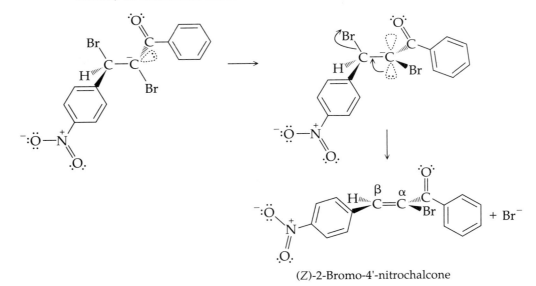

(Z)-2-Bromo-4'-nitrochalcone

Thus, the first stage of the reaction primarily leads to formation of an α-bromo-α,β-unsaturated ketone, (Z)-2-bromo-4'-nitrochalcone. This alkene could be easily isolated from the reaction medium, if required, as the second step in the reaction occurs at a considerably slower rate than the first step. Thus, appreciable concentrations of this unsaturated intermediate are obtained.

The *second stage* of the reaction involves 1,4 addition of the amine (in this case ammonia in equilibrium with the ammonium hydroxide) to the unsaturated ketone.

There are many examples of nucleophilic reagents that add to α,β-unsaturated aldehydes and ketones in a manner in which the addition is formally 1,4. This result is termed conjugate addition. Under basic conditions, these transformations involve initial attack by the nucleophile to the β-carbon atom, followed by electrophilic addition (normally of a proton) on the carbonyl oxygen; the nucleophile and electrophile add at the 1 and 4 positions relative to one another. The enolate formed in the early stages of the reaction is generally quickly protonated to give an enol. The enol will subsequently tautomerize to the ketone. A general mechanistic scheme is shown below for the 1,4 addition of water to an α,β-unsaturated carbonyl system.

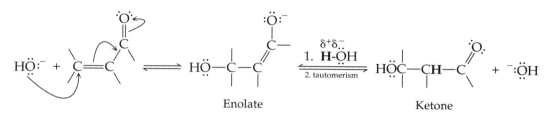

In the present case, conjugate addition has the potential to result in two diastereomers. As in the case of the starting dibromide, these stereoisomers can be either erythro or threo. The distribution between these two products is important as it ultimately determines the ratio of products in the next step (third) of the reaction.

From the geometry of the aziridine final product, and the fact that the internal nucleophilic substitution reaction will proceed with inversion of configuration, we can reliably postulate that conjugate addition must eventually result in the erythro diastereomer. In the conjugate addition of ammonia to 2-bromo-4'-nitrochalcone, the stereochemistry of the product is determined by which face of the resulting enolate receives the proton. The proton source may be the —NH$_3^+$ group resulting from the initial step of the conjugate addition reaction (as shown here), or it may be an ammonium ion or a molecule of ethanol or water hydrogen bonded to the amino group. A further consideration may be the geometry of the resulting enolate, as there is the potential to generate either the Z or the E enolate; the E enolate is illustrated here as it would appear to be the least hindered of the two diastereomeric enolates.

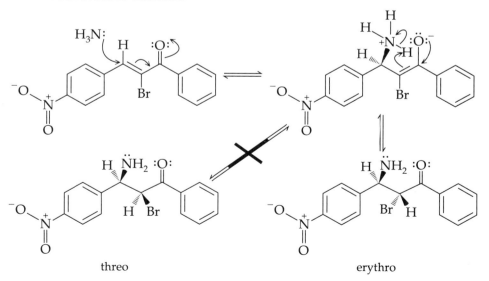

threo                                   erythro

It is also possible that both the threo and erythro diastereomers are produced at comparable rates and that the erythro is thermodynamically preferred by an equilibrium process between the two.

A comparison of the most-favored conformations, based on relative group sizes, for the erythro and threo diastereomers, indicates that the Conformation **A** for the erythro isomer would be less sterically crowed than Conformation **B** for the threo isomer, as shown here.

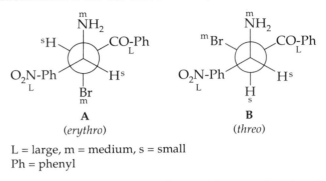

**A**
(*erythro*)

**B**
(*threo*)

L = large, m = medium, s = small
Ph = phenyl

The *final stage* on the route to the aziridine product involves an internal $S_N2$ ring closure in which the primary amine group attacks the $\alpha$-carbon atom holding the remaining bromine from the backside to close an aziridine ring and displace bromide with inversion of configuration at the $\alpha$ carbon. This displacement has been shown to go exclusively by this mechanism in the case of the $\beta$-amino-$\alpha$-bromoketones; an $S_N1$ reaction is less likely because of the instability of an $\alpha$-keto carbocation. For a detailed discussion of the mechanism of the classic $S_N2$ substitution reaction, refer to Experiment [22].

The stereochemistry of aziridine ring substitution, as pointed out above, is controlled by the product distribution in the $\beta$-amino-$\alpha$-bromoketone intermediate, which in this case favors the erythro configuration. Thus, inversion of configuration during ring formation leads to the trans-substituted aziridine ring in the present example.

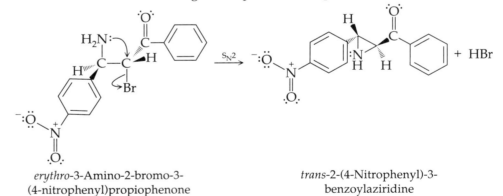

*erythro*-3-Amino-2-bromo-3-
(4-nitrophenyl)propiophenone

*trans*-2-(4-Nitrophenyl)-3-
benzoylaziridine

## EXPERIMENTAL PROCEDURE

Estimated time to complete the experiment: 3.5 hours.
Estimated time to complete Part A of the experiment: 1.5 hours.
Estimated time to complete Part B of the experiment: 2.0 hours.

### Physical Properties of Reactants and Product

| Compound | MW | Amount | mmol | mp(°C) | bp(°C) |
|---|---|---|---|---|---|
| *erythro*-2,3-Dibromo-3-(4-nitrophenyl)propiophenone | 411 | 450 mg | 1.10 | 151–153 | |
| Ethanol (95%) | | 4.4 mL | | | 78.5 |
| Ammonium hydroxide (concd) | | 1.3 mL | | | |
| *trans*-2-(4-nitrophenyl)-3-benzoylaziridine | 268.3 | | | 142–143 | |

## PART A

### Reagents and Equipment

Weigh and place 450 mg (1.10 mmol) of the *erytho*-2,3-dibromo-3-(4-nitrophenyl)propiophenone intermediate synthesized in Experiment [F2] in a labeled 25-mL round-bottom flask containing a magnetic stir bar. Add 4.4 mL (9.8 mL/g of chalcone) of 95% ethanol (graduated cylinder) and 1.3 mL (2.8 mL/g of chalcone) of concd ammonium hydroxide (automatic delivery pipet or 2-mL glass pipet) to the flask [**HOOD**]. Stopper the flask, swirl to mix the contents, and then seal it with Parafilm. (A polypropylene standard taper stopper is preferred for sealing the vessel for long periods in the presence of base.)

HOOD

---

> **CAUTION:**  Concentrated ammonium hydroxide is a strongly caustic reagent. You must handle it with care. Be particularly alert not to get this reagent on your skin, or breathe the vapors. All transfers of the reagent should be made in the *hood*.

HOOD

---

### Reaction Conditions

Stir the reaction mixture for 15 min. Then wrap the flask with aluminum foil and continue to stir the system for 24 hours. The reaction may be worked up at this point or stored in the dark for 1 week if necessary. Do not expect the chalcone to immediately dissolve in the reaction medium, it will do so over the course of several hours as the aziridine product begins to precipitate. If you wish, you can occasionally magnetically stir the reaction mixture for 15–20 min during this intervening period.

*This completes Part A of the experiment.*

## PART B

### Isolation of Product

Add 15 mL of ice cold water to the reaction mixture and swirl for 5 min. Collect the solid residue formed in the reaction by vacuum filtration on a Hirsch funnel. The product will appear as pale orange, very fine needles. There may also be a small amount of a more powdery yellow material that should dissolve in three 1.0-mL washes with ice cold water. Use the first wash to rinse the Erlenmeyer flask and transfer the rinse to the Hirsch funnel. Fine orange needles of the aziridine product will be deposited on the filter bed. (■)

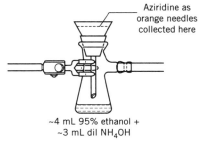

Aziridine as orange needles collected here

~4 mL 95% ethanol + ~3 mL dil NH$_4$OH

### Purification and Characterization

Collect the orange needles, dry them to constant weight (under reduced pressure), and determine an evacuated melting point. Recrystallize the crude product from hot methanol (~10–20 mL). After crystallization has begun, cool the system further in an ice bath for 10–15 min to complete the collection. The purified aziridine is obtained as long, shiny, pale orange needles via Hirsch filtration.

Weigh the *trans*-2-(4-nitrophenyl)-3-benzoylaziridine and calculate the percentage yield based on both the starting materials: *erythro*-2,3-dibromo-3-(4-nitrophenyl)propiophenone (Experiment [F3]), 4-nitrochalcone (Experiment [F2]), and on 4-nitrobenzaldehyde (Experiment [F1]). Recheck the evacuated melting point, to see if further purification is required, and obtain an IR spectrum using the KBr disk technique. Compare the spectrum to that of a reference standard shown in Figure 8.8.

***Optional.*** *Obtain the ¹H NMR spectrum of the aziridine. Establish from this spectrum and the NMR data obtained in Experiment [F2], if any unreacted dibromochalcone still contaminates the aziridine sample that has been purified for use in the preparation of the photochromic target molecule. Determine the diastereomeric purity of your product, and determine that the aziridine product is indeed trans substituted.*

**NOTE.**   *Approximately 50 mg of purified product with a melting point within 2–4 °C of the literature value is the minimum quantity of intermediate suggested for continuing the sequence on to Experiment [F4].*

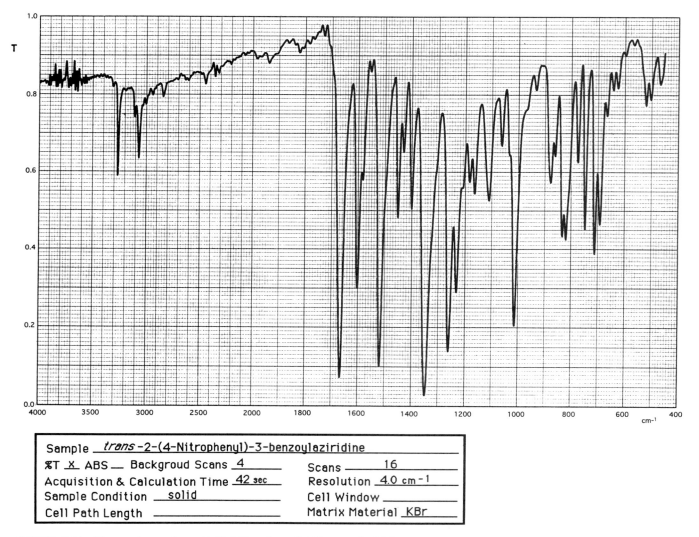

**FIGURE 8.8**   IR spectrum: *trans*-2-(4-Nitrophenyl)-3-benzoylaziridine.

**8-99.** The synthesis of 1-cyclohexyl-2-phenyl-3-(4-phenylbenzoyl)aziridine from cyclohexylamine and 4'-phenylchalcone dibromide, gives a mixture of 47% of the cis isomer and 44% of the trans isomer. Chromatography on activated alumina yielded, in the first eluates, a crystalline material, mp = 118–119 °C while the final eluates produced a higher melting material, mp = 144–146 °C. The lower melting substance exhibited a $\tilde{\nu}_{c=o}$ = 1656 cm$^{-1}$ while the higher melting compound had $\tilde{\nu}_{c=o}$ = 1686 cm$^{-1}$. Which product is the trans substituted aziridine and which is the cis-compound?

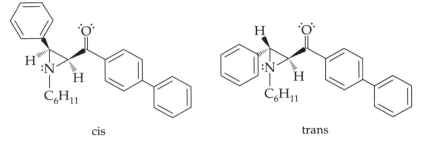

cis                    trans

**8-100.** *N*-Haloaziridines have been found to possess an extremely high nitrogen inversion barrier. The isolation of the cis (**I**) and trans (**II**) isomers of 1-chloro-2-methylaziridine has been accomplished and they represent the first isolated inversion isomers of trivalent nitrogen. These inversion isomers appear to have remarkable stability. For example, 1-chloro-2,2-dimethylaziridine (**III**) retains configurational stability at temperatures as high as about 135 °C! Make a list of the potential stereoisomerism available to structures **I**, **II**, and **III**. Use perspective drawings to illustrate each isomer and label the stereocenters by the *R* and *S* convention.

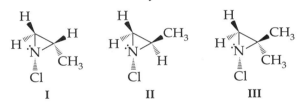

**8-101.** Another route to the effective synthesis of aziridines (**III**) is through the ring closure of the β-amino alcohol (**I**). The alcohol must first be converted to the β-amino hydrogen sulfate (**II**), which is the actual species that undergoes cyclization with strong base. Why is it necessary to convert the hydroxyl group into a hydrogen sulfate group prior to base treatment?

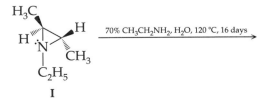

**8-102.** Aziridines are relatively reactive systems and undergo nucleophilic ring opening with the accompanying release of ring strain. Give a Fischer projection drawing of the expected product(s) of the reaction of *N*-ethyl-(2*S*,3*S*)-*trans*-2,3-dimethylaziridine (**I**) in aqueous ethyl amine.

I

## BIBLIOGRAPHY

Cromwell, N. H.; Barker, N. G.; Wankel, R. A.; Vanderhorst, P. J.; Olson, F. W.; Anglin, J. H. *J. Am. Chem. Soc.* **1951**, *73*, 1044.

Cromwell, N. H.; Cahoy, R. P.; Franklin, W. E.; Mercer, G. D. *J. Am. Chem. Soc.* **1957**, *79*, 922.

Cromwell, N. H.; Cram, D. J. *J. Am. Chem. Soc.* **1943**, *65*, *301*.

Cromwell, N. H.; Hudson, G. V.; Wankel, R. A.; Vanderhorst, P. J. *J. Am. Chem. Soc.* **1953**, *75*, 5384.

Cromwell, N. H.; Mercer, G. D. *J. Am. Chem. Soc.* **1957**, *79*, 3819.

Do Minh, T.; Trozzolo, A. M. *J. Am. Chem. Soc.* **1972**, *94*, 4046.

Heine, H. W.; Hanzel, R. P. *J. Org. Chem.* **1969**, *34*, 171.

Heine, H. W.; Smith, III, A. B.; Bower, J. D. *J. Org. Chem.* **1968**, *33*, 1097.

Heine, H. W.; Weese, R. H.; Cooper, R. A.; Durbetaki, A. J. *J. Org. Chem.* **1967**, *32*, 2708.

Padwa, A.; Clough, S.; Glazer, E. *J. Am. Chem. Soc.* **1970**, *92*, 1778.

## Experiment [F4]   A Photochromic Imine: 2-exo-6-exo-2,4-diphenyl-6-(4-nitrophenyl)-1,3-diazabicyclo[3.1.0]hex-3-ene

Common name: *exo*-2,4-Diphenyl-6-(*trans*-4-nitrophenyl)-1,3-diazabicyclo-[3.1.0]hex-3-ene
CA number: [36799-57-2]
CA name as indexed: 1,3-Diazabicyclo[3.1.0]hex-3-ene, 6-(4-nitrophenyl)-2,4-diphenyl-, (2α,5β,6β)-

### Purpose

To complete the synthesis of the photochromic imine, which is incorporated into the diazabicyclo[3.1.0]hex-3-ene skeleton. To obtain a rare molecular system in which an aziridine ring is fused to another heterocyclic ring. To explore the exceedingly interesting photochromic properties of the target molecule. To employ microscale techniques during the conversion and isolation of this light-sensitive material.

### *Prior Reading*

| | |
|---|---|
| *Technique 4:* | Solvent Extraction |
| | Liquid–Liquid Extraction (see pp. 80–82). |
| *Technique 5:* | Crystallization |
| | Craig Tube Crystallization (see pp. 95–96). |

## REACTIONS

### The Photosensitive Compound

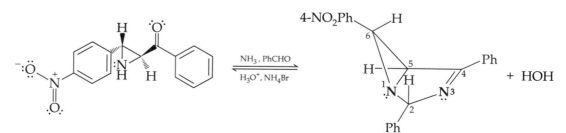

trans-2-(4-Nitrophenyl)-3-benzoylaziridine

2-*exo*-6-*exo*-2,4-Diphenyl-6-(4-nitrophenyl)-1,3-diazabicyclo[3.1.0]-hex-3-ene

## Photoproduct

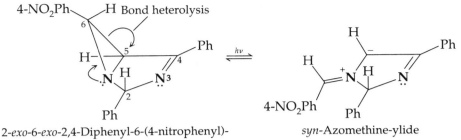

2-*exo*-6-*exo*-2,4-Diphenyl-6-(4-nitrophenyl)-
1,3-diazabicyclo[3.1.0.]hex-3-ene

*syn*-Azomethine-ylide

**DISCUSSION**

This reaction completes the synthesis of the photochromic target molecule, which possesses the 1,3-diazabicyclo[3.1.0]hex-3-ene ring system. The substituted aziridine ring system formed in Experiment [F3] is condensed with benzaldehyde and ammonia to yield this bicyclic system. The reaction may be viewed as proceeding under anhydrous conditions via an initial reaction between the aromatic aldehyde and the base, aided by the ammonium bromide catalyst, to generate an imine as indicated in the following scheme.

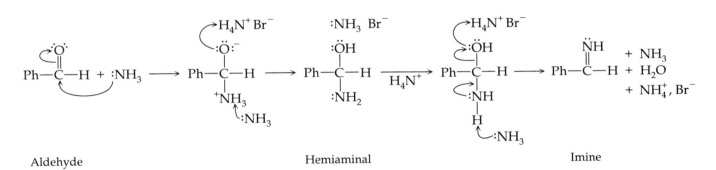

Aldehyde

Hemiaminal

Imine

The imine is subsequently attacked by the aziridine nucleophile to yield an aminal, again catalyzed by the ammonium bromide as shown in the following scheme.

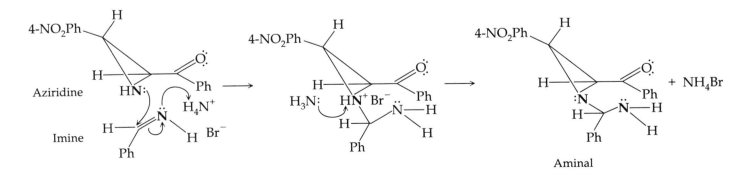

Finally, the aminal undergoes an internal condensation involving ring closure by nucleophilic attack of the primary amine group on the carbon of the carbonyl followed by dehydration of the hemiaminal to yield the diazabicyclo[3.1.0]hex-3-ene ring system, again catalyzed by ammonium bromide. A reasonable scheme is shown below.

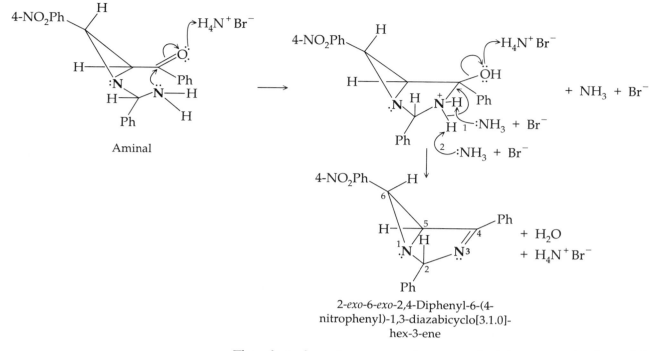

2-exo-6-exo-2,4-Diphenyl-6-(4-nitrophenyl)-1,3-diazabicyclo[3.1.0]-hex-3-ene

The photochromic compound appears to interconvert in the solid state to a bright blue azomethine–ylide (**I**) by an electrocyclic ring cleavage. An *ylide* is a neutral species whose Lewis structure contains opposite charges on adjacent atoms. The atoms involved are carbon, and an element from either Group 5A or 6A of the periodic table, such as N, P, or S. Of considerable interest is the fact that the photochemical process has been shown to give exclusively the syn isomer.

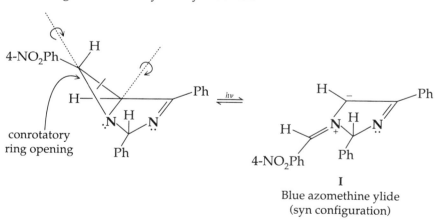

**I**
Blue azomethine ylide
(syn configuration)

This result is consistent with a *conrotatory ring opening* involving a symmetry-allowed concerted transformation in the *ground state*. As orbital symmetry requires a *disrotatory ring opening* from an *excited state* in these aziridines, the photochemically induced formation of the *syn*-azomethine–ylide (which is isoelectronic with the allyl anion) has been proposed to proceed via a dark reaction in which *electronically excited* states internally convert to *vibrationally excited* ground states. Indeed, evidence supporting this mechanism comes from the thermochromic behavior (a *ground-state* process) of the close relative 6-exo-2-dimethyl-4-diphenyl-6-(4-nitrophenyl)-1,3-diazabicyclo[3.1.0]-hex-3-ene (**II**), which when heated to 150 °C turns the same bright blue color (**III**).

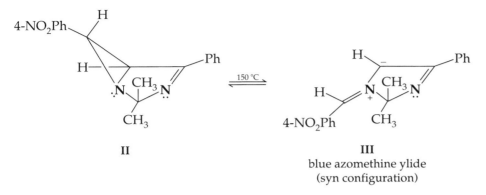

II

III
blue azomethine ylide
(syn configuration)

Exposure to light from tungsten lamps will not photochemically induce the ring-opening step. Most laboratories, however, are illuminated by fluorescent lights that emit small amounts of short wavelength radiation at the edge of the ultraviolet region (a wavelength long enough to not be absorbed by Pyrex glass), which will initiate the photochemical reaction. It is important to recognize that the azomethine–ylide photoproduct is a highly reactive species. Thus, if it is formed in solution where it can easily interact with other species, it rapidly decays to various byproducts, generally turning the solution yellow (not blue). It appears that if the ylide is produced in an environment that isolates it from other molecules the system is stable and, given enough time, it will slowly revert to the diazabicyclic starting material.

The crystalline state is the ideal solution to this problem. Thus, when the nearly colorless, solid imine is irradiated, it turns bright blue. The neighbors to any ylide in the crystal can be either the ylide itself or the diazabicyclo starting material. If colored material is then placed in the dark it will slowly revert to the colorless form once again. These molecular systems have been successfully cycled between the colorless and colored states many hundreds of times with little degradation of the crystalline material. Obviously, it is exceedingly important that (a) you do not expose the imine to light when it is in solution during recrystallization (or at least that contact with light is kept to a minimum; working in a red-light darkroom would be ideal conditions), and (b), recrystallizations should be carried out as quickly as possible, but obviously with care. Remember, you have a lot of time invested in this product, so work as quickly as is consistent with avoiding a costly spill.

---

Estimated time to complete the experiment: 2.5 hours.

## EXPERIMENTAL PROCEDURE

### Physical Properties of Reactants

| Compound | MW | Amount | mmol | mp(°C) | bp(°C) |
|---|---|---|---|---|---|
| *trans*-2-(4-Nitrophenyl)-3-benzoylaziridine | 268.3 | 40 mg | 0.15 | 142–143 | |
| Ethanol, absolute | | 800 mL | | | 78.5 |
| Benzaldehyde | 106.13 | 140 mL | 1.37 | | 178 |
| Ammonium bromide | 97.95 | 14 mg | 0.14 | 452 sub | |
| Ammonia | 17.03 | Excess | | | −33 |

## Reagents and Equipment

Weigh and place 40 mg (0.15 mmol) of the *trans*-2-(4-nitrophenyl)-3-benzoylaziridine synthesized in Experiment [F3] in a 15-mL screw-capped centrifuge tube containing a magnetic spin vane. Dissolve the aziridine (20 $\mu$L of EtOH/mg of aziridine) and benzaldehyde (140 $\mu$L, 1.37 mmol) in 800 $\mu$L of absolute ethanol. To this solution add 14 mg (0.14 mmol) of ammonium bromide (0.35 g of NH$_4$Br/g of aziridine). Stir this mixture for 1 min and saturate the system with anhydrous ammonia (NH$_3$ [**HOOD**]) (see your instructor for directions on this addition).

HOOD

> **CAUTION:  Anhydrous ammonia is a dangerous substance, particularly under pressure. The addition of this material to the reaction must be carried out in the *hood* under the *direct* supervision of the laboratory instructor.**

## Reaction Conditions

Gently bubble the ammonia gas through the reaction mixture until the system is saturated (the reaction mixture cools off, ~5 min). Tightly cap the tube (Teflon liner), wrap it in aluminum foil, and stir the mixture for a minimum of 24 hours. As in Experiment [F3], the reaction takes place at room temperature over a period of several hours. You can safely store the sealed reaction tube in your locker, protected from light with aluminum foil, for up to 1 week if necessary.

## Isolation of Product

After the 24-hour period, remove the supernatant liquid by centrifuging the tube and then transferring the liquid by means of a 9-in. Pasteur filter pipet to a Craig tube. Dry the remaining crystals in a stream of nitrogen gas. Remove a small sample of the solid material from the centrifuge tube on a glass rod or spatula and expose this material to direct sunlight or fluorescent light. If the solid material slowly turns blue, you have successfully synthesized the photochromic product. Concentration of the solution in the Craig tube may yield further quantities of the azomethine product. Weigh the tube and determine your percentage crude yield.

## Purification and Characterization

Recrystallize the crude product in the dark or red light (best) or with the laboratory lights off from hot 95% ethanol. Dissolution may require as much as 10 mL of solvent and is best carried out in the centrifuge tube. After cooling the centrifuge tube in ice and scratching the sides with a glass rod to induce crystallization, centrifuge the system, and remove the mother liquors with a Pasteur filter pipet. Dry the white (or near) crystals (should be white if the recrystallization is done without fluorescent lights or sunlight) using a stream of nitrogen gas.

**NOTE.**  *The photoproduct is an azomethine–ylide that is highly reactive. It is wise, therefore, to reduce the time that the material is in solution to a minimum, and while in solution to protect the contents of the capped centrifuge tube from light as much as possible. Once in the crystalline state, the large majority of the photochromic reactions takes place in an environment in which the azomethine–ylide is* *protected from further reaction and thus, given the opportunity (in the dark), the ylide will slowly recycle back to the diazabicyclic precursor with little loss.*

After removal of the mother liquors and drying of the purified photochromic product, determine an evacuated melting point. (The diazabicyclic intermediate substance is oxidatively sensitive and will decompose during atmospheric melting point measurements.) Be alert to color changes during the melting point determination.

Compare your results with the literature value of 169–172 °C. Obtain an infrared spectrum (KBr disk prepared in the dark or red light) of the white form of the product. Compare your spectral data to that of the reference standard shown in Figure 8.9. Then expose the disk to a bright fluorescent light for 1 min. and redetermine the infrared spectrum. What does a comparison of the two experimentally derived spectra tell you about this photochemically induced reaction?

Store the remainder of your azomethine–ylide in a clean, sealed vial, flushed with $N_2$, and protected from light. By your next laboratory period this material should be colorless.

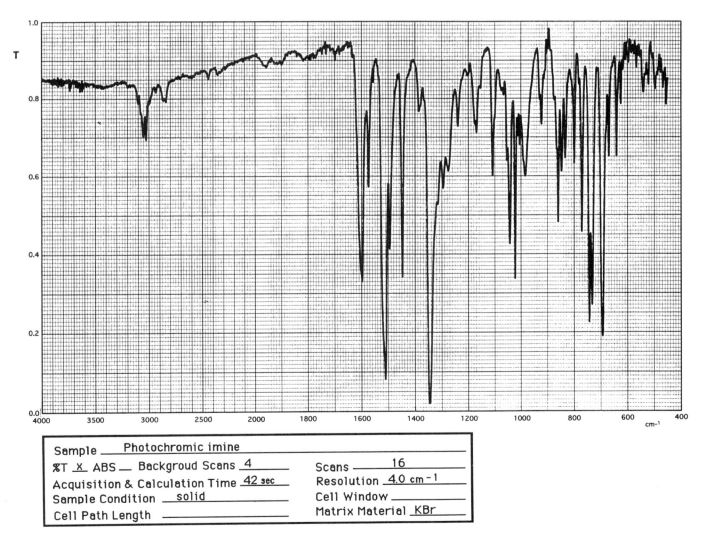

**FIGURE 8.9** IR spectrum: Photochromic imine.

## QUESTIONS

**8-103.** The heavily halogenated napthalenones were synthesized well over 100 years ago, but the structural details were sorted out less than 20 years ago. The compound, 2,3,4,4-tetrachloro-1-(4*H*)-naphthalenone (**I**), was discussed in the introduction to this sequence as an example of a substance that gains its photochromic activity by *bond homolysis*. In the synthesis of **I** two tetrachloro isomers were originally isolated. Compound **I** has photochromic activity while the second isomer (**II**) is simply a yellow colored material. The structure of **II** has been determined to be 2,2,3,4-tetrachloro-1-(2*H*)-naphthalenone. A key piece of physical evidence that allowed the assignment of the structures was the infrared frequencies of the carbonyl groups. These frequencies were found to be 1701 and 1675 cm$^{-1}$. Which wavenumber value belongs with which structure? Explain your reasoning.

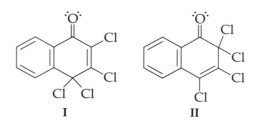

**8-104.** Explain why the neutral aziridine ring system is isoelectronic with the cyclopropyl carbanion.

**8-105.** As shown in Question 8-104, the aziridine ring system is isoelectronic with the cyclopropyl anion. Based on the theory of electrocyclic reactions these *even-number* $\pi$-electron-pair systems would be expected to undergo photochemical ring-cleavage in *disrotatory* fashion. Huisgen et al. have shown that in dimethyl 1-(4-methoxy-phenyl)aziridine-*trans*-2,3-dicarboxylate (**I**), upon photochemical excitation the ring opens to a 1,3-dipolar azomethine–ylide in a *disrotatory* cleavage that can be trapped by the addition of dimethyl acetylenedicarboxylate, which acts as a dipolarophile.

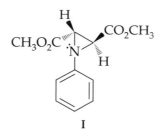

a. Give a perspective drawing of the expected adduct.
b. Give a perspective drawing of the expected adduct that would be formed from the blue azomethine ylide formed from our photochromic imine.
c. Is the relative stereochemistry the same in the two adducts in *a* and *b*? Explain.

**8-106.** The dramatic change in color that occurs when the diazabicyclo-[3.1.0]hex-3-ene ring system isomerizes on exposure to long wavelength UV radiation is related to what structural or electronic changes (or both) in the photochemically induced system?

**8-107.** Discuss the following:
  a. What role does ammonium bromide play in the conversion of the trans-substituted aziridine to the diazabicyclo[3.1.0]hex-3-ene derivative?
  b. Why is this particular bromide salt employed?

---

**BIBLIOGRAPHY**

DoMinh, T.; Trozzolo, A. M. *J. Am. Chem. Soc.* **1972,** *94,* 4046.

Heine, H. W.; Weese, R. H.; Cooper, R. A.; Durbetaki, A. J. *J. Org. Chem.* **1967,** *32,* 2708.

Hermann, H.; Huisgen, R.; Mäder, H. *J. Am. Chem.* **1971,** *93,* 1779.

Huisgen, R.; Scheer, S.; Huber, H. *J. Am. Chem. Soc.* **1967,** *89,* 1753.

Marckwald, W. *Z. Phys. Chem. (Leipzig)* **1899,** *30,* 140.

Padwa, A.; Clough, S; Glazer, E. *J. Am. Chem. Soc.* **1970,** *92,* 1778. *ibid.,* **1970,** *92,* 6997.

Padwa, A; Glazer, E. *J. Am. Chem. Soc.* **1972,** *94,* 7788.

Turner, A. B.; Heine, H. W.; Irving, J.; Bush, J. B. *J. Am. Chem. Soc.* **1965,** *87,* 1050.

Ullman, E. F.; Henderson, W. A. *J. Am. Chem. Soc.* **1964,** *86,* 5050.

Woodward, R. B.; Hoffmann, R. *Angew. Chem. Int. Ed. Engl.* **1969,** *8,* 781.

Zincke, T.; and Kegel, O. *Berichte* **1888,** *21,* 1030.

# 9

# Spectroscopic Identification of Organic Compounds

**C₉H₁₂, Triasterene** (*L. aster*, star)
Musso and Biethan (1964).

## INTRODUCTION TO INFRARED SPECTROSCOPY

The wavelike character of electromagnetic radiation can be expressed in terms of velocity $v$, frequency $\nu$, and wavelength $\lambda$ of sinusoidally oscillating electric and magnetic vectors traveling through space (Fig. 9.1). Frequency is defined as the number of waves passing a reference point per unit time, usually expressed as cycles per second ($s^{-1}$) or hertz (Hz). The velocity of the wave, therefore, equals the product of frequency and wavelength.

$$v = \nu\lambda$$

If the wavelength (the distance between each wave maxima or alternate nodes) is measured in centimeters, $v$ is expressed in centimeters per second (cm/s). For radiation traveling in a vacuum, $v$ becomes a constant, $c$ ($c \sim 3 \times 10^{10}$ cm/s), for all wavelengths. When electromagnetic radiation traverses other media, however, the velocity changes. The ratio of the speed in a vacuum, $c$, to the matrix velocity, $v$, is termed the *refractive index*, $n$, of the material.

$$n = c/v$$

Since $n$ is frequency dependent, the frequency at which the refractive index is measured must be specified. Frequency, however, has been shown to be independent of the medium and, therefore, remains constant. Wavelength thus varies inversely with $n$.

$$\lambda = c/n\nu$$

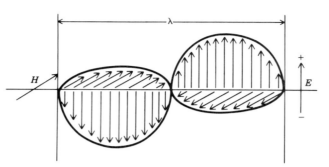

**FIGURE 9.1** Electromagnetic wave. *H* magnetic field, *E* electric field, $\lambda$ wavelength.

Since the velocity of electromagnetic radiation in a vacuum is normally greater than that in any other medium, $n$ will generally be greater than 1 at all frequencies. Thus, the wavelength must become shorter for a particular frequency when measured in any matrix.

Frequency can be considered to be a more fundamental property of radiation because it is independent of the medium. This property also requires that the energy $E$ associated with the radiation be matrix independent because $E$ is directly proportional to frequency by

$$E = h\nu$$

where $E$ equals the energy of a photon, which is related to frequency $\nu$ by Planck's constant ($h$) ($6.6 \times 10^{-27}$ erg s or $6.6 \times 10^{-34}$ J s).

The vibrational states present in molecules can be excited by absorption of photons. The nuclear masses and bond force constants determine the separation of these states and, therefore, the energies of the photons involved in the absorption process. The corresponding radiation frequencies fall predominantly in the IR region ($10^{14}$–$10^{12}$ Hz) of the electromagnetic spectrum.

The IR spectrum currently is measured in *wavenumbers* $\bar{\nu}$, which are units proportional to frequency and energy. The wavenumber is defined as

$$\bar{\nu} = \frac{\nu}{c} = \frac{E}{hc}$$

and as

$$\nu = \frac{c}{n\lambda} \qquad \text{then in air} \qquad \bar{\nu} = \frac{\sim 1}{\lambda}$$

The wavenumber, as expressed in units of reciprocal centimeters ($cm^{-1}$) (the number of waves per centimeter), offers several advantages.

**1.** Wavenumbers are directly proportional to frequency and are expressed in much more convenient numbers (in this region of the spectrum), 5000–500 $cm^{-1}$.

**2.** As shown above, wavenumbers are easily converted to wavelength values. The reciprocal of $\bar{\nu}$ and conversion of centimeters to wavelength units are all that is required (this is particularly handy because much early IR data were recorded linearly in wavelength). The wavelength unit employed in most of these spectra was the micron, $\mu$. The micron has been replaced by a unit expressed in meters, the micrometer, $\mu$m ($1\ \mu$m $= 1 \times 10^{-6}$ m).

**3.** Because the wavenumber is directly proportional to frequency and energy, the use of wavenumbers allows spectra to be displayed linear in energy, which is a distinct aid in sorting out related vibrational transitions.

**NOTE.** *It should also be pointed out that a reciprocal centimeter is not a unit of frequency. Wavenumbers are only proportional to frequency. Thus, it is not correct to refer to a vibrational absorption band as having a frequency of 3000 cm⁻¹ or to say that the vibration of the C—H bond possesses a frequency of 3000 cm⁻¹. The C—H oscillator, however, can be said to absorb radiation with an energy of 3000 cm⁻¹, or the C—H bond can be said to vibrate with a frequency of 9 × 10¹³ Hz.*

## MOLECULAR ENERGY

The total molecular energy $W$ may be expressed as the sum of the molecular translational, rotational, vibrational, and electronic energies:

$$W_{mol} = W_{trans} + W_{rot} + W_{vib} + W_{elec}$$

In this approximation, $W_{vib}$ is assumed to be independent of the other types of molecular energy. Translational and rotational motion involve much smaller energies, having little influence on the spectra under observation. The parameter $W_{elec}$ is the energy of the electrons. The energies of these latter transitions are very much larger than vibrational spacings, and their energy changes fall outside the IR. As a result, the IR region is active mainly to vibrational energy changes of ground electronic state molecules.

## MOLECULAR VIBRATIONS

Molecules can be characterized as being in constant vibrational motion. If we are to describe this motion for nuclei of a polyatomic system, we can utilize the Cartesian coordinates $x_m$, $y_m$, $z_m$ for a nucleus of mass $m$ referred to a fixed coordinate system. Then, for $n$ nuclei, we would generate $3n$ coordinates ($3n$ **degrees of freedom**) to describe the motion of all the atoms. Three of these coordinates, however, may be used to locate the center of mass of the system in space. These three coordinates define the translation of the entire system through space. Because translational energies have a small impact on vibrational spectra, the three coordinates of the center of mass can be dropped from the total required to determine the vibrational degrees of freedom. Therefore, $3n - 3$ coordinates are sufficient to determine the positions of the $n$ nuclei with respect to the center of mass. However, the molecular system is still free to rotate about the center of mass. For nonlinear molecules there are three additional coordinates that are required to fully describe rotational motion about the center of mass. For linear molecules only two coordinates are necessary to define rotation, as all the nuclei lie along one of the principal axes and are considered to be point groups. Thus, for nonlinear molecules $3n - 6$ coordinates fully define the vibrational motion of the nuclei. These coordinates are often referred to as the **vibrational degrees of freedom**. In linear systems one rotational degree of freedom can be considered to have been transformed into a vibrational degree of freedom ($3n - 5$).

*The number of vibrational degrees of freedom is directly related to the number of fundamental vibrational frequencies possessed by the molecular system.* These fundamental frequencies are often referred to as the **normal modes** of vibration.

To get a feel for the function of the normal modes in the vibrational pattern of a molecular system, let us consider a very simple arrangement of a single nucleus vibrating in two dimensions.[1] The nucleus of mass $m$ is

---

[1] Herzberg, G. *Molecular Spectra and Molecular Structure;* Van Nostrand: New York, 1945, Vol. 2, p. 62.

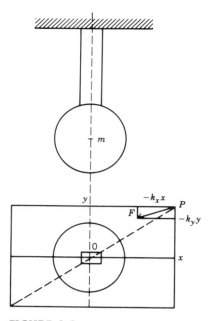

**FIGURE 9.2**   Harmonic vibration in two dimensions.

held by a rigid but elastic rectangular bar or rod. The mass $m$ can vibrate only in the plane perpendicular to the bar (Fig. 9.2). If the nucleus is displaced along the $x$ axis and then released, the system will oscillate with simple harmonic motion at a frequency given by

$$\nu_x = \frac{1}{2\pi} \sqrt{\frac{k_x}{m}}$$

where $k_x$ is the force constant of the bar in the $x$ direction where the restoring force $F = -k_x X$ for displacement $X$.

If displacement and release are carried out in the $y$ direction, a similar type of oscillation will occur with a frequency given by

$$\nu_y = \frac{1}{2\pi} \sqrt{\frac{k_y}{m}}$$

where $k_y$ is the force constant in the $y$ direction.

The frequency $\nu_x$ that results from displacement in the $x$ direction is different from the natural frequency $\nu_y$, which results from displacement in the $y$ direction, because the rectangular bar will possess different force constants $k_x$ and $k_y$. If the rectangular bar is replaced with a bar of square cross section, then the force constants that result from displacement in the $x$ and $y$ directions are equal. The frequencies of $x$ and $y$ motion also will be equal. In this situation the bar of rectangular dimensions can be referred to as having **degenerated** to a bar of square dimensions.

We now consider the displacement and release of the nucleus in a direction not along a principal axis; for example, the corner position $P$. Now on release, the motion performed by the nucleus is no longer simple harmonic. The restoring force $F$ will have components $-k_x X$ and $-k_y Y$, which are unequal and not directed toward the origin.

The complex motion of the nucleus in the $x$–$y$ plane, however, will still contain components that are simple harmonic in nature. The motion of the nucleus can be represented as the sum of these "normal modes" of vibration, which are perpendicular to each other. Thus, the position of the nucleus at any point in time after release can be expressed by the two coordinates,

$$x = x_0 \cos 2\pi\nu_x t$$
$$y = y_0 \cos 2\pi\nu_y t$$

where $x_0$ and $y_0$ are the coordinates of the initial position $P$, and $t$ is the time lapse from release.

The complicated pattern of motion performed by the nucleus on release from position $P$ is termed Lissajous motion. This type of motion is the superposition of two simple harmonic motions of differing frequency that are normal to each other. These are termed the *normal modes* or fundamental frequencies of the Lissajous motion of the nucleus of mass $m$. The $x$ and $y$ coordinates, thus, become the "normal coordinates."

In the case of a diatomic molecule the frequency (expressed as wavenumbers) is given by

$$\tilde{\nu} = \frac{1}{2\pi c} \sqrt{\frac{k}{\mu}}$$

where $k$ is the force constant, $m_1$ and $m_2$ are the atomic masses, and $\mu$ is the reduced mass.

$$\frac{1}{\mu} = \frac{1}{m_1} + \frac{1}{m_2} \quad \text{or} \quad \mu = \frac{m_1 m_2}{m_1 + m_2}$$

Nature has been kind in distributing vibrational energy in molecules! The vibrational (vib) states associated with a particular normal mode are not influenced, to a first approximation, by the energies of adjacent states. The portion of $W_{vib}$ contributed by a particular normal mode $\nu_i$ is given by

## QUANTIZED VIBRATIONAL ENERGY

$$W_{\text{vib}} = \left(v_i + \frac{1}{2}\right) h\nu_i$$

where $v_i$ is the vibrational quantum number for the normal mode and takes the values 0, 1, 2, . . . .

Each normal mode will possess a similar energy-quantum number relationship, and the total vibrational energy scheme can be obtained by summing over all 3$n$-6 fundamental vibrations:

$$W_{\text{vib}} = \sum_{i=1}^{3n-6} \left(v_i + \frac{1}{2}\right) h\nu_i$$

The characteristic energy level pattern for a normal mode as determined by the quantum relationship dictates that the level spacings will be equal with a value of $h\nu_i$. In the upper states, however, the potential energy curve begins to depart from the harmonic values and the vibration becomes *anharmonic*. In general, anharmonicity results in lower energy transitions or a contraction in the level spacing (Fig. 9.3). In rare instances the potential energy well develops steeper sides (quartic terms become important), and the spacing actually becomes greater at higher levels (*negative anharmonicity*). We will see a few of the more well-known departures of this type.

In addition to the equal spacing of the energy levels associated with each normal mode, the quantization of the vibrational energy requires that the lowest or zero vibrational level ($v_0$) does not occur at zero energy, but at $\frac{1}{2}h\nu$. Thus, the molecule retains, even at absolute zero, some small amount of vibrational energy. This energy is termed the "zero-point" energy, and its origin lies with the Heisenberg uncertainty principle.

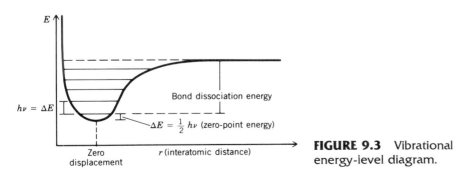

**FIGURE 9.3** Vibrational energy-level diagram.

## SELECTION RULES

The range of vibrational energy level transitions associated with the fundamental frequencies runs from somewhat under 5000 to approximately 100 $cm^{-1}$. Thus, the study of the absorption spectra of the normal modes centers on the infrared region of the spectrum. Although most molecules possess a large set of normal modes, the equal spacing of the vibrational levels and the operation of molecular selection rules greatly simplify what would otherwise be a very complex absorption pattern.

For infrared spectra the selection rules define the changes in the vibrational and rotational quantum numbers. The changes for the $3n - 6$ vibrational quantum numbers follow:

1. Only one quantum number can change during a transition.
2. The change $\Delta v$ is restricted to $+1$, $-1$.
3. For certain vibrations, $\Delta v$ must always be zero.

In a very large percentage of cases, an absorbed photon excites only a single normal mode (this rule also holds true for emission spectra). Thus, each frequency observed in the spectrum corresponds to a normal mode present in the molecule. The maximum number of frequencies observed corresponds directly to the $3n - 6$ (or $3n - 5$ for linear molecules) vibrational degrees of freedom.

In normal modes where $\Delta v = 0$, an incomplete set of frequencies will be observed. The symmetry elements present in the molecule largely determine whether $\Delta v = 0$ for a particular normal mode. *This results from the further requirement that there be a change in molecular dipole moment during the vibration for the absorption of a photon to occur.*

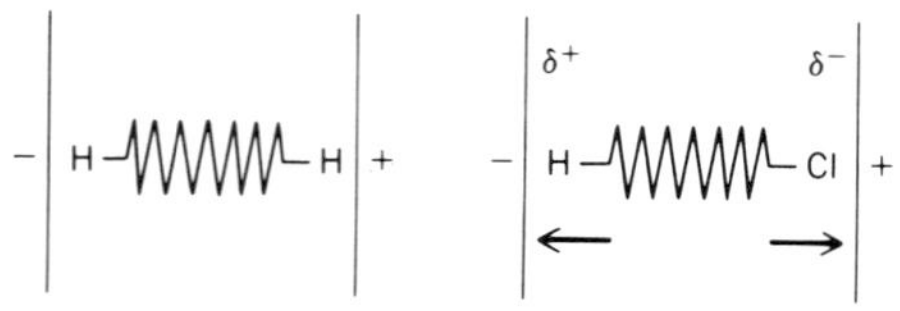

We can see why a variation in the magnitude of the dipole moment is essential to the absorption process by considering hydrogen and hydrogen chloride molecules placed between condenser plates (Fig. 9.4).

In the case of HCl, a permanent dipole moment exists along the molecular axis, with the negative pole closer to the chlorine atom and the positive pole closer to the hydrogen atom. When placed between condenser plates, as shown in the figure, the molecule will experience attractive forces at both ends, which will exert a stretching action. If the charge on the condenser plates is quickly reversed, the molecule will then experience repulsive forces at each end and be compressed. If it were possible to alternate the charge on the plates fast enough to match the natural frequency of the HCl molecule, the system would resonate. In the resonance or "tuned" condition, the oscillator will absorb energy from the condenser and expand its vibrational displacements. (If enough energy were absorbed, the molecule would dissociate.) The frequency remains constant, but the amplitude of the vibration increases (much the same way as the input of periodic energy with the correct phase into a child's swing increases the amplitude of the swing, but leaves the frequency constant). Within the band of infrared radiation (5000–100 $cm^{-1}$), oscillating electric fields of the radiation will act on molecules in a fashion similar to alternating condenser fields. The frequencies in this spectral region correspond to the natural vibrational frequencies present in the molecular systems. Thus, at the particular radiation frequency that matches the vibrational frequency of the HCl molecule, resonance will occur and the photon of corresponding energy will be absorbed as the molecule moves to the next higher vibrational state. In the case of the hydrogen molecule, there is no vibrating electric dipole present because of the symmetry of the system. Thus, no interaction with the oscillating electric vector of the radiation at the natural frequency can occur (this would correspond to no interaction with the

**FIGURE 9.4**   $H_2$ and HCl oscillators.

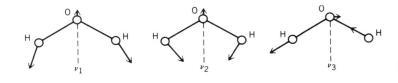

**FIGURE 9.5** Normal modes of water.

condenser plate fields). In this case the normal mode for the hydrogen molecule is not observed in the infrared spectrum, and the selection rule $\Delta v = 0$ applies.

Let us examine two simple examples to illustrate the preceding discussion.

## The Case of Water

Water is a nonlinear triatomic molecule. It has, therefore, three normal modes of vibration or fundamental frequencies ($3n - 6$, where $n = 3$). The displacements of the normal modes can be derived in much the same fashion as the two-dimensional case and are as shown in Figure 9.5.

All three vibrations are active in the infrared, and three absorption bands are observed. The arrows represent the relative atomic displacements involved in one phase of the vibration. The atoms all move in phase in simple harmonic motion in each fundamental mode. The high-frequency vibration $\nu_3$ at 3756 cm$^{-1}$ involves predominantly hydrogen motion, with one bond contracting while the other is stretching. This vibration is designated as the "antisymmetric stretching" mode. (Modes of this type are often incorrectly referred to as *asymmetric* vibrations. The vibration is not asymmetric, without symmetry, but a vibration of antisymmetric character, that is, opposite symmetry or opposed symmetry.) The other stretching vibration designated $\nu_1$ at 3652 cm$^{-1}$ involves in-phase and identical displacements of the two O—H bonds. It is termed the *symmetric stretching* mode. Finally, the low-frequency mode $\nu_2$ at 1595 cm$^{-1}$ corresponds to a bending of the molecule about the H—O—H bond angle. It occurs at lower frequencies than do the two stretching modes, as it takes less energy to bend a bond than it does to stretch one. Thus, the force constant $k_{bend}$ is considerably smaller than $k_{stretch}$. This bending vibration is often termed the *scissoring vibration* because the symmetric bending motion involved is similar to the action of scissors. The complex molecular vibrational pattern of the water molecule can be resolved into three simple harmonic components or normal modes that correspond directly to three absorption bands in the infrared spectrum. The atomic displacements of these normal modes can be used to characterize the type of fundamental vibration giving rise to the infrared absorption. In the water molecule we have two O—H stretching vibrations and one O—H bending mode. These modes represent the three vibrational degrees of freedom present in the water molecule.

## The Case of Carbon Dioxide

Carbon dioxide is a linear triatomic molecule. It possesses four vibrational degrees of freedom (one more than water). The displacements of the normal modes are given in Figure 9.6.

In the case of carbon dioxide only two absorption bands are observed in the infrared spectrum even though the molecule has four vibrational degrees of freedom. The high-frequency antisymmetric stretching mode $\nu_3$ occurs at 2350 cm$^{-1}$. The form of this mode is close to that of the antisymmetric stretching vibration found in water. The symmetric stretching vibration $\nu_1$ of carbon dioxide is similar to its counterpart in water; however, this

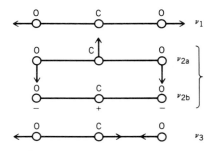

**FIGURE 9.6** Normal modes of carbon dioxide.

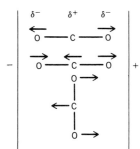

**FIGURE 9.7** Carbon dioxide oscillator.

vibration does not give rise to a change of dipole moment during the vibration. The arrangement of the atoms in the carbon dioxide molecule places a center of symmetry on the carbon atom. This symmetry element remains intact during the symmetric stretching vibration, and as a consequence no change in the dipole moment occurs. Thus, $\Delta v = 0$ and no absorption occurs at this frequency in the infrared. The normal mode still exists in the molecule, but it is infrared inactive. This fundamental has been located near 1334 cm$^{-1}$ in the Raman effect. (The Raman effect is another spectroscopic technique used to observe vibrational energy level transitions. It does not depend on a change in dipole moment.) There are two bending frequencies present in carbon dioxide $\nu_{2a}$ and $\nu_{2b}$. These two modes are identical except that one is rotated by 90° with respect to the other. The bending motion does produce a change in dipole moment, and as a result it is infrared active. Since the two modes have identical frequencies, however, only one absorption band will be observed at 667 cm$^{-1}$. Carbon dioxide can bend in two mutually perpendicular planes, and therefore two bending vibrations are required to fully characterize the vibrational motion of the molecule. The symmetry of the system dictates that the modes be identical. Vibrations possessing these characteristics are termed **degenerate vibrations** (remember the *degeneration* of the rectangular bar to a square that gave rise to two mutually perpendicular and identical modes). Thus, although carbon dioxide possesses four vibrational degrees of freedom, the observation of two absorption bands in the infrared spectrum can be satisfactorily explained. Indeed, if we examine the bending vibration present in water, we see that water can bend only in a single plane, and "bending" the molecule out of that plane results in rotation. It is this rotational degree of freedom of water that is translated into a vibrational degree of freedom in carbon dioxide as the nonlinear three-atom system is converted into a linear molecule.

If we place the carbon dioxide molecule between the plates of a condenser (Fig. 9.7), we can see why some of the carbon dioxide vibrations are infrared inactive, and other vibrations are active. As carbon dioxide does not possess a permanent dipole moment, it is important to note that the antisymmetric stretching mode and the degenerate bending vibration will develop dipole moments during these vibrations as a result of the nuclear displacements involved. These dipole moments obviously undergo changes during the different phases of the vibrations, and therefore these modes fit the requirements of the vibrational selection rules for infrared activity. In the symmetric stretching vibration, no dipole moment is developed during the vibration, and we have a case similar to that of the hydrogen molecule. Thus, no absorption band is observed in the infrared spectrum of carbon dioxide that corresponds to the symmetric stretching mode.

## VIBRATIONAL COUPLING

### Coupled Oscillators

Water and carbon dioxide are valuable examples of simple systems in which mechanical coupling between two oscillators is amply demonstrated. If we consider water to be constructed of two O—H oscillators, the individual diatomic systems would be expected to have identical frequencies. When welded together, however, the vibrations of one O—H oscillator interfere with the vibrations of the other O—H oscillator. The coupled oscillators generate two new vibrations, one at higher frequency and one at lower frequency (much the same as the resonance that develops between two coupled identical pendulums can be considered to involve two frequencies, one higher and one lower than the natural frequency of the pendulum). The coupling interaction is frequency dependent. The closer the frequencies of the two oscillators, the stronger the interaction. When there are identical frequencies and a direct mechanical connection, the coupling effect will be maximized. Under the conditions of strong interaction, the form of the new vibrations can be quite different from that of the isolated oscillators. Coupling is also angle dependent. Oscillators normal to each other couple poorly, whereas colinear oscillators will undergo maximum coupling. [Note the wavenumber separation in water (bent system) of $104$ $cm^{-1}$, as compared with $1016$ $cm^{-1}$ for carbon dioxide (back-to-back $C{=}O$ oscillators)].

**Second-Order Coupling**

The selection rules break down occasionally, particularly in condensed phases, to give overtone bands ($\Delta v > 1$). These departures may result from anharmonicities. The overtone frequencies are usually somewhat less than double that of the fundamental mode. The drop in expected frequency results from the compression of upper levels on the potential energy curve (Fig. 9.3). The absorption bands that result from these transitions are usually very weak, as the mode is formally forbidden.

One of the most spectacular of the second-order events is Fermi resonance. When the first overtone ($\Delta v = 2$) of a fundamental possesses very nearly the same energy as the $\Delta v = 1$ level of another normal mode, an interaction may occur in which the two close-lying levels are split into two new levels, one higher and one lower in frequency than the original modes. As a result of this mixing, the overtone often undergoes a dramatic intensity gain at the expense of the fundamental. The resulting doublet may even possess components of approximately equal intensity. The intensity distribution is dependent to a large extent on the value of the original frequency match. The classic example of the effect is the symmetric stretching frequency (Raman active only, see above) in carbon dioxide, which should occur near $1334$ $cm^{-1}$ but which, in fact, exists as a doublet ($1388$ and $1286$ $cm^{-1}$). The perturbation was explained by Fermi as the interaction of the overtone of the bending fundamental at $667$ $cm^{-1}$ with the first exited state of the symmetric stretching mode. For Fermi resonance to occur, (a) the oscillators involved must be so arranged that the anharmonic terms can interact (mechanical interaction can occur), and in addition, (b) the modes must meet certain symmetry restrictions. The large majority of all complex organic substances are of such low symmetry that the latter condition usually can be assumed to have been met.

In a few cases the overtone of a fundamental, although weak, will occur with higher than usual intensity, and in a region uncluttered by other absorptions. These bands can be utilized as confirming evidence in making assignments of fundamentals. In even rarer instances the first overtone will be observed to occur at slightly higher than double the fundamental values. These systems are considered to possess "negative anharmonicity" (see quantized vibrational energy, p. 589).

Another second-order effect is the "sum tone" or combination band. Although forbidden in the harmonic approximation, occasionally there will be absorbed a photon of the appropriate energy to simultaneously excite two normal modes. Combination bands occur as weak absorption bands that possess frequencies near the sum of the two fundamentals. If sum tones occur in regions open to observation, occasionally they can be of importance in group frequency interpretations (see out-of-plane C—H bending modes on aromatic rings: see also Fig. 9.21).

# INTRODUCTION TO GROUP FREQUENCIES: Interpretation of Infrared Spectra

Studies of the vibrational spectra of thousands of molecules have revealed that many of the normal modes associated with particular atomic arrangements may be transferred from one molecule to another. These vibrational frequencies are associated with small groups of atoms that are essentially uncoupled from the rest of the molecule. The absorption bands that result from these modes, therefore, are characteristic of the small group of atoms regardless of the composition of other parts of the molecule. These vibrations are known as the *group frequencies,* and interpretation of infrared spectra of complex molecules based on group frequency assignments is an extremely powerful aid in the elucidation of molecular structure. Four factors that make significant contributions to the development of a good group frequency from a molecular vibration are discussed in the section on carbonyl group frequencies, see p. 606.

We will discuss group frequencies in the following sequence:

*Group Frequencies of the Hydrocarbons*

Alkanes
Alkenes
Alkynes
Arenes

*Factors Affecting the Carbonyl Frequencies*

Mass effects
Geometric effects
Electronic effects (inductive and conjugative effects)
Interaction effects

*Group Frequencies of the Functional Groups*

| | |
|---|---|
| Hydrocarbons | Amines |
| Alcohols | Nitriles |
| Aldehydes | Amides, primary |
| Ketones | Amides, secondary |
| Esters | Isocyanates |
| Acid halides | Thiols |
| Carboxylic acids | Halogens |
| Anhydrides | Phenyl |
| Ethers | |

*Strategies for Interpreting Infrared Spectra*

## GROUP FREQUENCIES OF THE HYDROCARBONS

### Characteristic Group Frequencies of Alkanes

The saturated hydrocarbons and the alkane section of mixed structures contain only C—C and C—H bonds. The fundamental modes derived from the hydrocarbon portion of these molecules, therefore, are limited to C—C and C—H stretching and bending vibrations. These two types of oscillators are good examples of structural units that give rise to both excellent and very poor group frequencies. We have established (see p. 590) that a change in dipole moment during the vibration is essential for the absorption process to occur in the infrared. The exact relationship between the vibrational displacements and the observed band intensities is rather complex. It is related to the slope of the curve of the variation of dipole moment with the normal coordinate at the equilibrium point. The intensity of the fundamental is proportional to the square of the derivative of the dipole moment with respect to the normal coordinates. The absorptivity of infrared bands, therefore, usually can be gauged from a rough estimate of the magnitude of the oscillating dipole moment.

The C—C *group* is an oscillator that, at best, will possess a very small dipole moment because of the symmetry inherent in the bond. Because of the low bond polarization, the absorption bands associated with this system can be expected to be quite weak and difficult to identify. In addition, the C—C oscillator, in most cases, will be directly connected to other C—C oscillators with similar or identical frequencies. In such an arrangement mechanical coupling effects are to be expected. This coupling will give rise to a very complex absorption pattern unique to a particular compound. Although these highly coupled vibrations have little value as group frequencies, they are the most powerful means of identifying organic materials by modern chemical instrumentation. The region of the infrared spectrum where these frequencies predominate, 1500–500 cm$^{-1}$, is often referred to as the "fingerprint region."

The C—H *oscillator* is at the other extreme from the C—C case. It gives rise to excellent group frequencies. The light terminal H atom, which is connected to a relatively massive carbon atom by a strong bond (large force constant), possesses a high natural frequency. Because of the separation of this frequency from other frequencies, the only coupling that can influence the oscillator is that of other C—H groups connected to the same carbon. A rule of thumb to remember is "coupling generates as many modes as there are coupled oscillators." The methyl group, CH$_3$—, has three stretching modes; the methylene group (—CH$_2$—), two stretching modes; and the methine group (—CH—), a single stretching mode.

Since there is very little mechanical coupling of C—H oscillators beyond the local carbon atom, the natural frequencies of the methyl, methylene, and methine groups remain relatively constant when these groups are transferred from compound to compound or from group to group within the same system. The dipole moment of the C—H bond is not large, but its derivative is sufficient to give rise to reasonably identifiable absorption bands. Because many compounds will contain several C—H bonds of similar character, they will also possess an equal number of nearly equivalent C—H frequencies. These C—H stretching modes overlap, often to produce the most intense collection of absorption bands observed in the infrared spectrum of a material.

The three *coupled C—H stretching modes of methyl groups* can be described in terms of two antisymmetric vibrations that are degenerate, or nearly degenerate (p. 592), depending on the symmetry of the system, plus a symmetric mode (Fig. 9.8). In most cases, the two antisymmetric methyl stretching modes, although not rigorously degenerate, will give rise to a pair of very close-lying bands that are seldom resolved. The antisymmetric methyl modes are the highest frequency vibrations of the purely $sp^3$ hybridized C—H bonds. These fundamentals occur near 2960 cm$^{-1}$. The methyl symmetric stretching mode is found close to 2870 cm$^{-1}$ (Fig. 9.9).

The two *coupled stretching modes of the methylene group* are very similar in displacement pattern to the fundamental vibrations of the water molecule (see Fig. 9.5). The higher frequency mode, as in water, is the antisymmetric stretch. This fundamental occurs near 2925 cm$^{-1}$ in hydrocarbons. The symmetric stretching mode, which is particularly sensitive to adjacent heteroatoms bearing lone-pair electrons, in saturated hydrocarbons has the lowest frequency of the $sp^3$ hybridized coupled C—H oscillators. The symmetric stretch occurs close to 2850 cm$^{-1}$ (see Fig. 9.9).

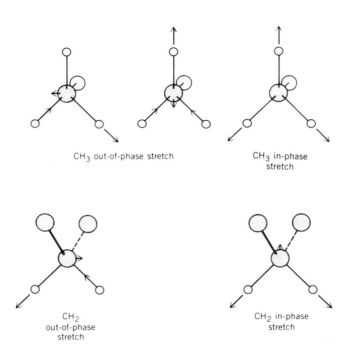

CH₃ out-of-phase stretch

CH₃ in-phase stretch

CH₂ out-of-phase stretch

CH₂ in-phase stretch

**FIGURE 9.8**   Stretching vibrations of methyl and methylene groups.

The *methine group* has a *single uncoupled mode*. As relatively few groups of this type are normally present in a structure, compared with methyl and methylene groups, this vibration gives rise to a weak fundamental, usually masked by the absorption of the other alkane groups. The absorption of the tertiary C—H bonds occurs in the 2900-wavenumber region.

With very few exceptions the $sp^3$ hybridized C—H bonds have their fundamental modes in the 3000–2800-wavenumber region.

Since the hydrogen atom is much lighter than other atoms, it undergoes most of the displacement. The mass term in the expression for frequency of a diatomic molecule is actually a reduced mass. In the case of hydrogen and carbon this is defined by

$$\frac{1}{\mu} = \frac{1}{m_H} + \frac{1}{m_C} \qquad \text{since } m_C \gg m_H, \ \mu \cong m_H$$

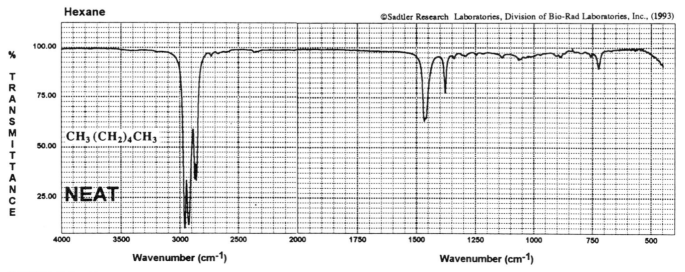

Hexane

©Sadtler Research  Laboratories, Division of Bio-Rad Laboratories, Inc., (1993)

$CH_3(CH_2)_4CH_3$

NEAT

% TRANSMITTANCE

Wavenumber (cm⁻¹)

Wavenumber (cm⁻¹)

**FIGURE 9.9**   IR spectrum: Hexane.

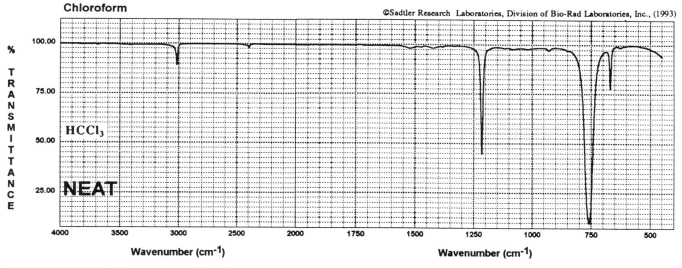

**FIGURE 9.10(a)** IR spectrum: Chloroform.

It is possible, therefore, to express to a very good first approximation the vibrational frequency of this system by a simple Hooke's law relationship, assuming an infinite mass for the carbon atom:

$$\tilde{\nu} = \frac{1}{2\pi c} \sqrt{\frac{k}{m_H}}$$

This expression makes it possible to predict frequency shifts on substitution by deuterium for hydrogen of $\tilde{\nu}_H/\tilde{\nu}_D = 1.41$. In practice these shifts are somewhat less than the theoretical values, usually falling in the range 1.32–1.38 (Fig. 9.10a and 9.10b).

If the $\tilde{\nu}_H/\tilde{\nu}_D$ ratio departs significantly from this range, the result can be taken as an indication that one or possibly both of the fundamentals are not behaving in harmonic fashion and that coupling is present. Chloroform is one of the rare exceptions to the 3000-wavenumber rule (see p. 596). The

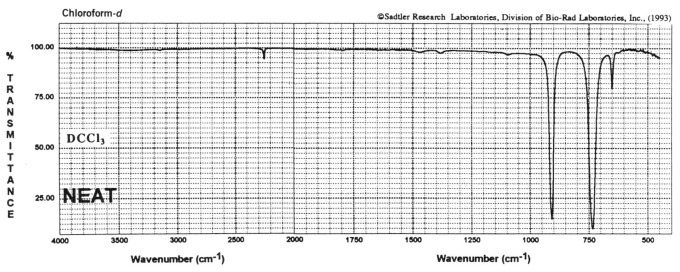

**FIGURE 9.10(b)** IR spectrum: Chloroform-*d*.

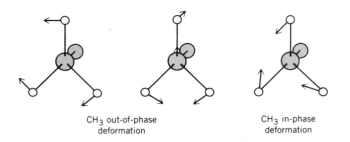

**FIGURE 9.11**    Bending vibrations of the methyl group.

single uncoupled C—H stretching mode occurs at 3022 cm$^{-1}$. The rise in frequency in this case results from extensive substitution by the strongly electronegative chlorine atoms.

The *bending frequencies of the C—H oscillator* will now be discussed. As in the case of stretching modes of the alkyl groups, the bending fundamentals are coupled only with those oscillators directly bonded to the carbon. Since there are three independent H—C—H bond angles, *there are three deformation or bending vibrations* associated with the *methyl group* (Fig. 9.11). As in the stretching fundamentals, two of these modes can be described in terms of antisymmetric degenerate or nearly degenerate, vibrations that occur near 1460 cm$^{-1}$. The third vibration is the symmetric bending mode (umbrella) close to 1375 cm$^{-1}$ (Figs. 9.9 and 9.11), which is easily identified.

The *methylene group* has a single H—C—H bond angle, and a *single deformation mode* (scissoring, Fig. 9.12) directly analogous to that of the water molecule. The symmetric bend of the —CH$_2$— group occurs near 1450 cm$^{-1}$ in hydrocarbons (Figs. 9.9 and 9.12).

*Three other bending modes* are available to the *methylene group* (Fig. 9.12). When the methylene group is fused into the molecule, three vibrational degrees of freedom develop, which are related to rotational motion of the isolated system, for example, the rotational motion of the water molecule (see p. 591). These fundamentals (wag, twist, and rock) are subject to significant coupling to adjacent methylene groups. The transitions associated with these modes are also rather weak in intensity. Thus, wag, twist, and rock are not useful group frequencies with the exception of a component of the rocking vibration (Fig. 9.12) in certain structures. In those molecules having four or more methylene groups in a row, the coupled mode corresponding to the *all-in-phase* rocking vibration develops a significant dipole moment change and a stable frequency. Thus, the in-phase rock that occurs near 720 cm$^{-1}$ (see Fig. 9.9) in the fingerprint region gives rise to an absorption band of sufficient intensity to allow for confident assignment. (Thus, we have all four bonds of the carbon associated with the methylene group involved in bending modes and four deformation vibrations are generated.)

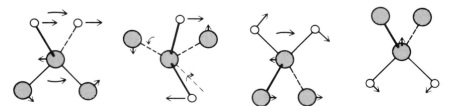

**FIGURE 9.12**    Bending vibrations of the methylene group.

The bending modes associated with the *methine hydrogen* are hard to identify in most hydrocarbons. (The in-plane bend of an isolated $sp^2$ C—H group can be particularly important, however, in certain heteroatom systems; see discussion of the aldehyde functional group.)

The C—H vibrational modes of the alkanes (or mixed compounds containing alkyl groups) that are characteristic and reliable group frequencies can be summarized as in Table 9.1.

**TABLE 9.1 Alkane Normal Modes**

| C—H Vibrational Modes | $\bar{\nu} \pm 10$ (cm$^{-1}$) |
| --- | --- |
| **Methyl Groups** | |
| Antisymmetric (degenerate) stretch | 2960 |
| Symmetric stretch | 2870 |
| Antisymmetric (degenerate) deformation | 1460 |
| Symmetric (umbrella) deformation | 1375 |
| **Methylene Groups** | |
| Antisymmetric stretch | 2925 |
| Symmetric stretch | 2850 |
| Symmetric deformation (scissor) | 1450 |
| Rocking mode (all-in-phase) | 720 |

Alkenes (olefins) possess the carbon–carbon double bond, C=C. The group frequencies of these molecules will be those of the alkanes for the saturated portion of the molecule, plus those modes contributed by the unsaturated group.

### C=C Stretching

The stretching fundamental of the C=C group is a useful group frequency. The increase in the force constant in going from the single bond to the double bond moves the frequency to sufficiently high values ($\bar{\nu}_{C=C} = 1616$ cm$^{-1}$, ethylene) to decouple this mode from adjacent C—C vibrations. Substitution of carbon for hydrogen on the double bond tends to raise the C=C stretching frequency, as the effective force constant has been shown to include an increased compression term (Fig. 9.13). It is possible to classify open-chain unsaturated systems into two groups as in Table 9.2.

The high-wavenumber set has quite weak bands unless conjugation occurs. Indeed, in the tetrasubstituted case, if the four groups are identical, the C=C band is formally forbidden in the infrared. When the alkyl groups are similar but not identical, the magnitude of the dipole moment is

**Characteristic Group Frequencies of Alkenes**

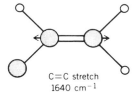

C=C stretch
1640 cm$^{-1}$

**FIGURE 9.13** Vinyl group double-bond stretching vibration.

**TABLE 9.2 Substitution Classification of C=C Stretching Frequencies**

| C=C Normal Modes | $\bar{\nu}$ (cm$^{-1}$) |
| --- | --- |
| trans-, tri-, tetrasubstituted | 1680–1665 |
| cis-, vinylidene- (terminal-1,1-), vinyl-substituted | 1660–1620 |

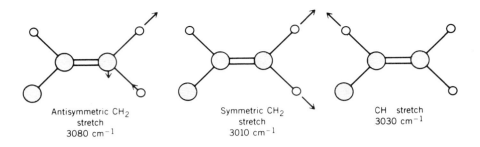

**FIGURE 9.14**   The C—H stretching vibrations of the vinyl group.

so small that it becomes difficult to detect an absorption band. In this situation the molecule is considered to have a "pseudocenter of symmetry." In the low-wavenumber set more intense absorption bands occur, with the average intensity falling in the medium-to-strong range.

### Alkene C—H

Several fundamental modes associated with the alkene C—H groups are good group frequencies.

The C—H stretching frequencies occur above 3000 cm$^{-1}$ and are localized in two regions defining three groups as follows: (1) If two C—H groups are present on an $sp^2$ carbon, two coupled (antisymmetric and symmetric) vibrations occur near 3080 and 3010 cm$^{-1}$. (2) If a single C—H group is attached to an $sp^2$ hybridized carbon, it gives rise to a single mode near 3030 cm$^{-1}$. (3) A vinyl group will have both sets of bands, but the lower wavenumber modes are seldom resolved from the saturated C—H stretching modes (Figs. 9.14 and 9.15).

The alkene C—H bending frequencies fall into two categories: (1) There are bending modes that occur in the plane of the double bond. These fundamentals are not useful group frequencies. (2) There are bending modes that occur out of the plane of the double bond. These fundamentals are useful group frequencies.

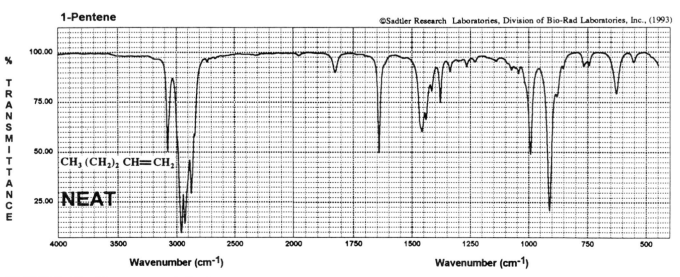

**FIGURE 9.15**   IR spectrum: 1-Pentene.

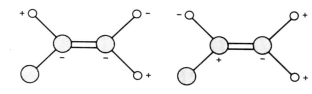

**FIGURE 9.16** Out-of-plane C—H bending vibrations of the vinyl group.

## Out-of-Plane Deformation Modes

**Vinyl Groups:** The vinyl group has three C—H bonds and therefore three out-of-plane bending modes. Two of these normal modes are good group frequencies. The fundamental with the two trans hydrogen atoms bending in-phase occurs near 990 cm$^{-1}$ (Fig. 9.16). A second vibration involving primarily the two hydrogen atoms attached to the terminal carbon, bending in-phase together (wag), is located close to 910 cm$^{-1}$ (Fig. 9.16). The absorption bands resulting from both of these fundamentals are strong and easily detected in the fingerprint region.

**Vinylidene Groups:** The 1,1-substituted system will have two bending frequencies. One of these bending modes is a good group frequency. When both hydrogen atoms wag out of the C=C plane together, a strong dipole moment change develops and gives rise to an intense absorption band near 890 cm$^{-1}$. This mode is related to the low-frequency mode found in the vinyl group at 910 cm$^{-1}$ (Figs. 9.16 and 9.17).

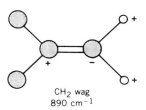

CH$_2$ wag
890 cm$^{-1}$

**FIGURE 9.17** Out-of-plane C—H bending vibration of the vinylidene group.

**Trans Alkenes:** A trans alkene has two out-of-plane bending modes, but again only one gives rise to a good group frequency. The mode that involves both hydrogen atoms moving in phase together occurs close to 965 cm$^{-1}$. This mode is directly related to the high-frequency mode of the vinyl group near 990 cm$^{-1}$ (Fig. 9.16).

**Cis Alkenes:** A cis-substituted C=C group does not possess a very good out-of-plane group frequency. The only mode with reasonable intensity involves the *in-phase bend* of the two hydrogen atoms. This fundamental is derived from a rotational type motion that couples to the rest of the system (Fig. 9.18). Therefore, the cis mode is not localized, but occurs in a broad region near 700 cm$^{-1}$.

**Trisubstituted Alkenes:** In the case of R$_2$C=CHR systems we have a single out-of-phase bending vibration that occurs near 820 cm$^{-1}$. The mode is uncoupled and gives rise to a medium-intensity band.

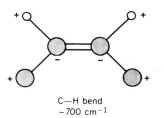

C—H bend
~700 cm$^{-1}$

**FIGURE 9.18** Out-of-plane C—H bending vibration of cis-substituted HC=CH groups.

**Tetrasubstituted Alkenes:** These groups have no C—H bending modes.

## Overtones

The overtones of the fundamentals that involve wagging of the terminal hydrogen atoms in the vinyl and vinylidene groups occur with unusual intensity. These bands are observed at 1825 and 1785 cm$^{-1}$, slightly more than double the fundamental frequency. Thus, these harmonics exhibit *negative anharmonicity* in addition to unusual intensity for forbidden vibrations.

The group frequencies of the alkene C—H modes can be summarized as in Table 9.3.

**TABLE 9.3 Alkene Normal Modes**

| C—H Vibrational Modes | $\bar{\nu} \pm 10$ (cm$^{-1}$) |
|---|---|
| **Stretching Modes** | |
| Antisymmetric stretch (=CH$_2$) | 3080 |
| Symmetric stretch (=CH$_2$) | 3020 |
| Uncoupled stretch (=CH) | 3030 |
| **Out-of-plane Bending Modes** | |
| *Vinyl group* | |
|     Trans hydrogen atoms (in-phase) | 990 |
|     Terminal hydrogen atoms (wag) | 910 |
| *Vinylidene group* | |
|     Terminal (wag) | 890 |
| *Trans group* | |
|     Trans hydrogen atoms (in-phase) | 965 |
| *Cis group* | |
|     Cis hydrogen atoms (in-phase) | ~700 |
| *Trisubstituted group* | |
|     Uncoupled hydrogen atom | 820 |
| *Tetrasubstituted group: no modes* | |

## Characteristic Group Frequencies of Alkynes

### C≡C Stretching Vibration

Triple-bond formation further increases the force constant involved in the C—C stretching vibration. Thus, alkynes possess the highest of all observed C—C stretching frequencies. The triple-bond group frequency is located near 2120 cm$^{-1}$ in monosubstituted alkynes. Compression effects on the force constant similar to those observed in substituted alkenes raise the stretching mode into the 2225-wavenumber region in disubstituted alkynes. The alkyne stretching vibration involves a relatively small dipole moment change. The resulting bands, therefore, are particularly weak in the infrared. In the disubstituted case, if the two groups are identical, the vibration is infrared inactive. Here, as in the alkenes, a pseudocenter of symmetry can operate to significantly suppress the intensity of the mode. Even though these bands occur in a region that is essentially devoid of other absorptions, their inherent weakness and variable intensity present significant problems in making confident band assignments. The high frequency of the mode effectively decouples the vibration from the rest of the system. Thus, triple bonds show little evidence of any first-order coupling. On the other hand, these vibrations are prone to second-order effects that complicate their interpretation. Triple-bond assignments must be handled with care. These vibrations constitute a set of group frequencies relatively difficult to deal with, if they can be observed at all.

### Alkyne C—H Vibrations

The stretching of alkynyl C—H bonds gives rise to the highest carbon–hydrogen vibrations observed. These modes are relatively intense uncoupled single sharp modes that occur near 3300 cm$^{-1}$. Although this normal mode occurs in the same region as O—H and N—H fundamental vibrations, alkynyl C—H stretches usually can be distinguished by the sharpness of the band. These stretches are highly reliable group frequencies.

The bending modes of the C—H alkynyl group do not give rise to reliable group frequencies.

The group frequencies of the alkynes are summarized in Table 9.4.

**TABLE 9.4   Alkyne Normal Modes**

| C≡C, C—H Vibrational Modes | $\tilde{\nu} \pm 10$ (cm$^{-1}$) |
|---|---|
| Triple-bond stretch (monosubstituted) | 2120 |
| Triple-bond stretch (disubstituted) | 2225 |
| ≡C—H bond stretch (monosubstituted) | 3300 |

**Characteristic Group Frequencies of Arenes**

Aromatic ring systems represent the final class of hydrocarbons to be considered in this section. The discussion will center on the benzene ring, but many of the more complicated systems have been examined in detail.

The infrared spectra of aromatic compounds possess many needle-sharp bands. This characteristic sets these spectra apart from the spectra of aliphatic compounds. It arises because aromatic systems are tightly bound rigid molecules having little opportunity for rotational isomerism. With aliphatic compounds the observed spectrum, is in reality, often the spectrum of a complex mixture of rotamers. These isomers all exhibit very similar but not identical spectra that overlap and result in band broadening.

**Group Frequencies of the Phenyl Group**

The group frequencies of the phenyl group can be classified as carbon–hydrogen vibrations consisting of stretching and out-of-plane bending modes, plus carbon–carbon ring stretching and out-of-plane bending modes. The in-plane bending modes in both cases are not effective group frequencies.

### C—H Stretching Modes

The C—H stretching vibrations occur as a series of weak bands in the region 3100–3000 cm$^{-1}$. This observation is consistent with $sp^2$ hybridization of the carbon atom. These modes directly overlap the alkene C—H stretching fundamentals. Substitution of heteroatoms into the ring can significantly perturb these frequencies (oxygen raises the mode into the 3200–3100-wavenumber region). As the bands are generally weak, they may be masked by strong aliphatic absorption in mixed compounds if they lie close to 3000 cm$^{-1}$. Care must be taken in the assignment of these modes.

### C=C Stretching Modes

The phenyl ring modes, which possess excellent group frequency properties, involve two pairs of closely related C=C stretching vibrations. These vibrations are related to degenerate fundamentals in unsubstituted benzene. On ring substitution the degeneracy is removed because of the lowered symmetry. The ring vibrations $\nu_{8a}$ and $\nu_{8b}$ (the numbering has been carried over from the benzene fundamental assignments by Wilson) involve displacements in three-carbon units at each end of the ring, which are analogous to the symmetric and antisymmetric stretching modes of water. The two modes result from the two sets of displacements, which are in-phase (Fig. 9.19).

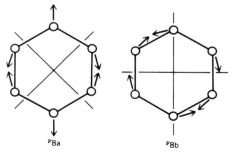

**FIGURE 9.19**   Ring stretching vibrations of benzene, $\nu_{8a}$, $\nu_{8b}$.

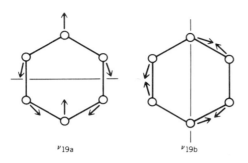

$\nu_{19a}$  $\nu_{19b}$

**FIGURE 9.20** Ring stretching vibrations of benzene, $\nu_{19a}$, $\nu_{19b}$.

These vibrations are degenerate and inactive in the infrared spectrum of the unsubstituted ring. In general, the higher frequency vibration, $\nu_{8a}$, is the more intense of the pair. If a substituent is conjugated to the ring system, however, the lower frequency component gains in intensity and sometimes becomes the most intense member of the pair. These two modes are also substituent independent with para disubstitution; however, if the groups are identical or nearly identical, the modes are infrared forbidden or active with greatly suppressed intensity. These modes, $\nu_{8a}$ and $\nu_{8b}$, occur near 1600 and 1580 cm$^{-1}$, respectively.

The second pair of vibrations corresponds to the identical displacements of the first pair of modes with the sets now out-of-phase (Fig. 9.20). Substitution on the ring removes the degeneracy, giving two bands corresponding to $\nu_{19a}$ and $\nu_{19b}$ in benzene.

These fundamentals, as with $\nu_{8a}$ and $\nu_{8b}$, are substituent independent on mono substitution or para substitution. The dipole moment change associated with these normal modes generally gives rise to rather intense absorption bands. The high-frequency component, however, is sensitive to electron-withdrawing substituents that can significantly suppress the intensity of this mode. These fundamentals occur near 1500 and 1450 cm$^{-1}$, respectively.

### C—H Bending Vibrations

The C—H bending normal modes of group frequency value are the out-of-plane vibrations. These fundamentals are useful guides to the substitution pattern on the ring system. There will be as many out-of-plane vibrations as there are C—H groups. The modes of interest, however, are those fundamentals in which all the hydrogen atoms move in phase. These vibrations have substantial dipole moment changes and, thus, give rise to intense absorption bands (Fig. 9.21). The very strong intensity of the out-of-plane deformation modes plays a key role in our ability to make confident assignments for these fundamentals, as they fall in the heart of the fingerprint region. The five all-in-phase bending vibrations are as presented in Figure 9.21 and Table 9.5.

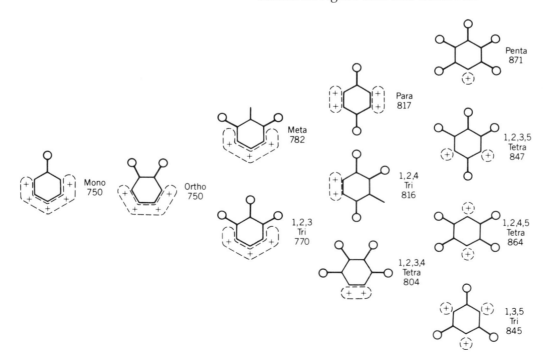

**FIGURE 9.21** Arene out-of-plane C—H bending vibrations.

**TABLE 9.5    Arene Out-of-Ring Plane C—H Deformation Modes**

| Arene Fundamentals | $\bar{\nu}$ Range (cm$^{-1}$) |
|---|---|
| 5 | 770–730 |
| 4 | 770–735 |
| 3 | 810–750 |
| 2 | 860–800 |
| 1 | 900–860 |

Although there is considerable overlap of the ranges, the uncertainty of the assignment can often be reduced by the identification of an additional strong band in the 690-wavenumber region. This band results from a carbon–carbon out-of-plane ring deformation, $\nu_4$, of benzene (Fig. 9.22). This mode is substituent insensitive to mono-, meta-, and 1,3,5-substitution. Thus, a band will seldom occur in this region in ortho-disubstituted systems.

**FIGURE 9.22**    Phenyl out-of-plane bending vibrations, $\nu_4$.

### Sum Tone Patterns

Some of the out-of-plane C—H bending modes can be excited simultaneously with other low-frequency fundamentals. As these sum tone transitions are formally forbidden, the resulting absorption bands are very weak. With optically thick samples, however, weak bands are observed in the 2000–1650-wavenumber region. The pattern of these bands is highly characteristic of the substitution arrangement on the ring, since they have their origin in the out-of-plane C—H bending frequencies. Unless carbonyl groups are present in the molecule, the 2000–1650-wavenumber region will be open for observation. These combination band patterns can be used to remove ambiguity about the ring substitution pattern based on assignments of the out-of-plane fundamentals (Fig. 9.23).

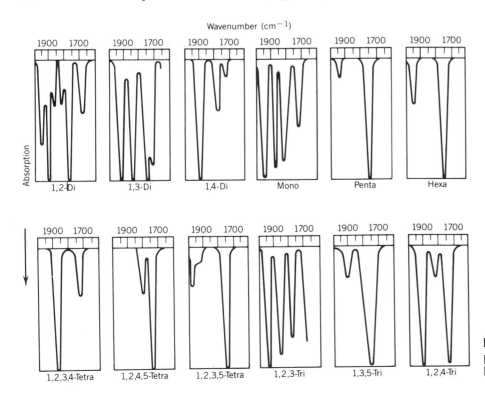

**FIGURE 9.23**    Combination band patterns in the 2000–1650-wavenumber region.

The group frequencies of the arenes can be summarized as in Table 9.6.

**TABLE 9.6  Arene Group Frequencies**

| Arene Fundamentals | $\bar{\nu}$ Range (cm$^{-1}$) |
| --- | --- |
| C—H stretch | 3100–3000 |
| C=C ring stretch ($\nu_{8a}$) | 1600 ± 10 |
| C=C ring stretch ($\nu_{8b}$) | 1580 ± 10 |
| C=C ring stretch ($\nu_{19a}$) | 1500 ± 10 |
| C=C ring stretch ($\nu_{19b}$) | 1450 ± 10 |
| C—H out-of-plane bend (1H) | 900–860 |
| C—H out-of-plane bend (2H) | 860–800 |
| C—H out-of-plane bend (3H) | 810–750 |
| C—H out-of-plane bend (4H) | 770–735 |
| C—H out-of-plane bend (5H) | 770–730 |
| C—C ring out-of-plane bend (1; 1,3; 1,3,5) | 690 ± 10 |
| C—H out-of-plane bend sum tones | 2000–1650 |

## FACTORS AFFECTING THE CARBONYL GROUP FREQUENCIES

The carbonyl group is perhaps the single most important functional group in organic chemistry. It is certainly the most commonly occurring functionality. Infrared spectroscopy can play a powerful role in the characterization of the carbonyl because this group possesses all of the properties that give rise to an excellent group frequency.

1. The carbonyl group has a large dipole moment derivative, which gives rise to very intense absorption bands.
2. As a result of the large force constant, it has a stretching frequency that occurs at high values outside the fingerprint region. In addition, this portion of the spectrum is devoid of most other fundamentals.
3. Its stretching fundamental occurs in a range that is reasonably narrow (little coupling), 1750 ± 150 cm$^{-1}$, but sensitive enough to the local environment to allow for considerable interpretation of the surrounding structure.
4. The range of frequencies is determined by a number of factors that are now well understood in terms of the effects outlined below.

### Mass Effects

The mode of principal interest is the stretching vibration. In this oscillator the C and O atoms undergo comparable displacements; thus, we must replace the simplified mass expression in the Hooke's law approximation that was applied to the C—H stretching fundamentals by the reduced mass $\mu$ (where $\mu = m_C m_O / (m_C + m_O)$, see above). On substitution of the isotopes $^{13}$C and $^{18}$O, the predicted small frequency shifts (~30–40 cm$^{-1}$) are observed. These results are consistent with a relatively low degree of mechanical coupling to the rest of the system. This lack of coupling is expected, because the force constant of the multiple bond is significantly different from the values of the force constants of the bonds that connect the carbonyl to the rest of the molecule.

Geometric effects can play a major role in determining the location of the carbonyl frequency within the 1750-wavenumber region. Although the displacement of the oxygen atom involves simply stretching or compressing the C=O bond, the displacement of the carbon atom is more complex. This latter movement also contains a compression component of the force constants in the two connecting single bonds as the carbonyl carbon is being stretched. In the opposite phase of the vibration, a stretching component of the force constants in the two connecting bonds is required, as the carbonyl carbon is being compressed. The magnitude of these additional force constant components is angle dependent. As the angle between the single bonds (C—CO—C angle) decreases, the contribution of the single-bond components to the effective C=O stretching force constant will be raised. Since the frequency of the vibration is directly proportional to the square root of the force constant, a decrease in the internal carbonyl bond angle will raise the frequency (Table 9.7). Alternatively, an increase in the internal carbonyl bond angle will lower the carbonyl frequency.

**TABLE 9.7  Variation of Carbonyl Frequency (cm⁻¹) Versus Bond Angle**

| Ring Size: | 7 | 6 | 5 | 4 | 3 | (2) |
|---|---|---|---|---|---|---|

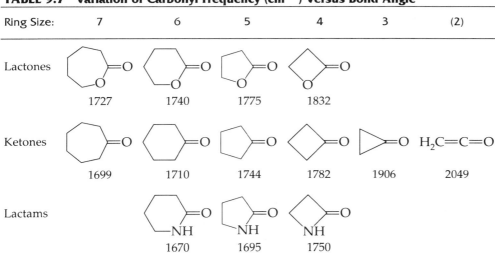

Lactones — 1727, 1740, 1775, 1832

Ketones — 1699, 1710, 1744, 1782, 1906, 2049

Lactams — 1670, 1695, 1750

Resonance and inductive effects can profoundly influence the vibrational frequency of the carbonyl group. In the following discussion the carbonyl stretching mode for acetone ($\bar{\nu}_{C=O} = 1715$ cm⁻¹, liq) will be used as a reference frequency representative of simple alkyl substitution on the carbonyl group. Effects that perturb this reference fundamental to either higher or lower values will be examined.

### Electronic Effects That Raise the Carbonyl Frequency
When an alkyl substituent is replaced by a more electronegative system, the balance of contributing resonance forms in the carbonyl is slightly shifted away from dipolar forms by strong inductive effects. This shift results in a larger effective C=O force constant and higher frequencies. (e.g., in hexanoyl chloride, $\bar{\nu}_{C=O} = 1805$ cm⁻¹, Fig. 9.28).

### Electronic Effects That Lower the Carbonyl Frequency

Direct conjugation of the carbonyl via $\alpha,\beta$-unsaturation will introduce new dipolar carbonyl resonance forms that lower the effective force constant values and, thus, decrease the carbonyl stretching frequency (e.g., consider cyclohexenyl methyl ketone, **I**, $\bar{\nu}_{C=O} = 1685$ cm$^{-1}$, and acetophenone, **II**, $\bar{\nu}_{C=O} = 1687$ cm$^{-1}$).

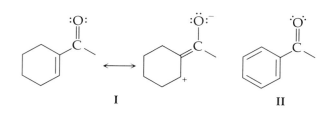

       **I**                    **II**

### The Case of Ester Carbonyl Vibrations (Competing Inductive and Resonance Effects)

The substitution of the more electronegative oxygen for carbon in going from ketones to esters will raise the carbonyl frequency in esters as a result of inductive influences (see Electronic Effects That Raise the Carbonyl Frequency, above). On the other hand, the lone-pair electrons present on the ether oxygen of the ester will be in direct conjugation with the carbonyl. This latter interaction will generate dipolar resonance forms that will tend to drop the C=O frequency. The balance between these two competing effects in esters, which might be difficult to anticipate, appears to favor the inductive effects. The ester carbonyls commonly are located 20–40 cm$^{-1}$ higher than the simple aliphatic ketones, in the range 1755–1735 cm$^{-1}$.

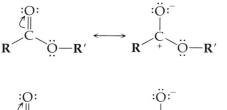

    Now let us examine ester carbonyl frequencies in somewhat greater detail.

    If the ester carbonyl is directly involved with $\alpha,\beta$ unsaturation, the normal ester frequency is lowered by 20–30 cm$^{-1}$. Thus, unsaturated ester carbonyl frequencies occur very nearly in the same region as simple aliphatic ketone frequencies (e.g., in ethyl benzoate, $\bar{\nu}_{C=O} = 1720$ cm$^{-1}$).

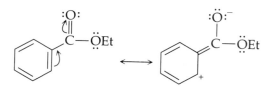

If the ester is conjugated, but the conjugation is located adjacent to the ether oxygen rather than alpha to the carbonyl group, then the carbonyl frequency is raised. The higher $\bar{\nu}_{C=O}$ results from resonance competition for the lone-pair electrons of the ether oxygen by the carbonyl and the new conjugating group (e.g., phenyl acetate, $\bar{\nu}_{C=O} = 1769$ cm$^{-1}$). These frequency shifts support the arguments concerning competition between inductive and resonance effects within the ester ether group.

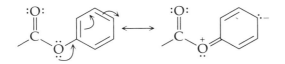

If the ester group is directly conjugated on both sides, then the resonance effects should cancel, and we would expect this type of system to exhibit near normal carbonyl frequencies (e.g., phenyl benzoate, $\bar{\nu}_{C=O} = 1743$ cm$^{-1}$) as compared with ethyl acetate, $\bar{\nu}_{C=O} = 1742$ cm$^{-1}$.

**Interaction Effects**

Interaction effects vary from those that have a dramatic impact on the spectra to those that are barely detectable. Our understanding of these terms completes the discussion of the major factors affecting the carbonyl group frequencies. We can roughly divide this area of discussion into intramolecular and intermolecular types of interactions.

### Intramolecular Carbonyl Interactions

*First-order coupling effects* are rarely observed as this oscillator is rather effectively decoupled from the rest of the molecule by differences in frequency. The most spectacular example (see p. 591) is the case of carbon dioxide. In $CO_2$ the two coupled oscillators are aligned for maximum interaction and possess identical frequencies. The splitting between the antisymmetric and symmetric levels is very large, approximately 1000 cm$^{-1}$. A second, much less dramatic, example is the case of anhydrides. In this instance the two oscillators are joined through a central oxygen. Delocalization across the connecting atom operates to maintain planarity of the system and thereby to increase the coupling. Even so, since the carbonyls are no longer held at the optimum angle and are vibrationally insulated by an intervening atom, the first-order coupling drops to about 70 cm$^{-1}$, less than 10% of the $CO_2$ value (e.g., hexanoic anhydride, $\bar{\nu}_{C=O} = 1817, 1750$ cm$^{-1}$). The uncoupled vibration would be expected to occur near 1770 cm$^{-1}$. This latter value is consistent with an oxygen-substituted carbonyl in which conjugation of the lone-pair electrons on the ether oxygen has been nearly canceled. The full inductive effect of the ether oxygen atom on the carbonyl stretching vibration can be inferred from these data. Equalized resonance competition by both carbonyl systems of the anhydride for the ether lone-pair electrons might be expected to bring about just such an effect.

*Intramolecular hydrogen bonding* of carbonyls can be enhanced by resonance interactions and, thus, significantly perturb the stretching frequency. For example, let us consider the anthraquinone series shown below. In Structure **I**, hydrogen bonding is not present and a single stretching frequency for the quinone is observed at 1675 cm$^{-1}$. This value reflects direct conjugation with the aromatic ring and with the methoxyl groups conjugated in equivalent fashion with both carbonyls. Little coupling between the carbonyls is observed through the ring. In Structure **II** a phenol group replaces one of the methoxyl groups. Resonance forms operating through a tightly hydrogen-bonded six-membered ring act to reduce the effective force constant of the carbonyl, with the result that the observed frequency, $\bar{\nu}_{C=O} = 1636$ cm$^{-1}$, is lowered nearly 40 cm$^{-1}$. In Structure **III** both methoxyl groups have been replaced by phenolic groups. This change results in both carbonyls undergoing strong hydrogen bonding, which involves resonance forms similar to those found in Structure **II**. Thus, a single band is observed at $\bar{\nu}_{C=O} = 1627$ cm$^{-1}$. This downward shift approaching 50 cm$^{-1}$ can be ascribed primarily to strong internal hydrogen bonds present in the anthraquinone system.

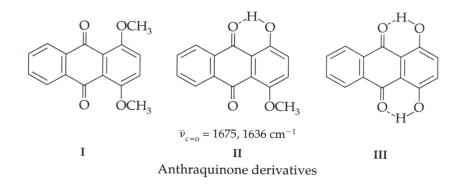

$$\bar{\nu}_{c=o} = 1675, 1636 \text{ cm}^{-1}$$

I                     II                     III

Anthraquinone derivatives

Several other examples of very strong intramolecular hydrogen bonding are known. In tropolone, where the hydrogen appears to be essentially equidistant from the two oxygen atoms, the absorption occurs at 1605 cm$^{-1}$.

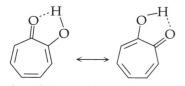

Tropolone (H is equidistant from oxygens)

*Second-order coupling (Fermi resonance)* often occurs with carbonyl vibrations in complex organic molecules. In the large majority of cases, the frequency match of overtone and fundamental is relatively poor so that the frequency of the fundamental is not affected. The main evidence for the interaction in these cases will be weak shoulders associated with the main carbonyl peak. An example of the coupling not being trivial is the case of cyclopentanone. The Fermi interaction involves an overtone or combination of a C—H level, as the splitting collapses to a singlet in 2,2,5,5-tetra-$d_4$-cyclopentanone.

*Field effects* will also perturb the carbonyl frequency. The classic case is that of the chloroacetones. In those rotamers, in which the chlorine atom is in the eclipsed position with respect to the oxygen, repulsive lone-pair interactions occur. Field effects result in suppression of the contribution of the dipolar carbonyl resonance form, and therefore cause a rise ($\sim$30 cm$^{-1}$) in the stretching frequency (Table 9.8).

**TABLE 9.8   Field Effect in Chloroacetone**

| Compound | $\bar{\nu}_{C=O}$(cm$^{-1}$) |
|---|---|
| Acetone | 1715 |
| Chloroacetone | 1752,1726 |
| 1,1-Dichloroacetone | 1743,1724 |
| 1,1,1-Trichloroacetone | 1729 |

These arguments are supported by the observation that two carbonyl frequencies are present in the mono- and dichloroacetones, and single frequencies in acetone and trichloroacetone. *The small frequency rise observed in the low-frequency component is attributed to inductive effects,* which, therefore, can have only a minimal influence on the high-frequency component.

*Transannular interactions* occur when cyclic carbonyl groups are sterically positioned so that the carbon atom of the carbonyl is oriented toward an electron-rich center lying across the ring. The interaction can greatly enhance the dipolar resonance form of the carbonyl and result in a significant drop in the stretching frequency. The effect has a major impact on the carbonyl frequency of the alkaloid protopine ($\bar{\nu}_{C=O} = 1660$ cm$^{-1}$).

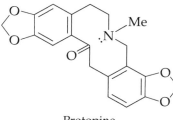

Protopine

Most interesting, however, are the results obtained from a number of model compounds (e.g., the cyclooctaaminoketone, **I**, $\bar{\nu}_{C=O} = 1666$ cm$^{-1}$, and its perchlorate salt, **II**, which exhibits no carbonyl absorption band at all!

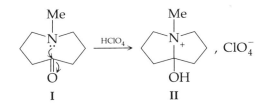

**I**          **II**

## Intermolecular Carbonyl Interactions

*Strong intermolecular hydrogen bonding* can significantly perturb carbonyl frequencies. It is known in the case of aliphatic carboxylic acids that these substances form strongly hydrogen-bonded dimers when neat or in highly concentrated solutions. Association through the carbonyl groups leads to the formation of a symmetric eight-membered ring containing two hydrogen bonds. Coupling through the tightly bonded ring results in a splitting of the carbonyl levels of approximately the same magnitude as found in anhydrides ($\bar{\nu}_{C=O} = $ ca. 70 cm$^{-1}$). As the dimer possesses a center of symmetry, the in-phase mode will not be active in the infrared ($\bar{\nu}_{C=O} = $ ca. 1650 cm$^{-1}$). The out-of-phase stretch of the carbonyls, however, will be active. The antisymmetric C=O stretch gives rise to a strong band in the infrared ($\bar{\nu}_{C=O} = $ ca. 1720 cm$^{-1}$). In very dilute solution, it is sometimes possible to observe these systems in the monomeric state. Under these conditions the carbonyl frequencies return to expected values ($\bar{\nu}_{C=O} = $ ca. 1770 cm$^{-1}$). As the discussion of group frequencies expands, we will see several other examples of the effect of intermolecular hydrogen bonding on the carbonyl group frequency.

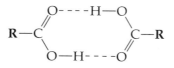

**TABLE 9.9  Carbonyl Dipolar Interactions**[a]

| Compound | $\Delta\bar{\nu}$ (cm$^{-1}$) |
|---|---|
| Acetyl chloride | 15 |
| Phosgene | 13 |
| Acetone | 21 |
| Acetaldehyde | 23 |
| N,N-Dimethylformamide | 50 |

[a] Shift measured between dilute nonpolar solution and neat sample.

The interaction of weak hydrogen bonds is relatively hard to detect in the infrared, as the shifts are measured in terms of a few wavenumbers. One of the better examples is the effect on the carbonyl stretch of acetone ($\bar{\nu}_{C=O} = 1722$ cm$^{-1}$) as measured in hexane solution. When the hydrocarbon solvent is replaced by chloroform, weak hydrogen bonds (O$\cdots$H—C) develop, and the carbonyl mode drops 12 wavenumbers to 1710 cm$^{-1}$.

Weak dipolar interactions between carbonyls can also be observed in the infrared. The frequency shifts caused by these interactions parallel the development of polarization in the carbonyl group, as can be judged by the data in Table 9.9.

**The major factors perturbing carbonyl frequencies can be summarized as follows:**

*Factors That Raise the C=O Frequency*

**1.** Electronegative substitution
**2.** Decrease in C—CO—C internal bond angle

*Factors That Lower the C=O Frequency*

**1.** Conjugation
**2.** Hydrogen bonding

**TABLE 9.10  Carbonyl Group Frequencies**

| Compound | $\bar{\nu}$ (cm$^{-1}$) |
|---|---|
| Ketones, aliphatic, open chain (R$_2$CO) | 1725–1700 |
| Ketones, conjugated | 1700–1675 |
| Ketones, ring | (see Table 9.7) |
| Acid halides | >1800 |
| Esters, aliphatic | 1755–1735 |
| Esters, conjugated | 1735–1720 |
| Esters (conjugated to oxygen) | 1780–1760 |
| Lactones | (see Table 9.7) |
| Anhydrides. aliphatic, open chain | 1840–1810 and 1770–1740 |
| Acids, aliphatic | 1725–1710 |
| Amides | (see Tables 9.22–9.24) |
| Lactams | (see Table 9.7) |
| Aldehydes | 1735–1720 |

As several of these factors may be operating simultaneously, careful judgment as to the contribution of each individual effect must be exercised in predicting carbonyl frequencies. This judgment develops rapidly with practice at interpretation.

This completes the discussion of factors affecting carbonyl group frequencies. A number of additional examples will be discussed in detail in the section on Characteristic Frequencies of Functional Groups (see below), in which a survey of the infrared spectra of functional groups is considered. Carbonyl frequencies are summarized in Table 9.10.

---

Now that we have examined the major group frequencies associated with the common hydrocarbon platforms (platform = hydrocarbon structural unit supporting a functional group) and the principal parameters affecting the carbonyl group, let us consider the vibrations associated with the common functional groups that lead to good group frequency correlations. For the most part we will use, in these discussions, a series of infrared spectra derived from straight-chain aliphatic $C_6$ compounds.

## CHARACTERISTIC FREQUENCIES OF FUNCTIONAL GROUPS

### Hexane

The spectrum of *n*-hexane (see Fig. 9.9) obtained with the pure liquid, as expected, contains simply the group frequencies of an aliphatic hydrocarbon. The antisymmetric and symmetric methyl stretching modes occur below 3000 cm$^{-1}$ at $\bar{\nu} = 2960$ and 2876 cm$^{-1}$. The antisymmetric and symmetric methylene stretching fundamentals occur near $\bar{\nu} = 2938$ and 2860 cm$^{-1}$. The antisymmetric methyl deformation ($\bar{\nu} = 1467$ cm$^{-1}$) overlaps the symmetric methylene scissoring vibration, which is found as a difficult-to-identify shoulder at $\bar{\nu} = 1455$ cm$^{-1}$. The symmetric methyl bend (umbrella mode) is easily assigned to the sharp band at 1379 cm$^{-1}$. Finally, the all-in-phase rocking mode of a sequence of four or more methylene groups can be identified by its intensity in the fingerprint region near 725 cm$^{-1}$. We often will be able to identify this collection of platform group frequency bands as we progress through the infrared spectra of the following series of compounds (see Fig. 9.9 and Table 9.1).

### Hexanol

If an oxygen atom is inserted across one of the terminal C—H bonds, we obtain the alcohol, 1-hexanol. The change in the infrared spectrum obtained with a sample path length of less than one half that used to obtain the spectrum of hexane is remarkable (Fig. 9.24). A very intense band appears at 3350 cm$^{-1}$, which is assigned to the stretching mode of the single O—H group (Table 9.11). The very broad and intense properties of this absorption are characteristic of the stretching of hydrogen-bonded hydroxyl groups. The increase in intensity of this mode also reflects an increase in the polarity of the bond involved in the vibration over that of the C—H bond in hexane. A second strong band in the spectrum is located near 1058 cm$^{-1}$. This absorption has been identified as the C—O stretching mode. The vibrational displacements of this fundamental are similar to the antisymmetric stretch of water. Since the vibration involves significant displacement of the adjacent C—C oscillator, the vibration will be substitu-

**TABLE 9.11    Normal Modes of the Hydroxyl Group**

| $\bar{\nu}$ (cm$^{-1}$) | Intensity | Mode Description |
|---|---|---|
| 3500–3200 | Very strong | O—H stretch (only strong when hydrogen bonded) |
| 1500–1300 | Medium to strong | O—H in-plane bend (overlap $CH_2$, $CH_3$ bend) |
| 1260–1000 | Strong stretch | C—C—O antisymmetric |
| 650 | Medium | O—H out-of-plane bend |

**TABLE 9.12    Substitution Effects on C—O Stretch of Alcohols**

| Type of —OH Substitution | $\bar{\nu}_{C—O}$ (cm$^{-1}$) |
|---|---|
| $RCH_2$—OH | 1075–1000 |
| $R_2CH$—OH | 1150–1075 |
| $R_3C$—OH | 1200–1100 |
| $C_6H_5$—OH | 1260–1180 |

tion sensitive. These latter shifts can be of value in determining the nature of the alcohol (primary, secondary, or tertiary, see Table 9.12).

The only other new modes observed in going from hexane to 1-hexanol are the O—H bending vibrations. Two types of bending vibrations would be expected, the in-plane and out-of-plane displacements. The in-plane bend of the O—H oscillator is not a very good frequency because it is coupled to adjacent methylene group bending vibrations (wagging). It can be found because of its breadth (O—H hydrogen bonding) as an underlying absorption running across the 1500–1300-wavenumber region (Table 9.11). The out-of-plane O—H bending fundamental occurs at lower frequencies as a broad band (H-bonding) near 650 cm$^{-1}$. The group frequencies of the hydrocarbon portion of the molecule are easily identified, including the rocking fundamental (727 cm$^{-1}$), which is superimposed on the broad O—H out-of-plane bending mode. The normal modes of the hydroxyl group possess many of the characteristics that lead to excellent group frequency correlations (Fig. 9.24).

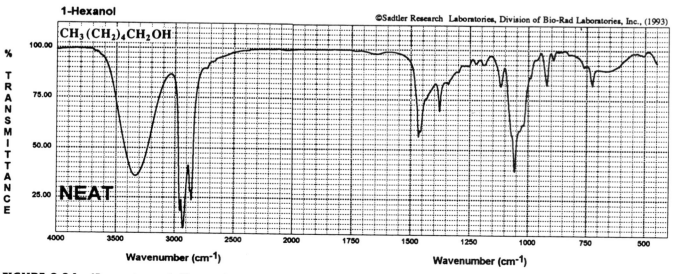

**FIGURE 9.24**    IR spectrum: 1-Hexanol.

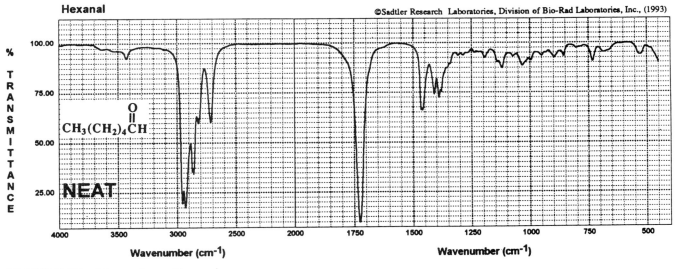

**FIGURE 9.25**   IR spectrum: Hexanal.

## Hexanal

If two terminal hydrogen atoms of hexane are replaced by a single oxygen atom, we have hexanal. The aldehyde functional groups gives rise to several good group frequencies (Fig. 9.25 and Table 9.13). The system has strong bonds and a large dipole moment, and it is essentially decoupled from the rest of the molecule by the low-frequency C—C connecting bond. A component of the C—H stretching mode of the aldehyde group can be assigned to a band of weak-to-medium intensity at $\bar{\nu} = 2723$ cm$^{-1}$. The low frequency of this mode is interpreted on the basis of a Fermi resonance interaction. The aldehyde in-plane C—H bending fundamental found at 1390 cm$^{-1}$ would be expected to generate an overtone very close to the aldehyde group C—H stretching mode, which must occur near 2775 cm$^{-1}$. The two levels interact and split to give two components. The more easily identified low-frequency component is very characteristic of the aldehyde group and is located near 2730 cm$^{-1}$. The higher frequency component

**TABLE 9.13   Normal Modes of the Aldehyde Group**

| $\bar{\nu}$ (cm$^{-1}$) | Intensity | Mode Description |
|---|---|---|
| 2750–2720 | Weak to medium | C—H stretch, in Fermi resonance with C—H bend |
| 1735–1720 | Very strong | C=O stretch |
| 1420–1405 | Medium | CH$_2$ symmetric bend, —CH$_2$— $\alpha$ to —CHO |
| 1405–1385 | Medium | C—H in-plane bend |

often is masked by other aliphatic C—H stretching absorptions in the 2900–2800-wavenumber region. For hexanal, the upper band is observed as a distinct shoulder occurring at 2823 cm$^{-1}$. The carbonyl stretching frequency is the most intense band in the spectrum and is located at 1728 cm$^{-1}$. The structural change in going from ketone to aldehyde will produce mass and inductive effects that will lower the frequency and bond angle and hyperconjugative effects that will raise the carbonyl frequency. The outcome of this competition is that aliphatic aldehyde carbonyl stretching modes generally occur at slightly higher values than do those of the saturated ketones. The only other identifiable group frequency associated with the aldehyde system is the C—H in-plane bending fundamental ($\bar{\nu} = 1390$ cm$^{-1}$) responsible for the overtone that undergoes Fermi resonance with the aldehyde C—H stretching mode. Note that the aldehyde group perturbs one of the aliphatic chain group frequencies. Thus, the frequency of the symmetric deformation (scissoring) of the methylene group alpha to the carbonyl is lowered ($\bar{\nu} = 1408$ cm$^{-1}$), and intensified by hyperconjugation with the carbonyl. It is also evident that the vibrational modes of the aliphatic portion of the molecule now contribute a much smaller fraction of the overall absorption by the sample. Hence, the spectrum is obtained with a path length even shorter than that of 1-hexanol.

### 3-Heptanone

If we insert a carbonyl group between the first two methylene groups of hexane, we have the ketone 3-heptanone. The only group frequency mode associated with this group is the stretching frequency ($\bar{\nu}_{C=O} = 1718$ cm$^{-1}$), which occurs within the expected region for an aliphatic ketone (Fig. 9.26 and Table 9.14). As in the case of aldehydes having $\alpha$-methylene groups, the symmetric bending modes (scissoring) of the adjacent methylene groups at carbon atoms 2 and 4 (C-2 and C-4) in 3-heptanone are perturbed to lower frequencies ($\bar{\nu} = 1425$ cm$^{-1}$) by hyperconjugation with the carbonyl. The shift of the methylene bending frequency is a useful indication of the substitution surrounding the ketone. The very weak band at 3425 cm$^{-1}$ can be confidently assigned as the overtone of the carbonyl stretching frequency ($2 \times \bar{\nu}_{C=O} = 3444$ cm$^{-1}$). The drop in intensity from that of the

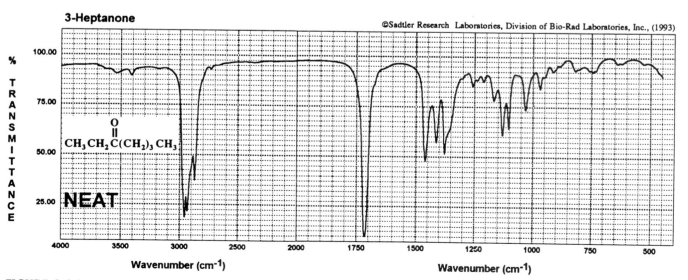

**FIGURE 9.26**   IR spectrum: 3-Heptanone.

**TABLE 9.14    Normal Modes of the Ketone Group**

| $\bar{\nu}$ (cm$^{-1}$) | Intensity | Mode Description |
|---|---|---|
| 3430–3410 | Very weak | Not fundamental, overtone of carbonyl stretch |
| 1725–1700 | Very strong | C=O stretch |
| 1430–1415 | Medium | —CH$_2$— symmetric bend, —CH$_2$— $\alpha$ to ketone C=O |

fundamental and the frequency contraction are typical of these forbidden transitions. Note that in this molecule the sequence of methylene groups has dropped below 4, and the rocking vibration ($\sim$720 cm$^{-1}$) is no longer easily detectable.

### n-Hexyl Acetate

Replacing the O—H hydrogen of hexanol with CH$_3$CO— gives $n$-hexyl acetate. The very strong band found at 1743 cm$^{-1}$ is typical of the carbonyl frequency of an aliphatic ester, particularly acetate esters (Fig. 9.27 and Table 9.15). Two very intense bands occur in the spectra of acetate esters in the 1250–1000-wavenumber region. In primary acetates these bands are found near 1250 and 1050 cm$^{-1}$ (hexyl acetate, 1242, 1042 cm$^{-1}$). The higher frequency mode is assigned to the antisymmetric C—CO—O stretch (similar to that in water), and the lower frequency mode to the antisymmetric O—CH$_2$—C stretch. Although there is some coupling of these vibrations to the adjacent structure, resonance through the carbonyl group by the ether oxygen tends to localize the higher of the two vibrations. The lower mode, which would be expected to be more highly coupled, is in fact more subject to substitution effects. The upper mode is found in most saturated esters at frequencies (1210–1160 cm$^{-1}$) slightly lower than those that occur in acetates. The only other absorption to note in acetate esters is the methyl

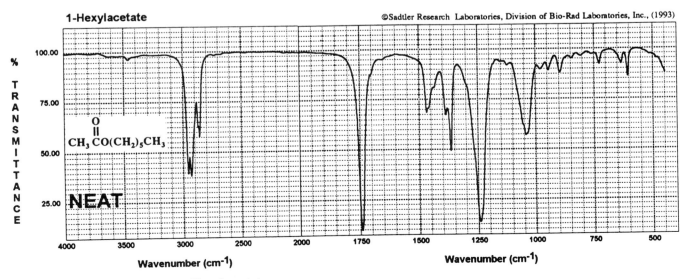

**FIGURE 9.27**   IR spectrum: $n$-Hexyl acetate.

**TABLE 9.15**   **Normal Modes of the Ester Group**

| $\bar{\nu}$ (cm$^{-1}$) | Intensity | Mode Description |
|---|---|---|
| 1755–1735 | Very strong | C=O stretch |
| 1370–1360 | Medium | CH$_3$ symmetric bend $\alpha$ to ester carbonyl |
| 1260–1230 | Very strong | C—CO—O antisymmetric stretch—acetates |
| 1220–1160 | Very strong | C—CO—O antisymmetric stretch—higher esters |
| 1060–1030 | Very strong | O—CH$_2$—C antisymmetric stretch—1° acetates |
| 1100–980 | Very strong | O—CH$_2$—C antisymmetric stretch—higher esters (may overlap with upper band) |

symmetric bending mode. Here the umbrella deformation of the methyl adjacent to the carbonyl occurs at a slightly lower (hyperconjugation) frequency ($\bar{\nu} = 1366$ cm$^{-1}$), and the band is significantly intensified by the interaction with the carbonyl as compared with the deformation of the methyl at the end of the hexyl chain ($\bar{\nu} = 1384$ cm$^{-1}$).

### Hexanoyl Chloride

Hexanoyl chloride can be formed from hexane by the exchange of three terminal methyl hydrogen atoms for an oxygen and a chlorine. The carbonyl stretching mode dominates the spectrum (Fig. 9.28 and Table 9.16). It is an extremely intense band occurring near 1802 cm$^{-1}$. The high frequency and intensity result from inductive effects of chlorine substitution directly on the carbonyl group. The chlorine modes, on the other hand, are not easy to identify, and these vibrations do not develop good group frequencies. The symmetric deformation (scissoring) of the methylene group adjacent to the carbonyl group again shows the same frequency

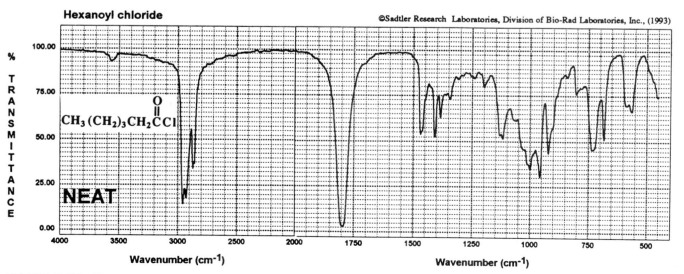

**FIGURE 9.28**   IR spectrum: Hexanoyl chloride.

**TABLE 9.16   Normal Modes of the Acid Halide Group**

| $\bar{\nu}$ (cm$^{-1}$) | Intensity | Mode Description |
|---|---|---|
| 1810–1800 | Very strong | C=O stretch, acid chlorides |
| 1415–1405 | Strong | —CH$_2$— symmetric bend, $\alpha$ to —COCl carbonyl |

decrease resulting from hyperconjugation as observed with the ketones and aldehydes. In this case the mode ($\bar{\nu} = 1408$ cm$^{-1}$) gains considerable intensity via interaction with the highly polarized carbonyl group.

**Hexanoic Acid**

Hexanoic acid is obtained from hexane by substituting a carbonyl oxygen and an —OH group for the terminal hydrogen atoms. The acid possesses a very intense band with a width at one-half peak height of about 1000 cm$^{-1}$, which covers the region 3500–2200 cm$^{-1}$ (Fig. 9.29 and Table 9.17). This absorption is characteristic of very strongly hydrogen-bonded carboxylic acid groups. The relatively weak C—H stretching absorption of the aliphatic chain is superimposed on the O—H stretch between 3000 and 2800 cm$^{-1}$. Also occurring along this broad absorption are a characteristic set of weak overtone and combination bands running from about 2800 to 2200 cm$^{-1}$. The spectrum of the neat material will be that of the hydrogen-bonded dimer, which as noted earlier (see p. 611), has a center of symmetry. The carbonyl mode (out-of-phase stretch) is found at 1709 cm$^{-1}$. Two rather broad and intense bands located between 1450 and 1400 and between 1300 and 1200 cm$^{-1}$ are associated with the in-plane O—H bend and the antisymmetric CH$_2$—CO—O stretch. These modes show evidence of considerable mixing, a situation quite different from the case of alcohols. A strong broad absorption band at 930 cm$^{-1}$ is assigned to an out-of-plane bending mode of the hydrogen-bonded dimer ring. Thus, this latter band is present only when detectable concentrations of the dimers exist. In dilute solution the dimer band vanishes. This low-wavenumber absorption is termed the acid dimer band. It is worth noting that even in carboxylic

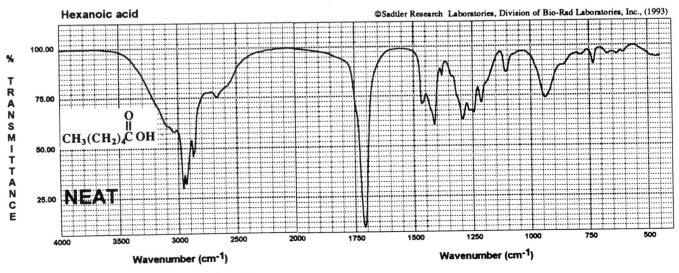

**FIGURE 9.29**   IR spectrum: Hexanoic acid.

**TABLE 9.17   Normal Modes of the Carboxylic Acid Group**

| $\tilde{\nu}$ (cm$^{-1}$) | Intensity | Mode Description |
|---|---|---|
| 3500–2500 | Very very strong | O—H stretch intensified by hydrogen bonding |
| 2800–2200 | Very weak | Overtone and sum tones |
| 1725–1710 | Very strong | C=O antisymmetric hydrogen-bonded dimer stretch |
| 1450–1400 | Strong | CH$_2$—CO—O antisymmetric stretch mixed with O—H bend |
| 1300–1200 | Strong | CH$_2$—CO—O antisymmetric stretch mixed with O—H bend |
| 950–920 | Medium | Out-of-plane O—H bend, acid dimer |

acids, in which the normal modes of a large, highly polarized functional group dominate the spectrum, the group frequencies of the aliphatic molecular backbone are still identifiable.

### Hexanoic Anhydride

Hexanoic anhydride can be formed from two molecules of hexanoic acid by removing the elements of water. Coupling of the carbonyls through the ether oxygen splits the carbonyls ($\tilde{\nu}_{C=O} = 1831, 1761$ cm$^{-1}$) by about 70 cm$^{-1}$ (see Fig. 9.30 and Table 9.18). In this instance the higher frequency mode is the in-phase vibration. Strong bands occur in aliphatic anhydrides, near 1050 cm$^{-1}$, which are directly related to C—O stretching modes. The scissoring deformation of the —CH$_2$— groups alpha to the carbonyls is assigned to the band near 1415 cm$^{-1}$.

### Dihexyl Ether

Dihexyl ether is obtained by removing the elements of water from two molecules of 1-hexanol. Long-chain aliphatic ethers have many physical properties similar to those of the hydrocarbons, and the infrared spectrum

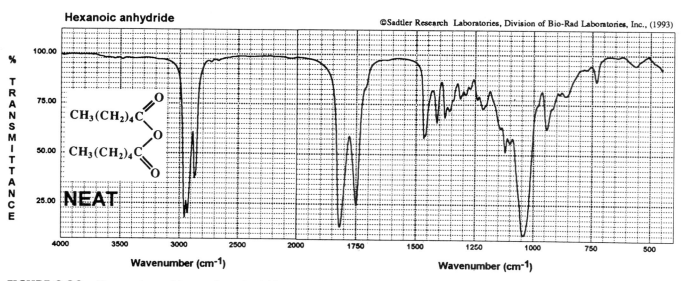

**FIGURE 9.30**   IR spectrum: Hexanoic anhydride.

**TABLE 9.18    Normal Modes of the Anhydride Group (Open Chain)**

| $\bar{\nu}$ (cm$^{-1}$) | Intensity | Mode Description |
|---|---|---|
| 1840–1810 | Very strong | C=O in-phase stretch |
| 1770–1740 | Very strong | C=O out-of-phase stretch |
| 1420–1410 | Strong | —CH$_2$— symmetric bend $\alpha$ to carbonyls |
| 1100–1000 | Very strong | C—O stretch, mixed modes |

of dihexyl ether (Fig. 9.31) is not very different from that of tridecane. The single major departure in the spectrum is the presence of a very strong band near 1100 cm$^{-1}$. In dihexyl ether this absorption is found at 1130 cm$^{-1}$. The vibration responsible for this absorption band must involve a considerable amount of antisymmetric C—O—C stretch. Heavy coupling, however, is also involved in this level. Because of the extensive mechanical coupling with the chain carbon atoms, substitution adjacent to the ether linkage can rather significantly shift this frequency. Fortunately, the large intensity associated with this fundamental relative to the other bands occurring in this region makes it possible, in most cases, to assign with confidence the antisymmetric C—O—C stretching mode (see Table 9.19).

**TABLE 9.19    Normal Mode of the Ether Group**

| $\bar{\nu}$ (cm$^{-1}$) | Intensity | Mode Description |
|---|---|---|
| 1150–1050 | Strong | C—O—C antisymmetric stretch, mixed mode |

### *n*-Hexylamine

If an NH group is inserted across one of the terminal C—H bonds of *n*-hexane, we obtain 1-hexylamine. Compare the infrared spectrum of 1-hexylamine (Fig. 9.32) with those of 1-hexanol (Fig. 9.24) and *n*-hexane (Fig. 9.9). The difference in the region above 3310 cm$^{-1}$ is quite remarkable. Although *n*-hexane is essentially devoid of absorption in this region and 1-hexanol exhibits a very strong band, 1-hexylamine possesses two bands ($\bar{\nu}_{N-H}$ = 3380, 3290 cm$^{-1}$) of medium-to-weak intensity. These latter bands are the antisymmetric and symmetric N—H stretching modes, respectively, of the primary amino group (Table 9.20). Hydrogen-bond intensifi-

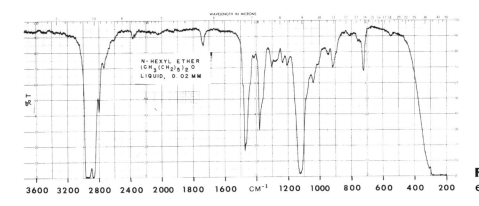

**FIGURE 9.31**    IR spectrum: Dihexyl ether. *(Courtesy of Bowdoin College.)*

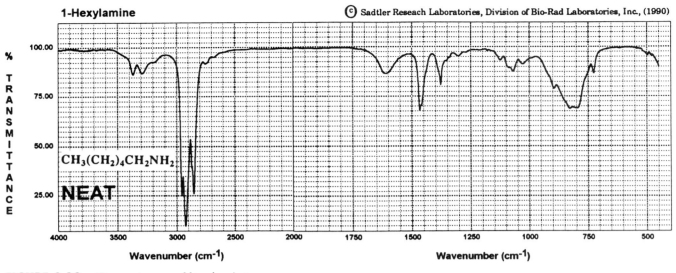

**FIGURE 9.32**    IR spectrum: *n*-Hexylamine.

cation of the N—H group in simple primary amines does not equal that of the hydroxyl system. A band of medium intensity just above 1600 cm$^{-1}$ is assigned to the symmetric (scissoring) deformation of the amino group. This vibration is directly related to the methylene mode near 1450 cm$^{-1}$. The occurrence of this fundamental requires that the amino group be unsubstituted, as the two hydrogen atoms undergo major displacements during the vibration. A second bending mode of primary amino groups can sometimes be observed. This vibration resembles the methylene wagging motion in which the hydrogen atoms are displaced more or less parallel to the molecular axis. As is usual for hydrogen-bonded bending fundamentals, the mode occurs as a fairly broad and quite strong band near 800 cm$^{-1}$. Frequencies related to the normal modes of single bonded C—N systems are highly coupled to the surrounding C—C oscillators and cannot be assigned with any confidence.

**TABLE 9.20    Normal Modes of the Primary Amine Group**

| $\bar{\nu}$ (cm$^{-1}$) | Intensity | Mode Description |
|---|---|---|
| 3400–3200 | Weak to medium | NH$_2$ stretch, (antisymmetric and symmetric) |
| 1630–1600 | Medium | NH$_2$ symmetric bend |
| 820–780 | Medium | NH$_2$ wag |

### Hexanenitrile

Replacement of three terminal hydrogen atoms of hexane by a nitrogen gives hexanenitrile. The nitrile group is a very simple two-atom oscillator. The very strong triple bond (as in the case of the alkynes) contributes to an unusually high stretching frequency (Fig. 9.33 and Table 9.21), and the polar character of the group gives rise to very strong bands. These two factors allow for easy distinction of nitrile bands from alkynyl absorption. The stretching fundamental of saturated nitriles falls in the region 2260–2240 cm$^{-1}$. As expected, conjugation lowers this fundamental (2240–2210 cm$^{-1}$). In hexanenitrile the normal mode occurs at 2255 cm$^{-1}$. Primary nitrile groups can interact through hyperconjugation with the adjacent

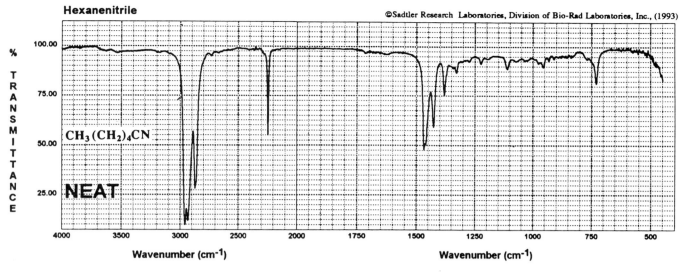

**FIGURE 9.33** IR spectrum: Hexanenitrile.

methylene groups so that the symmetric deformation (scissoring) mode is lowered into the 1425-wavenumber region. In hexanenitrile this bending vibration is assigned to a band at 1430 cm$^{-1}$. The remaining absorption bands are the aliphatic group frequencies.

**TABLE 9.21    Normal Modes of the Nitrile Group**

| $\bar{\nu}$ (cm$^{-1}$) | Intensity | Mode Description |
|---|---|---|
| 2260–2240 | Strong | C≡N stretch, aliphatic |
| 2240–2210 | Strong | C≡N stretch, conjugated |

## Hexanamide

Hexanamide is obtained by replacing the terminal hydrogen atoms of hexane with the elements $ONH_2$. The highly polar amide group leads to very strong hydrogen bonding, which in turn leads to greatly intensified N—H antisymmetric and symmetric stretching modes ($\bar{\nu}_{N-H}$ = 3375, 3200 cm$^{-1}$; Fig. 9.34 and Table 9.22). The carbonyl stretch occurs at low values (1675 cm$^{-1}$) for a saturated system substituted with an electronegative atom.

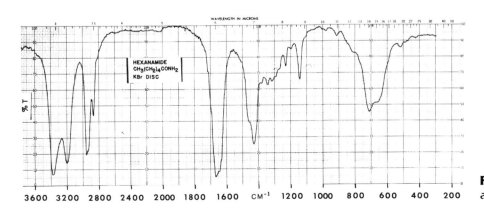

**FIGURE 9.34** IR spectrum: Hexanamide. *(Courtesy of Bowdoin College.)*

**TABLE 9.22    Normal Modes of the Primary Amide Group**

| $\bar{\nu}$ (cm$^{-1}$) | Intensity | Mode Description |
|---|---|---|
| 3400–3150 | Very strong | —NH$_2$ antisymmetric and symmetric stretch, hydrogen bonded |
| 1680–1650 | Very strong | C=O stretch, hydrogen bonded |
| 1660–1620 | Strong | —NH$_2$ symmetric bend (overlap with C=O stretch) |
| 1430–1410 | Strong | —CH$_2$— symmetric bend $\alpha$ to amide carbonyl |
| 750–650 | Medium | —NH$_2$ wag |

Resonance between the carbonyl and the nitrogen lone pair, plus strong hydrogen bonding, appears to overcome the inductive effect. The carbonyl band is *accidentally degenerate* (two fundamentals occurring at the same frequency by chance rather than being required to have the same frequency by symmetry restrictions), but does not interact to any appreciable extent with the symmetric bending mode of the —NH$_2$ group in hexanamide (symmetry constraints restrict the interaction in this case). In some situations both bands can be resolved, but they will often occur as a single band. The wagging vibration of the amino group in which the hydrogen atoms are displaced parallel to the chain axis is found in the 700-wavenumber region as a strong broad band. The symmetric deformation of the methylene group adjacent to the amide carbonyl undergoes the conventional frequency drop to the 1425-wavenumber region. The perturbed scissoring fundamental, however, gains considerable intensity from the interaction with the highly polarized amide carbonyl. Because of the presence of very strong intermolecular hydrogen bonds, the overall spectra of solid or pure liquid amides exhibit broad, rather ill-defined bands.

### *N*-Methylhexanamide

If we substitute on the amide group by replacing a hydrogen with a methyl group, we obtain the secondary amide, *N*-methylhexanamide. The single N—H group gives rise to a very strong band at about 3300 cm$^{-1}$, which is very indicative of strong hydrogen bonding (Fig. 9.35). A medium intensity band near 3100 cm$^{-1}$ is the overtone of the N—H bending mode

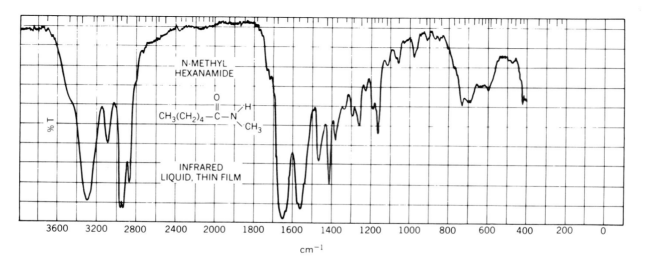

**FIGURE 9.35**    IR spectrum: *N*-Methylhexanamide. *(Courtesy of Bowdoin College.)*

identified at 1570 cm$^{-1}$ in Fermi resonance with the N—H stretching fundamental. The overtone does not match the fundamental particularly well, but it is close enough to acquire substantial intensity enhancement. The carbonyl stretching mode at 1650 cm$^{-1}$ exhibits the very intense and broad characteristics of the amide C=O system. The N—H in-plane bend of the single oscillator occurs near 1570 cm$^{-1}$. The drop in frequency from that of the primary scissoring mode near 1600 cm$^{-1}$ allows for confident assignment of the substitution on secondary amide groups (Table 9.23). The band at 1570 cm$^{-1}$, although often referred to as the N—H bending mode, is in reality a heavily mixed mode. The pure in-plane N—H bend naturally falls near 1450 cm$^{-1}$. Resonance between the carbonyl and the nitrogen lone pair results in a stiffening of the C—N bond. The resulting C—N stretch is raised into the 1400-wavenumber region. Mechanical coupling between the N—H bend and the C—N stretch results in first-order interaction and a splitting of the levels. The upper level occurs near 1570 cm$^{-1}$, whereas the lower level often can be identified near 1300 cm$^{-1}$ as a weak-to-medium band. The out-of-plane bend of the N—H group is identified as a broad, medium-intensity band centered near 700 cm$^{-1}$. The symmetric deformation of the methylene group alpha to the carbonyl is located at 1410 cm$^{-1}$. The unusual intensity of this mode results from interaction with the heavily polarized C=O system.

**TABLE 9.23  Normal Modes of the Secondary Amide Group**

| $\bar{\nu}$ (cm$^{-1}$) | Intensity | Mode Description |
|---|---|---|
| 3350–3250 | Strong | N—H stretch, intensified by hydrogen bonding |
| 3125–3075 | Medium | Overtone N—H bend in Fermi resonance with N—H stretch |
| 1670–1645 | Very strong | C=O stretch, hydrogen bonded |
| 1580–1550 | Strong | N—H in-plane bend mixed with C—N stretch |
| 1415–1405 | Strong | —CH$_2$— symmetric bend $\alpha$ to amide C=O |
| 1325–1275 | Medium | C—N stretch mixed with N—H in-plane bend |
| 725–680 | Medium | N—H out-of-plane bend |

Studies of amide carbonyl frequencies in dilute nonpolar solution indicate that hydrogen-bonding effects are largely responsible for the low frequencies observed with primary and secondary amides, but play no role in tertiary amides.

The data (Table 9.24) indicate that when hydrogen bonding is removed in primary amides, the inductive effect of the nitrogen dominates over the influence of conjugation, but not as much as in the case of esters. This result is consistent with the relative electronegativities involved in esters and amides. In secondary amides with an electron-releasing $N$-alkyl

**TABLE 9.24  Amide Carbonyl: Solution and Solid Phase Data**

| Amide | Dilute Solution (cm$^{-1}$) | Solid (cm$^{-1}$) |
|---|---|---|
| R—CO—NH$_2$ | ~1730 | ~1690–1650 |
| R—CO—NHR | ~1700 | ~1670–1630 |
| R—CO—NR$_2$ | ~1650 | ~1650 |

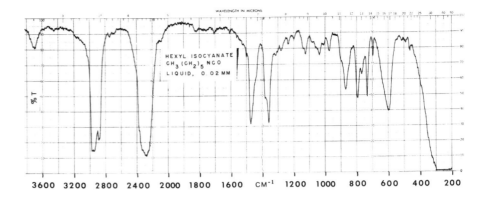

**FIGURE 9.36** IR spectrum: Hexyl isocyanate. *(Courtesy of Bowdoin College.)*

group replacing a hydrogen, conjugation involving the nitrogen lone pair with the carbonyl begins to overcome the inductive effect. In tertiary amides with two N-alkyl substituents present, conjugation now dominates the inductive effect. Under these conditions polarized resonance forms make large contributions to the character of the carbonyl, and the tertiary amide C=O frequencies are observed to decrease to values in the same range as the frequency shifts generated via hydrogen bonding.

### 1-Hexyl Isocyanate

1-Hexyl isocyanate is obtained by replacement of a terminal hydrogen atom with an —N=C=O group. The out-of-phase stretching mode of the isocyanate group attached to the hexyl chain occurs at 2275 cm$^{-1}$, as a broad and very strong band (Fig. 9.36). This functional group is representative of a number of cumulated double-bond systems that possess vibrations mechanically identical to that of carbon dioxide ($\bar{\nu} = 2350$ cm$^{-1}$). The range of stretching frequencies observed for alkyl-substituted isocyanates is very narrow, $\bar{\nu} = 2280$–$2260$ cm$^{-1}$, which implies little coupling to the rest of the system (Table 9.25). Interestingly, conjugation appears not to have any significant effect on the mode. The symmetric stretching fundamental is not easily observed in the infrared as it is a weak band occurring in the fingerprint region. The remaining group frequencies in the spectrum of 1-hexyl isocyanate are those of the alkyl group.

**TABLE 9.25  Normal Mode of the Isocyanate Group**

| $\bar{\nu}$ (cm$^{-1}$) | Intensity | Mode Description |
|---|---|---|
| 2280–2260 | Very strong | —N=C=O antisymmetric stretch |

### 1-Hexanethiol

Insertion of a sulfur atom in a terminal C—H bond of *n*-hexane gives 1-hexanethiol. The spectrum of this material resembles that of hexane itself except for small changes in the fingerprint region and a weak band near 2570 cm$^{-1}$ (Fig. 9.37). The latter absorption is assigned to the S—H stretching fundamental (Table 9.26). Although this mode is quite weak, it is not

**TABLE 9.26  Normal Mode of the Thiol Group**

| $\bar{\nu}$ (cm$^{-1}$) | Intensity | Mode Description |
|---|---|---|
| 2580–2560 | Weak | S—H stretch |

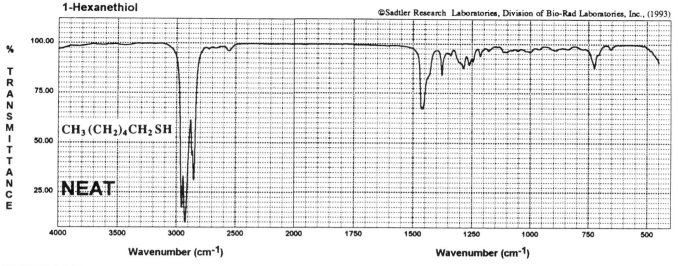

**FIGURE 9.37**   IR spectrum: 1-Hexanethiol.

involved in any significant coupling, and it occurs in a region of the spectrum sparsely populated by other absorption bands. The S—H stretch, therefore, can be considered a reliable group frequency. The S—H bending and C—S stretching modes also are weak, and as they fall in the fingerprint region they are not useful as group frequencies. The remaining bands of 1-hexanethiol that can be assigned belong to the alkyl portion of the molecule.

## 1-Chlorohexane

Replacement of a terminal hydrogen atom of $n$-hexane by a chlorine atom gives 1-chlorohexane. The massive chlorine atom is connected to the alkyl section by a fairly weak but highly polarized bond, which dictates that the C—Cl stretching frequency appears as an intense band at low frequencies (Fig. 9.38 and Table 9.27). The spectrum of 1-chlorohexane does possess a number of moderately intense absorption bands in the low-frequency region (800–600 cm$^{-1}$). Some coupling to the main structure adjacent to the C—Cl bond is expected, since the carbon atom will be carrying out the

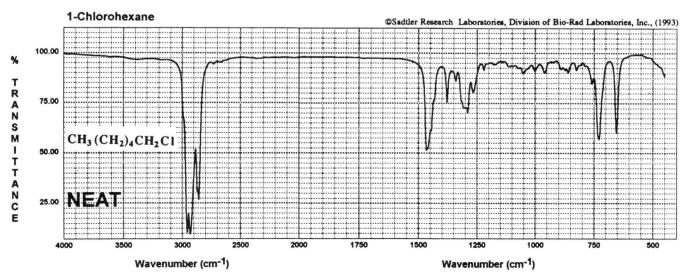

**FIGURE 9.38**   IR spectrum: 1-Chlorohexane.

**TABLE 9.27    Normal Mode of the Chloro Group**

| $\bar{\nu}$ (cm$^{-1}$) | Intensity | Mode Description |
|---|---|---|
| 750–650 | Strong | C—Cl stretch, rotamers, and mixed modes occur |

majority of the displacement. Reliable assignment of the halogen stretching mode, therefore, is not easy, because the surrounding C—C modes will pick up intensity from the polar C—Cl bond. In the case of 1-chlorohexane, it is possible to assign two C—Cl stretching modes ($\bar{\nu}_{C-Cl} = 731$, $658$ cm$^{-1}$), based on Raman spectral data. The presence of two modes is attributed to the presence of rotamers. The higher frequency is assigned to the **anti** (trans) conformer, and the lower frequency to the **gauche** conformer. Note that the stretching frequency of the **anti** isomer, 731 cm$^{-1}$, falls at the same frequency as the methylene rocking vibration of the hexyl chain. Thus, without additional data it would have been difficult to assign the C—Cl stretching modes, even in these fairly simple systems. The carbon–halogen stretching vibration must be employed with care as a group frequency.

The bending modes of the halogens usually occur at such low frequencies as to be of little use as conventional group frequencies. The remaining bands are related to the hydrocarbon portion of the molecule.

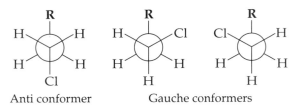

Anti conformer            Gauche conformers

### Chlorobenzene

The final spectrum to be considered in this section is the spectrum of the C$_6$ compound chlorobenzene (Fig. 9.39). Here we have introduced a new hydrocarbon platform bearing the functional group. Chlorobenzene is a

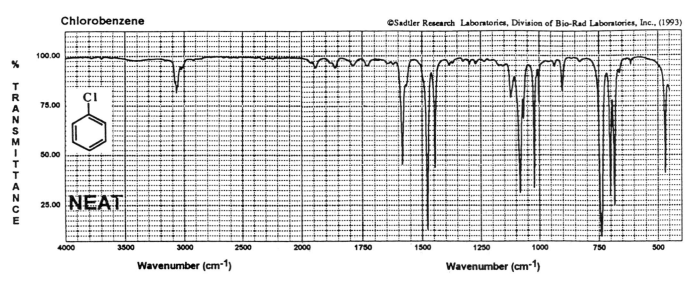

**FIGURE 9.39**   IR spectrum: Chlorobenzene.

material in which the oscillators are tightly bound together as a single conformer. The spectrum contains many needle-sharp bands as compared, for example, with the spectrum of 1-chlorohexane. Thus, it is often possible to detect the presence of either aliphatic or aromatic systems simply on the basis of overall appearance of the spectrum. In this spectrum the group frequencies of the hydrocarbon portion can be assigned as in Table 9.28.

**TABLE 9.28  Group Frequency Assignments for Chlorobenzene**

| $\tilde{\nu}$ (cm$^{-1}$) | Intensity | Mode Description |
| --- | --- | --- |
| 3080 | Medium | C—H stretch, C—H bonded to $sp^2$ carbon |
| 1585 | Strong | $\nu_{8a}$ ring stretching |
| 1575 | Weak | $\nu_{8b}$ ring stretching |
| 1475 | Strong | $\nu_{19a}$ ring stretching |
| 1450 | Strong | $\nu_{19b}$ ring stretching |
| 747 | Strong | C—H all in-phase, out-of-plane bend |
| 688 | Strong | Ring deformation |
| 1945, 1865, 1788, 1733 | All weak | Sum tones, out-of-plane C—H bends, pattern matches monosubstitution of ring |

**1.** Divide the spectrum at 1350 cm$^{-1}$.

**2.** Above 1350 cm$^{-1}$, absorption bands have a high probability of being good group frequencies. The interpretation is usually reliable and free from ambiguities. We can be much more confident of our assignments in this region even with rather weak bands.

**3.** Because of the reliability of the high-wavenumber region, we always begin the interpretation of a spectrum at this end.

**4.** Bands below 1350 cm$^{-1}$ may be either group frequencies or fingerprint frequencies. (The fingerprint region is often considered to begin at slightly higher wavenumber values, close to 1500 cm$^{-1}$, but for interpretation purposes we will consider the region to be that lying below 1350 cm$^{-1}$.)

**5.** Below 1350 cm$^{-1}$, group frequencies are less easily assigned. In addition, even if a reliable group frequency occurs in this region, absorption at that frequency is not necessarily a result of that mode.

**6.** To make more confident assignments below 1350 cm$^{-1}$, it is helpful to be able to associate a secondary property, such as band shape, with the particular mode. For example, the band is very intense, broad, sharp, occurs as a characteristic doublet, gives the correct frequency shift on isotopic substitution, or the like.

**7.** A good rule to remember is that in the fingerprint region the **absence** of a band is more important than the presence of a band.

**8.** Before beginning the interpretation, establish the sampling conditions and as much other information about the sample as possible (such as molecular weight, melting point, boiling point, color, odor, elemental analysis, solubility, and refractive index).

**9.** In the interpretation try to assign the most intense bands first. These bands very often will be associated with a polar functional group.

**10.** Do not try to assign all the bands in the spectrum. Fingerprint bands are unique to a particular system. Occasionally, intense bands will

**STRATEGIES FOR INTERPRETING INFRARED SPECTRA**

be fingerprint type absorptions; these bands, generally, will be ignored in the interpretation. Fingerprint bands do, however, play an extremely important role when infrared data are employed for identification purposes.

**11.** The correlation chart (back endpaper) can act as a helpful quick aid for checking potential assignments. It is not a substitute for understanding the theory and operation of group frequency logic. *The use of the correlation chart without a good knowledge of group frequencies is the shortest path to disaster!*[2]

**12.** Try to utilize the so-called *macro group frequency* approach. That is, if the functionality or molecular structural group requires the presence of more than a single group frequency mode, make sure that all modes are correctly represented. The *macro frequency train* represents a very powerful approach to the interpretation of relatively complex spectral data. This technique is at the core of current work on the automatic computer interpretation of infrared spectra. Contained in the product characterization section of the example given at the end of this chapter is a detailed discussion that demonstrates the operational use of the *macros*. Careful reference to this discussion will be helpful in the initial stages of learning this interpretation technique. It should become relatively easy to extend this interpretive approach to other reactions by reference to common infrared library files. This last suggestion is perhaps the most important strategy to master in learning to interpret infrared spectra. Practice on *macro group frequencies* will pay big dividends in the laboratory.

---

**INFRARED SPECTROSCOPY**    **Instrumentation and Sample Handling**

---

The infrared spectrometer is a complex and expensive instrument that you will encounter on a regular basis in the organic laboratory. TREAT THE IR SPECTROMETER WITH RESPECT. The instrument is particularly adapted to the characterization of microscale products. Data obtained from the IR spectrometer will be used many times throughout the semester. You will become more proficient at obtaining this type of spectral information as you gain experience with the instrument. Practice in preparing the sample for instrumental analysis can significantly improve the quality of the spectra.

The late Robert B. Woodward, Nobel Laureate, one of the most outstanding synthetic organic chemists of this century, once stated: "But no single tool has had more dramatic impact upon organic chemistry than infrared measurements. The development [of easily] operated machines for the determination of infrared spectra has permitted a degree of immediate and continuous analytical and structural control in synthetic organic work which was literally unimaginable. . . . The power of the method grows with each day, and further progress may be expected for a long time to come. Nonetheless, its potential is even now greater than many realize."[3]

Although the preceding statement was made some time ago (when NMR spectroscopy was in its infancy), and the more dramatic impact has shifted to high-field NMR and high-resolution mass spectrometry (MS) for structural elucidation, Woodward's final point still holds true when it comes to substance identification. Infrared data and particularly spectra obtained from gas chromatographic (GC) coupled IR interferometers, continue to play a dominant role in compound identification and characterization (where nanogram, $10^{-9}$ g, and even picogram, $10^{-12}$ g, sensitivities are routinely required in these measurements). The infrared spectrum is a reflection of the vibrational energy levels present in a molecule. As no two

---

[2] Bellamy, L. J. *The Infrared Spectra of Complex Molecules,* 3rd ed., Chapman and Hall: London, 1975, p. 3.

[3] Woodward, R. B. In *Perspectives in Organic Chemistry;* A. Todd, Ed.; Interscience: New York, 1956, p. 157.

substances have the same exact set of vibrational frequencies, no two substances have identical infrared spectra. There are at present a number of excellent collections of infrared spectra that may be used for reference comparisons.[4]

The use of IR spectra to identify the presence of particular functional groups in a molecule is one of the principal applications of this technique. Many of the absorption bands present in an IR spectrum can be related to specific arrays of atoms in that material. These "group frequencies" can be extremely helpful in the interpretation of experimental results. With the recent coupling of $^{13}C$ NMR data and IR group frequency arguments, the interpretation of the functional group environment has become extremely powerful.

## INSTRUMENTATION

The workhorse infrared instrument used for routine characterization of materials in the undergraduate organic laboratory is still the optical null double-beam grating spectrometer (Fig. 9.40). For a discussion of double-beam spectrometers, see UV–vis Instrumentation discussions, p. 688. Although most undergraduate instructional laboratories still utilize this type of instrumentation, the winds of change are blowing. Over the last decade the development of interferometric nondispersive spectrometers, which depend on high-speed computer manipulation of interferograms to generate the data in spectral form, have invaded the research laboratory. It is now clear that "low-cost" infrared interferometers will become the next generation's infrared instructional instrumentation. The spectra utilized in the interpretive discussions that follow were generated on a prototype of this kind of infrared instrumentation, the Perkin–Elmer model 1600.

This instrument will acquire 16 scans and carry out the required calculations in 42 s. While the spectrum is being printed out (~40 s) the data on a second sample can be acquired. The 42-s acquisition data are significantly superior to those currently recorded by dispersive instruments that take from 5 to 8 min to scan a sample from 4000 to 600 $cm^{-1}$.

[4] For example, see the Bibliography.

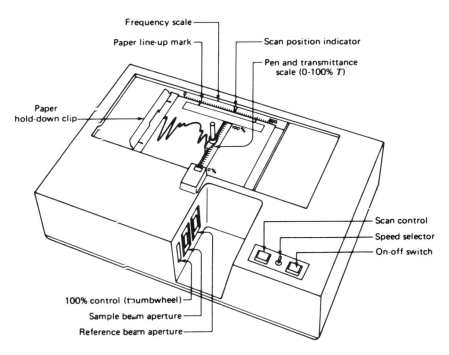

**FIGURE 9.40** The Perkin–Elmer model 710B IR spectrometer. *From Zubric, James W. The Organic Chem Lab Survival Manual, 3rd ed.; Wiley: New York, 1992. (Reprinted by permission of John Wiley & Sons, Inc., New York.)*

Thus, as infrared interferometers are now beginning to enter the instructional laboratory in significant numbers, a short description of this instrumentation becomes important for understanding how modern infrared data are acquired.

The design of the majority of current interferometers, employed in infrared measurements, is based on adaptations of a two-beam system originally developed by Michelson for photon velocity measurements (see Fig. 9.41).

The principal function of the optical system of the two-beam interferometer is (1) to divide (beamsplitter) the source radiation into two components traveling separate paths; (2) to introduce an oscillating path difference between these two beams (by moving one of the mirrors back and forth while keeping the other mirror fixed; this is however, not the case in the PE 1600, which introduces a slight wrinkle); and (3) to create an interference condition at the intersection of the beams during their second pass through the beamsplitter. Thus, the path difference is varied in a cyclic fashion and the resulting intensity variation of the recombined beams is measured as a function of the moving mirror or changing path length (see Fig. 9.41). The intensity fluctuation with time generates an interferogram that ultimately yields the desired spectral information when it undergoes Fourier transformation (FT).

The path difference between the two beams where the mirror travel is $x$ is, therefore, $2x$. This difference is defined as the retardation ($\delta$). When the two paths are equivalent, there is zero retardation and both beams will be in phase when they intersect at the beam splitter the second time. Under these conditions constructive interference occurs, and the intensity of the beam reaching the detector is the sum total of both beams. When the moving mirror displacement equals $\frac{1}{4}\lambda$ (wavelength), the retardation is $\frac{1}{2}\lambda$ and destructive interference takes place during recombination of the beams at the beamsplitter. Under these conditions, no signal reaches the detector.

If we now displace the movable mirror at a constant velocity, the signal at the detector will vary sinusoidally with a maximum occurring

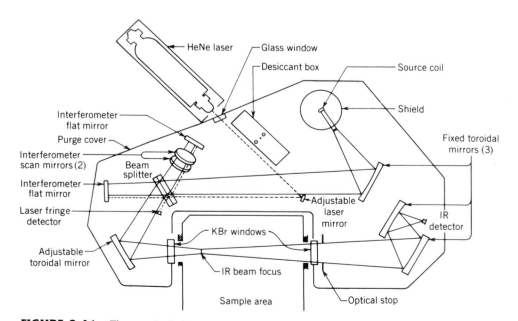

**FIGURE 9.41**   The optical diagram of the Perkin–Elmer model 1600 infrared interferometer. *(Courtesy of the Perkin–Elmer Corp., Norwalk, CT.)*

every time retardation is an integral multiple of $\lambda$. The source intensity $[I\bar{\nu}]$ at the detector, is therefore a function of the retardation, $\delta$, and $I'(\delta)$.

At other values of $\delta$, the intensity of the exit beam is given by:

$$I'(\delta) = 0.5\ I(\bar{\nu})\{1 + \cos 2\pi\delta/\lambda\}$$
$$= 0.5\ I(\bar{\nu})\{1 + \cos 2\pi\bar{\nu}\delta\}$$

The parameter $I'(\delta)$ is a function that involves a constant or *dc* term $[0.5\ I(\bar{\nu})]$ and a modulated *ac* term $[0.5\ I(\bar{\nu}) \cos 2\pi\bar{\nu}\delta]$. It is the latter part of the expression that is usually referred to as the interferogram. If we employ a monochromatic (single-frequency) source, the intensity at the detector is simply defined as

$$I(\delta) = 0.5\ I(\bar{\nu})\ \cos 2\pi\bar{\nu}\delta$$

Spectral information is extracted from the interferogram by calculating the cosine Fourier transform of $I(\delta)$. Normally, as the moving mirror is displaced at constant velocity, $v$ (cm/s), we consider how the interferogram varies with time, $I(t)$, rather than as a function of retardation, $I(\delta)$. In this form at $t$ (s) from the zero point the retardation ($\delta$) may be expressed as $\delta = 2vt$ in centimeters.

Substituting in the above, we obtain

$$I(t) = 0.5I(\bar{\nu})\ \cos 2\pi\bar{\nu}(2vt)$$

The units of the abscissa of the interferogram will be the inverse of the units of the spectrum, in this case reciprocal centimeters ($\text{cm}^{-1}$).

The amplitude of the signal at time $t$ for the cosine wave frequency $f$ may be expressed as

$$A(t) = A_0 \cos 2\pi f t \qquad (A_0 = \text{maximum amplitude})$$

Thus, the frequency $f$ of the radiation in terms of $\bar{\nu}$ ($\text{cm}^{-1}$) for the elementary case of a monochromatic interferogram, $I(t)$, corresponds to

$$f\bar{\nu} = 2\bar{\nu}v.$$

Thus, in this example, in which we have obtained the interferogram of a monochromatic source, the determination of the "spectrum" via Fourier transform is simple, as the amplitude and wavelength can both be measured directly (see Fig. 9.42).

The extrapolation to the case of sources that emit continuous bands of radiation, however, presents a considerably more complex interferogram, which requires the use of digital computers to compute the Fourier transform and generate the data in terms of an absorption spectrum. With the advent of high-speed computation, and mathematical tricks that have shortened the calculation routine, it is now possible to generate infrared spectral data very quickly and accurately via interferometric instrumentation.

The older generation optical null, double-beam arrangement, although not nearly as sensitive and quick as the interferometric instruments, does have some distinct advantages. For example, absorption signals from any substances present in both beams in equal quantities are automatically canceled. Although atmospheric water and carbon dioxide have particularly strong absorption bands in the infrared, the double-beam instrument automatically subtracts the absorption of these substances from

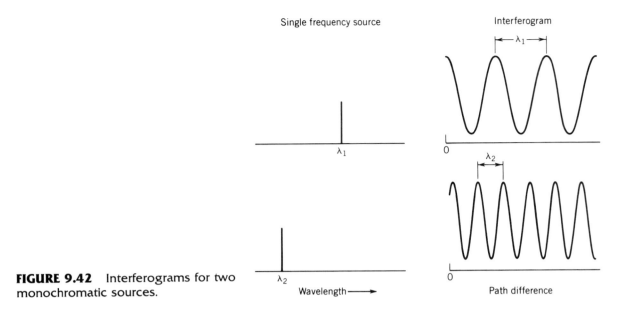

**FIGURE 9.42** Interferograms for two monochromatic sources.

the recorded spectrum. In the case of the FT systems this problem is resolved by subtracting a background spectrum, which can only be carried out sequentially with the acquisition of the sample spectrum.

## SAMPLE HANDLING IN THE INFRARED

For a spectrum to be obtained in the infrared region, the sample must be mounted in a cell that is transparent to the radiation. Glass and quartz absorb in this spectral region and cells constructed of these materials cannot be employed. Alkali metal halides have large regions of transmission in the infrared as do silver halides. Of these materials, sodium chloride, potassium bromide, and silver chloride are most often used as cell windows in infrared sampling.

### Liquid Samples

For materials boiling above 100 °C, the procedure is very simple. Using a syringe or Pasteur pipet, place 3–5 $\mu$L of sample on a polished plate of sodium chloride or silver chloride. Then cover it with a second plate of the same material and clamp it in a holder that can be mounted vertically in the instrument. Be sure that the plates are clean when you start and when you are through! Obviously, the sodium chloride cannot be cleaned with water. Silver chloride is very soft and scratches easily; it also must be kept in the dark when not in use because it darkens quickly in direct light. Spectra obtained in this fashion are referred to as *capillary film spectra* (see Fig. 9.43).

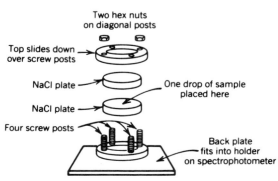

DO NOT OVERTIGHTEN

Two hex nuts on diagonal posts

Top slides down over screw posts

NaCl plate — One drop of sample placed here

NaCl plate

Four screw posts

Back plate fits into holder on spectrophotometer

**FIGURE 9.43** IR salt plates and holder. *From Zubric, James W. The Organic Chem Lab Survival Manual, 3rd ed.; Wiley: New York, 1992. (Reprinted by permission of John Wiley & Sons, Inc., New York.)*

These samples generally require a sealed cell constructed of either sodium chloride or potassium bromide. Such cells are expensive, they need careful handling and maintenance, and are assembled as shown in Figure 9.44.

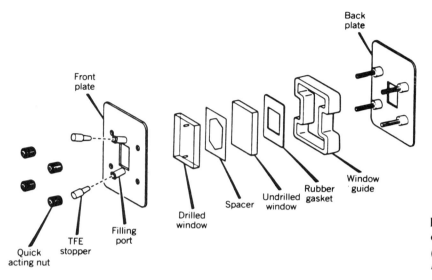

**FIGURE 9.44** Sealed demountable cell or demountable cell with ports. *(Courtesy of the Perkin-Elmer Corp., Norwalk, CT.)*

**Solid Samples**

Solid powders can be mounted on horizontal sodium chloride plates, and the beam diverted through the sample by mirrors. This procedure would make sample preparation very easy for solids. Unfortunately, powders tend to scatter the entering radiation very efficiently by reflection, refraction, and molecular scattering. Some of these effects become rapidly magnified at higher frequencies since they vary as the fourth power of the frequency. The result of solid-sample scattering is that large amounts of energy are removed from the sample beam. This results in poor absorption spectra, as the instrument is forced to operate at very low energies. The detector cannot differentiate between a drop in energy from absorption or one derived from scattering.

For materials melting below 80 °C the simplest technique, however, is to mount the sample between two salt plates and **gently** apply a heat lamp until melting occurs. With the fast acquisition times of interferometers, the melting point range is now as high as 100 °C (i.e., the spectrum can be obtained so rapidly that the sample does not have time to cool and crystallize – heated cells are used in research laboratories, but they are rather expensive and difficult to maintain). For substances melting above 100 °C the sampling routine most often employed to avoid scattering problems is the potassium bromide (KBr) disk. The sample (2–3 mg) is finely ground in a mortar, the finer the better (lower reflection or refraction losses). Then, 150 mg of previously ground and dried KBr is added to the mortar and quickly mixed by stirring, not grinding, it with the sample. (Potassium bromide is very hygroscopic and will rapidly pick up water while being ground in an open mortar.) When mixing is complete the mixture is transferred to a die and pressed into a solid disk. Potassium bromide will flow under high pressure and seal the solid sample in the alkali metal halide matrix. Potassium bromide is transparent to infrared radiation in the region of interest. Most important, however, the KBr makes a much better match of the refractive indexes between the sample and its matrix than does air. Thus, reflection and refraction effects at the crystal faces of the sample are greatly suppressed. Several styles of dies are commercially

GENTLY

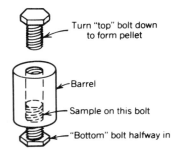

**FIGURE 9.45** The KBr pellet mini-press. *From Zubric, James W. The Organic Chem Lab Survival Manual, 3rd ed.; Wiley: New York, 1992. (Reprinted by permission of John Wiley & Sons, Inc., New York.)*

available. For routine use a die consisting of two stainless steel bolts and a barrel is the simplest to operate. The ends of the bolts are polished flat to form the die faces. The first bolt is seated to within a turn or two of the head. Then the sample mixture is added (avoid breathing over the die while adding the sample). The second bolt is firmly seated in the barrel, and then the clamped assembly is tightened by a torque wrench to 240 in.-lb. After standing for 1.5 min, the two bolts are removed, leaving the KBr disk mounted in the center of the barrel, which can then be mounted in the instrument. After the spectrum of the sample is run, the disk can be retrieved and the sample recovered if necessary (Fig. 9.45). *Always clean the die immediately after use. KBr is highly corrosive to steel.*

The standard techniques of sample preparation employed to obtain infrared spectra of microscale laboratory products are the use of capillary films with liquids on NaCl or AgCl plates and the use of KBr disks and melts in solids.

When infrared spectra are obtained, it is important to establish that the wavenumber values have been accurately recorded. Successful interpretation of the data often depends on very small shifts in these values. Calibration of the frequency scale is usually accomplished by obtaining the spectrum of a reference compound, such as polystyrene film. To save time, record absorption peaks only in the region of particular interest.

**NOTE.** *It should be noted that most of the infrared spectra found in the experimental sections of the text which are Fourier transform derived (Perkin-Elmer 1600) have been plotted on a slightly different scale than the spectra presented in Chapter Nine. The former spectra utilize a 12.5-cm$^{-1}$/mm format below 2000 cm$^{-1}$ and undergo a 2:1 compression above 2000 cm$^{-1}$ (25 cm$^{-1}$/mm).*

## QUESTIONS

**9-1.** The form of the C—H out-of-plane bending vibrations of the vinyl group are shown below:

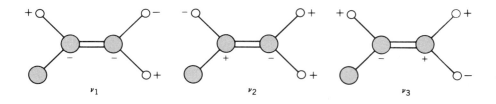

The first two modes give rise to excellent group frequencies, while the third fundamental does not lend itself to these correlations.

    a. Explain the factors that lead to the third mode being such a poor group frequency.

    b. Predict the location in the spectrum of the third fundamental.

**9-2.** In the figure below, the mass of the terminal hydrogen atoms on the acetylene is hypothetically varied from zero to infinity. The response of the C—H symmetric stretching (3374 cm$^{-1}$) and triple-bond stretching (1974 cm$^{-1}$) modes to the change in mass is shown.

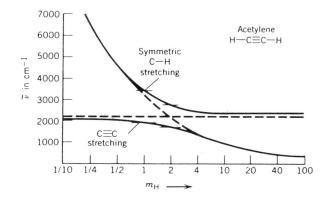

a. Calculate the expected D-isotopic shift for the C—H symmetric stretching. Is the hypothetical value close to the calculated value? Explain.

b. Explain why the triple bond stretching frequency is approximately 100 cm$^{-1}$ higher for high-mass terminal isotopes as compared to the low-mass terminal isotopes.

**9-3.** Acetylene has two C—H groups. It will have two C—H stretching frequencies, the in-phase and out-of-phase stretching modes. The in-phase (symmetric) stretch occurs at 3374 cm$^{-1}$ and the out-of-phase stretch at 3333 cm$^{-1}$. Explain why the in-phase vibration is located at a higher frequency than the out-of-phase stretch in the case of acetylene.

**9-4.** The carbonyl stretching frequencies of a series of benzoyl derivatives are listed below:

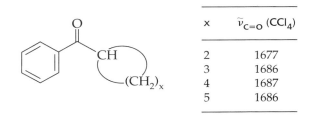

| x | $\tilde{\nu}_{C=O}$ (CCl$_4$) |
|---|---|
| 2 | 1677 |
| 3 | 1686 |
| 4 | 1687 |
| 5 | 1686 |

If we consider the $\nu_{C=O}$ of acetone at 1715 cm$^{-1}$ as a reference frequency, then identify the factors affecting $\nu_{C=O}$ in the series of compounds listed.

**9-5.** Explain how mass effects act to lower the carbonyl frequency, as well as how inductive and hyperconjugation effects act to raise the carbonyl frequency of aldehydes relative to ketones.

**9-6.** The carbonyl stretching frequency of aliphatic carboxylic acids in dilute solution is located near 1770 cm$^{-1}$. This frequency is much higher than the carbonyl frequency of these substances when measured neat (~1720 cm$^{-1}$). Also, it is considerably higher than the corresponding simple aliphatic ester (1745 cm$^{-1}$) value. Explain.

**9-7.** In a number of cases, dipolar interactions control the frequency shifts found in carbonyl stretching vibrations. Table 9.9 lists wavenumber shifts in going from neat to dilute nonpolar solutions.
Explain the observed values.

**9-8.** The antisymmetric —CH$_2$—CO—O— stretching vibration in carboxylic acids is heavily mixed with the in-plane bending mode of the O—H group. In alcohols these two vibrations seldom show evidence of mechanical coupling. Explain.

**9-9.**  Conjugation of the functional group in alkyl isocyanates has little impact on the antisymmetric —N=C—O stretching vibration located near 2770 cm$^{-1}$. Explain.

**9-10.**   In the infrared spectrum of 2-aminoanthraquinone (**I**) two carbonyl stretching frequencies are observed at 1673.5 and 1625 cm$^{-1}$.

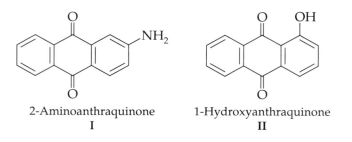

2-Aminoanthraquinone        1-Hydroxyanthraquinone
         **I**                           **II**

a.  Assign carbonyl bands in the infrared spectrum to the carbonyl groups in Structure **I** and explain your reasoning.

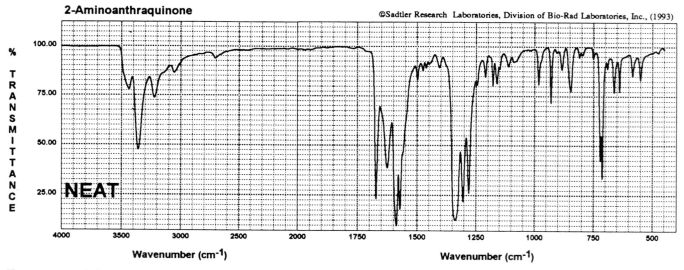

IR spectrum: 2-Aminoanthraquinone.

b.  The infrared spectrum of 1-hydroxyanthraquinone (**II**) also exhibits two carbonyl frequencies, which are located at 1675 and 1637 cm$^{-1}$. Assign the carbonyl groups to the related absorption bands. Explain your reasoning.

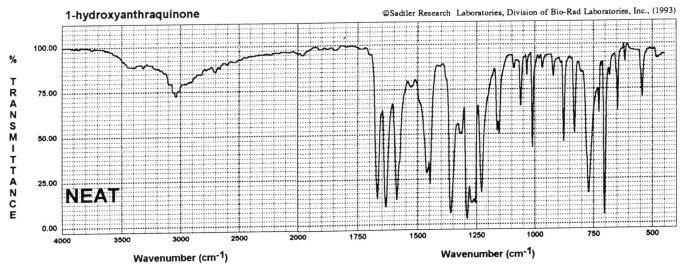

IR spectrum: 1-Hydroxyanthraquinone.

c. The spectrum of 2-hydroxyanthraquinone exhibits a single carbonyl stretching frequency near 1673 cm$^{-1}$. Explain why a single carbonyl band would be expected in the system and why this vibration is located at 1673 cm$^{-1}$.

**9-11.** Suggest a possible structure for the hydrocarbon $C_6H_{14}$, which has the infrared spectrum shown below:
Is there more than one correct structure?

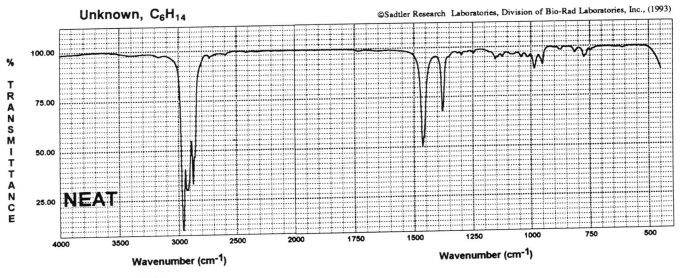

IR unknown spectrum: $C_6H_{14}$.

**9-12.**   The hydroxylamine **I** can be oxidized by $MnO_2$ to the amide oxo-haemanthidine (**II**). In dilute solution the carbonyl absorption band of **II** occurs at 1702 cm$^{-1}$. Explain this observation.

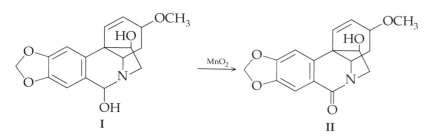

**9-13.**   Identify the following olefinic hydrocarbons. All samples were obtained from distillation cuts in the $C_6$ boiling range.

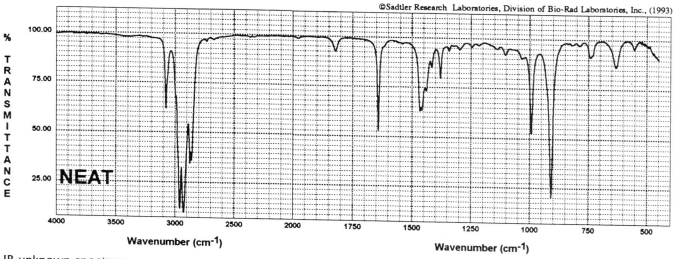

IR unknown spectrum *a*.

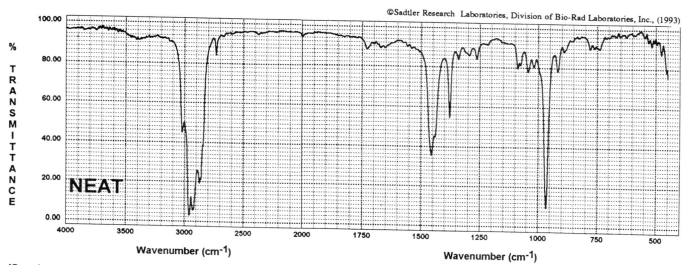

IR unknown spectrum *b*.

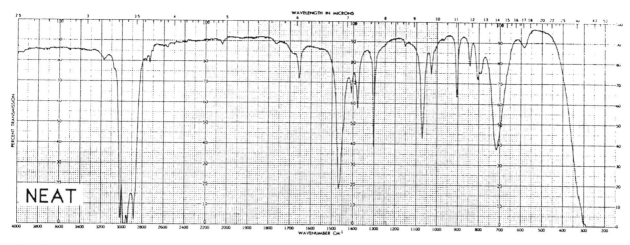

IR unknown spectrum *c.* *(Courtesy of Bowdoin College)*

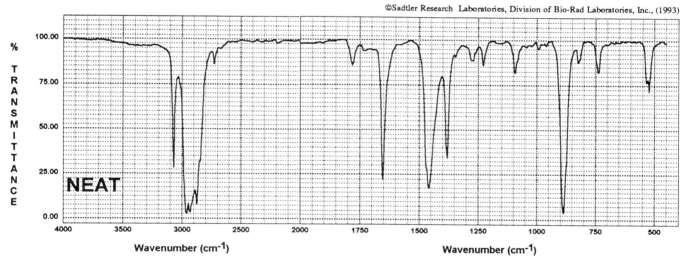

©Sadtler Research Laboratories, Division of Bio-Rad Laboratories, Inc., (1993)

IR unknown spectrum *d.*

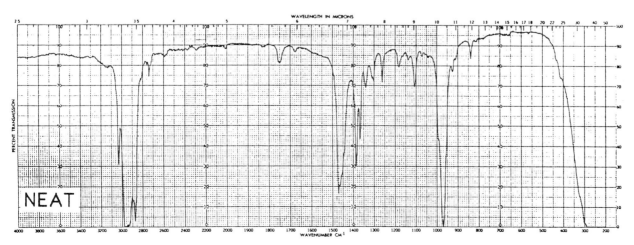

IR unknown spectrum *e.* *(Courtesy of Bowdoin College)*

**9-14.**  The infrared spectra of the xylene isomers, and an additional aromatic hydrocarbon, are given below. Assign the spectra and suggest a potential structure for the remaining unknown substance.

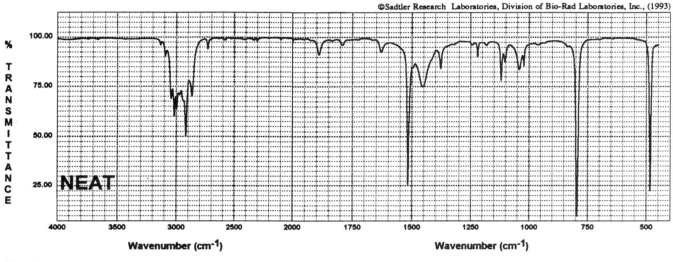

IR unknown spectrum *a*.

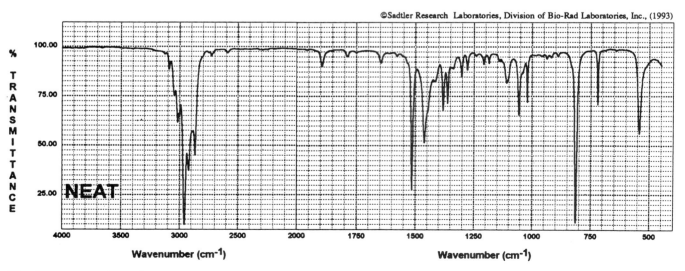

IR unknown spectrum *b*.

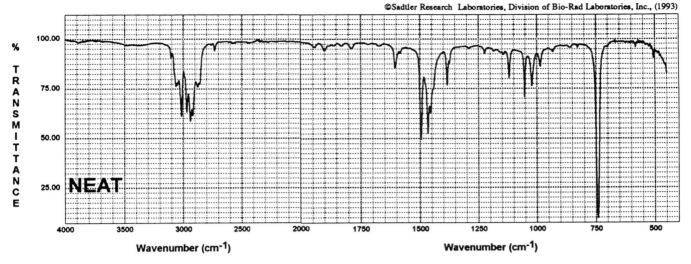

IR unknown spectrum *c*.

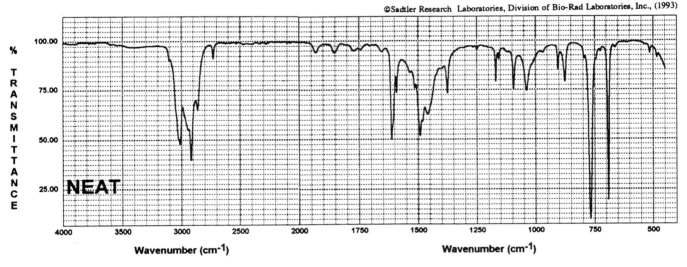

IR unknown spectrum *d*.

**9-15.** The C—H stretching mode of chloroform ($CHCl_3$) which occurs at 3022 cm$^{-1}$, is one of the rare exceptions to the 3000-cm$^{-1}$ rule. What is the rule? Suggest an explanation for this exception.

*Aldrich Library of Infrared Spectra*, Aldrich Chemical Co., Inc., 940 West Saint Paul Avenue, Milwaukee, WI 53233. 3rd ed., 1981, 10,000 spectra arranged by chemical type.

API collection (American Petroleum Institute). About 4500 spectra. M.C.A. collection (Manufacturing Chemists' Association). About 3000 spectra. Chemical Thermodynamics Property Center, Texas A&M College, Department of Chemistry, College Station, TX 77843.

Coblentz Society 10,000 spectra. Marketed through Sadtler Research Labs, 3316 Spring Garden Street, Philadelphia, PA 19104.

**BIBLIOGRAPHY**

D.M.S. System (Documentation of Molecular Spectra). About 15,000 spectra. IFI/Plenum Data Corp., 227 West l7th Street, New York, 10011.

Grasselli, J. G. *Atlas of Spectral Data and Physical Constants for Organic Compounds;* CRC Press: Cleveland, OH, 1973.

Japanese collection. About 17,000 spectra. Good quality, but labeled in Japanese. Infrared Data Committee of Japan. Nankodo Co., Haruki-Cho, Tokyo.

Sadtler Library. About 80,000 spectra of single compounds; about 12,000 spectra of commercial products. Sadtler Research Labs, 3316 Spring Garden Street, Philadelphia, PA 19014.

# NUCLEAR MAGNETIC RESONANCE SPECTROSCOPY:
## Introduction to Nuclear Magnetic Resonance

**NUCLEAR SPIN**

Nuclear spin is an energy property intrinsic to a nucleus, and analogous to the electron spin that plays such an important role in determining electron configurations. Nuclear spin values are quantized, as are electron spins, and are represented by $I$, the nuclear spin quantum number. Nuclear spin quantum numbers range from 0 through $\frac{7}{2}$, in increments of $\frac{1}{2}$. For nuclei of greatest interest to organic chemists, the $^1H$, $^{13}C$, $^{19}F$, and $^{31}P$ nuclei have spins of $\frac{1}{2}$; the $^{12}C$, $^{16}O$, and $^{32}S$ nuclei have spins of 0 (and thus cannot be observed by nuclear magnetic resonance spectroscopy, NMR); the $^2H$ (deuterium, D) and $^{14}N$ nuclei have a spin of 1. Since any spinning charged particle (or body) produces a magnetic moment, a nucleus with a nonzero spin quantum number has a magnetic moment, $\mu$.

Nuclear spin values are quantized because the nuclear angular momentum, and thus the nuclear magnetic moment, is quantized. When placed in an external magnetic field, nuclei orient their magnetic moments in certain ways with respect to the magnetic field, which is assumed to be aligned with the $z$ axis of a Cartesian coordinate system. These orientations are referred to as the $z$ components of the nuclear magnetic moment, $\mu_z$. For a nucleus with a spin of $\frac{1}{2}$, $\mu_z$ may be $+\frac{1}{2}$ or $-\frac{1}{2}$. In general, for a nucleus of spin $I$, the $\mu_z$ takes quantized values from $[-I, -I + 1, \ldots, I - 1, I]$; or $(2I + 1)$ different values in all. For this discussion we will limit ourselves to nuclei with spin $\frac{1}{2}$, since this is easier to describe, and since most nuclei of interest in organic chemistry are of spin $\frac{1}{2}$.

When placed in a static magnetic field of strength $H_0$, the magnetic moment, $\mu_z$, of the spinning nucleus precesses about the magnetic field at a frequency, $\nu$, such that $\nu = \gamma H_0/2\pi$, where $H_0$ is the strength of the applied magnetic field, and $\gamma$ is a characteristic property of the nucleus known as the gyromagnetic ratio. When a nucleus of spin $\frac{1}{2}$ is placed in a magnetic field, the energies of the $\mu_z = +\frac{1}{2}$ and $-\frac{1}{2}$ states are separated, since in one spin state the nuclear magnetic moment is aligned with the applied magnetic field; in the other spin state the nuclear magnetic moment is opposed to the applied magnetic field.

The amount of separation of the two energy states, $\Delta E$, is proportional to the magnetic field, and is given by the following expression,

$$\Delta E = \frac{h\gamma H_0}{2\pi} = h\nu$$

When nuclei in the magnetic field are exposed to radiation of the proper frequency, transitions between the two energy states are stimulated, and the nucleus is said to be in *resonance*, or to *resonate*. This transition occurs when the frequency and the energy difference are related by the Planck relation, $\Delta E = h\nu$, and thus the sample will absorb energy of frequency $\nu$. The study of these energy changes is known as <u>n</u>uclear <u>m</u>agnetic <u>r</u>esonance, or NMR, spectroscopy.

---

**INSTRUMENTATION**

In an NMR spectrometer, the magnetic field is provided by a large permanent magnet, electromagnet, or superconducting electromagnet. Commercially available NMR spectrometers have magnets with field strengths that range from 1.4 to 16.3 Tesla (the earth's magnetic field at its surface is roughly $5 \times 10^{-5}$ Tesla), and thus operate at frequencies from 60 to 700 MHz for protons. In general, most spectrometers with an operating frequency above 100 MHz use a superconducting electromagnet.

Traditionally, NMR spectra were acquired either by holding the applied magnetic field constant and sweeping the radio frequency (rf), or by holding the rf constant and sweeping the applied magnetic field. Energy absorption by the sample was detected, and the result was the NMR spectrum, a plot of intensity (of energy absorption) versus frequency (or field). This instrumental technique is referred to as continuous wave, or CW, spectroscopy (Fig. 9.46). Over the last 15 years, however, it has been commonly replaced by pulsed, or Fourier transform (FT), NMR spectroscopy. Among many other benefits, FT-NMR spectroscopy allows very rapid acquisition of spectral data, which permits analysis of small samples and of rare nuclei, such as $^{13}C$.

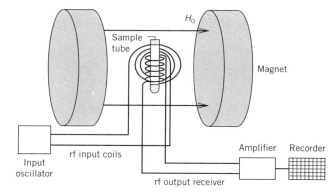

**FIGURE 9.46** Schematic of NMR spectrometer. *(Reprinted with permission of John Wiley & Sons, New York.)*

The basic principles of FT-NMR spectroscopy can be qualitatively explained as follows. Take, for example, an NMR spectrum that contains a single peak at a given frequency. The graph of this spectrum (Fig. 9.47) is a plot of intensity versus frequency. The same information can be conveyed, however, by a plot of intensity versus time that shows a cosine wave at the frequency described by the graph of the usual NMR spectrum. This is shown in Figure 9.48, and for a spectrum with a single frequency, this plot of intensity versus time is almost as easy to interpret as the usual NMR spectrum shown in Figure 9.47.

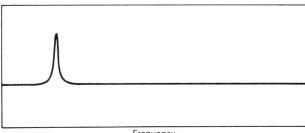

**FIGURE 9.47** Intensity versus frequency (usual NMR spectrum).

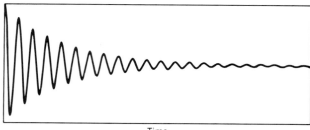

**FIGURE 9.48**    Intensity versus time.

Of course, it would be very difficult to determine the frequencies of many superimposed cosine waves from this kind of plot, and it would be at best awkward to interpret a complex NMR spectrum presented in such a fashion (Fig. 9.49). The use of the Fourier transform allows us mathematically to interconvert these time domain (Fig. 9.49) and frequency domain spectra (Fig. 9.50). Fourier transform of the apparently complex spectrum in Figure 9.49 gives the spectrum in Figure 9.50, and it is then easy to see that there are actually only three different resonance signals contained in the time domain of the data of Figure 9.49.

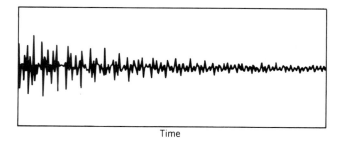
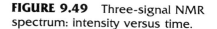

**FIGURE 9.49**    Three-signal NMR
spectrum: intensity versus time.

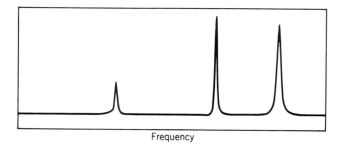

**FIGURE 9.50**    Three-signal NMR
spectrum: intensity versus frequency.

Fourier transform NMR spectra are obtained by applying a short (~1–10 μs), high-powered pulse of rf energy to the sample (Fig. 9.51). This pulse affects all the nuclei to be observed. Before the pulse is applied, the equilibrium net nuclear magnetization is aligned with the applied magnetic field, along the $z$ axis. The coordinate system is presumed to be rotating about the $z$ axis at the frequency of the rf pulse. The pulse, applied down

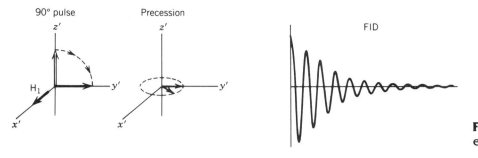

**FIGURE 9.51** Basic pulsed NMR experiment.

the $x$ axis, applies a torque to the nuclear magnetic moments and rotates them into the $xy$ plane. At this point the pulse is turned off and the nuclear magnetic moments return to their equilibrium alignments. In the process, they precess about the $z$ axis (applied magnetic field) in the $xy$ plane and induce a current in a detector coil, which can be thought of as being aligned with the $y$ axis. This current varies in a sinusoidal manner, and the observed frequency will be the difference between the resonance frequency of the nuclei and the frequency of the rf pulse.

The detected signal, which is called the free induction decay (FID), is digitized and stored. For small organic molecules in a nonviscous solution, the FID will disappear after a few seconds, which corresponds to the time it takes the nuclei to regain equilibrium alignments after the rf pulse. Thus, an entire $^1$H NMR spectrum can be obtained in approximately 2 seconds, in contrast to the 10 or 15 min usually needed to obtain a CW spectrum. A major advantage of FT-NMR is that many spectra of a sample can be rapidly obtained and added together to increase the signal-to-noise (S/N) ratio. Noise is presumably random about some zero level, so when many spectra are added together, the noise level is reduced, while real signals are reinforced when added. The S/N ratio is proportional to the square root of the number of spectra added together. Thus, one can obtain the $^1$H spectrum of a 10-$\mu$mol sample in a few minutes. With FT-NMR, it is possible to obtain spectra of isotopes that are insensitive and/or of low natural abundance, as well as spectra of large biological molecules in dilute solution. By adding a few hundred spectra together, an adequate $^{13}$C NMR spectrum of a 60-mg sample of molecular weight 200 can be obtained in about 20 min.

## CHEMICAL SHIFT

In a molecule, the magnetic field at a nucleus depends not only on $H_0$, the field generated by the instrument (the external field), but also on the magnetic fields associated with the electron density near the nucleus. Electrons are influenced by the external field in such a way that their motion generates a small magnetic field that opposes the applied field, and reduces the actual field experienced at the nucleus. This reduction is very small (relative to the external field) and is on the order of 0.001%, 10 ppm, for most protons, and about 200 ppm for $^{13}$C nuclei. This reduction of the external field is known as *shielding*, and it gives rise to differences in the energy separation for nuclei in different electronic environments in a molecule. The differences in the energy separation are known as *chemical shifts*.

The magnitude of the chemical shift depends on the nature of the valence and inner electrons of the nucleus and also even on electrons that are not directly associated with the nucleus. Chemical shifts are influenced by inductive effects, which reduce the electron density near the nucleus

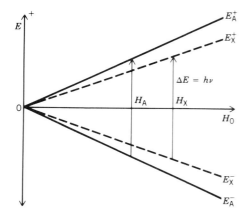

**FIGURE 9.52** The energy splitting for two chemically different protons. The differences between the $A$ energy levels (solid lines) and the $X$ levels (dashed lines) have been amplified for illustrative purposes. At 60 MHz, nucleus $A$ absorbs energy at field $H_A$ and nucleus $X$ absorbs energy at field $H_X$. Nucleus $X$ is said to be more strongly shielded than $A$. The resonance for $X$ is said to occur upfield of that for $A$.

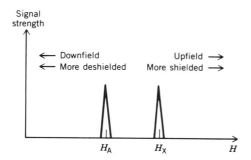

**FIGURE 9.53** The spectrum for the system in Figure 9.52 as it would be displayed. It is conventional to display the spectrum with magnetic field strength increasing to the right so that upfield (and more strongly shielded) is toward the right and downfield (and deshielded) is toward the left.

and reduce the shielding. The orientation of the nucleus relative to $\pi$ electrons also plays an important role in determining the chemical shift. A proton located immediately outside a $\pi$-electron system (as in the case of the protons on benzene rings) will be significantly deshielded. In most molecules the chemical shift is determined by a combination of these factors; chemical shifts are difficult to predict using theoretical principles, but have been well studied and can usually be easily predicted empirically upon comparison to reference data.

In an NMR spectrum, the absorption of rf energy is detected, and this is illustrated in Figure 9.52, where the energy absorption is shown for increasing frequency. In this example we illustrate the case with two different nuclei, $A$ and $X$. Since $A$ and $X$ are different, they absorb energy at different frequencies while in the same applied magnetic field.

The spectrum would be displayed as shown in Figure 9.53. The difference in the resonances is known as a *chemical shift* and is expressed in parts per million (ppm). The use of frequency units is cumbersome, and is complicated because NMR spectrometers of different magnetic field strengths (and thus operating frequencies) are used. The use of ppm units allows direct comparisons of spectroscopic data obtained on different instruments, when chemical shifts are referenced relative to a reference compound whose chemical shift is arbitrarily defined as 0 ppm. The accepted reference standard for $^1H$ and $^{13}C$ NMR in organic solutions is tetramethylsilane [$(CH_3)_4Si$, TMS]. The chemical shift relative to TMS is symbolized by $\delta$, which is defined below.

$$\delta = \frac{10^6(\nu - \nu_{TMS})}{\nu_{TMS}}$$

Tetramethylsilane is used as a reference substance for a number of reasons. It is more strongly shielded (Si is more electropositive than C) than most other protons and carbon atoms, and its resonance is thus well removed from other areas of interest in the NMR spectrum. Tetramethylsilane is inert and thus unlikely to react with the compound being analyzed, it is volatile (bp 26 °C) and thus easily removed after a sample has been analyzed, and its 12 identical protons per molecule provide a strong signal per molecule of TMS.

## SPIN–SPIN COUPLING

In a molecule with several protons, the exact frequency at which a proton resonates depends not only on the chemical shift of that proton, but also on the spin states of nearby protons. This occurs because the magnetic moments of the nearby protons can either shield or deshield the proton in question from the applied magnetic field, depending on the orientation of the nearby magnetic moments relative to the applied magnetic field. The extent of this perturbation is independent of the applied magnetic field strength. The effect of the spin state of one nucleus on another's resonance is known as *coupling* or *splitting*.

The spectra resulting from spin–spin coupling depend on the types of nuclei, the distance and geometry between the nuclei, the nature of the bonding, the electronic environment, and the total number of spin states possible. The latter may be illustrated by looking at the spectrum of an imaginary compound that has protons $H_A$ and $H_X$ on adjacent carbons, and connected by three bonds: $H_A$—C—C—$H_X$ (Fig. 9.54). In the first approximation we would expect one resonance for $H_A$ and one resonance for $H_X$, and the spectrum would resemble that shown in Figure 9.53. In the presence of coupling, the resonance for $H_A$ splits into two signals, one of which corresponds to $H_X$ having $\mu_z = +\frac{1}{2}$ and the other to $\mu_z = -\frac{1}{2}$. The coupling effect is symmetric in that the $H_X$ resonance also splits into two resonances, one for each spin state of $H_A$. The magnitude of the separation of the $H_A$ pair (a doublet) or the $H_X$ pair (also a doublet) is known as the coupling constant, or $J$. It is usually expressed in frequency units (Hz), since $J$ is independent of the magnetic field strength.

A simple way to explain this is to consider the effect the two possible spin states of $H_X$ have on the resonance frequency of $H_A$. The equilibrium population distribution of the two spin states in $H_X$ is very close to $1:1$, since $\Delta E$ is only about $10^{-3}$ cal/mol. Since $H_X$ has a magnetic moment, there are then two slightly different magnetic fields at $H_A$. We thus see two signals for $H_A$, one for those $H_A$ nuclei adjacent to $H_X$ nuclei with $\mu_z$ aligned with the applied magnetic field, and one signal for those $H_A$ nuclei adjacent to $H_X$ nuclei with $\mu_z$ aligned opposed to the applied field. While coupling between protons connected by more than three bonds does occur, its magnitude, $J$, is usually small and often not directly observed in a usual NMR spectrum.

The splitting becomes more interesting when there are several nuclei of one type. 1,1-Dibromoethane ($CH_3CHBr_2$) has three equivalent protons in the methyl group and one proton on the C-1 atom. The methyl group in this case exhibits rapid internal rotation so that its three protons are equivalent. The chemical shift for the C-1 proton is 5.86 ppm and that for the methyl protons is 2.47 ppm. Here we can see an example of decreased shielding resulting from the presence of electronegative substituents. Equivalent protons do not couple with one another (this is an important rule in interpreting spectra), but the methyl protons will affect the proton on C-1, and vice versa.

To analyze the splitting pattern, we need to consider the orientations of the nuclear magnetic moments, with respect to the applied magnetic field, for all three methyl protons. Since each of the three protons may have two spin states that are of nearly equal probability, there are $2^3 = 8$ possible combinations of spin states in all for the methyl protons. The net sums of these may have only four different values, as shown in Figure 9.55. The symbol $(+)$ is used to represent $\mu_z = +\frac{1}{2}$ for a single proton and

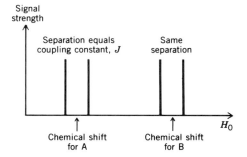

**FIGURE 9.54**   Spectrum of two chemically different protons that are coupled.

| Individual $\mu_z$ | SUM |
|---|---|
| $(+)(+)(+)$ | $+\frac{3}{2}$ |
| $(+)(+)(-)$ or $(+)(-)(+)$ or $(-)(+)(+)$ | $+\frac{1}{2}$ |
| $(+)(-)(-)$ or $(-)(+)(-)$ or $(-)(-)(+)$ | $-\frac{1}{2}$ |
| $(-)(-)(-)$ | $-\frac{3}{2}$ |

**FIGURE 9.55**   Possible Combinations of Spin States for a Methyl Group

$(-)$ is used to represent $\mu_z = -\frac{1}{2}$. Thus $(+)(+)(-)$ means that protons 1 and 2 have $\mu_z = +\frac{1}{2}$, while proton 3 has $\mu_z = -\frac{1}{2}$

The number of different $\mu_z$ states is $(2N + 1)$, where $N$ is the number of equivalent nuclei (of spin $\frac{1}{2}$). Thus the three methyl protons can generate four slightly different magnetic fields, and the proton on the C-1 of $CH_3CHBr_2$ sees (in different molecules) four different magnetic fields. Since these different magnetic fields are not of equal probability, but rather populated in a ratio of $1:3:3:1$, the four signals we see for the proton on the C-1 when coupled to the methyl group are of intensities $1:3:3:1$, and are referred to as a *quartet*. This is shown schematically in Figures 9.56a–9.56c.

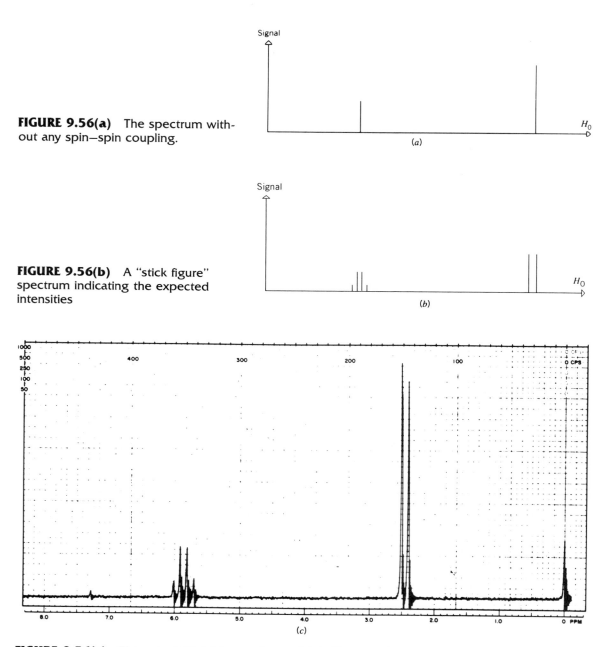

**FIGURE 9.56(a)**  The spectrum without any spin–spin coupling.

**FIGURE 9.56(b)**  A "stick figure" spectrum indicating the expected intensities

**FIGURE 9.56(c)**  The actual 60-MHz spectrum of 1,1-dibromoethane. The TMS signal at 0 ppm is seen as well as a weak signal at 7.3 ppm, which is not from this molecule.

Since the proton on C-1 has two possible spin states of nearly equal probability, the protons of the methyl group experience two slightly different magnetic fields, and are observed in the spectrum as two slightly separated signals of equal intensity, or a *doublet*. The separation between each of the C-1 proton signals is the coupling constant, *J*, and will equal the *J* of the methyl signal. The proposed spectrum is shown in Figure 9.56*b*. The coupling constant in this case is about 7 Hz. The 60 MHz NMR spectrum of 1,1-dibromoethane is shown in Figure 9.56*c*.

**NOTE.**    *The net effect of spin–spin coupling is that a proton (or group of equivalent protons) adjacent to N other protons, will be observed as a multiplet with (N + 1) lines.*

A proton, or group of equivalent protons, may be coupled to more than one group of nuclei. The spectrum of 1-nitropropane ($CH_3CH_2CH_2NO_2$) is shown in Figure 9.57. The signal from the central methylene ($CH_2$) group is seen at about 2.0 ppm. Because the methylene group is adjacent to (and thus coupled to) five protons, its signal is a (5 + 1) or six-line multiplet; a *sextet*.

Nuclei with spins of 1 or greater exhibit more complex spin–spin coupling, since they can exist in more than two different spin states. For a nucleus coupled to *N* nuclei of spin *I*, a multiplet of $2\,I \cdot N$ lines will be observed. Nuclei of spin zero do not couple.

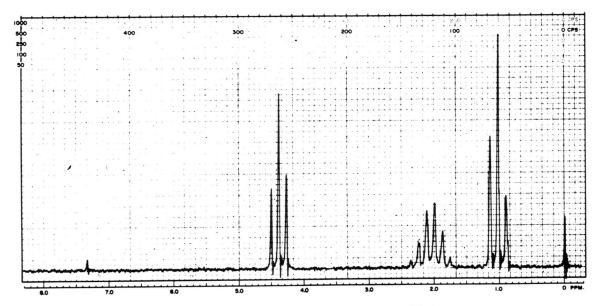

**FIGURE 9.57**    The 60-MHz spectrum of 1-nitropropane. *(Courtesy of Varian Associates, Palo Alto, CA.)* Starting from the right, the TMS signal at 0 ppm is seen. Next is a 1:2:1 triplet at 1.03 ppm. This triplet results from the protons on C-3 and their coupling with the two protons on C-2. Next is the sextet centered at 2.07 ppm. This multiplet is from the protons on C-2 and their coupling with the protons on C-1 and C-3. Finally we have the signal from the protons closest to the nitro group centered at 4.38 ppm. These protons appear as a 1:2:1 triplet due to their coupling with the protons on C-2.

## INTENSITIES

The area under an NMR peak is proportional to the number of nuclei giving rise to that signal. The intensity of a resonance is thus best determined by the *integral* of the NMR spectrum over a resonance, or group of resonances. Nuclear magnetic resonance spectrometers have the capability of measuring the integral, though integration data from a FT spectrometer are less reliable than those from a CW spectrometer. In more complex spectra the intensities are useful as a measure of the number of protons of a given type. For instance, in the above case the integral over both peaks of the methyl group doublet will be three times the integral over the quartet of the proton on C-1. Integration can thus often provide useful information when used in determining the identity of a compound.

## SECOND-ORDER EFFECTS

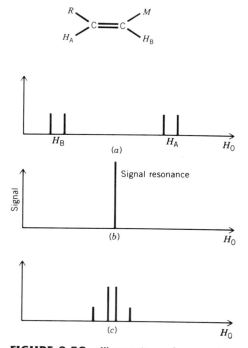

**FIGURE 9.58**  Illustration of second-order effects. (*a*) The chemical shift difference is much larger than the coupling constant and a first-order spectrum is observed. (*b*) Protons A and B are equivalent and a single resonance is observed. (*c*) The chemical shift difference is of the same order of magnitude or less than the coupling constant. Note the "leaning in" of the peak intensities in this spectrum relative to that in part *a*.

The examples so far considered have all consisted of first-order spectra. First-order spectra are those multiplets interpretable through elementary coupling analysis, such as that above; second-order spectra are those that are not interpretable in this manner. These highly symmetric and fairly simple first-order spectra are generally observed when the chemical shift differences (expressed as a frequency) are much greater than the coupling constant. Second-order effects occur when the coupling constants become comparable to or greater than the chemical shift differences. Thus, spectra obtained on instruments with higher magnetic fields are more likely to be first order, since the frequency differences between given signals increase with increasing magnetic fields. However, the chemical shift differences (in ppm) remain the same regardless of magnetic field strength.

Second-order effects may be understood in qualitative terms by considering the limiting cases. Let us consider the hypothetical disubstituted ethylene shown in Figure 9.58, where R and M are substituents that might be identical or may have very different effects on the alkenyl protons. In Figure 9.58*a* the spectrum is shown for the situation in which R and M have very different effects. In this case we will observe a first-order spectrum consisting of two doublets. The coupling constant is the separation in the doublets and the chemical shift of each nucleus is the geometric midpoint of each doublet.

In Figure 9.58*b*, groups R and M are identical. $H_A$ and $H_B$ are identical in this case and only a single resonance is observed (coupling between equivalent nuclei is not observed).

In Figure 9.58*c* the difference in the chemical environment of $H_A$ and $H_B$ is very slight. The spectrum shown may be seen as intermediate between the limiting cases in Figure 9.58*a* and Figure 9.58*b*. It should be noted that there is a "leaning in" of the doublets as the central members increase in intensity at the expense of the outer members. A full continuum of behavior may be expected with cases observed in which the outer members are lost in the noise and the central members take the appearance of a doublet. This would be one example of a class of spectra known as "deceptively simple spectra."

The second-order spectra of systems with more than two protons are difficult to describe even in qualitative terms. Second-order spectra may well display more lines than one would predict from simple coupling theory. It should also be noted that in second-order spectra, the coupling constants and the chemical shift differences may not be obtainable as simple differences in the positions of spectral lines. Thus, spectra obtained at high frequencies (and magnetic fields) are often more useful. As the operating frequency of the instrument is increased, the chemical shift differences (in frequency terms) increase while the spin–spin coupling remains constant. Thus, the complicating second-order effects are likely to be less noticeable in high-field spectra. The reader is referred to more extensive treatments of NMR for a discussion of second-order cases.

## INTERPRETATION OF ¹H NMR SPECTRA

The first issues that must be addressed are molecular symmetry and the magnetic equivalence or nonequivalence of protons or other functional groups. Even if two protons, or groups, are chemically equivalent, they may or may not be magnetically equivalent. Although molecular symmetry can often simplify NMR spectra, one must be able to discern which protons or groups are equivalent by symmetry. The two most useful symmetry properties (or symmetry operators) are the plane of symmetry and the axis of symmetry.

A plane of symmetry is simply a mirror plane such that one half of the molecule is the mirror image of the other half, such as in *meso*-pentane-2,4-diol.

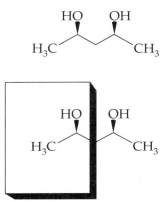

*meso* pentane-2,4-diol

The methyl groups are identical by symmetry, and one would expect this stereoisomer to show one methyl doublet in its ¹H NMR spectrum, and one signal for a methyl group in its ¹³C spectrum.

Consider the other diastereomer of pentane-2,4-diol, the chiral *d,l* isomer. This isomer has an axis of symmetry. If the molecule is rotated 180° about an axis in the plane of the paper passing through the central carbon, the molecule can be converted into itself.

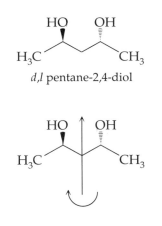

*d,l* pentane-2,4-diol

Here too the methyl groups are identical by symmetry, and one would expect this stereoisomer to show one methyl doublet in its $^1$H NMR spectrum, and one signal for a methyl group in its $^{13}$C spectrum.

It is, however, relatively simple to use NMR spectroscopy to distinguish between these stereoisomers. To do this, look at the two methylene protons on the central carbon, C-3, of each isomer.

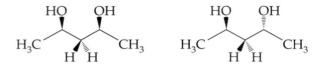

The plane of symmetry in the meso isomer bisects each of the two protons. In the $d,l$ isomer, the axis of symmetry interconverts the two protons. Thus, in the $d,l$ isomer, the two methylene protons are equivalent by symmetry, but they are not equivalent in the meso isomer. This can also be seen by inspecting the molecule. On the left, one H is syn to both —OH groups and the other is anti to both —OH groups. On the right, each H atom is syn to one —OH group and anti to the other.

The more rigorous way to determine equivalence or nonequivalence is to determine whether the two protons (or groups) are homotopic (identical), diastereotopic, or enantiotopic. To compare two protons, one uses the usual Cahn–Ingold–Prelog system for the nomenclature of stereoisomers. One artificially distinguishes the relative priority of two protons by a method such as drawing them in different colors or by pretending that one is deuterium (as long as the molecule does not contain D). One draws the two possibilities (i.e., the first H as D and then the second H as D) and then determines the stereochemical relationship between the two molecules drawn.

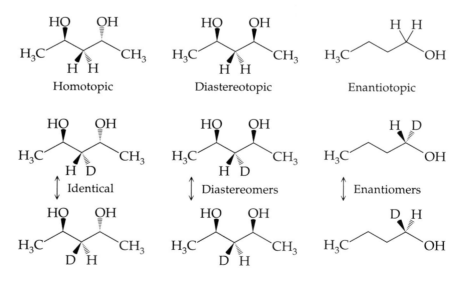

If the two are identical, the two protons are identical, or *homotopic*. If the two structures are diastereomers, the two protons are *diastereotopic*, and if the two structures are enantiomers, the two protons are *enantiotopic*. Diastereotopic protons, or groups, will be magnetically nonequivalent. Enantiotopic protons, or groups, will be magnetically equivalent only in an achiral environment and may appear nonequivalent in a chiral environment, such as a chiral solvent or in a biological sample. Homotopic protons may or may not be magnetically equivalent. Of course, it is possible for magnetically nonequivalent signals to be so close to one another in the NMR spectrum as to overlap (accidentally degenerate).

Homotopic protons may be magnetically nonequivalent if the two protons have different coupling constants to the same third proton. The most common example of this occurs in para-substituted benzenes.

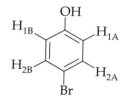

By symmetry, $H_{1A}$ and $H_{1B}$ are equivalent. These protons are not, however, magnetically equivalent because $H_{1A}$ and $H_{1B}$ have different coupling constants to, for example, $H_{2A}$, and the spectrum of this molecule may well be more complex than one would at first expect.

The equivalence or nonequivalence of functional groups, as well as protons, can easily be determined. The ¹H NMR spectrum of menthol shows three methyl doublets, as the two methyls in the isopropyl group are diastereotopic. The ¹³C spectrum of menthol shows three distinct resonances for the three different methyl groups.

Menthol

---

## ¹H CHEMICAL SHIFTS

Figure 9.59 summarizes the chemical shifts of protons in a large range of chemical environments. It is, however, a bit dangerous to use figures such as this one, without understanding some of the factors that underlie shielding and the chemical shift. To give some flavor of the factors that determine chemical shifts and the range of values observed, we will briefly examine chemical shifts in methyl groups and chemical shifts for protons on $sp^2$ carbon atoms.

Methyl groups bonded to an $sp^3$ carbon generally have chemical shifts in the range 0.8–2.1 ppm as long as there is no more than one electron-withdrawing group attached to the carbon. The shifts generally increase as the strength of the electron withdrawing group increases, or as more electron-withdrawing groups are added. Groups that inductively withdraw electrons reduce the electron density near the methyl group protons. This results in less shielding and a downfield shift of the methyl resonance. This effect is clearly seen in the spectra of 1-nitropropane and 1,1-dibromo-ethane (Figure 9.56c and Figure 9.57), respectively. The chemical shifts for methyl groups bonded to $sp^2$ carbon atoms fall in the range 1.6–2.7 ppm.

In the case of a proton bonded to an $sp^2$ carbon, the location of the proton relative to the $\pi$ cloud plays an important role in determining the chemical shift. In unconjugated alkenes the chemical shifts fall in the range 5–6 ppm. Where more than one proton is bonded to an alkene, complex second-order spectra can be expected at low operating frequencies since the coupling constants are usually fairly large relative to the difference in resonance frequencies. In aldehydes, RCHO, the increased electronegativity of the oxygen increases the deshielding and the chemical shift falls in the range 9–10.5 ppm.

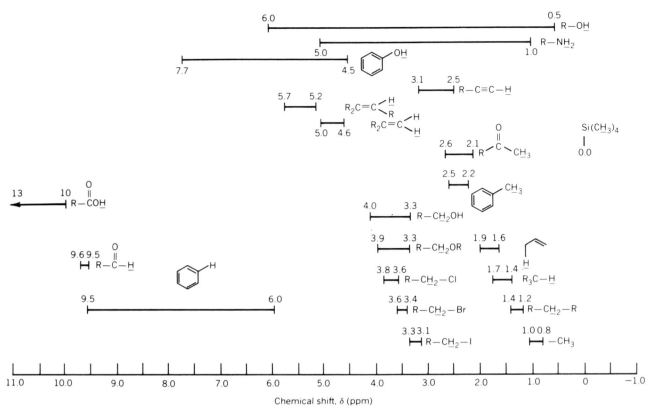

**FIGURE 9.59** NMR $^1$H chemical shifts. *From Zubrick, J. W. The Organic Lab Survival Manual, 3rd ed.; Wiley: New York, 1992. (Reprinted by permission of John Wiley & Sons, New York.)*

The chemical shift in an aromatic system is generally greater than that for alkenes. For example, the chemical shift of benzene is 7.37 ppm, substantially greater than the 5.6 ppm for the alkenyl protons of cyclohexene. Much of this difference results from the "ring current" effect and the orientation of the proton relative to the aromatic $\pi$ electrons. If the ring substituents are not strongly electron withdrawing or electron donating, such as alkyl groups, the chemical shift for ring protons will not be shifted greatly from that of benzene itself. Furthermore, these substituents generate only small chemical shift differences among the ring protons. Thus, the 60-MHz spectra for toluene (methylbenzene) appears to have a single resonance in the aromatic region at about 7.1 ppm. If, on the other hand, the substituents are electron withdrawing, the ortho and para ring protons will be somewhat deshielded relative to benzene. Pi electron-donating substituents, such as a methoxy group, will increase the shielding of groups ortho and para to it.

## SPIN–SPIN COUPLING

Coupling information is the primary reason that $^1$H NMR is such a powerful tool for organic structure determination. Since coupling information is transmitted through bonds, coupling provides information about nearby protons and can often be used to deduce stereochemistry.

The sign of the coupling constant (usually symbolized as *J*) may be positive or negative. However, first-order spectra are not sensitive to the sign of the coupling constant. In second-order cases, the sign of *J* may be determined by a detailed analysis of the spectrum, though the sign of *J* is generally of little value for organic structure determination.

Nonequivalent protons attached to the same carbon (geminal protons) will couple with one another. These coupling constants tend to be large (> 10 Hz) for $sp^3$ carbon atoms and small (< 4 Hz) for $sp^2$ carbon atoms. Geminal coupling constants tend to decrease with decreasing ring size, because of hybridization changes at carbon, and with the increasing electronegativity of the substituents on a given methylene group.

**Geminal Coupling**

Vicinal coupling describes the coupling over three bonds observed between protons attached to two bonded carbon atoms, H—C—C—H. Vicinal coupling constants (*J values*) can range from near 0 to greater than 15 Hz, depending on the stereochemical relationship (dihedral angle) between the coupled protons, the hybridization of the carbon atoms, and the electronegativity of other substituents. For vicinal $sp^3$ protons, the coupling constant is related to the dihedral angle and is expressed graphically by the Karplus curve (Fig. 9.60).

**Vicinal Coupling**

Though the magnitude of vicinal coupling is very sensitive to the angle of rotation about the central bond, in many simple cases nearly all coupling constants are equal. This situation is often the case if internal rotation about a C—C single bond can occur on a time scale that is very short relative to the NMR time scale, such as in acyclic systems. In these cases, the effect of internal rotation is completely blurred as far as NMR is concerned, and only an average coupling constant is observed. Vicinal coupling constants in freely rotating alkyl groups are usually observed in the 6.5–8 Hz range.

When the central C—C bond between two coupled protons is a double bond, rotation is restricted and separate coupling constants for cis and trans protons may be observed. Cis coupling constants fall in the range 5–12 Hz, whereas trans coupling constants range from 12–20 Hz. As a result of these large coupling constants, second-order effects are often observed in substituted alkenes in instruments of lower field strengths.

When rotation about carbon–carbon single bonds is restricted, or when stereochemistry dictates significant conformational preferences, nonaveraged coupling constants may be observed that can complicate the appearance of the NMR spectrum. For example, two diastereotopic protons of a methylene (CH$_2$) group may well each couple to a given third proton with different coupling constants. The familiar coupling explained

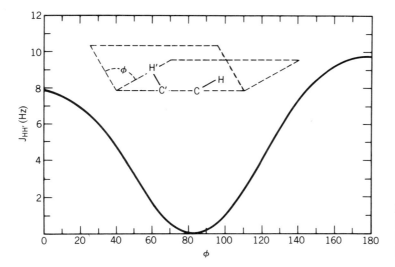

**FIGURE 9.60** The vicinal Karplus correlation. Relationship between dihedral angle and coupling constants for vicinal protons.

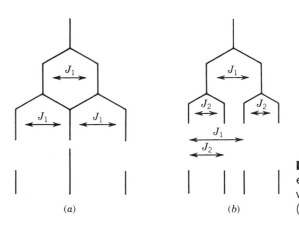

**FIGURE 9.61**   (*a*) Triplet equals doublet of doublets with equal *J* values (*b*) Doublet of doublets.

at the elementary level suggests that if one proton is adjacent to, for example, two others, the NMR signal of the first proton will be a triplet. This simplification will be true only if the two coupling constants are identical. A triplet is merely a doublet of doublets with equal coupling constants, which gives rise to the familiar triplet with intensities of $1:2:1$ (Figure 9.61*a*). A doublet of doublets, on the other hand, gives rise to a four-line multiplet with peaks of roughly equal intensity (Figure 9.61*b*).

## Long-range Coupling

Longer range coupling involving four or more bonds is common in allylic systems, and in aromatic rings and other conjugated $\pi$ systems. These coupling constants are generally smaller than the values considered above (i.e. < 3 Hz).

## EXAMPLES OF COMPLEX, YET FIRST-ORDER, COUPLING

## Ethyl Vinyl Ether

The coupling constants of even a seemingly complex multiplet can be discerned in a relatively simple manner. First, the total width (outside peak to outside peak) of a first-order multiplet is equal to the sum of all the coupling constants, keeping in mind that, for example, a triplet of $J = 7$ Hz is really a doublet of doublets with both $J$ *values* equal to 7 Hz. The expansion of the proton spectrum of ethyl vinyl ether is presented as an example in Figures 9.62*a*–9.62*c*. Integration data are displayed between the spectrum and the horizontal axis in Figure 9.62*a*.

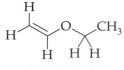

ethyl vinyl ether

Consider the multiplet centered at 6.45 ppm (Figure 9.62*b*). By measuring the distance (in Hz) from either outside line to the next inner line, which is 6.8 Hz, the first coupling constant is determined. Then, by measuring from the outside line to the second line in, the second coupling constant is found to be 14.4 Hz. We know that this is the last coupling constant to be found for several reasons. First, if we measure from the outside line to the third line in, we get a value, 21.2 Hz, which is equal to

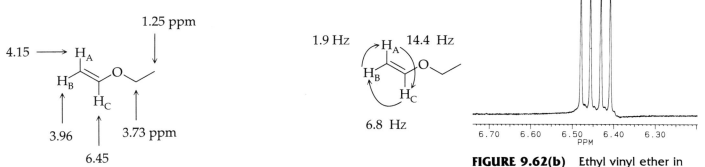

**FIGURE 9.62(a)** Ethyl vinyl ether in CDCl$_3$.

the sum of the previously determined coupling constants. Second, the width of the multiplet (the same measurement in this simple case), is equal to our two coupling constants. Thus, the NMR signal at 6.45 ppm is a doublet of doublets with $J = 14.4$ and 6.8 Hz.

The two doublets of doublets at 4.15 and 3.96 ppm (Figure 9.62c) must be coupled to one another as they both have the coupling constant of 1.9 Hz in common. This geminal coupling constant is typical of the terminal methylene of an alkene. Since the proton at 3.96 is coupled to the proton at 6.45 ppm by $J = 6.8$ Hz, and the proton at 4.15 ppm is coupled to the one at 6.45 ppm by $J = 14.4$ Hz, the proton at 3.96 ppm must be cis, and the proton at 4.15 ppm must be trans, to the alkene proton at 6.45 ppm.

The simple coupling observed for the ethyl group in ethyl vinyl ether can be readily assigned. The triplet at about 1.25 ppm, which integrates for three protons, is due to the methyl group; it is a triplet because the equivalent protons of the methyl group are coupled to the two protons on the adjacent carbon with equal coupling constants. The O—CH$_2$ protons are observed in the NMR spectrum as the quartet at about 3.75 ppm; they are a quartet because they are coupled equally to the three equivalent protons of the methyl group.

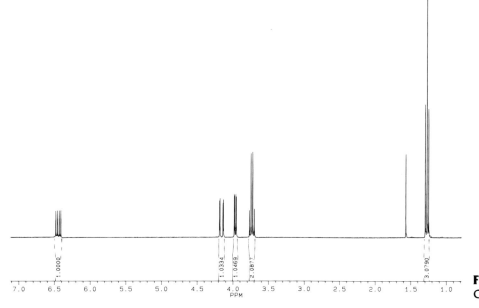

4.15 ⟶ H$_A$

H$_B$

H$_C$

1.25 ppm

3.96        3.73 ppm

6.45

Chemical shifts

Ethyl Vinyl Ether

1.9 Hz    H$_A$    14.4 Hz

H$_B$

H$_C$

6.8 Hz

Coupling constants

HERTZ

1944.32
1937.48
1929.95
1923.13

**FIGURE 9.62(b)** Ethyl vinyl ether in CDCl$_3$ (expansion).

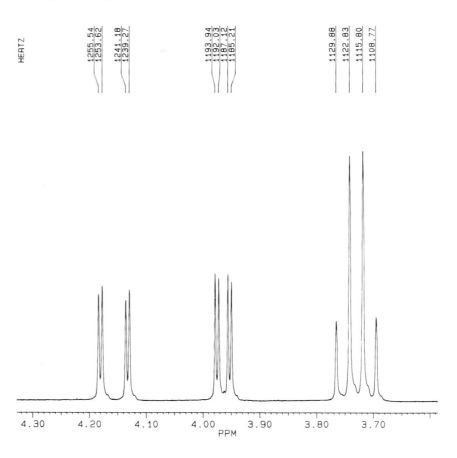

**FIGURE 9.62(c)**  Ethyl vinyl ether in CDCl₃ (expansion).

## Allyl Acetate

For a more complex example, refer to the ¹H NMR spectrum of allyl acetate (the NMR signal for the methyl group has been omitted) in Figures 9.63a–9.63d.

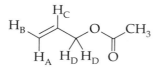

Protons *A*, *B*, and *C* are all chemically distinct, and the two protons labeled *D* are equivalent to one another by symmetry (the plane of the paper). The multiplet at 4.58 ppm (Figure 9,63d) corresponds to H$_D$ and is a doublet of triplets. The coupling constant for the triplet is 1.4 Hz, and the coupling constant for the doublet is 5.8 Hz, which can be measured between any two corresponding peaks in the two triplets.

The four quartets around 5.3 ppm (Figure 9.63c) are actually two doublets of quartets at 5.32 and 5.24 ppm and correspond to H$_A$ and H$_B$ in the structure above. At 5.32 ppm, the multiplet is a doublet of quartets, *J* = 17.2, 1.5 Hz. At 5.24 ppm, we have another doublet of quartets, *J* = 10.4, 1.3 Hz. We see quartets because the long-range allylic coupling to the two H$_D$ signals gives a triplet that has a coupling constant *J* that is approximately equal to the geminal coupling constant (~1.4 Hz) between H$_A$ and H$_B$. Since NMR line widths are naturally several tenths of a hertz, it is not possible to distinguish between coupling constants such as these that differ only by 0.2 Hz. We can unambiguously distinguish H$_A$ and H$_B$ by the magnitude of their coupling constants to H$_C$, which are 17.2 and 10.4 Hz. Since trans coupling constants are larger than cis ones, H$_A$ must have the 17.2-Hz coupling constant to H$_C$ and is thus assigned to the signal centered

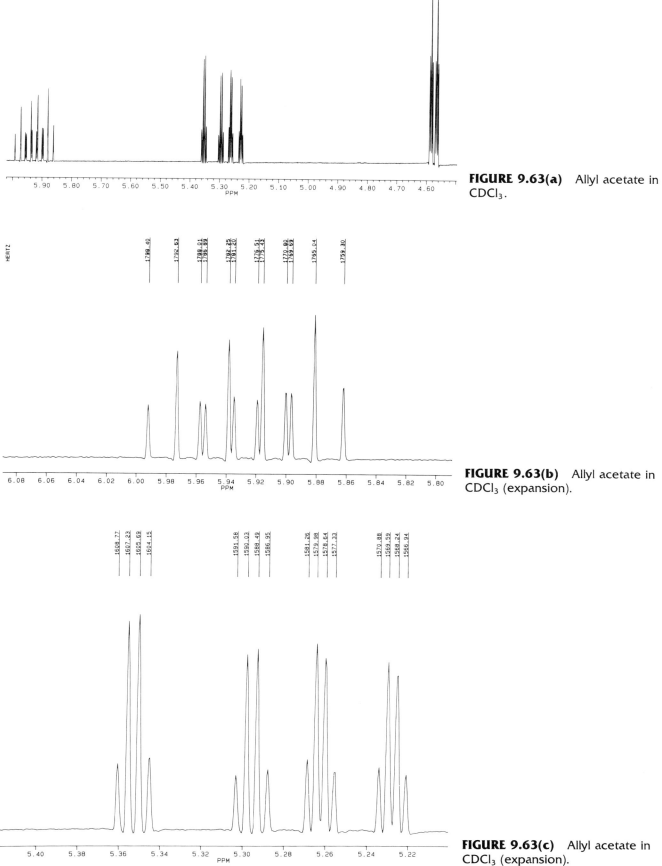

**FIGURE 9.63(a)**  Allyl acetate in CDCl$_3$.

**FIGURE 9.63(b)**  Allyl acetate in CDCl$_3$ (expansion).

**FIGURE 9.63(c)**  Allyl acetate in CDCl$_3$ (expansion).

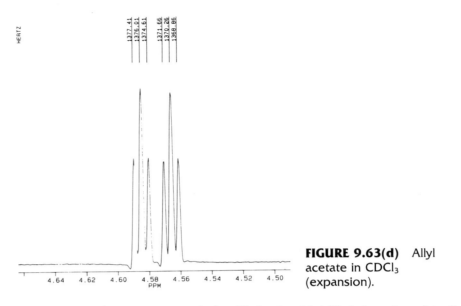

**FIGURE 9.63(d)** Allyl acetate in CDCl₃ (expansion).

at 5.32 ppm. Since H$_B$ is coupled to H$_C$ by $J$ = 10.4 Hz it is assigned to the signal centered at 5.24 ppm.

Finally, we already know what the multiplet for H$_C$ should look like, since we know all of its coupling constants. It is coupled to the two H$_D$ protons with a coupling constant of 5.8 Hz, to H$_A$ with $J$ = 17.2 Hz, and to H$_B$ with $J$ = 10.4 Hz. The multiplet for H$_C$ at 5.93 ppm should be, therefore, a doublet of doublets of triplets with $J$ = 17.2, 10.4 and 5.8 Hz, respectively. There should be 2 × 2 × 3 = 12 lines and the width should be 17.2 + 10.4 + (2 × 5.8) = 39.2 Hz. There are indeed 12 lines (Figure 9.63b) and the distance between the outside peaks is 39.1 Hz, which is a perfectly reasonable deviation from the ideal.

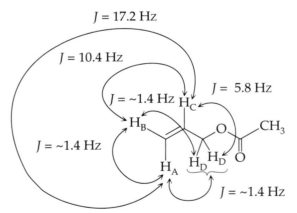

allyl acetate, proton–proton coupling constants

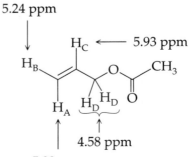

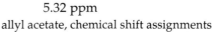

allyl acetate, chemical shift assignments

Since cyclohexane rings are often held in at most two potential conformations (both chairs), coupling constants may allow the determination of relative stereochemistry. On cyclohexane rings in chair conformations, the axial–axial coupling constants for vicinal protons (180° dihedral angle) are on the order of 9–12 Hz. Equatorial–equatorial and equatorial–axial coupling constants (60° dihedral angles) are on the order of 2–4 Hz. Thus it is often a relatively simple matter to determine stereochemical relationships on a six-membered ring using NMR spectroscopy.

Take, for example, the two diastereomers of 4-*tert*-butylcyclohexanol.

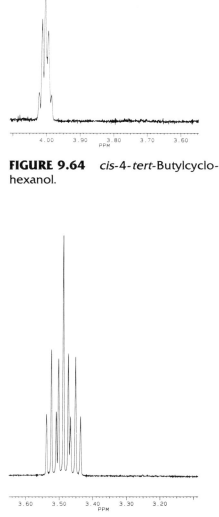

**FIGURE 9.64**    *cis*-4-*tert*-Butylcyclohexanol.

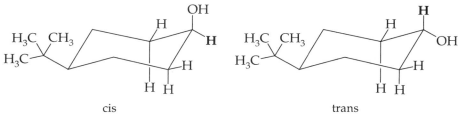

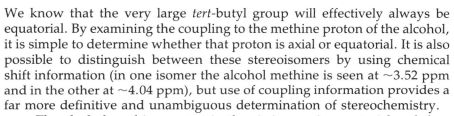

cis                                      trans

We know that the very large *tert*-butyl group will effectively always be equatorial. By examining the coupling to the methine proton of the alcohol, it is simple to determine whether that proton is axial or equatorial. It is also possible to distinguish between these stereoisomers by using chemical shift information (in one isomer the alcohol methine is seen at ~3.52 ppm and in the other at ~4.04 ppm), but use of coupling information provides a far more definitive and unambiguous determination of stereochemistry.

The alcohol methine proton in the cis isomer is equatorial and thus has a 60° dihedral angle to all four adjacent protons that give rise to a pentet (which is really a doublet of doublets of doublets of doublets with equal coupling constants) with a coupling constant $J = $ ~3 Hz, which is seen in Figure 9.64.

The alcohol methine in the trans isomer is axial and thus has a 180° dihedral angle to each of the two adjacent axial protons and a 60° dihedral angle to each of the two adjacent equatorial protons. This arrangement gives rise to a triplet of triplets with $J = $ ~13 and 3 Hz, which is shown in Figure 9.65.

**FIGURE 9.65**    *trans*-4-*tert*-Butylcyclohexanol.

# ¹³C NMR SPECTROSCOPY

With the advent of Fourier transform (FT) NMR spectrometers, ¹³C NMR spectroscopy is now available as a simple and routine tool for the structure determination of organic molecules. Since ¹³C is of low natural abundance (1.1%), addition of many spectra is required in order to obtain acceptable signal-to-noise (S/N) levels. With modern spectrometers, ¹³C spectra can often be acquired simply by issuing software commands; in some instruments a different probe is inserted into the magnet. Since ¹³C resonates at roughly 25% of the proton operating frequency of a spectrometer system, an instrument that acquires ¹H spectra at 300 MHz will be reset to about 75 MHz for ¹³C work.

Generally, ¹³C NMR spectra are acquired while the entire ¹H frequency range is irradiated by a second rf coil inside the probe assembly. These spectra are referred to as broadband-decoupled ¹³C spectra and they do not show the effect of spin–spin coupling to ¹H nuclei. Such decoupling is done because $^1H-^{13}C$ coupling constants can be quite large (a few hundred Hz) relative to chemical shift differences, which leads to multiplets split over a large portion of the spectrum and subsequent confusion (Figure 9.66). It is often simpler to see a single line for each distinct carbon atom in a molecule. Furthermore, irradiation of the $^1H$ spectrum results in signal enhancement of the $^{13}C$ signals of the attached carbon atoms. This enhancement is the nuclear Overhauser effect (NOE).

The $^{13}C$ NMR chemical shifts follow the same rough trends as seen in $^1H$ chemical shifts. It should be noted, however, that $^{13}C$ chemical shifts are not nearly as amenable to prediction based on the electronegativity of substituents as are $^1H$ chemical shifts. The $^{13}C$ chemical shifts are, in general, less sensitive to substituent electronegativities, and are far more sensitive to steric effects than are $^1H$ chemical shifts. A brief listing of approximate $^{13}C$ chemical shifts is provided in Table 9.29; a more extensive and thorough listing is available in the Silverstein et al. reference (Recommended reading). As in $^1H$ NMR spectroscopy, TMS [Si(CH$_3$)$_4$] is used as the internal reference and the chemical shift of TMS is defined as zero. Except for functional groups such as acetals and ketals, $sp^3$ hybridized carbon atoms appear upfield (to the right) of 100 ppm; and $sp^2$ hybridized carbon atoms appear downfield of 100 ppm. Common carbonyl-containing functional groups appear downfield of 160 ppm. Aldehydes and ketones appear at 195–220 ppm; esters, amides, anhydrides, and carboxylic acids appear at 165–180 ppm.

Typical $^{13}C$ NMR spectroscopy provides an NMR spectrum that is not amenable to integration because of the NOE, and insufficient relaxation delays, and therefore the number of carbon atoms giving rise to a given signal cannot generally be determined by these techniques. It is possible to obtain $^{13}C$ NMR spectra that can be accurately integrated (inverse-gated decoupling), but this experiment requires a great deal of acquisition time to achieve adequate signal-to-noise levels.

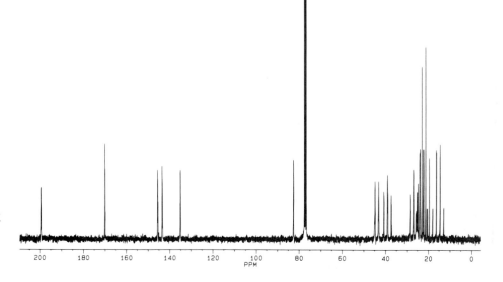

**FIGURE 9.66** Fully $^1H$-coupled $^{13}C$ NMR spectrum of 5-(1-acetoxy-1-methylethyl)-2-methyl-2-cyclohexenone in CDCl$_3$.

**TABLE 9.29    Approximate ¹³C NMR Chemical Shifts**

| Functional Group | Carbon[a] | Chemical Shift/ $\delta$ (ppm) |
|---|---|---|
| Alkyl carbon atoms | | ~5–45 |
| | 1° R—CH₃ | ~5–30 |
| | 2° R—CH₂—R′ | ~15–35 |
| | 3° R—CHR′R″ | ~20–40 |
| | 4° RCR′R″R‴ | ~25–45 |
| Alkenyl carbon atoms | | ~110–150 |
| | H₂C=C | ~100–125 |
| | HRC=C | ~125–145 |
| | RR′C=C | ~130–150 |
| Aromatic carbon atoms | | ~120–160 |
| Alkynyl carbon atoms | C≡C | ~65–90 |
| Nitriles | R—C≡N | ~115–125 |
| Alcohols and Ethers | C—OH(R) | ~50–75 |
| | C—O (epoxides) | ~35–55 |
| Amines | C—N | ~30–55 |
| Alkyl Halides | C—X | ~0–75 |
| Carbonyl groups | C=O | ~165–220 |
| Ketones, aldehydes | RCOR′, RCHO | 195–220 |
| Carboxylic acids, esters | RCO₂H, RCO₂R′ | 165–180 |
| Amides, anhydrides | RCON, RCO₂OCR′ | 160–175 |

[a] R = Alkyl group

Information about C—H coupling can be readily obtained, however. Fully coupled ¹³C NMR spectra are not very useful for structure determination because C—H couplings are large (~120–270 Hz, depending mainly on the hybridization at carbon) and multiplets tend to overlap. Furthermore, when the hydrogen atoms are not irradiated, there is no NOE, and the signal-to-noise ratio suffers significantly. The most common use for coupling information is to determine the number of protons attached to a given carbon atom, and this can be determined in a variety of ways, some of which do not actually display the carbon signals as multiplets due to coupling to attached protons.

<u>S</u>ingle <u>f</u>requency <u>o</u>ff-<u>r</u>esonance <u>d</u>ecoupling (SFORD) is a useful technique for determining the number of hydrogen atoms attached to a given carbon. The decoupler is tuned off to one side of the proton spectrum and the sample irradiated at a single frequency giving rise to ¹³C spectra that show C—H couplings as a fraction of their actual values and that show a partial NOE. The apparent C—H coupling is dependent on both the actual coupling constant, as well as the difference between the decoupler frequency and the resonance frequency of the hydrogen in question. The major disadvantage of SFORD is its low signal-to-noise ratio, which is due to two factors. First, there is only a partial NOE. Second, when NMR signals are split into multiplets, the signal intensity becomes distributed among several peaks. Thus, SFORD spectra require significantly more spectral acquisitions than do fully decoupled ¹³C NMR spectra, and to some extent have been replaced with <u>D</u>istortionless <u>E</u>nhancement by <u>Po</u>larization <u>T</u>ransfer (DEPT) spectra.

Distortionless enhancement by polarization transfer $^{13}$C NMR spectroscopy provides a rapid way of determining the number of hydrogen atoms attached to a given carbon atom. DEPT spectra result from a multiple pulse sequence which terminates in a "read pulse," which can be varied according to the spectrum desired. In DEPT spectra, all peaks are singlets and quaternary carbon atoms (without attached hydrogen atoms) are not seen in any DEPT spectra. In DEPT-135° spectra, CH and CH$_3$ groups appear as singlets of positive intensity, and CH$_2$ groups appear as negative peaks. The DEPT-90° spectra show only CH groups, and thus allow CH$_3$ and CH groups to be distinguished. In combination with a routine fully decoupled spectrum, DEPT spectra allow unambiguous assignment of the number of hydrogen atoms attached to each carbon. In practice, such spectral editing techniques are not perfect, and one often sees small residual peaks where, in principle, there should be none; these are usually small enough to be readily distinguished from the "real" peaks.

The fully coupled $^{13}$C NMR spectrum of the acetoxy-enone (**I**) is shown in Figure 9.66, the broadband decoupled spectrum in Figure 9.67, and the SFORD spectrum in Figure 9.68. The DEPT-135° spectrum is shown in Figure 9.69, and the DEPT-90° spectrum in Figure 9.70. The 1:1:1 triplet centered at 77 ppm is due to the solvent, CDCl$_3$. Interpretation and assignment of the $^{13}$C NMR spectrum are much easier when one unambiguously knows how many protons are attached to each carbon.

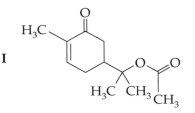

I

The $^{13}$C NMR is often better than $^1$H NMR for distinguishing functional groups because typical $^{13}$C chemical shifts are in the range from 0–

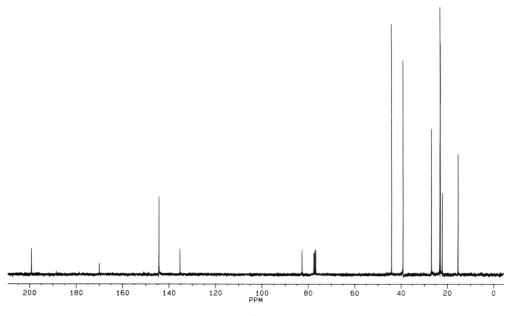

**FIGURE 9.67** Broadband $^1$H-decoupled $^{13}$C NMR spectrum of 5-(1-acetoxy-1-methyl-ethyl)-2-methyl-2-cyclohexenone in CDCl$_3$.

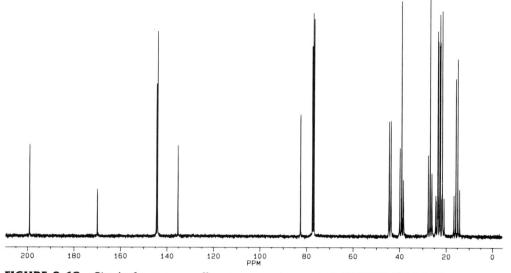

**FIGURE 9.68**  Single frequency off-resonance decoupled (SFORD) ¹³C NMR spectrum of 5-(1-acetoxy-1-methylethyl)-2-methyl-2-cyclohexenone in CDCl₃.

200 ppm relative to TMS, as compared to 0–10 ppm for proton chemical shifts. Coupling between adjacent ¹³C nuclei is not observed (except in isotopically enriched samples) because the probability of having two rare isotopes adjacent to one another is very small. Because of the absence of decoupling, ¹³C spectra are less complex than ¹H spectra, and ¹³C spectra are often better suited for the detection and identification of isomeric or

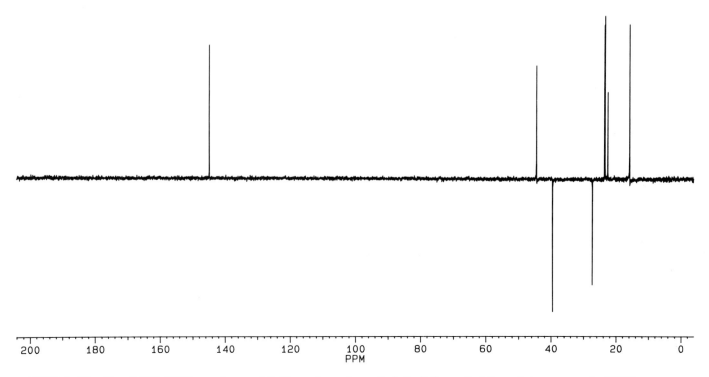

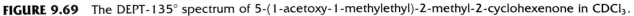

**FIGURE 9.69**  The DEPT-135° spectrum of 5-(1-acetoxy-1-methylethyl)-2-methyl-2-cyclohexenone in CDCl₃.

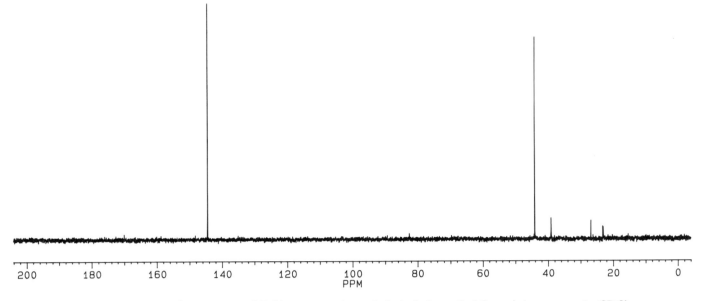

**FIGURE 9.70**   The DEPT-90° spectrum of 5-(1-acetoxy-1-methylethyl)-2-methyl-2-cyclohexenone in CDCl$_3$.

other impurities in a sample; it is easy for small peaks to be concealed underneath a complex second-order multiplet in the $^1$H NMR spectrum.

The 300-MHz $^1$H spectrum of 4-cyclohexene-*cis*-1,2-dicarboxylic acid anhydride is shown in Figure 9.71 . Owing in part to the presence of two stereocenters, as well as to long-range coupling through the $\pi$ system of the alkene, the entire $^1$H spectrum is second order at this field strength, and no information is available from the coupling constants because the spectrum is too complex. Limited assignments to peaks could be made on the basis of chemical shift, but it would be difficult to make any statements regarding purity of your sample based on the $^1$H NMR spectrum, as an impurity could easily be hidden underneath any of the complex signals.

On the other hand, the $^{13}$C spectrum of 4-cyclohexene-*cis*-1,2-dicarboxylic acid anhydride (Figure 9.72) is much less complex. Because of the mirror plane of symmetry in the compound, there are only four different carbon atoms and thus only four lines are seen in the fully decoupled $^{13}$C NMR spectrum. This simplicity makes it easy to detect the presence of isomeric or other impurities, which are present in this sample. These impurities were not as easy to detect in the $^1$H NMR spectrum. The 1 : 1 : 1 triplet centered at 77 ppm is due to the solvent, CDCl$_3$. Although no $^1$H–$^{13}$C

**FIGURE 9.71**   The $^1$H NMR spectrum of 4-cyclohexene-*cis*-1,2-dicarboxylic acid anhydride in CDCl$_3$.

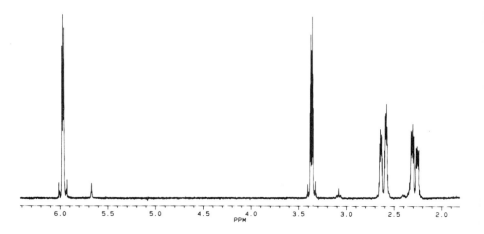

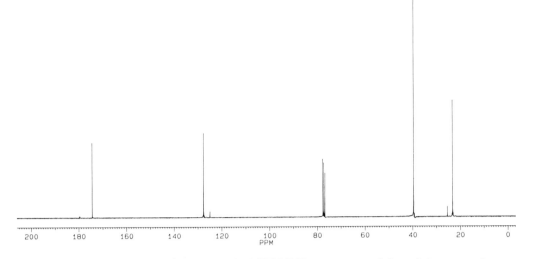

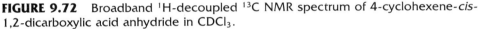

**FIGURE 9.72**   Broadband $^1$H-decoupled $^{13}$C NMR spectrum of 4-cyclohexene-*cis*-1,2-dicarboxylic acid anhydride in CDCl$_3$.

coupling is observed because of the $^1$H broadband decoupling, $^2$H–$^{13}$C coupling *is* observed because $^1$H and $^2$H resonate at different frequencies.

## TWO-DIMENSIONAL NMR SPECTROSCOPY

The two most significant recent developments in NMR spectroscopy are probably the use of Fourier-transform techniques, and the development of two-dimensional (2D) NMR spectroscopy. Two-dimensional spectra are obtained using a sequence of rf pulses that includes a variable delay or delays. A set of FIDs is acquired and stored. The variable delay is incremented by a small amount of time and a new set of FIDs are obtained and stored, and so on. At the end, the resulting matrix of FID data is Fourier transformed twice: once with respect to the acquisition time (as in normal FT-NMR) and second with respect to the time of the variable delay in the pulse sequence. The resulting data represent a surface and are presented as a contour plot of that surface.

The most useful sort of 2D spectra for organic compound identification are called correlation spectra. Correlation Spectroscopy (COSY) spectra are presented as a contour plot with routine proton spectra along both of the axes, as shown in the COSY spectrum of ethyl vinyl ether in Figure 9.73. The spectra along the axes are low-digital-resolution spectra and appear to be a bit different from those generated as usual NMR spectra. Note that the 2D spectrum is symmetric about the diagonal that runs from the lower left corner to the upper right corner. Every peak is represented by a peak along the diagonal and, in fact, the diagonal *is* the normal proton spectrum. Where the contour plot indicates a peak other than on the diagonal, the interpretation is that the corresponding peaks on the two axes represent protons coupled to one another.

For example, draw a line down from the signal at 1.2 ppm on the horizontal axis. This line encounters the diagonal at 1.2 ppm and then there is a peak at 3.7 ppm. By drawing a horizontal line over to the spectrum on the vertical axis, we can see that the peaks at 1.2 and 3.7 ppm are

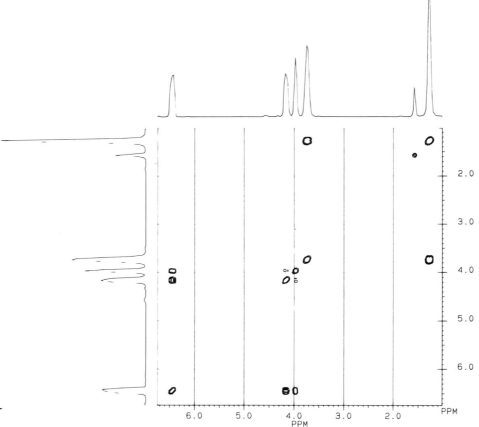

**FIGURE 9.73** The COSY NMR spectrum of ethyl vinyl ether.

coupled to one another; these are the signals from the methyl triplet and methylene quartet of the ethyl group. We can also see that neither of these is coupled to the remainder of the spectrum, which of course is the vinyl group. Each proton in the vinyl group is coupled to each of the others, as we can see in the COSY spectrum. For example, the signal at 4.2 ppm is coupled to both the signal at 4.0 ppm and the one at 6.4 ppm.

It is possible to obtain 2D spectra that correlate the spectra of different nuclei, such as $^1H$ and $^{13}C$. The heteronuclear correlation spectrum of ethyl vinyl ether is shown in Figure 9.74. The $^{13}C$ spectrum is along the horizontal axis and the $^1H$ spectrum is along the vertical axis. The peaks of the 2D spectrum indicate that the corresponding peaks on the axes represent a carbon and a proton (or protons) that are directly bonded. By using this spectrum we can easily verify the $^{13}C$ and $^1H$ chemical shift assignments below.

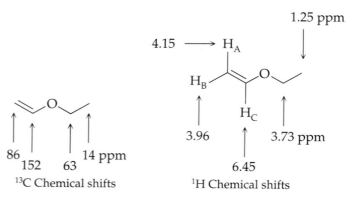

$^{13}C$ Chemical shifts

$^1H$ Chemical shifts

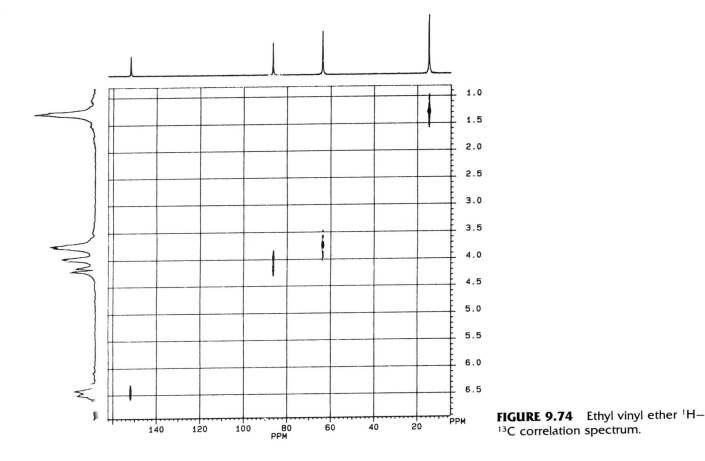

**FIGURE 9.74**  Ethyl vinyl ether $^1$H—$^{13}$C correlation spectrum.

There are many other powerful 2D NMR spectroscopic techniques that can provide a wealth of information about molecular structure, even in organic molecules as large as proteins and nucleic acids. A few of the good texts that provide further information about these powerful tools are listed at the end of the chapter in the Bibliography.

## NUCLEAR MAGNETIC RESONANCE SAMPLING

It is usually simple to prepare a sample of a small organic molecule (MW < 500) for NMR analysis. For $^1$H NMR spectroscopy, the sample size compatible with CW spectrometers is in the range 30–50 mg dissolved in about 0.5 mL of solvent. Fourier-transform spectrometers require only 2–3 mg of sample in the same volume of solvent. Most samples are measured in solution in thin-walled tubes 5 mm in diameter and 18–20 cm long. The NMR sample tubes are expensive and delicate; they must be as perfectly straight, and have as perfectly concentric walls, as possible. The sample tube is filled to a depth of about 3 cm. Filling the tube to this depth maximizes sample concentration in the active part of the NMR probe, and thus the strength of the signal. The addition of more solvent just wastes sample by dilution. The sample tube is spun about its axis in the instrument to average out small changes in the tube walls and in the magnetic field strength over the sample volume. There must be enough solvent in the tube to ensure that the vortex, or whirlpool, created when the tube is spun, does not extend down into the portion of the tube where the rf coils in the NMR probe are active. Many instruments use a depth gauge that shows exactly where this area is. Glass microcells are now commercially available. An inexpensive microcell technique for CW-NMR spectrometers using ordinary 5-mm NMR tubes has been described.[5]

HOOD

The most practical NMR solvent is deuterochloroform ($CDCl_3$), as it is relatively cheap and dissolves many different compounds. Handle this solvent with care, in the **hood**, as it is **toxic**! Many other deuterated solvents are commercially available, including acetone, methanol, and water. The universally accepted internal reference compound employed in making these measurements is tetramethylsilane, $Si(CH_3)_4$ (TMS). The most convenient source of TMS is commercially available $CDCl_3$ which contains about 1% TMS for use with CW spectrometers (commercially available 0.03% solutions are more appropriate for FT spectrometers).

The most significant problem in sample preparation is the exclusion of small pieces of dust and dirt, because they may contain magnetic material, which will result in poor spectra. Scrupulously clean samples of liquids can often simply be added to the NMR tube, followed by the solvent. Liquids containing visible impurities, and solids, are best prepared by dissolving the sample in about 0.3–0.4 mL of solvent in a small vial. The solution can then be filtered into the NMR tube through a Pasteur pipet plugged with a small piece of cotton, and the pipet can be rinsed with the NMR solvent to achieve the appropriate volume in the NMR tube. Since TMS is volatile (bp = 26.5 °C), the tube should be capped following addition, or the TMS will evaporate. If several people in a laboratory section are going to be obtaining NMR spectra, you will find that the spectrometer will be easier to tune with each new sample if all the sample tubes are filled to exactly the same level.

At this point, all specific instructions are dependent on the NMR spectrometer available to you. In general, the sample is inserted into the magnet and the magnetic field is adjusted very slightly (called shimming or tuning) to obtain a magnetic field that is as homogeneous as possible throughout the sample volume observed by the spectrometer. These adjustments are accomplished by energizing a collection of small electromagnetic coils around the sample, a process that is often done by the spectrometer's computer. Symptoms of a poorly shimmed spectrometer include broad and/or asymmetric peaks. The best place to check for this is on the TMS peak as it is the one peak in the spectrum you *know* should be narrow and symmetric. Once the spectrometer is tuned, the spectrum is obtained and plotted, and the sample is removed from the magnet.

Your NMR sample can be easily recovered by emptying the NMR tube into a small vial, rinsing the tube once or twice with (a nondeuterated) solvent, and then evaporating the solvent in a hood under a gentle stream of dry nitrogen.

**QUESTIONS**

Several  60-MHz [1]H NMR spectra are given below along with the molecular formula of the compound.[6] You should be able to account for at least one acceptable structure and for all of the observed resonances.

---

[5] Yu, S. J. *J. Chem. Educ.* **1987,** *64,* 812.

[6] From Pouchert, C. J. *The Aldrich Library of NMR Spectra,* 2nd ed., Aldrich Chemical Co.: Milwaukee, WI, 1983.

**9-16.** $C_4H_8O$. Spectrum *a*.

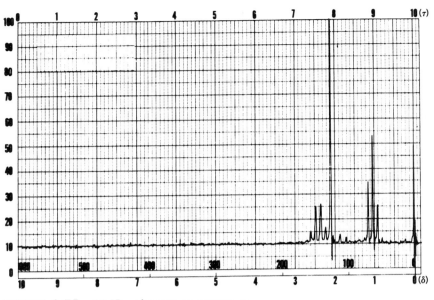

**FIGURE 9.75**    NMR unknown spectrum *a*.

**9-17.** $C_3H_6O_2$. Spectra *b* and *c*. Two compounds with the same empirical formula.

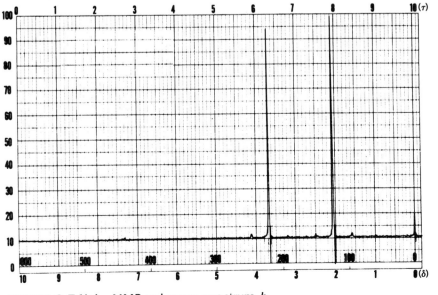

**FIGURE 9.76(a)**    NMR unknown spectrum *b*.

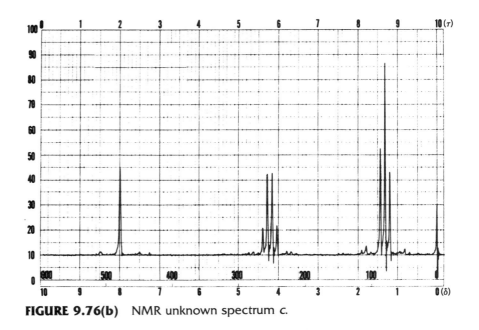

**FIGURE 9.76(b)** NMR unknown spectrum c.

**9-18.** $C_4H_8O$. Spectrum d. Also give some thought to the weak resonances at 0.5 and 1.8 ppm.

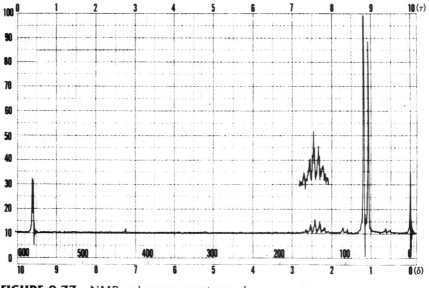

**FIGURE 9.77** NMR unknown spectrum d.

**9-19.**  $C_7H_7Cl$. Spectrum *e*.

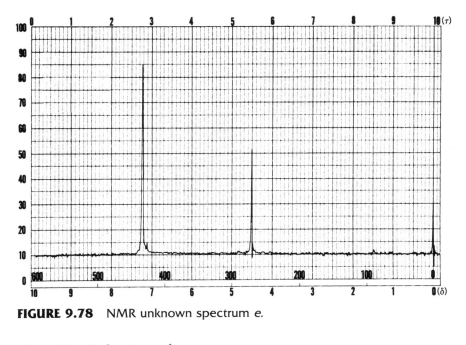

**FIGURE 9.78**    NMR unknown spectrum *e*.

**9-20.**  $C_8H_{10}O$. Spectrum *f*.

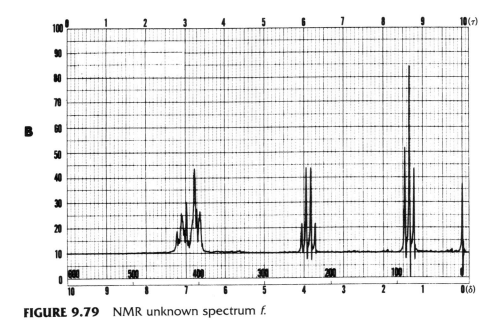

**FIGURE 9.79**    NMR unknown spectrum *f*.

---

Nuclear magnetic resonance theory and principles of interpretation:

**RECOMMENDED READING**

Abraham, R. J.; Fisher, J.; Loftus, P. *Introduction to NMR Spectroscopy*; Wiley: London, 1988.

Cooper, J. W. *Spectroscopic Techniques for Organic Chemists*; Wiley: New York, 1980.

Richards, S. A. *Laboratory Guide to Proton NMR Spectroscopy*; Blackwell Scientific Publications: London, 1988.

Silverstein, R. M.; Bassler, G. C.; Morrill, T. C. *Spectrometric Identification of Organic Compounds*, 5th ed., Wiley: New York, 1991.

Sorrell, T. N. *Interpreting Spectra of Organic Molecules*; University Science Books: Mill Valley, CA, 1988.

Sternhell, S.; Kalman, J. R. *Organic Structures from Spectra*; Wiley: London, 1986.

Advanced theory and spectroscopic techniques:

Atta-ur-Rahman, *Nuclear Magnetic Resonance. Basic Principles*; Springer-Verlag: New York, 1986.

Derome, A.E. *Modern NMR Techniques for Chemistry Research*; Pergamon Press: Oxford, 1987.

Duddeck, H.; Dietrich, W. *Structure Elucidation by Modern NMR. A Workbook*; Springer-Verlag: New York, 1989.

Sanders, J. K. M.; Constable, E. C.; Hunter, B. K. *Modern NMR Spectroscopy. A Workbook of Chemical Problems*; 2nd ed., Oxford University Press: Oxford, 1993.

Sanders, J. K. M.; Hunter, B. K. *Modern NMR Spectroscopy. A Guide for Chemists*; 2nd ed., Oxford University Press: Oxford, 1993.

Libraries of NMR spectra:

Bhacca, N. S. ; Hollis, D. P.; Johnson, L. F.; Pier, E. A.; Shoolery, J. N. *NMR Spectra Catalog*; Varian Associates: Palo Alto, CA, 1963.

Pouchert, C. J. *The Aldrich Library of NMR Spectra*, 2nd ed.; Aldrich Chemical Co.: Milwaukee, WI, 1983.

Pouchert, C. J.; Behnke, J. *The Aldrich Library of $^{13}C$ and $^1H$ FT-NMR Spectra*; Aldrich Chemical Co.: Milwaukee, WI, 1992.

## ULTRAVIOLET–VISIBLE SPECTROSCOPY: Introduction to Absorption Spectroscopy

In an atom, molecule, or ion, there are a limited number of electronic energy states available to the system because of the quantized nature of the energies involved. The absorption of a photon by the system can be interpreted as corresponding to the occupation of a new energy state by an electron. The difference in energy between these two states may be expressed as $\Delta E$.

————Upper state (Excited Electronic State, $E_1$)

$\Delta E$

————Lower State (Ground Electronic State, $E_0$)

where the energy of the photon, $E$, is related to the frequency of the radiation by the Planck equation,

$$E = h\nu_i$$

where $h$ is Planck's constant, $6.626 \times 10^{-34}$ J s and $\nu_i$ is the frequency in hertz. In the case above $\Delta E = E_1 - E_0 = h(\nu_1 - \nu_0) = h\nu_i$.

Thus, when a frequency match between the radiation and an energy gap ($\Delta E$) in the substance occurs, a transition between the two states involved may be induced. The system can either absorb or emit a photon corresponding to $\Delta E$ depending on the state currently occupied (emission would occur if the system relaxed from an upper-level excited state to a lower state). All organic molecules absorb photons with energies corresponding to the visible or ultraviolet regions of the electromagnetic spectrum, but to be absorbed, the incident energy in this frequency range must correspond to an available energy gap between an electronic ground state and an upper-level electronic excited state. The electronic transitions of principal interest to the organic chemist are those that correspond to the excitation of a single electron from the <u>h</u>ighest <u>o</u>ccupied <u>m</u>olecular <u>o</u>rbital (HOMO) to the <u>l</u>owest <u>u</u>noccupied <u>m</u>olecular <u>o</u>rbital (LUMO). As we will see, this will be the molecule's absorption occurring at the longest wavelength in the electronic absorption spectrum and is therefore, the most easily observed.

Electromagnetic radiation can be defined in terms of a frequency, $\nu$, which is inversely proportional to a wavelength $\lambda$ times a velocity $c$ ($\nu = c/\lambda$, where $c$ is the velocity of light in a vacuum, $2.998 \times 10^8$ m/s and $c = \nu\lambda$ the wave velocity). Thus,

$$\Delta E = h\nu = \frac{hc}{\lambda} = hc\bar{\nu}$$

where $\bar{\nu}$ is the wavenumber, defined as the reciprocal of the wavelength $(1/\lambda) \times$ the velocity of light.

Most ultraviolet and visible (UV and vis) spectra are recorded linear in wavelength, rather than linear in frequency or in units proportional to frequency (the wavenumber) or in energy values. Wavelength in this spectral region is currently expressed in nanometers (nm, where 1 nm = $10^{-9}$ m) or angstrom units (Å, where 1 Å = $10^{-10}$ m). The older literature is full of UV–vis spectra in which wavelength is plotted in millimicrons (m$\mu$), which are also equivalent to $10^{-9}$ m. For a further discussion of the relationship between frequency, wavelength, wavenumber, and refractive index, see Introduction to Infrared Spectroscopy.

It is unfortunate that because of instrumentation advantages this region of the spectrum is most often plotted in units that are nonlinear in energy (note the inverse relationship of $E$ to $\lambda$). A convenient formula for expressing the relationship of wavelength and energy in useful values is:

$$E = 28{,}635/\lambda \text{ kcal/mol} \quad (\lambda \text{ in nm})$$

or in terms of wavenumbers:

$$E = (28.635 \times 10^{-4})\bar{\nu} \quad (\bar{\nu} \text{ in cm}^{-1})$$

The electromagnetic spectrum, and the wavelength ranges corresponding to a variety of energy state transitions are listed in Table 9.30. Infrared, UV–vis, and rf are of particular interest to the organic chemist as the excitation of organic substances by radiation from these regions of the spectrum can yield significant structural information about the molecular system being studied.

The absorption of rf energy by organic molecules immersed in strong magnetic fields involves exceedingly small energy transitions ($\sim$0.05 cal/mol), which correspond to nuclear spin excitations and result in NMR spectra.

When a molecule absorbs microwave radiation, the energy states available for excitation correspond to molecular rotations and involve energies of roughly 1 cal/mol. With relatively simple molecules (in the gas

**TABLE 9.30** **Spectroscopic Wavelength Ranges**

| Region | Wave-length (m) | Energy | Change Excited |
|--------|-----------------|--------|----------------|
| Gamma ray | less than $10^{-10}$ | $>10^6$ kJ/mol | Nuclear transformation |
| X-ray | $10^{-8} - 10^{-10}$ | $10^4 - 10^6$ kJ/mol | Inner-shell electron transitions |
| Ultraviolet (UV) | $4\times10^{-7} - 1\times10^{-8}$ | $10^3 - 10^4$ kJ/mol | Valence shell electrons |
| Visible (vis) | $8\times10^{-7} - 4\times10^{-7}$ | $10^2 - 10^3$ kJ/mol | Electronic transitions |
| Infrared (IR) | $10^{-4} - 2.5\times10^{-6}$ | $1 - 50$ kJ/mol | Bond vibrations |
| Microwave | $10^{-2} - 10^{-4}$ | $10 - 1000$ J/mol | Molecular rotations |
| ESR | $10^{-2}$ | $10$ J/mol | Electron spin transitions |
| NMR | $0.5 - 5$ | $0.02 - 0.2$ J/mol | Nuclear spin transitions |

phase) possessing a dipole moment (required for the absorption process) the analysis of the microwave spectrum can yield highly precise measurements of the molecular dimensions (bond lengths and angles). The number of organic systems which exhibit pure rotational spectra that can be rigorously interpreted is, unfortunately, a relatively slender collection.

Absorption of radiation in the infrared region of the spectrum involves the excitation of vibrational energy levels and corresponds to energies in the range of about 1–12 kcal/mol.

The excitation of electronic states requires considerably higher energies, from a little below 40 to nearly 300 kcal/mol. The corresponding radiation wavelengths would fall across the visible (400–800 nm), the near-UV (200–400 nm), and the far- (or vacuum) UV (100–200 nm) regions. The long-wavelength visible and near-UV regions of the spectrum hold the information of particular value to the organic chemist. Here the energies involved correspond to the excitation of loosely held bonding ($\pi$) or lone-pair electrons. The far-UV region, however, involves high-energy transitions associated with the inner-shell and $\sigma$-bond electronic energy transitions. This region is difficult to access as atmospheric oxygen begins to absorb UV radiation below 190 nm, which requires working in evacuated or purged instruments (thus, the reason this region is often referred to as the vacuum UV).

## UV–VIS SPECTROSCOPY

As we have seen, the application of electronic absorption spectroscopy in organic chemistry is restricted largely to excitation of ground state electronic levels in the near-UV and vis regions. When photons of these energies are absorbed, the excited electronic states that result have bond strengths appreciably less than their ground-state values, and the internuclear distances and bond angles will be altered within the region of the molecules where the electronic excitation occurs (see Figure 9.80). It is normally reasonable to assume that nearly all of the molecules are present in the ground vibrational state within the ground electronic state. The upper electronic state also contains a set of vibrational levels and any of these may be open to occupation by the excited electron (see Figure 9.80). Thus, an electronic transition from a particular ground-state level can be to any number of upper-level vibrational states on the excited electronic state, as shown here for a diatomic molecule:

Thus, the shape of an electronic absorption band will be determined to a large extent by the spacing of the vibrational levels and the distribution of band intensity over the vibrational sublevels. In the large majority of cases these effects lead to broad absorption bands in the UV–vis region.

The wavelength maximum at which an absorption band occurs in the UV–vis region is generally referred to as the $\lambda_{max}$ of the sample (where wavelength is determined by the band maximum).

The quantitative relationship of absorbance (the intensity of a band) to concentration is expressed by the Beer–Lambert equation.

$$A = \log \frac{I_0}{I} = \varepsilon \cdot c \cdot \ell$$

Where $A$ = absorbance, expressed as $\frac{I_0}{I}$, $I_0$ is the intensity of the incident light, and $I$ is the intensity of the light transmitted through the sample.

$\varepsilon$ = molar absorbtivity, or the extinction coefficient, **(a constant characteristic of the specific molecule being observed)**. Values for conjugated dienes typically range from 10,000 to 25,000.

$c$ = concentration (mol/L) and $l$ = length of sample path (cm).

The calculated extinction coefficient and solvent are usually listed with the wavelength at the band maximum. For example, data for methyl vinyl ketone (3-buten-2-one) would be reported as follows:

$\lambda_{max}$ 219 nm ($\varepsilon$ = 3600, ethanol)
$\lambda_{max}$ 324 nm ($\varepsilon$ = 24, ethanol)

Typical UV–vis spectra are shown in Experiments [5], [19D], [A2$_a$], and [A3$_a$]. As part of the characterization data, UV–vis information is also given in Experiments [16], [A1$_a$], [19A], [2$_{adv}$], and [33A].

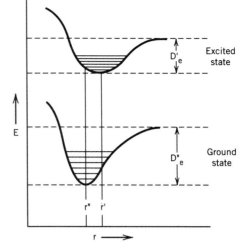

**FIGURE 9.80**  Two electronic energy levels in a diatomic molecule.

# APPLICATION TO ORGANIC MOLECULES

In organic compounds containing *conjugated* systems of $\pi$ electrons, a particular *chromophore* present can often be identified by the use of UV–vis spectroscopy. A *chromophore* is, in this case, a group of atoms able to absorb light in the UV–vis region of the spectrum. As the electronic transitions involved are limited primarily to $\pi$-electron (and lone-pair) systems this type of spectroscopy is less commonly used than the other modern spectroscopic techniques which, in fact, it predates by several decades. Ultraviolet-visible spectroscopy, however, can play a very valuable role in certain situations. For example, if a research problem involves synthesizing a series of derivatives of a complex organic molecule that possesses a strong chromophore, the UV-vis spectrum will be highly sensitive to structural changes involving the arrangement of the $\pi$-electron system (see, e.g., Experiment [5]).

In a conjugated alkene, such as 1,3-butadiene, the long-wavelength photon absorbed corresponds to the energy required for the excitation of a $\pi$ electron from the HOMO, $\pi_2$ to the LUMO, $\pi_3^*$. For these alkenes, this transition is represented as $\pi \longrightarrow \pi^*$; that is, an electron is promoted from a $\pi$ (bonding) molecular orbital to a $\pi^*$ (antibonding) orbital. This type of excitation is depicted below for both ethylene and 1,3-butadiene. Note that as a consequence of extending the chromophore and raising the energy of the highest occupied level in butadiene, the energy gap between the HOMO and LUMO levels of ethylene is larger than that in the conjugated system. Thus, the photon required for excitation of ethylene has a higher energy (higher frequency → shorter wavelength, $\lambda_{max} = 171$ nm) than the photon absorbed by 1,3-butadiene ($\lambda_{max} = 217$ nm).

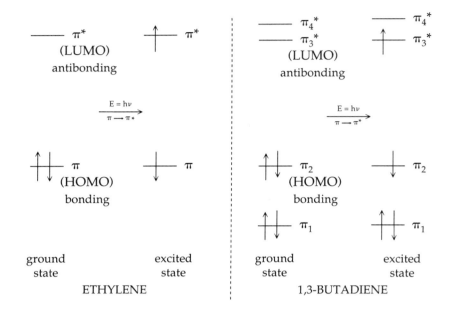

If we continue to extend the chromophore, the decrease of the energy gap between the HOMO and LUMO levels also continues. This drop in $\Delta E$ is then reflected in a drop in energy of the photon required to excite the $\pi \rightarrow \pi^*$ transition. This effect is illustrated in Table 9.31.

**TABLE 9.31    Absorption Maxima of Conjugated Alkenes**

| Name | Structure | $\lambda_{max}$ (nm) |
|---|---|---|
| Ethylene | $CH_2{=}CH_2$ | 165 |
| 1,3-Butadiene | $CH_2{=}CH{-}CH{=}CH_2$ | 217 |
| 1,3,5-Hexatriene | $CH_2{=}CH{-}CH{=}CH{-}CH{=}CH_2$ | 268 |
| 1,3,5,7-Octatetraene | $CH_2{=}CH{-}CH{=}CH{-}CH{=}CH{-}CH{=}CH_2$ | 290 |

As the extension of the chromophore continues, the $\lambda_{max}$ of the $\pi \rightarrow \pi^*$ transition will eventually shift into the visible region. At this point the substance exhibits color. As the absorbed wavelength is coming from the blue end of the visible spectrum these compounds will appear yellow. The color will become deeper and become red as the energy of the photon required for electronic excitation continues to drop. For example, tetraphenylcyclopentadienone is purple (Experiment [A3ₐ]), the dye, Methyl Red, is deep red (Experiment [26]), and *trans*-9-(2-phenylethenyl)anthracene is golden yellow (Experiment [19D]).

Compounds that contain a carbonyl chromophore $\overset{\backslash}{\underset{/}{C}}{=}\ddot{O}$ , also absorb radiation in the UV region. A $\pi$ electron in this unsaturated system undergoes a $\pi \rightarrow \pi^*$ transition. However, unless the carbonyl is part of a more extended chromophore, such as an $\alpha,\beta$-unsaturated ketone system, the $\pi \rightarrow \pi^*$ transition requires a fairly high-energy photon for excitation, usually below 190 nm in the far-UV and similar to the energy required for excitation of a carbon–carbon double bond. The edge of the $\pi \rightarrow \pi^*$ absorption band may just barely be observed on instrumentation designed for near-UV studies. This partially observed absorption band is generally referred to as *end absorption*. In the case of carbonyls, however, the heteroatom also loosely holds two pairs of nonbonding electrons that are often termed *lone-pair* electrons. These nonbonding electrons reside in orbitals ($n$) which are higher in energy than the bonding $\pi$ orbital, but lower in energy than the antibonding $\pi^*$ orbital. Thus, while a transition from an $n$ level to a $\pi^*$ level is formally forbidden, in fact, weak bands are observed at $\lambda_{max}$ in the near-UV that have their origin in the excitation of a lone-pair electron by an $n \rightarrow \pi^*$ transition. An energy diagram of a typical carbonyl system follows:

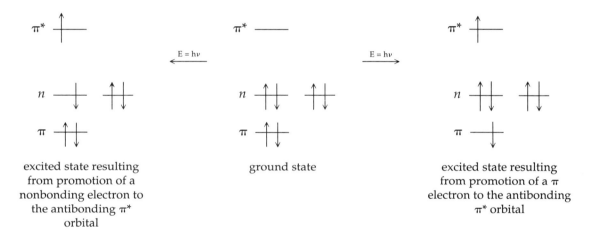

excited state resulting from promotion of a nonbonding electron to the antibonding $\pi^*$ orbital

ground state

excited state resulting from promotion of a $\pi$ electron to the antibonding $\pi^*$ orbital

Thus, those substances that contain the carbonyl chromophore absorb radiation of wavelengths that correspond to both the $n \rightarrow \pi^*$ and the $\pi \rightarrow \pi^*$ transitions. For a simple ketone, such as acetone ($CH_3COCH_3$), the $\pi \rightarrow \pi^*$ transition is found in the far-UV and the $n \rightarrow \pi^*$ in the near-UV. When the carbonyl becomes part of an extended chromophore, such as in methyl vinyl ketone (3-buten-2-one), the spectra reveal that these two transitions have shifted to longer wavelengths (a bathochromic shift, see Fig. 9.81 for

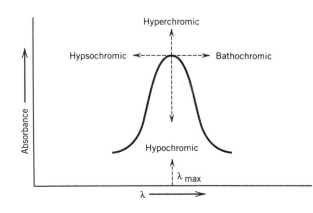

**FIGURE 9.81** Terms describing direction of wavelength and intensity shifts.

the definition of terms used in UV-vis spectra to indicate the direction of wavelength and intensity shifts).

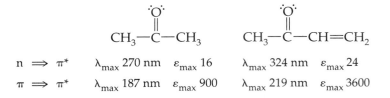

| | | | |
|---|---|---|---|
| $n \Rightarrow \pi^*$ | $\lambda_{max}$ 270 nm  $\varepsilon_{max}$ 16 | $\lambda_{max}$ 324 nm  $\varepsilon_{max}$ 24 |
| $\pi \Rightarrow \pi^*$ | $\lambda_{max}$ 187 nm  $\varepsilon_{max}$ 900 | $\lambda_{max}$ 219 nm  $\varepsilon_{max}$ 3600 |

Saturated systems containing heteroatoms with nonbonded electrons also exhibit weak absorption bands, often as end absorptions, which have their origin in forbidden $n \rightarrow \sigma^*$ transitions. When these hetero groups are attached to chromophores, both the wavelength and the intensity of the absorption can be altered. These are often referred to as *auxochromes* and *auxochromic shifts*.

Often, model compounds containing a chromophore of interest are referred to as an aid in the interpretation of the UV–vis spectrum of a new structure. Substantial collections of data have been developed for a wide variety of chromophores as an aid to this type of correlation. A number of empirical correlations, such as the Woodward–Fieser rules, of substituent effects on $\lambda_{max}$ values (see Silverstein et al. and Pasto et al. in the Bibliography section).

The Woodward–Fieser rules are a set of empirical correlations derived from studies of UV–vis spectral data. Using these rules it is possible to predict with reasonable accuracy the $\lambda_{max}$ for *new* systems containing various substituents on *known* chromophores. The rules are summarized in Table 9.32.

**TABLE 9.32    Woodward–Fieser Rules for Conjugated Dienes**

| Functionality | Increment (nm) |
|---|---|
| Base value for homoannular diene | 253 |
| Base value for heteroannular diene | 214 |
| Add: | |
| for each double bond extending conjugation | +30 |
| for double bond outside of ring (exocyclic) | +5 |
| for alkoxy groups | +6 |
| for S-alkyl groups | +30 |
| for Cl, Br groups | +5 |
| for dialkylamino groups | +60 |
| for parts of rings attached to butadiene fragment | +5 |

Examples of homoannular and heteroannular dienes are shown below.

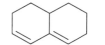        

a heteroannular diene        a homoannular diene

An example illustrating the use of these rules follows. Calculate the wavelength at which the following steroidal methyl sulfide will absorb.

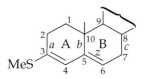

The base value for the diene is 214 nm, as the system is heteroannular (if a homoannular diene were present it would take precedence over the heteroannular diene; see the following example). There are three ring residues (or alkyl substituents) attached to the chromophore. Through hyperconjugation, the $\pi$ system is slightly extended by this type of substitution. The residues are labeled $a$, $b$, and $c$. Each of these substituents are assumed to add 5 nm to the $\lambda_{max}$ of the parent heteroannular diene, for a total of 15 nm. The 5,6-double bond in the B ring marked $z$ is exocyclic to the A ring, so empirically we add an additional 5 nm. Finally, for the thiomethyl substituent at the 3 position we add 30 nm. The total is $214 + 15 + 5 + 30 = 264$ nm.

Thus we have:

Predicted value    $\lambda_{max}$ (calcd) = 264 nm
Observed value    $\lambda_{max}$ (obsd) = 268 nm    ($\varepsilon = 22{,}600$)

As another example, consider ergosta-3,5,7,9-tetraene-3-acetate (**I**):

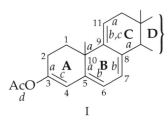

I

*Prediction of $\lambda_{max}$ for a substituted homoannular diene*

| | | |
|---|---|---|
| Parent homoannular diene in ring **B** | | 253 nm |
| Increments for | | |
| Double bond extending conjugation | $c$ [2 × 30] | 60 |
| Alkyl substituent or ring residue | $a$ [5 × 5] | 25 |
| Exocyclic double bond | $b$ [3 × 5] | 15 |
| Polar substituents | $d$ [0] | 0 |
| | $\lambda_{calcd}$ | 353 |

Predicted value    $\lambda_{max}$ (calcd) = 353 nm
Observed value    $\lambda_{max}$ (obsd) = 355 nm    ($\varepsilon = 19{,}700$)

There are additional rules for carbonyl-containing compounds, such as ketones, aldehydes, carboxylic acids, and so on, and for aromatic compounds. For example, Table 9.33 lists the parameters for conjugated carbonyl systems. Note that in contrast to the conjugated diene compounds, in which we are observing $\pi \rightarrow \pi^*$ transitions, the $n \rightarrow \pi^*$ transitions of the carbonyl $\lambda_{max}$ chromophore are often solvent dependent. Thus, solvent effects will have to be considered when predicting $\lambda_{max}$ values in these systems.

### TABLE 9.33  Conjugated Carbonyl Systems

| $\alpha,\beta$-Unsaturated Functionality | Base Value (nm) |
|---|---|
| Acyclic or six-membered or higher cyclic ketone | 215 |
| Five-membered ring ketone | 202 |
| Aldehydes | 210 |
| Carboxylic acids and esters | 195 |

| | Increment (nm) |
|---|---|
| Extended conjugation | +30 |
| Homoannular diene | +39 |
| Exocyclic double bond | +5 |

| | Substituent Increments (nm) | | |
|---|---|---|---|
| Substituent | $\alpha$ | $\beta$ | $\delta$ |
| Alkyl | +10 | +12 | +18 ($\gamma$ and higher) |
| Hydroxyl | +35 | +30 | +50 |
| Alkoxy | +35 | +30 | +31 ($\gamma$ + 17) |
| Acetoxy | +6 | +6 | +6 |
| Dialkylamino | | +95 | |
| Chloro | +15 | +12 | |
| Bromo | +25 | +30 | |
| Alkylthio | | +85 | |

| Solvent | Solvent Increments (nm) |
|---|---|
| Water | −8 |
| Ethanol | 0 |
| Methanol | 0 |
| Chloroform | +1 |
| Dioxane | +5 |
| Ether | +7 |
| Hexane | +11 |
| Cyclohexane | +11 |

An example of $\lambda_{max}$ (ethanol) calculation for a carbonyl system is presented here.

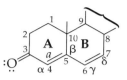

The base value for the $\alpha,\beta$-unsaturated six-membered ring ketone system is 215 nm. Extended conjugation adds an additional 30 nm. The presence of an exocyclic double bond, marked $a$, extends the $\lambda_{max}$ another + 5 nm. There is a substituent on the $\beta$-carbon atom (+12 nm) and on the $\delta$-carbon atom (+ 18 nm). There is no solvent effect as the spectrum was obtained in ethanol (0 shift). The total is 215 + 30 + 5 + 12 + 18 = 280 nm.

Predicted value      $\lambda_{max}$ (calcd) = 280 nm

Observed value      $\lambda_{max}$ (obsd) = 284 nm

The Woodward–Fieser rules work well for systems with four or fewer double bonds. For more extensively conjugated systems, $\lambda_{max}$ values are more accurately predicted using the Fieser–Kuhn equation:

$$\text{Wavelength} = 114 + 5M + n(48.0 - 1.7n) - 16.5 R_{endo} - 10 R_{exo}$$

where

$n$ = number of conjugated double bonds
$M$ = number of alkyl substituents in the conjugated system
$R_{endo}$ = number of rings with endocyclic double bonds in the system
$R_{exo}$ = number of rings with exocyclic double bonds in the system

Sample calculation: Find the UV $\lambda_{max}$ of $\beta$-carotene.

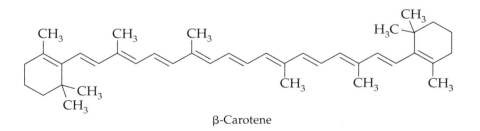

β-Carotene

In the structure there are 11 conjugated double bonds, $n = 11$. There are 6 alkyl groups and 4 ring residues on the conjugated system, $M = 10$. Both rings have an endocyclic double bond, $R_{endo} = 2$. Neither ring has any exocyclic double bonds, therefore $R_{exo} = 0$. Substituting in the equation,

$$\text{Wavelength} = 114 + 5(10) + 11[48 - 1.7(11)] - 16.5(2) - 10(0)$$
$$= 114 + 50 + 322.3 - 33 - 0 = 453 \text{ nm}$$

Predicted value      $\lambda_{max}$ (calcd) = 453 nm

Observed value      $\lambda_{max}$ (obsd) = 455 nm

**TABLE 9.34   The Benzoyl Chromophore[a]**

| Parent Chromophore $C_6H_5$—CO—**R** | | | |
|---|---|---|---|
| Function | | | Wavelength (nm) |
| **R** = alkyl or ring residue | | | 246 |
| **R** = H | | | 250 |
| **R** = OH, O-alkyl | | | 230 |

| Increment for Each Substituent | o- | m- | p- |
|---|---|---|---|
| Alkyl or ring residue | 3 | 3 | 10 |
| —OH, —OCH₃, —O-alkyl | 7 | 7 | 25 |
| —O⁻ (p-sensitive to steric effects) | 11 | 20 | 78 |
| —Cl | 0 | 0 | 10 |
| —Br | 2 | 2 | 15 |
| —NH₂ | 13 | 13 | 58 |
| —NHAc | 20 | 20 | 45 |
| —NHCH₃ | | | 73 |
| —N(CH₃)₂ | 20 | 20 | 85 |

[a] Spectra obtained in alcohol solvents.

Two examples of this correlation scheme (see Table 9.34) are

**1.** 6-Methoxytetralone

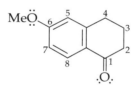

Predicted $\lambda_{max}$ is calculated by taking:
Parent value, 246 nm + one o-ring residue, 3 + one p-OMe, 25 = 274 nm

Predicted value      $\lambda_{max}$ (calcd) = 274 nm
Observed value      $\lambda_{max}$ (obsd) = 276 nm   ($\varepsilon$ = 16,500)

**2.** 3-Carboethoxy-4-methyl-5-chloro-8-hydroxytetralone

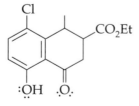

Parent value, 246 nm + one o-ring residue, 3 + one o-OH, 7 + 0 m-Cl = 256 nm

Predicted value      $\lambda_{max}$ (calcd) = 256 nm
Observed value      $\lambda_{max}$ (obsd) = 257 nm   ($\varepsilon$ = 8000)

For further examples of these types of calculations, see Silverstein et al. and Pasto et al. in the Bibliography section.

In summary, UV–vis spectra can make substantial contributions to understanding the molecular structure of organic substances that possess *chromophores*.

**1.** Interpretation of ultraviolet-visible spectra often can be a powerful approach for identifying the molecular structure of that section of a new substance that contains the chromophore.

**2.** The $\lambda_{max}$ increases within a series of compounds that contain a common chromophore that is lengthened (increased conjugation) over the series. The intensity of the absorption ($\varepsilon_{max}$) also generally becomes greater as conjugation increases, but can be very sensitive to steric effects (see Experiment [5]).

**3.** The $\lambda_{max}$ is sensitive to hyperconjugation by alkyl substituents, conformational changes that restrict $\pi$-system overlap, configurational, or geometric isomerization in which $\pi$ systems are perturbed, and structural changes, such as the isomerization of a double bond from an *exocyclic* to an *endocyclic* position and changes in ring size.

**4.** In many instances, accurate prediction of the $\lambda_{max}$ of a new molecular system can be made based on empirical correlations of the parent chromophore giving rise to the absorption.

Table 9.35 lists the $\lambda_{max}$ values of a number of common organic molecules.

**TABLE 9.35   Absorption Maxima of Several Unsaturated Molecules**

| Compound | Structure | $\lambda_{max}$ (nm) | $\varepsilon_{max}$ |
|---|---|---|---|
| Ethylene | $CH_2{=}CH_2$ | 171 | 15,530 |
| 1,3-Butadiene | $CH_2{=}CH{-}CH{=}CH_2$ | 217 | 21,000 |
| Cyclopentadiene | | 239 | 3,400 |
| 1-Octene | $CH_3(CH_2)_5CH{=}CH_2$ | 177 | 12,600 |
| *trans*-Stilbene | | 295 | 27,000 |
| *cis*-Stilbene | | 280 | 13,500 |
| Toluene | | 189 / 208 / 262 | 55,000 / 7,900 / 260 |
| 4-Nitrophenol | $HO{-}C_6H_4{-}NO_2$ | 320 | 9,000 |
| 3-Penten-2-one | $CH_3CH{=}CHCCH_3$ | 220 / 311 | 13,000 / 35 |

## INSTRUMENTATION

The acquisition of UV–vis absorption spectra for use in the elucidation of organic molecular structure is now carried out with instrumentation that is typically an automatic-recording photoelectric spectrophotometer. The optical components of one of the classic spectrophotometers is given in Figure 9.82. This system is typical of a high-quality double-beam double-monochromator instrument.

The instrument consists of a number of components:

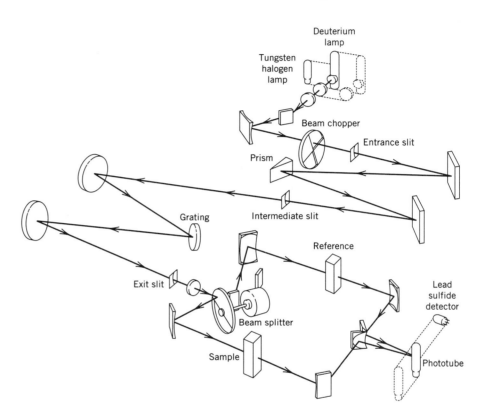

**FIGURE 9.82** Schematic optical diagram of a double beam-in-time spectrophotometer with double monochromation (Cary Model 17D). *(Courtesy of Varian Associates, Inc.)*

### The Source of Radiation

Radiant energy may be generated by either a deuterium discharge lamp or a tungsten–halogen lamp depending on the spectral region to be observed. Deuterium is generally preferred over hydrogen as the intense radiating ball of plasma is slightly larger in the case of deuterium, and therefore source brightness is enhanced by a factor of about 4. Below 360 nm, deuterium gas emits an intense continuum band that covers a major portion of the UV region. With special windows the short wavelength cutoff can be extended down to about 160 nm well out in to the vacuum-UV. Emission line spectra limit the long wavelength use of these lamps to about 380 nm. The lamps of choice for the region above 350 nm (the visible) are incandescent filament lamps, as they emit a broad band of radiation from 350 nm on the short wavelength end all the way to about 2.5 $\mu$m (the near-IR) on the long wavelength side. The majority of the radiation emitted falls, in fact, outside the visible, peaking at about 1$\mu$m in the near-IR. Nevertheless, tungsten lamps are *the* choice for measurements in the visible region, as they are extremely stable light sources.

Thus, the sources are required to possess two basic characteristics: (1) they must emit a sufficient level of radiant energy over the region to be studied so that the instrument detection system can function, and (2) they must maintain constant power during the measurement period. Source power fluctuations can potentially result in spectral distortion.

As the name implies, a monochromator (making a single color or hue) functions to isolate a single frequency from the source band of radiation. In practice we settle for isolating a small collection of overlapping frequencies surrounding the monochrome we wish to observe. Thus, the monochromator section of the instrument takes all the source radiation in at one end, and releases a very narrow set of bands of radiation at the other end. This function is accomplished, as shown in Figure 9.82, by focusing the entering radiation on an entrance slit that forms a narrow image of the source. After passing through the entrance slit, the spreading radiation is collimated by being reflected off a parabolic mirror, and is converted into parallel light rays (just as in a search light). The collimated radiation is then directed to the dispersing agent that is usually a quartz prism (quartz is transparent to UV, glass is not) or diffraction grating. The dispersing device functions to spread the different wavelengths of collimated light out in space. After emerging from the prism the dispersed radiation is redirected to either the same, or a new, collimator mirror and refocused as an image of the source on the exit slit of the monochromator. The exit slit has only a small fraction of the original radiation focused on it, and allows it to pass through in the image of the source. The remaining frequencies lie at different angles on either side of the exit slit. By mechanically turning the prism or grating, and thus changing the angle of the dispersing device with respect to the exit slit, all of the narrowly dispersed bands of radiation can be passed out of the monochromator in sequential fashion.

Instruments that are designed to reduce unwanted radiation to an absolute minimum will place two monochromators in tandem with an intermediate slit connecting the dispersing systems. In the case illustrated in Figure 9.82 the first monochromator uses a prism, while the second uses a grating. The two monochromators, however, have to be in perfect synchronization or no light at all will be transmitted.

After leaving the monochromator the radiation is directed to the sample compartment by a rotating sector mirror, where it is alternately focused on the substance to be examined (which is contained in a cell with quartz windows) and a reference cell (which holds the pure solvent used to dissolve the sample). The system now has two beams, hence the name *double-beam spectrophotometer*. After passing through the sample where the absorption of radiation may occur, the beams are recombined.

It should be noted that the sampling position could be placed either before or after the monochromator. In infrared instruments (such as the PE Model 710B, Fig. 9.40) it was generally often found before the monochromator until the introduction of interferometers. In UV systems, the sampling area is most often placed after the monochromator, and for good reasons. If the sample were placed before the monochromator, it would be exposed to the entire band of high-energy UV radiation being emitted by the source over the entire sampling period. By positioning the sample after the monochromator, at any one time the sample sees only the very small fraction of the dispersed radiation passed by the exit slit. Thus, sample stability is greatly protected by this arrangement. Remember that near-UV radiation carries photons with energies that approach those of the bond energies of organic molecules.

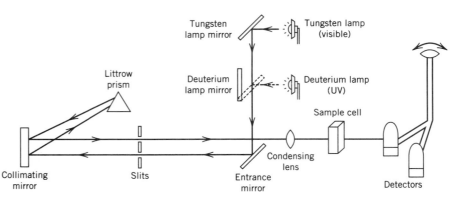

**FIGURE 9.83**  UV–Visible single beam spectrophotometer.

## The Detector

The recombined beams are then focused on the detector. Detectors function as transducers as they convert electromagnetic radiation into electrical current. There are a number of radiation-sensitive transducers available as detectors for these instruments. One such detector is the photomultiplier tube. These detectors operate with photocathodes that emit electrons in direct proportion to the number of photons striking the photosensitive surface and possess very large internal amplification. Thus, these detectors operate at low power levels. One particular advantage of the photomultiplier is the ability to adjust their sensitivity over a wide range simply by adjusting the supply voltage

## The Electronics: The Amplifier and Recorder

In the case of the double-beam instrument, the two signals generated by the sample and reference beams (each referenced against a dark current) in the detector are amplified and the ratio of the sample signal to the reference signal is plotted on a recorder.

The simplest of the absorption spectrometers are the single-beam instruments (see Fig. 9.83). These spectrometers are generally employed for problems involving simple one-component analyses. The photometric accuracy of scanned spectra should not be of paramount importance with these systems. Single-beam spectrometers require extremely stable sources and detectors.

## SAMPLE PREPARATION

Ultraviolet spectra are usually obtained on samples in solution using quartz cells. Quartz is used because it is transparent to both UV and visible light. For spectra restricted to the visible region, Pyrex® cells are satisfactory (and a good deal less expensive), but as Pyrex absorbs UV radiation, these cells cannot be employed for measurements in this region.

Solution cells usually have a horizontal cross section of 1 cm$^2$ and require about 3 mL of sample solution. Cells must be absolutely clean, and it is advisable to rinse the cell several times with the solvent used to dissolve the sample. A background spectrum of the solvent-filled cell (*without* a reference sample) can easily be obtained at this time and used as a check against contamination of either the cell or the solvent or both.

As the intensities of electronic transitions vary over a very wide range, the preparation of samples for UV–vis spectra determination is highly concentration dependent. Intense absorption can result from the high molecular extinction coefficients found in many organic chromophores. The sampling of these materials requires preparing very dilute solutions (on the order of $10^{-6}$–$10^{-4}$ $M$). These solutions can be conveniently obtained by the technique of *serial dilution*. In this method a sample of the material to be analyzed is accurately weighed, dissolved in the chosen solvent, and diluted to volume in a volumetric flask. Sample weights of 4–5-mg in 10-mL volumetric flasks are typical. Once dissolved and brought to volume an aliquot is then taken from this original solution, transferred to a second volumetric flask, and diluted as before. This sequence is repeated until the desired concentration is obtained.

Numerous choices of solvent are available and a list is given in Table 9.36, most of them are available in "spectral grade." The most commonly used solvents are water, 95% ethanol, methanol, and cyclohexane.

**TABLE 9.36  Solvents used in the Near-UV[a]**

| Solvent | Cut-Off Wavelength (nm) |
|---|:---:|
| Acetonitrile | 190 |
| Chloroform | 245 |
| (Toxic, substitute $CH_2Cl_2$) | 235 |
| Cyclohexane | 205 |
| 1,4-Dioxane | 215 |
| (Toxic, substitute EtOEt) | 218 |
| 95% Ethanol | 205 |
| *n*-Hexane | 195 |
| Methanol | 205 |
| Isooctane | 195 |
| Water | 190 |

[a] As these solvents have no color, they are transparent in the visible.

When considering a solvent:

- The most important factor is solubility of the sample. As UV–vis spectra can be very intense, even low solubility may be quite acceptable in sample preparation.
- The wavelength cutoff for the solvent may become important if the sample absorbs below about 250 nm.
- Sample–solvent molecular interactions. An example of these effects would be hydrogen bonding of protic solvents with carbonyl systems. Hydrocarbon chromophores are less influenced by solvent character than are the more polar chromophores.

**BIBLIOGRAPHY**

American Petroleum Research Institute Project 44 *Selected Ultraviolet Spectral Data* Vols. I–IV, Thermodynamics Research Center, Texas A&M University: College Station, TX, 1945–1977 (1178 compounds).

Friedel, R. A.; Orchin, M. *Ultraviolet Spectra of Aromatic Compounds*; Wiley: New York, 1951.

Grasselli, J. G.; Ritchey, W. M. *Atlas of Spectral Data and Physical Constants for Organic Compounds,* 2nd ed., CRC Press: Cleveland, OH, 1975.

Jaffe, H. H.; Orchin, M. *Theory and Application of Ultraviolet Spectroscopy*; Wiley: New York, 1962.

Lambert, J. B.; Shurvell, H. F.; Verbit, L.; Cooks, R. G.; Stout, G. H. *Organic Structural Analysis*; Macmillan: New York, 1976.

Lang, L., Ed.; *Absorption Spectra in the Ultraviolet and Visible Region*, Vols. 1–20, Academic Press: New York, 1961–1975; Vols. 21–24, Kreiger: New York, 1977–1984.

Pasto, D. J.; Johnson, C. R.; Miller, M. J. *Experiments and Techniques in Organic Chemistry*; Prentice Hall: New Jersey, 1992.

Silverstein, R. M.; Bassler, G. C.; Morrill, T. C. *Spectrometric Identification of Organic Compounds*, 5th ed., Wiley: New York, 1991, Chapter 7.

*Standard Ultraviolet Spectra*; Sadtler Research Laboratories: Philadelphia, PA.

Stern, E. S.; Timmons, T. C. J. *Electronic Absorption Spectroscopy in Organic Chemistry*; St. Martin Press: New York, 1971.

*UV Atlas of Organic Compounds*, Vols. I–IV, Butterworths: London, 1966–1971.

# 10

# Qualitative Identification of Organic Compounds

**C₁₀H₁₀, Housane**
Eaton, Or, Branca and Shankar (1968).

## ORGANIC QUALITATIVE ANALYSIS

One of the exciting challenges that a chemist faces on a regular basis is the identification of organic compounds. To the introductory student this challenge is an excellent way to be initiated into the arena of chemical research. Millions of organic compounds are recorded in the chemical literature. At first glance it may seem a bewildering task to attempt to identify one certain compound from this vast array. However, it is important to realize that the majority of these substances can be grouped, generally by functional groups, into a comparatively small number of classes. In addition, chemists have an enormous database of chemical and spectroscopic information, which has been correlated and organized over the years, at their disposal. Determination of physical properties, of the functional groups present in a molecule, and of the reactions it undergoes, have allowed the chemist to establish a systematic, logical identification scheme.

Forensic chemistry, the detection of species causing environmental pollution, the development of new pharmaceuticals, progress in industrial research and development of polymers, to name a few, all depend to a large extent on the ability of the chemist to isolate, purify, and identify specific chemicals. The objective of organic qualitative analysis is to place a given compound, through screening tests, into one of a number of specific classes, which in turn greatly simplifies the *identification* of the compound. This screening is usually done by using a series of preliminary observations and chemical tests, in conjunction with the instrumental data that developments in spectroscopy have made available to the analyst. The advent of infrared (IR) and nuclear magnetic resonance (NMR) spectroscopy, and mass spectrometry (MS), have had a profound effect on the approach taken to identify a specific organic compound. Ultraviolet (UV) spectra may also be utilized to advantage with certain classes of materials.

The systematic approach taken in this text for the identification of an unknown organic compound is as follows:

**1.** Preliminary tests are performed to determine the physical nature of the compound.

**2.** The solubility characteristics of the unknown species are determined. This identification can often lead to valuable information related to the structural composition of an unknown organic compound.

**3.** Chemical tests, mainly to assist in identifying elements other than C, H, or O, may also be performed.

**4.** Classification tests to detect *common functional groups* present in the molecule, are carried out. The majority of these tests may be done using a few drops of a liquid or milligrams of a solid. There is an added benefit, especially in relation to the chemical detection of functional groups, that an introductory student in the organic chemistry laboratory may obtain. That is, a vast amount of chemistry can be *observed* and *learned* in performing these tests. The successful application of these tests requires that you develop the ability to think in a logical manner and based on your observations, that you learn to interpret the significance of each result. Later, as the *spectroscopic techniques* are introduced, *the number of chemical tests performed are usually curtailed.*

**5.** The spectroscopic method of analysis is utilized. As you develop further in your knowledge of chemistry, you will appreciate more and more the revolution that has taken place in chemical analysis over the past 25–30 years, and the powerful tools now at your disposal for the identification of organic compounds. For the introductory laboratory, the techniques of *IR, NMR,* and *UV–vis* spectroscopy are generally developed.

It is important to realize that *negative* findings are often as important as *positive* results in identifying a given compound. Cultivate the habit of following a *systematic pathway or sequence* so that no clue or bit of information is lost or overlooked along the way. It is important also to develop the *attitude* and *habit* of planning ahead. Outline a logical plan of attack, depending on the nature of the unknown, and follow it. As you gain more and more experience in this type of investigative endeavor, the planning stage will become easier.

At the *initial* phase of your training, the unknowns to be identified will be relatively pure materials and will all be known compounds. The properties of these materials are recorded in the literature, and/or in the tables in Appendix A. Later perhaps, mixtures of compounds or samples of commercial products will be assigned for separation, analysis, and identification of the component compounds.

*Record all observations and results of the tests in your laboratory notebook.* Review these data as you execute the sequential phases of your plan. This method serves to keep you on the path to success.

A large number of texts have been published on organic qualitative analysis. Several references are cited here.

---

## BIBLIOGRAPHY

Cheronis, N. D.; Entrikin, J. B.; Hodnett, E. M. *Semimicro Qualitative Organic Analysis*, 3rd ed., Interscience: New York, 1965.

Cheronis, N. D.; Ma, T. S. *Organic Functional Group Analysis by Micro and Semimicro Methods;* Interscience: New York, 1964.

Feigl, F.; Anger, V. *Spot Tests in Organic Analysis,* 7th ed., Elsevier: New York, 1966.

Kamm, O. *Qualitative Organic Analysis,* 2nd ed., Wiley: New York, 1932.

Pasto, D. J.; Johnson, C. R.; Miller, M. J. *Experiments and Techniques in Organic Chemistry;* Prentice Hall, New Jersey, 1992.

Schneider, F. L. In *Qualitative Organic Microanalysis,* Vol. II of *Monographien aus dem Gebiete der qualitativen Mikroanalyse;* Benedetti-Pichler, Ed.; Springer-Verlag: Vienna, Austria, 1964.

Shriner, R. L.; Fuson, R. C.; Curtin, D. Y.; Morrill, T. C. *The Systematic Identification of Organic Compounds* 6th ed., Wiley: New York ,1980.

Vogel, A. I. *Qualitative Organic Analysis,* Part 2 of *Elementary Practical Organic Analysis,* Wiley: New York, 1966.

---

## PRELIMINARY TESTS

### Overview

The objective of preliminary tests is to assist you in selecting a route to follow in order to ultimately identify the unknown material at hand. It must be emphasized, however, that these tests frequently consume material. Given the amounts of material generally available at the micro- or semimicroscale level, judicious selection of the tests to perform must be made; however, in some tests, the material analyzed may be recovered. Each preliminary test that can be conducted with *little expenditure of time and material* can offer valuable clues as to which class a given compound belongs.

### Nonchemical Tests

#### Physical State

If the material is a *solid,* a few milligrams of the sample may be viewed under a magnifying glass or microscope, which may give some indication as to the homogeneity of the material. Crystalline shape often is an aid in classifying the compound.

Determine the melting point, using a small amount of the solid material. If a narrow melting point range (1–2 °C) is observed, it is a good indication that the material is quite pure. If a broad range is observed, the compound must be recrystallized from a suitable solvent before proceeding. If the material undergoes decomposition on heating, it is worthwhile to try an evacuated (sealed-tube) melting point. If any evidence indicates that sublimation is occurring, an evacuated melting point should be run. Furthermore, this result indicates that sublimation might be used to purify the compound, if necessary.

If the material is a *liquid,* the boiling point is determined by the ultra-micro method. If sufficient material is on hand, and the boiling point reveals that the material is relatively pure (narrow boiling point range), the *density* and the *refractive index* can provide valuable information for identification purposes.

#### Color

Since the majority of organic compounds are colorless, examination of the color can occasionally provide a clue as to the nature of the sample. Use caution, however, since tiny amounts of some impurities can color a substance. Aniline is a classic example. When freshly distilled it is colorless, but on standing a small fraction oxidizes and turns the entire sample a reddish-brown color.

Colored organic compounds contain a *chromophoric group,* usually indicating extended conjugation in the molecule. For example, 1,2-dibenzoylethylene (Experiments [3A] and [6]) is yellow; 5-nitrosalicylic acid

(Experiment [29C]) is light yellow; azobenzene (Experiment [35]) is red; tetraphenylcyclopentadienone (Experiment [A3ₐ]) is purple.

Can you identify the chromophores that cause these compounds to be colored? Note that a colorless liquid or white solid would *not contain* these units. Thus, compounds containing these groupings would be excluded from consideration as possible candidates in identification of a given substance.

### Odor

Detection of a compound's odor can occasionally be of assistance, since the vast majority of organic compounds have no definitive odor. You should become familiar with the odors of the common compounds or classes. For example, aliphatic amines have a fishy smell; benzaldehyde (like nitrobenzene and benzonitrile) has an almond odor; esters have fruity odors (Experiments [8A–C]). Common solvents, such as acetone, diethyl ether, and toluene all have distinctive odors. Butyric and caproic acids have rancid odors. Low molecular weight mercapto (—SH) containing compounds have an intense smell of rotten eggs. In many cases, extremely small quantities of certain relatively high molecular weight compounds can be detected by their odor. For example, a $C_{16}$ unsaturated alcohol released by the female silk worm moth elicits a response from male moths of the same species at concentrations of 100 molecules/cm³. Odors are one extremely important facet of chemical communication between plants and animals and they can often result in a spectacular behavioral response (see also Experiment [8]).

Odor detection in humans involves your olfactory capabilities and thus can be considered a helpful lead, *but very rarely can this property be used to strictly classify or identify a substance.* As mentioned above, contamination by a small amount of an odorous substance is always a possibility.

CAUTION

> **CAUTION:  You should be very cautious when detecting odors. Any odor of significance can be detected several inches from the nose. Do not place the container closer than this to your eyes, nose, or mouth. Open the container of the sample and gently waft the vapors toward you.**

## Ignition Test

> **CAUTION:   Make sure you are wearing safety glasses.**

Valuable information can be obtained by carefully noting the manner in which a given compound burns. The ignition test[1] is carried out by placing 1–2 mg of the sample on a spatula, followed by heating and ignition with a microburner flame. Do not hold the sample directly in the flame; heat the spatula about 1 cm from the flat end and move the sample slowly into the flame (see Fig. 10.1).

Important observations to be made concerning the ignition test are summarized in Table 10.1.

As the heating of the sample takes place, you should make the following observations.

---

[1] For an extensive discussion on examination of ignition residues see Feigl, F.; Anger, V. *Spot Tests in Organic Analysis,* 7th ed., Elsevier: New York; 1966; p. 51.

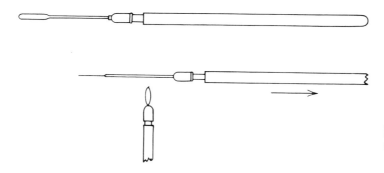

**FIGURE 10.1**   Heating on the micro-spatula *(Courtesy of Springer-Verlag, Vienna, Austria.)*

**1.** Any melting or evidence of sublimation: This observation gives an approximate idea of the melting point by the temperature necessary to cause melting.

**2.** Color of the flame as the substance begins to burn (see Table 10.1).

**3.** Nature of the combustion (flash, quiet, or an explosion). Rapid, almost instantaneous combustion indicates high hydrogen content. Explosion indicates presence of nitrogen, or nitrogen–oxygen, containing groups. For example, azo groups (Experiment [35]) or nitro groups (Experiment [29]).

**4.** Nature of the residue, if present, after ignition.

a. If a black residue remains and disappears on further heating at higher temperature, the residue is carbon.

b. If the residue undergoes swelling during formation, the presence of a carbohydrate or similar compound is indicated.

c. If the residue is black initially but still remains after heating, an oxide of a heavy metal is indicated.

d. If the residue is white, the presence of an alkali or alkaline earth carbonate, or $SiO_2$ from a silane or silicone is indicated.

**TABLE 10.1   Ignition Test Observations**[a]

| Type of Compound | Example | Observation |
|---|---|---|
| Aromatic compounds, unsaturated, or higher aliphatic compounds | Toluene | Yellow, sooty flame |
| Lower aliphatic compounds | Hexane | Yellow, almost nonsmoky flame |
| Compounds containing oxygen | Ethanol | Clear bluish flame |
| Polyhalogen compounds | Chloroform | Generally do not ignite until burner flame applied directly to the substance |
| Sugars and proteins | Sucrose | Characteristic odor |
| Acid salts or organometallic compounds | Ferrocene | Residue |

[a] Cheronis, N. D.; Entrikin, J. B. *Semimicro Qualitative Analysis*; Interscience: New York, 1947, p. 85.

**SEPARATION OF IMPURITIES**

If the preliminary tests outlined above indicate that the unknown in question contains impurities, it may be necessary to carry out one of several purification steps. These techniques are discussed in earlier chapters and are summarized below for correlation purposes.

1. For a liquid, distillation is generally used (see Techniques 2 and 3).
2. For a solid, recrystallization is generally used (see Technique 5).
3. Extraction is used especially if the impurity is insoluble in a solvent in which the compound itself is soluble (see Technique 4).
4. Sublimation is a very efficient technique, if the compound sublimes (see Technique 9).
5. Chromatography (gas, column, and thin-layer) is often used (see Techniques 1 and 6A).

It should be realized that these techniques may be applied to the separation of mixtures as well.

**DETECTION OF ELEMENTS OTHER THAN CARBON, HYDROGEN, OR OXYGEN**

The elements, other than C, H, and O, that are most often present in organic compounds are nitrogen, sulfur, and the halogens (F, Cl, Br, or I). To detect the presence of these elements, the organic compound is generally fused with metallic sodium. This reaction converts these heteroatoms to the water-soluble inorganic compounds, NaCN, $Na_2S$, and NaX. Inorganic qualitative analysis tests enable the investigator to determine the presence of the corresponding anions.

$$\text{Organic compound containing} \begin{Bmatrix} C \\ H \\ O \\ N \\ S \\ X \end{Bmatrix} \xrightarrow[\Delta]{Na} \begin{Bmatrix} NaCN \\ Na_2S \\ NaX \end{Bmatrix}$$

**Sodium Fusion[2]**

HOOD

**IMPORTANT.** *The fusion reaction is carried out in the **hood**. Make sure you are wearing safety glasses. All reagents must be of analytical grade, and deionized water must be used.*

CAUTION

**CAUTION: Sodium metal can cause serious burns and it reacts *violently* with water.**

In a small (10 x 75-mm) test tube (soft glass preferred), supported in a transite board (see Fig. 10.2), is placed about 25–30 mg of clean sodium metal (about one half the size of a pea).

[2] See Campbell, K. N.; Campbell, B. K. *J. Chem. Educ.* **1950,** *27,* 261 for a discussion of the procedure.

CAUTION: Use forceps to make this transfer, *never* touch sodium metal with your fingers.

CAUTION

Heat the tube with a flame until the sodium melts and sodium vapor is observed rising in the tube.

Mix a small sample of your unknown compound (1–2 drops of a liquid; 6–10 mg if a solid) with about 15–25 mg of *powdered* sucrose.[3] Gentle mixing of solids may be done on filter paper or glassine weighing paper and liquids on a watch glass. Then carefully add this mixture to the tube. During the transfer, be careful not to get any material on the sides of the test tube.

**NOTE.** *The addition of sucrose to the sample aids in the reduction of various nitrogen or sulfur compounds. Also, it absorbs volatile materials so that they may undergo the desired reaction before significant vaporization can occur.*

Now heat the tube gently to initiate the reaction with sodium. Remove the flame until the reaction subsides, and then heat to redness for 1–2 min. Allow the tube and contents to cool to room temperature. **Then,** *and only then,* **cautiously** add several drops of methanol (Pasteur pipet) to decompose any unreacted metallic sodium. Gently warm to drive off the excess methanol.

Reheat the tube to a bright red. While the tube is still red hot, lift the transite board and test tube from the iron ring and place the tube in a small beaker (30 mL) containing about 15 mL of deionized water (the transite board acts as a cover on the beaker).

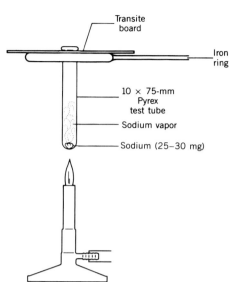

**FIGURE 10.2** Apparatus for sodium fusion.

CAUTION: The soft-glass tube usually cracks and breaks during this operation.

CAUTION

Break up the tube with a glass rod, heat the solution to boiling and filter it by gravity into a clean 50-mL Erlenmeyer flask. Wash the filter paper with an additional 2.0 mL of distilled water and combine this wash with the original filtrate.

**NOTE.** *If a Pyrex test tube is used, after the unreacted sodium metal is completely destroyed by adding methanol, add 2 mL of deionized water directly to the tube and contents. Place a glass stirring rod in the tube and heat the solution to boiling with stirring and then filter as described above. Dilute the filtrate with deionized water to about 5 mL.*

### Using the Fusion Solution

The clear, colorless fusion solution is used to test for the presence of $CN^-$ (nitrogen), $S^{-2}$ (sulfur), and $X^-$ (halogens except $F^-$) as described in the following sections.

a. Place 2–3 drops (Pasteur pipet) of the fusion solution on a white spot plate, followed by 2 drops of water. Now add 1 drop of dilute

**Sulfur**

---

[3] Confectioner's sugar is used and can be purchased at the supermarket.

(2%) aqueous sodium nitroprusside solution. The formation of a deep blue-violet color is a positive test for sulfur.

$$Na_2S + Na_2Fe(CN)_5NO \quad \rightarrow \quad Na_4[Fe(CN)_5NOS]$$

Sodium nitroprusside      Blue-violet complex

b. Place 3–4 drops (Pasteur pipet) of the fusion solution on a white spot plate followed by 1–2 drops of acetic acid. Now add 1 drop of 1% lead(II) acetate solution. The formation of a black precipitate (lead sulfide) indicates the presence of sulfur.

## Nitrogen[4]

*Reagents*

**1.** A 1.5% solution of *p*-nitrobenzaldehyde in 2-methoxyethanol.
**2.** A 1.7% solution of *o*-dinitrobenzene in 2-methoxyethanol.
**3.** A 2.0% solution of NaOH in distilled water.

**NOTE.** *All reagent drops are dispensed using Pasteur pipets.*

On a white spot plate, place together: 5 drops of Reagent **1,** 5 drops of Reagent **2,** and 2 drops of Reagent **3.** Stir this mixture gently with a glass rod.

Now add 1 drop of the fusion solution. The formation of a deep-purple color is a positive test for the presence of $CN^-$ ion; a yellow or tan coloration is negative. If a positive result is obtained, nitrogen is present in the sample.

The test is valid in the presence of halogens (NaX) or sulfur ($Na_2S$). It is much more sensitive than the traditional Prussian Blue test.[5]

### The Soda Lime Test

In a 10 × 75-mm test tube, mix about 50 mg of soda lime and 50 mg of $MnO_2$. Add 1 drop of a liquid unknown or about 10 mg of a solid unknown. Place over the mouth of the tube a moist strip of Brilliant Yellow paper (moist, red litmus paper is an alternative). Using a test tube holder, hold the tube at an incline (*pointing away from you and others*) and heat the contents gently at first and then quite strongly. Nitrogen-containing compounds will usually evolve ammonia.

A positive test for nitrogen is the deep red coloration of the brilliant yellow paper (or blue color of the litmus paper).

## The Halogens (Except Fluorine)

### Using the Fusion Solution

In a 10 × 75-mm test tube containing a boiling stone, place 0.5 mL (calibrated Pasteur pipet) of the fusion solution. Carefully acidify this solution by the dropwise addition of dilute $HNO_3$ acid, delivered from a Pasteur pipet (test acidity with litmus paper). If a positive test for nitrogen or sulfur was obtained, heat the resulting solution to a gentle boil (stir with a micro-

---

[4] Adapted from Guilbault, G. G.; Kramer, D. N. *Anal. Chem.* **1966,** *39,* 834. *Idem. J. Org. Chem.* **1966,** *31,* 1103. See also Shriner, R. L.; Fuson, R. C.; Morrill, T. C. *The Systematic Identification of Organic Compounds,* 6th ed., Wiley: New York, 1980, p. 80.

[5] See Vogel, A. I. *Elementary Practical Organic Chemistry,* Part 2, 2nd ed., Wiley: New York, 1966, p. 37.

spatula to prevent boilover) for 1 min over a microburner **[HOOD]** to expel    HOOD
any HCN or H$_2$S that might be present. Then cool the tube to room temperature.

To the resulting cooled solution, add 2 drops (Pasteur pipet) of aqueous 0.1 $M$ AgNO$_3$ solution.

A heavy curdy-type precipitate is a positive test for the presence of
Cl$^-$, Br$^-$, or I$^-$ ion. A faint turbidity is a negative test.

AgCl precipitate is white.
AgBr precipitate is pale yellow.
AgI precipitate is yellow.
AgF is not detected by this test since it is relatively soluble in water.

The silver halides have different solubilities in dilute ammonium hydroxide solution.

Centrifuge the test tube and contents and remove the supernatant
liquid using a Pasteur filter pipet. Add 0.5 mL (calibrated Pasteur pipet) of
dilute ammonium hydroxide solution to the precipitate and stir with a
glass rod to determine whether the solid is soluble.

AgCl is soluble in ammonium hydroxide due to the formation of the com-
plex ion, [Ag(NH$_3$)$_2$]$^+$.
AgBr is slightly soluble in this solution.
AgI is insoluble in this solution.

## Further Test

Once the presence of a halide ion has been established, a further test[6] is
available to aid in distinguishing between Cl$^-$, Br$^-$, and I$^-$ ion.

As described above, acidify 0.5 mL of the fusion solution with dilute
HNO$_3$. To this solution, add 5 drops (Pasteur pipet) of a 1.0% aqueous
KMnO$_4$ solution and shake the test tube for about 1 min.

Now add 10–15 mg of oxalic acid, enough to decolorize the excess
purple permanganate, followed by 0.5 mL of methylene chloride solvent.
The test tube is stoppered, shaken, vented, and the layers allowed to
separate. Observe the color of the CH$_2$Cl$_2$ (lower) layer.

A clear methylene chloride layer indicates Cl$^-$ ion.
A brown methylene chloride layer indicates Br$^-$ ion.
A purple methylene chloride layer indicates I$^-$ ion.

The colors may be faint and should be observed against a white
background.

## The Beilstein Test[7]

Organic compounds that contain chlorine, bromine, or iodine, and hydrogen decompose on ignition in the presence of copper oxide, to yield the
corresponding hydrogen halides. These hydrogen halides react to form the
volatile cupric halides that impart a green or blue-green color to a nonlumi-
nous flame. It is a very sensitive test, but some nitrogen-containing compounds and some carboxylic acids also give positive results.

---

[6] For a further test to distinguish between the three halide ions see Shriner, R. L.;
Fuson, R. C.; Morrill, T. C. *The Systematic Identification of Organic Compounds*, 6th ed., Wiley:
New York, 1980, p. 81. Also see this reference (p. 85) for a specific test for the F$^-$ ion.
[7] Beilstein, F. *Berichte* **1872,** *5,* 620.

Pound the end of a copper wire to form a flat surface that can act as a spatula. The other end of the wire (~4 in. long) is stuck in a cork stopper to serve as an insulated handle.

Heat the flat tip of the wire in a flame until coloration of the flame is negligible.

On the **cooled** flat surface of the wire, place a drop (Pasteur pipet) of liquid unknown, or a few milligrams of solid unknown. Gently heat the material in the flame. The carbon present in the compound will burn first, and thus the flame will be luminous, but then the characteristic green or blue-green color may be evident. It may be fleeting, so watch carefully.

It is recommended that a known compound containing a halogen be tested so that you become familiar with the appearance of the expected color.

Fluoride ion is not detected by this test, since copper fluoride is not volatile.

## SOLUBILITY CHARACTERISTICS

Determination of the solubility characteristics of an organic compound can often give valuable information as to its structural composition. It is especially useful when correlated with spectral analysis.

Several schemes have been proposed that place a substance in a definite group according to its solubility in various solvents. The scheme presented below is similar to that outlined in Shriner et al.[8]

There is no sharp dividing line between soluble and insoluble, and an arbitrary ratio of solute to solvent must be selected. We suggest that a compound be classified as soluble if its solubility is greater than 15 mg/500 $\mu$L of solvent.

Carry out the solubility determinations, at ambient temperature, in $10 \times 75$-mm test tubes. Place the sample (~15 mg) in the test tube and add a total of 0.5 mL of solvent in three portions from a graduated or calibrated Pasteur pipet. Between addition of each portion, stir the sample vigorously with a glass stirring rod for 1.5–2 min. If the sample is water soluble, test the solution with litmus paper to assist in classification according to the solubility scheme that follows.

**NOTE.**  *To test with litmus paper, dip the end of a small glass rod into the solution and then gently touch the litmus paper with the rod. DO NOT DIP THE LITMUS PAPER INTO THE TEST SOLUTION.*

In doing the solubility tests follow the scheme in the order given. *Keep a record of your observations.*

**Step I** Test for water solubility. If soluble, test the solution with litmus paper.

**Step II** If water soluble, determine the solubility in diethyl ether. This test further classifies water-soluble materials.

**Step III** Water-insoluble compounds are tested for solubility in a 5% aqueous NaOH solution. If soluble, determine the solubility in 5% aqueous $NaHCO_3$. The use of the $NaHCO_3$ solution aids in distinguishing between strong (soluble) and weak (insoluble) acids.

**Step IV** Compounds insoluble in 5% aqueous NaOH are tested for solubility in a 5% HCl solution.

[8] Shriner, R. L.; Fuson, R. C.; Morrill, T. C. *The Systematic Identification of Organic Compounds*, 6th ed., Wiley: New York, 1980.

**Step V** Compounds insoluble in 5% aqueous HCl are tested with concentrated $H_2SO_4$. If soluble, further differentiation is made using 85% $H_3PO_4$, as shown in the scheme.

**Step VI** Miscellaneous neutral compounds containing sulfur or nitrogen are normally soluble in strong acid solution.

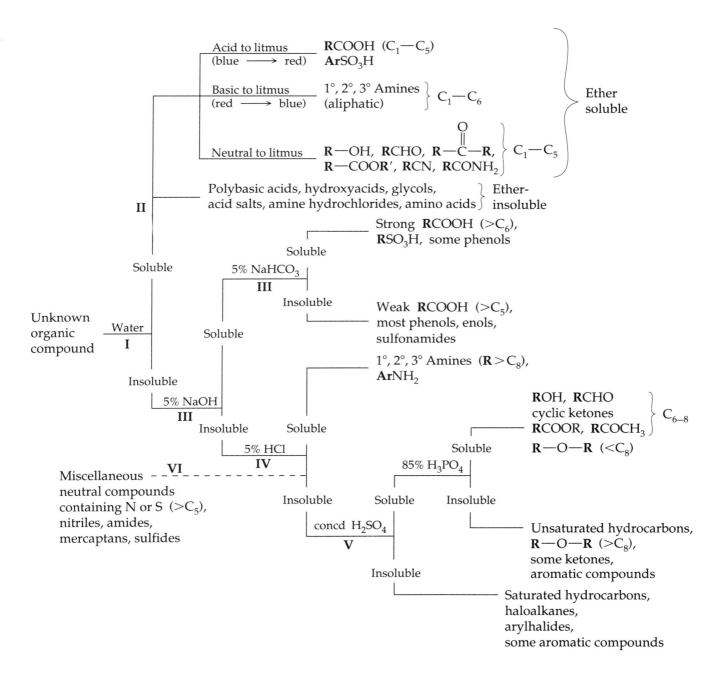

Note that to classify a given compound, it may not be necessary to test its solubility in every solvent. *Do only those tests that are required to place the compound in one of the solubility groups.* Make your observations with care, and proceed in a logical sequence as you make the tests.

# THE CLASSIFICATION TESTS[9]

**NOTE.** *For all tests given in this section, drops of reagents are measured out using Pasteur pipets.*

## Alcohols

### Ceric Nitrate Test

**INSTRUCTOR PREPARATION.** *The reagent is prepared by dissolving 4.0 g of ceric ammonium nitrate [$(NH_4)_2Ce(NO_3)_6$] in 10 mL of 2 N $HNO_3$. Warming may be necessary.*

Primary, secondary, and tertiary alcohols having fewer than 10 carbon atoms give a positive test as indicated by a change in color from *yellow* to *red*.

$$(NH_4)_2Ce(NO_3)_6 + RCH_2OH \rightarrow [\text{alcohol + reagent}]$$
$$\text{Yellow} \qquad\qquad\qquad (\text{Red complex})$$

Place 5 drops of test reagent on a white spot plate. Add 1–2 drops of the unknown sample (5 mg if a solid). Stir with a thin glass rod to mix the components and observe any color change.

**1.** If the alcohol is water insoluble, 3–5 drops of dioxane may be added, but run a blank to make sure the dioxane is pure. Efficient stirring gives positive results with most alcohols.
**2.** Phenols, if present, give a brown color or precipitate.

### Chromic Anhydride Test: The Jones Oxidation

**INSTRUCTOR PREPARATION.** *The reagent is prepared by slowly adding a suspension of 1.0 g of $CrO_3$ in 1.0 mL of concentrated $H_2SO_4$ to 3 mL of water. Allow the solution to cool to room temperature before using.*

The Jones oxidation test is a rapid method to distinguish primary and secondary alcohols from tertiary alcohols. A positive test is indicated by a color change from *orange* (the oxidizing agent, $Cr^{6+}$) as the oxidizing agent is itself reduced to the *blue green* ($Cr^{3+}$).

$$\left.\begin{array}{c} RCH_2OH \\ \text{or} \\ R_2CHOH \end{array}\right\} + H_2Cr_2O_7 \xrightarrow{H_2SO_4} Cr_2(SO_4)_3 + \begin{array}{c} RCO_2H \\ \text{or} \\ R_2C{=}O \end{array}$$
$$\qquad\qquad\quad \text{Orange} \qquad\qquad \text{Green}$$

The test is based on oxidation of a primary alcohol to an aldehyde or acid, and of a secondary alcohol to a ketone.

On a white spot plate, place 1 drop of the liquid unknown (10 mg if a solid). Add 10 drops of acetone and stir the mixture with a thin glass rod. Add 1 drop of the test reagent to the resulting solution. Stir and observe any color change within a 2-second time period.

**1.** Run a blank to make sure the acetone is pure.
**2.** Tertiary alcohols, unsaturated hydrocarbons, amines, ethers, and ketones give a negative test within the 2-second time frame for observing

---

[9] For a detailed discussion of classification tests see (a) footnote 8, p. 138 or (b) Pasto, D. J.; Johnson, C. R.; Miller, M. J. *Experiments and Techniques in Organic Chemistry;* Prentice Hall: New Jersey, 1992.

the color change. Aldehydes, however, give a positive test since they are oxidized to the corresponding carboxylic acids.

## The HCl/ZnCl₂ Test: The Lucas Test

**INSTRUCTOR PREPARATION.** *The Lucas reagent is prepared by dissolving 16 g of anhydrous ZnCl₂ in 10 mL of concd HCl while cooling in an ice bath.*

The Lucas test is used to distinguish between primary, secondary, and tertiary monofunctional alcohols having fewer than six carbon atoms.

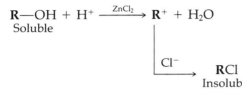

The test requires that the alcohol initially be soluble in the Lucas test reagent solution. As the reaction proceeds, the corresponding alkyl chloride is formed, which is insoluble in the reaction mixture. As a result, the solution becomes cloudy. In some cases a separate layer may be observed.

**1.** Tertiary, allyl, and benzyl alcohols react to give an immediate cloudiness to the solution. You may be able to see a separate layer of the alkyl chloride after a short time.
**2.** Secondary alcohols generally produce a cloudiness within 3–10 min. The solution may have to be heated to obtain a positive test.
**3.** Primary alcohols having less than six carbon atoms dissolve in the reagent but react very, very slowly. Those having more than six carbon atoms do not dissolve to any significant extent, no reaction occurs, and the aqueous phase remains clear.
**4.** A further test to aid in distinguishing between tertiary and secondary alcohols is to run the test using concentrated hydrochloric acid. Tertiary alcohols react immediately to give the corresponding alkyl halide, whereas secondary alcohols do not react under these conditions.

In a small test tube prepared by sealing a Pasteur pipet off at the shoulder (■), place 2 drops of the unknown (10 mg if a solid) followed by 10 drops of the Lucas reagent.
Shake or stir the mixture with a thin glass rod and allow the solution to stand. Observe the results. Based on the times given above, classify the alcohol. Additional points to consider:

**1.** Certain polyfunctional alcohols also give a positive test.
**2.** If an alcohol having three or fewer carbons is expected, a 1-mL conical vial equipped with an air condenser should be used to prevent low molecular weight alkyl chlorides (volatile) from escaping and thus remaining undetected.

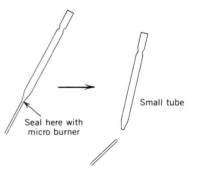

Making a small test tube.

## The Iodoform Test
This test is positive for compounds that on oxidation generate methyl ketones (or acetaldehyde) under the reaction conditions. For example, methyl carbinols (secondary alcohols having at least one methyl group attached to the carbon atom to which the —OH is attached), acetaldehyde, and ethanol give positive results.
For the test see Methyl Ketones and Methyl Carbinols (p. 715).

### Periodic Acid: Vicinal Diols

**INSTRUCTOR PREPARATION.** *This reagent solution is prepared by dissolving 250 mg of periodic acid ($H_5IO_6$) in 50 mL of deionized water.*

Vicinal diols (1,2 diols) are differentiated from the simple alcohols by the characteristic reaction below. Metaperiodic acid ($HIO_4$) selectively oxidizes 1,2 diols to give carbonyl compounds.

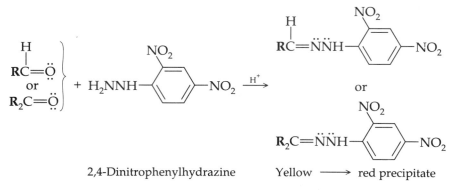

The test is based on the *instantaneous* formation of a white precipitate of silver iodate ($AgIO_3$) following addition of silver nitrate.

$$HIO_3 + AgNO_3 \rightarrow HNO_3 + AgIO_3 \downarrow$$

Place 2 mL of the periodic acid reagent solution in a small test tube.
Add 2 drops of concentrated nitric acid and mix the solution thoroughly. Add 2 drops of a liquid (~2–5 mg of a solid) and mix again. Now add 2–3 drops of 5% aqueous silver nitrate solution. An *instantaneous white precipitate* constitutes a positive test.
$\alpha$-Hydroxyaldehydes, $\alpha$-hydroxyketones, $\alpha$-hydroxyacids, 1,2-diketones, and $\alpha$-aminoalcohols, also give a positive test.

## Aldehydes and Ketones

### The 2,4-Dinitrophenylhydrazine Test

**INSTRUCTOR PREPARATION.** *The reagent solution is prepared by dissolving 1.0 g of 2,4-dinitrophenylhydrazine in 5.0 mL of concentrated sulfuric acid. This solution is slowly added, with stirring, to a mixture of 10 mL of water and 35 mL of 95% ethanol. After mixing, filter the solution.*

Aldehydes and ketones react rapidly with 2,4-dinitrophenylhydrazine to form 2,4-dinitrophenylhydrazones. These derivatives range in color from *yellow* to *red*, depending on the degree of conjugation in the carbonyl compound.

2,4-Dinitrophenylhydrazine        Yellow ⟶ red precipitate

On a white spot plate place 7–8 drops of 2,4-dinitrophenylhydrazine reagent solution.
Then add 1 drop of a liquid unknown. If the unknown is a solid, 1 drop of a solution prepared by dissolving 10 mg of the material in 10 drops

of ethanol is added. The mixture is stirred with a thin glass rod. The formation of a red-to-yellow precipitate is a positive test.

**NOTE:** *The reagent, 2,4-dinitrophenylhydrazine, is orange-red and melts at 198 °C (dec). Do not mistake it for a derivative!*

Reactive esters or anhydrides react with the reagent to give a positive test. Allylic or benzylic alcohols may be oxidized to aldehydes or ketones, which in turn give a positive result. Amides do not interfere with the test. Be sure that your unknown is pure and does not contain aldehyde or ketone impurities.

Phenylhydrazine and *p*-nitrophenylhydrazine are often used to prepare the corresponding hydrazones. These reagents also yield solid derivatives of aldehydes and ketones.

## Silver Mirror Test for Aldehydes: Tollens Reagent

This reaction involves the oxidation of aldehydes to the corresponding carboxylic acid, using an alcoholic solution of silver ammonium hydroxide. A positive test is the formation of a *silver* mirror, or a black precipitate of finely divided silver.

$$\underset{\overset{|}{\text{RC}}}{\overset{\text{H}}{|}}=\ddot{\text{O}}\colon \;+\; 2\,\text{Ag(NH}_3)_2\ddot{\text{O}}\colon\!\text{H} \;\longrightarrow\; 2\,\text{Ag}\!\downarrow \;+\; \text{R}\!-\!\overset{\overset{\displaystyle\ddot{\text{O}}\colon}{\|}}{\underset{\underset{\displaystyle\colon\!\ddot{\text{O}}\colon^{-},\;\text{NH}_4^{+}}{}}{\text{C}}} \;+\; \text{H}_2\ddot{\text{O}} \;+\; 3\,\ddot{\text{N}}\text{H}_3$$

The test should only be run after the presence of an aldehyde or ketone has been established.

In a small test tube prepared from a Pasteur pipet (see the Lucas test) place 1.0 mL of a 5% aqueous solution of $AgNO_3$, followed by 1 drop of aqueous 10% NaOH solution. Now add concentrated aqueous ammonia, drop by drop (2–4 drops) with shaking, until the precipitate of silver oxide just dissolves. Add 1 drop of the unknown (10 mg if a solid), with shaking, and allow the reaction mixture to stand for 10 min at room temperature. If no reaction has occurred, place the test tube in a sand bath at 40 °C for 5 min. Observe the result. Additional points to consider:

**1.** Avoid a large excess of ammonia.
**2.** Reagents must be well mixed. Stirring with a thin glass rod is recommended.
**3.** *This reagent is freshly prepared for each test. It should not be stored since decomposition occurs with the formation of $AgN_3$, which is explosive.*
**4.** This oxidizing agent is very mild and thus alcohols are not oxidized under these conditions. Ketones do not react. Some sugars, acyloins, hydroxylamines, and substituted phenols do give a positive test.

## Chromic Acid Test

**INSTRUCTOR PREPARATION.** *The reagent is prepared by dissolving 1 g of chromium trioxide in 1 mL of concd $H_2SO_4$, followed by 3 mL of $H_2O$.*

Chromic acid in acetone rapidly oxidizes aldehydes to carboxylic acids. Ketones react very slowly, or not at all.

In a 3 mL vial or small test tube, place 2 drops of a liquid unknown (~10 mg if a solid) and 1 mL of spectral grade acetone. Now add several drops of the chromic acid reagent.

A green precipitate of chromous salts is a positive test. Aliphatic aldehydes give a precipitate within 30 seconds; aromatic aldehydes take 30–90 seconds.

The reagent also reacts with primary and secondary alcohols (see Chromic Anhydride Test: Jones Oxidation, p. 704).

### Bisulfite Addition Complexes

**INSTRUCTOR PREPARATION.** *The reagent is prepared by mixing 1.5 mL of ethanol and 6 mL of a 40% aqueous solution of sodium bisulfite. Filter the reagent before use, if a small amount of the salt does not dissolve.*

Most aldehydes react with a saturated sodium bisulfite solution to yield a crystalline bisulfite addition complex.

The reaction is reversible and thus the carbonyl compound can be recovered by treatment of the complex with aqueous 10% $NaHCO_3$ or dilute HCl solution.

Place 50–75 $\mu$L of the liquid unknown in a small test tube and add 150 $\mu$L of the sulfite reagent and mix thoroughly.

A crystalline precipitate is a positive test.

Alkyl methyl ketones and unhindered cyclic ketones also give a positive test.

## Alkanes and Cycloalkanes: Saturated Hydrocarbons

### Iodine Charge-Transfer Complex

Alkanes exhibit a *negative* iodine charge-transfer complex test. Species containing $\pi$ electrons or nonbonded electron pairs produce a brown colored solution. This color formation is due to the charge-transfer complex between iodine and the available electrons.

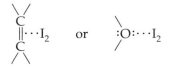

Solutions of iodine and nonparticipating compounds are violet in color.

On a white spot plate, place a small crystal of iodine. Now add 2–3 drops of a liquid unknown. Alkanes give a *negative* test (violet color).

The test is run only on liquid unknowns. Saturated hydrocarbons, fluorinated and chlorinated saturated hydrocarbons, and aromatic hydrocarbons and their halogenated derivatives all give violet colored solutions. All other species give a positive test (brown colored solution).

### Concentrated Sulfuric Acid

Saturated hydrocarbons, halogenated saturated hydrocarbons, simple aromatic hydrocarbons, and their halogenated derivatives are insoluble in *cold* concentrated sulfuric acid.

CAUTION In a small test tube place 100 $\mu$L of *cold* concentrated sulfuric acid **[CAUTION]**. Now add 50 $\mu$L of an unknown. A resulting heterogeneous solution (the unknown does *not* dissolve) is a positive test for a saturated hydrocarbon.

Alkenes, and compounds having a functional group containing a nitrogen or oxygen atom, are soluble in cold, concentrated acid.

### Bromine in Methylene Chloride

Unsaturated hydrocarbons readily add bromine ($Br_2$). An example of this reaction is given in Experiment [F2].

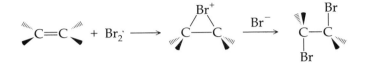

The test is based upon the decolorization of a red-brown bromine–methylene chloride solution.

---

**CAUTION:  Bromine is highly toxic and can cause burns.**

---

In a $10 \times 75$-mm test tube, or in a small tube prepared from a Pasteur pipet (see Lucas test), place 2 drops of a liquid unknown ($\sim 15$ mg if a solid) followed by 0.5 mL of methylene chloride. Add dropwise, with shaking, a 2% solution of bromine in methylene chloride solvent **[HOOD]**. The presence of an unsaturated hydrocarbon will require 2–3 drops of the reagent before the reddish-brown color of bromine persists in the solution. Additional points to consider:

**1.** Methylene chloride is used in place of the usual carbon tetrachloride ($CCl_4$) as it is less toxic.

**2.** Phenols, enols, amines, aldehydes, and ketones interfere with this test.

### Permanganate Test: Baeyer Test for Unsaturation

Unsaturation in an organic compound can be detected by the decolorization of permanganate solution. The reaction involves the cis hydroxylation of the alkene to give a 1,2 diol (glycol).

$$\overset{}{\underset{}{>}}C{=}C\overset{}{\underset{}{<}} + 2\,MnO_4^- + 4\,H_2O \longrightarrow \overset{}{\underset{:\underset{\cdot\cdot}{O}H}{C}}{-}\overset{}{\underset{:\underset{\cdot\cdot}{O}H}{C}} + 2\,MnO_2 + 2\,OH^-$$

On a white spot plate, place 0.5 mL of *alcohol-free* acetone, followed by 2 drops of the unknown compound ($\sim 15$ mg if a solid). Now add dropwise (2–3 drops), with stirring, a 1% aqueous solution of potassium permanganate ($KMnO_4$). A positive test for unsaturation is the discharge of purple permanganate color from the reagent and the precipitation of brown manganese oxides.

Any functional group that undergoes oxidation with permanganate interferes with the test (phenols, aryl amines, most aldehydes, primary and secondary alcohols, etc.).

**Alkenes and Alkynes: Unsaturated Hydrocarbons**

## Alkyl Halides

### Silver Nitrate Test

Alkyl halides that undergo the $S_N1$ substitution reaction react with alcoholic silver nitrate ($AgNO_3$) to form a precipitate of the corresponding silver halide.

Secondary and primary halides react slowly or not at all at room temperature. However, they do react at elevated temperatures. Tertiary halides react immediately at room temperature.

In a 1.0-mL conical vial place 0.5 mL of 2% ethanolic $AgNO_3$ solution and 1 drop of unknown (~10 mg if a solid). A positive test is indicated by the appearance of a precipitate within 5 min. If no reaction occurs, add a boiling stone and equip the vial with an air condenser. Heat the solution at *gentle* reflux for an additional 5 min using a sand bath. Cool the solution.

If a precipitate is formed, add 2 drops of dilute $HNO_3$. Silver halides will not dissolve in nitric acid solution. Additional points to consider:

**1.** The order of reactivity for R groups is allyl $\cong$ benzyl > tertiary > secondary >>> primary. For the halide leaving groups: I > Br > Cl.

**2.** Acid halides, $\alpha$-haloethers, and 1,2-dibromo compounds also give a positive test at room temperature. Only activated aryl halides give a positive test at elevated temperatures.

### Sodium Iodide in Acetone

**INSTRUCTOR PREPARATION.**   *The reagent is prepared by dissolving 3 g of sodium iodide (NaI) in 25 mL of acetone. Store in a dark bottle.*

Primary alkyl chlorides and bromides can be distinguished from aryl and alkenyl halides by reaction with sodium iodide in acetone (Finkelstein reaction).

$$R\text{—}X + NaI \xrightarrow{\text{acetone}} R\text{—}I + NaX \downarrow$$
$$X = Cl, Br$$

Primary alkyl bromides undergo an $S_N2$ displacement reaction within 5 min at room temperature, primary alkyl chlorides only at 50 °C.

In a 1.0-mL conical vial, place 1 drop of a liquid unknown (~10 mg if a solid) and 3 drops of acetone. To this solution add 0.5 mL of sodium iodide–acetone reagent.

A positive test is the appearance of a precipitate of NaX within 5 min. If no precipitate is observed, add a boiling stone and equip the vial with an air condenser. Warm the reaction mixture in a sand bath at about 50 °C for 5 min. Cool to room temperature and determine whether a reaction has occurred. Additional points to consider:

**1.** Benzylic and allylic chlorides and bromides, acid chlorides and bromides, and $\alpha$-haloketones, $\alpha$-haloesters, $\alpha$-haloamides, and $\alpha$-halonitriles also give a positive test at room temperature.

**2.** Primary and secondary alkyl chlorides, and secondary and tertiary alkyl bromides, react at 50 °C under these conditions.

**3.** If the solution turns red brown in color, iodine is being liberated.

## Amides, Ammonium Salts, and Nitriles

### Hydroxamate Test for Amides

Unsubstituted (on nitrogen) amides, and the majority of substituted amides, will give a positive hydroxamate test.

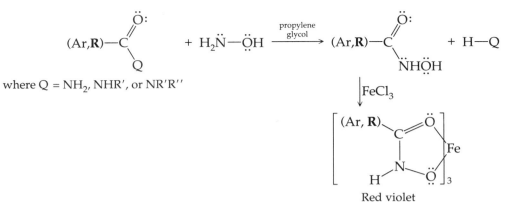

Red violet

The hydroxamic acid is identified by formation of a red-to-purple color in the presence of $Fe^{3+}$ ion, as for the test with esters (see below).

In a 3.0-mL conical vial containing a boiling stone and equipped with an air condenser place 1 drop of a liquid unknown ($\sim$10 mg if a solid), followed by 0.5 mL of 1 $M$ hydroxylamine hydrochloride–propylene glycol solution. Heat the resulting mixture to reflux temperature ($\sim$190 °C) using a sand bath, and reflux for 3–5 min. Cool the solution to room temperature, and add 2 drops of 5% aqueous $FeCl_3$ solution. The formation of a red-to-purple color is a positive test.

Ammonium salts, amides, and nitriles undergo hydrolysis in alkaline solution to form ammonia gas, or an amine.

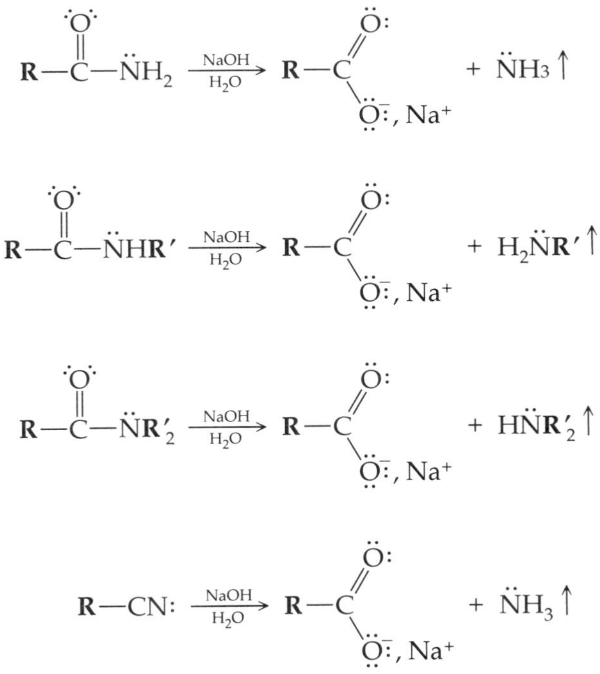

**Alkaline Hydrolysis**

Detection of ammonia from ammonium salts, primary amides, and nitriles, by use of a color test using copper sulfate solution, constitutes a positive test for these functional groups. The same test may also be used for secondary and tertiary amides that can generate low molecular weight (volatile) amines upon hydrolysis.

In a 1.0-mL conical vial containing a boiling stone, and equipped with an air condenser, place 1–2 drops of the unknown liquid (~10 mg if a solid) and 0.5 mL of 20% aqueous NaOH solution. Heat this mixture to *gentle* reflux on a sand bath. Moisten a strip of filter paper with 2 drops of 10% aqueous copper sulfate solution and place it over the top of the condenser. Formation of a *blue* color [copper ammonia (or amine) complex] is a positive test.

The filter paper may be held in place using a small test tube holder or other suitable device.

## Amines

### Copper Ion Test

Amines will give a blue-green coloration or precipitate, when added to a copper sulfate solution. In a small test tube, place 0.5 mL of a 10% copper sulfate solution. Now add 1 drop of an unknown (~10 mg if a solid). The blue-green coloration or precipitate is a positive test. Ammonia will also give a positive test.

### Hinsberg Test

The Hinsberg test is useful for distinguishing between primary, secondary, and tertiary amines. The reagent used is *p*-toluenesulfonyl chloride in alkaline solution.

*Primary* amines with fewer than seven carbon atoms form a sulfonamide that is soluble in the alkaline solution. Acidification of the solution results in the precipitation of the insoluble sulfonamide.

$$H_3C-\langle \rangle-SO_2Cl + R-NH_2 \xrightarrow{NaOH}$$

$$H_3C-\langle \rangle-SO_2\bar{N}R, Na^+ \underset{\substack{\text{excess} \\ \text{base}}}{\overset{\substack{\text{excess} \\ \text{acid}}}{\rightleftarrows}} H_3C-\langle \rangle-SO_2NHR + NaCl + H_2O$$

(soluble)                    (insoluble)

*Secondary* amines form an insoluble sulfonamide in the alkaline solution.

$$H_3C-\langle \rangle-SO_2Cl + R_2NH \xrightarrow{NaOH} H_3C-\langle \rangle-SO_2NR_2 + NaCl + H_2O \xrightarrow{\substack{\text{excess} \\ \text{base}}} \text{no change}$$

(insoluble)

*Tertiary* amines normally give no reaction under these conditions.

$$H_3C-\langle \rangle-SO_2Cl + R_3N \xrightarrow{NaOH} H_3C-\langle \rangle-SO_3^- + NR_3 + 2 Na^+ + Cl^- + H_2O$$

(soluble)       (oil)

In a 1.0-mL conical vial containing a boiling stone, and equipped with an air condenser, place 0.5 mL of 10% aqueous sodium hydroxide solution, 1 drop of the sample unknown (~10 mg if a solid), followed by 30 mg of *p*-toluenesulfonyl chloride [**HOOD**]. Heat the mixture to reflux for 2–3 min on a sand bath, and then cool it in an ice bath. Test the alkalinity of the solution using litmus paper. If it is not alkaline, add additional 10% aqueous sodium hydroxide dropwise.

HOOD

Using a Pasteur filter pipet, separate the solution from any solid that may be present. Transfer the solution to a clean 1.0-mL conical vial [**SAVE**].

SAVE

**NOTE.** *If an oily upper layer is obtained at this stage, remove the lower alkaline phase* [**SAVE**] *using a Pasteur filter pipet. To the remaining oil add 0.5 mL of cold water and stir vigorously to obtain a solid material.*

SAVE

If a solid is obtained, it may be (1) the sulfonamide of a secondary amine; (2) recovered tertiary amine, if the original amine was a solid; or (3) the insoluble salt of a primary sulfonamide derivative (if the original amine had more than six carbon atoms). Additional points to consider:

1. If the solid is a tertiary amine, it is soluble in aqueous 10% HCl.
2. If the solid is a secondary sulfonamide, it is insoluble in aqueous 10% NaOH.
3. If no solid is present, acidify the alkaline solution by addition of 10% aqueous HCl. If the unknown amine is primary, the sulfonamide will precipitate.

## Bromine Water

Aromatic amines, since they possess an electron-rich aromatic ring, can undergo electrophilic aromatic substitution with bromine, to yield the corresponding arylamino halide(s). Therefore, if elemental tests indicate that an aromatic group is present in an amine, treatment with the bromine water reagent may indicate that the amine is attached to an aromatic ring.

For the test, see Phenols and Enols (p. 717).

## Fuming Sulfuric Acid

**Aromatic Hydrocarbons with NO Functional Groups**

Simple aromatic hydrocarbons are insoluble in sulfuric acid ($H_2SO_4$) but are soluble in fuming sulfuric acid. If these hydrocarbons contain more than two alkyl substituents, they may be sulfonated under these conditions.

In a small test tube place 100 $\mu$L of fuming sulfuric acid [**CAUTION**]. Now add 50 $\mu$L of the unknown suspected to be aromatic. A resulting homogeneous solution is a positive test.

CAUTION

## Azoxybenzene and Aluminum Chloride

This color test is run only on those aromatic compounds that are insoluble in sulfuric acid (see previous test). The color produced in this test results from the formation of a complex of $AlCl_3$ and a *p*-arylazobenzene derivative.

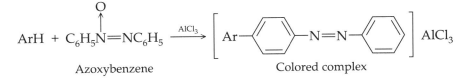

Azoxybenzene                    Colored complex

In a small dry test tube, place 250 $\mu$L of the aromatic unknown. Add a small crystal of azoxybenzene and about 12 mg of anhydrous aluminum chloride. If a color is not produced immediately, warm the mixture for a few min.

Aryl halides and other simple aromatic hydrocarbons give a deep-orange to dark-red color or precipitate. Polynuclear aromatic hydrocarbons, such as naphthalenes and anthracenes, give brown colors. Aliphatic hydrocarbons give no color, or at most a light yellow tint.

## Carboxylic Acids

The presence of a carboxylic acid is detected by its solubility behavior. An aqueous solution of the acid will be acidic to litmus paper (or pH paper may be used). Since a sulfonic acid would also give a positive test, the test for sulfur (sodium fusion) is used to distinguish between the two types of acids. A water-soluble phenol is acidic toward litmus paper but also would give a positive ferric chloride test.

Carboxylic acids also react with a 5% solution of sodium bicarbonate (see Experiment [4B]).

Place 1–2 mL of the bicarbonate solution on a watch glass and add 1–2 drops of the acid (~10 mg if a solid). Gas bubbles of $CO_2$ constitute a positive test.

## Esters

### Hydroxamate Test

Carboxylic esters can be identified by conversion to hydroxamic acid salts. Acidification of this salt produces the corresponding hydroxamic acid (RCONHOH), which is identified by formation of a red-to-purple color in the presence of $Fe^{3+}$ ion.

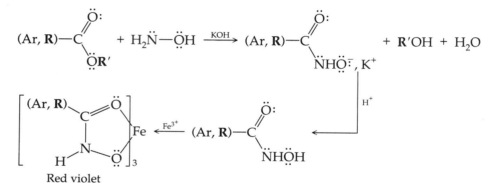

In a 3.0-mL conical vial containing a boiling stone, and equipped with an air condenser, place 1 drop of the liquid unknown (~10 mg if a solid) followed by 0.5 mL of 1.0 M ethanolic hydroxylamine hydrochloride solution. Add 10% methanolic KOH to this solution (dropwise) until the resulting solution has pH ~10 (pH paper). Heat this mixture to reflux temperature using a sand bath for 5 min, cool to room temperature, and acidify to pH = 3–4 by dropwise addition of 5% aqueous HCl solution. Now add 2 drops of 5% aqueous $FeCl_3$ solution. The formation of a red-to-purple color is a positive test. Additional points to consider:

1. It is suggested that a blank be run for comparison purposes.
2. Acid chlorides, anhydrides, lactones, and imides also give a positive test.

### Saponification

This well-known reaction of esters can often be used to classify these compounds. It also may lead to a useful derivative if the corresponding carboxylic acid is isolated.

In a 3.0-mL conical vial containing a magnetic spin vane, place 100 $\mu$L of the liquid unknown (~150 mg if a solid) and add 1 mL of 6 M NaOH solution. Attach the vial to a reflux condenser. Now place the vial in a sand bath on a magnetic stirring hot plate and, with stirring, heat the mixture at reflux for 0.5 hour, or until the solution becomes homogeneous.

A positive test is the disappearance of the organic layer (if the original unknown was water insoluble) or the lack of the usually pleasant aroma of the unknown ester.

High-boiling esters (bp > 200 °C) are usually not saponified under these conditions due to their low solubility in the aqueous solvent.

**Ethers**

**NOTE.**    *Upon standing, ethers may form peroxides. Peroxides are very explosive. To test for the presence of these substances, use starch–iodide paper that has been moistened with 6 M HCl. Peroxides cause the paper to turn blue. To remove peroxides from ethers, pass the material through a short column of highly activated alumina (Woelm basic alumina, active grade 1, see p. 318 of the reference cited in footnote 9b).*

### Ferrox Test

The ferrox test is a color test sensitive to oxygen, which may be used to distinguish ethers from hydrocarbons that, like most ethers, are soluble in sulfuric acid.

In a dry 10 × 75-mm test tube using a glass stirring rod, grind a crystal of ferric ammonium sulfate and a crystal of potassium thiocyanate. The ferric hexathiocyanatoferrate that is formed adheres to the rod.

In a second clean 10 × 75-mm test tube, place 2–3 drops of a liquid unknown. If dealing with a solid, use about 10 mg and add toluene until a saturated solution is obtained. Now using the rod with the ferric hexathio-cyanatoferrate attached, stir the unknown. *If the unknown contains oxygen, the ferrate compound dissolves and a reddish-purple color is observed.*

Some high molecular weight ethers do not give a positive test.

### Bromine Water

Aromatic ethers, since the aromatic ring is electron rich, can undergo electrophilic aromatic substitution with bromine to yield the corresponding aryl ether–halide(s). Therefore, if elemental tests indicate that an aromatic group is present in an ether, treatment with the bromine water reagent may substantiate the presence of an aryl ether.

For the test see Phenols and Enols (p. 717).

**Methyl Ketones and Methyl Carbinols**

### Iodoform Test

**INSTRUCTOR PREPARATION.**    *Dissolve 3 g of KI and 1 g I$_2$ in 20 mL of water.*

The iodoform test involves hydrolysis and cleavage of methyl ketones to form a yellow precipitate of iodoform (CHI$_3$).

$$R-\overset{\overset{\ddot{\text{O}}}{\|}}{C}-CH_3 + 3\,I_2 + 3\,KOH \longrightarrow R-\overset{\overset{\ddot{\text{O}}}{\|}}{C}-CI_3 + 3\,KI + 3\,H_2O$$

$$\downarrow \text{KOH}$$

$$R-\overset{\overset{\ddot{\text{O}}}{\|}}{C}-\overset{-}{O},\overset{+}{K} + CHI_3 \downarrow$$

Yellow

It is also a positive test for compounds that, upon oxidation, generate methyl ketones (or acetaldehyde) under these reaction conditions. For example, methyl carbinols (secondary alcohols having at least one methyl group attached to the carbon atom to which the OH unit is linked), acetaldehyde, and ethanol give positive results.

In a 3.0-mL conical vial equipped with an air condenser, place 2 drops of the unknown liquid (10 mg if a solid), followed by 5 drops of 10% aqueous KOH solution.

HOOD **NOTE.** *If the sample is insoluble in the aqueous phase, either mix vigorously or add dioxane* **[HOOD]** *or bis(2-methoxyethyl) ether to obtain a homogeneous solution.*

Warm the mixture on a sand bath to 50–60 °C and add the KI–$I_2$ reagent dropwise until the solution becomes dark brown in color (~1.0 mL). Additional 10% aqueous KOH is now added (dropwise) until the solution is again colorless.

---

**CAUTION: Iodine is highly toxic and can cause burns.**

---

After warming for 2 min, cool the solution and determine whether a yellow precipitate ($CHI_3$, iodoform) has formed. If a precipitate is not observed, reheat as before for another 2 min. Cool and check again for the appearance of iodoform. Additional points to consider:

**1.** The iodoform test is reviewed elsewhere.[10]
**2.** An example of the general haloform reaction, using bleach to oxidize a methyl ketone, is given in Experiment [34].

## Nitro Compounds

### Ferrous Hydroxide Test

Many nitro compounds give a positive test based on the following reaction.

$$R—NO_2 + 4 H_2O + 6 Fe(OH)_2 \rightarrow R—NH_2 + 6 Fe(OH)_3 \downarrow$$
$$\text{Red-brown}$$

The nitro derivative oxidizes the iron(II) hydroxide to iron(III) hydroxide, the latter is a red-brown solid.

In a 1.0-mL conical vial place 5–10 mg of the unknown compound, followed by 0.4 mL of freshly prepared 5% aqueous ferrous ammonium sulfate solution. After mixing, add 1 drop of 3 $N$ sulfuric acid followed by 10 drops of 2 $N$ methanolic KOH. Cap the vial, shake vigorously, vent, and then allow it to stand over a 5-min period. The formation of a red-brown precipitate, usually within 1 min, is a positive test for a nitro group.

### Sodium Hydroxide Color Test

Treatment of an aromatic nitro compound with 10% sodium hydroxide solution may often be used to determine the number of nitro groups present on the aromatic ring system.

Mononitro compounds produce no color (a light yellow may be observed).
Dinitro compounds produce a bluish-purple color.
Trinitro compounds produce a blood-red color.

The color formation is due to formation of Meisenheimer complexes (for a discussion, see p. 321 of the text cited in footnote 9b).

---

[10] Fuson, R. C.; Bull, B. A. *Chem. Rev.* **1934**, *15*, 275.

To run the test, dissolve 10 mg of the unknown (1–2 drops if a liquid) in 1 mL of acetone in a small test tube. Now add about 200 $\mu$L of 10% NaOH solution and shake. Observe any color formation.

If amino, substituted amino, or hydroxyl groups are present in the molecule, a positive color test is not obtained.

### Ferric Ion Test

Most phenols and enols form colored complexes in the presence of ferric ion, $Fe^{3+}$.

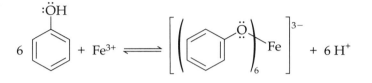

Phenols give red, blue, purple, or green colors. Sterically hindered phenols may give a negative test. Enols generally give a tan, red, or red-violet color.

On a white spot plate place 2 drops of water, or 1 drop of water plus 1 drop of ethanol, or 2 drops of ethanol, depending on the solubility characteristics of the unknown. To this solvent system add 1 drop (10 mg if a solid) of the substance to be tested. Stir the mixture with a thin glass rod to complete dissolution. Add 1 drop of 2.5% aqueous ferric chloride ($FeCl_3$) solution (light yellow in color). Stir and observe any color formation. If necessary, a second drop of the $FeCl_3$ solution may be added. Additional points to consider:

1. The color developed may be fleeting or it may last for many hours. A slight excess of the ferric chloride solution may or may not destroy the color.
2. An alternate procedure using $FeCl_3$–$CCl_4$ solution in the presence of pyridine is available.[11]

### Bromine Water

Phenols, substituted phenols, aromatic ethers, and aromatic amines, since the aromatic rings are electron rich, undergo aromatic electrophilic substitution with bromine to yield substituted aryl halides. For example,

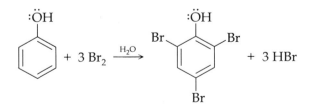

---

**CAUTION:    The test should be run in the *hood*.**          HOOD

---

[11] Soloway, S.; Wilen, S. H. *Anal. Chem.* **1952**, *4*, 979.

In a small test tube, place 1–2 drops of the unknown (~20 mg if a solid) and add 1–2 mL of water. Check the pH of the solution with pH paper. In the **hood** add saturated bromine water dropwise until the bromine color persists. A precipitate generally forms.

A positive test is the decolorization of the bromine solution, and often the formation of an off-white precipitate. If the unknown is a phenol, this should cause the pH of the original solution to be less than 7.

## PREPARATION OF DERIVATIVES

Based on the preliminary and classification tests carried out to this point, you should have established the type of functional group (or groups) present (or lack of one) in the unknown organic sample. The next step in qualitative organic analysis is to consult a set of tables containing a listing of known organic compounds sorted by functional group and/or by physical properties or by both. Using the physical properties data for your compound, you can select a few possible candidates that appear to "fit" the data you have collected. On a chemical basis, the final step in the qualitative identification sequence is to prepare one or two *crystalline derivatives* of your compound. Selection of the specific compound, and thus final confirmation of its identity, can then be made from the extensive derivative tables that have been accumulated. With the advent of spectral analysis, the preparation of derivatives is often not necessary, but the wealth of chemistry that can be learned by the beginning student in carrying out these procedures is extensive and important. The preparation of selected derivatives for the most common functional groups are given below. Condensed tables of compounds and their derivatives are summarized in Appendix A. For extensive tables and alternative derivatives that can be utilized, see the following Bibliography.

**IMPORTANT.**   *In each of the procedures outlined below, drops of reagents are measured using Pasteur pipets.*

**BIBLIOGRAPHY**

Pasto, D. J.; Johnson, C. R.; Miller, M. J. *Experiments and Techniques in Organic Chemistry*; Prentice Hall: Englewood Cliffs, NJ, 1992.

Rappoport, Z.; *Handbook of Tables for Organic Compound Identification*, 3rd ed., CRC Press: Boca Raton, FL, 1967.

Shriner, R. L.; Fuson, R. C.; Curtin; D. Y.; Morrill, T. C. *The Systematic Identification of Organic Compounds*, 6th ed., Wiley: New York, 1980.

## CARBOXYLIC ACIDS[12]

### Preparation of Acid Chlorides

$$\underset{\ddot{O}}{R-\overset{\overset{\ddot{O}}{\|}}{C}}-\ddot{O}H + Cl-\overset{\overset{\ddot{O}}{\|}}{S}-Cl \xrightarrow{DMF} R-\overset{\overset{\ddot{O}}{\|}}{C}-Cl + HCl\uparrow + SO_2\uparrow$$

[12] See Tables A.1 and A.2.

Weigh and place 20 mg of the unknown acid in a dry 3.0-mL conical vial containing a boiling stone and fitted with a cap. Now add **[HOOD]** 4 drops of thionyl chloride and 1 drop of *N,N*-dimethylformamide (DMF). Immediately attach the vial to a reflux condenser that is protected by a calcium chloride drying tube.

HOOD

---

**CAUTION:   This reaction is run in the *hood* since hydrogen chloride and sulfur dioxide are evolved. Thionyl chloride is an irritant and is harmful to breathe. Immediately recap the vial after each addition until the vial is attached to the reflux condenser.**

HOOD

---

Allow the mixture to stand at room temperature for 10 min, heat it at gentle reflux on a sand bath for 15 min, and then cool it to room temperature. Dilute the reaction mixture with 5 drops of methylene chloride solvent.

The acid chloride is not isolated but is used directly in the following preparations.

**Amides**

$$\text{R} - \overset{\overset{\displaystyle \cdot \cdot}{\overset{\displaystyle O}{\|}}}{\text{C}} - \text{Cl} + 2\,\ddot{\text{N}}\text{H}_3 \longrightarrow \text{R} - \overset{\overset{\displaystyle \cdot \cdot}{\overset{\displaystyle O}{\|}}}{\text{C}} - \ddot{\text{N}}\text{H}_2 + \text{NH}_4\text{Cl}$$

Cool the vial in an ice bath and add 10 drops of concentrated aqueous ammonia **[HOOD]** via Pasteur pipet, *dropwise*, with stirring. *It is convenient to make this addition down the neck of the air condenser.* The amide may precipitate during this operation. After the addition is complete, remove the ice bath and stir the mixture for an additional 5 min. Now add methylene chloride (10 drops) and stir the resulting mixture to dissolve any precipitate. Separate the methylene chloride layer from the aqueous layer using a Pasteur filter pipet and transfer it to another Pasteur filter pipet containing 200 mg of anhydrous sodium sulfate. Collect the eluate in a Craig tube containing a boiling stone. Extract the aqueous phase with an additional 0.5 mL of methylene chloride. Separate the methylene chloride layer as before and transfer it to the same column. Collect this eluate in the same Craig tube. Evaporate the methylene chloride solution using a warm sand bath **[HOOD]** under a gentle stream of nitrogen gas. Recrystallize the solid amide product using the Craig tube. Dissolve the material in about 0.5 mL of ethanol, add water (dropwise) until the solution becomes cloudy, cool the Craig tube in an ice bath, and collect the crystals in the usual manner. Dry the crystalline amide on a porous clay plate and determine the melting point.

HOOD

HOOD

**Anilides**

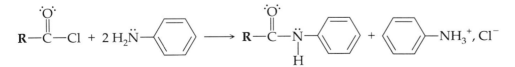

In a 3.0-mL conical vial containing a magnetic spin vane, and equipped with an air condenser, place 5 drops of aniline and 10 drops of methylene chloride. Cool the solution in an ice bath and transfer the acid chloride

HOOD

solution (prepared above) via Pasteur pipet, *dropwise*, with stirring, to the aniline solution **[HOOD]**. *It is convenient to make this addition down the neck of the condenser.* After the addition is complete, remove the ice bath and stir the mixture for an additional 10 min.

Transfer the methylene chloride layer to a $10 \times 75$-mm test tube, and wash it with 0.5 mL of $H_2O$, 0.5 mL of 5% aqueous HCl, 0.5 mL of 5% aqueous NaOH, and finally with 0.5 mL of $H_2O$. For each washing, shake the test tube and remove the top aqueous layer by Pasteur filter pipet. Transfer the resulting wet methylene chloride layer to a Pasteur filter pipet containing 200 mg of anhydrous sodium sulfate. Collect the eluate in a Craig tube containing a boiling stone. Rinse the original test tube with an additional 10 drops of methylene chloride. Collect this rinse and pass it through the same column. Both eluates are combined.

HOOD

Evaporate the methylene chloride solvent on a warm sand bath under a gentle stream of nitrogen gas **[HOOD]**. Recrystallize the crude anilide from an ethanol–water mixture using the Craig tube. Dissolve the material in about 0.5 mL of ethanol, add water (dropwise) to the cloud point, cool in an ice bath, and collect the crystals in the usual manner. Dry the purified derivative product on a porous clay plate, and determine its melting point.

**Toluidides**

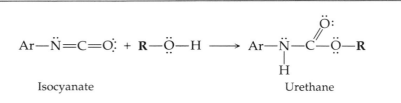

The same procedure described for the preparation of anilides is used except that *p*-toluidine replaces the aniline.

---

# ALCOHOLS [13]

**Phenyl- and α-Naphthylurethanes (Phenyl- and α-Naphthylcarbamates)**

$$Ar-\ddot{N}=C=\ddot{O} + R-\ddot{O}-H \longrightarrow Ar-\underset{\underset{H}{|}}{\ddot{N}}-\overset{\overset{\ddot{O}:}{\|}}{C}-\ddot{O}-R$$

Isocyanate                                         Urethane

**NOTE.**  *For the preparation of these derivatives, the alcohols must be anhydrous. Water hydrolyzes the isocyanates to produce arylamines that react with the isocyanate reagent to produce high-melting, disubstituted ureas.*

In a 3.0-mL conical vial containing a boiling stone and equipped with an air condenser protected by a calcium chloride drying tube place 15 mg of an anhydrous alcohol or phenol. Remove the air condenser from the vial and add 2 drops of phenyl isocyanate or α-naphthyl isocyanate. Replace the air condenser immediately. If the unknown is a phenol, add 1 drop of pyridine in a similar manner.

HOOD

---

**CAUTION:  This addition must be done in the *hood*. The isocyanates are lachrymators! Pyridine has the characteristic strong odor of an amine.**

---

[13] See Table A.3.

If a spontaneous reaction does not take place, heat the vial at about 80–90 °C, using a sand bath, for a period of 5 min. Then cool the reaction mixture in an ice bath. It may be necessary to scratch the sides of the vial to induce crystallization. Collect the solid product by vacuum filtration, using a Hirsch funnel, and purify it by recrystallization from ligroin. For this procedure, place the solid in a 10 × 75-mm test tube and dissolve it in 1.0 mL of warm (60–80 °C) ligroin. If diphenyl (or dinaphthyl) urea is present, (formed by reaction of the isocyanate with water), it is insoluble in this solvent. Transfer the warm ligroin solution to a Craig tube using a Pasteur filter pipet. Cool the solution in an ice bath and collect the resulting crystals in the usual manner. After drying the product on a porous clay plate, determine the melting point.

### 3,5-Dinitrobenzoates

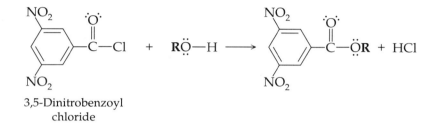

3,5-Dinitrobenzoyl
chloride

**NOTE.**   *The dinitrobenzoyl chloride reagent tends to hydrolyze on storage to form the corresponding carboxylic acid. Check its melting point before use (3,5-dinitrobenzoyl chloride, mp = 74 °C; 3,5-dinitrobenzoic acid, mp = 202 °C).*

In a 3.0-mL conical vial containing a boiling stone, and equipped with an air condenser protected by a calcium chloride drying tube, place 25 mg of pure 3,5-dinitrobenzoyl chloride and two drops of the unknown alcohol. Heat the mixture to about 10 °C below the boiling point of the alcohol (but not over 100 °C) on a sand bath for a period of 5 min. Cool the reaction mixture, add 0.3 mL of water, and then place the vial in an ice bath to cool. Collect the solid ester by vacuum filtration, using a Hirsch funnel, and wash the filter cake with three 0.5-mL portions of 2% aqueous sodium carbonate ($Na_2CO_3$) solution, followed by 0.5 mL of water. Recrystallize the solid product from an ethanol–water mixture using a Craig tube. Dissolve the material in about 0.5 mL of ethanol, add water (dropwise) until the solution is just cloudy, cool in an ice bath, and collect the crystals in the usual manner. After drying the product on a porous clay plate, determine the melting point.

### ALDEHYDES AND KETONES[14]

### 2,4-Dinitrophenylhydrazones

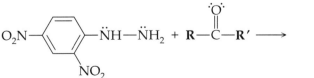

2,4-Dinitrophenylhydrazine

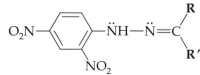

A 2,4-dinitrophenylhydrazone

[14] See Tables A.4 and A.5.

The procedure outlined in the Classification Test Section for aldehydes and ketones (p. 706) is used. Since the derivative to be isolated is a solid, it may be convenient to run the reaction in a 3-mL vial, or in a small test tube. Double the amount of the reagents used. If necessary, the derivative can be recrystallized from 95% ethanol.

The procedure is generally suitable for the preparation of phenylhydrazone and *p*-nitrophenylhydrazone derivatives of aldehydes and ketones.

## Semicarbazones

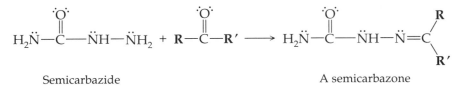

Semicarbazide                             A semicarbazone

In a 3.0-mL conical vial place 12 mg of semicarbazide hydrochloride, 20 mg of sodium acetate, 10 drops of water, and 12 mg of the unknown carbonyl compound. Cap the vial, shake vigorously, vent, and allow the vial to stand at room temperature until crystallization is complete (varies from a few min to several hours). Cool the vial in an ice bath if necessary. Collect the crystals by vacuum filtration, using a Hirsch funnel, and wash the filter cake with 0.2 mL of cold water. Dry the crystals on a porous clay plate. Determine the melting point.

---

## AMINES[15]

### Primary and Secondary Amines: Acetamides

In a 3.0-mL conical vial equipped with an air condenser, place 20 mg of the unknown amine, 5 drops of water, and 1 drop of concentrated hydrochloric acid.

In a small test tube, prepare a solution of 40 mg of sodium acetate trihydrate dissolved in 5 drops of water. Stopper the solution and set it aside for use in the next step.

Warm the solution of amine hydrochloride to about 50 °C on a sand bath. Then cool it, and add 40 $\mu$L of acetic anhydride in one portion **[HOOD]** through the condenser by aid of a 9-in. Pasteur pipet. In like manner, *immediately* add the sodium acetate solution (prepared previously). Swirl the contents of the vial to ensure complete mixing.

HOOD

Allow the reaction mixture to stand at room temperature for about 5 min, and then place it in an ice bath for an additional 5–10 min. Collect the white crystals by vacuum filtration, using a Hirsch funnel, and wash the filter cake with two 0.1-mL portions of water. The product may be recrystallized from ethanol–water using the Craig tube, if desired. Dry the crystals on a porous clay plate and determine the melting point.

[15] See Tables A.6 and A.7.

$$R-\overset{\cdot\cdot}{N}H_2 + \text{(benzoyl chloride)} \xrightarrow{\text{NaOH}} \text{(benzamide)} + NaCl + H_2O$$

In a 3.0-mL conical vial place 0.4 mL of 10% aqueous NaOH solution, 25 mg of the amine, and 2–3 drops of benzoyl chloride **[HOOD]**. Cap and shake the vial over a period of about 10 min. Vent the vial periodically to release any pressure buildup.  HOOD

Collect the crystalline precipitate by vacuum filtration, using a Hirsch funnel, and wash the filter cake with 0.1 mL of dilute HCl followed by 0.1 mL of water. It is generally necessary to recrystallize the material from methanol or aqueous ethanol using the Craig tube. Dry the product on a porous clay plate and determine the melting point.

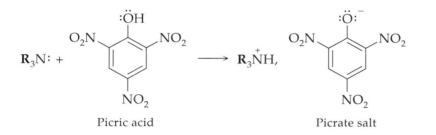

Picric acid        Picrate salt

In a 3.0-mL conical vial containing a boiling stone, and equipped with an air condenser, place 15 mg of the unknown amine and 0.3 mL of 95% ethanol.

**NOTE.** *If the amine is not soluble in the ethanol, shake the mixture to obtain a saturated solution and then transfer this solution, using a Pasteur filter pipet, to another vial.*

Now add 0.3 mL of a saturated solution of picric acid in 95% ethanol.

---

**CAUTION:** Picric acid explodes by percussion or when rapidly heated.

---

Heat the mixture at reflux, using a sand bath, for about 1 min and then allow it to cool slowly to room temperature. Collect the yellow crystals of the picrate by vacuum filtration, using a Hirsch funnel. Dry the material on a porous clay plate and determine the melting point.

---

$$R-\overset{\cdot\cdot}{\underset{\|}{C}}-Cl + 2\,NH_3 \longrightarrow R-\overset{\cdot\cdot}{\underset{\|}{C}}-NH_2 + NH_4Cl$$

In a 10 × 75-mm test tube, place 0.4 mL of ice cold, concentrated ammonium hydroxide solution. To this solution is added slowly, **[HOOD]** with  HOOD

[16] See Table A.8.

shaking, about 15 mg of the unknown acid chloride or anhydride. Stopper the test tube and allow the reaction mixture to stand at room temperature for about 5 min. Collect the crystals by vacuum filtration, using a Hirsch funnel, and wash the filter cake with 0.2 mL of ice cold water. Recrystallize the material using a Craig tube, from water or an ethanol–water mixture. Dry the purified crystals on a porous clay plate and determine the melting point.

## AROMATIC HYDROCARBONS[17]

### Picrates

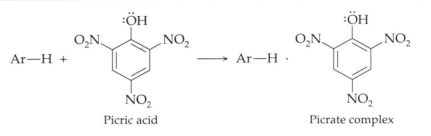

Picric acid                    Picrate complex

The procedure outlined on page 723 is used to prepare these derivatives.

## NITRILES[18]

### Hydrolysis to Amides

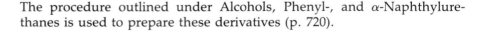

Conversion of nitriles to water-insoluble amides, by hydrolysis with alkaline hydrogen peroxide, is a possible method of characterization for these compounds. It is especially useful for aromatic nitriles.

In a 5-mL conical vial containing a magnetic spin vane, weigh and place about 50 mg of the nitrile and 500 $\mu$L of a 1 $M$ NaOH solution. Cool the mixture in a water bath and, with stirring, add dropwise 500 $\mu$L of 12% $H_2O_2$ solution. Attach the vial to an air condenser and warm the solution on a sand bath while stirring at 50–60 °C for approximately 45 min. To the cooled reaction mixture, add 1–2 mL of cold water, and then collect the solid amide by vacuum filtration. Wash the product with two 1-mL portions of cold water, and recrystallize the amide from aqueous ethanol using the Craig tube. Dry the solid and determine the melting point.

## PHENOLS[19]

### α-Naphthylurethanes (α-naphthylcarbamates)

### Bromo Derivatives

The procedure outlined under Alcohols, Phenyl-, and α-Naphthylurethanes is used to prepare these derivatives (p. 720).

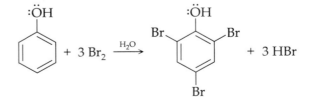

[17] See Table A.9.
[18] See Table A.13.
[19] See Table A.10.

**INSTRUCTOR PREPARATION.** *The brominating reagent is prepared by adding 1.0 mL (3 g) of bromine [HOOD] to a solution of 4.0 g of KBr in 25 mL of water.*    HOOD

In a 1.0-mL conical vial, place 10 mg of the unknown phenol followed by 2 drops of methanol and 2 drops of water. To this solution, add 3 drops **[HOOD]** of brominating agent from a Pasteur pipet.    HOOD

Continue the addition (dropwise) until the reddish-brown color of bromine persists. Now add water (4 drops), cap the vial, shake, vent, and then allow it to stand at room temperature for 10 min. Collect the crystalline precipitate by vacuum filtration using a Hirsch funnel and wash the filter cake with 0.5 mL of 5% aqueous sodium bisulfite solution. Recrystallize the solid derivative from ethanol, or from an ethanol–water mixture, using a Craig tube. Dissolve the material in about 0.5 mL of ethanol, add water until it becomes cloudy, cool in an ice bath, and collect the crystals in the usual manner. Dry the purified product on a porous clay plate and determine the melting point.

---

These compounds do not give derivatives directly, but are usually converted into another material that can then be derivatized. The procedures are, for the most part, lengthy, and frequently give mixtures of products. It is recommended that compounds belonging to these classes be primarily identified using spectroscopic methods. Measurement of their physical properties is also of utmost importance.

### ALIPHATIC HYDROCARBONS, HALOGENATED HYDROCARBONS, AMIDES, NITRO COMPOUNDS, ETHERS, AND ESTERS[20]

### QUESTIONS

**10-1.** The following six substances have approximately the same boiling point and are all colorless liquids. Suppose you were given six unlabeled bottles, each of which contained one of these compounds.

Explain how you would use simple chemical tests to determine which bottle contained which compound.

| | |
|---|---|
| Ethanoic acid | Toluene |
| Propyl butanoate | Diisobutylamine |
| 1-Butanol | Styrene |

**10-2.** A colorless liquid ($C_4H_6O$) with a boiling point of 81 °C, was found to be soluble in water and also in ether. It gave a negative test for the presence of halogens, sulfur, and nitrogen. It did, however, give a positive test with the Baeyer reagent and also gave a positive test with the 2,4-dinitrophenylhydrazine reagent. It gave negative results when treated with ceric nitrate solution and with Tollens reagent. Treatment with ozone followed by hydrolysis in the presence of zinc gave formaldehyde as one of the products.

What is the structure and name of the colorless liquid?

**10-3.** A colorless liquid **A** ($C_3H_6O$) was soluble in water and ether, and had a boiling point of 94–96 °C. It decolorized a $Br_2$–$CH_2Cl_2$ solution and gave a positive ceric nitrate test. On catalytic hydrogenation it formed Compound **B** ($C_3H_8O$), which did not decolorize the above bromine solu-

---

[20] See Tables A.11, A.12, and A.14–A.17.

tion, but did give a positive ceric nitrate test. Treatment of Compound **A** with ozone, followed by hydrolysis in the presence of zinc, gave formaldehyde as one of the products. Compound **A** formed an $\alpha$-naphthylurethane with a melting point of 109 °C.

What are the names and structures of Compounds **A** and **B**?

**10-4.** A compound of formula $C_{14}H_{12}$ gave a positive Baeyer test and burned with a yellow, sooty flame. Treatment with ozone followed by hydrolysis in the presence of zinc gave formaldehyde as one of the products. Also isolated from the ozonolysis reaction was a second compound, $C_{13}H_{10}O$, which burned with a yellow, sooty flame, and readily formed a semicarbazone with a melting point of 164 °C. The $^1H$ NMR spectrum of this compound ($C_{13}H_{10}O$) showed only complex multiplets near 7.5 ppm; the fully $^1H$-decoupled $^{13}C$ NMR spectrum showed only 5 peaks.

What are the structures and names of the two compounds?

**10-5.** Compound **A** ($C_7H_{14}O$) burned with a yellow, nonsooty flame and did not decolorize a bromine–methylene chloride solution. It did give a positive 2,4-dinitrophenylhydrazine test, but a negative Tollens test. Treatment of the compound with lithium aluminum hydride followed by neutralization with acid, produced a Compound **B**, which gave a positive Lucas test in about 5 min. Compound **B** also gave a positive ceric nitrate test. The $^1H$ NMR spectrum for Compound **A** gave the following data:

| 1.02 ppm | 9H, singlet |
|----------|-------------|
| 2.11 ppm | 3H, singlet |
| 2.31 ppm | 2H, singlet |

Give suitable structures for Compounds **A** and **B**.

**10-6.** A friend of yours who is a graduate student attempting to establish the structure of a chemical species from field clover, isolated an alcohol that was found to have an optical rotation of +49.5°. Chemical analysis gave a molecular formula of $C_5H_{10}O$. It was also observed that this alcohol readily decolorized $Br_2$–$CH_2Cl_2$ solution. On this basis, the alcohol was subjected to catalytic hydrogenation and it was found to absorb 1 mol of hydrogen gas. The product of the reduction gave a positive ceric nitrate test, indicating that it too was an alcohol. However, the reduced compound was optically inactive.

Your friend has come to you for assistance in determining the structures of the two alcohols. What do you believe the structures are?

**10-7.** An unknown compound burned with a yellow, nonsmoky flame and was found to be insoluble in 5% sodium hydroxide solution but soluble in concentrated sulfuric acid. Measurement of its boiling point gave a range of 130–131 °C. Combustion analysis gave a molecular formula of $C_5H_8O$. It was found to give a semicarbazone with a melting point of 204–206 °C. However, it gave a negative result when treated with Tollens reagent and it did not decolorize the Baeyer reagent. It also gave a negative iodoform test.

Identify the unknown compound.

**10-8.** An unknown organic carboxylic acid, mp = 139–141 °C, burned with a yellow, sooty flame. The sodium fusion test showed that nitrogen was present. It did not react with $p$-toluenesulfonyl chloride, but did give a positive test when treated with 5% aqueous ferrous ammonium sulfate solution, acidified with 3 $N$ $H_2SO_4$, and then followed by methanolic KOH solution. A 200-mg sample of the acid neutralized 12.4 mL of 0.098 $N$ sodium hydroxide solution.

Identify the acid.

Does your structure agree with the calculated equivalent weight?

**10-9.** An unknown organic liquid, **A,** was found to burn with a yellow, sooty flame and give a positive Lucas test (~5 min). Upon treatment with sodium dichromate–sulfuric acid solution it produced Compound **B,** which also burned with a yellow, sooty flame. Compound **B** gave a positive 2,4-dinitrophenylhydrazine test, but a negative result when treated with the Tollens reagent. However, **B** did give a positive iodoform test. The $^1$H NMR spectrum for Compound **A** showed the following:

| 1.4 ppm | 3H (doublet) |
| 1.9 ppm | 1H (singlet) |
| 4.8 ppm | 1H (quartet) |
| 7.2 ppm | 5H (complex multiplet) |

Give the structures and suitable names for Compounds **A** and **B.**

**10-10.** A hydrocarbon **A** ($C_6H_{10}$) burned with a yellow, almost nonsmoky flame. On catalytic hydrogenation over platinum catalyst it absorbed 1 mol of hydrogen to form Compound **B.** It also decolorized a $Br_2$–$CH_2Cl_2$ solution to yield a dibromo derivative, **C.** Ozonolysis of the hydrocarbon gave only one compound, **D.** Compound **D** gave a positive iodoform test when treated with iodine–sodium hydroxide solution. On treatment of Compound **D** with an alcoholic solution of silver ammonium hydroxide, a silver mirror was formed within a few min.

Identify the hydrocarbon **A** and Compounds **B–D.**

**10-11.** A high-boiling liquid, bp = 202–204 °C, burns with a yellow, sooty flame. Sodium fusion indicates that halogens, nitrogen, and sulfur are not present. It is not soluble in water, dilute sodium bicarbonate solution, or dilute hydrochloric acid. However, it proved to be soluble in 5% aqueous sodium hydroxide solution. The compound gives a purple color with ferric chloride solution and a precipitate when reacted with bromine–water. Treatment with hydroxylamine reagent did not give a reaction, but a white precipitate was obtained when the compound was treated with $\alpha$-naphthyl isocyanate. On drying, this white, solid derivative had a mp = 127–129 °C.

Identify the original liquid and write a structure for the solid derivative.

After identifying the unknown liquid, can you indicate what the structure of the precipitate obtained on reaction with bromine might be?

**10-12.** A colorless liquid, bp = 199–201 °C, burns with a yellow, sooty flame. The sodium fusion test proved negative for the presence of halogens, nitrogen, and sulfur. It was not soluble in water, 5% aqueous sodium hydroxide, or 5% hydrochloric acid. However, it dissolved in sulfuric acid with evolution of heat. It did not give a precipitate with 2,4-dinitrophenylhydrazine solution, and it did not decolorize bromine–methylene chloride solution. The unknown liquid did give a positive hydroxamate test and was found to have a saponification equivalent of 136.

Identify the unknown liquid.

**10-13.** Your friend of Question 10-5 still needs your help. A week later a low-melting solid **A** was isolated, which combustion analysis showed had composition $C_9H_{10}O$. The substance gave a precipitate when treated with 2,4-dinitrophenylhydrazine solution. Furthermore, when reacted with iodoform reagent, a yellow precipitate of $CHI_3$ was observed. Acidification of the alkaline solution from the iodoform test produced a solid material, **B.**

Reduction of Compound **A** with $LiAlH_4$ gave Compound **C** ($C_9H_{12}O$). This material, **C,** also gave Compound **B** when treated with iodoform reagent.

Vigorous oxidation of **A, B,** or **C** with sodium dichromate–sulfuric acid solution gave an acid having a mp = 121–122 °C.

Your friend needs your assistance in determining the structures for Compounds **A**, **B**, and **C**. Can you identify the three compounds?

**10-14.**   An organic compound ($C_9H_{10}O$) showed strong absorption in the IR spectrum at 1735 cm$^{-1}$ and gave a semicarbazone having a melting point of 198 °C. It burned with a yellow, sooty flame and also gave a positive iodoform test. The $^1$H NMR spectrum of the compound provided the following information:

| | |
|---|---|
| 2.11 ppm | 3H (singlet) |
| 3.65 ppm | 2H (singlet) |
| 7.20 ppm | 5H (complex multiplet) |

Identify the unknown organic compound.

**10-15.**   An unknown compound (**A**) was soluble in ether but only slightly soluble in water. It burned with a clear blue flame and combustion analysis showed it to have the molecular formula of $C_5H_{12}O$. It gave a positive test with the Jones reagent producing a new compound (**B**) with a formula of $C_5H_{10}O$. Compound **B** gave a positive iodoform test and formed a semicarbazone. Compound **A** on treatment with sulfuric acid produced a hydrocarbon (**C**) of formula $C_5H_{10}$. Hydrocarbon **C** readily decolorized a $Br_2$–$CH_2Cl_2$ solution, and on ozonolysis, produced acetone as one of the products.

Identify the structure of each of the lettered compounds.

**10-16.**   Compound **A** ($C_7H_{14}$) decolorized a $Br_2$–$CH_2Cl_2$ chloride solution. It reacted with BH$_3$•THF reagent, followed by alkaline peroxide solution, to produce Compound **B**. Compound **B**, on treatment with chromic acid–sulfuric acid solution, gave carboxylic acid **C**, which could be separated into two enantiomers. Compound **A**, on treatment with ozone, followed by addition of hydrogen peroxide, produced Compound **D**. Compound **D** was identical to that material isolated from the oxidation of 3-hexanol with chromic acid–sulfuric acid reagent.

Identify the structures of Compounds **A**, **B**, **C**, and **D**.

**10-17.**   Compound **A** ($C_8H_{16}$) decolorized a bromine–methylene chloride solution. Ozonolysis produced two compounds, **B** and **C**, which could be separated easily by GC. Both **B** and **C** gave a positive 2,4-dinitrophenylhydrazine test. Carbon–hydrogen analysis and molecular weight determination of **B** gave a molecular formula of $C_5H_{10}O$. The $^1$H NMR spectrum revealed the following information for **B**:

| | |
|---|---|
| 0.92 ppm | 3H, triplet |
| 1.6  ppm | 2H, pentet |
| 2.17 ppm | 3H, singlet |
| 2.45 ppm | 2H, triplet |

Compound **C** was a low-boiling liquid (bp 56 °C). The $^1$H NMR of this material showed only one singlet.

Identify Compounds **A**, **B**, and **C**.

# Tables of Derivatives

**TABLE A.1  Derivatives of Carboxylic Acids (Liquids)**

| Acid | bp (°C) | Melting Point of Derivative (°C)[a] | | |
|---|---|---|---|---|
| | | Amide | Anilide | p-Toluidide |
| Methanoic (formic) | 101 | — | 50 | 53 |
| Ethanoic (acetic) | 118 | 82 | 114 | 153 |
| Propenoic (acrylic) | 141 | 84 | 104 | 141 |
| Propanoic | 141 | 81 | 106 | 126 |
| 2-Methylpropanoic (isobutyric) | 155 | 128 | 105 | 109 |
| Butanoic (butyric) | 163 | 115 | 96 | 75 |
| 2-Methylpentenoic (methacrylic) | 163 | 102 | 87 | — |
| Pyruvic | 165 | 124 | 104 | 109 |
| 3-Methylbutanoic | 177 | 135 | 110 | 106 |
| Pentanoic (valeric) | 186 | 106 | 63 | 74 |
| 2-Methylpentanoic | 186 | 79 | 95 | 80 |
| 2,2-Dichloroethanoic | 194 | 98 | 118 | 153 |
| Hexanoic (caproic) | 205 | 100 | 94 | 74 |
| Heptanoic (enanthic) | 223 | 96 | 65; 70 | 81 |
| Octanoic (caprylic) | 239 | 106; 110 | 57 | 70 |
| Nonanoic (pelargonic) | 254 | 99 | 57 | 84 |

[a] Two values are given for those derivatives that may exist in polymorphic forms.

**TABLE A.2  Derivatives of Carboxylic Acids (Solids)**

| Acid | bp (°C) | Melting Point of Derivative (°C)[a] | | |
| --- | --- | --- | --- | --- |
| | | Amide | Anilide | p-Toluidide |
| Decanoic | 31–32 | 108 | 70 | 78 |
| Lauric | 43–45 | 87 | 78 | 100 |
| Myristic | 54 | 103 | 84 | 93 |
| Trichloroacetic | 54–58 | 141 | 97 | 113 |
| Chloroacetic | 61 | 121 | 137 | 162 |
| Palmitic | 62 | 106 | 90 | 98 |
| Octadecanoic (stearic) | 70 | 109 | 95 | 102 |
| Crotonic | 72 | 158 | 118 | — |
| 3,3-Dimethyl acrylic | 69 | 107 | 126 | — |
| Phenylethanoic | 77 | 156 | 65 | 117 |
| 2-Benzoylbenzoic | 128 | 165 | 195 | — |
| Pentandioic (glutaric) | 97 | 175 | 223 | 218 |
| Ethanedioic (oxalic) | 101 | 219 | 148 | 169 |
| 2-Methylbenzoic (o-toluic) | 105 | 143 | 125 | 144 |
| 3-Methylbenzoic (m-toluic) | 112 | 94 | 126 | 118 |
| Benzoic | 122.4 | 130 | 160 | 158 |
| Sebacic | 131–134 | 170 (mono) | 122 (mono) | 201 |
| | | 210 (di) | 200 (di) | |
| trans-Cinnamic | 133 | 147 | 153 | 168 |
| 2-Acetoxybenzoic (aspirin) | 135 | 138 | 136 | — |
| cis-Butenedioic (maleic) | 137 | 172 | 198 (mono) | 142 (di) |
| | | | 187 (di) | |
| Malonic | 137 | — | 132 (mono) | 86 (mono) |
| | | | 230 (di) | 253 (di) |
| 2-Chlorobenzoic | 140 | — | 118 | 131 |
| 3-Nitrobenzoic | 140 | 143 | 154 | 162 |
| 2-Nitrobenzoic | 146 | 176 | 155 | — |
| Diphenylacetic | 148 | 168 | 180 | 172 |
| 2-Bromobenzoic | 150 | 155 | 141 | — |
| Benzilic | 150 | 153 | 175 | 190 |
| Hexanedioic (adipic) | 153 | 125 (mono) | 151 (mono) | — |
| | | 230 (di) | 241 (di) | |
| 2-Hydroxybenzoic (salicylic) | 158 | 142 | 136 | 156 |
| 2-Iodobenzoic | 162 | 110 | 141 | — |
| 4-Methylbenzoic (p-toluic) | 179 | 160 | 144 | 160; 165 |
| 4-Methoxybenzoic (p-anisic) | 185 | 167 | 170 | 186 |
| 2-Naphthoic | 186 | 192 | 171 | 192 |
| Succinic | 190 | 157 (mono) | 143 (mono) | 180 (mono) |
| | | 260 (di) | 230 (di) | 255 (di) |
| Phthalic | 211 | 149 (mono) | 170 (mono) | 150 (mono) |
| | | 220 (di) | 254 (di) | 201 (di) |
| 3,5-Dinitrobenzoic | 205 | 183 | 234 | — |
| 4-Nitrobenzoic | 241 | 198 | 204; 211 | 192; 204 |

[a] Two values are given for those derivatives that may exist in polymorphic forms.

## TABLE A.3 Derivatives of Alcohols

| Alcohol | bp (°C) | Melting Point of Derivative (°C) | | |
|---|---|---|---|---|
| | | Phenyl-urethan | α-Naphthyl-urethan | 3,5-Dinitro-benzoate |
| Methyl (methanol) | 65 | 47 | 124 | 108 |
| Ethyl (ethanol) | 78 | 52 | 79 | 93 |
| Isopropyl (2-propanol) | 82 | 88 | 106 | 122 |
| tert-Butyl (tert-butanol) | 83 | 136 | 101 | 142 |
| Allyl | 97 | 70 | 109 | 49 |
| n-Propyl (1-propanol) | 97 | 51 | 80 | 74 |
| sec-Butyl (2-butanol) | 99 | 65 | 97 | 76 |
| tert-Pentyl (2-methyl-2-butanol) | 102 | 42 | 71 | 116 |
| Isobutyl (2-methyl-1-propanol) | 108 | 86 | 104 | 87 |
| 3-Pentanol | 116 | 48 | 71 | 101 |
| n-Butyl (1-butanol) | 118 | 63 | 71 | 64 |
| 2,3-Dimethyl-2-butanol | 118 | — | — | 111 |
| 2-Pentanol | 119 | — | 76 | 61 |
| 2-Methyl-2-pentanol | 121 | 239 | — | 72 |
| 3-Methyl-3-pentanol | 123 | 50 | — | 97 |
| 2-Methoxyethanol | 125 | — | 113 | — |
| 2-Methyl-1-butanol | 129 | — | — | 70 |
| 2-Chloroethanol | 131 | 51 | 101 | 95 |
| 4-Methyl-2-pentanol | 132 | 143 | 88 | 65 |
| 3-Methyl-1-butanol | 132 | 55 | — | 61 |
| 2-Ethoxyethanol | 135 | — | 67 | 75 |
| 3-Hexanol | 136 | — | — | 77 |
| 2,2-Dimethyl-1-butanol | 137 | — | — | 51 |
| 1-Pentanol | 138 | 46 | 68 | 46 |
| 2-Hexanol | 139 | — | — | 39 |
| 2,4-Dimethyl-3-pentanol | 140 | — | — | — |
| Cyclopentanol | 141 | 132 | 118 | 115 |
| 2-Ethyl-1-butanol | 148 | — | — | 52 |
| 2-Methyl-1-pentanol | 148 | — | — | 51 |
| 4-Heptanol | 156 | — | 80 | 64 |
| 1-Hexanol | 158 | 42 | 59 | 58 |
| 2-Heptanol | 159 | — | 54 | 49 |
| Cyclohexanol | 161 | 82 | 128 | 113 |
| 2-Furfuryl | 172 | 45 | 129 | 81 |
| 1-Heptanol | 177 | 68 | — | 47 |
| Tetrahydrofurfuryl | 178 | 61 | — | 84 |
| 2-Octanol | 179 | 114 | — | 32 |
| 1-Octanol | 195 | 74 | — | 61 |
| Benzyl | 205 | 78 | — | 113 |
| 2-Phenylethanol | 221 | 79 | — | 108 |
| 1-Decanol | 231 | 59 | — | 57 |
| Cinnamyl | (mp 35) | 90 | — | 121 |
| Benzohydrol | (mp 67) | 139 | — | 141 |
| Cholesterol | (mp 147) | 168 | — | — |

**TABLE A.4  Derivatives of Aldehydes**

| Aldehyde | bp (°C) | Melting Point of Derivative (°C)[a] | |
| --- | --- | --- | --- |
| | | Semi-carbazone | 2,4-Dinitrophenyl-hydrazone |
| Acetaldehyde | 21 | 162 | 168 |
| Propionaldehyde | 50 | 89 (154) | 154 |
| Isobutyraldehyde | 64 | 125 | 187 (183) |
| n-Butyraldehyde | 74 | 104 | 123 |
| Isovaleraldehyde | 92 | 107 | 123 |
| n-Valeraldehyde | 103 | 108 | 107 |
| Crotonaldehyde | 104 | 199 | 190 |
| n-Hexaldehyde | 131 | 106 | 104; 107 |
| n-Heptaldehyde | 153 | 109 | 108 |
| 2-Furaldehyde | 161 | 202 | 212 (230) |
| Benzaldehyde | 179 | 222 | 237 |
| Salicylaldehyde | 197 | 231 | 252 dec |
| p-Tolualdehyde | 204 | 221 | 239 |
| 2-Chlorobenzaldehyde | 215 | 146 (229) | 213 |
| Citral | 228 | 164 | 116 |
| 4-Anisaldehyde | 248 | 210 | 253 |
| trans-Cinnamaldehyde | 252 | 215 | 255 |
| 4-Chlorobenzaldehyde | (mp 47) | 230 | 254 |

[a] Two values are given for those derivatives that may exist in polymorphic forms or as syn and anti geometrical isomers.

**TABLE A.5   Derivatives of Ketones**

| Ketone | bp (°C) | Melting Point of Derivative (°C)[a] | |
| | | Semi-carbazone | 2,4-Dinitro-phenylhydrazone |
|---|---|---|---|
| Acetone | 56 | 187 | 126 |
| 2-Butanone | 80 | 146 | 117 |
| 3-Methyl-2-butanone | 94 | 113 | 120 |
| 2-Pentanone | 102 | 112 | 143 |
| 3-Pentanone | 102 | 139 | 156 |
| 3,3-Dimethyl-2-butanone | 106 | 157 | 125 |
| 4-Methyl-2-pentanone | 119 | 134 | 95 |
| 2,4-Dimethyl-3-pentanone | 124 | 160 | 88 (94) |
| 2-Hexanone | 129 | 122 | 106 |
| Cyclopentanone | 131 | 205 | 142 |
| 4-Heptanone | 145 | 133 | 75 |
| 3-Heptanone | 149 | 101 | — |
| 2-Heptanone | 151 | 127 | 89 |
| Cyclohexanone | 155 | 166 | 162 |
| 2-Octanone | 173 | 122 | 58 |
| Acetophenone | 200 | 198 | 240 |
| Benzalacetone | (mp 41) | 187 | 223 |
| Benzophenone | (mp 48) | 164 | 239 |
| Benzalacetophenone | (mp 58) | 168; 180 | 245 |
| Benzil | (mp 95) | 175 (182) | 189 |
| Benzoin | (mp 133) | 206 (dec) | 245 |

[a] Two values are given for those derivatives that may exists in polymorphic forms or as syn and anti geometrical isomers.

**TABLE A.6   Derivatives of Primary and Secondary Amines**

| Amine | bp (°C) | Melting Point of Derivative (°C)[a] | | |
|---|---|---|---|---|
| | | Acetamide | Benzamide | Picrate |
| Methylamine | −6 | — | 80 | — |
| Ethylamine | 17 | — | 71 | — |
| Isopropylamine | 33 | — | 71 | 165 |
| *tert*-Butylamine | 45 | 98 | 134 | 198 |
| *n*-Propylamine | 49 | 47 | 84 | 135 |
| Allylamine | 53 | — | — | 140 |
| Diethylamine | 55 | — | 42 | 155 |
| *sec*-Butylamine | 63 | — | 76 | 139 |
| Isobutylamine | 69 | — | 57 | 150 |
| *n*-Butylamine | 77 | — | 42 | — |
| Diisopropylamine | 84 | — | — | 140 |
| Di-*n*-propylamine | 109 | — | — | 75 |
| Piperidine | 106 | — | 48 | 152 |
| Ethylenediamine | 116 | 172 (di) | 244 (di) | 233 |
| Cyclohexylamine | 134 | 104 | 149 | — |
| Diisobutylamine | 139 | 86 | — | 121 |
| Di-*n*-butylamine | 159 | — | — | 59 |
| Benzylamine | 185 | 60 | 105 | 199 |
| Aniline | 185 | 114 | 163 | 198 |
| *N*-Methylaniline | 196 | 102 | 67 | 145 |
| 2-Methylaniline | 199 | 110 | 144 | 213 |
| 4-Methylaniline | 200 (mp 45) | 148 | 158 | — |
| 3-Methylaniline | 203 | 65 | 125 | 200 |
| *N*-Ethylaniline | 205 | 111 | 147 | 194 |
| 2-Chloroaniline | 208 | 87 | 99 | 134 |
| 2,5-Dimethylaniline | 215 | 139 | 140 | 171 |
| 2,6-Dimethylaniline | 216 | 177 | 168 | 180 |
| 2,4-Dimethylaniline | 217 | 133 | 192 | 209 |
| *N*-Ethyl-3-methylaniline | 221 | — | 72 | — |
| 2-Methoxyaniline | 225 | 85 | 60 (84) | 200 |
| 4-Chloroaniline | 232 (mp 70) | 179 | 192 | 178 |
| 4-Methoxyaniline | 243 (mp 57) | 130 | 154 | 170 |
| 2-Ethoxyaniline | 229 | 79 | 104 | — |
| 4-Ethoxyaniline | 254 | 135 | 173 | — |
| Diphenylamine | (mp 54) | 101 | 180 | 182 |
| 3-Nitroaniline | (mp 114) | 152 | 155 | — |
| 4-Nitroaniline | (mp 147) | — | 199 | — |

[a] Two values are given for those derivatives that may exist in polymorphic forms.

**TABLE A.7    Derivatives of Tertiary Amines**

|  |  | Melting Point of Derivative (°C)[a] |
|---|---|---|
| Tertiary Amine | bp (°C) | Picrate |
| Trimethylamine | 3 | 216 |
| Triethylamine | 89 | 173 |
| Pyridine | 116 | 167 |
| 2-Methylpyridine (2-picoline) | 129 | 169 |
| 2,6-Dimethylpyridine (2,6-lutidine) | 142 | 168 (161) |
| 3-Methylpyridine (3-picoline) | 143 | 150 |
| 4-Methylpyridine (4-picoline) | 143 | 167 |
| Tripropylamine | 157 | 116 |
| N,N-Dimethylaniline | 193 | 163 |
| Tributylamine | 216 | 105 |
| N,N-Diethylaniline | 216 | 142 |
| Quinoline | 237 | 203 |
| Triisopentylamine | 245 | 125 |

[a] Two values are given for those derivatives that may exist in polymorphic forms.

**TABLE A.8    Derivatives of Acid Chlorides and Anhydrides**

|  |  |  | Melting Point of Derivative (°C) |
|---|---|---|---|
| Acid Chloride or Anhydride | bp (°C) | mp (°C) | Amide |
| Acetyl chloride | 52 | — | 82 |
| Propionyl chloride | 77–79 | — | 81 |
| Butyryl chloride | 102 | — | 115 |
| Acetic anhydride | 138–140 | — | 82 |
| Propionic anhydride | 167 | — | 81 |
| Butyric anhydride | 198–199 | — | 115 |
| Benzoyl chloride | 198 | — | 130 |
| 3-Chlorobenzoyl chloride | 225 | — | 134 |
| 2-Chlorobenzoyl chloride | 238 | — | 142 |
| cis-1,2-Cyclohexanedicarboxylic anhydride | — | 32 | 192d (acid) |
| Benzoic anhydride | — | 39–40 | 130 |
| Maleic anhydride | — | 54–56 | 181 (mono) 266 (di) |
| 4-Nitrobenzoyl chloride | — | 72–74 | 201 |
| Succinic anhydride | — | 119–120 | 157 (mono) 260 (di) |
| Phthalic anhydride | — | 131–133 | 149 (mono) 220 (di) |

**TABLE A.9   Derivatives of Aromatic Hydrocarbons**

| Aromatic Hydrocarbon | bp (°C) | mp (°C) | Melting Point of Derivative (°C)[a] |
|---|---|---|---|
| | | | Picrate |
| Benzene | 80 | — | 84 |
| Toluene | 111 | — | 88 |
| Ethylbenzene | 136 | — | 96 |
| p-Xylene | 138 | — | 90 |
| m-Xylene | 138–139 | — | 91 |
| o-Xylene | 143–145 | — | 88 |
| Mesitylene | 163–166 | — | 97 |
| 1,2,4-Trimethylbenzene | 168 | — | 97 |
| 1,2,3,4-Tetramethylbenzene | 205 | — | 92 |
| 1-Methylnaphthalene | 242 | — | 142 |
| 2-Methylnaphthalene | — | 35 | 116 |
| Pentamethylbenzene | — | 51 | 131 |
| Naphthalene | — | 81 | 149 |
| Acenaphthene | — | 94 | 161 |
| Phenanthrene | — | 100 | 144 (133) |
| Anthracene | — | 216 | 138 |

[a] Two values are given for those derivatives that may exist in polymorphic forms.

**TABLE A.10  Derivatives of Phenols**

| Phenol | mp (°C) | Melting Point of Derivative (°C) | |
|---|---|---|---|
| | | Bromo | α-Naphthylurethan |
| 2-Chloro- | 7 (bp 175) | 48 (mono) 76 (di) | 120 |
| Phenol | 42 | 95 (tri) | 133 |
| 4-Methyl- (p-cresol) | 35 | 49 (di) 108 (tetra) | 146 |
| 3-Methyl- (m-cresol) | 203 (bp) | 84 (tri) | 128 |
| 3,4-Dimethyl- | 229 (bp) | 171 (tri) | 141 |
| 2-Methyl- (o-cresol) | 33 | 56 (di) | 142 |
| 4-Ethyl- | 45 | — | 128 |
| 2-Nitro- | 45 | 117 (di) | 113 |
| 2,6-Dimethyl- | 48 | 79 | 176 |
| 2-Isopropyl-5-methyl- (thymol) | 50 | 55 | 160 |
| 3,5-Dimethyl- | 64 | 166 (tri) | — |
| 4-Bromo- | 66 | 95 (tri) | 168 |
| 2,5-Dimethyl- | 73 | 178 (tri) | 173 |
| 1-Naphthol | 95 | 105 (di) | 152 |
| 3-Nitro- | 96 | 91 (di) | — |
| 4-tert-Butyl- | 98 | 50 (mono) 67 (di) | 110 |
| 1,2-Dihydroxy- (catechol) | 105 | 192 (tetra) | 175 |
| 1,3-Dihydroxy- (resorcinol) | 110 | 112 (tri) | 275 |
| 4-Nitro- | 112 | 142 (di) | 150 |
| 2-Naphthol | 123 | 84 | 157 |
| Pyrogallol | 134 | 158 (di) | 173 |
| 1,4-Dihydroxy- (hydroquinone) | 171 | 186 (di) | — |

**TABLE A.11  Aliphatic Hydrocarbons**

| Compound | bp (°C) | Compound | bp (°C) |
|---|---|---|---|
| **Alkanes** | | | |
| Pentane | 36 | 2,2,4-Trimethyl-pentane | 99 |
| Cyclopentane | 49 | | |
| 2,2-Dimethylbutane | 50 | trans-1,4-Dimethyl-cyclohexane | 119 |
| 2,3-Dimethylbutane | 58 | | |
| 2-Methylpentane | 60 | Octane | 126 |
| 3-Methylpentane | 63 | Nonane | 151 |
| Hexane | 69 | Decane | 174 |
| Cyclohexane | 81 | Eicosane | 343 (mp 37) |
| Heptane | 98 | Norbornane | (mp 87, subl) |
| | | Adamantane | (mp 268, sealed) |
| **Alkenes and Alkynes** | | | |
| 1-Pentene | 30 | 3-Hexyne | 82 |
| 2-Methyl-1,3-butadiene (isoprene) | 34 | Cyclohexene | 84 |
| | | 2-Hexyne | 84 |
| trans-2-Pentene | 36 | 1-Heptene | 94 |
| cis-2-Pentene | 37 | 1-Heptyne | 100 |
| 2-Methyl-2-butene | 39 | 2,4,4-trimethyl-1-pentene | 102 |
| Cyclopentadiene | 41 | | |
| 1,3-Pentadiene (piperylene) | 41 | 2,4,4-Trimethyl-2-pentene | 104 |
| 3,3-Dimethyl-1-butene | 41 | 1-Octene | 123 |
| 1-Hexene | 63 | Cyclooctene | 146 |
| cis-3-Hexene | 66 | 1,5-Cyclooctadiene | 150 |
| trans-3-Hexene | 67 | d,l-α-Pinene | 155 |
| 1-Hexyne | 71 | (−)-β-Pinene | 167 |
| 1,3-Cyclohexadiene | 80 | Limonene | 176 |
| | | 1-Decene | 181 |

**TABLE A.12   Halogenated Hydrocarbons**

| Compound | bp (°C) | Compound | bp (°C) |
|---|---|---|---|
| **Alkyl Halides** | | | |
| **Chlorides** | | **Bromides** | |
| n-Propyl | 47 | Ethyl | 38 |
| tert-Butyl | 51 | Isopropyl | 60 |
| sec-Butyl | 68 | Propyl | 71 |
| Isobutyl | 69 | tert-Butyl | 72 |
| n-Butyl | 78 | Isobutyl | 91 |
| Neopentyl | 85 | sec-Butyl | 91 |
| tert-Pentyl | 86 | Butyl | 101 |
| Cyclohexyl | 143 | tert-Pentyl | 108 |
| Hexachloroethane | 185 (mp 187, subl) | Neopentyl | 109 |
| Triphenylmethyl | (mp 113) | 1-Bromoheptane | 178–179 |
| | | **Iodides** | |
| | | Methyl | 43 |
| | | Ethyl | 72 |
| | | Isopropyl | 90 |
| | | Propyl | 102 |

| Compound | bp (°C) | mp (°C) |
|---|---|---|
| **Aryl Halides** | | |
| Chlorobenzene | 132 | — |
| Bromobenzene | 156 | — |
| 2-Chlorotoluene | 157–159 | — |
| 4-Chlorotoluene | 162 | — |
| 1,3-Dichlorobenzene | 172–173 | — |
| 1,2-Dichlorobenzene | 178 | — |
| 2,4-Dichlorotoluene | 196–203 | — |
| 3,4-Dichlorotoluene | 201 | — |
| 1,2,4-Trichlorobenzene | 214 | — |
| 1-Bromonaphthalene | 279–281 | — |
| 1,2,3-Trichlorobenzene | — | 51–53 |
| 1,4-Dichlorobenzene | — | 54–56 |
| 1,4-Bromochlorobenzene | — | 66–68 |
| 1,4-Dibromobenzene | — | 87–89 |
| 1,2,4,5-Tetrachlorobenzene | — | 138–140 |

**TABLE A.13  Nitriles**

| Compound | bp (°C) | Compound | mp (°C) |
|---|---|---|---|
| Acrylonitrile | 77 | 4-Chlorobenzylcyanide | 30.5 |
| Acetonitrile | 81 | Malononitrile | 34 |
| Propionitrile | 97 | Stearonitrile | 40 |
| Isobutyronitrile | 108 | 2-Chlorobenzonitrile | 41 |
| n-Butyronitrile | 117 | Succinonitrile | 48 |
| Benzonitrile | 191 | Diphenylacetonitrile | 75 |
| 2-Methylbenzonitrile | 205 | 4-Cyanopyridine | 80 |
| 3-Methylbenzonitrile | 212 | | |
| 4-Methylbenzonitrile | 217 | | |
| Benzylcyanide | 234 | | |
| Adiponitrile | 295 | | |

**TABLE A.14  Amides**[a]

| Compound | bp (°C) | mp (°C) |
|---|---|---|
| N,N-Dimethylformamide | 153 | — |
| N,N-Diethylformamide | 176 | — |
| N-Methylformamide | 185 | — |
| N-Formylpiperidine | 222 | — |
| N,N-Dimethylbenzamide | — | 41 |
| N-Benzoylpiperidine | — | 48 |
| N-Propylacetanilide | — | 50 |
| N-Benzylacetamide | — | 54 |
| N-Ethylacetanilide | — | 54 |
| N,N-Diphenylformamide | — | 73 |
| N-Methyl-4-acetotoluidide | — | 83 |
| N,N-Diphenylacetamide | — | 101 |
| N-Methylacetanilide | — | 102 |
| Acetanilide | — | 114 |
| N-Ethyl-4-nitroacetanilide | — | 118 |
| N-Phenylsuccinimide | — | 156 |
| N-Phenylphthalimide | — | 205 |

[a] Also see Tables A.1 and A.2 for amides prepared as derivatives of carboxylic acids.

**TABLE A.15 Nitro Compounds**

| Compound | bp (°C) | mp (°C) |
|---|---|---|
| Nitrobenzene | 211 | — |
| 2-Nitrotoluene | 225 | — |
| 2-Nitro-*m*-xylene | 225 | — |
| 3-Nitrotoluene | 231 | — |
| 3-Nitro-*o*-xylene | 245 | — |
| 4-Ethylnitrobenzene | 246 | — |
| 2-Chloro-6-nitrotoluene | — | 36 |
| 4-Chloro-2-nitrotoluene | — | 38 |
| 3,4-Dichloronitrobenzene | — | 42 |
| 1-Chloro-2,4-dinitrobenzene | — | 50 |
| 4-Nitrotoluene | — | 54 |
| 1-Nitronaphthalene | — | 56 |
| 1-Chloro-4-nitrobenzene | — | 84 |
| *m*-Dinitrobenzene | — | 90 |

**TABLE A.16 Ethers**

| Compound | bp (°C) | mp (°C) |
|---|---|---|
| Furan | 32 | — |
| Ethyl vinyl ether | 33 | — |
| Tetrahydrofuran | 67 | — |
| *n*-Butyl vinyl ether | 94 | — |
| Anisole | 154 | — |
| 4-Methylanisole | 174 | — |
| 3-Methylanisole | 176 | — |
| 4-Chloroanisole | 203 | — |
| 1,2-Dimethoxybenzene | 207 | — |
| 4-Bromoanisole | 215 | — |
| Anethole | 234–237 | — |
| Diphenyl ether | 259 | — |
| 2-Nitroanisole | 273 | — |
| Dibenzyl ether | 298 | — |
| 4-Nitroanisole | — | 50–52 |
| 1,4-Dimethoxybenzene | — | 56–60 |
| 2-Methoxynaphthalene | — | 73–75 |

## TABLE A.17 Esters

| Compound | bp (°C) | Compound | bp (°C) |
|---|---|---|---|
| **Liquids** | | | |
| Methyl formate | 32 | Pentyl formate | 132 |
| Ethyl formate | 54 | Ethyl 3-methylbutanoate | 135 |
| Methyl acetate | 57 | Isobutyl propanoate | 137 |
| Isopropyl formate | 68 | Isopentyl acetate | 142 |
| Ethyl acetate | 77 | Propyl butanoate | 143 |
| Methyl propanoate | 80 | Ethyl pentanoate | 146 |
| Methyl propenoate | 80 | Butyl propanoate | 147 |
| Propyl formate | 81 | Pentyl acetate | 149 |
| Isopropyl acetate | 91 | Isobutyl 2-methylpropanoate | 149 |
| Methyl 2-methylpropanoate | 93 | Methyl hexanoate | 151 |
| sec-Butyl formate | 97 | Isopentyl propanoate | 160 |
| tert-Butyl acetate | 98 | Butyl butanoate | 165 |
| Ethyl propanoate | 99 | Propyl pentanoate | 167 |
| Propyl acetate | 101 | Ethyl hexanoate | 168 |
| Methyl butanoate | 102 | Cyclohexyl acetate | 175 |
| Allyl acetate | 104 | Isopentyl butanoate | 178 |
| Ethyl 2-methylpropanoate | 110 | Pentyl butanoate | 185 |
| sec-Butyl acetate | 112 | Propyl hexanoate | 186 |
| Methyl 3-methylbutanoate | 117 | Butyl pentanoate | 186 |
| Isobutyl acetate | 117 | Ethyl heptanoate | 189 |
| Ethyl butanoate | 122 | Isopentyl 3-methylbutanoate | 190 |
| Propyl propanoate | 122 | Ethylene glycol diacetate | 190 |
| Butyl acetate | 126 | Tetrahydrofurfuryl acetate | 194 |
| Diethyl carbonate | 127 | Methyl octanoate | 195 |
| Methyl pentanoate | 128 | Methyl benzoate | 200 |
| Isopropyl butanoate | 128 | Ethyl benzoate | 213 |

| Compound | mp (°C) | Compound | mp (°C) |
|---|---|---|---|
| **Solids** | | | |
| d-Bornyl acetate | 29 (bp 221) | Ethyl 3,5-dinitrobenzoate | 93 |
| Ethyl 2-nitrobenzoate | 30 | Methyl 4-nitrobenzoate | 96 |
| Ethyl octadecanoate | 33 | 2-Naphthyl benzoate | 107 |
| Methyl cinnamate | 36 (bp 261) | Isopropyl 4-nitrobenzoate | 111 |
| Methyl 4-chlorobenzoate | 44 | Cyclohexyl 3,5-dinitrobenzoate | 112 |
| 1-Naphthyl acetate | 49 | Cholesteryl acetate | 114 |
| Ethyl 4-nitrobenzoate | 56 | Ethyl 4-hydroxybenzoate | 116 |
| 2-Naphthyl acetate | 71 | tert-Butyl 4-nitrobenzoate | 116 |
| Ethylene glycol dibenzoate | 73 | Hydroquinone diacetate | 124 |
| Propyl 3,5-dinitrobenzoate | 74 | tert-Butyl 3,5-dinitrobenzoate | 142 |
| Methyl 4-bromobenzoate | 81 | Hydroquinone dibenzoate | 204 |

# B

# Chapters 6, 7, and 8: Experiments Classified by Mechanism

## A. ELIMINATION REACTIONS

*Acid-Catalyzed Dehydration of Alcohols:*
The E1 Elimination Reaction: Dehydration of 2-Butanol to Yield 1-Butene, *trans*-2-Butene, and *cis*-2-Butene
Experiment [9], 210

*Base-Promoted Dehydrohalogenations:*
The E2 Elimination Reaction: Dehydrohalogenation of 2-Bromobutane to Yield 1-Butene, *trans*-2-Butene, and *cis*-2-Butene
Experiment [10], 219

Dehydrohalogenation of *meso*-Stilbene Dibromide: Diphenylacetylene
Experiment [$A3_b$], 490

An Elimination Reaction with *erythro*-2,3-Dibromo-3-phenylpropanoic Acid: 2′-Bromostyrene
Experiment [D3], 530

*trans*-2-(4-Nitrophenyl)-3-benzoylaziridine
Experiment [F3], 568

*Base-Promoted Elimination of 2,4-Dinitrophenol:*
Piperonylonitrile
Experiment [E3], 548

## B. ADDITION TO CARBON–CARBON MULTIPLE BONDS

*Hydrogenation of an Alkene:*
Reductive Catalytic Hydrogenation of an Alkene: Octane
Experiment [12], 246

*Hydroboration–Oxidation of an Alkene:*
Octanol
Experiment [13], 252

## C. ADDITION TO CARBONYL GROUPS (CARBON–HETERO MULTIPLE BONDS)

## D. ALIPHATIC ELECTROPHILIC SUBSTITUTION

## E. ALIPHATIC NUCLEOPHILIC SUBSTITUTION AT TETRAHEDRAL CARBON

## F. NUCLEOPHILIC SUBSTITUTION AT TRIGONAL CARBON

**G. AROMATIC ELECTROPHILIC SUBSTITUTION**

## H. AROMATIC NUCLEOPHILIC SUBSTITUTION

## I. AROMATIC FREE RADICAL SUSTITUTION

## J. OXIDATION AND REDUCTION REACTIONS

*Reductive Hydrogenation of an Alkene:*
Octane
Experiment [12], 246

*Hydroboration–Oxidation of an Alkene:*
Octanol
Experiment [13], 252

*Diborane Reductions:*
Thioxanthene
Experiment [1A$_{adv}$], 419
Xanthene
Experiment [1B$_{adv}$], 420

*Hypochlorite Oxidation of an Alcohol:*
Cyclohexanone
Experiment [32], 396

*Chromium Trioxide-Resin Oxidation of Alcohols:*
9-Fluorenone
Experiment [33A], 402

Piperonal
Experiment [E1], 538

*Copper(II) Ion Oxidation of Benzoin:*
Benzil
Experiment [A2$_a$], 473

*Ferric Chloride Oxidative Coupling of 2-Naphthol:*
1,1′-Bi-2-naphthol
Experiment [5$_{adv}$], 442

*Hypochlorite Oxidation of Methyl Ketones by the Haloform Reaction:*
Benzoic Acid
Experiment [34A], 407
*p*-Methoxybenzoic Acid
Experiment [34B], 408

*Oxidation of Cyclohexanol:*
Adipic Acid
Experiment [B1], 499

# K. OTHER EXPERIMENTAL TRANSFORMATIONS

*Photochemical Isomerization of an Alkene:*
cis-1,2-Dibenzoylethylene
Experiment [6], 169

*Photochemical Isomerization:*
cis-Azobenzene
Experiment [35], 411

*Preparation of an Arene Sulfonamide:*
Sulfanilamide
Experiment [C3], 515

---

# L. REARRANGEMENTS

# Glossary

**Absorb** To take up matter (to dissolve), or to take up radiant energy.

**Active methylene** A methylene group with hydrogen atoms rendered acidic due to the presence of an adjacent ($\alpha$) electron withdrawing group, such as a carbonyl.

**Activity (of alumina)** A measure of the degree to which alumina adsorbs polar molecules. The activity (adsorbtivity) of alumina may be reduced by the addition of small amounts of water. Thus the amount of water present in a sample of alumina determines the activity grade. Alumina of a specific activity can be prepared by dehydrating alumina at 360 °C for about 5–6 hours and then allowing the dehydrated alumina to absorb a suitable amount of water. The Brockmann scale of alumina activity is based on the amount of water (weight percent) that the alumina contains: Grade I = 0%, Grade II = 3%, Grade III = 6%, Grade IV = 10%, and Grade V = 15% . For further information, see Brockmann, H.; Schodder, H. *Chem. Ber.* **1941**, *74*, 73.

**Adsorb** The process by which molecules or atoms (either gas or liquid) adhere to the surface of a solid.

**Aliquot** A portion.

**Anilide** A compound that contains a $C_6H_5NHCO$ group. An amide formed by acylation of aniline (aminobenzene).

**Capillary action** The action by which the surface of a liquid, where it contacts a solid, is elevated or depressed because of the relative attractions of the molecules of the liquid for each other and for the solid. It is particularly observable in capillary tubes, where it determines the ascent (descent) of the liquid above (below) the level of the liquid in which the capillary tube is immersed.

**Characterize** To conclusively identify a compound by the measurement of its physical, spectroscopic, and other properties.

**Condensation reaction** A condensation reaction is an addition reaction that produces water (or another small neutral molecule such as $CH_3OH$ or $NH_3$) as a byproduct.

**Deliquescent** Liquefying by the absorption of water from the surrounding atmosphere.

**Dihedral angle** The angle between two intersecting planes. In organic chemistry the term *dihedral angle* (or *torsional angle*) is used to describe the angle between two atoms (or groups) bonded to two adjacent atoms, such as H—C—C—H, and can be determined from a molecular model by looking down the axis of the bond between the two central atoms.

**Eluant** A mobile phase in chromatography.

**Eluate** The solution that is eluted from a chromatographic system.

**Elute** To cause elution.

**Elution** The flow, in chromatography, of the mobile phase through the stationary phase.

**Emulsion** A suspension composed of immiscible drops of one liquid in another liquid (e.g., oil and vinegar in salad dressing).

**Enol** A functional group composed of a hydroxyl group bonded to an alkene.

**Enolate** The conjugate base of a enol, that is, a negatively charged oxygen atom bonded to an alkene. An enolate results from deprotonation of a carbon $\alpha$ to a carbonyl group.

**Filter cake** The material that is separated from a liquid, and remains on the filter paper, after a filtration.

**Glacial acetic acid** Pure acetic acid containing less than 1% water.

**Heterolysis** Cleavage of a covalent bond in a manner such that both the bond's electrons end up on one of the formerly bonded atoms.

**Homogeneous** Consisting of a single phase.

**Homolysis** Cleavage of a covalent bond in a manner such that the bond's electrons are evenly distributed to the formerly bonded atoms.

**Hygroscopic** Absorbs moisture.

**In situ** In chemistry, the term usually refers to a reagent or other material generated directly in a reaction vessel and not isolated.

**Lachrymator** A material that causes the flow of tears.

**Ligroin** A solvent composed of a mixture of alkanes.

**Metabolites** The compounds consumed and produced by metabolism.

**Metabolism** The chemical processes performed by a living cellular organism.

**Methine** A CH group (with no other hydrogen atoms attached to the carbon atom).

**Methylene** A $CH_2$ group (with no other hydrogen atoms attached to the carbon atom).

**Mother liquor** The residual, and often impure, solution remaining from a crystallization.

**Olefin** An older term for an alkene.

**Oxonium ion** A trivalent oxygen cation with a full octet of electrons (e.g., $H_3O^+$).

**Polymer** A large molecule constructed of repeating smaller (monomer) units.

**Racemic** Consisting of an equimolar mixture of two enantiomers.

**Reagent** A chemical or solution used in the laboratory to detect, measure, react with, or otherwise examine other chemicals, solutions, or substances.

**Reflux** The process by which all vapor evaporated or boiled from a vessel is condensed and returned to that vessel.

**Rotamers** Conformational isomers that can be interconverted by rotation about one or more single bonds (e.g., gauche and anti butane).

**Tare** A tared container is one whose weight has been measured. The term may also refer to the process of zeroing a balance after a container has been placed on the weighing platform.

**Triturate** To grind to a fine powder.

**Zwitterion** A neutral molecule containing separated opposite formal charges.

# Index

## A

Abderhalden vacuum drying apparatus, 94 *illus.*
Absorption spectroscopy, 676–678
Accidents, reporting of, 7
Acetamides, as amine derivatives, 722
Acetanilide:
  bromination, 371–374
  synthesis, 339, 342–344
Acetic acid, Fischer esterification, 203–206
  with acidic resins, 207–209
Acetic anhydride:
  in amide synthesis, 339, 342–346
  anhydride exchange reactions of, 354–357
  in imide synthesis, 351–352
Acetone, aldol reaction with benzaldehyde, 313, 315–320
Acetophenone:
  aldol condensation with 4-nitrobenzaldehyde, 557–560
  hypochlorite oxidation, 406, 407–408
Acetyl coenzyme A, 191–192
1-Acetylferrocene, synthesis, 364, 366–369
Acid anhydrides:
  carbonyl vibrational frequency, 612
  cyclic, 354–357
  derivative preparation, 723–724
  derivatives, 735 *table*
  group frequencies, 620, 621
Acid–base solvent extraction, 87–88
  experiments, 150–154
Acid chlorides:
  as carboxylic acid derivatives, 718–719

derivative preparation, 723–724
derivatives, 735 *table*
group frequencies, 618–619
synthesis by acid reaction with thionyl chloride, 502–504
Acid halides, carbonyl vibrational frequency, 612
Acid hydrolysis, of esters, 201
Acidic resins, for Fischer esterification, 207–209
Acids, extraction separation from bases, 87–88
Actual yield, 41
Acylation, Friedel–Crafts, 365, 366–369
Acylium ions, 454
Addition–elimination two-step sequences, 389
Addition reactions, 248
  concerted, 254, 262
Adipic acid:
  reaction with thionyl chloride, 502–504
  synthesis, 499–501
Adipoyl chloride, polymerization with 1,6-hexanediamine, 505–506
Air condensers, 24, 25 *illus.*
Alcohols, 73–74
  dehydration, 213–218
  derivative preparation, 720–721
  derivatives, 731 *table*
  in esterification reactions, 191–193
  Grignard synthesis, 279–280
    secondary alcohols, 286–290
    tertiary alcohols, 281–284
  group frequencies, 613–614

oxidation with chromium trioxide resin, 401–404
oxidation with hypochlorite, 397–399
selective oxidation to aldehydes, 538–539
tests for, 704–706
water solubility, 75
Aldehydes:
  carbonyl vibrational frequency, 612
  derivative preparation, 721–722
  derivatives, 732 *table*
  group frequencies, 615–616
  reaction with Grignard reagents, 286–290
  reduction by metal hydrides, 155–158
  tests for, 706–708
Alder, Kurt, 273. *See also* Diels-Alder reaction
Aldol condensation:
  benzil and 1,3-diphenylacetone, 478–481
  crossed, 321
  dibenzalacetone synthesis, 313, 315–320
  Knoevenagel reaction, 520–523
  4-nitrobenzaldehyde and acetophenone, 557–560
  retro-aldol reaction, 314
  theory, 314–315, 521, 558–559
Aldoximes, 549
*Aldrich Catalog Handbook of Fine Chemicals*, 121
Aliphatic hydrocarbons, 725, 738 *table*
  tests for, 708–709
Aliquat 336, 302, 305–306, 309–311
Alkaline hydrolysis test, 711–712